Austin & Winfield, Publishers

PRINCIPLES OF INTEGRATIVE ENVIRONMENTAL PHYSIOLOGY

G. Edgar Folk Jr.
Department of Physiology, University of Iowa

Marvin L. Riedesel
Department of Biology, University of New Mexico

Diana L. Thrift
Department of Physiology, University of Iowa

Principles of Integrative Environmental Physiology

Ann Bancroft in Antarctica. Leader of the American Women's Expedition, Bancroft and four other women skied over 660 miles pulling 200-pound sleds. They pulled uphill to reach 12,000 feet, consuming 5,500 calories a day, 45% of which was animal fat. Bancroft lost 20 pounds. The women's average height was 5'4", weight, 120 pounds. The expedition reached the South Pole on January 14, 1993. (Photo courtesy of Ann Bancroft.)

Principles of Integrative Environmental Physiology

G. Edgar Folk, Jr.
Department of Physiology
University of Iowa

Marvin L. Riedesel
Department of Biology
University of New Mexico

Diana L. Thrift
Department of Physiology
University of Iowa

Austin & Winfield, Publishers
San Francisco London Bethesda
1998

Library of Congress Cataloging-in-Publication Data

Folk, G. Edgar (George Edgar), 1914-
Principles of integrative environmental physiology / G. Edgar Folk, Jr., Marvin L. Riedesel ; Diana L. Thrift, editor.
p. cm.
Updated, expanded, and revised from Introduction to environmental physiology (1966) and Textbook of environmental physiology (1974)
Includes bibliographical references and index.
ISBN 1-57292-109-9 (hc : alk. paper). -- ISBN 1-57292-108-0 (pb alk. paper)
1. Animal ecophysiology. I. Riedesel, Marvin LeRoy, 1925- . II. Thrift, Diana L., 1948- . III. Title.
QP82.F62 1997
571.1'9—dc21
97-43186
CIP

Editorial Inquiries:
Austin & Winfield, Publishers
7831 Woodmont Avenue, #345
Bethesda, MD 20814
(301) 654-7335

To Order: (800) 99-AUSTIN

Book design by Lisa Parker, Parker Davis Graphics, Iowa City.
Cover photo by Tom and Pat Leeson.

To our parents

Rev. G. Edgar Folk and May Davis Folk

Elmer V. Riedesel, M.D. and Augusta Stankee Riedesel

William B. Thrift and Betty Wright Thrift

Contents

LIST OF ENVIRONMENTAL DIAGRAMS AND NOMOGRAMS

ENVIRONMENTAL DIAGRAMS
Introducing Each Chapter

NOMOGRAMS
Last Five Pages of Book

LIST OF FIGURES

PREFACE

This is a book about the integrative environmental physiology of responses to challenging physical and chemical environments. In it we explore the physiological and anatomical adaptations by which mammals and other animals combat these extreme ambient conditions. The word *integrative* in the title reflects the widely accepted viewpoint that all responses to demanding environments represent coordinated and balanced contributions from many widely separated tissues and organs of the body, as well as cellular compartments and genetic and molecular systems (Boyd and Noble 1993).

While the emphasis of the book is on vertebrates and mammals, we have included a few examples of the diverse adaptations of plants, invertebrate, and unicellular organisms as they adjust to specific features of the environment (Diamond 1994). Note that the book is also about that part of *comparative physiology* that relates to extremes of the physical environment.

Much of the discussion throughout the book is devoted to a detailed description of the changes that occur as whole-body organisms adjust to environmental extremes. This adjustment amounts to a redesign of the organism; it can occur within minutes at the molecular or cellular level, or take weeks to years at the higher levels of organization. We call this adjustment *acclimatization*, which is the major theme of this book.

In each chapter we have attempted to illustrate the *principles* of environmental physiology. Because the book serves several purposes, it was not feasible to cover the field completely; more detailed material can be found in modern review articles, such as those published in the *Handbook of Physiology: Environmental Physiology* (Fregly and Blateis 1995).

The book presents facts about extreme climates along with the concern expressed by many that these extremes will become worse. How does climate relate to the theme of this book, *acclimatization?* One example is the climate of India, where people must tolerate air temperatures that reach 48.8°C (120°F). What happens if the temperature there increases to 54.4°C (130°F)? In the book's final chapter, we'll raise these and other questions as we examine today's changing climate—one that may require degrees of acclimatization never before experienced.

In one sense, this book presents an oversimplified view as we select single factors of the physical environment to analyze for their effects. For example, one chapter examines responses to *cold* while another examines *altitude*, yet there is little examination of the *combined effects* of these two factors. Perhaps there should be a separate book on cross-acclimatization. But as long as the reader recognizes the significance and failings of this single-factor approach, we have found it valuable to compile single environmental influences into one volume.

Our intended audience for this book includes both undergraduate and graduate students, teachers, and researchers. (Students should already have had at least an introductory course in physiology.) We particularly direct the book to those readers interested in the following:

- comparative mammalogy,
- the outdoor environment,
- indigenous people enduring hostile environments,
- explorers and others drawn to wild and remote places where they face challenges posed by extreme environments,
- an interdisciplinary approach including meteorology, physiology, ecology, biology, agriculture, and medicine.

This book has been updated, expanded, and revised from two previous editions, *Introduction to Environmental Physiology*, (Folk 1966a), and *Textbook of Environmental Physiology*, (Folk 1974). As with those volumes, we intend that this new book serve as a convenient reference. To that end, we have updated the extensive bibliography to reflect recent developments in the field. At the same time, however, we wanted to retain descriptions of many of the outstanding and classic papers in environmental physiology in the hopes that it will prevent unnecessary duplication of experiments. The earlier references—especially those of Krogh, Kuno, Dill, Scholander, Irving, Schmidt-Nielsen, and Griffin—provide a record of the development of the field.

Another useful feature we have retained and expanded in this edition is the series of Environmental Diagrams that introduce each chapter. In addition to information on environmental extremes and variations of climate, these diagrams present physiological data, including tolerance of mammals to extreme temperatures, physiological effects of light, high pressure, and more.

We have also retained and extended the Nomograms, which are located at the very back of the book for easy reference. We urge our readers to become familiar with them, particularly the Temperature and Length Scales conversions on the inside back cover.

A note about temperature scales and other measurements used in this book: despite worldwide support for the metric system known as Système International d'Unités (SI), several countries continue to use other forms of measure. In the U.S., for example, forecasters present meteorological data to the public in degrees Fahrenheit; road signs show speed limits and distances in miles; and doctors discuss patients' temperatures in Fahrenheit. Thus, in an attempt to address all our readers, we usually present temperatures in degrees Centigrade (C°) followed by the degrees Fahrenheit (F°) in parentheses; similarly, distances are most often given first in meters, then feet. For examples of of SI units and their conversions to other units of measure, we refer the reader to Appendices D and E.

ACKNOWLEDGEMENTS

This book was inspired and completed only because of the editing and illustrations in two earlier editions by the late Mary Arp Folk. For this volume, direct contributions were made by Beth Dennis, H.T. Hammel (Ted), Paul R. Kerkof, Witold F. Krajewski, Stephen Loring, Timothy P. Lyons, L. David Mech, Louise Murray, Evelyn Steffanson Nef, and Lisa Parker. In many cases we telephoned a friend asking, "Do you have a reference to . . .?" The following produced one or more critical references, or made similar contributions: Ann Bancroft, Richard Bovbjerg, Barbara Boyle, Mike Castellini, John Coady, Peter Colony, David S. Curry, Robin Davidsson, William Gannon, Carl Gisolfi, Donald R. Griffin, June Helm, Louise Janes, Jack A. Loeppky, Robin Looft-Wilson, Gary Malvin, Robert Merlino, Francisco Mora, Connie Mutel, Jerry Schnoor, Holmes Semken, Chuck Swenson, Thomas A. Potts, Robert Tomanek, and Charles Wunder.

We also thank our families for their patience, understanding, support, and help, especially Rosemary and David, Witek, Joanna, Sophia, and Sean, Victoria, Christopher, and last, but not least, Bess.

Principles of Integrative Environmental Physiology

1. ENVIRONMENTAL PHYSIOLOGY: HISTORY AND TERMINOLOGY

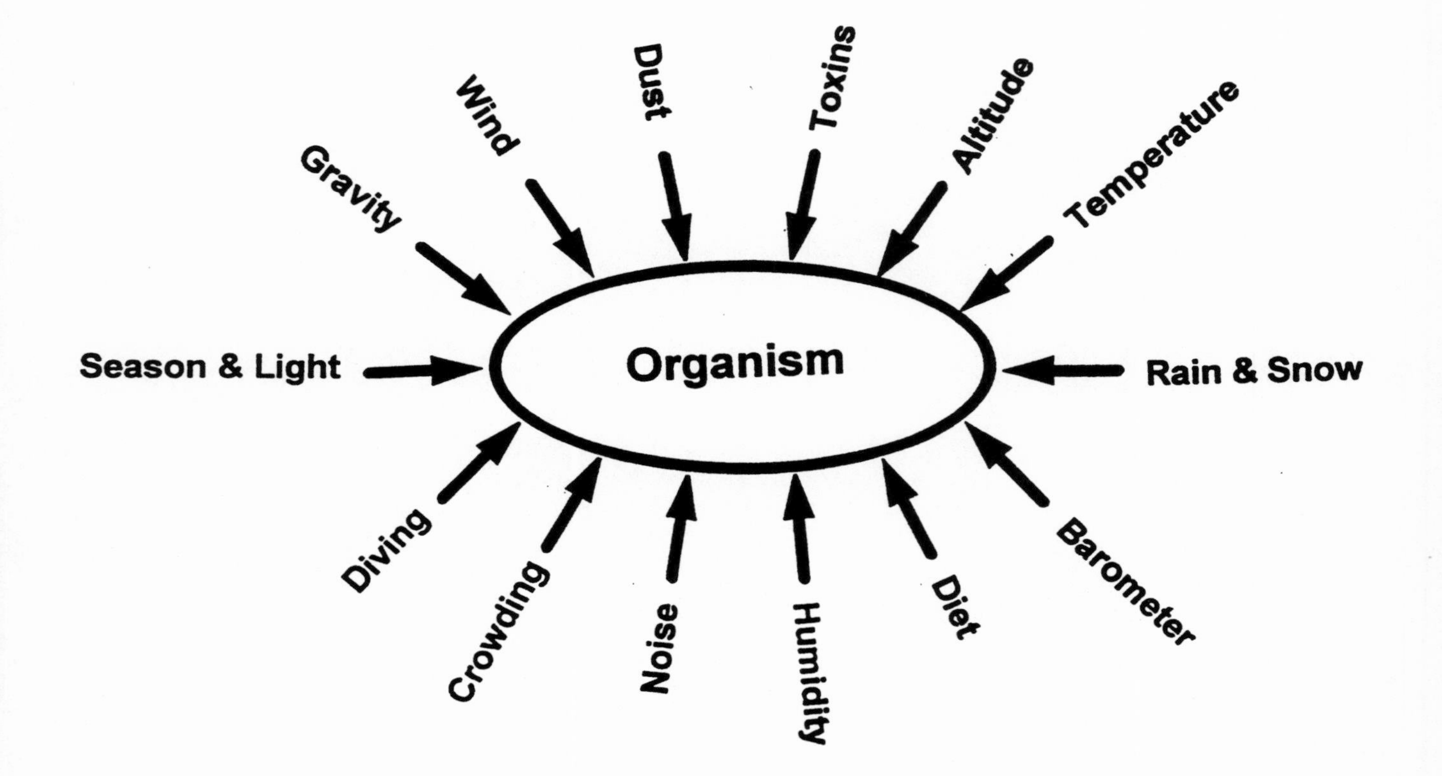

Environmental Diagram: Chapter 1

1. Environmental Physiology: History and Terminology

1. ENVIRONMENTAL PHYSIOLOGY: HISTORY AND TERMINOLOGY

INTRODUCTION

The noted physiologist, Walter B. Cannon, used an integrative approach to physiological function early in the twentieth century when he described a principle he termed *homeostasis*. According to this principle, dynamic small changes occur continuously within the mammalian organism to enable it to maintain a relatively constant condition. Thus, homeostasis depends upon the integrative function of all aspects of an organism's physiology. Some examples of the maintenance of constancy are blood pressure, the pH of the blood and its chemical components, body temperature, and in a certain percentage of the human population, body weight. With some individuals, whether human or other mammal, there are departures from this constancy to a new level as there is a gradual change in the individual; an example is the change that occurs with regular exercise. It is these regulated departures from carefully controlled homeostasis that represent the theme of this book, namely, the adjustments of mammals to hostile environments.

HISTORY

Hippocrates investigated the effects of weather and climate on human conditions. Darwin's theoretical views, found in *The Origin of Species,* stressed the interrelations of life and the physical environment. Further study of this area was demarked by Haeckel in 1866 as "Oecology." This science, "Ecology," was eventually roughly subdivided for practical reasons into "Autecology," the reactions between a species and its physical environment, and "Synecology," the causes behind the successions and fluctuations in plant and animal communities. The latter term could also be described as "Population Ecology."

During the period when the science of ecology was developing, parallel groups of environmentalists were organizing independently. For example, some of these maintained an applied medical interest, and each new area of interest was named by an authority in the field. The overall theme reflected a new orientation toward climatic effects upon humans and the other mammals.

As early as 1902, physiological measurements in extreme environments were made by August Krogh (Schmidt-Nielsen 1995)(see

Fig. 1-1). He was followed by Per Scholander, who named their activities "Expeditionary Physiology," though the term was seldom used. In his autobiography, Scholander (1990) described experiments he conducted in many parts of the world, often on the ship he designed, the *Alpha Helix*. The work of Krogh and Scholander is presented in detail in later chapters of this book. Building upon their work, a larger group of cross-discipline scientists, led by D.B. Dill, developed the concepts and discipline of *Environmental Physiology.* (See sidebar.)

In the 1960's, *Biometeorology*, a further expansion of Environmental Physiology, was established by Tromp (1963a, 1963b), Sargent (1963, 1972), Lee (1953), and Folk (1969).

However, whether these and other key workers were known as "ecologists," "biometeorologists," or "environmental physiologists," they were all working with essentially the same material, doing similar experiments, and teaching similar courses.

Figure 1-1. Early Environmental Physiologists August Krogh (Nobel Laureate) and Lawrence Irving. Krogh studied the effect of a high fat diet on the energy metabolism of the Inuit in Greenland. Irving's work on mammals in the cold and diving mammals is presented throughout this book. Photo courtesy of B. Schmidt-Nielsen.

INTEGRATIVE PHYSIOLOGY

By 1990 a new approach to physiology arose. Some physiologists who liked to work with tissues and organs (organismic physiology) felt that their field had been fragmented into areas with a narrow focus (membrane and molecular physiology). They also felt that they had been at fault for not integrating their field with a cellular approach. Thus, integrative physiology was born. Several symposia offered coordination on several fronts: from molecular to whole organism function; from biochemical to biophysical processes; across species; and between organs (Wagner 1995). Engelberg (1995) writes that integration does not involve cementing together areas of specialization, but instead confronts living systems as wholes (cells, multicellular organisms, societies) of all kinds. To relate our particular textbook to the idea, we add that Animal Behavior is a discipline to meld into Integrative Physiology.

THE STIMULI

Since this approach to physiology (i.e. studying climatic effects) is essentially the study of stimuli and responses, at this point let us look at the physical environmental factors that represent the stimuli or impacts on the organism. See also Environmental Diagram, and sidebar (this page).

Natural Physical Factors

1. Heat
2. Cold
3. Humidity
4. Water effects on skin
5. Air movement
6. Barometric and deep water pressure
7. Visible light
8. Ultraviolet light
9. Substratum
10. Dust

EVER-PRESENT IMPACTS OF THE ENVIRONMENT

In addition to the environmental impacts depicted in the Environmental Diagram and listed in this text, consider that at some time each day, a climatic disaster is occurring somewhere on the globe. For example, in one newspaper on November 12, 1997, the following Associated Press reports appeared:

- In Somalia, the Juba River flooded; 300,000 people are without homes, 130 people drowned, and the crop of the staple, sorghum, was destroyed.
- In southern Vietnam, the October typhoon destroyed the homes of many thousands of families; 435 people were killed, and several thousand more were missing.

Thus, the physiology of the discomfort of populations receives the focus of our attention in this book.

11. Gravity
12. Subtle factors: cosmic radiations, atmospheric electricity, and terrestrial magnetism

This classification is practical, but it is not yet certain whether those environments listed as "subtle factors" have an important influence upon the behavior or health of any mammals.

THE RESPONSES

Physiological responses may be measured quantitatively in environmental chambers, or somewhat less specifically in the free environment. The responses that are measured may result from acute, short, or chronic exposures to hostile environments. This measurement of physiological responses needs no explanation, but what about observations of animal behavior? Often, behavior is an initial response, followed by the physiological compensations. When this happens, the physiological changes represent a permanent response, and the behavior can cease. Although in this book we emphasize the physiological responses of the whole animal to environmental extremes, in some cases it will be difficult to separate the physiological from the behavioral response.

The difference between these two responses can be understood by observing cats and white rats in a cold environment. In the cat, the piloerector muscles contract, and every hair stands out straight—a physiological response. White rats, as described by Richter (1927), increase their use of nesting material (in terms of length of ticker tape) in direct proportion to the decrease in environmental temperature—a behavioral response. The contribution of behavior to temperature regulation is also observed when the cold mammal curls itself into a ball, and when the hot mammal stretches out to increase surface area. Many such illustrations demonstrate that behavior warrants inclusion in the analyses concerned with integrative environmental physiology.

TERMINOLOGY OF THE RESPONSES

All organisms have a series of functional and structural adaptations that allow them to reside where environmental extremes are found. These physical environments seem to us to be harsh and alien.

Various workers have described different kinds of responses to extreme environments. These will be presented in the following sections.

Cannon's *Emergency Syndrome*

Walter B. Cannon described the release of adrenergic substances from sympathetic nerve endings and called the resulting reaction "the flee, fright, or fight syndrome." He emphasized that two types of stimuli bring about this reaction: 1) the condition of danger, and 2) exposure to environmental extremes. Thus, when a mammal is challenged by either of these there is a rapid total body response referred to as the *Emergency Syndrome*. This syndrome is usually dramatically revealed to a person who has nearly had a violent or

D.B. DILL (1891–1979) AND THE HARVARD FATIGUE LABORATORY

When the Harvard Fatigue Laboratory was established by Lawrence J. Henderson in 1927, its interdisciplinary, collaborative approach to research in the field of human physiology was unique; the goal was to look at the interrelatedness of the systems and organs which comprise a biological organism. Its staff and associates included physiologists, biochemists, biologists, medical clinicians, sociologists, anthropologists, and psychologists, all of whom brought with them "an interest in learning more about physiology as an integrated system of parts, a Gestalt, as it were" (Horvath & Horvath 1973). In the beginning, techniques were perfected for blood analysis of subjects while exercising; later, these techniques were applied to samples from subjects exposed to heat, cold, and mountain altitude (Folk 1969). The Fatigue Laboratory also provided a workplace for field scientists from its early years up to 1947.

In 1927, Henderson appointed David Bruce Dill as Director of Research at the Fatigue Laboratory; Dill held this position until the Laboratory closed 20 years later. Dill had earned his Ph.D. from Stanford, and later received their Honorary Degree. He received the Legion of Merit and other awards for his military contributions during World War II. He published 300 papers, mostly from the Fatigue Laboratory, and three of his four books were on environmental matters: *Life, Heat, and Altitude; Adaptation to the Environment;* and *The Hot Life of Man and Beast.*

Through his dynamic leadership at the Harvard Fatigue Lab, Dill encouraged the early careers of others in the field: Sid Robinson, Steve Horvath, Robert Darling, Robert Johnson, Fred traumatic accident: one to two minutes later the knees of the individual may shake violently with an accompanying overall weakness due to the release of adrenergic substances. The delayed response is considered evidence for the endocrine component of the syndrome. The importance of this environmental physiology is illustrated when a cat with a sympathetic nerve chain removed on one side is exposed to extreme cold. In the cold room, the hair is not erected to increase insulation on the side without the sympathetic nerve chain. If the sympathetic chains are removed on both sides, the cat is completely unable to tolerate severe exposure.

Selye's *General Adaptation Syndrome*

Also confusingly called the "Stress Syndrome," the *General Adaptation Syndrome* occurs when there is a gradual or prolonged exposure to some of the extremes of the physical environment. It is also induced by other factors, such as internally introduced agents typified by toxins, or by holding a low position in the "peck-order of dominance" in an animal population. The Adaptation Syndrome is recognized by physiological changes, especially in the adrenal and thymus glands. If exposure to an extreme environment is prolonged, there is first a gradual increase in weight of the adrenal gland, then a successive loss of lipid in the adrenal cortex, followed by a regaining of the control level of lipid in the enlarged gland. In addition, there is enlargement of the thymus. Selye (1950) defines the new condition as a "state manifested by a specific syndrome which consists of all non-specifically induced changes within a biological system." To Selye, burned skin on a rat had, first, a local effect (which was not his main interest), and second, a non-specific induced hormonal change in the entire body of the rat, which was Selye's special interest. It was the second effect which he named the General Adaptation Syndrome.

The response mechanisms described by Cannon and Selye have laid special emphasis on the function of individual endocrine organs. Their descriptions must be kept in mind, but their usefulness in describing how mammals respond to environments is limited because they do not go far enough. This is the central theme in a book published in 1965 by Lissák and Endroczi entitled, *The Neuroendocrine Control of Adaptation*. They have approached the response problem from a wider angle and tried to take into consideration the entire neural and humoral systems. They showed that environmental challenge regulates behavior not only by a direct nervous influence but also by a humoral action on the nervous system. Lissák and Endroczi emphasize that the changes in the quantity and nature of hormones released in response to various environmental stimuli are so numerous that the whole subject is one of great complexity.

Turning again to Selye, it has been difficult to detect in human subjects those changes which are characteristic of the General Adaptation Syndrome in lower mammals. It is partly for this reason that physiologists have had to use an additional expression to describe adjustments of all animals to environmental changes. Their conventional term has been *acclimatization*, since there are many expres-

Sargent, and Donald Griffin. The Harvard Fatigue Laboratory gained an international reputation, and Dill used it to provide a home base for many European scientists such as F.J.W. Roughton, famous as the co-discoverer of carbonic anhydrase, and Per Scholander, the great field biologist and master of micro-gasometric techniques.

D.B. Dill could be called the first "environmental physiologist." Throughout his long and productive career, he often served as his own first experimental subject, sometimes exercising to the point of fatigue, and measuring his own responses to exercise, heat, altitude, and finally, aging. He made eleven expeditions to altitude, his last at age 87. It is not surprising that many symposia have been dedicated to him; the latest was entitled *Environmental Physiology: Aging, Heat, and Altitude* (Horvath and Yousef 1981).

The Fatigue Laboratory was dissolved in 1947, a result of lack of funding and support. However, the dispersal of its excellent staff led to the establishment or strengthening of many other research laboratories around the world. Harwood Belding, former head of the Fatigue Laboratory's Cold Division, left Harvard to hold the first chair of Environmental Physiology, established at the University of Pittsburgh in 1951. Dill continued to hold professorships at Harvard until 1961. He later established the Desert Research Institute at the University of Nevada, where he was Professor of Environmental Physiology from 1966 to 1979. (See Fig. 7-19 in Heat chapter.) He also continued to collect data on some of the same individuals who served as subjects in the Fatigue Laboratory. Often referred to as a "scientist's scientist," no one has had a greater influence on environmental physiology than D.B. Dill.

sions of special types of acclimatization which are not part of the Adaptation Syndrome. An example of these special acclimatization effects is the change, after prolonged exposure to high altitude, in red cell count from about 5 million/mm^3 to 8 million/mm^3.

We should avoid referring to the "Stress Syndrome" of Selye because in physics the word *stress* is the "force," and *strain* is the "result." Working definitions sometimes used by physiologists are as follows: 1) *stress* denotes the magnitude of forces external to the bodily system that tend to displace the system from its resting or ground state; 2) *strain* is the internal displacement from the resting or ground state brought about by the application of the stress (Lee 1965; Yousef 1988). The word *stress* does occur in common physiological usage to denote discomfort, such as "exposed to heat stress" or "exposed to cold stress."

ACCLIMATIZATION AND ACCLIMATION

Our English words *acclimatize* and *acclimate* were both derived from the French *acclimater;* Folk (1974) has found examples of these words in common usage in Europe as early as the late eighteenth century. A dictionary published in 1888 defines these terms as follows: "Acclimatization, like acclimation, is the process of being habituated or inured to a new climate, or to one not natural." The terms have been used and accepted in technical physiological literature from the mid-1940s.

Our definition of acclimatization is: *The functional compensation over a period of days or weeks in response to a complex of environmental factors, as in seasonal or climatic changes.* An example can be found in acclimatization to heat; after a human subject has worked for one week outdoors in the heat, there are three conspicuous physiological changes: an increase in sweating, and a decrease in both body temperature and heart rate from the control values at the start of the experiment. Thus, we are concerned with a phenotypic change which could be referred to as a redesign of the animal body.

But what about changes that occur when the experiments were done in an environmental chamber and only one variable was studied? This situation, of course, differs vastly from the free environment where an animal is subject to a battery of effects, which include radiation, temperature, wind, and humidity changes. To differentiate between such situations, a group of physiologists (Hart 1957; Dill 1964; Eagan 1963) agreed that a separate term was needed, and suggested applying the term *acclimate* to exposures in environmental chambers. They were later supported by a committee of The American Physiological Society. Therefore, we define acclimation as: *The functional compensation over a period of days to weeks in response to a single environmental factor only, as in controlled experiments.* Examples of the single factors to be varied are heat, cold, humidity, or wind.

DARWINIAN "GENETIC" ADAPTATION

This type of response essentially speaks for itself; one need only think of breeds of cattle, such as the Hindu Brahman cow, which is tolerant to heat, and the Highland cow, which is tolerant to cold. Clearly the genotype has been selected to produce the two breeds. Thus, Darwinian genetic adaptation is used for alterations which favor survival of a species or of a strain in a particular environment, which alterations have become part of the genetic heritage of the particular species or strain.

To illustrate the difference between Darwinian genetic adaptation and acclimatization, let us consider the differences between lowland natives in Peru and those who had lived for many generations at high altitude. Those living at high altitude have what is referred to as a "barrel-chest," while those living at sea level do not. Apparently, no one knows whether this is an example of Darwinian genetic adaptation, which would be inherited, or whether it is an example of acclimatization from birth. To our knowledge, no one has measured the chests of newborn babies delivered by these high-altitude natives to see whether they *inherit* the barrel-chest or *develop* it. Although blood samples have been taken from near-term fetuses of human subjects at high altitude, the results could be interpreted in terms of acclimatization through the mother's condition and through the fetal circulation.

Some authors write about "evolution observed," and they are referring to Darwinian genetic adaptation. Thomson objects to their vocabulary, saying that these authors are observing "local adaptation" (acclimatization) and "differentiation of populations," not evolution. He believes that true evolution is the same as "speciation" (the occurrence of new species) (Thomson 1997).

Regardless of debate about the details of the origin of speciation, we must consider what happens when it occurs. As evolution proceeds, animals seek out niches which best provide the needs of the specific species. Competition among organisms can lead to anatomical and physiological adjustments and changes, which we call Darwinian genetic adaptation. Within a given organism, tissues and organs develop to meet the needs of the cells that make up the tissue or organ. In a similar manner, specific cells develop features which must meet the needs of the organelles within the cell. In 1981, Taylor and Weibel coined the word *symmorphosis,* which postulates that the quantity of structure an animal builds into a functional system is matched to what is needed: enough but not too much.

HABITUATION

One of the early studies on the topic of habituation was done by Glaser and colleagues in 1959. He defined habituation as the process of *forming into a habit* or *accustoming*, the implication being that it depends on the mind, is reversible, and may involve the diminution of normal responses or sensations.

This is illustrated in the following example. A team of physiologists wanted to study tolerance to cold exposure, especially of the feet and legs, in the Alacaluf Indians of Tierra del Fuego (Hammel 1963). One of their measures was to compare the Alacaluf natives with the investigators themselves. The measurements consisted of plethysmography of the foot and lower leg. The natives left their

FEMALE SUBJECTS AND ENVIRONMENTAL PHYSIOLOGY

Before 1960, most work in environmental physiology (as well as other human physiology) recorded *male* physiology. Then, in the early 1960s, Hertig and Sargent published *Acclimatization of Women to Heat*, and Rahn released his landmark paper, *Physiology of Breath-Hold Diving and the Ama of Japan*. Those publications made an unusual contribution to the field because they studied the physiology of human *females*.

Environmental physiology is by no means the only area to have neglected the study of female physiology. Until fairly recently, the medical profession's studies of heart disease focused almost exclusively on males, thereby ignoring subtle physiological differences that could have significant implications for half the human population.

As environmental physiologists, we must ask whether, in addition to the obvious differences in male and female physiology, there are also certain hormonal differences that may (or may not) influence responses to environmental extremes. More studies comparing differences in the physiological responses of women and men would provide useful information.

UNCOMFORTABLE INVESTIGATORS AND SUBJECTS

Although environmental physiologists conduct many experiments under controlled laboratory conditions, they also often work in the field under uncomfortable conditions as they study hostile environments and their effects on living organisms. Indeed, many researchers have, of necessity, used themselves as subjects in these unfriendly environments. For example, Blagden and Fordyce in 1775 took a dog and some obliging associates into a room heated to 198°F; they carried the dog in a basket so its feet would not be burned. Hammel and colleagues immersed their feet and legs in painfully cold icebaths to compare their reactions with the indigenous people at the tip of South America. In the Arctic, Laurence Irving (see Fig. 1-1) drove dog teams and lived under conditions where it was hard to employ even simple procedures to obtain physiological information from the mammals whose body temperatures he was studying. Harold Lewis also became an excellent dog team driver as he studied sled dogs in both the Arctic and Antarctic. Ian Redmond studied herds of salt-mining elephants deep inside an African cave; K. Schmidt-Nielsen studied the blood volume of the camel in the desert; and H. Reynolds and others have tracked wild grizzly bears to their dens to insert radio-capsules in the animals for long-range studies of hibernation.

limbs in ice water, considering it a game, and were able to continue this exposure indefinitely. For the scientists, who were cold-acclimatized, the exposure to the icebath for a twenty-minute period was exceedingly painful. Nevertheless, the heat given off into the calorimeter in both cases came out to be the same, indicating that there were not differences in vasomotor function in the limbs of the Alacaluf compared with the limbs of the scientists. The difference was in the men's reaction to the potential pain of the extreme cold. The experiment indicates that there are differences to be found within the nervous system, and we may call this *habituation*.

Another illustration may help: Eagan, with a series of subjects, had them hold one finger in an icebath for several minutes every day for one week. On the first day the subjects found this extremely painful, but by the end of seven days they did not report pain. It was suggested that the term *habituation* probably applies here, although circulatory measurements were not made.

ADAPTATION

We have been discussing acclimatization, which is an adaptation involving integrated physiological and structural mechanisms. Some workers use the term *adapted* to express an increase in heat or cold tolerance. Prosser has stated that the term *adapt* means something different to different biologists; the term is used by both geneticists and experimental cell biologists. For example, we say: 1) the rods of the retina *adapt* in darkness; 2) the ear *adapts* to a repeated constant frequency; 3) a single proprioceptive axon *adapts* by decreasing frequency of firing, with a constant stimulus.

It seems unnecessary to replace the words *to acclimatize*, which have had so many years of usage, with the nonspecific term *to adapt*. A simple solution for the beginner is to ignore the umbrella word *adaptation*, which has so many applications, and to substitute specific vocabulary; even the word *adjust* is useful. We direct the reader's attention to the more advanced textbook, *Adaptational Biology, Molecules to Organisms,* by C. Ladd Prosser (1986). This is a book for the specialist. When he considers acclimatization, Prosser uses the following important words: "Environmentally induced variations are possible only within genetically fixed limits." He goes on to say that the phenotype is not completely described even when both genotype and environment have been specified. His approach to the analysis of the impact of the physical environment is that of population genetics. His book is a successful treatise on molecular evolution, and on page 8, Prosser accepts the terms *acclimation* and *acclimatization.*

We will, in later chapters, turn our attention to individual examples of habituation, acclimatization, and acclimation. First we will consider the environment of light and its effect upon the biological clock.

2. THE INFLUENCE OF THE RADIANT ENVIRONMENT

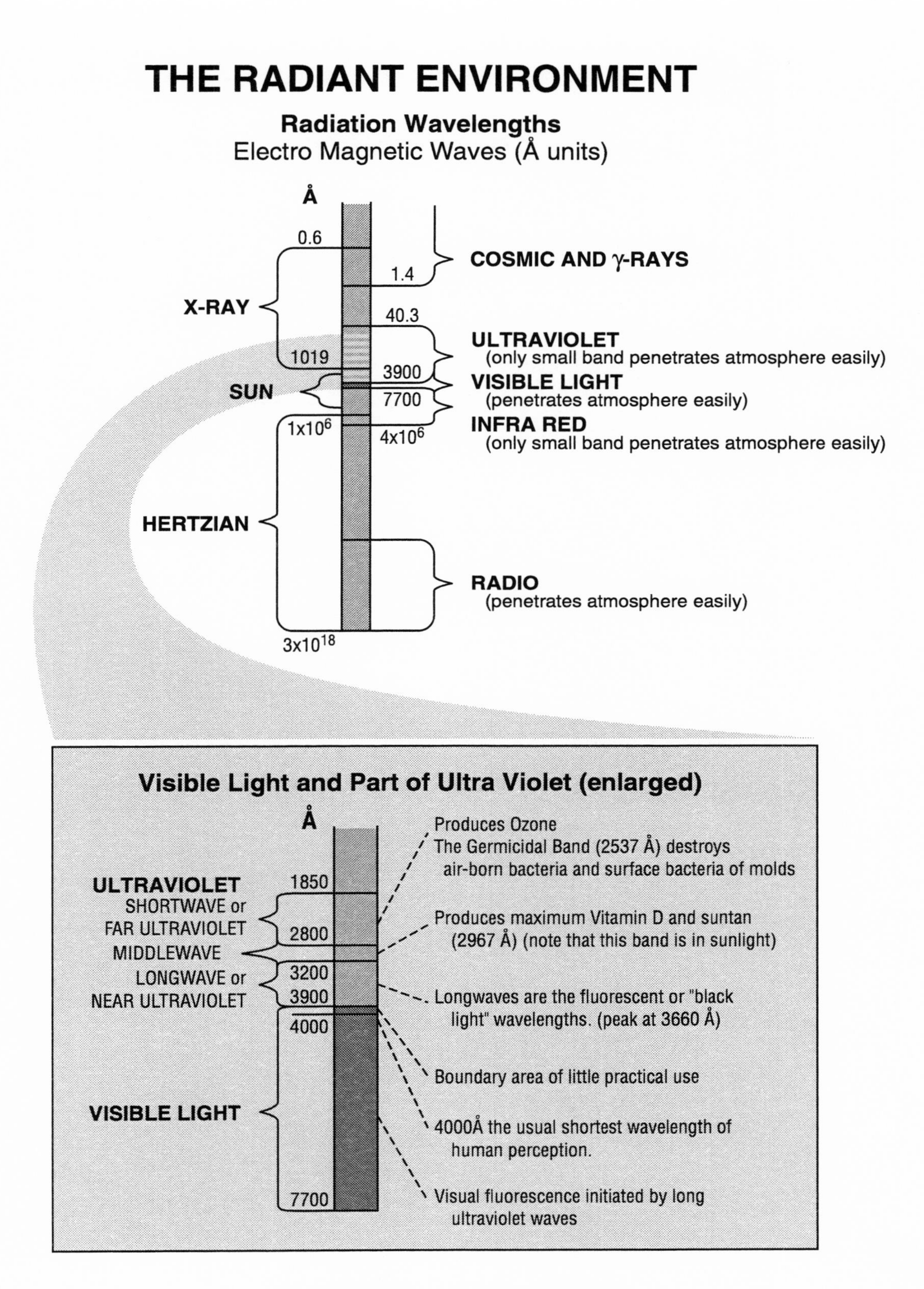

Environmental Diagram A: Chapter 2

EXTREME SEASONAL PHOTOPERIODS
Arctic & Antarctic

There are three Arctic cities of 10,000 inhabitants in Norway and Sweden alone. Example:

Tromsø, 69°39' N, is 200 miles north of the Arctic Circle, and 1,400 miles from the North Pole. It has the following light cycle:

WINTER	SPRING	SUMMER	AUTUMN
Continuous darkness	Short days becoming long days	Continuous light	Short nights becoming long nights
Nov. 21 - Jan. 23	Jan. 23 - May 21	May 21 - July 23	July 23 - Nov. 21

The Development of Continuous Darkness
The "Shortest Day of the Year" at Different Latitudes

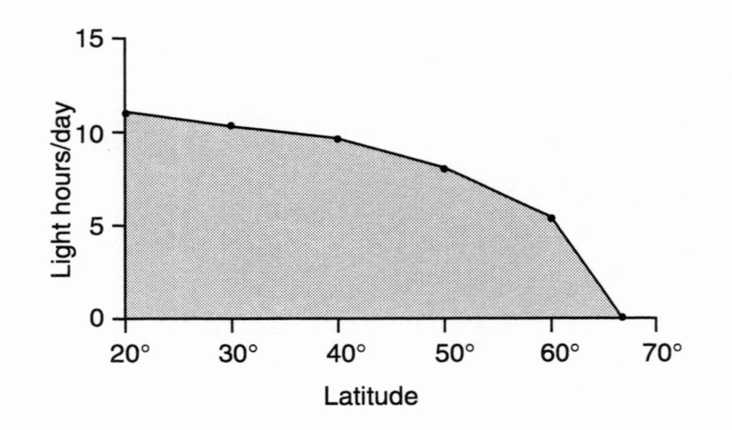

Environmental Diagram B: Chapter 2

2. The Influence of The Radiant Environment

- Introduction
- Radiation Descriptions
- Physiological Effects of Radio Waves
- Overview of Effects of Light
  - Physics of Light
  - Quality of the Light Stimulus
  - Laboratory Studies on Exposed Cells
- Physiological Effects of Ultraviolet Radiation
  - Origins of Ultraviolet Radiation
  - Three Bands of Ultraviolet Radiation
  - Effects on Sub-Primate Mammals
  - Effects on Humans
  - Ultraviolet and Vitamin D
- Physiological Effects of Visible Light: Daylength
  - Duration and Intensity Compared
  - Responses of Plants to Daylength
- Mechanism of Plant Photoperiodism
  - Natural Photoperiods and Geographical Variations
  - Varied Effects of Photoperiod on Birds and Mammals
  - Examples of Breeding Cycles
  - Effects and Applications of Natural Photoperiods
  - Exceptions to Photoperiodic Control
- The Mechanism of Photoperiodicity
  - A Sequence of Physiological Events
  - The Primary First-Action
  - Pigments and the Mechanism of Photoperiodicity
  - Anatomical Pathway of the Light Stimulus
  - Light and Pineal Function
  - Anatomy and Innervation of the Pineal
  - The Pineal, Daily Time of Activity, and Seasonal Events
- Human Photoperiodicity
- The Influence of Moonlight
- The Special Case of a White Coat Color
  - Photoperiodic Control
  - White Coats, Latitude, and Photoperiod
- Experimental Continuous Light
- Natural Continuous Light
- Natural Continuous Darkness
  - Solar Absence
  - Ultrasonic Sounds and Hearing
- Experimental Continuous Darkness

2. THE INFLUENCE OF THE RADIANT ENVIRONMENT

INTRODUCTION

In earlier times, students and scientists limited their consideration of the radiant environment to the effects of ultraviolet, visible, and infrared light. Now, the longer and shorter electromagnetic waves must also be considered: Cosmic, X-ray, Hertzian, and Radio (see Environmental Diagram A, *The Radiant Environment,* on a preceding page). Although this book emphasizes natural environmental extremes, this chapter will also include some mention of artificially produced electromagnetic waves. First we will discuss some radiation physics; then we will consider the effects of electromagnetic wavelengths, which are far removed from the visible spectrum. Finally, we will present the major topic of this chapter: the visible and near-visible spectrum and its effect on mammals and birds. We will consider it as both an *experimental* stimulus, and more importantly, a *natural* stimulus acting upon the life history of both migratory and non-migratory higher vertebrates. Included will be such matters as the control by light over the seasonal laying down of body fat, breeding, and nest-building.

RADIATION DESCRIPTIONS

Radiations are described in terms of both wavelength, expressed as Ångström units (Å), and frequency, expressed in hertz (Hz). As wavelengths decrease, frequencies increase. In our Environmental Diagram, all measurements are given in wavelengths. The frequencies of some common and useful radiations are presented in Table 2-1 (Thom 1966).

PHYSIOLOGICAL EFFECTS OF RADIO WAVES

The dramatic physiological effects of X-ray frequencies (10^{19} Hz) or visible light (10^{14} Hz) are obvious. However, the responses are not as obvious if the frequency is lowered by a factor of about 10,000 (10^{4} Hz) to the microwave region. At high microwave power levels, heat is generated in tissues by thermal degradation of the absorbed energy. At low power levels, no heat is generated. For years, scientists in the former Soviet Union claimed the presence of physiological effects, such as the stimulation of nerve impulses, with use of even this low power. It was not until the late 1970's that some American scientists agreed upon the existence of this non-thermogenic effect (Dodge and Glaser 1977).

Are radar (microwave) radiations safe? Biological effects from the wavelength are not apparent until at least 0.1 watt of energy per square cm of tissue is applied. Large radar antennas usually have a maximum capability of 0.01 watt per square cm in the center of the radar beam at a distance of 152 meters (500 feet). It is unlikely that such beams will heat biological tissue to a danger point. The larger radar antennas to be used in the future will not have a linear increase in intensity of energy, since the greater energy output will be spread over a larger area. This information is important to biologists who use radar equipment to study the migration of birds.

X-rays will be discussed in a later chapter. While there is a natural source of X-rays in space, human-produced X-rays pose a problem below the atmosphere. What is our tolerance to these artificial X-rays? Some authorities say that a safe dose is 5 roentgens per year (see Chapter 9, *Aerospace Physiology*).

OVERVIEW OF EFFECTS OF LIGHT

Although our theme in this chapter includes the effects of light in the outdoor environment, we can nevertheless obtain a preliminary understanding of this influence by considering (1) the physics of light, and (2) the effects of light upon cells and tissues under experimental circumstances. This information will be useful later when we consider the effects of light (photoperiodicity) on the brains of higher vertebrates in the free environment. Mammals will be used as examples when possible; however, most of the progress in understanding photoperiodicity has been made by studying plants and birds.

Physics of Light

There are three essential points about the radiant energy of visible and ultraviolet light: (1) when a molecule (or atom) absorbs a photon, it becomes excited; (2) the absorption of a quantum of light produces only one activated molecule; and (3) the energy absorbed may be reradiated at the same wavelength (resonance), at longer wavelengths (fluorescence), or degraded to heat.

Table 2-1. Frequencies and Wavelengths of Common Types of Radiation

Radiation	Approx. Wavelength	Approx. Frequency	Uses
X-ray	1.000 Å	1019 Hz	X-ray diagnosis
Ultraviolet	0.001 mm	1015 Hz	Room Air Sterilization
Visible Light	0.010 mm	1014 Hz	Vision
Ultrashortwave Radio	10.000 cm	109 Hz	Microwave Oven, Radar
Mediumwave Radio	100.000 m	106 Hz	Diathermy

Shortwaves such as light and X-rays are expressed in Ångström units: 1 nm (nanometer) = 10^{-9} meters, 1 Å (one Ångström) = 10 nm, 1 Å = 10^{-10} meters or 10^{-8} cm (centimeters). (See Environmental Diagram.)

As we mentioned earlier, wavelengths of radiations are measured in Ångström units: 1 Å = 10^{-8} cm (10 Å = 1 millimicron). Light affects living cells by the impact of photons; these are quantums of light energy analogous to the electron. So important are present-day applications of the harnessing of photons that some say we are leaving the "Century of the Electron," and entering the "Century of the Photon."

Quality of the Light Stimulus

What do biologists mean by *light?* In the laboratory they usually study the effects of *visible white light*, while in the field they observe the effects of *natural daylight*. Seldom do researchers examine the *quality* of light (wavelength or color). For example, in insects, it is light in the blue region of the spectrum that advances or delays the response to a changing sunrise (Frank and Zimmerman 1969). Such research on animals is rare.

Laboratory investigators sometimes believe they are using visible white light when, in fact, they are not. Incandescent bulbs provide primarily the red part of the spectrum, while fluorescent bulbs produce the yellow-green portion. However, commercial light bulbs that mimic the total visible spectrum of the sun are available.

We know that UV light poses a hazard to living tissues, and there is also danger from blue light. A tubular quartz mercury light is germicidal in the UV region without being hazardous in the blue region; a quartz halogen bulb is hazardous in the blue region but not the UV.

Laboratory Studies on Exposed Cells

The effect of light on single cells and exposed tissues is important because this will be related to our later discussions of the penetration of light through the skulls of birds and mammals.

The history of botanical blue light experiments began when a worker in the 19th century placed a flask of red port wine between a growing plant and light from a window; the plant grew as well as before, but no longer bent toward the light. Later, in 1864, von Sachs demonstrated that the bending of plants toward light is stimulated by the blue end of the spectrum. When weakly protected cells are exposed to light, the cells are harmed or stimulated; there may be changes in protoplasmic viscosity and streaming, in electrical charges, in permeability, and in the colloidal behavior of proteins (Clayton 1970, 1971). These effects are so powerful that we can easily accept the following reactions to light of intact plants and animals: in plants, photosynthesis; in animals, the stimulation of photoreceptors and the discharge of nerve impulses. The responses of both to hours of daylight (photoperiodicity) is a topic that will receive much of our attention in this chapter.

The above results from the disciplines of cellular and molecular physiology were obtained by studying exposed cells and tissues; this information causes us to wonder how transparent, free-living animals such as some fish, crustacea, and nudibranchiae (transparent mollusks) are able to control the environmental stimulus of light falling on their internal organs.

Before elaborating upon the major topic of this chapter—the influence of hours of daylight on the life histories of mammals and birds—we must consider separately the powerful ultraviolet component of sunlight.

PHYSIOLOGICAL EFFECTS OF ULTRAVIOLET RADIATION

Origins of Ultraviolet Radiation

Animals in the free environment and people on vacation bask in sunshine. The sun's rays are a combination of ultraviolet, visible light, and infrared radiations. In this section we will discuss the ultraviolet; later we will consider visible light and infrared. These three radiations that originate from the sun must pass through our upper atmosphere. This upper layer contains the ozone protective shield, which is known to have been weakened in some areas, especially over the Antarctic (Fig. 2-1).

Ozone Protective Shield. The presence of ozone influences the quantity of ultraviolet light that reaches the earth. The layer of ozone (O_3) is found at about 25 to 50 km (16 to 30 miles) in the stratosphere. This layer absorbs or filters ultraviolet energy which, in excess, could be dangerous and even lethal. Should large amounts of this energy pass through, there may be effects on weather and climate, increases in mutation rates, and an increase in the rate of skin cancer in humans. Indeed, the dramatic increases seen recently in three forms of skin cancer (Glass and Hoover 1989; Gallagher et al. 1990) may be due, in part, to the diminishment of the ozone layer. This is discussed further in "Effect on Humans." Ozone also absorbs some of the visible and infrared wavelengths.

The ozone layer is about five km (three miles) thick, but several influences have reduced the amount of ozone. Some calculations suggest that the reduction could be as much as 7% due to a reaction with nitrogen oxides (Gleason 1993). These oxides are produced by SST airplanes and atmospheric nuclear explosions. In 1976, Maugh proposed that fluorocarbons from aerosol sprays were responsible for mopping up some of the ozone. In the early 1980s, many governments banned or limited the use of ozone-depleting chemicals, and by 1995 their steady accumulation in the atmosphere appeared to have halted; levels of most chemicals peaked in 1994 and declined 1.5% in 1995 (Montzka et al. 1996). These chemicals are chlorine and bromine, found in the chlorofluorocarbons, hydrochlorofluorocarbons, halons, and chlorinated solvents. One hundred and fifty nations, including the United States, have agreed to continue to limit the manufacture and use of these chemicals; if successful, we may one day see the ozone hole close.

The above information refers to the stratospheric ozone protective shield. The presence of stratospheric ozone tends to make the global climate cooler and more variable. A reduction in this ozone shield means an increase in CO_2, which in turn means a warmer glo-

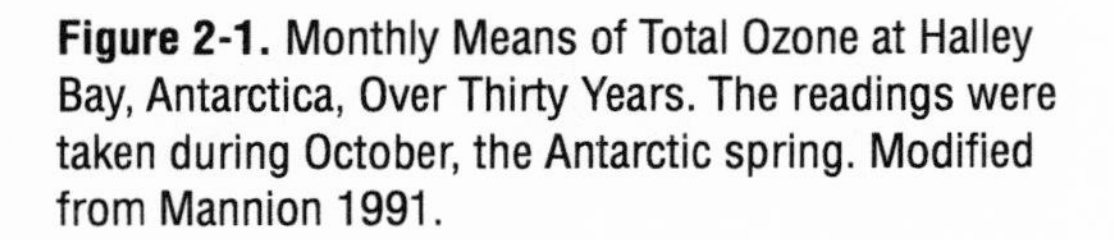

Figure 2-1. Monthly Means of Total Ozone at Halley Bay, Antarctica, Over Thirty Years. The readings were taken during October, the Antarctic spring. Modified from Mannion 1991.

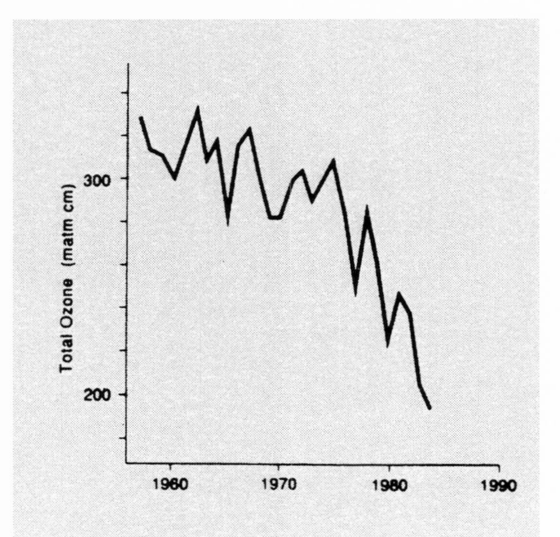

bal climate. Thus, the subtraction of ozone should add to the CO_2 effect. (For a further discussion of this topic, see Chapter 11, *Human Evolution, Resources, and Pollution.*)

In addition to stratospheric ozone, there is also tropospheric ozone, produced by human activity and found relatively near the earth's surface in all regions of the world. Human health is adversely impacted by this "sea level" ozone in the summer (Horvath and McKee 1994). Ozone at 1 ppm narrows airways in the lung, resulting in increased airway resistance. Urban air usually has 0.1 to 0.2 ppm ozone (Seinfeld 1986).

Variable Composition of the Sun's Radiation. The sunlight reaching the animal varies, depending on several factors: clouds, dust, the altitude of the sun, and the thickness of the ozone layer (which is thinner at the North and South Poles). All of these change with time of day and season. Cloudy days do not offer much protection from the sun's radiation; more than 70% of the burning power of the UV ray can penetrate through clouds to earth, even reaching three feet into water. At high terrestrial altitude, the thinner atmosphere is less effective as a radiation filter. Additionally, snow can reflect 85% of the sun's rays—all greatly increasing the chances of sunburn on a skier's face.

Window glass also influences passage of the sun's radiation. Of the three constituent bands of ultraviolet light only one, the longwave UVA, penetrates window glass. Yet many people in homes, offices, and hospitals, as well as experimental animals in laboratories, receive all their daylight through this glass. Thus, they may have a slight ultraviolet deficit, which may, in turn, influence their level of vitamin D.

Three Bands of Ultraviolet Radiation

Ultraviolet radiation is especially interesting because of its effects upon human skin: it produces *erythema* (sunburn), is responsible for tanned skin, and builds Vitamin D in the skin. These effects arise not only from the natural radiation of the sun, but from many human-generated sources as well.

There are three distinct bands of ultraviolet in natural radiation, all of which affect the animal *simultaneously*; however, these bands can be partitioned experimentally for research and clinical therapy by use of special bulbs and filters. These three UV radiation bands differ markedly in biological effectiveness: UVA (320 to 400 nm), UVB (280 to 320 nm), and UVC (185 to 280 nm) (320 nanometers = 3200 Ångström units). We will refer to UVA as *longwave*, UVB as *middlewave*, and UVC as *shortwave* or *far ultraviolet.* These ultraviolet components are shown in this chapter's Environmental Diagram A.

Each of the bands has different physiological effects and different photobiological "targets." These targets are all in the skin: they are the stratum corneum (the outer layer), the epidermis, the dermis, and (to be considered separately) the melanocytes, which are in all layers including the stratum corneum (see Fig. 2-2). It is important to emphasize that melanocytes can be completely without pigment.

Longwave UVA and Tanning. Let us first consider the longwave UVA. This is the most useful band, while UVB is very harmful. Fortunately UVA is many times more plentiful; when the radiation conditions are perfect, approximately 92% of our global ultraviolet radiation is UVA. This ultraviolet portion is approximately 5% of the total global radiation.

The primary target of the longwave UVA is the dermis layer, though UVA also affects other layers of the skin, as we shall see. There are three main types of physiological reactions: erythema, genetic damage, and tanning.

With overexposure, the effects of UVA may begin with the production of *erythema*; it takes a 500 to 5,000 times greater dose of UVA to produce this burn than it takes for UVB. This sunburn from UVA reaches a peak in two to three hours, while it takes seven to eight hours for UVB to burn. Skin burned by UVA does not blister, but skin burned by even low doses of UVB will blister.

UVA causes some *genetic damage* in the cells of the dermis, especially in the DNA; but the damage from UVB is much greater. Human skin, however, has an adaptation for protection against both of these radiations called "photoreactivation." This process consists of an enzyme-mediated, light-dependent repair of the DNA that has been damaged by UV.

On the beneficial side, UVA radiation affects the immune system

Figure 2-2. A Section of Human Skin. This illustration shows the several layers and other structures important in its function. Modified from Schottelius and Schottelius 1973.

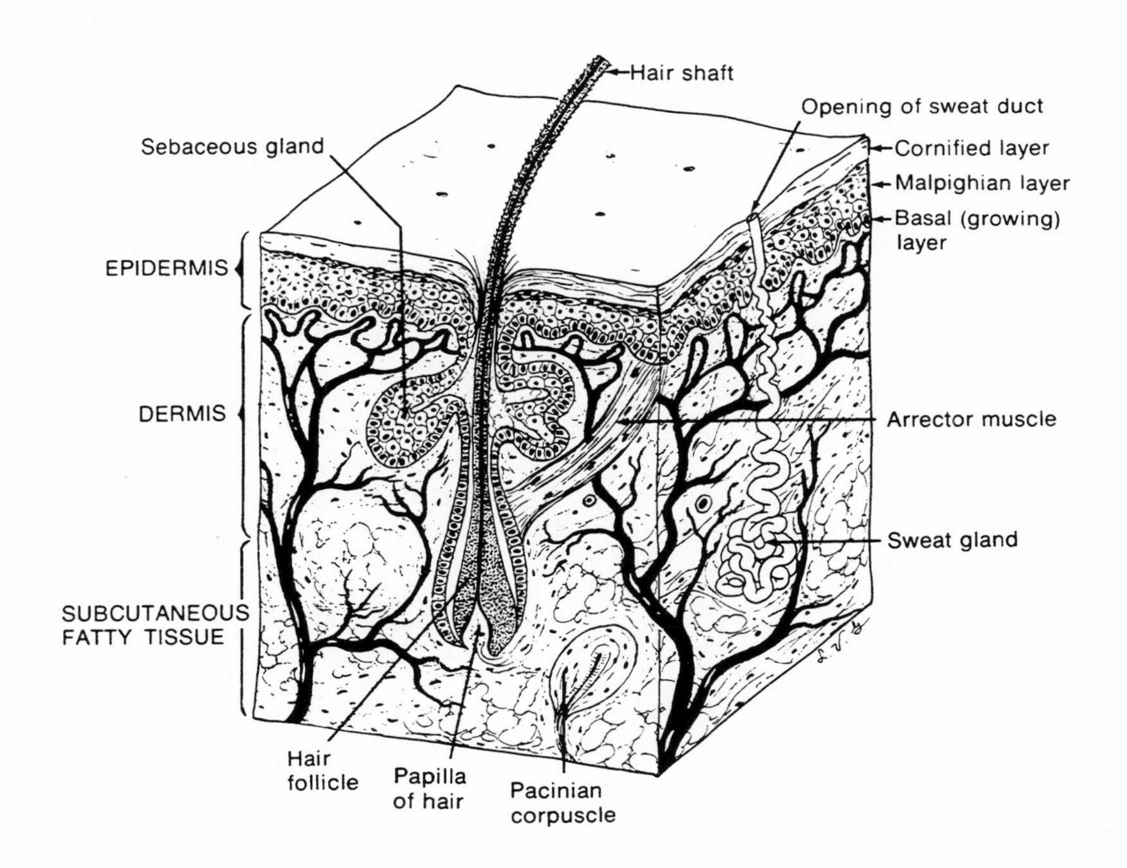

THE THREE BANDS OF ULTRAVIOLET RADIATION

UVA—Longwave (320 to 400 nm)

- Can tan without burning; the tan offers some protection against UVB
- Causes some genetic damage in dermis
- The most prevalent band; comprises 92% of total global ultraviolet

UVB—*Middlewave (280 to 320 nm)*

- Appears to raise blood levels of vitamin D
- Blisters skin
- Burns before it tans
- Damages epidermal cells
- Most carcinogenic band

UVC—*Shortwave (185 to 280 nm)*

- Germicidal band; destroys airborne bacteria
- Very little natural UVC occurs; together with UVB comprises only 8% of total global ultraviolet

by producing antigens within the epidermis. Furthermore, it produces changes in the superficial vasculature of the upper dermis, probably by inducing immediate release of vasodilators such as prostaglandins.

If the exposure to sunlight is long enough, tissues not protected by hair convert pre-existent colorless pigment granules into brown melanin. This *tanning* process happens in all races of humans, whether light-skinned or dark.

The conversion of pigment by UVA does not have to be preceded by sunburn. The tanning process is due to melanocytes in all three layers of the skin, including the stratum corneum. When more melanin is produced, less sunburn occurs. UVA can also cause delayed tanning by stimulating melanocytes to activate the L-dopa reaction which produces more melanin; this effect has a typical four to five day delay. UVA not only turns pigments brown, but causes new pigments to become visible. This pigment in the three layers protects the skin against the more harmful UVB. It is possible that even more protection is provided by a thickening of the stratum corneum.

Note that tanning is a form of acclimatization since it gradually protects the skin against sunburn.

Middlewave UVB and Shortwave UVC. With the above background on UVA, we now turn our attention to UVB and UVC. Only 8% of the natural global ultraviolet radiation is in the form of B and C. This is fortunate, since both are harmful and powerful. The shortwave UVC band (185 nm to 280 nm) is the germicidal band; this band can be produced clinically with special lamp bulbs, and is used to destroy bacteria. The reason that living cells are affected very little by natural sources of UVC is that most of the UVC is absorbed not only by ozone (as discussed above), but also by oxygen in the atmosphere.

UVB is another matter; this much more harmful band has its target effect upon the epidermis. The sunburn it produces is more serious and forms blisters. UVB cannot tan without previously causing sunburn. Fortunately, the tan from UVA provides some protection against UVB. In humans, UVB causes considerable damage to cells in the epidermis since its wavelengths span the photoabsorption spectrum of DNA. As a result, the carcinogenically effective dose of UVB is three times that of UVA (Sutherland 1981).

Effects on Sub-Primate Mammals

At first thought, one would consider that there is very little tanning or sunburn among mammals because of the protective effect of fur. Although there are quite a few species of hairless mammals, most of these are either fossorial (living within the soil) or nocturnal. However, even in the free environment, basking mammals are susceptible to sunburn and tanning on their ears, which are often white and hairless. In fact, many of the experiments that separated the effect of UVA and UVB have been done using the ears of both white and black guinea pigs as models because of their different pigment distribution. In the laboratory environment we find many hairless mammals, especially in the mutant strains of rodents. These strains would be useful for future work that is much needed in this field; there is very little information about the effects of UV upon immune systems, blood vessels, endothelial cells, perivascular cells, and connective tissue.

Effects on Humans

Many more people are being exposed to UV than ever before. Though the main reason seems to be an increase in voluntary sun exposure, others include a decrease in the amount of clothing worn, the popularity of tanning parlors (most of which use UVA), the use of liquid sunscreens that block only part of the UV spectrum, and the use of UV in clinical therapy. One result of all this exposure has been a dramatic increase in the three kinds of skin cancer attributable to the sun: basal cell carcinoma, squamous cell carcinoma, and malignant melanoma. The American Cancer Society (1996) predicts more than 850,000 new cases of sun-related skin cancer each year through the end of this decade. Though the majority of these projected cancers are of the highly curable basal cell and squamous cell forms, nearly 10,000 people per year will die from all forms of skin cancer. Malignant melanoma, the least common and most dangerous skin cancer, accounts for 7,300 of these projected deaths, with 38,300 new cases expected. Since 1973, incidence of malignant melanoma has increased at the rate of about 4% per year.

In the northwestern United States, investigators found that the incidence of sun-induced malignancies for all three forms of skin cancer has nearly tripled since 1960 (Glass and Hoover 1989). Another study in British Columbia showed similar increases for these cancers (Gallagher et al. 1990). Both these studies are significant for another reason: the geographical areas studied—Oregon, Washington, and British Columbia—are relatively low-sunlight areas, not "sunspots" where people are typically considered to be at risk for excessive sun exposure. Perhaps these increases are related to changes observed in the past decade in the ozone layer over North America.

Although most medical authorities consider that skin cancer is caused primarily by radiation from the sky, others believe that some of the increase must be from chemical carcinogens and human-produced radiation. We must consider as well that some of the dramatic increase seen in skin cancer rates may be attributable to a growing awareness of the problem and to improved detection methods.

Light-skinned people are at higher risk for skin cancer than are dark-skinned people: the incidence rates for malignant melanoma, for example, are over 40 times higher for whites than for blacks. But in addition to skin type, cultural factors that increase UV exposure may also contribute to increased risk of sun-induced cancers in people of all skin colors. Let us compare the exposure of people living in so-called industrialized societies with that of people of the Kalahari Desert. At first it might be supposed that the normal ancestral condition of the desert-dweller is a daily exposure to UVA and UVB, an exposure even greater than the sun-worshipers on our beaches. A sunbather will soon become tanned enough to be protected in the same way as the desert native; neither group, protected by this "natural barrier," should be especially vulnerable to cancer. However, the industrialized populations experience far more UVA exposure than less-developed, desert-living populations because of the use of cosmetic-tanning parlors and the application of sunscreen preparations that block only part of the UV spectrum. Both the tan from the tanning parlor, and the filter provided by the sunscreen protect against UVB, with the result that the users stay in the sun much longer than the Kalahari native. After a few decades, the cumulative dose of UVA can be enormous. This may be a factor in understanding the epidemiology of skin cancer in developed countries.

Ultraviolet and Vitamin D

Another effect of UV is its influence on the production of vitamin D, with a maximum effect at the UVB wavelength of 2967Å. Interest in the blood levels of vitamin D in mammals goes back to the early part of this century, when it was shown that rickets (a vitamin D deficiency) in experimental animals could be cured by exposing them to UV light. The vitamin D blood level of experimental animals depends upon exposure to UV light, whether from direct effect upon the skin or from the licking of fur by the animal. Perhaps there has been an overemphasis on the contribution of the skin; some workers suggest that because vitamin D is also manufactured in the fur, some animals may obtain the vitamin by licking it off their fur.

Early experiments of Quarterman and his associates supported the concept that the UV light acts on the skin to change blood levels of vitamin D. Their experiment showed that if sheep were without fleece (shorn), as sunshine (measured in 10 day units) changed from an average of 2 hours/day to 10 hours/day, the blood level of vitamin D increased by 350%. Specifically, when sunlight was low, the blood value was 20 i.u./100 gm, and when high it was 70 i.u./100 gm. However, at the same times, the blood value of unshorn sheep hardly increased at all. The process is that 7-dehydrocholesterol in the skin is converted to previtamin D_3, which is then converted by a transport protein to vitamin D_3. This vitamin is known to occur in the body in several forms; one form is referred to as a hormone instead of a vitamin. We know that this hormone acts upon at least eight organs of the mammal because specific receptors for the hormone have been found in all eight.

In an earlier section, we discussed the effect of UV on free-living mammals; the question arises as to the source of vitamin D in fossorial and nocturnal mammals. They must obtain it from food or manufacture it, since UV does not fall upon their skin.

The use of UV-irradiation has received considerable attention in the former Soviet Union, presumably because of its relation to vitamin D. Prophylactic UV-irradiation has been used there in industry and in institutions, and experimental work was done to describe the difference between continuous and intermittent use of UV-irradiation. Intermittent irradiation had a more pronounced biological effect than continuous. The types of indices used concerned plasma coagulation time and phagocyte activity.

PHYSIOLOGICAL EFFECTS OF VISIBLE LIGHT: DAYLENGTH

We must now turn our attention to visible sunlight because it usually accompanies UV radiation. Of all the sun's shortwave (visible) radiation, only 47% reaches the earth as direct, scattered, or diffuse radiation; the rest is reflected back into space from the earth, clouds, and atmosphere. This earthly light component (the 47%) drives the food cycle of the world and acts as a stimulus to living tissue.

Duration and Intensity Compared

Biologists define *photoperiod* as "hours of daylight." The light stimulus, whether outdoors or in experimental environments, varies in two dimensions: *duration* (length of daylight) and *intensity* (actual quantity of energy measured in gram-calories or foot-candles). For studies with either duration or intensity, there are two terms applied for the same light-stimulus phenomenon: *photoperiodism* for plant experiments, and *photoperiodicity* for animals. Thus, any biological phenomenon is *photoperiodic* if it changes systematically with an incremental increase or decrease in either duration or intensity of light. Note that during seasonal changes in daylight, both duration and intensity are modified.

Changes in duration of daylight affects mammals in a variety of ways, including changes in body fat, hair coat length, and, in some, reproductive cycles. Photoperiodic changes in intensity (quantity of energy instead of changes in exposure time) have been shown to alter sequences of activity. Consider experiments involving birds or mammals maintained in continuous light: their time of starting daily 24-hour sequences of locomotor activity was altered systematically by incremental changes in light intensity (see Aschoff's rule in a later discussion).

Plant experiments serve as a model for illustrating changes in duration of photoperiod; actually they were begun long before any animal experiments were undertaken in this field. Furthermore, many investigators of animal photoperiodicity refer to plants in their interpretations for two reasons: (1) the analysis of the *mechanism* of light-effects depends upon those data from plants that were obtained using flashes of light and the determination of pigment changes; and (2) the students of evolution who interact with the photoperiodists use the power of photoperiodic phenomena in their arguments, lumping together support from both plant and animal data (Giese 1979).

Responses of Plants to Daylength

Length of daylight is all-important in photoperiodic control of the life history of outdoor plants, while intensity of light and variable ambient air temperature have little effect. Thus, plants rely on a stimulus reproducible each year, while avoiding complete dependence upon *sun hours* (i.e., hours of sunshine), which vary more than 16% from year to year, causing light intensity to vary. Plants respond to predictable *daylight hours* rather than to unpredictable sunshine.

Early theorizing about the influence of daylength began in 1852.

By 1912, Ternois had shown that hemp and hops will flower if daylight is shortened to 6 hours. The first two authoritative papers were written by Cajlachjan (1936), and Garner and Allard (1920). Today, knowledge of plant photoperiodism permits the following classification:

- long-day plants which bloom with more than 12 hours of light per day;
- short-day plants which bloom with less than 12 hours; and
- plants of wide tolerance with light not a limiting factor.

Duration of daylight may control not only blooming, but also (1) germination of seeds, (2) growth of entire plant, (3) formation of bulbs, (4) shape of plant (rosette or spreading), (5) autumn falling of leaves, and (6) sex of individual plant.

Daylength is not the only factor in both animal and plant photoperiodism, but the *night length* (either alone or with a changed daylength) may be important; an easy experiment is to interrupt the dark period by a flash of light. Experiments with the cocklebur showed that after one exposure to a *long night* (LN) the plants would flower, but a single flash of light in the dark period would *inhibit* flowering (Fig. 2-3). In other species, the flash of light may promote flowering (Melchers 1952).

MECHANISM OF PLANT PHOTOPERIODISM

This sensitive control over the growth process of plants by use of a flash of light reminds one of the chemistry of photographic processes. Photochemistry is certainly involved in the growth process, since we know that one artificially darkened leaf on a lighted plant can control the flowering that occurs at some distance from the leaf.

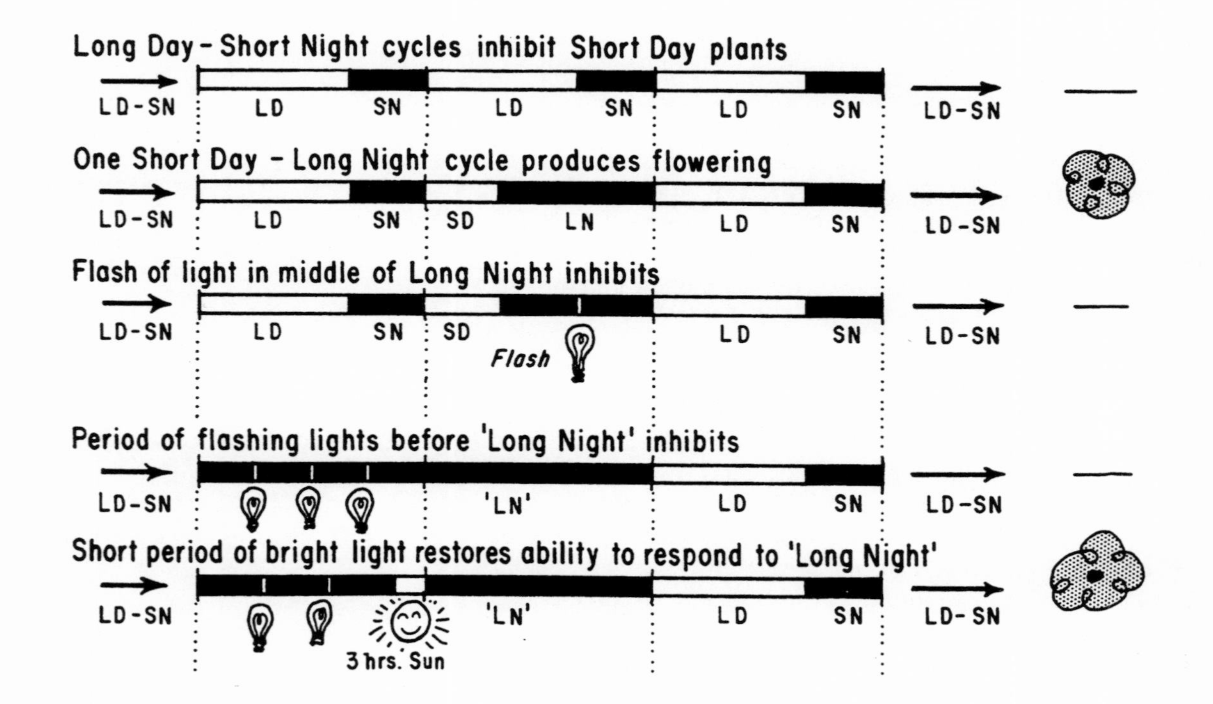

Figure 2-3. Photoperiod and Short-Day Plants. The kinetics of the interplay of light and darkness with short-day plants. These experiments on the cocklebur show the persistent after effects of one long night of photoperiodic treatment, resulting in subsequent flowering under formerly unfavorable conditions. The single long-night period must be linked with an appropriate preceding period of high intensity light. Modified from Hamner 1940.

Due primarily to experiments by Hendricks, we now know the initial photocontrols through which changes in light cause plants to respond, but some of the compounds (pigments) and enzymes involved are still to be found. Hendricks' description of this photoreaction is found in Table 2-2.

The pigment (P) is now called phytochrome; apparently it is a protein which exists in a growth- and flowering-inhibiting form which is converted to a noninhibiting form. For short-day plants, if there is sufficient darkness, then a flowering hormone is produced stimulated by phytochrome. If there is inadequate darkness, then a modified phytochrome *inhibits* flowering. This photoreaction is compared to that found in birds and mammals later in this section.

The contribution of classic experiments on plant pigments to our understanding of animal photoperiodism is presented as an example of interdisciplinary sharing of information; the use of a flash of light during the dark period is a technique employed in both plant and animal studies. It is not possible here to review the modern experiments with plant pigments. The student may want to read further about the xanthophyll cycle, zeaxanthin, and lutein.

Natural Photoperiods and Geographical Variations

As we turn our attention from plants back to animal responses, recall that the theme of this book includes the impact of extreme environments. Is a change in daylength an environmental extreme? Actually this contrast between daylight and darkness is an extreme which causes free-living mammals to either avoid light (nocturnal types), or to conceal themselves during darkness (diurnal types). The term *daylight* refers to daylengths which vary from 6 to 18 hours or longer. Some appropriate vocabulary to apply to these hours of daylight in the free environment follows.

The two most conspicuous events of winter and summer are the longest day and the shortest day of the year. They occur at the two times of the year when the sun is at its greatest distance from the celestial equator; they are the *summer solstice* (about June 21) and the *winter solstice* (about December 21). An example of the summer solstice for a region which is fairly far north, Scotland, would be

Table 2-2. The Reactions of Plant Photochemistry.

P	+	RX	RED RADIATION ↔	PX	+	R
Pigment absorption maximum 6400-6600Å		Another reactant, called "flowering hormone" by some	FAR-RED RADIATION OR DARKNESS	Pigment absorption maximum 7200-7400Å		Changed Reactant

when the sunrise is at 0400 hours and the sunset is at approximately 2200 hours (an 18-hour day).

Another conspicuous event occurs twice each year at all latitudes: a day with 12 hours of daylight and 12 hours darkness. These two occasions happen when the sun's distance from the equator causes night and day to be of equal length in all parts of the earth (Fig. 2-4). These two times are called the *vernal equinox* (about March 21), and the *autumnal equinox* (about September 22). This 12-hour photoperiod is the one most frequently used in laboratory experiments.

We have been discussing modest quotas of daylight as observed by field biologists in the temperate zones; these photoperiods are amplified near the two geographic poles. As a result, in the northern hemisphere, millions of people and their domestic animals experience two or more months of continuous light, and then, several months later, a similar period of continuous darkness. A detailed description of the daylight changes in one such community, Tromsø, is found in Environmental Diagram B at the beginning of this chapter. The stages of the photoperiodic change from the equator to 70° of latitude are also shown.

Although we often refer to the "continuous darkness of winter" in the polar regions, a more accurate term would be *solar absence.* Near noontime on many winter days north of the Arctic Circle and in Antarctica, the amount of light is more like that of twilight in the temperate zone.

The intensity of the daily photoperiod varies with the latitude. In midsummer at noon in the shade, the light under a north temperate evergreen forest, a tropical rain forest, or under 100 meters of ocean water measures 40 foot-candles, while at the same time in direct sunlight in a north temperate area, the measurement is 9000 foot-candles.

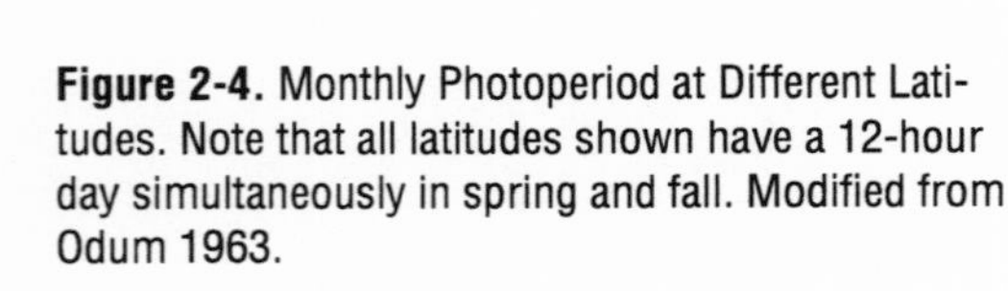

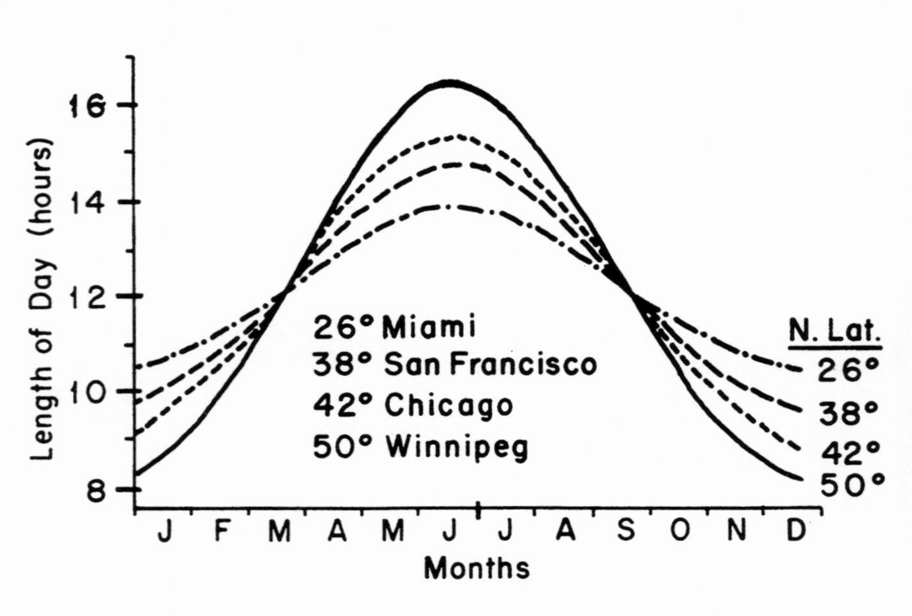

Figure 2-4. Monthly Photoperiod at Different Latitudes. Note that all latitudes shown have a 12-hour day simultaneously in spring and fall. Modified from Odum 1963.

Varied Effects of Photoperiod on Birds and Mammals

The influence of hours of daylight on higher vertebrates has two aspects: one is the effect of photoperiod on the migration, seasonal physiological changes, and reproduction of vertebrates; the other relates to daily physiological rhythms, also referred to as circadian rhythms (which means "almost 24-hour rhythms"). Light is one (but only one) of the factors that can influence and change these circadian rhythms. This is such an important and specialized matter that the entire next chapter will be devoted to the environment-rhythm relationship: *Biological Clocks and Rhythms* (Chapter 3). Nevertheless, some attention must be given in the present chapter to the extremes of continuous light and darkness as unrelated to biological rhythms.

The various effects of photoperiod on birds and mammals include (1) seasonal color and density changes in hair or feather coat, (2) initiation of migration, (3) readiness to reproduce, (4) physiological preparation for winter, and (5) changes in type of nest built. Some mammalian photoperiodic responses have been confirmed by experiments (see Table 2-3).

An experiment with ferrets in Australia included six physiological variables which either increased or decreased in magnitude with an increase in photoperiod (Fig. 2-5).

There have been surprisingly few studies of day-active rodents, such as the gray squirrel. McElhinney and co-workers (1996) showed, by activity recording and body temperature measurements, that the Nile grass rat was day active; mating took place in daylight, but most litters were born in darkness. In contrast, the Ethiopian white-footed rat is strictly nocturnal. Interestingly, and rare among free-living mammals, this rat breeds at all months of the year (Bekele 1996).

Different environmental signals influence reproduction in mammals inhabiting tropical regions. In a long photoperiod and high temperature, the palm squirrel is expected to show active testes, but when these variables are combined with high humidity, then the pineal-hypothalamus-hypophyseal axis calls for complete testicular regression (Haldar 1996).

To give an example from birds, starlings maintained under 15 hours of light had testes four times the size of those kept under 8 hours of light (Fig. 2-6). A few investigators have studied the specific wavelength of light that is effective, beginning with the classic work of Bartholomew on English sparrows (1949). In insects, light in the blue region of the spectrum advances or delays the response to a changing sunrise (Frank and Zimmerman 1969).

An unusual effect of a natural photoperiod is described in the control over torpor in hummingbirds. When in daily torpor (blood tem-

Table 2-3. Some Mammalian Reproductive Responses to Photoperiod.

SPECIES	RESPONSES: Short Days and Long Nights	RESPONSES: Long Days and Short Nights
Mink	Mating and delayed implantation	Non-reproductive growth
Northern weasel	Non-reproductive growth and ermine coat	Mating
Domestic cat	....	Non-seasonal estrus
White rats on poor diet	....	Higher incidence of mating
Sheep	Mating	Bearing of young (6 mo. gestation)
Horse	Non-reproductive growth and winter coat formation	Mating and bearing of young (11 mo. gestation)

perature near environmental temperature), they awakened when light stimulated them, and did not awaken when maintained in darkness (Warncke 1994).

Examples of Breeding Cycles

Comments on the normal physiology of reproduction are included here because this system is unlike other systems of the mammalian body in that it shows great variation in design. Some physiological systems are consistent. For example, if you stimulate the vagus nerve of either an elephant or a 20-gram shrew, in both cases the heartbeat will slow down or stop. The physiology of reproduction does not show this type of consistency. For example, some bats will ovulate one egg from one ovary one year, and one egg from the other ovary the next year. Some insectivores, however, will ovulate twenty or more eggs simultaneously from each ovary.

We must discuss the word *estrus* (oestrus), which has its origin from the Greek *gadfly* or *frenzy*. It is used in the sense that the female comes into estrus or "heat," which means sexual receptivity. There may be an extremely large expenditure of energy at this time by the female, resulting in a running reaction. For example, a female white rat may run 500 revolutions per day in a wheel for three or four days, but on the night of estrus may run 10,000 revolutions (34 km/21 mi). This mechanism results in the female covering a good deal of ground in the free environment, thereby increasing the possibility of coming in contact with a male so that breeding can take place. Indeed, many horseback riders have experienced the exploits of the mare in heat. The occurrence of estrus has great variability; in the white rat, it is every four or five days, while in very large mammals it may be once a year or less often.

Figure 2-5. Photoperiod and Seasonal Changes in Ferrets. Correlation of changes in daylight with those attributes of male ferrets which are governed by daylight. Note that in Australia, day length increases from June to December. Modified from Harvey and Macfarlane 1958.

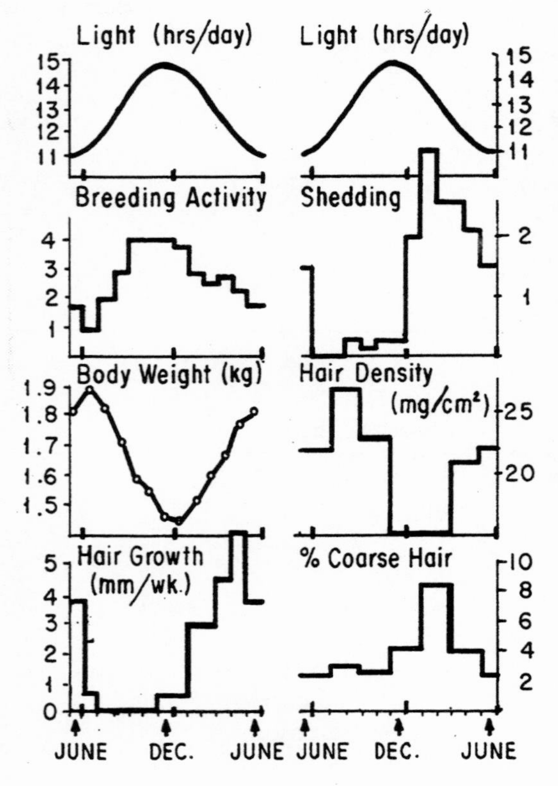

Other examples of unusual reproduction are found in colonial or troop mammals. The hairless mole-rat of Africa inhabits a subterranean climate; in a large colony only one female bears young (as in the honeybee colony). The rest of the colony gathers food and does the digging. Likewise, in a troop of common marmosets, one female bears the young, and the other females help to raise this dominant female's offspring. The neuroendocrine systems of the female companion marmosets suppress their own fertility. Yet in the case of the golden lion tamarin, the fertility of the subordinate females is not suppressed; it is aggression by the dominant female that prevents them from mating. In the marmosets, the mechanism of suppression is as follows: the hypothalamus normally produces gonadotrophine-releasing hormones (GnRH). These in turn cause the pituitary to release other hormones that foster the development of eggs in the ovary. It is GnRH which is suppressed by the behavior of the dominant female (Black 1995).

Even more unusual is reproduction in the armadillo, which also is controlled by photoperiod. Storrs reports that this species delays implantation of the fertilized ovum; however, if the female is stressed, such as by being brought to captivity from a free environment, she may delay implantation for the uniquely long time of two years. The armadillo ovulates only one ovum, which always divides in four to provide identical quadruplets.

The effect of natural photoperiods on horses seems to be unusual. In most breeds, the mare's estral cycle is every 21 days from March through September, with estrus (heat) lasting 5 to 7 days. The periods of estrus in August and September are shorter (3 to 5 days), and the last estrus in September is often characterized by behavioral receptivity to the stallion, but without ovulation. Mares in temperate regions do not have estral cycles from about September through February, but estrus can be induced in January and February by increasing the daily photoperiod.

There are two types of ovulation: spontaneous and induced. An embryologist who wants to collect ova from the oviducts of mice, for example, can usually find them spontaneously present every four or five days. In other species, such as rabbits and cats, ovulation must be induced by the breeding process itself. Breeding cycles and types of ovulation for eight species are found in Table 2-4.

One species of lemur will breed on only one day a year. Most species mate when days are shortest, but if the light does not vary, they will not breed at all.

Light controls breeding cycles and the preparation of the uterus for pregnancy. In monkeys there is elaborate preparation of the uterus to receive the fertilized egg, with a thickened bed of tissue and ample circulation. It can be said that the normal condition for monkeys is pregnancy; in the free environment the female will usually become pregnant, and this preparation of the uterus is not ordinarily wasted.

This same mechanism has been inherited by humans to the extent that there is a monthly preparation for pregnancy, although most of the time, pregnancy does not occur. This preparation is the phenomenon of menstruation whereby in a somewhat inefficient process the soft bed that has been prepared for the fertilized egg is sloughed off. In an extreme political response, one of today's civilizations has attempted to control reproduction in the human female so that there is only one child per family: one fertilization per lifetime compared to, in the monkey, a pregnancy for every ovulation.

Nursing by mammals controls the timing of the next pregnancy after the light cycle determines breeding and pregnancy. Is nursing a socially constructed attitude? *Social constructionists* view maternal care as a learned and indoctrinated drive, while *essentialists* assume that females are genetically programmed to do essential nurturing (Hrdy 1995).

The composition of mother's milk varies with the habitat of the mother and the demands of survival (Hayssen 1995; Iverson et al. 1995). The black bear has the highest milk fat content of any land mammal. Note the variations among some species:

Figure 2-6. Photoperiod and Starling Reproduction. The condition of the testes of male starlings after maintaining two groups from December 1st to April 1st under two different photoperiods. The testes of four birds from the short daylight group and three birds from the long daylight group are shown.

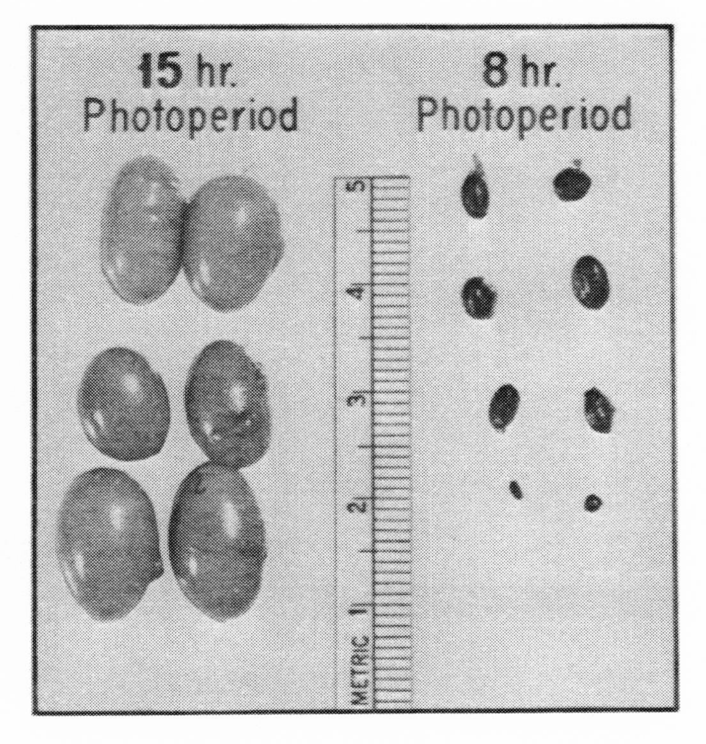

COMPOSITION OF MILK

	Water	Fat	Sugar	Protein
Cow	88%	4%	2%	6%
Human	88%	4%	6%	2%
Black Rhino	-	0.2%	high	low
Hooded Seal	-	61%	-	low
Black Bear	-	64%	1-3%	24%

Effects and Applications of Natural Photoperiods

The natural photoperiod varies with latitude so that in Miami, Florida the time of sunset has a maximum annual change of about three hours, while in Winnipeg, Canada the change is closer to eight hours (refer again to Fig. 2-4). Thus, Florida's local mammal species will have a different response to photoperiod compared to those in Canada.

It is sometimes difficult to extend laboratory observations on photoperiodicity to animals in the free environment. Extensive and fundamental work has been completed by Fuller and Stebbins (1969) at Hay River in northern Alberta on several species of rodents living freely in the forest or in enclosures in the forest. The photoperiodic stimuli on these animals do not fit our ordinary expectations: these animals had high intensity photoperiodic stimuli in winter from the lack of leaves and the presence of snow, but had low stimulation in summer from the presence of leaves and lack of snow. The situation is further complicated because one must measure the influence of light upon the animal when it is living much of the time under the snow.

Other complications include the type of reproduction of the animals in the free environment. For example, photoperiodic stimuli are known to control delayed implantation (also called seasonal embryonic diapause) in the weasel and deer families. The armadillo, black bear, and seal are also members of this exclusive list. As an example consider the European badger (of the weasel family), which

Table 2-4. Sexual Receptivity (Estrus) and Ovulation in Mammals.

		BREEDING CYCLES	
	Estrus	Gestation* (days)	Ovulation
Mice	4 1/2-5 days	17-21 (19)	Spontaneous
Rat	4-5 days	20-22 (21)	"
Hamsters	4 days	15-18 (16)	"
Guinea Pigs	16 days	59-67 (63)	"
Rabbits	Induced	29-33 (31)	Induced
Cats	15-28 days	52-69 (63)	"
Dogs	Biannually	58-67 (63)	Spontaneous
Monkeys	28 days	155-180 (164)	"

*The average figure is shown in parentheses. There will be some variation according to species and breeding systems.

mates during an estrus that occurs right after giving birth to the young. It then keeps the blastocyte free for about 300 days before implantation; thus its total pregnancy lasts almost one year (Canivenc 1968). This is a difficult area of study because the long gestation period requires much more sophisticated techniques of studying the physiology of reproduction. (Several mammals of different species are mentioned above; the reader is urged to read for interest the groups or orders of mammals listed in Appendix A.)

Natural photoperiods can be markedly fractionated and will still attain the desired effect (see "Skeleton photoperiods" in the next chapter). For example, what light cycle produces the most eggs in a commercial hennery of 60,000 birds producing 47,000 eggs per day? After the hens are mature, the lights need to be on for only 15 minutes per hour during the 16-hour long "waking" day.

Similar work was done with domestic pigs in Siberia (Klotchkov et al. 1971). In one experiment, a prolonged 17-hour day in autumn or winter was used, both before breeding and through gestation; in another, the prolonged day lasted through gestation alone. In both cases, the prolonged day resulted in (1) increased litter sizes and weights at birth, (2) decreased numbers of stillborn, and (3) increased weight during growth of young pigs. The explanations are two: under the increased illumination there were more and bigger corpora lutea in the pregnant pigs, and fewer embryonic deaths and resorptions. In contrast, the control pigs had a greater than 50% mortality of embryos.

The light cycle affects the increased metabolism associated with cold-exposure. Lynch (Lynch and Folk 1970) has reported that he could obtain with warm-acclimated deer mice "cold acclimatization without cold" by means of a light cycle of LD 9:15 (Light 9 hr/Darkness 15 hr per day). He also studied their nesting while varying the T_a (T_a = environmental or air temperature) and photoperiod; when he decreased photoperiod for warm-acclimated deer mice at room temperature, they built nests 80% larger (Lynch 1972). In another study, a short photoperiod induced greater cold resistance for domestic mice than did a long photoperiod (Ferguson 1979). The endocrine influence for this resistance was studied by Pistole and Cranford (1982); thyroid activity was lower in wild meadow mice maintained under a short photoperiod than in those kept in a long photoperiod.

Above we mentioned "cold acclimatization without cold." Under some circumstances, this cold resistance is not acquired by photoperiod alone. Zegers and Merritt (1988) found that deer mice needed both low temperature and short photoperiod to experience an elevated metabolism.

Some of the photoperiodic influences on free-living mammals are very subtle. For example, several decades ago, Pinter and Negus reported that mean litter size of *Microtus* was significantly larger with 18-hour illumination compared with 6-hour.

We must now ask whether artificial and natural short photoperiods have the same effect. When two groups of hamsters were both maintained under 7 hours of daylight, with one group outdoors and the other in the laboratory, both groups showed reproductive collapse, but the outdoor group exhibited this significantly sooner (Vaughan, Brainard, and Reiter 1984)

Exceptions to Photoperiodic Control

With some mammals the photoperiodic response is independent of temperature (weasel), but with others the response can be prevented by cold (sheep). Some mammals do not respond to light changes, but to other environmental factors such as temperature or diet, which may trigger growth of reproductive organs and mating. Under what conditions is it an advantage to a species to abandon photoperiodic control? Some of the 200 ungulate species now have breeding and gestation periods that are timed only for parturition at the time of clement weather (Fraser 1968).

One must also take into account an *internal* reproductive rhythm—perhaps lasting about 12 months—that would follow its course even if the physical environment were unchanging. This inherent annual rhythm may be accelerated or retarded by changes in environmental factors, or it may be unaffected by them. The term, *circannual*, is now applied to this annual rhythm.

Examples of independent inherent rhythms are not difficult to find. One need only keep male and female frogs in hibernation in a refrigerator at 3° to 4°C all winter. When spring comes 8 months later, the frogs, still in the constant conditions of the refrigerator, will spawn. As early as the mid-thirties, Baker and Bird listed many species of plants with inherent rhythms: the most interesting is a carefully studied specimen of the plant, *Breynia cornus,* which flowered about every 5.5 months for 11 years.

Figure 2-7. Photoperiod and Rate of Testes Development. The graph shows the effect of day length on the growth of the male reproductive organs of migratory birds. The fully functional state for production of mature sperm cells occurs much sooner if the bird is in a long-day situation. Modified from Wolfson 1959.

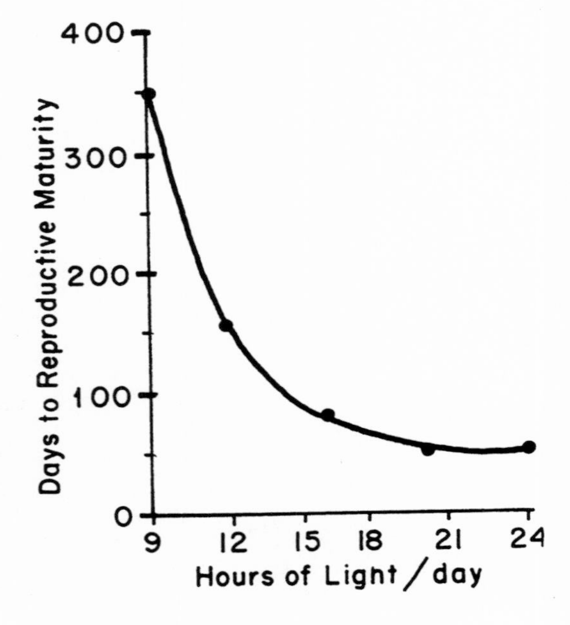

At about that same time, Marshall was the first to suggest that more birds than mammals have freed themselves from the external seasonal influences. The classic example is that of the sooty tern, which has a breeding season about every 9 months. On the other hand, this internal reproductive rhythm is quite weak in some higher animals. The white-crowned sparrow, with which Farner developed some of the principles of animal photoperiodicity, is among the most completely photoperiodic of any species. When this bird is maintained on short daily photoperiods of 8 hours or less, there is no measurable gonadal development for at least a year. Wolfson contributed similar advances using a bird with a stronger internal reproductive rhythm, the slate-colored junco (Fig. 2-7). In the junco, testicular regression and refractoriness can be brought to a maximum by 18 hours of darkness per day. Interestingly, Wolfson was able to bring this about in both July and September.

The internal reproductive rhythm is found in both Syrian and Turkish hamsters. When exposed to short days, the gonads of both sexes become nonfunctional due to a decrease in serum LH (luteinizing hormone) and FSH (follicle stimulating hormone). After 20 to 25 weeks of short days, gonadal function is reinstated "spontaneously" under the demand of the internal reproductive rhythm (Goldman 1980).

Two examples in mammals of reproductive cycles that do not depend upon light now follow. First, a plant factor (designated as 6-MBOA) triggers reproductive activity in a vole, *Microtus montanus*, in the free environment. In the laboratory, this plant derivative also increases fertility of this species when made available in its diet (Sanders et al. 1981).

A second example is found in another rodent that lives entirely underground in several countries in Africa: the naked mole-rat *Heterocephalus glaber,* which is a rodent and not an insectivore (see Appendix A). Apparently the inherent rhythm of reproduction is important here because breeding takes place four times over the year without a specific season. Surprisingly, in a colony of 40 or 60 of these rats, there is only one breeding female. There are, however, three castes referred to as "frequent workers," "infrequent workers," and "non-workers." It is suggested that this rodent is the only known vertebrate which qualifies as a eusocial animal (Jarvis 1981); other examples are bees and termites (Sherman 1991).

Even when photoperiodicity is important in the life cycle of a species, there is individual variation in response. For deer mice in the free environment, Desjardins states, "seasonal reproductive performance is probably occasioned by one or more external cues, depending on photoperiod, temperature, food availability, and the individual's sensitivity to these cues."

THE MECHANISM OF PHOTOPERIODICITY

A Sequence of Physiological Events

The numerous photoperiodic responses tabulated earlier are the superficially evident ones. Let us look at some less conspicuous ones which form a chain of events in migratory North American animals. Typical North American migratory birds demonstrate a sequence, first identified by Wolfson, which seems to be triggered by changes in daylight:

1. fat is deposited in wintering birds in the south (experimentally it is possible within five days to change the composition by weight of the bird from 25% fat to 50% fat);
2. the day-active bird becomes nocturnal;
3. northward migration occurs;
4. spring breeding takes place;
5. there is a period of molt and inactivity;
6. fat is deposited again; and
7. southward migration occurs.

These events have been reproduced in the laboratory. When migrant German birds were kept in a large circular cage in a short photoperiod, the birds huddled at the south side of the cage; when kept in a long photoperiod, they fluttered at the north side of the cage (McMillan, Gauthreaux, and Helms 1970).

Most of the seven steps of the sequence mentioned above probably apply to migratory mammals, although at this point the effect of light is conjecture. Examples of such mammals are sea otters, seals, caribou, and whales. During the longest migration of any mammal, the gray whale travels 6,000 miles from its cold summer feeding ground to the warm Baja California lagoons for breeding and parturition. A few mammals add an interesting element into the sequence: spontaneous fasting (hypophagia). The humpback whale stops feeding for 4 to 5 months (Folk 1980). The male fur seal fasts for 6 weeks. In some mammalian species, fasting is preceded by an increase to 150% of summer weight.

The Primary First-Action

To understand the internal mechanism responsible for the photoperiodic responses in birds and mammals, we must decide which of these physiological "events" could be primary. For example, perhaps it is the depositing of fat. This event was referred to by Hendricks as the "first-action in the animal arising from a recognized property of the environment." The question also arises whether the hypothetical primary internal event alone is light-triggered, or whether the same event is temperature-triggered in nonphotoperiodic animals.

The mystery of the "primary first-action" must now be within our power to solve, because models of bird migration are being successfully developed. Above we described an experiment in which wild birds were kept in circular cages; these experiments have also been done under an artificial planetarium sky. When a long or short photoperiod was used under these circumstances, the birds did not orient to true north or south, but to an artificial north or south star pattern, no matter where the pattern was placed.

Nevertheless, several investigators have suggested that the migratory restlessness does not entirely depend upon photoperiods. When researchers produced a changed internal physiological state, the birds changed their direction of migratory restlessness (McMillan, Gauthreaux, and Helms 1970). How then do we decide what is the "primary first-action"?

Evidence for one type of primary first-action has now been provided by Meier (Meier and Fivizzani 1980). He was working with prolactin, which we may consider an ancient hormone since it can change the amount of body fat, even in fish. When Meier worked in the laboratory of Farner, he injected a pulse of prolactin into white-crowned sparrows. This resulted in a triad of effects including migratory activity, increase in body fat, and increase in gonadal size. When Meier moved to Kentucky, he repeated the experiment with white-throated sparrows and obtained opposite results, namely a quiescence in behavior, a loss of body fat, and regression of gonads. He then obtained some white-crowned sparrows and found exactly the same results. After about two years of puzzling over the matter, he found that the difference was due to the time of day of the injection. If the birds were injected four hours after dawn, one effect was obtained, but if they were injected eight hours after dawn, the opposite effect was obtained. He then tried pulses of corticosterone and found that this had an important relationship to the pulse of prolactin: the amount of time between the two injections controlled the physiology and behavior of the bird. Experiments were done with four hours, eight hours, and twelve hours between injections, and the results were different in every case. It is possible that the corticosterone is relatively stable, and that throughout the year the moving, or floating, or variable timing is found in the pulse of prolactin. Apparently these pulses in nature are spaced correctly by changes in daylight. The most exciting part of the story involves the search for whether these pulses were found in birds in the free environment. They *were* found to be there. These results not only clarify our knowledge of the migratory urge, but they help us to understand the importance of pulses of hormones in regulating the physiology and behavior of mammals as well as birds.

Similar pulses are important in nocturnal rodents; early pregnancy in rats is maintained by a surge of prolactin at "lights-on" and another surge at "lights off."

Even more important than prolactin in the mechanism of photoperiodism is *melatonin* (N-acetyl-5-methoxytryptamine). This hormone in the pineal gland pulses each day. There is a ten-fold variation in production of melatonin in chickens and rats from day to night (2ng in light:20ng in darkness) (Ralph 1976). The melatonin pulses are known to influence the time of starting activity each day in birds, and in mammals the pulses also have a role in controlling the annual sequence of breeding. Important though these hormone pulses are, however, it is probably the mechanism that *initiates* them which is the primary first-action.

Melatonin has also been studied in men and local mammals living in the Antarctic winter (75°S). The men adjusted by shifting the start of the pulse only (i.e. shifting the phase), but the outdoor animals changed their *duration* of melatonin secretion (Broadway and Arendt 1990).

Pigments and the Mechanism of Photoperiodicity

In the search for the mechanism of the photoperiodic response, it is reasonable to compare this phenomenon in plants and mammals. At first this seems to be straining a point, until we realize that the same photochemical-like experiments which were done with plants have also been duplicated on some species of birds and sheep. Some varieties of sheep come into estrus only when the days are short and nights are long, but this effect can be prevented by interrupting the night period with a flash of light as described earlier in plant experiments by Withrow. In hamsters, artificial short days and long nights inhibit reproduction, but if a flash of light is introduced in the middle of the dark period, the effect is to stimulate reproduction (Hoffman et al. 1981).

Even more specifically, Farner provided evidence from birds that an essential compound (perhaps a pigment) in the response mechanism is formed rapidly in light and disappears slowly in darkness. The supporting data were obtained by dividing the daily photoperiod into a number of short segments. Hendricks explains these photoreactions in terms of pigment activity. He postulates that the energy received by one pigment in the brains of mammals and birds induces the oxidation of another molecule; this chemistry is different from the reversible reaction found in plants. This oxidation-reaction is the cause of pathological sensitivity to sunlight (klamath weed toxicity in sheep and photosensitization by porphyrins in humans).

Anatomical Pathway of the Light Stimulus

Let us consider again the two types of photoperiodic action: it may involve (1) changes in annual reproductive and migratory activity; or (2) the ability of the mammal or bird to alter its time of beginning daily activity according to a change in dawn or sunset. The incoming stimuli of light and the action of pigments would begin along the same pathway regardless of the behavior which is triggered. The organs involved are the eyes, the hypothalamus, the pituitary, and the pineal gland. Probably all of these are part of the path of a *feedback system*. Perhaps some of them merely temper certain physiological reactions without actually exerting maximum control (Nalbandov 1970). The contribution of the hypothalamus to the pathway has been shown in birds by removing their hypothalamic portal system; there was then no further photoperiodic reproductive response (Farner 1961). A significant part of the hypothalamus, the *suprachiasmatic nuclei* (SCN) will be discussed later.

The contribution of the eye to the feedback system must now be analyzed. It is not clear whether it is nerve impulses initiated by light, or the direct action of light on cells, which is responsible for the accelerated or reduced responses due to changes in light. Probably neither rods nor cones are involved, and the photoperiodic control is effected by a low level of radiant energy compared to full daylight. The action spectra usually have maxima in the red-orange. In some cases the light stimulus does not have to travel through the eye or the optic nerve. Benoit, and then Underwood and Menaker, demonstrated that blinded sparrows and ducks are fully capable of mediating the gonadal response to photoperiodic stimuli. Light reaches the brain of the bird through the skull; Benoit confirmed with photographic plates that light can indeed reach the hypothalamus via this route. There may be rhodopsin-based receptors in the bird hypothalamus.

In at least one case, not only reproductive cycles but also the change in time of activity due to a change in sunrise can be controlled by light through the skull. Underwood and Menaker (1970) studied one lizard species after removal of eyes, pineal organ, and parietal eyes. These lizards appear still to have an extra-retinal photoreceptor which can mediate the change in the pattern of activity. Perhaps in some way the recent information on electrical responses from skin under the stimulus of intense flashes of light may be related to the pathway of light to the central nervous system.

The parietal eye mentioned above is of interest. This structure is located in the mid-forehead or on

Figure 2-8. An Enlargement of the Synapse Where One Neuron Makes Contact with Another. Impulses are transmitted to the receiving neuron by release of transmitter substance from the synaptic vesicles. This arrangement is called a *tight junction*. Modified from Schmidt-Nielsen 1997.

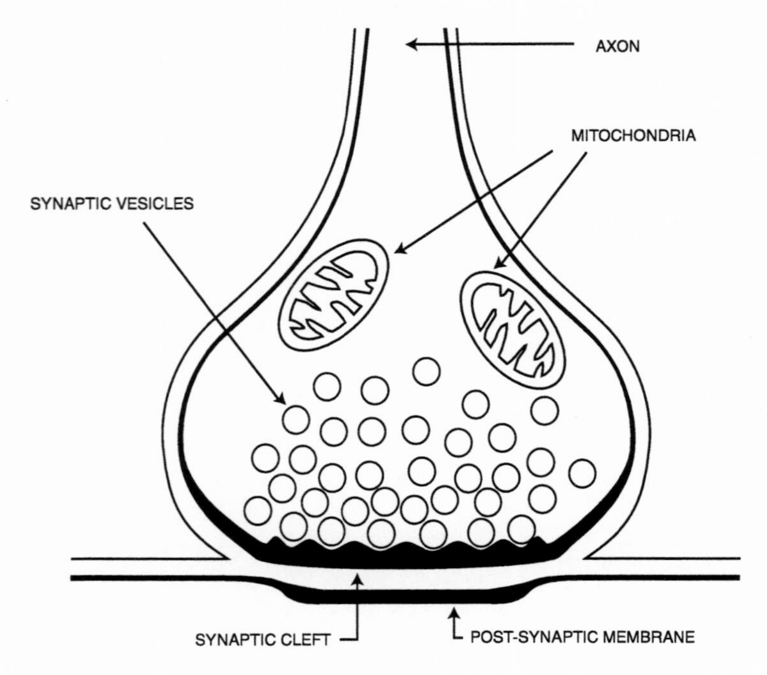

top of the skull of many reptiles. According to Phillips and Harlow (1981), when this eye was covered in the horned lizard, the reptiles changed their temperature regulation mechanism by panting and bladder discharge.

On the other hand, it is possible that the white rat does not receive light stimuli without eyes. Halberg recorded the time of starting activity of the wild Norway rat blinded and exposed to a 12-hour light-dark cycle (LD 12:12); the light cycle did not influence the time of beginning activity of the blinded rat. The experiment was repeated on other species of rodents, and there was still no evidence of extraocular photoreception (Nelson and Zucker 1981). Perhaps it is only in birds and reptiles that visible light penetrates the skull.

It was stated above that the hypothalamus is part of the feedback system that responds to light. Benoit was a pioneer in such studies: after removal of the eyes of ducks, he inserted quartz rods through the empty orbit to the hypothalamus. By conducting different wavelengths of light to the hypothalamus, the ducks were stimulated to full reproductive activity.

Light and Pineal Function

Regardless of how light first acts as a stimulus, we must consider the pineal gland as a key organ in the feedback chain. This long-neglected gland is now a favorite area of study by environmental and reproductive physiologists. The function of the pineal gland is to store potent biological transmitter materials that influence seasonal reproductive activity and the time of starting motor activity. The materials stored are serotonin and melatonin (see Fig. 2-9 below, and sidebar on next page). Also stored is a neuron-specific phosphoprotein, which is especially abundant at synapses (Fiske 1965; Fiske and Huppert 1968). Note that these materials are not produced evenly throughout the 24-hour period, but in large and variable pulses, mostly during the night (Klein 1985).

Early work on the influence of the pineal gland on reproduction was done by Wurtman, Axelrod, and Kelly (1968), and others (Hoffman and Reiter 1965). Their work with the white rat, and the definitive work by Reiter (1978) on the Syrian hamster, provide the following generally accepted conclusion: In a number of mammals, the pineal organ is a neuroendocrine gland producing an antigonadotropic hormone, of which the synthesis and release is stimulated by darkness and inhibited by light. Thus, short days suppress reproductive function via the pineal. (The pineal system of the Turkish hamster is more complex.)

The dominant hormone is melatonin, and it is studied in tissues by identifying the enzyme (HIOMT) responsible for its production. By searching for this hormone in various tissues, Ralph has shown that melatonin is formed and stored in the retina as well as in the pineal gland. The enzyme can be detected in the retina of fetal rats, but it cannot be detected in the pineal gland until a few days after birth (Ralph 1976). In the adult mole (an insectivore), the eyes produce two to ten times more melatonin than the pineal gland (Ralph 1980). The results from mice show that the ability to synthesize

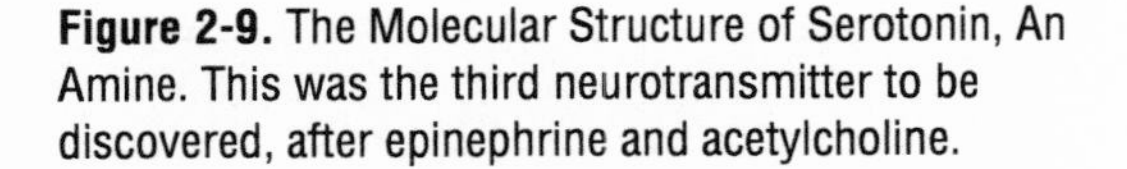

Figure 2-9. The Molecular Structure of Serotonin, An Amine. This was the third neurotransmitter to be discovered, after epinephrine and acetylcholine.

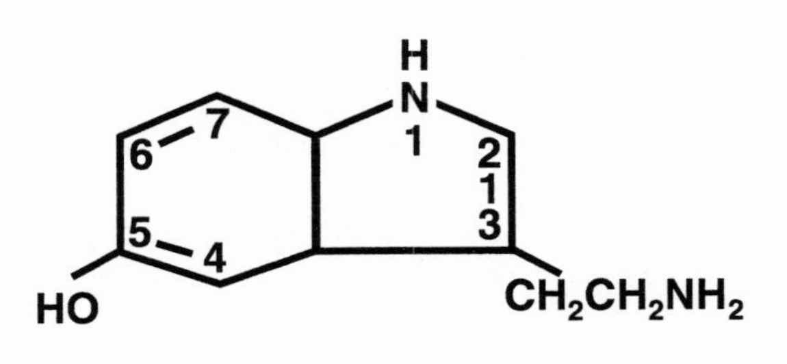

Melatonin and Serotonin

The importance of *melatonin* has been described in the text. It originates from *serotonin*, which in addition to being important in photoperiodism, is also a key hormone in control over biological clocks and temperature regulation (see the next two chapters). In vertebrates, melatonin is produced by the pineal and parietal glands and by the retina. In invertebrates and plants, it is produced by cells, and the amount is three times the concentration found in vertebrate blood (Hardeland 1997).

Most endocrine glands store large amounts of hormones, but the pineal stores very little serotonin or melatonin. Production in the pineal results from an adrenergic neuron which releases a neurotransmitter, *norepinephrine;* this stimulates β (beta) receptors which enhance N-acetyltransferase activity, which in turn converts serotonin into melatonin. This occurs during dark hours but not light hours (Devlin 1992).

Serotonin is one of the neurotransmitters also called *neurohumors*. By definition, such a material is produced from synaptic vesicles in the axon at the time of the arrival of an action potential at the presynaptic ending (Fig. 2-8). In other words, nerves act like glands. In the 1950's, there were vigorous debates as to whether such transmission was electrical, or only by neurotransmitters (Cook 1986); professors were often overheard asking each other, "Are you one of the "spark boys" or "soup boys"? Until 1953, there were only two transmitters known: acetylcholine and epinephrine. In that year, John Welsh discovered a third substance which acts as a transmitter, serotonin (also referred to as 5-hydroxytryptamine or 5-HT) (Fig. 2-9) (Welsh 1953; Folk and Long 1988).

melatonin depends upon two independently assorting Mendelian genes, both of which are homozygous-recessive in C57 mice and homozygous-dominant in field mice (Birau and Schloot 1980).

The seasonal reproductive changes are influenced by melatonin via changes in secretion of LH, FSH, and prolactin (Goldman and Darrow 1983). This melatonin can be produced from the pineal or the retina. The independence of the retinal melatonin biological clock has been shown by the culturing of the hamster retina by Tosini and Menaker (1996).

Although there is a tendency to call the eyes a *second* pineal gland, this is not always the case, especially in mammals. In one species, the daily pulse of circulating melatonin is derived from both the pineal and the retina; in other species, circulating melatonin may come exclusively from the pineal, or only from the eyes. This is the case with some lizards: either retinas or pineal glands—but not both in the same species—have important regulatory influences on reproduction. Melatonin from non-pineal sources in a variety of species is described in detail by Ralph (1980) and Roth et al. (1980).

The relation between light and the pineal gland was studied on the Arctic brown lemming, which experiences over 80 days of continuous light and 80 days of continuous darkness each year. When the pineal glands of these animals were removed, the normal reproductive responses to photoperiod ceased (Folk, Hagelstein, and Ringens 1977).

Anatomy and Innervation of the Pineal

In studies of the pineal gland, it is difficult to relate structure to function. This is because the cells of the gland contain relatively few secretory granules. Most of the gland consists of one cell type: the pinealocytes. The gland is richly innervated by cervical post-ganglionic sympathetic nerve fibers, most of which are found in the perivascular spaces of capillaries. We realize that in the lower vertebrates, pineal photoreceptors are similar to retinal cones. Thus, it is generally accepted that mammalian pinealocytes are phylogenetically derived from ancestral pineal photoreceptor cells. The activity of the pineal gland in respect to both synthesis of its chemical messengers and its control over reproduction apparently is influenced by nerve action, but does not always depend upon it.

The distribution of nerves is understood most thoroughly in the frog. From the pars intermedia of the frog, two types of axons are distributed to the pineal gland; one type of axon has its firing inhibited by light and the other does not (Oshima and Gorbman 1969). In the bird there is sympathetic innervation of the pineal; cutting of this innervation does not prevent the pineal from forming melatonin, nor does it interfere with the control by light over the frequency of egg laying (Ralph 1980).

The situation is very different in the mammal. In this group the pineal gland is not innervated exclusively from the sympathetic system, but also by fibers of central origin from the commissures. The sympathetic fibers originate in the superior cervical ganglia of the sympathetic trunk, continue in the internal carotid nerve, and then enter the pineal gland. Their importance for the regulation of melatonin synthesis has been demonstrated by sympathectomy and by electrical stimulation. Electrical stimulation of the superior cervical ganglion can lead to an approximately 50-fold increase of pineal serotonin activity. The central innervation reaches the pineal gland via two of the commissures, and then continues with an uneven distribution in the gland, some lying in the periphery and others in the center. These central fibers contain as many as five peptides such as oxytocin and vasopressin. It is not known whether these neuropeptides are released into the pineal blood vessels or whether they act locally to regulate pinealocyte function.

Neural connections from the pineal to the brain have recently been described in the Syrian hamster; the tracts travel to the brain regions associated with sleep and behavior (Evered and Clark 1985).

To be more specific, we can now generalize that in the mammal, the light stimulus (which may be inhibitory) begins with the retina and travels through seven discreet nuclei (aggregations of cell bodies) of which the SCN is in some cases the most important, and finally, impulses arrive at the pineal gland. Here, the main control over function in the pineal is through the release of norepinephrine (NOR) from the terminals of nerves whose cell bodies lie in the SCN. The NOR probably acts through beta adrenoreceptors. Serotonin is converted to melatonin. This change is accomplished by dramatically synchronized enzyme pulses. The shapes of these curves in the rat gland are different from those of the sheep, monkey, and hamster.

Further information on the control of pineal activity by sympathetic innervation in a mammal has been reported by an Australian team: when the superior cervical ganglion which serves the pineal gland is removed, female wallabies no longer show delayed implantation (seasonal embryonic diapause). This experiment not only shows the control over the pineal gland by the cervical ganglion, but also provides the first evidence that the pineal gland is involved in delayed implantation in mammals (Renfree et al. 1981).

To conclude, we now see the pineal as a locus of conversion of neural input; thus we can identify it as a neuro-endocrine transducer (Tamarkin, Curtis, and Almeida 1985).

Now there are some 50 or 60 known neurotransmitters, for example nitric oxide and GABA. For nerve-to-nerve transmission outside the brain and spinal cord, *acetylcholine* is the ubiquitous transmitter substance. The many other newer substances act within the central nervous system. Note that some neurons contain near the synapse as many as three neurotransmitters.

For convenience, we call some of the alike neurotransmitters *amines*; examples are serotonin, norepinephrine (or "noradrenalin," the British term) and dopamine. What was the result of the long-term debate as to whether transmission at the synapse is by electrical or by neurohumoral means? Both sides were right, although electrical transmission is less common. This last type depends on two adjoining neuronal membranes, so thin that they are almost "open," one to the other; such a synapse is called a *gap junction* and provides rapid transmission. This release of neurotransmitters depends on a large increase of intracellular calcium around the transmitter vesicles. It occurs occasionally in four Phyla, including invertebrates, and in the central nervous system of some vertebrates. However, most synapses are chemical and are called *tight junctions.* Their transmission is slower than that of gap junctions (Schmidt-Nielsen 1997).

The Pineal, Daily Time of Activity, and Seasonal Events

We must now summarize the relation between two of the functions controlled by the pineal: (1) *the beginning of daily activity* under the influence of dawn or sunset, and (2) *seasonal events* such as reproduction or the molting of a hair coat.

We have described the pathway of the light-initiated stimulus as extending through the retina, seven discrete nuclei, the pineal, and the pituitary. The critical experiments on photoperiodic control were done by severing parts of the neuronal path, recording normal pulses of hormones, and injecting pulses of hormones.

The pineal gland in the bird serves a different function from that of the mammal. In both cases the neuronal control over the pineal is exquisitely sensitive to environmental illumination, and melatonin is not evenly secreted over 24 hours; there is a pulse of the hormone in darkness which changes its length with photoperiod. Birds do not use a melatonin seasonal message for control of reproduction, but it is essential for their control of daily activity: the start of the pulse is the sunset stimulus (Gaston and Menaker 1968). This signal is especially needed when there is no regular signal from the environment (such as dawn); the irregularity of this signal would occur during migration.

Mammals have an opposite mechanism from birds. Quay (1972) and others found that in the white rat, the pineal gland is not needed to signal time of daily activity, but only to adjust the activity when sunset changes; this has been confirmed in other rodents. On the other hand, mammals are able to determine the progression of seasons by "reading" the daily pulses of melatonin; or, to state it differently, they are sensing the *duration* of nocturnal elevation in the antigonadotropic hormone (Evered and Clark 1985). Most of these experiments were done on rodents, mink, and sheep.

Mammals exploit these pineal changes to control the annual sequence of breeding and other events. The pineal exerts its control partly by modulating the frequency with which the pituitary secretes gonadotropin; this luteinizing hormone level is determined by the *duration* of nightly melatonin pulses rather than by the photoperiod itself. The mammalian pineal is essential for this seasonal timing of the annual cycle, but the gland does not modify the events themselves once they are initiated.

Another function of the pineal gland is control of body temperature and thermogenesis (Ralph et al. 1979; Ralph, Firth, and Turner 1979). It is not the function of this chapter to discuss this; details can be found in Chapter 4, *Temperature Regulation and Energy Metabolism*.

HUMAN PHOTOPERIODICITY

Is there a photoperiodic effect on people? Most speculation concerning the cause of "spring fever" centers around changes in vasomotor tone due to the effects of temperature rather than the possible influence of light. However, the cause of the "spring in the step" of the person who starts the day in bright sunlight after a period of cloudy weather may have its roots in the powerful photoperiodic effects seen in the lower mammals.

Indeed, light appears to play a strong role in human health, affecting not only levels of fatigue and resistance to illness, but reproductive and mental health as well. Research began several decades ago to uncover the effects of light on humans, as well as to develop ways it can be used to correct malfunctions. As early as 1950, Kovacs wrote a book entitled *Light Therapy,* and in the former Soviet Union, experiments were carried out to assess and improve the amount of daylight for industrial and other workers. Considerable attention was given to the intensity of illumination: some combinations appeared to produce less fatigue. Soviet investigators also studied workers in film production plants and other laboratories where employees work under conditions of either complete or partial darkness. In these places, the rate of illness was one and a half to two times higher than for people working under normal light conditions. The researchers recommended that workers be given doses of oxygen inhalation and ultraviolet irradiation. These early findings encourage us to bear in mind that the human animal evolved as a day-active mammal; industry often overlooks the facts of evolution.

In 1968, Wurtman and colleagues developed the hypothesis that melatonin from the pineal gland is related to, or controls, the timing of the menstrual cycle. Other early experiments supported the concept of the powerful influence of light upon the human reproductive cycle. In one (Dewan 1967), fourteen women suffering from infertility were instructed to sleep with a 100-watt bulb burning all night about 10 feet from the head of the bed, but only on the 14th, 15th, and 16th nights of the menstrual cycle. All other nights of the month were to be spent sleeping in total darkness. There was a significant increase in the regularity of the menstrual cycle when this procedure with light was used, and the length of the cycles peaked prominently at 29 days.

More recently, researchers have investigated why certain people are elated in the summer and depressed in the winter; this kind of depression is called *Seasonal Affective Disorder (SAD),* or sometimes *winter blues* (Rosenthal 1993a). People in the northern regions have long referred to such winter depressions as "cabin fever." SAD differs from non-seasonal depressive disorders in several ways, one being that instead of experiencing insomnia as a symptom of the depression, those who are *winter-depressed* experience *hypersomnia*, desiring to sleep at least 12 hours a day. (This disease-state may explain stories of Siberian natives who appear to enter a form of hibernation.) Other winter-symptoms noted by Regan (1982) were extreme overeating, constant fatigue, and a suicidal tendency. Not surprisingly, Rosen et al. (1989) found that the incidence of SAD increased with latitude, ranging from 1.4% in Florida to 9.7% in New Hampshire. Researchers suggest that seasonal affective disorder may not be uncommon in the general population.

Much research over the past decade points to the importance of identifying patients who suffer from winter depression as distinguished from other forms of depression; this is because these people can be helped by phototherapy. In one of the early studies at the National Institute of Mental Health, Rosenthal (1985) controlled the depression of 11 winter-depressed patients by adding three additional hours of bright artificial light each day to their wintertime environment. In 1990, Terman showed that exposures of only 30 minutes/day were effective when the intensity of the lights was increased to 10,000 lux. Typically, patients are instructed to sit either in front of, or under, a bank of bright (at least 2,500 lux) florescent lights. Indirect or diffused light must reach the patients' eyes, indicating the importance of a peripheral retinal mechanism. Studies of headmounted visors delivering as little as 400 lux suggest that lower illuminance might be effective when the source is within a few inches of the eye (Moul et al. 1990). Some researchers have also observed that exposure to an artificial dawn (created by connecting a bedside lamp to a timer) can have an antidepressant effect on those with Seasonal Affective Disorder, even though the patient's eyes are closed and the light intensity is much less (Terman et al. 1989; Avery et al. 1993). The efficacy of phototherapy has been proven; studies continue in order to determine optimum intensity, duration, time of day, placement, and spectrum of light.

In humans, one effect of a bright light is to suppress nocturnal melatonin secretion. Formerly, it was considered necessary to use 2,500 lux; Honma (1995) has succeeded in suppressing nocturnal melatonin secretion with 500 lux. Rosenthal and Wehr (1992) have concluded that the secretion and modification of melatonin does not appear to play a critical role in the etiology and treatment of SAD, although it was the recognition that bright light suppresses nocturnal melatonin secretion that first inspired the use of phototherapy in treating this disorder (Lewy et al. 1980).

Regan has coined the term *therapeutic photomedicine* to apply to these and other medical techniques that make use of light. Its beneficial effect may be through the changes in melatonin, although its influence on human subjects is little understood. There is a 75% decrease in the melatonin of children from age 5 to puberty at 11. Considering what is now known about melatonin, this correlation (low melatonin vs. puberty) probably means that melatonin suppresses puberty up to an appropriate age (Evered and Clark 1985).

Plasma melatonin in human subjects is derived exclusively from the pineal gland, as it is in rats. As we discussed above, the nighttime level in the blood can be manipulated by photoperiod or bright lights; these observations led to the finding that there is a pineal malfunction in some mental disorders. Also, there is a relation between melatonin and hormone-dependent tumors; melatonin seems to protect against tumors by suppressing prolactin secretion or by acting on estrogen receptors.

THE INFLUENCE OF MOONLIGHT

A much neglected facet of visible light radiation is the influence of moonlight. The amount of moonlight is extended in northern regions; the early Arctic pioneer, Charles Brower, and also Vilhjalmur Stefansson, often successfully hunted by moonlight to provide food for their expeditions. Much of the exploration by Commander Robert E. Peary in northern Greenland was done by moonlight; he made extensive sledding journeys during the two-week periods when the moon was continuously traveling in an ellipse overhead.

Surely this strong light during the nighttime period must influence so-called nocturnal animals in the free environment—perhaps as a navigational aid—yet experimental work concerned with this is still rare. One experiment on the field activity of *Microtus* is reported by Doucet and Bider (1969); after meticulously correcting for extraneous factors, they concluded that the new moon phase is associated with higher activity.

Bünning and Moser (1969) went further and compared *Euglena*, short-day plants, and long-day plants for the interference of moonlight with the normal photoperiod of the plant. Bünning took into account the fact that the usual threshold value of light intensity in photoperiodic influences on plants corresponds approximately to the intensity of moonlight; specifically, intensities as low as 0.1 lux may influence photoperiodicity of plants and animals. The most interesting finding was that the introduction during the night of a light intensity similar to that of moonlight promoted the flowering of long-day plants.

THE SPECIAL CASE OF A WHITE COAT COLOR

Photoperiodic Control

The control of coat color by light is of special interest. We have already mentioned that light controls shedding, changes in coarseness of hair, and hair density. Some mammals attain a permanent white coat in the north (which must be associated with protective coloration) and other mammals change from a white coat in winter to a dark coat in summer. Any analysis of this conspicuous physiological change must involve a blend of information from the fields of photoperiodicity, endocrinology, biological clocks, and cold acclimatization. (See also Golger's rule, discussed later.)

An outstanding contribution to this area was made in the early 1940's by the noted biologist, Charles Lyman, when he studied the mechanism of photoperiodic control in the varying hare (*L. americanus*). Long days induce a moult from winter white fur to summer brown fur, a high concentration of gonadotropic hormone in the blood, estrus in the female, and enlarged testes in the male. Short days in the fall induce a white coat, a low level of gonadotropic hormone, and a suppression of reproduction. If these light cycles are experimentally reversed in spring and fall, the physiological responses are prevented. These results depended upon light stimuli through the eyes. The photoperiodic effects were not affected by air temperature.

White Coats, Latitude, and Photoperiod

These color changes are related to that situation when a small mammal ventures onto the surface of snow; for example, the collared lemming turns white in winter and is frequently active on the snow surface. The brown lemming stays brown all year and does not travel on the snow. Sometimes a particular species does not change color (perhaps remaining brown) in part of its range, but farther north it turns white in winter; the New York weasel is an example. The difference, in an evolutionary sense, is probably due to the decreased fall photoperiod found farther north.

The mechanism of color change in the collared lemming (*Dicrostonyx*) has been studied. This color change is controlled simply by photoperiod; any short days bring about white coats. However, maximal *gonadal growth* depends upon photoperiod history. Raising lemmings on long days does not produce gonadal enlargement; they must be primed by preceding short days to produce the long-day effect (Gower et al. 1996).

White species that never change color include the narwhal and the Dall mountain sheep. Perhaps the narwhal has an adaptive advantage: conspicuous visibility in Arctic waters may aid in keeping the school of whales together. Note also that the Arctic beluga whale is also a bright white (although the subadults are gray). Dall sheep are extremely conspicuous in summer; many striking photographs have been taken of this magnificent species standing white against green or gray mountain slopes.

The white coat picture is not always simple; sometimes latitude, insular isolation, and photoperiod work together to change the rule of *brown-in-summer, white-in-winter.* The noted Arctic ornithologist, David Parmalee, observed coat color of animals at three locations, each farther north, in the Arctic and high-Arctic. His observations showed an incremental change to a permanent white coat:

	Bathurst Island (All seasons)	Victorialand and Elsmereland (In Summer)*
Wolf	all white	black, gray, or brown
Arctic Hare	"	brown
Caribou	"	brown, all seasons
Gyrfalcon	"	black or gray
Ptarmigan	"	brown
Collared Lemming	—	brown

* All white only in Winter.

Robert E. Peary observed that the caribou of northern Greenland are all white, instead of the brown of the lower Arctic, and there are four similar species on Bathurst Island. More work needs to be done on this variation in color with latitude. For example, does the arctic fox or the collared lemming (mentioned above) remain white in summer even farther north at Ellesmere Island? Which is more important, extreme photoperiod or genetic isolation, in explaining the above observations?

One variation of the "all white" color in winter is found in large white weasels (ermine) which have a black tail tip (*Mustela frenata*); as we will see, photoperiod must program this black tip as well as the white coat. The least weasel (*M. nivalis*) also changes to a white coat for the winter but does not have the black tail tip. It was first considered possible that the black tail tip may help young weasels to follow their mother, or that the black tip may be used in intraspecific communication. However, an ingenious series of experiments by Roger Powell (1982) strongly supports the hypothesis that the black tail tip on white weasels serves to reduce capture of weasels by "confusing" avian predators.

EXPERIMENTAL CONTINUOUS LIGHT

With white rats and mice (nocturnal mammals) maintained in continuous light (LL), the daily block (a 9- to 12-hour unit) of locomotor activity shows a regular and constant delay in time of starting; thus the *activity block* travels around the clock (Aschoff 1953; Folk 1955). Although this is now called the *Aschoff effect,* it was originally described by Johnson in 1939. The rate of change of time of activity is faster with an increase in the intensity of the light. When animals in continuous light are then placed in continuous darkness, the daily delay in time of starting ceases. Behavior modification such as this should be called a *photoperiodic response.* Although we will discuss this effect of light again in the next chapter, it is helpful here to develop a definition: "Photoperiodicity is a quantitative biological principle in which a systematic and incremental variation in the stimulus of light produces a graded and sequential response. One assumes that a threshold must be reached." Of course, the same definition applies as much to *daylength* as to *intensity;* the rate of testes development depends incrementally on photoperiod (refer again to Fig. 2-7).

In studies on rats, the delay resulting from constant light (5.1 foot-candles) was 45 minutes per day, and their activity block traveled around the clock in 16 days (Folk 1959). Day-active rodents have not been satisfactorily tested for their type of response to continuous light, but with day-active birds, Aschoff found the opposite of this regular delay in time of starting activity, namely an acceleration (Aschoff 1963). This *early starting* of activity increased with the intensity of light and is thus a photoperiodic response. The biologists who are interested in photobiology now describe these delays and accelerations of time of activity as applications of Aschoff's rule, stated as follows: *In continuous light with increasing intensity of illumination, light-active animals increase their daily spontaneous frequency, while dark-active animals decrease it* (Pittendrigh 1960).

This rule is illustrated in a simplified composite diagram (Fig. 2-10) and in a copy of an actual record (Fig. 2-11). Aschoff's rule does not apply to plants.

This mechanism of incrementally changing time of starting activity depending upon the light intensity is used by animals in the free environment when there is a change of season. As sunset comes later, nocturnal animals delay the time of becoming active. It is the reverse with day-active birds as sunrise comes earlier. This will be discussed in more detail in the next chapter.

There are practical applications of continuous light; considerable attention has been given to its effect upon the domestic chicken. In an extensive study Kadono has even interrelated egg-laying, body temperature, and food-intake in the hen under the influence of continuous light (Kadono, Besch, and Usami 1981). Fewer eggs are laid during the lifetime of the hen in continuous light, although there is an initial increase in egg-laying.

Artificial constant light (LL) affects the reproduction of mammals. To understand this, let us consider for a moment the rodents of the Arctic, where there are in nature long periods of this environmental challenge (LL). Is there an effect on reproduction? At Point Barrow, about every four years vast numbers of lemmings breed in continuous light (as well as in other light conditions); the lemmings are so numerous and ubiquitous that they even run into the laboratory from the tundra. Thus, breeding does take place in LL.

To study this phenomenon of breeding under LL, it was reasonable to use the laboratory white rat for a comparison; in artificial continuous light of several months' duration, all rats went into permanent estrus and still could be bred. The resulting young were not harmed by this extreme environment; in fact, at birth and at 20 days of age, the experimental young were over 30% heavier than controls. A similar effect of LL on estrus and litter-size is also found in pigs and mink (Klotchkov et al. 1971).

What is continuous estrus? The typical rat and rodent cycle under a normal light cycle includes spontaneous ovulation and a complete endocrine cycle every 4 or 5 days. If there is no breeding, the cycle has two parts:

- stage one, the *follicular* (proestrus and estrus), lasts about 24 hours, is controlled by FSH and LH and characterized by sexual receptivity, ovulation, a LH surge, high blood levels of estrogens, and extreme 1ocomotor activity;
- stage two, the *luteal* (metestrus 12 hr and diestrus 65 hr), is controlled by the combination of LH and prolactin and characterized by high blood progesterone, corpora lutea in the follicles, and later a lowered progesterone which stimulates FSH/LH-RH to start growth of follicles and a low secretion of estrogen.

Continuous estrus under continuous light may last for 6 months (Klotchkov and Belyaev 1977); the rodents remain in stage one as described above with all of its characteristics except for ovulation and the LH surge, unless breeding takes place. Here the comparison with lemmings ends because lemmings always breed in nature in continuous light, and we know nothing about the endocrine control of their ovulatory cycle.

Figure 2-10. Daily Running Activity Rhythms of White-Footed Mouse. (A) Rhythm of mouse under short-day photoperiod (12 hours of light). (B) Rhythm of mouse under long-day photoperiod (16 hours of light). (C) Free-Running rhythm of mouse under continuous darkness. (D) Rhythm showing Aschoff effect of mouse under continuous light. Modified from Rawson 1959.

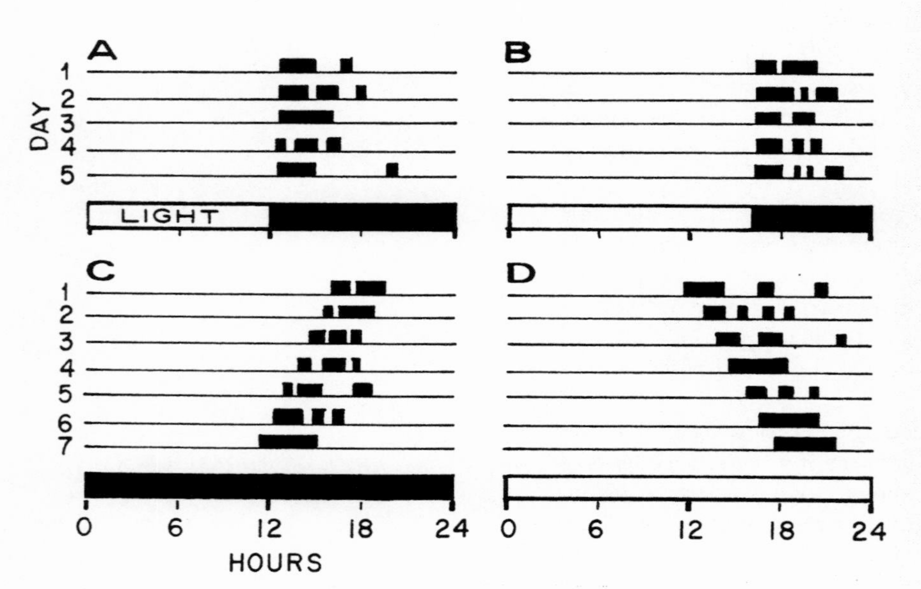

NATURAL CONTINUOUS LIGHT

Up to this point, the use of continuous light in experimental conditions has been discussed. Some years ago it occurred to us that the 82 days of continuous light each summer in the Arctic and Antarctic regions might cause animal activity to change according to Aschoff's rule (Folk 1964). The intensity of this continuous light will vary with time and location; at noontime it will measure 40 foot-candles (fc) under 100 meters of ocean water, but 9000 fc in direct sunlight. After many summers of studying (via radio-telemetry) Arctic ground squirrels, porcupines, foxes, wolverines, and wolves, we reached a tentative conclusion that natural continuous light does not induce a period consistently longer or shorter than 24 hours in Arctic mammals (Folk 1961a); in other words, Aschoff's rule does not apply. We studied four individuals of each species in most cases. Swade reached this same conclusion several years later.

To determine what environmental signal could keep Arctic animals in phase with solar time, we attempted to set up the following hypothesis: the most likely regular event is the position of the sun when it sinks low to the horizon, even though it is actually going in a circle overhead (Folk 1964). Near midnight each day the sun is closer to the northern horizon than it is at noontime to the southern horizon (Fig. 2-12). This explanation would not do for animals in north Greenland or on the islands near the North Pole since the sun is much higher above the horizon during the period of continuous light at midnight.

To pursue the hypothesis further for application to mammals on continental North America, we must ask whether birds or mammals can sense celestial objects with sufficient resolution to use such a phenomenon as the distance of the sun from the horizon. Penney, who released penguins in a featureless snow field in the Antarctic, indicated that by celestial navigation, these birds headed directly home with an error of only 1% (Emlen and Penney 1966). Emlen (1969) found a similar remarkable resolution with birds in a simulated natural sky; he reported that only 8.9% of the sky was needed for his indigo buntings to orient to the night sky in appropriate fashion. It would be of interest to determine whether *mammals* show the same resolution in interpreting position of celestial objects. A reasonable experiment would be to have young geese or mammals imprint on large cylinders or cubes, then the cubes could be reduced in size to determine the resolution of recognition of these young animals.

In considering environmental effects on the life history of a species, one must know the *duration* of the natural continuous light and the *rate of change* from one daylight season to another. The time of year for solar absence and for continuous light is shown in Figure 2-13. The rate of change between these two conditions means that the adjustment on the part of mammals and birds is difficult because there are not regular changes each day. In January, the time of sunrise changes rapidly from day to day, by March the change is constant and slower, then by May there is once again a very rapid change each day in both sunrise and sunset. For some animals, this must mean irregular feeding habits. Ptarmigan, for example, always begin activ-

Figure 2-11. Actual Running Activity Record of Deer Mouse. Twenty-day record (May 16-June 4) of running wheel activity recorded in constant darkness showing the effect of hunger on the level of activity. Two weeks supply of food was presented at time of feeding. Each vertical pen excursion equals 100 revolutions per minute. Modified from Rawson 1960.

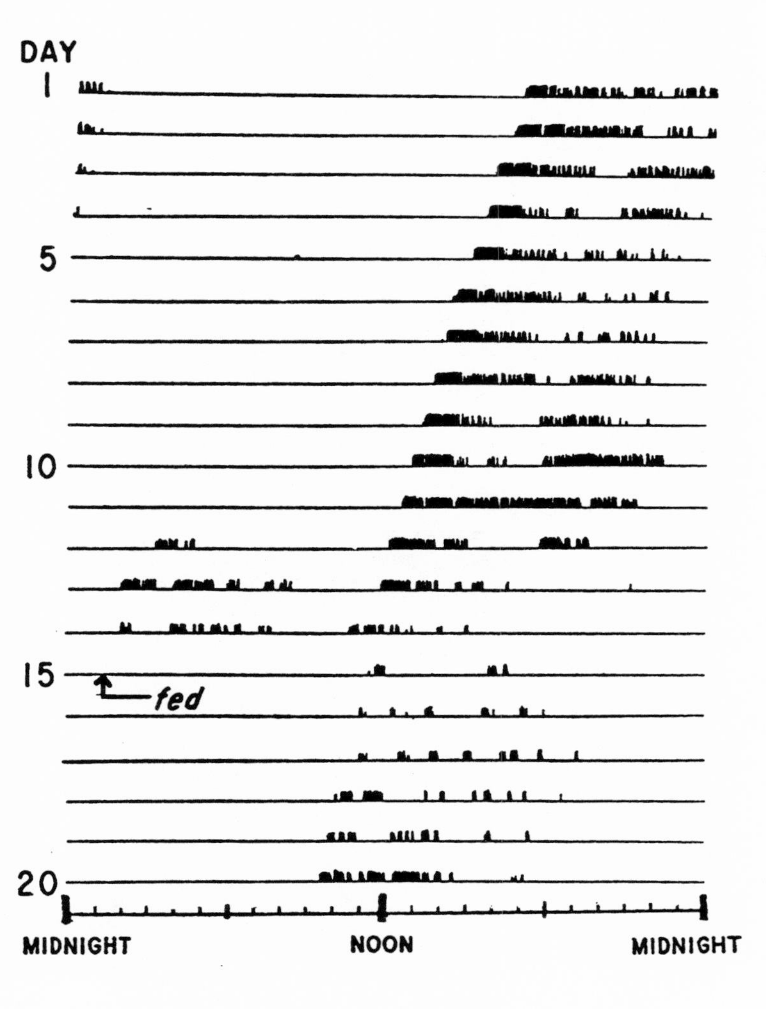

ity at the first light of day. The result is that by the time of continuous daylight they are apt to show approximately 24 hours of phasic activity.

NATURAL CONTINUOUS DARKNESS

Solar Absence

Continuous darkness is not merely an experimental environment. As stated earlier, many thousands of people and their domestic animals experience this condition for 2 or 3 months each winter (see Environmental Diagram B). The domestic animals (dogs, rabbits, cattle, sheep, and reindeer) are provided with little artificial light, and the countless wild animals have practically no light at all.

Our use of the phrase "countless wild animals" appears justified, since it applies to the following Arctic land animals: lemmings, mice, ground squirrels, hare, reindeer, caribou, seals, musk-oxen, and the carnivores. Some estimates put the total number of individual land mammals in the Arctic life-zone at roughly the same number as there are in all of the other five major life-zones combined.

The question naturally arises whether the seasonal lack of daylight quota has any adverse physiological effect on animals adapted to Arctic life. It is useless to ask whether these animals are essentially diurnal or nocturnal, because during some part of the year they are

Figure 2-12. Pathway of Sun at High Latitudes. A person in the center of the diagram (at Barrow) would see the sun overhead or on the horizon on different dates as indicated. Barrow is at approximately 72° N.

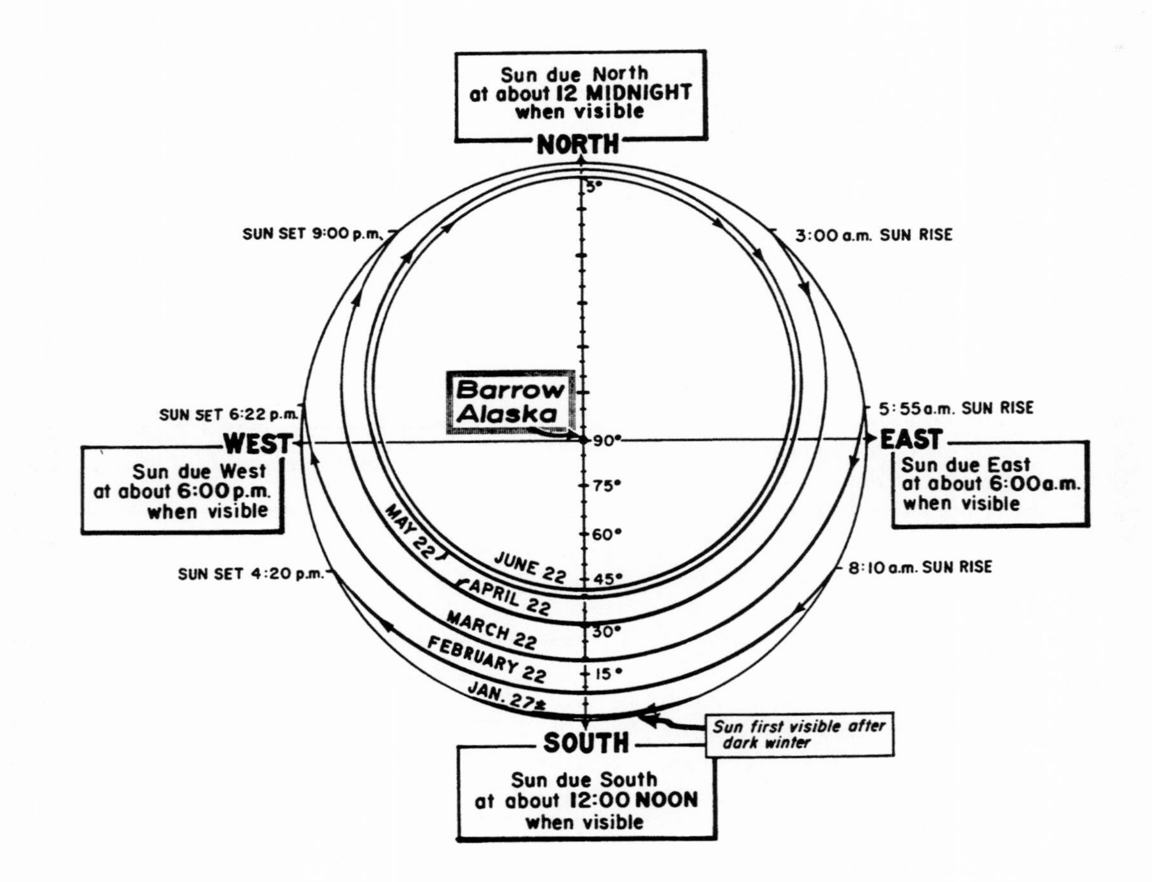

forced to have both of these behavior patterns. The arctic seal has a special challenge to survive in winter because it must orient in darkness to its breathing hole in the ice, and must locate and catch fish in darkness under the ice. Although it was earlier supposed that seals must ecolocate their prey in darkness, it has now been shown that they have nowhere near the sonar ability found in whales and dolphins. Seals are assisted in near darkness by an incredibly sensitive visual sense; also the bioluminescence of food prey may be a cue used by seals in winter (Schusterman 1997).

We should not overlook the ability of other terrestrial animals to move in total darkness. Redmond (1982) studied herds of elephants, sometimes as many as 19 in a herd, that penetrated over 150 meters (500 feet) into a rocky cave at night in order to obtain salt from the wall at the back of the cave. They did not even have the benefit of starlight. He writes: "To look down into the chamber and see, in the dim light of a tiny pen torch, an adult elephant shuffling along in slow motion was an incomparable experience. Feeling her way around boulders bigger than herself, she carefully placed each foot down before lifting the next." We do not know what senses these elephants use to master this environment of total darkness.

One mechanism for finding food in darkness is olfaction. Experiments were done by deploying a greenhouse gas, dimethyl sulfide (DMS), in sub-Antarctic waters. This gas is given off when zooplankton graze on phytoplankton. DMS attracted three species of petrels in darkness, but not albatrosses at anytime (Nevitt 1996). Petrels feed on zooplankton at night; albatrosses feed in daylight.

Mammals living underground have made various adaptations to an environment with little or no daylight. Hildebrand (1985) summarized the anatomy, senses, and energetics of mammals underground. For example, a pocket gopher may require 3400 times more energy to burrow than to move the same distance on the surface. In moles, the kidneys perform differently from those of surface-mam-

Figure 2-13. Annual Hours of Sunlight in the High Latitudes. The examples given hold for Point Barrow, Alaska. The diagram demonstrates the changes from continuous lack of sunlight to continuous sunlight that begins in May. One can read both the annual hours of sunlight and more importantly, the daily rate of change or increases in sunlight. For example, in January the change adds about two hours per day, but in April only about 20 minutes per day.

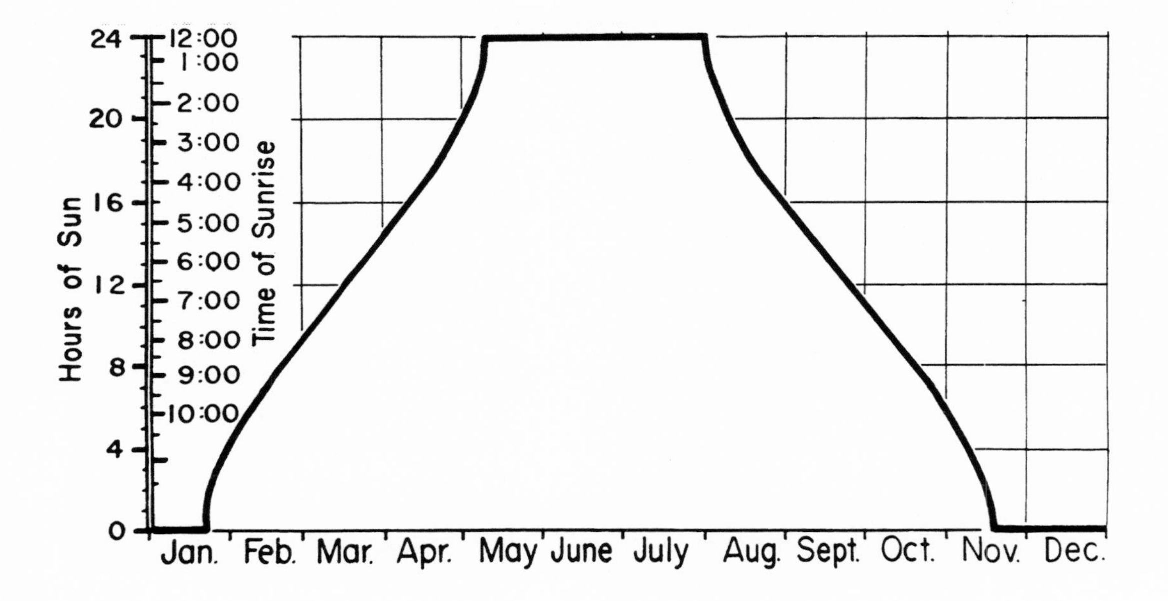

mals. When compared with the urine of rats on a carnivorous diet, mole urine contained bicarbonate, while that of the rats did not. This is an adaptation of the mole for unloading CO_2 without increasing its concentration in the burrow (Haim et al. 1987).

Some 150 species of mammals spend most or all of their lives underground. Representatives of different evolutionary groups have converged independently to the underground life. Ar (1987) describes various physiological adaptations made by these underground-dwelling mammals.

Ultrasonic Sounds and Hearing

One method of sensing in darkness is echolocation. Consider the use of echolocation by flying bats to avoid obstacles and to catch prey, as first described by Griffin (1959). Table 2-5 shows types of sounds made by one species of bat (*Eptesicus*), including their frequency, pulse intervals, and durations.

Other bat species have also been studied: 15 species had a pulse duration of ultrasonic bursts varying from about 0.5 to 65 milliseconds. This variation is because the echolocating bat, as it searches for and closes in on its prey, changes the duration and frequency of the call to help locate the prey. A characteristic increase in the rate of pulse production appears to tell the bat of last-millisecond changes in the position of the prey. Some bats may also alter the harmonic structure of their orientation calls to increase precision of target resolution (Fenton and Fullard 1981).

Can the prey hear the sounds made by bats? The flying cricket does hear ultrasound by its tympanal organ or "ear" on its leg. These frequencies, up to 100,000 Hz (100 kHz) are sensed by a neuron receptive to ultrasound (Hoy et al. 1982).

Maternal-infant communication by ultrasonic frequencies has also been studied. The most challenging and interesting situation is when thousands of adult bats and their young are calling simultaneously, which results in tremendous cacophony and confusion. Other species, including tenrecs, shrews, and infant rodents, communicate by ultrasonics between parents and young in the darkness of the nest (Geyer and Barfield 1979).

It is well known that whales and dolphins navigate by echolocation. Many bowhead whales, gray whales, and orcas winter on the Siberian Coast where they experience months of continuous dark-

Table 2-5. The Duration and Frequency of Ultrasonic Bursts for Ecolocating*

Phase	PULSE Interval msec	PULSE Duration msec	PULSE Number sec	Frequency Sweep kHz
Search	50-100	12	10	77 to 30
Approach	Intermediate between Search and Terminal phase			67 to 26
Terminal	5	0.4	200	23 to 19

*(From Gould 1970)

ness, although some of the gray whales also migrate as far south as California's Baja peninsula. The whale echolocation includes low-frequency sounds that travel farther in water, and high frequencies that allow smaller-size discrimination. Sound passes by objects that are smaller than the wavelengths of the sound. Some frequencies produced and used are:

Species	Frequencies in Hz	Size of Object Detected
Gray whale	80 to 2,000	75 cm
Killer whale	up to 24,000	8 cm
Dolphins	up to 96,000	2 cm

The gray whale does not use echolocation to find food, since krill are smaller than the wavelength of the gray whale's sounds.

EXPERIMENTAL CONTINUOUS DARKNESS

The effects of artificial continuous darkness on vision must now be considered, although the evidence is fragmentary. A few mammals have been raised in darkness. When kittens are raised in darkness, considerable damage is done to the visual pathways (Dews and Wiesel 1970). When chimpanzees (which are day-active) are also raised in darkness, there is optic nerve degeneration (Riesen 1960). In nocturnal rodents like the rat, there is little effect of continuous darkness upon vision. There are, of course, several dozen species of animals, mostly blind, which have become adapted to life deep in caves.

As far as reproduction is concerned, it is evident that many generations of wild rats breed (too successfully) in sewers, where some of them receive little daylight. As for more completely nocturnal rodents such as flying squirrels, it is also possible that their entire life is spent in continuous darkness, except for moonlight. Yet when Eayers studied white rats in the fifties, he found that those raised in continuous darkness grow more slowly and mature more slowly. This is probably because lactation by the nursing mother decreased during this period. Darkness has a similar effect on the seasonal milk yields of cows; there is a decrease in early winter followed by an increase in late winter months. This was tested experimentally; when 12 to 15 hours of darkness per day was decreased to 8 hours of darkness, milk yield increased 10% to 15% without increased consumption of feed.

This relates to a topic we can call "the appetite for light." This has been studied by operant conditioning. Calves choose illumination for about 67% of the day if only a single action is necessary to turn the lights on and off, but for only 0.5% of the day if continuous work is required to maintain illumination. Darkness does not appear to have an adverse effect if it is within reasonable limits; it is well known that cattle and horses feed at night.

As far as mechanism is concerned, there has been little analysis (either physiological or anatomical) of mammals raised in darkness. In one study, the thyroid gland was found to be more active in rodents raised in continuous darkness compared with those raised in continuous light.

The recording of functions of animals in continuous darkness has been an essential part of the experimental work of biologists interested in biological rhythms of nocturnal animals. In spite of possible detrimental effects of continuous darkness, the use of these environmental conditions has been necessary to test the degree of control of light cycles over biological rhythms.

A rhythm which continues in darkness is said to *persist* or to be *free-running*. Experimenters in this field have not reported adverse effects of continuous darkness on activity; on the contrary, the running of rats in continuous darkness usually increases over control values. Other effects will be described in detail in the chapter on Biological Rhythms.

3. BIOLOGICAL CLOCKS AND RHYTHMS

THE DAILY AND ANNUAL GEOGRAPHICAL INFLUENCE ON BIOLOGICAL RHYTHMS

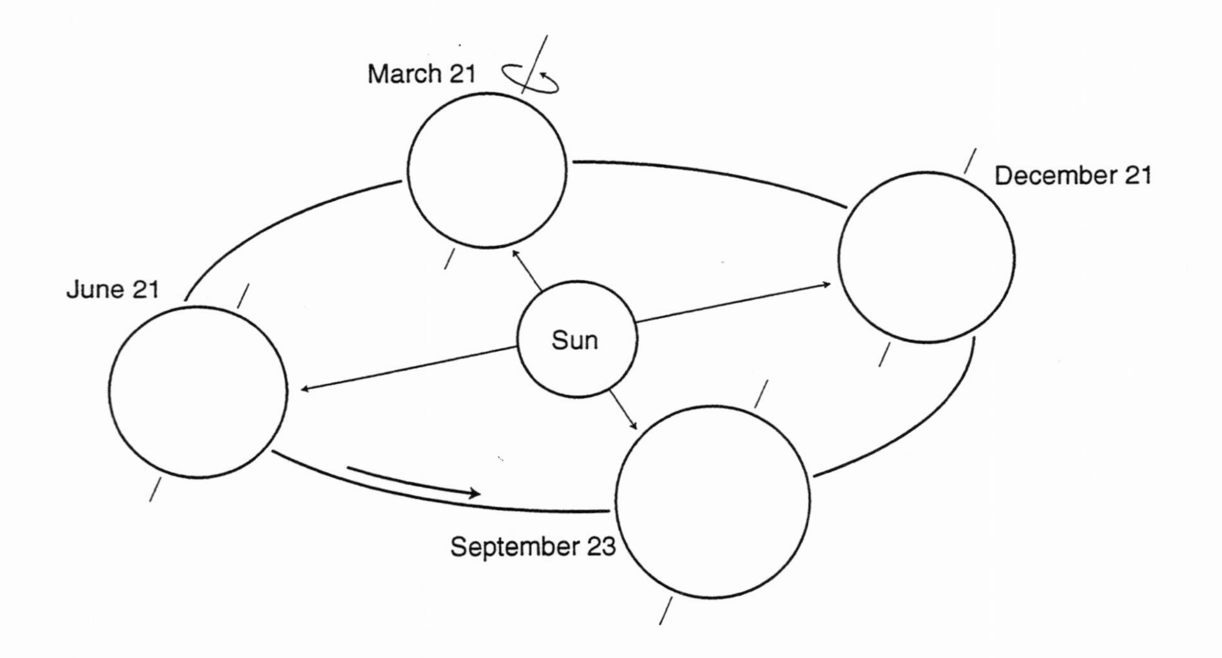

The Earth and the Sun

Environmental Diagram: Chapter 3

3. Biological Clocks and Rhythms

3. BIOLOGICAL CLOCKS AND RHYTHMS

INTRODUCTION

This chapter presents the principles of a topic studied and reviewed by many scientists and medical doctors, that of *biological clocks*. The present writing is designed for students and those scientists who need the basics of "clockmanship" or "clockwomanship." To cover the vast literature of the 1980s and 1990s would take an entire book. However, along with a review of the principles of this field, we also present quite a few examples of experiments from the eighties and nineties.

LIGHT, EVOLUTION, AND CIRCADIAN RHYTHMS

The temperate zone light cycle of day and night represents a natural environmental extreme. This cycle can be expanded to an even greater extreme in the form of daily continuous light or daily continuous darkness under both natural (Arctic and Antarctic) and artificial circumstances. The extremely sensitive mammalian retina can detect a lighted match in *darkness* a mile away; yet the eye in *daylight* must protect itself from 10,000 foot-candles of direct sunlight. Some mammals counteract this stimulus by being nocturnal; some spend their days in forests under deciduous trees where the low light value of less than 100 foot-candles in summer rises to 3000 foot-candles in winter. Regardless of habitat, there has been a powerful 24-hour cycle of light and darkness that has continued to influence the mammalian group since its origin 160 million years ago (end of Triassic). Even the invertebrate ancestors of the vertebrates were influenced by this light cycle 500 million years ago. This influence is unlike those of localized environments such as temperature, water, or mountains, in that *all* mammalian ancestors at some time have had to experience the light-darkness extreme. This is one factor of evolutionary pressure that is common to all mammals, and one they share with other forms of life including the plant kingdom. The result has been the 24-hour biological rhythm, a composite coupled with the light-darkness cycle; it can be detected in the physiology or behavior of most animals and green plants. Since this daily biological rhythm usually does not last for exactly 24 hours, its timing is now called *circadian*, meaning "about 24-hour." It is not only coupled with the light-darkness cycle but was probably caused by it. The specialized field of such rhythms is *Chronobiology.*

The circadian rhythm of physiology or behavior is the most common type of the several biological rhythms, and it will be the only one discussed in full in this chapter. Although the light-darkness cycle is coupled with this rhythm, this physical cycle is also reinforced by lesser factors such as temperature. For purposes of explanation, our attention must be limited at first to the relationship between light and rhythms, although it is evident that in the free environment other factors may be present. For example, in many parts of the globe, the seasonal variation often results in warm long-days contrasted with cold short-days, and the 24-hour period itself may consist of a humid and cold darkness compared to a drier and warmer daylight. *Thus, the natural occurrences of these multiple factors force the investigator to interpret cautiously the results of single-factor studies from environmental chambers.*

The circadian rhythm deserves close inspection and analysis by all students of biological science because it is ubiquitous, and because several important and intriguing biological problems are implicit in all observations of this rhythm. Two representative questions are 1) is the rhythm inherited or learned; and 2) why, in most cases, does the rhythm represent a biological process or mechanism that is independent of temperature? To answer such questions we begin with examples of circadian rhythms.

TYPICAL MAMMALIAN CIRCADIAN RHYTHMS

Activity Rhythms

We can find typical circadian rhythms by recording the locomotor activity of rodents (Fig. 3-1). Soon after "sunset," nocturnal rodents will spontaneously begin to exercise on a running wheel, often running 6 to 10 miles a night. The pattern of running for a day-active rodent may be the same, but will occur during daylight. The time of starting activity is quite regular; Fig. 3-2 gives examples typical of three species. All of these records were made using early classic instruments; today such records are usually made with self-graphing recorders or on-line computers.

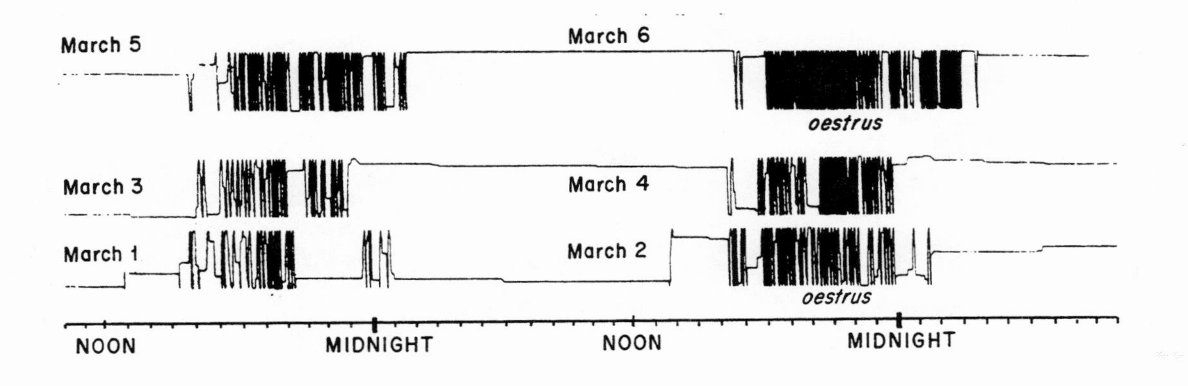

Figure 3-1. Running Activity of the Rat. Six-day record on paper of persistent spontaneous running activity of the white rat. During this period, the rat was in a soundproof room in continuous darkness. Each complete excursion of the pen represents 200 revolutions of the running wheel. The drum turned once in 48 hours.

Advantages of Running Wheel Records

The student of mammalian biological rhythms is faced with deciding which physiological measurement is the best index to the rhythmical status of the animal. The best technique for small mammals would be to measure three physiological functions continuously, or at least as frequently as possible. Practical measurements might be heart rate or body temperature by radio capsule, volume and chemistry of voided urine, and locomotor activity. Continuous measurements are technically difficult, and many investigators settle upon recording a single function. Should one select a behavior measurement (locomotor activity) or a physiological measurement? Locomotor activity, whether recorded by tambour cage, photoelectric cell, or running wheel, measures the time of the greatest metabolic expenditure almost as well as if oxygen-consumption were being recorded.

If the activity is measured by running wheel, the amplitude of daily spontaneous running is of interest as a feat of "athletic accomplishment." Activity records frequently prove that the animal does not stop to rest throughout a period of running, and it travels 8 or 10 miles. Female white rats in estrus, although running more intermittently, frequently complete 15 miles in 12 hours. The white rat used as an illustration in Fig. 3-3 would run 21 miles in one night; Richter and Sloanaker recorded astonishing distances in this species of 27 miles and 38 miles in 24 hours. These efforts were more strenuous than the distances alone represent, since the rats were running *uphill* on the inclined side of a running wheel.

The accomplishment of these short-legged animals compares favorably to that of the best runner among some trained dogs at the Harvard Fatigue Laboratory; on an inclined treadmill this dog ran in one 17-hour period to 2 times the height of Mount Everest (reaching the equivalent altitude of 14 miles, or 23 km) (Dill 1960). While accomplishing this test, the dog willingly ran the 125 miles with only a 10-minute break each hour for food and rest. In contrast, the running of the rodents just described was spontaneous. The occurrence of these prolonged rodent "marathons" increases our interest in the *running period,* which is but one sequence in a variety of daily behaviors.

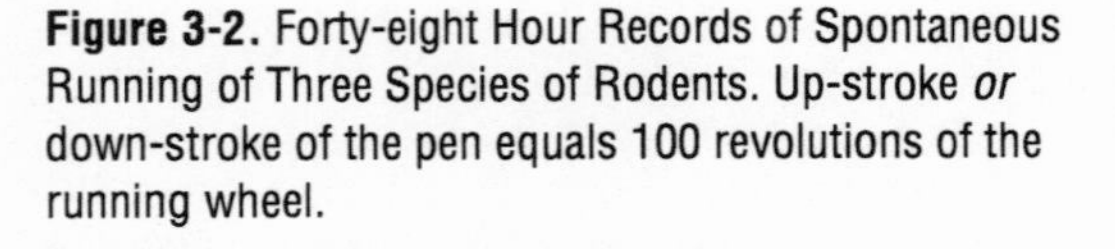

Figure 3-2. Forty-eight Hour Records of Spontaneous Running of Three Species of Rodents. Up-stroke *or* down-stroke of the pen equals 100 revolutions of the running wheel.

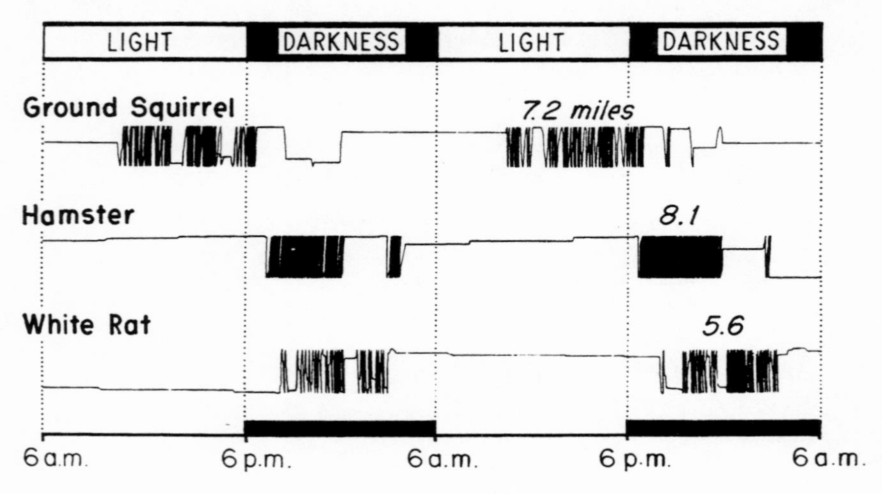

What is the meaning of this nightly running period? Richter, a pioneer in the field, called it a *hunger reaction* or *regulatory behavior.* Random hunger running, in the free environment, increases the probability that the mammal will encounter food. Some mammals in cages run vigorously away from available food because running is the "correct" sequence in the *daily series of events.* A test of the hunger reaction was made by George Wald by simply removing food from rats. Hunger running started at 1,000 revolutions a day and increased for each of 6 days by 1,000 or 2,000 revolutions. As expressed by Wald (1944), "a hungry animal stakes its metabolic reserves against the chance of finding food." However, until starvation is near, the amplified hunger reaction occurs at the same time of day as the normal hunger running reaction.

Another cause of increased running is the *gonadal-endocrine component* superimposed on the basic hunger reaction. One expression of this component is the *estrus peak of high running,* which is observed every 4 or 5 days in the records of female rats (Figs. 3-3) and hamsters. *Estrus* means sexual receptivity; which is usually coupled with hyperactivity (see Chapter 2, Examples of Breeding Cycles). Again, as we can see in this figure, the amplified running comes at the usual time in the 24-hour sequence of behavior. This excessive running increases the probability that the estrous female will cross the path of a male and will be bred. It is seen in many other mammals such as dogs, horses, and cattle. At times the circadian biological clock is necessary for the condition of estrus; if rats in continuous light do not have a regular daily high-low light cycle, the four-day endocrine cycle is lost (Weber and Adler 1979).

One warning: Not all *individuals* of a species are good runners, and some mammalian *species* show remarkably little total activity each day. If an animal is quiet for 24 hours, one cannot assume that there is no 24-hour rhythm. A marked biochemical rhythm will still be present.

The Day-Animal and the Night-Animal

During the period of the prolonged running activity of the nocturnal animal, the *basal body temperature* or *core temperature* of the resting animal is higher during the night than it is during the day; likewise, resting nighttime *heart rate* is higher than resting daytime heart rate. Similar figures have been obtained for oxygen-consumption, total activity, bladder activity, kidney activity, and many other functions and mechanisms. It will be shown later that these higher nighttime values are *not merely the result of exercise,* although an exercise rise may be superimposed upon them.

Let us now accept the advantage of recording a variety of physi-

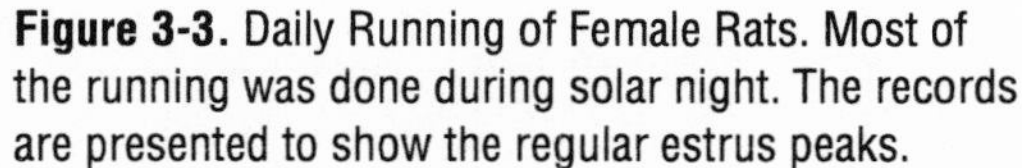

Figure 3-3. Daily Running of Female Rats. Most of the running was done during solar night. The records are presented to show the regular estrus peaks.

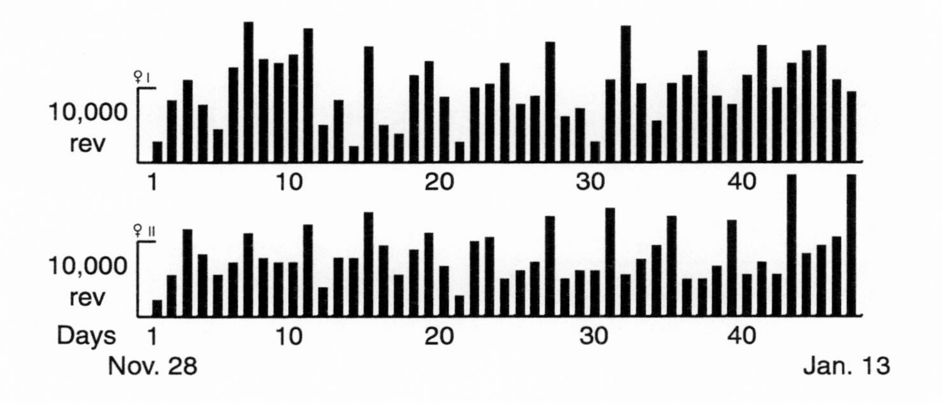

ological functions. Some of these functions can be measured by radio-telemetry with little disturbance to the animal (Folk and Folk 1979; Folk et al. 1995).

Of all the physiological measures, body temperature is the most popular. The circadian temperature rhythm has been studied in humans of all ages. It does not exist in the three week old baby (see Figure 3-4), but is well established at 10 weeks (Guilleminault et al. 1996). It is such a strong component of homeostasis that it is even present when a human has been declared brain-dead (Orita et al. 1995).

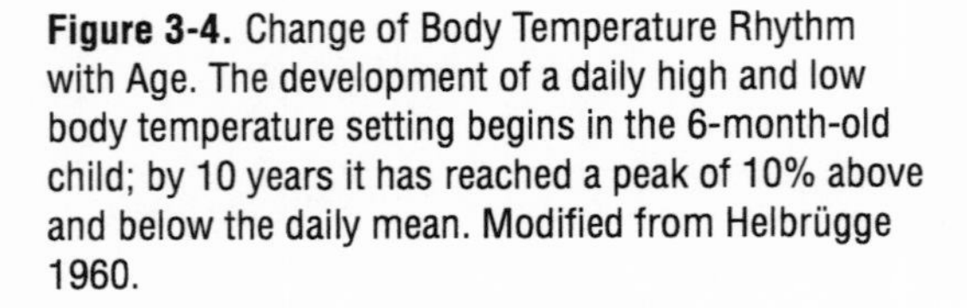

Figure 3-4. Change of Body Temperature Rhythm with Age. The development of a daily high and low body temperature setting begins in the 6-month-old child; by 10 years it has reached a peak of 10% above and below the daily mean. Modified from Helbrügge 1960.

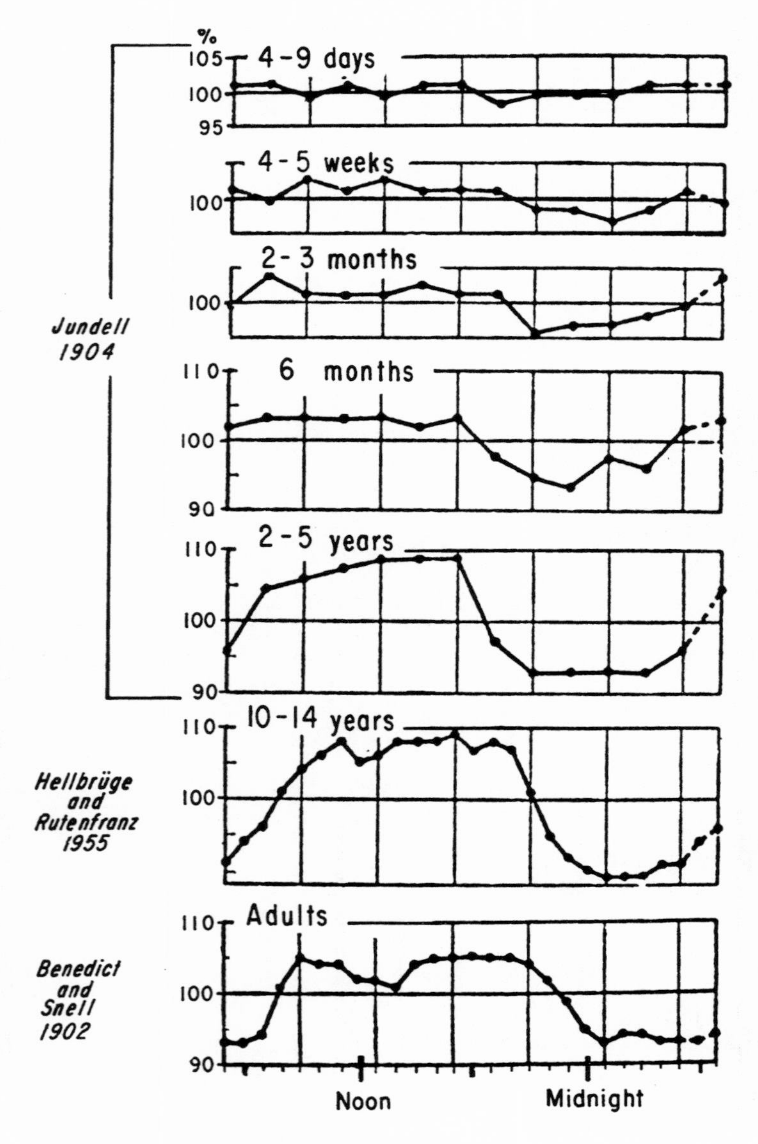

Cellular measurements have not been neglected. The peak and trough of circadian mitotic activity in the hamster cheek pouch and ear epithelium are the reverse of body temperature highs and lows (Folk and Folk 1979). Hormones in plasma are rhythmic; one with a particularly high amplitude is the growth hormone (Krieger 1979). A circadian rhythm of *serotonin* (see sidebar, Chapter 2) has been found even in spinal fluid (Taylor et al. 1985). The circadian nature of cholesterol biosynthesis has been well documented (Biali et al. 1995).

Simultaneous recordings of physiological rhythms are valuable for determining whether the operating mammalian "machine" consists of separate rhythms, perhaps slightly out of phase, or closely synchronized rhythms involving the entire physiology of the animal. Picture these simultaneous rhythms coming to a peak at approximately the same time of day; we can refer to the whole pattern as a *rhythm profile.* The important concept shown in this profile is that each animal is made up of two parts: the *day-animal* and the *night-animal.* The animal is thus committed to a continuous sequence of Jekyll-Hyde changes throughout its lifetime. This idea must be carried through the interpretation of all circadian rhythms, whether they concern the responses of Protozoa, or the abrupt changing of day-shift industrial workers to a new night-shift program. In the latter case, it is especially helpful to think in terms of the *day-person* and the *night-person.* The meaning of these terms will be discussed in more detail later.

TERMINOLOGY OF RHYTHMS

Free-running, Endogenous, Exogenous. The typical mammalian rhythms that have been described are synchronized with the powerful solar light-darkness cycle. One approach to the study of any rhythm is to test for the dependence of the rhythm upon the light cycle by placing the animal under constant and continuous conditions of darkness (or light), temperature, and sound. Using rodents as an example, the results of this test show with remarkable clarity that the amplitude and period (length) of the activity rhythm does not depend upon a light-darkness cycle. Nocturnal rodents in continuous darkness make activity records for many months which can hardly be distinguished from records made with a light cycle. Most investigators today refer to such activity rhythms as *free-running* (Moore-Ede et al. 1982). (Note that use of this term is not to be confused with our use of the expression, *free environment.*) The period of the rhythm in darkness is never 24 hours, and may be about 23 or 25 hours; under these circumstances the designation *circadian* rhythm is especially appropriate. It is certainly preferable to the ambiguous term *diurnal* rhythm, which is found in some of the literature and which has several meanings.

The period of a free-run varies with individuals within a species; in one 40-day study of eight deer mice, three mice had a period of about 23.9 hours, two about 24.05 hours, and three varied from 24.1 to 24.4 hours (Pittendrigh 1957, 1981).

Rhythms that continue in constant darkness can be found in groups as diverse as crayfish, crabs, fruit flies and plants. However, all rhythms do not continue when cyclical environmental conditions are replaced by constant conditions. Various terms have been applied to these two types of rhythms, first by botanists and later by zoologists. Most investigators use either *free-running* or *persistent rhythm* for any biological periodicity (such as running or body temperature) that continues in a constant environment; in other words, light, sound, and temperature cycles are absent. If the rhythm does not persist, it is usually called *exogenous*; if still present, it is called *endogenous*. It is conceivable that a mammal may show both exogenous and endogenous rhythms; perhaps in continuous darkness its locomotor rhythm ceases (exogenous), but its biochemical rhythms continue (endogenous).

The approach to testing for free-running rhythms must differ with nocturnal and day-active animals. With nocturnal animals the rhythm will persist in darkness for months with minor changes in period length because darkness is not a positive environmental factor (Figs. 3-3 and 3-5). One experiment had a standard deviation of only three minutes (Pittendrigh and Daan 1976; Pittendrigh 1981). This persistence is not present in continuous light, which may affect the animal so that its activity shows a regular and constant amount of alteration which varies with the intensity of the light. As discussed in Chapter 2, this effect is recognized as an application of Aschoff's rule. In view of this alteration of the rhythm, how can day-active mammals be tested? Continuous darkness cannot be used since it alters the activity of such mammals. For example, some species of day-active ground squirrels, showing a distinct running pattern of 7 miles daily, cease running completely in total darkness. According to Aschoff, the solution is to use continuous dim light, which alters very little the normal activity pattern of day-active animals.

Cycles. Another term, *cycle,* has gradually attained specific meaning among students of biological rhythms. It is applied to *physical* phenomena (for example, seasons, light cycles) and to *physiological* periodicities which are short-term and have no relation to solar or other cosmic events. Examples of such physiological periodicities are heart cycles, menstrual cycles, and the 4- or 5-day estrus cycles of rats. Note in Fig. 3-3 the regularity of not only the estrus running of the control female, but especially that of the reversed cycle female; reversal did not alter the estrus regularity.

Figure 3-5. Induced Metabolic Rhythms in *Euglena*. Artificial rhythms were succcessfully induced by light-darkness cycles, and these rhythms were tested for persistence by using continuous darkness. The *Euglena* always reverted to its free-running (natural) period. Above each horizontal rhythmical record the experimental light-dark cycle is indicated. Each vertical line represents a twenty-four hour period. Modified from Schnabel 1968.

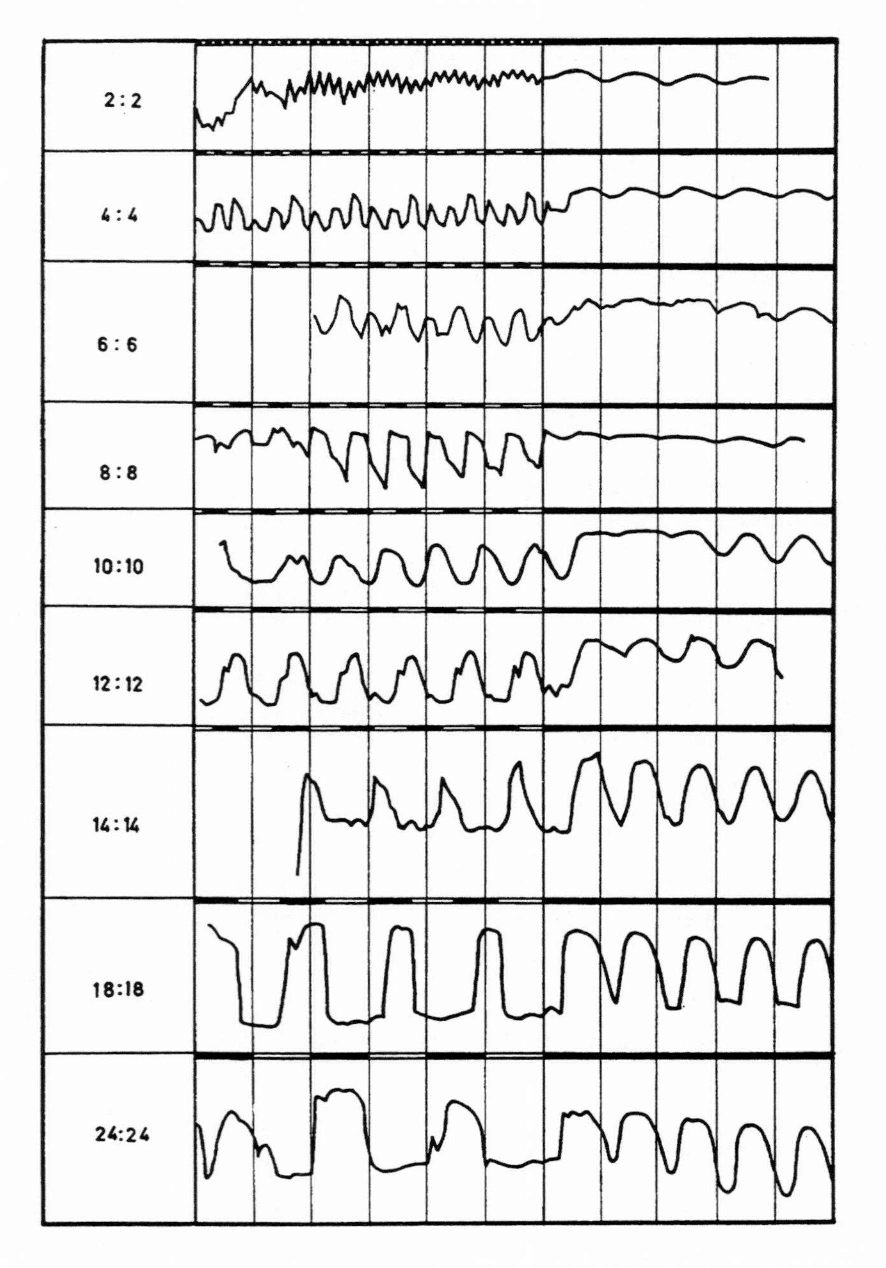

Period, Frequency, Amplitude, and Phase. The most common term in use by chronobiologists is *period* (tau, signified by the Greek letter τ). This is very familiar because we are used to speaking of the 24-hour period (i.e., 24-hour day). It is immediately clear when we say that an animal in darkness has changed its activity to a rhythm with a period of 22 hours or 26 hours. The period is the time interval between two identical points or defined events in a rhythm; these events might be the start of the activity block, its peak, or its end.

The next most common term is *frequency*; this is the reciprocal of the *period*. The frequency is the number of cycles occurring per time unit. Rhythms with circadian periods are *low frequency* rhythms, while rhythms shorter than circadian periods (2.5 to 8 hours) are *high frequency*.

Another term used to describe a rhythm is *amplitude*. Technically, the calculated amplitude is the difference between the maximum (or minimum) and the mean value in a sinusoidal rhythm. The maximum is also called the *crest* or peak. However, some rhythms do not describe a sinusoidal curve; instead the curves have a flat bottom or are more in the shape of a square wave (see Figs. 3-4 and 3-5). A plot of the hourly running of rodents often describes a square wave.

The amplitude of any rhythm may vary systematically; an illustration is seen in the effect of the 4- or 5-day estrus cycle on the daily rhythm (see Fig. 3-3).

The *phase* of a rhythm is somewhat vaguely defined as the *instantaneous condition of the rhythm within a period* (Moore-Ede et al. 1982). Essentially it is the placement of a rhythm in time. A clue to the phase is a *phase point* which might be the onset of locomotor activity of a rodent each night. If by light manipulation, the onset and the activity comes later, we say there is a *phase shift* in the direction of a *phase delay*. If two rhythms such as heart rate and body temperature have identical timing, we say they are *in phase*. If their peaks are 12 hours apart, they are *out of phase* by 12 hours. The *acrophase* is the lag between the reference time (noon or midnight in a 24-hour day) and the time of the rhythm's crest (Cornelissen et al. 1989).

The waveform from rhythm data can often be quantified by fitting a cosine curve; calculating this best-fit curve is like obtaining the deviations from a straight line in a linear regression analysis (Halberg and Howard 1958).

Biological Clocks. In about 1940, the concept was introduced that the physiological or behavior rhythm, especially when free-running, was used by the animal to measure time accurately. Then, between 1940 and 1960, more and more biologists began writing about biological "clocks." This at first appeared to be a term too loosely used, but numerous papers since 1950 have justified these statements: *Many animals can measure time accurately,* and *a clock is an integral part of animal physiology.* However, there is a tendency to overapply this concept. The more useful term is *rhythm,* because if a tissue has a rhythmical function, it is a clock; i.e., every rhythm measures time. Nevertheless, it is useful for us to consider what might be a component of a biological clock.

The Hands of the Clock. One goal of this chapter is to analyze the *mechanism* of the biological clock. In some cases it is useful to consider that observed rhythms such as the locomotor rhythm are not *the* clock mechanism, but are the *result* of the mechanism. The locomotor activity represents the *hands* of the clock. Moore-Ede uses as an example an animal with its limbs tied. It will not show its activity rhythm (i.e., the hands of the clock), but its underlying circadian clock keeps on ticking. Once the animal is released, it will resume activity at the circadian time determined by the mechanism of the clock (Moore-Ede et al. 1982).

Oscillators. A well-accepted term which adds concreteness to the concept of the biological clock is *oscillator*; it is basic to the working theory to explain the *mechanism* of biological rhythms. Pittendrigh (1957) first described the oscillator model and developed appropriate related terminology; apparently the model is adequate for interpretation of most rhythms studied so far. His model could be called a *biological pendulum*. He suggested that the basic element in all biological rhythms is an *endogenous self-sustaining oscillation* (ESSO). This concept, that animals and plants are composed of interacting oscillators that *result* in circadian rhythms and ultradian rhythms (2 or more cycles within 24 hours), has received firm acceptance. As a development of this concept, there has been a tendency to substitute the term *oscillation* for *biological rhythm;* this approach, intended to improve theorizing, has been helpful. For example, as early as 1973, Britton Chance and others edited a large volume entitled "Biological and Biochemical Oscillators" (Chance et al. 1973).

As an extension of the use of an oscillator for the mechanism controlling a biological rhythm, the concept was next developed of a coupled oscillator system consisting of primary and secondary oscillators (Pittendrigh and Daan 1976). We will cover this in more detail later. However, the oscillator model does not cover all cases; Bowen describes two species of insects that make use of an hourglass mechanism for time measurement, rather than an oscillator (Bowen and Skopik 1976).

Polar Plots and the Acrophase. Before considering a common term, the *acrophase*, we will begin to think of rhythms in terms of circular plots. Any time period, whether designated in hours, days, or months, can be conceived as a circular process; thus a rhythm being analyzed can conveniently be depicted as a 360° rotation. The periphery of the circle can have designated time units, but it is usually marked off in degrees totaling 360° (Fig. 3-6). These *polar plots* will be described in detail later. Any two selected radii on the circle will set apart or designate a part of the rhythm; these radii will describe a phase angle. One type of phase angle is the acrophase. It is a description of the crest of an oscillation in terms of time or degrees. For example, Aschoff (1981) discusses a series of seasonally timed events over a yearly time scale, including suicides, conceptions, and mortality. He considers the month of maximal value to be the acrophase.

In an earlier section we discussed treatment of data in the form of a sinusoid instead of a circular plot. On the waveform of such a continuing sinusoidal curve, if it has a time base, the acrophase is the time of the peak or maximum.

Figure 3-6. Example of a Polar Plot. Cyclicity of leukocytes in the blood of Arabian horses. Acrophases: 1. Group I—pregnant mares; 3. Group II—barren mares; 3. Group III—foals born in 1981; 4. Group IV—foals born in 1982. Dotted lines indicate error field. (Halberg 1990)

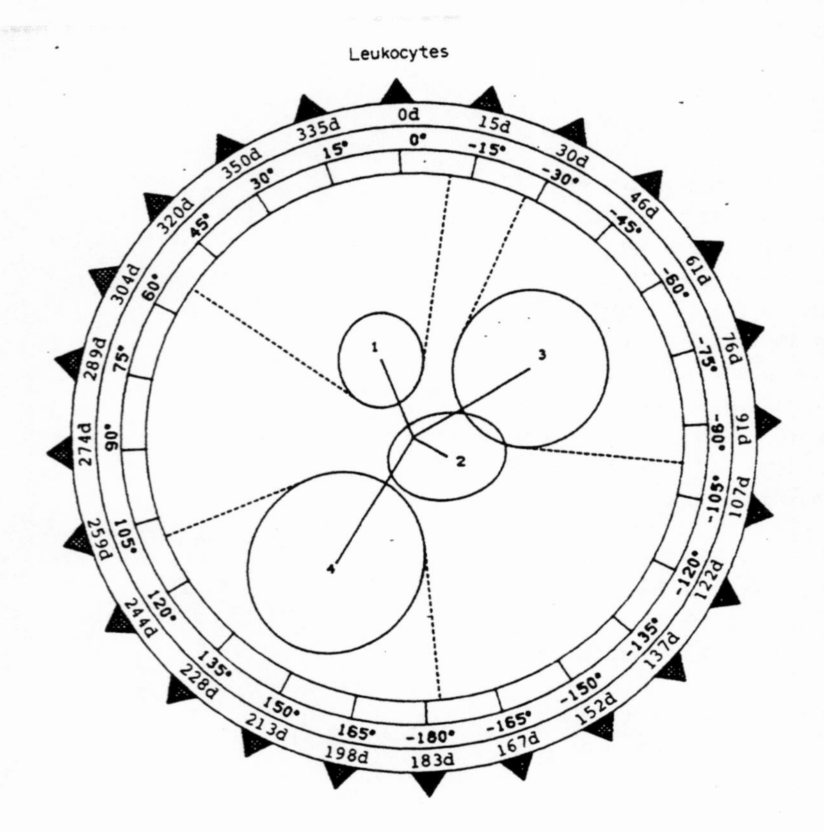

Non-Circadian Rhythms. Biological rhythms with a period shorter than that of circadian rhythms are called *ultradian*; those with a period longer than that of circadian rhythms are *infradian*. Examples of ultradian rhythms (more frequent than 24 hours) are found in the unusual activity rhythms of insectivores and some types of wild mice, which have short endogenous rhythms (2.5 to 8 hours) uninfluenced by any light-darkness cycle. However, some fossorial (*underground-living*) rodents are completely arrhythmic (Hickman 1984). There is a one-hour rhythm in the internal and external temperature of the udder of cows (Lefcourt et al. 1982). An example of a 4-hour rhythm is found in the core temperature rhythm of an arctic ground squirrel (Fig. 3-7); the temperature peaks correlate with activity peaks. At the opposite extreme there are 2-day rhythms (infradian) in human functions, and incongruously, in the flight activity of mosquitoes (Clopton 1985).

Ultradian rhythms of growth are also found in tissue cultures with periods from 1 to 7 hours and up to 34 oscillations, and also with a period of 12 hours, with 3 oscillations (Chance et al. 1973; Kadle and Folk 1983).

Another ultradian rhythm is called *circatidal*

(period 12.4 hours). The fiddler crab shows two free-running rhythms superimposed: a circadian rhythm of color change and a circatidal rhythm of activity of 12.4 hours. Many marine worms also have a tidal rhythm (Saunders 1977). Two types of infradian rhythms are called *circalunar* (period 20 days) and *circannual* (period one year). Circalunar rhythms are apt to be oceanic; there are five species of marine worms which swarm to the water surface in a lunar rhythm. One species appears during a few nights in the last quarter of the lunar cycle; another during the third quarter of the cycle. Some of the species were tested in an unchanged photoperiod; their lunar rhythm was found to be endogenous with a free-running period longer than the lunar cycle. In a later section, a human lunar rhythm will be described.

An example of a long-range annual cycle is the sooty tern which breeds every nine months (see Chapter 2). Another case involves ground squirrels which have been blinded; they repeated their annual reproductive and hibernation rhythms with a period close to one year (circannual rhythm) (Pengelley 1974; Pengelley et al. 1978). The mechanism behind the circannual rhythm is believed to be circadian; in other words, the physiological status of the animal at the time of the summer solstice (longest day of the year) is the climax of an accumulation of slight, daily circadian changes.

Abbreviations. Symbols for test environments for biological rhythm are now standardized (Folk 1974):

L = artificial daylight

D = darkness

If "dawn" is at 0800 (8 AM) and "sunset" at 2000 (8 PM), the conditions are described as: LD 12:12 (8-20).

LL designates conditions of constant light

DD designates conditions of constant darkness.

New symbols should be added to this scheme:

Figure 3-7. Daily Body Temperatures of Arctic Ground Squirrels. Circadian body temperature rhythms and activity records by observation; animal was in an outdoor enclosure in continuous Arctic daylight. Body temperatures taken by radio-capsule.

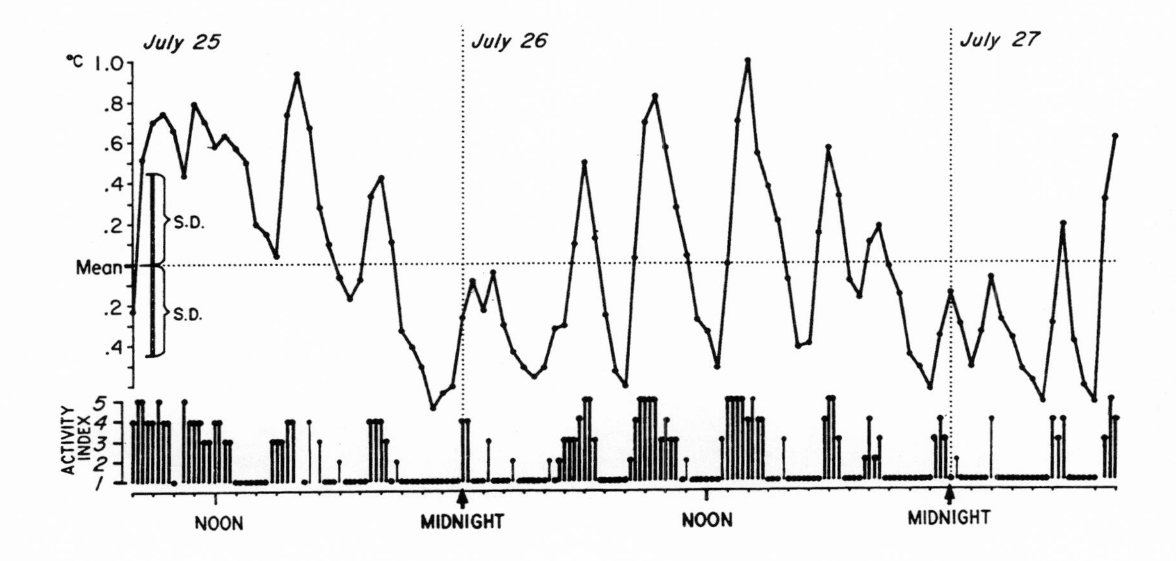

CT = controlled* temperature
CS = controlled sound
CH = controlled humidity
* *controlled* means constant or non-periodic.

A test environment can be concisely described as follows: (DD, CT, CS) or (LL, CT, CH).

STATISTICAL ANALYSIS

There are varied techniques for the application of statistics to biological rhythms (Luce 1970). One approach is to prove the data are rhythmic by use of spectral analysis, and then by the *least squares* procedure. This will tell whether a cycle is regular, and the procedure will specify the amplitude and phase of the rhythm. If the cycle is stable and relatively nonvariable, then one may make a polar plot as described earlier. Taking a 24-hour cycle as an example, one may record this in the form of the circumference of a clock. One of the clock hands might represent the phase at which one daily peak event occurs, while another hand may indicate the peak phase of some other physiological occurrence. Or one could use this polar display to compare physiological rhythms of several individuals. Such a plot would not show exact peaks, but rather an average estimate or typical phase representing the timing of all peaks that occurred during the collection of the data. Also, Halberg has made use cosinor analysis in his "cosinor polar plot," which expresses amplitude by the length of the hand of the clock. By this procedure it is possible to quantify aspects of rhythms that are very cumbersome in graphs, charts, and tables.

ZEITGEBERS AND ENTRAINMENT

Modification of Free-Running Rhythms

We discussed earlier some free-running rhythms of rodents which persist in continuous darkness or continuous dim light; these rhythms can be modified by clues or signals from the physical environment called *zeitgebers* (timegivers). We say that the free-running rhythms are composed of a series of events following each other in a chain-pattern: sleeping-grooming-feeding-running-defecating-sleeping-grooming, etc. Perhaps this free-running rhythm has a period of 25 hours, but if we turn on a 24-hour light-dark cycle (for example LD 12:12), the chain pattern will take on this new period of 24 hours. Dawn can be programmed at 0600; this process is called *entrainment*. Next, if we experimentally move the sunset backwards by 6 hours to noon, then the entire sequence as a unit is caused to rotate "like a dial" backwards to match the new light cycle (Fig. 3-8). This process is another example of entrainment.

Entrainment by Light in the Free Environment. The illustration described above made use of a laboratory light cycle of 12 hours. In the free environment, such a regular light cycle seldom exists because the time of dawn or dusk changes each day or at least each week. Thus, slight entrainment occurs continuously as the seasons change. The environmental change "taps the circadian pendulum."

We say that the rhythm shifts to match the period and phase of the environmental clues. The fact that the environment does change each day in most parts of the globe shows the importance of thinking of the activity sequence as a dial which is normally and sensitively shifted or readjusted by regular changes in the natural light cycle. Light reversal experiments of 12 hours are really not unbiological, but are accelerated cases of a natural process. Some details of this natural process were recorded on flying squirrels and primitive primates in outdoor enclosures; all species showed that environment had a regulatory effect on activity (De Coursey 1960; Kavanau and Peters 1976).

We should consider some of the extremes of this environmental control by light. In the Arctic and Antarctic there is a seasonal variation from constant light to constant darkness; in an area such as southern England, the length of daylight varies seasonally from about 8 to about 16 hours. In such an environment, the daily zeitgeber (dawn or sunset) must change the period of animal activity by a small amount each day or every few days.

Does an Entrained Rhythm Persist? In Fig. 3-8 we observed the gradual entrainment of a hamster to an early sunset (6 hour change). What happens to this new pattern of activity (which has a new relationship to solar time) when the animal is then placed in continuous darkness? Let us take a more extreme case of an entrainment by 12 hours (complete light reversal). This is a common procedure endocrinologists use so that they may work with the night-animal during their own bankers' hours (0800 to 1700). If the reversed animal is placed in continuous darkness it will free-run; but does the activity block revert to the former nighttime (1800 to 0600), or to the entrained nighttime (0600 to 1800) as it free-runs? In one experiment, a reversed-cycle female rat limited her activity for 31 days to her entrained nighttime; she did not revert by 12 hours to the former nighttime. Many other experiments show that the entrained rhythm begins to free-run in darkness at the new entrained activity time, not the original control activity time.

Limits of Entrainment. Up to now our discussion has emphasized components of light and darkness which total 24 hours. The next reasonable experiment is to try to entrain mammals or other organisms to light-darkness cycles with a 28-hour period or an 18-hour period. The results depend upon both the kind of mammal and the rhythm being studied. In some cases, mammals will take on these rhythms with no effect on health or growth (Ingram et al. 1985): however, this entrainment does not persist if a nocturnal animal that has been entrained is placed in continuous darkness. Furthermore, the limits of any entrainment are quickly discovered; for ex-

Figure 3-8. Example of Phase Change. Rate of adjustment of female hamster to a sunset moved earlier by 6 hours.

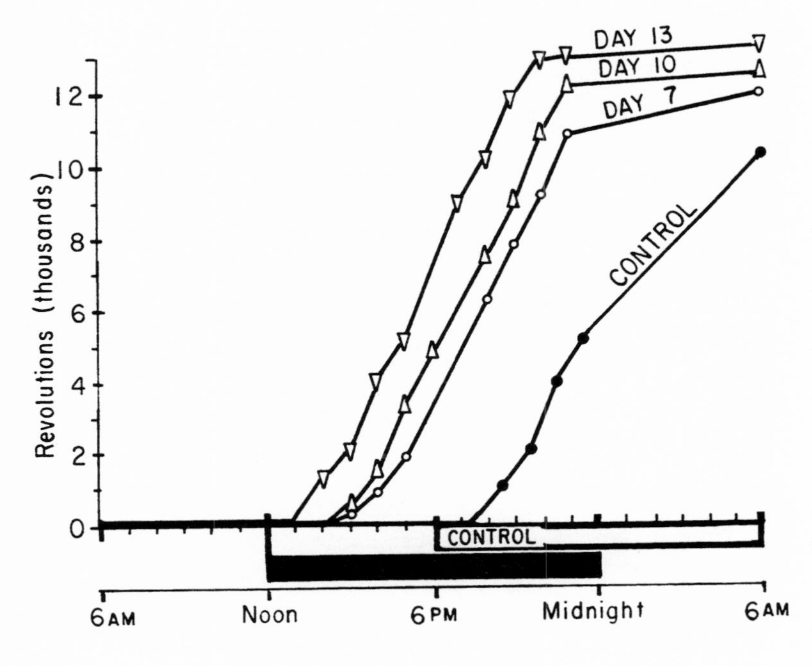

ample, if a light-darkness cycle of 15 hours is imposed on a naturally nocturnal animal, then the animal breaks up its activity into two components, or may even become arrhythmic.

Light and 16-Hour Animals. A number of investigators have attempted to convert rodents from their circadian pattern of activity to one of 16 hours (LD 8:8) by entrainment with a light cycle. Using the standard laboratory light cycles made by abruptly turning a light bulb on and later turning it off, no one succeeded in obtaining so-called "16-hour rodents" (Hoffman 1968; 1969). However, when Kavanau simulated dawn and dusk by gradually lowering the lights or gradually raising the light intensity, his mice responded to a pattern of LD 8:8. Two peaks of activity were noted, one during the dusk of the 16-hour day, and one during the dawn (Kavanau and Peters 1976). As in so many similar experiments, Kavanau did not test for free-running after the 16-hour experiments.

Changing Sensitivity to the Zeitgeber. Single brief pulses of light can cause shifts in the steady-state phase of the free-running circadian locomotor rhythm. When a hamster, free-running in constant darkness, was exposed to light for 15 minutes at the start of the activity block, there was an immediate phase delay; when the 15 minutes of light fell on the second portion of the locomotory pattern, there was a gradual phase advance (Fig. 3-9). In the latter case there were several so-called "transients" before a new steady-state pattern was obtained. From this experiment we find that a prerequisite for entrainment is a *periodically changing sensitivity to the stimulus* provided by a zeitgeber (Pittendrigh 1964).

Figure 3-9. Locomotor Activity Rhythm of the Golden Hamster *Mesocricetus.* The hamster was free-running in DD and perturbed by 12-hour pulses of white light. At (a) the pulse falls early in the subjective night and causes a phase delay which is accomplished almost immedicately; at (b) the pulse falls late in the subjective night and causes a phase advance which passes through 5-6 transient cycles before achieving steady state. Modified from Pittendrigh 1964.

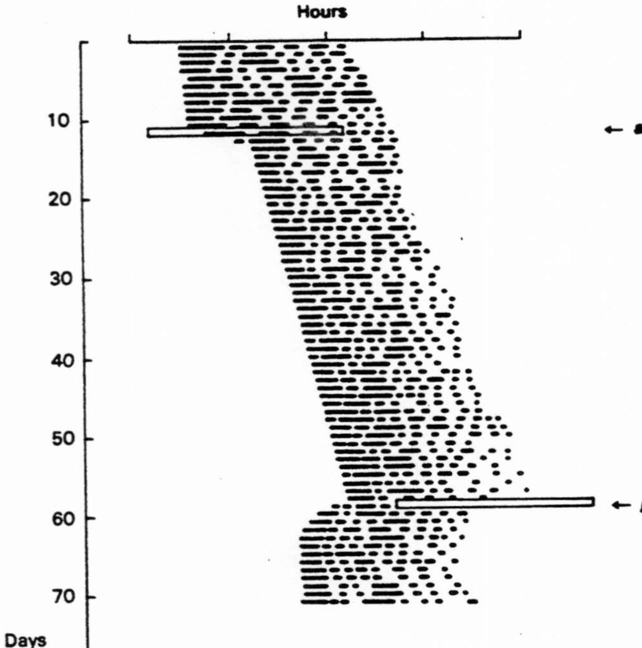

By applying light pulses to all portions of the free-running circadian rhythm of activity, it is possible to generate a phase response curve upon which *phase advances* are plotted as positive and *phase delays* are plotted as negative. These data are obtained in reference to what is called the "objective day of the hamster" (the first half of the free-running rhythm) compared with the "subjective night of the hamster" (the second half of the circadian rhythm) (Takahashi and Zatz 1982). Here is a specific example about the phase response curve: if a flash of light is delivered to a free-running hamster in constant darkness at subjective night, there is no shift the next day; if given at 1300, there is a delay of 1 hour the next day; if at 1700 there is an *advance* of 2 hours the next day; if at 2200 there is an *advance* of 20 minutes the next day.

In a phase response experiment on the rhythm of the isolated eye of an invertebrate, the sea hare *Aplysia*, an inhibitor of protein synthesis was applied to the eye. The result was that the rhythm was advanced or delayed, depending on the phase at which the pulse was applied (Jacklett 1969; 1977). The eye showed an ability to carry out circadian activity in a culture medium, maintaining a period of about 27 hours. The phase of this rhythm appears to be under the control of adenylate cyclase acting upon serotonin in the pacemaker of this eye of *Aplysia*

(Eskin 1983). Injected serotonin shifts the phase of this circadian rhythm (Corrent, McAdoo, and Eskin 1978).

We must now consider Aschoff's rule and its relationship to this changing sensitivity phenomenon.

Aschoff's Rule and the Phase Response Curve. In Chapter 2 we saw that when a nocturnal mammal is placed in continuous light, the time of beginning activity is delayed a specific amount depending upon the intensity of the light (Aschoff 1955). The peak of activity of such an animal travels around the clock (Folk 1959). The observation is referred to as *Aschoff's rule.*

In the preceding section we stated that, in DD, if a light stimulus falls on the first part of the activity period, then the activity is delayed the next day, but if the light falls on the second part of the activity period, the activity is accelerated the next day. When the animal is in continuous light, both of these stimuli occur at the same time; yet the activity of the animal is only *delayed* and travels around the clock. We deduce that this is an example of a statement of Aschoff's to the effect that the animal obeys the strongest of the zeitgebers. Apparently, the stimulus producing the delay in time of activity is stronger than the stimulus producing the advance in time of activity. The explanation of why this occurs is clarified by consideration of the two-oscillator theory, first described in an earlier section.

The Two-Oscillator Theory. The above discussion of phase delay (peak activity comes later each day) and of phase advance (peak activity comes earlier each day) depends upon some details of events concerned with changing sensitivity to the zeitgeber: the rhythm reaches a new steady-state almost immediately after causing a delay-type of phase shift; however, in the case of an advanced phase shift, the system passes through several non-steady-state cycles called *transients* before achieving entrainment (see Fig. 3-9). Some theoretical models for the circadian biological clock suggest that a single oscillator can explain these phase shifts, but the transients just described can best be explained by the two-oscillator model proposed by Pittendrigh. It is as if the first oscillator is responsible for phase delay and the second oscillator is responsible for phase advance. We will refer to the phase delay oscillator as the light-sensitive *A*-oscillator; it is the pacemaker, and it is responsible not only for phase delay, but also for another oscillator, the *B*-oscillator, which governs phase advance. The *A*-oscillator is immediately reset (perhaps within minutes) by a light pulse, but the *B*-oscillator must work through the pacemaker (*A*-oscillator), and it requires several cycles before it obtains a steady-state relationship to the driver oscillator; hence the series of transients.

ASCHOFF'S RULE

This phenomenon is described in Chapter 2. Although the effect on day active animals is different from that on night-active animals, we will discuss only the latter. In an application of Aschoff's rule, the block of activity in the night-active animal comes in darkness. When light falls on this block, the animal does not instantly switch its activity to a dark period. For a while it runs in the light, but gradually moves its activity block to fit the daily darkness. The rate of moving the block to darkness depends on the intensity of the light; a quantitative photoperiodic phenomenon.

This rule applies to numerous groups of mammals including bats (Bay 1978), and also the invertebrates.

Fiddler crabs (*Uca subcylindrica*) from south Texas also showed the Aschoff effect in continuous light. Under this regimen, their locomotor rhythm took on a frequency of 24.6 ± 1 h (N=19). Curiously, this species of crab did not show the tidal rhythm seen in other members of the genus (Broghammer and Thurman 1997).

Non-Photoperiodic Entrainment

What components of the physical environment can "impose their will" on the free-running rhythm? They must be events that occur approximately every 24 hours. They might be a) the start of a light cycle, b) the end of a light cycle, c) a regular artificial feeding time, d) the regular noise of other animals in cages, or e) a change in environmental temperature or barometric pressure. Two flashes of light simulating the start and end of a photoperiod (referred to as a skeleton photoperiod) will suffice (Pittendrigh 1964). Such physical events today are called zeitgebers, although earlier they were called synchronizers and clues. Let us discuss several of the physical events from the above list.

Temperature has been used to initiate the start or end of activity (in other words for entrainment). Hoffman (1968) influenced the circadian periodicity of lizards and field mice by varying ambient temperature in sinusoidal fashion. He was able to use temperature as a replacement for the beginning of a light cycle with the lizards, but not with the field mice. However, with the field mice, high or rising temperatures could suppress locomotor activity without actually synchronizing as a zeitgeber the circadian oscillation.

Aschoff (1979a) provides a more specific example. He maintained a pig-tailed macaque (*Macaca nemestrina*) in conditions of constant illumination (450 lux), and found that its time of starting activity phase advanced by more than 1 hour a day, thereby following Aschoff's rule for day-active animals. The next experiment was also in continuous light; when 12 hours of 17°C ambient *air* was alternated with 33°C air, the animal entrained perfectly, after some transients, to the time of temperature *change*, acquiring a 24-hour rhythm. Entrainment overrode the Aschoff effect. The same results were obtained with squirrel monkeys (Aschoff 1981). However, other workers using other species found that temperature does not act *alone* as a zeitgeber, but it does influence (change) the effect of light as a zeitgeber (Tokura and Oishi 1985).

Another illustration of an unusual zeitgeber is an experiment in which a 24-hour cycle of *pressure* was used, varying only from 1.0 to 1.09 atmospheres. Under constant conditions of environmental temperature and light, the pocket mouse (*Perognathus*) entrained its circadian rhythm of body temperature to the 24-hour cycle of pressure (Hayden and Lindberg 1969).

An even more unusual zeitgeber was demonstrated by Remmert (1980); an artificial sun was rotated once in 24 hours around a caged bird. The bird showed a strict 24-hour rhythm. When the same bulb was stationary, the bird's own free-running frequency took over. Probably the shadows cast by the moving light acted as zeitgeber.

Meal timing can be substituted for the light zeitgeber (Folk 1954). As is usually the case, the exact time of applying the zeitgeber is important; food must be given only during the *early* part of the original light-span (Caradente et al. 1982).

BIOLOGICAL CLOCKS

Advantages

The previous discussion of biological rhythms has emphasized the basic free-running rhythm, which in the free environment is continuously entrained to an almost 24-hour period by environmental factors, especially light. Regardless of the result (the rhythm might have some variation of a 23.5- or 24.5-hour period), the rhythm serves to measure time fairly accurately. After all, it *is* a clock.

There are several advantages to such a clock for animals in free environments. One advantage has been described in relation to Cannon's concept of homeostasis. He emphasized the constancy of the internal environment of animals. Moore-Ede now points out that the biological clock *predicts* an oncoming environmental challenge to constancy, and initiates appropriate corrective response *in advance* (Moore-Ede et al. 1982). Other more specific advantages are the action of rhythms to stand in for zeitgebers, and their assistance to orientation and navigation.

Replacement Zeitgebers. In the free environment, zeitgebers are not regular, and there is known to be "competition" between them. Sunrise may be completely hidden for a number of days; at that same time the barometer may be unusually high and the environmental temperature may show a large change. When such disturbed environmental conditions happen, the internal clock of the mammal maintains its habitual program. This persistence of regularity is of advantage to the species. Perhaps the necessity for the mechanism can be illuminated thus: the ground squirrel who is off in his burrow sleeping when the rest of the colony is active, will miss the breeding activities of the colony.

Navigation by Vertebrates. Birds, fishes, reptiles, and numerous invertebrates are known to navigate by the sun. They travel to various goals by maintaining an appropriate angle to the sun, or to polarized light when the sun is not visible. They must change this angle according to the time of day. Thus, they must measure or know time accurately in order to make appropriate corrections. Either with or without a zeitgeber as a starter, they appear to measure off personal time units; perhaps we can think of these as bird-units, fish-units, bee-units, or spider-units. In Chapter 2, we mentioned Emlen's experiments on the celestial navigation by birds during which they need to visualize only 8.9% of the area of the night sky with visible stars in order to demonstrate orientation toward the path of migration. To interpret the star pattern, they must have an internal clock.

The ultimate demonstration of the ability of birds to use sun or celestial navigation came in two experiments: a Laysan albatross returned to Midway Island from Washington state, a flight of 3,200 miles which took 10 days; a shearwater returned to Scotland from New York, a flight of 3,600 miles (Matthews 1968). The story of the penguins is even more unusual: in the mid-1960's, Soviet researchers who released penguins at the Antarctic Base in Mirnyj documented one bird who returned to its nest area after traveling some 2,800 miles along Antarctica's eastern shore. Penguins need the sun to navigate: when they have sunlight, their return path on the featureless Antarctic snowfields leads directly to their nests, varying from a straight line only by a factor of less than 1% (Emlen and Penney 1966). This course consistency is one of the most remarkable examples of accuracy of sun orientation in the animal kingdom.

Pigeons require a full day of experience with the sun on a particular day in order to use it as a compass; without the sun they use a magnetic compass (Wiltschko et al. 1981).

This relationship between biological rhythms and biological clocks is so important that we will describe an imaginary experiment to assist the student in understanding sun navigation. Suppose you remain in a constant environment (DD, CT, CS) for 2 days. You are then taken blindfolded in an ocean boat out of sight of land. Your blindfold is removed and you are told to row to shore. You know that shore is "to the west," but if you do not know the time of day (by a good physiological clock), then you cannot tell the difference between various positions of the sun, such as at 1000 or 1400, and you would not know which direction to go. Numerous experimental animals knew the time, even when previously kept in darkness for several days.

Animal navigators have another challenge; the sun's azimuth changes each day. Thus they must memorize this change in azimuth by day, by date, and in some cases, by latitude.

Navigation by Insects. Experiments with insects are of two classes, those which show that these animals *measure time* and those which show that they *keep time* by an internal clock. Ants and bees can learn certain feeding periods such as every 3 or 5 hours (they measure time), and they continue to come for food for 6 to 9 days after food is discontinued. Bees can also orient to food at some distance from the hive by traveling at an appropriate *angle to the sun.* Northern hemisphere bees were released under circumstances where they would fly in a southern direction using the sun as a reference, regardless of the time of day. When the bees were taken to the southern hemisphere and released, they might again have traveled south to attain their goal; but as the sun was in the inappropriate position from what they had learned, instead of traveling south, they traveled with a remarkable bias toward the east. They used the wrong angle to get to the experimental goal by employing the sun formula of the northern hemisphere. Cloudy days do not pose a navigational problem for bees; they seem to remember from previous days the sun's position at each time of day (Dyer and Gould 1981).

Navigation by Mammals. Mammals have not been shown to navigate by polarized light, yet this ability is found in so many other diverse groups that it is unlikely that they are exceptions. We may assume that mammals orient themselves, perhaps to recognize their territories or to escape to burrows, by seeing "streaks" of polarized light that change their angle to the observer as the day progresses. Even humans can learn to see polarized light.

The migration distances of birds and mammals are well-known. Norris points out that fur seals migrate south as far as the Baja California coast, and then return to the small cluster of islands in the middle of the Bering Sea, a location which is usually fog-shrouded (Norris 1967). Among the longest and most regular migrations of any animal are those taken by the great whales, such as the finback, the blue, and the gray whale. In general, the giant whales bear young during winter in warm tropical waters and then feed during summer in waters at high latitudes. There are many points in this migratory path that require the use of an orientation and migration system. Norris suggests that these migrating sea mammals will use whatever relevant sensory data they receive during the long trip. One sense may predominate, and although vision is the primary sense, smell, hearing, touch, and thermal senses may also enter into navigational decisions under certain circumstances. Undoubtedly, celestial navigation is used during part of the migration by these seals and whales. Recent observations suggest that these sea mammals can follow magnetic lines of flux.

The humpback whale covers several thousand miles between north latitudes 15° and 70°; one individual sperm whale was recorded to have migrated 4,000 nautical miles (1 naut. mile = 6076 ft.). Perhaps cetacea orient visually using the sun's position.

The experiments just discussed were selected to illustrate the biological principle that many animals have an accurate sense of time. Field and planetarium studies on birds have shown their sense of time and their navigational ability; other studies on birds using feeding experiments have also revealed their sense of time (Menaker 1969; Hoffman 1968). Mammalian feeding experiments to determine time measurements are rare; students may find an excellent opportunity to do interesting and simple experiments in this area.

Types of Biological Clocks

The acceptance of internal biological clocks as used in navigation developed to the point that Pittendrigh (1957) named three types:

Continuously Consulted Clocks. The mechanisms used in sun navigation would come under this category, as well as the ability to learn and anticipate feeding periods, as shown by bees and ants. Some remnant of the time sense is also present in humans: perhaps you have, on some occasion, decided to wake up without an alarm clock an hour before your usual awakening time—and found that you did wake up exactly at the selected time.

Interval Timers. When *Drosophila* are about to leave the pupal case to take adult form, the creatures demonstrate awareness of "forbidden periods" for emergence, relative to a light stimulus at egg laying. Most emergence occurs only for a specific 3-hour period out of each 24 hours according to an internal interval-timer. The adaptive function of this interval-timer is a variation of what was previously called *standing-in for a zeitgeber.* We might very well call it *anticipating the zeitgeber.* As Pittendrigh expressed it:

> The advantage of hatching from pupal cases (eclosing) at dawn is clear: this is the time when the evaporating power of the atmosphere is at its lowest. It is what might be called a target-time. The functional problem of restricting eclosion to this recurrent target-time includes the problem of identifying it; but since the fly initiates processes leading to eclosion some hours prior to the act, the identification problem is not the simple one of directly recognizing dawn as the appropriate time. Indeed, the real problem is identifying an appropriate point in time

some hours before dawn, and this is difficult or impossible to do by the recognition of external cues. In fact, it is done *by a time measurement from the previous dawn.*

Pure Rhythms: The free-running rhythms of activity, heart rate, and body temperature in hamsters serve as examples of mammalian pure rhythms. Each organ is constantly taking part in the development and progress of the rhythm.

The above classification does not include annual clocks. There are two migratory species of birds that arrive in the north on the same specific dates each year: the swallows of Capistrano, California, and the turkey vultures of Hinckley, Ohio.

BIRDS MEASURE TIME

Ringed turtle-dove pairs alternate brooding the eggs; the female broods for 18 hours, and the male reports promptly and precisely at 10 AM for his shift, and the female leaves. He broods for six hours, and she reports back at 4 PM. Have they measured how much time has passed, or do they read the sun's position? To answer this question, an experimenter, Dr. Rae Silver, confined the male away from the nest for the first three hours of his usual six-hour stint. When the male showed up three hours late, the female vacated the nest as usual, but returned promptly at 4 PM to resume her duties. The male, however, refused to budge. The female was going constitutionally by *sun time,* judged under the condition of being away from the nest. The male, under the condition of being in the shade on the nest, apparently had carefully *measured time* and seemed to say, "I don't care how late in the day it is, I still have three hours to go." The result was a pecking, shoving, and pushing attack by the female in her attempts to dislodge the brooding male.

Adaptive Significance

Behavioral Advantages. Let us now take a closer look at the relationship between evolutionary pressure and biological rhythms. We have made the point that both animals and plants have been conditioned by light and warmth for many millions of years; it is not surprising that the physiology of the day-organism and the night-organism in the same individual is so different. Unless this difference, this internal rhythm, were advantageous, it would probably not have survived. We have discussed the value of the rhythm as a stand-in for a zeitgeber, and to make navigation possible. Menaker adds to this the predictive value of measuring time. For example, if the survival of a wide-ranging animal depends upon its being back in its burrow by daylight, then it must begin its return trip well before it sees the first light of dawn. Another value is to keep species together socially; as stated earlier, if one member of a species is nocturnal while all others in the population are day-active, then that one member will not take part in the breeding activities of the colony.

Clocks can keep apart two similar competing species which are *sympatric* (this means occupy and hunt the same territory). An example is the ocelot (*Leopardus*) and the jaguarundi (*Herpailurus*). Caso determined by radio-tracking nineteen specimens of each species that the ocelots were predominantly nocturnal (75% of the time) and the jaguarundis were diurnal (85% of the time). They seldom came into contact with each other (Caso et al. 1997).

Economy of Energy Reserves. What is the advantage to the animal of the nadir (the lowest point) of the daily rhythm? In some circadian rhythms, this low level is maintained for several hours; the convenient term *low setting* can be applied to the function. As will be shown later in this chapter, this low setting is by no means associated only with sleep. It will be demonstrated that human subjects have a 25% reduction at certain times of the day in their resting heart rate and other functions; this reflects a reduction in activity of the entire body. Apparently this low setting is an energy-saving mechanism; less daily food is required because of it. Let us carry this idea to the cell. Johnson and Hastings, who studied the circadian biochemistry of *Gonyaulax*, suggest that during its rhythm nadir, this organism conserves nitrogen by degrading enzymes whose function is *temporarily* unnecessary (1986).

UBIQUITY OF CIRCADIAN UNITS IN BIOLOGICAL PHENOMENA

We have been discussing a clear-cut circadian locomotor rhythm associated with a composite group of internal physiological rhythms. This composite of rhythms persists in a constant environment (DD or LL, CT, CS). Such rhythms as these have been demonstrated in at least six orders of mammals, in all groups of vertebrates, and in eight phyla of the Animal Kingdom. It is significant that the Protozoa are included in the list. The presence of persistent circadian periodicity in Protozoa (and also in one-celled plants) complicates theorizing about biological clocks. Although we can postulate for the Metazoa that some organ *acts* as a pacemaker to control the entire animal, in the Protozoa the mechanism would have to be much simpler. Even then there is the possibility of further breakdown, as indicated by the questions of Pittendrigh as to whether these one-celled organisms *are* a clock, or *contain* a clock.

Botanists have contributed much of the early proof of periodic biological occurrences that take place in units of about 24 hours, and of the environmental independence of these units. Bünning's bean plant (*Phaseolus*) shows a rhythm of leaf movement of 25 hours in constant conditions. It is difficult to relate a rhythm of 25 hours to any periodic cosmic event. In the free environment, the addition of a zeitgeber can convert this inherent rhythm to one *near* 24 hours. Circadian rhythms of plant growth have even been shown in darkness (Wilkins 1960). As many as three consecutive circadian periods could be measured with individual plants. These rhythms needed a brief period of light to initiate them, but the sequence of circadian periods could be started at any time of the day. This last observation, one of the more significant findings of the experiment, suggests that cosmic periodic events with a 24-hour frequency are not the cause of the observed plant rhythms. To summarize, the circadian units in biological functions occur at both the cellular and organ level.

The broad distribution through the animal and plant kingdom of the same pattern of rhythms as in the mammals, shows that these rhythms must be related to the daily revolution of the earth on its axis. Below we will examine the three schools of thought that seek to explain the exact nature of the relationship to cosmic events.

SCHOOLS OF RHYTHM CAUSATION

Three theories as to the origin of 24-hour rhythms may be found in the literature. They may informally be referred to as 1) a Conditioning Theory, 2) an Inherited Circadian-Clock Theory, and 3) a Cosmic Receiving-Station or External Timing Theory.

Conditioning Theory

This theory assumes that as animals develop from the zygote, they are arrhythmic; they then *learn* what is approximately a 24-hour rhythm from environmental conditioning or from the behavior of parents. This condition is reinforced and made accurate by the environment as the animal develops, so that when tested in a constant environment the rhythm persists. There is not yet sufficient direct evidence to justify discarding this theory for mammals; no mammals have been raised in constant darkness without the possibility of being conditioned by the circadian rhythm of the mother through the placental circulation or by her circadian locomotor activity.

The learning of a rhythm pattern by young animals such as white rats is complicated by the fact that some such youngsters are arrhythmic for a period of time. It is not until the white rat is over a month old that one can record a distinct daily behavior that is circadian. The question must be asked whether the animal could learn a pattern at an early age, but not express it until later; in other words, is there any justification for the possibility that a physiological event or a behavior could express itself long after the stimulus is given? The answer is yes; we find this with imprinting (Sluckin 1965). Many examples of imprinting are provided by Lorenz (1957). In one experiment, young ducks were hatched by geese; after a short period, they were separated from the geese and raised by ducks. When the young ducks came to breeding age, they would court and attempt to mate only with geese. The early imprint on the young ducks was that they are *geese,* and it had survived without reinforcement through the period of growth to sexual maturity. Perhaps the conditioning of young rodents by adults works in a similar fashion: the conditioned rhythm of about 24 hours survives a growth period without reinforcement and expresses itself later as a crisp circadian rhythm.

An Inherited Clock Theory

The second theory assumes the presence of an inherited circadian clock; in other words, the rhythm is born, not made. Direct evidence exists using birds as experimental material. Chicks were hatched and raised under constant dim lights and without any known periodic environmental factor. They developed a rhythm of slightly less than 24 hours (Menaker 1969). If subtle daily environmental factors had been an influence, these would have acted as zeitgebers and set up a more exact 24-hour rhythm in these birds. Much more evidence has been obtained using invertebrates, many of which show genetically determined rhythms of approximately 24 hours. The clocks of all members of an invertebrate population (raised under constant conditions of darkness and temperature) are not necessarily synchronized, especially if several generations are raised under these conditions. *Drosophila* clocks could be set in the same phase by a flash of light acting as a daybreak signal (Pittendrigh 1964). Pittendrigh unified the concept of a genetic self-sustaining rhythm by use of an oscillator analogy (or model):

Our working hypothesis, which is wholly consistent with the available facts, assumes that all organisms are capable of time measurements; that their clocks have a common and ancient basic mechanism like the hairspring and balance wheel of diverse human timepieces from wrist watches to clocks; and that this basic element is an oscillatory system with a natural period evolved to match, approximately, the period of those environmental variables that are ecologically significant (day, month, etc.).

In other words, Pittendrigh and many other investigators believe that a physiological interval-timer or innate physiological periodicity of about 24 hours is as much a fundamental characteristic of animals as certain oft-repeated organ patterns such as nervous systems and brains, or circulatory systems and hearts. This theory is supported by experiments that demonstrated three mutations on the chromosome of fruit flies, all of which altered the biological rhythms of the flies in different ways (Konopka 1985).

Further experiments are in progress to demonstrate conclusively the genetic nature of circadian rhythms in *mammals*. An early attempt was made by Wolf to test for the presence of circadian rhythms in Japanese dancing mice raised from birth under constant conditions of darkness and temperature. The young mice developed a circadian rhythm which was not synchronized with external environmental changes. Similar results have been obtained by other investigators who studied rodents blinded at birth. Aschoff raised several generations of mice under constant conditions of temperature and darkness. The *mean* free-running period of their daily rhythms remained at about 23.5 hours, even to the last generation of the series, and the *range* of values for these periods with succeeding litters remained approximately the same. Aschoff interpreted his results as proving the genetic origin of a natural period on the grounds that if the rhythms were learned, there would be an accumulative error that would cause the third generation to have a period other than 23.5 hours (1979).

This interpretation and the experiment itself are so critical to the field of biological rhythms that we must offer an alternative interpretation for consideration. Perhaps the experiment merely proves the remarkable susceptibility of young mice to conditioning by adults to a rhythm of about 24 hours. In an experiment where a mother rodent raises her young in constant darkness, is there a regular zeitgeber approximately once in 24 hours to influence the young as they mature? This zeitgeber would establish the conditioning of circadian periodicity, not its inheritance. Such a zeitgeber exists! It is well known by keepers of rat colonies that mother rats nurse their young almost continuously in daylight, however, they do most of their own feeding at night, and are away from the litter much of this time to carry out other activities. The white crescent-shaped area on the young rats indicating milk-filled stomachs is not visible most of the night. It is not the length of time the mother is away that is important, but the fact that at a specific time of day she regularly leaves the nest, the nest cools, the young cool, and the stomachs of the young rats begin to empty. Records have been made of the regular feeding (eating) periods of white rats (Folk 1954). In the complete set of records for one of these animals (Fig. 3-10), note the regularity of eating (especially the 4

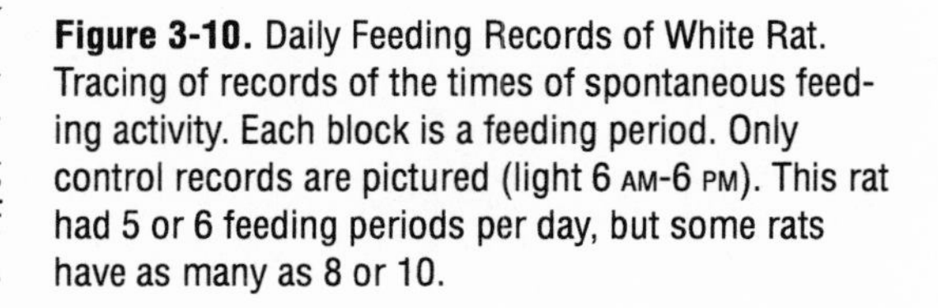
Figure 3-10. Daily Feeding Records of White Rat. Tracing of records of the times of spontaneous feeding activity. Each block is a feeding period. Only control records are pictured (light 6 AM-6 PM). This rat had 5 or 6 feeding periods per day, but some rats have as many as 8 or 10.

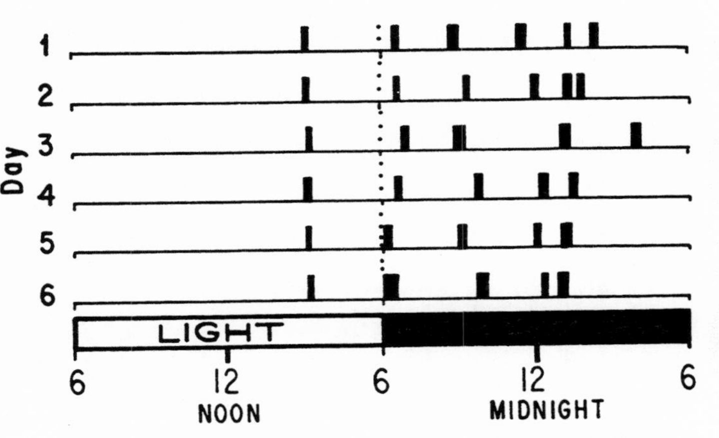

o'clock "tea time"), which might act as zeitgeber to young rats left in the nest by the mother. Such a zeitgeber in the case of the rat would be accidental; according to Barnett, young rats do not receive maternal guidance, as is the case in the Carnivora, Proboscidea, and Primates (Barnett 1963). In other words, the mother rat would not actively push or guide the young to eat at a specific time. Her own activity, however, would influence them.

Do any factors exist which would facilitate the synchronization of activities of a litter of young rats or mice according to a single zeitgeber? Mammalian experiments on this problem, like the ones above, are rare; of other animals, such as the fruit fly, Pittendrigh says, "The likelihood that individuals in a population *entrain* (synchronize) each other in their activity cycles is the...first obvious possibility." We will cite two examples: 1) Stevens (1962) reports that in the fiddler crab, *Uca,* the endogenous rhythm is less stable and accurate in isolated individuals than in large groups; 2) Folk (1974) studied a reversed-cycle female rat (active during solar day) that had persisted in this new activity rhythm as punctually as a timepiece in a continuous environment (DD, CT, CS) for 24 days. Two other females in another soundproof chamber persisted in their usual night activity rhythm (DD for 60 days) with regularity, thus following a time sequence 12 hours different from the first female (Fig. 3-3). The first female was moved into a room with the other two rats (DD, CT, CS). For 14 days the reversed cycle rat resisted change, and regularly ran up to 10,000 revolutions during the period when her two neighbors were resting. Then the single female ran very little or not at all for 4 days and at that point changed her activity pattern, now running at the same time as the other two females (Fig. 3-11). This experiment can only be interpreted as evidence favoring synchronization of rodent activity. Evidence from adult animals may not necessarily apply to baby rats, but no other pertinent information is available.

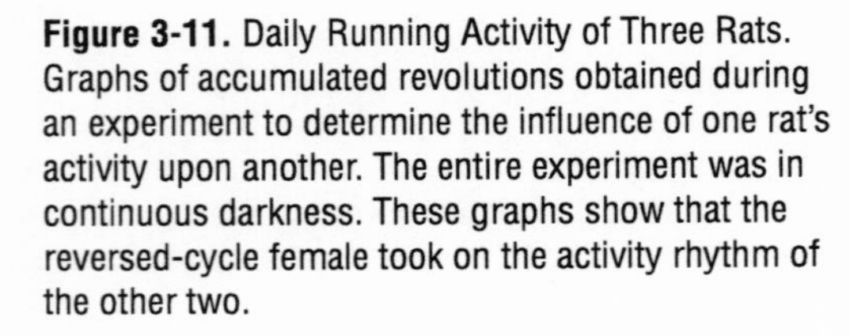
Figure 3-11. Daily Running Activity of Three Rats. Graphs of accumulated revolutions obtained during an experiment to determine the influence of one rat's activity upon another. The entire experiment was in continuous darkness. These graphs show that the reversed-cycle female took on the activity rhythm of the other two.

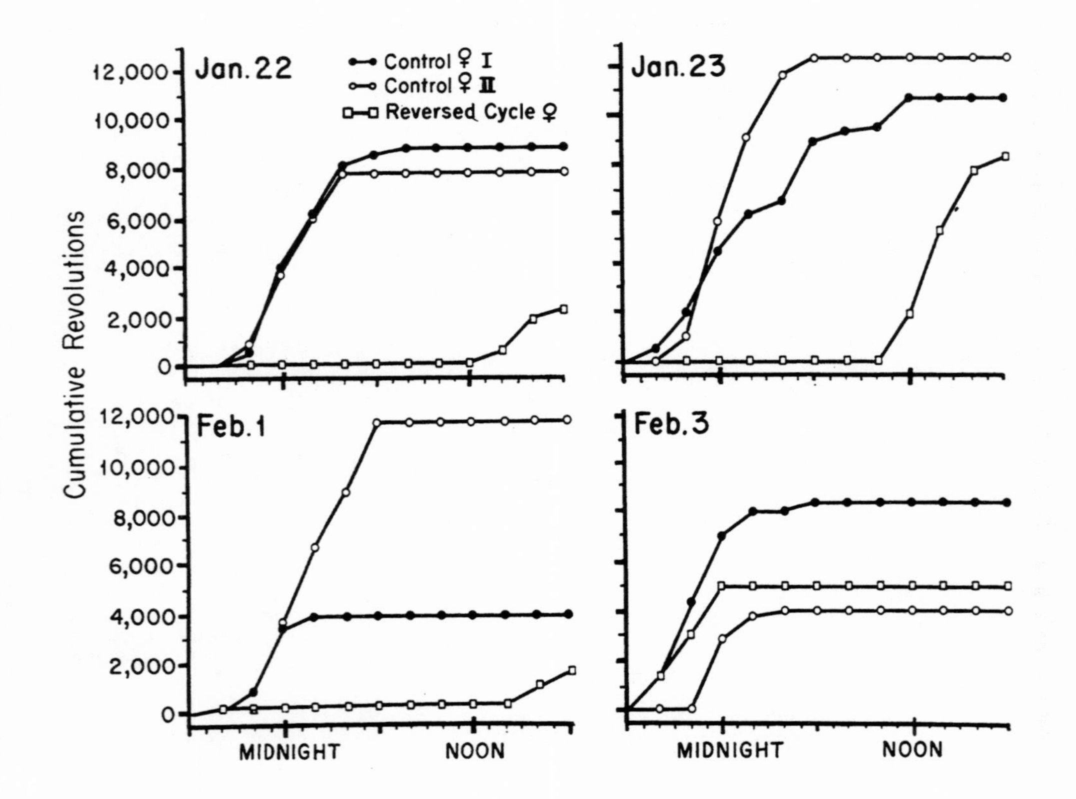

A way was found to prevent the mother rat from providing zeitgebers to her nursing young. A litter of rats was raised under constant conditions with a foster mother in addition to the natural mother. The mothers had different daily rhythms; they were exchanged at random times each day, and it was observed that the young had suckled from both mothers equally over each 24 hours. By chance the litter consisted of six males. Their mean weight at weaning was 39 g compared to the usual mean of 33.6 g for a litter of this size. After they were 95 days old, the three best runners were studied in running recorders in a sub-basement room, which was insulated from regular environmental noise. The free-running period of the rhythms of these three animals turned out to vary from 24.5 to 25 hours. They had one extended period of running (8 to 12 hours) in each of these periods. This evidence appears to support the theory of a genetic 23- to 25-hour rhythm. The behavior of "a mother" during the period of nursing could not have made an imprint. Of course this experimental approach must now be continued backward in the history of the young rodent *into the period of gestation*, since some periodic factor from the mother acting through the placental circulation of the fetus could act as zeitgeber. It would not be difficult to prevent all periodic factors from the mother from acting as zeitgebers to her fetuses, but the experiment has not been done. What *has* been done is to study the fetuses of rats and monkeys directly; there is a circadian rhythm of metabolism in pacemaker cells of the fetal hypothalamus. The rhythm was shown to be "conditioned" by the mother (Reppert 1985).

The above represents experiments on the reproduction and behavior of whole animals to analyze the inherited circadian clock theory. This assumes a genetically endowed, fundamental, pure rhythm with a free-running period of about 23 to 25 hours. In the free environment, zeitgebers set this inherent or endogenous rhythm at about 24 hours, permitting it to be the clock in animal-navigation. Occasionally, experimenters can set up abnormal rhythms as small as 16 hours, but these are temporary and the animals revert to the fundamental rhythm when zeitgebers are removed. Most of the direct evidence supports this theory of inheritance.

Clearly the next stage in our knowledge of a presumptive inherited clock is finding a gene that controls the circadian timing of some species. This was first accomplished in the fruit fly in the 1970s; the gene is on the locus called *period (per)*. Mutations here either abolished rhythm expression or lengthened or shortened its period (Fass 1993). For 10 years, the biochemistry of the gene was little understood, nor was a comparable gene in the mammal found. Neither the protein product of the fruit fly gene nor its function were known, nor were the other molecules known to complete the oscillatory loop. Now the *period protein* (PER) has been found in nuclei of the eyes and brain of the fruit fly; the protein level and associated RNA shows a circadian fluctuation response to 12-hour light/dark cycles (Vosshall et al. 1994).

The next finding by Sehgal was even more important, another clock gene in the fruit fly, called *timeless (tim)*. She showed (by study of mutations) that *tim* regulates *per*. Thus, *tim* affects the central timekeeping mechanism of the biological clock through molecular control of *per* (Sehgal et al. 1994).

Are there comparable genes in mammals? Ingenious work in the laboratory of Takahashi, which involved recording the activity rhythm of 300 mice automatically, showed a circadian gene on mouse chromosome 5 that would correspond to a region on human chromosome 4. A mouse with this mutation on this *clock* gene had a circadian period four hours longer than normal mice (Vitaterna et al. 1994).

Then Takahashi went on to chemically induce clock mutations in mice. When screening such mice, the team found another mutant gene that profoundly changed the circadian period of the mouse strain when studied in darkness (King et al. 1997). Normal mice have a precise 23.7 hour period in darkness. The mutant clock gene stretched the period to 25 hours with one copy of the mutant gene, and stretched the period to 27 to 28 hours in a strain of mice with two copies of the mutant gene. The gene discovered by Takahashi's team is called CLOCK, and it is a transcriptional activator. When the team sequenced the gene, they found that it codes for a protein that has PAS domains—an element first identified in PER. PAS may be a "signature motif" appearing in clock genes (Barinaga 1997).

External-Timing Theory

The Hypothesis. Not all clock-scientists accepted the concept of an autonomous endogenous timer or clock. Brown (1959, 1970) and associates believed that animals and plants respond to periodic environmental changes other than light or temperature; some of these periodic variables include cosmic ray showers, magnetic lines of flux, air ionization, pressure, and humidity. We will call these factors *subtle environmental zeitgebers*; Pittendrigh uses the term *residual periodic variables* (RPV). Brown's experiments soon convinced him that organisms in standard laboratory constant conditions were using subtle, rhythmic, geophysical forces—those that easily permeated the barriers of an experimental setup—as informational input to time overt rhythmic processes. More specifically, he said: "A good working hypothesis for the present appears to be that there exists in organisms an endogenous clock-system which maintains its regular frequencies through some kind of an external pacemaking signal which continues to be effective under what is deemed *constant conditions.*" He seemed to have believed that once a day, several subtle environmental zeitgebers—perhaps changes in cosmic ray count, terrestrial magnetism, or air ionization—all come to a peak in phase, and this "bundle" could act as a zeitgeber. This theory no longer has supporters because of the discovery of "clock genes." However, Brown had made some important discoveries.

Lunar Frequency of Activity Peaks

Some of Brown's experiments were in an important area we will call *Subtle Environmental Influences.* These experiments were his greatest contribution; one could say they were studies in "stimulus and response physiology."

His first experiment involved a rat kept in continuous illumination (Brown 1969). Under these circumstances, the activity phase of the rhythm of the rat lasted 9 to 12 hours but showed a regular, constant, and definite amount of alteration or delay in time of starting, so that the activity block traveled around the clock. In Brown's experiment (LL) the amount of activity did not remain constant, but increased whenever the activity peak of the rat coincided with high tide, or with certain phases of the moon. These effects are difficult to visualize until the daily amplitude of running is graphed along the same axis as the time of high tide, and the phases of the moon. To summarize, the experiment describes persisting effects of exogenous factors acting upon a moving endogenous rhythm. The influence shown here upon the white rat's activity can only be said to have a lunar frequency of 27 days. The environmental factor is unidentified.

There have been numerous other papers that show activity peaks of animals correlated with phases of the moon (Brown 1965; Bennett and Huguenin 1969; Klinowska 1972a). Perhaps the most thorough is that of Klinowska (1972b) who used several statistical techniques to establish correlations with the activity peaks of male hamsters; she made comparisons with 40 environmental factors including phases of the moon. As Brown had pointed out previously, Klinowska found a number of lag and lead correlations between some of the meteorological parameters and high activity of the hamsters when maintained under natural daylight conditions. Brown provided the hypothesis that perhaps the phases of the moon are correlated with changes in the *magnetic lines of flux,* and it is this physical change that might possibly influence animals.

In a later section of this chapter, we discuss the circadian rhythms of a blind man: all of his four rhythms displayed a lunar periodicity.

Polar Experiments. Experiments were also done to study the possible subtle effects of the earth's rotation. Hamner and colleagues (1962) took hamsters, fruit flies, and a fungus (*Nemospora*) to the South Pole; all these organisms were maintained on a surface rotating counter to the earth's rotation. All displayed crisp circadian rhythms.

Gamma Radiation

The next experiments to show the response to subtle environments were again done by Brown. He justified his use of very weak gamma radiation in his experiments as follows: he wanted to expose one group of mice and a group of snails to a gamma field five times higher than the control mice and snails. He felt that this increase in background radiation would be well within the natural range for the field, since it varies more than this as one travels over the earth. As an example, he stated that the strength of the natural background of gamma radiation is doubled when one travels from Illinois to Massachusetts; in places in South America it is reportedly 50 times higher, and in some places in Asia, as much as 100 times higher than in Illinois. Thus, it is reasonable to test mice for their response to a 5-fold change in background radiation (Brown 1970). Using this increase in radiation, Brown and colleagues found that under the influence of a 5-fold change in gamma radiation both mice and snails change their spontaneous activity (1966, 1968).

Magnetic Lines of Flux

For 16 years experts have suspected, but not proven, a link between electromagnetic fields and cancer. Let us look at some of the experiments on magnetic lines of flux. In many laboratories the stimulus used has been 1.5 gauss, which is about 9 times stronger than the earth's magnetic field. The history of the influence of magnetism on living things involves information that goes back 2.5 million years. At Columbia University's LaMonte Geological Observatory, Hayes (1972) studied eight species of microscopic marine animals from remains buried in hundreds of feet of sediment. He found that six of the eight species suffered mass extinction at the time of distinct magnetic field reversal so that the magnetic poles which formerly caused the compass needle to point north were reversed. Some scientists have speculated that there is an indirect effect of this reversal of the magnetic field, but Hayes states that the effect could be directly upon the animal.

The designs of experiments using applied magnetic fields have varied greatly; Bennett and Huguenin (1969) used earthworms and an experimental apparatus in which the magnetic field had an intensity of almost zero. There was an effect upon the ***light-withdrawal reflex*** of earthworms exposed in this apparatus. *Tetrahymena* populations are sensitive to low intensity, low frequency magnetic fields; they show morphologic changes and reduced cell division (Tabrah et al. 1978).

In some way the presence of aggregations of magnetite ($Fe_3 0_4$) in the snouts of some marine vertebrates relates to sensing magnetism (Maugh 1982); magnetite is strongly attracted by a magnet. Perhaps the signal is at the cellular level; there are changes in cell sodium pump activity following whole mice exposure to weak magnetic fields (Batkin 1978).

Bee behavior is influenced by external magnetic fields (Hsu and Li 1994). The authors found specialized cells containing magnetite that could relay changes in magnetic fields to the nervous system.

Another model is the nudibranch mollusk, *Tritonia*. This species orients in a geomagnetic field, and it has large identifiable neurons. Only one neuron showed enhanced electrical activity when stimulated by ambient, earth-strength, magnetic fields. No other neurons responded (Lohman and Willows 1991).

Articles from eight different Russian laboratories described experiments in which magnetic fields were used on humans and animals. The exposure varied from less than 1 gauss to 300. Each article described an effect of this exposure. Typical results are described by Chizhenkova (1966); in her experiments, the effect of a permanent field and the effects of switching a field on and off were measured on the EEG. She attempted to speculate concerning the specific receptor of the effect of these fields. Kholodov (1966) has reviewed much of this field of work, especially that of American researchers. He stated that magnetic fields have been shown to strongly affect all organisms up to the primate level; the engineering attempts to protect astronauts from ionizing radiations by using magnetic fields must take into account this influence upon tissues.

Brown (1966) used graded exposures, varying from 0.05 gauss up to 4 gauss. When changes were made in the magnetic fields, there were immediate effects and after-effects upon planarians.

A magnetic influence was demonstrated on a green turtle in a definitive experiment done by Archie Carr at his field station in the Carribean (Gessaman, p. 74, 1973). A displaced green turtle, carrying a brass bar, returned directly to its nesting beach. When a small magnet was substituted for the brass bar, the displaced turtle followed a disoriented path (Fig. 3-12).

Most exciting, however, is the influence of magnetic fields upon birds. The most outstanding European investigator in this area was Wiltschko; a sample of his laboratory findings are as follows: 1) field intensity determines whether the magnetic field can be used for migratory orientation; 2) the normal intensity range can be enlarged by having the bird become accustomed to weaker or stronger fields; and 3) the birds can perceive the axial direction of the lines but not their polarity (Wiltschko and Wiltschko 1972). Keeton (1972) obtained additional evidence by attaching magnets to homing pigeons. He tentatively concluded that magnetic clues for these birds rank higher than familiar landmarks, at least at distances of 32 km or more. Salamanders can be trained to variable magnetic fields, showing that they use magnetic compass orientation to find their home pond (Phillips 1986).

Figure 3-12. Interferring with Magnetic Lines of Flux. Two experiments displaced a green turtle due east of its nesting beach and tracked its return to the beach. When carrying a magnet, the turtle became disoriented. Modified from Baldwin in Gessaman 1973.

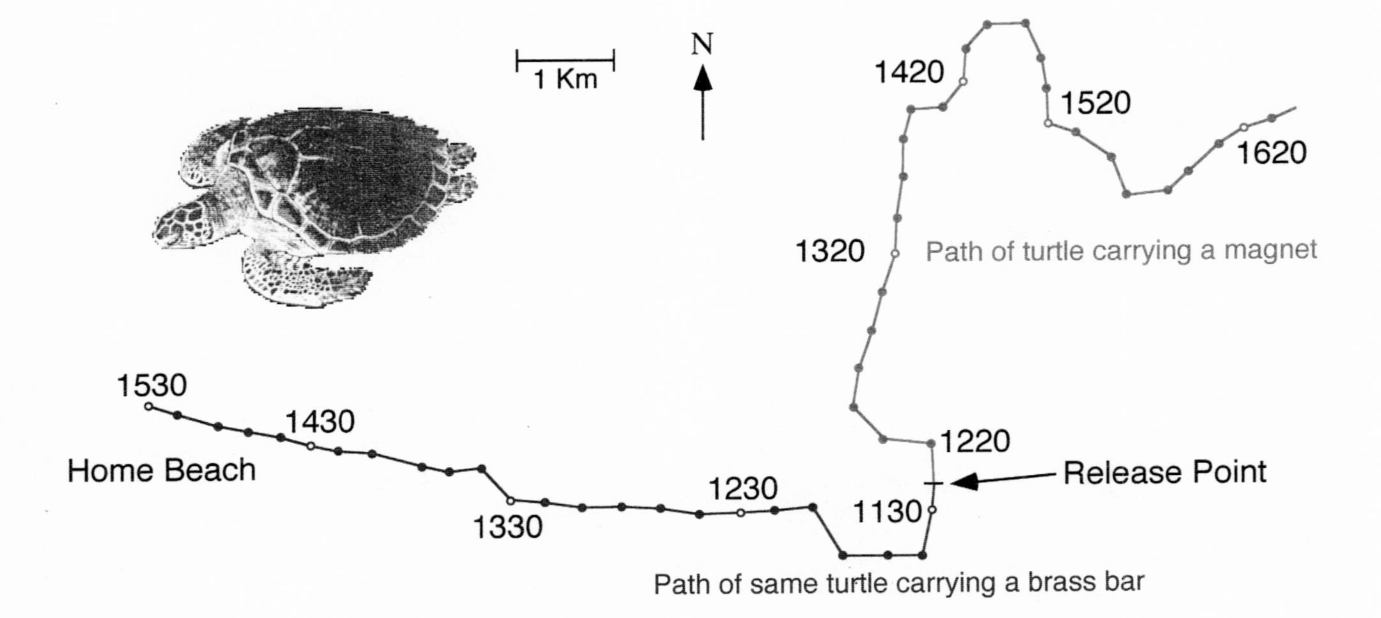

Very few of the investigators using magnetic fields have speculated about how these animals perceive the stimulus. Lindauer and Martin (1972) have done so, although in their studies on magnetic effects they used insects. Perhaps their theory may be applicable to higher animals: they believe it is improbable that there are any specific receptors. They suggest that the effect is by electric induction in sense organs or in the nerve fibers, as well as para-magnetic effects in the sense organs which could be gravity receptors; alternatively, perhaps the effect is upon protein molecules acting as dipoles during locomotion so that the force of the magnetic field could act in a specific way. They conclude that a mechanism based on paramagnetic effects seems most probable.

We have been discussing the effects of magnetic fields. As a special case, we find that humans, livestock, and wild animals live and reproduce in the magnetic fields caused by voltage transmission lines. While no effects have been found on reproduction, there is a magnetic field effect upon a critical component of circadian rhythm control, namely the daily flood of melatonin which is produced from the pineal gland, the retina, and other tissues (Harlow et al. 1981). The all-important suprachiasmatic pacemaker (to be discussed later), connected to the pineal by nerve tracts, has an enormous number of melatonin receptors, an indication of the close coordination of the two organs. In experimental animals, high-intensity magnetic fields suppress the valuable daily bursts of melatonin (Preston-Martin 1996). As we have discussed, this can influence daily circadian rhythms. But melatonin has another task: it mops up free radicals, thus preventing oxidative damage (Reiter 1996). Therefore, any reduction in melatonin should be avoided. In fact, it has been used as a supplement by military forces to prevent sleep disruption at times of rapid deployment across time zones (Comperatore et al. 1996).

Developers and builders are advised to take these findings into consideration when locating houses and schools in areas where there are high voltage transmission lines.

Electric Fields

There is considerable interest in the influence of electric fields on circadian rhythms of mammals. A convenient example is the work of Wever with human subjects (1971). Using as many as 100 subjects in carefully controlled experiments, Wever first determined the free-running rhythms of activity in human subjects. Then, when a weak electric field of 10 cps was turned on, all subjects altered their free-running rhythm (Fig. 3-13). The investigator did not present any theory as to how

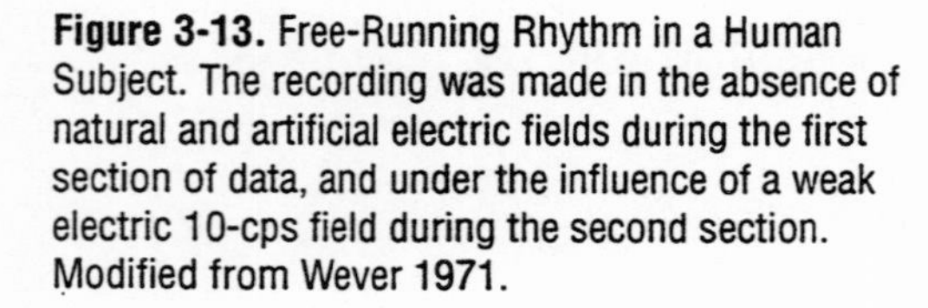
Figure 3-13. Free-Running Rhythm in a Human Subject. The recording was made in the absence of natural and artificial electric fields during the first section of data, and under the influence of a weak electric 10-cps field during the second section. Modified from Wever 1971.

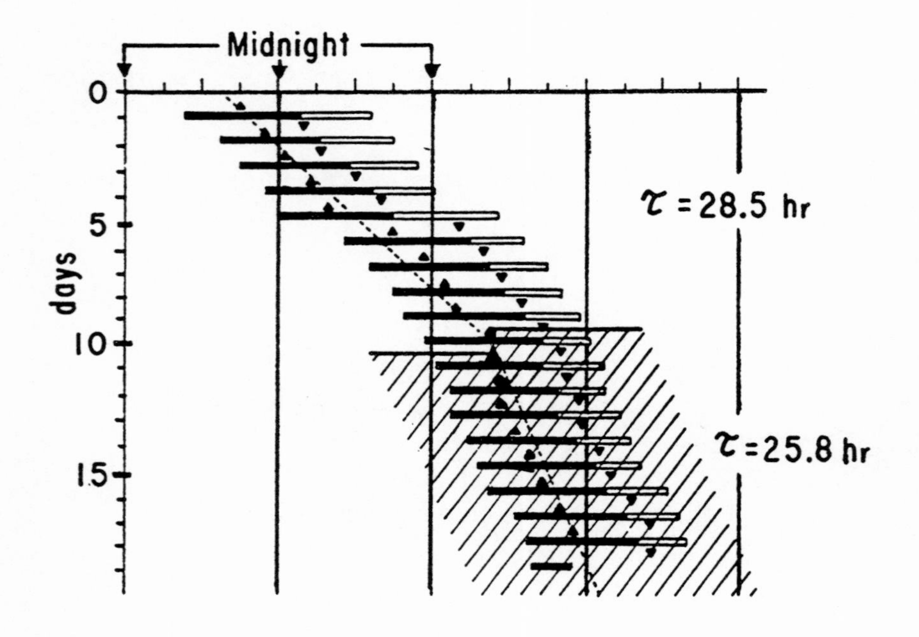

this electric field is sensed. Mice also entrain their circadian rhythms to electric fields.

Air Ions

There is much evidence, not universally accepted, that air ions of either charge are biologically active; usually the counts of ambient ions are 200/cm^3 at nightime, increasing to 1000/cm^3 at midday, negative ions prevailing at both times (Hinsull 1986). One author records that negative air ions can improve various cardiovascular functions as well as subjective feelings during physical effort (Inbar 1982). Sulman (1980) has measured high serotonin levels in the blood in the presence of positive air ions. Air ions can protect guinea pigs against anaphylactic shock (Deleanu 1986).

Although it will require many more years of investigation, let us hope that these experiments with the subtle environmental factors will eventually help us understand the nature of and the control over the biological clock.

ANATOMICAL LOCATION OF THE PACEMAKER

Extirpation Experiments

A number of investigators over the years have done experiments which were designed to locate a hypothetical biological timer or clock. In the past ten years these efforts have been called the search for a "master synchronizer," or "pacemaker" (Moore-Ede et al. 1980). One approach is to use operative procedures. Probably the most important results are those where rhythms are present after the operations, since the absence of a rhythm in a post-operative (perhaps sick) mammal is difficult to interpret. Modern experiments in this field are based on a new premise, but for the sake of history, some experiments of interest are listed below (the references for each experiment can be found in Harker's 1958 review):

1. Removal of stomachs: Rats showed no alteration in maze testing and running activity due to the operation (Tsang 1938).
2. Removal of adrenals: The rhythm of liver storage of glycogen ceases, but the activity rhythm in rats continues (Bacq 1931).
3. Hypopituitary: Genetic hypopituitary mice (dwarf mice) are rhythmic (Osborne 1940).
4. Hypophysectomy: Rats show a circadian rhythm after removal of the pituitary (Richter and Wislocki 1930; Levinson, Welsh, and Abromowitz 1941).
5. Removal of thyroid: The circadian rhythm of rats is upset but not eliminated (Richter 1933).
6. Lesions in frontal brain tissue and corpus striatum: Some rats retain a rhythm (Beach 1941).
7. Removal of cortex: No loss of circadian rhythm in decorticate dogs (Rothmann 1923).

These experiments have been oversimplified for presentation here; these abstracts do show, however, that at an early stage in the experimental approach to mammalian circadian rhythms, it was evident that the control of these rhythms must be deep-seated and complicated. For this reason, it is valuable to refer to animals of simpler architecture than mammals for information on the anatomical location of the biological clock. One experiment on invertebrates demonstrated that a clock can be approximately located, but it was a 40-minute timer: Wells (1955) showed that the very regular activity of a marine worm is controlled by a discrete portion of the head anatomy. Three-minute bursts of feeding movements (with or without food) and 1-minute rests are followed by locomotion movements every 40 minutes (Fig. 3-14). The clock is the esophagus or even a slice of this structure in vitro. The rhythm of this tissue can be likened to the activity of the mammalian heart. Unlike the heart, the rhythm from the esophagus is transferred to the entire nervous system; the esophagus drives the proboscis which drives the rest of the body. Wells continued making contributions through such thorough and fundamental experiments over a 30-year period.

A pioneer in the field, Harker (1964, page 64), introduced similar experiments with the cockroach; she reported a hormonal clock in the subesophageal ganglion of the cockroach that could be transplanted to another cockroach. The recipient cockroach took on the daily rhythm of the donor. Since Harker's early experiments, three workers have surgically transferred portions of the nervous system of insects, and thereby transferred the phase or period of what is now called the pacemaker from one animal to another; the experiments were successful in silk moths, *Drosophila*, and cockroaches (Page 1982; Takahashi and Zatz 1982).

It is evident from the preceding experiments that the thinking of investigators of circadian rhythms was similar to that of Wells and Harker: the search was for a localized center of control over the circadian rhythm. This point of view was shared by Pittendrigh; he suggested that the elaborate chronometers of the higher animals (birds and insects) depend upon and are in the nervous systems

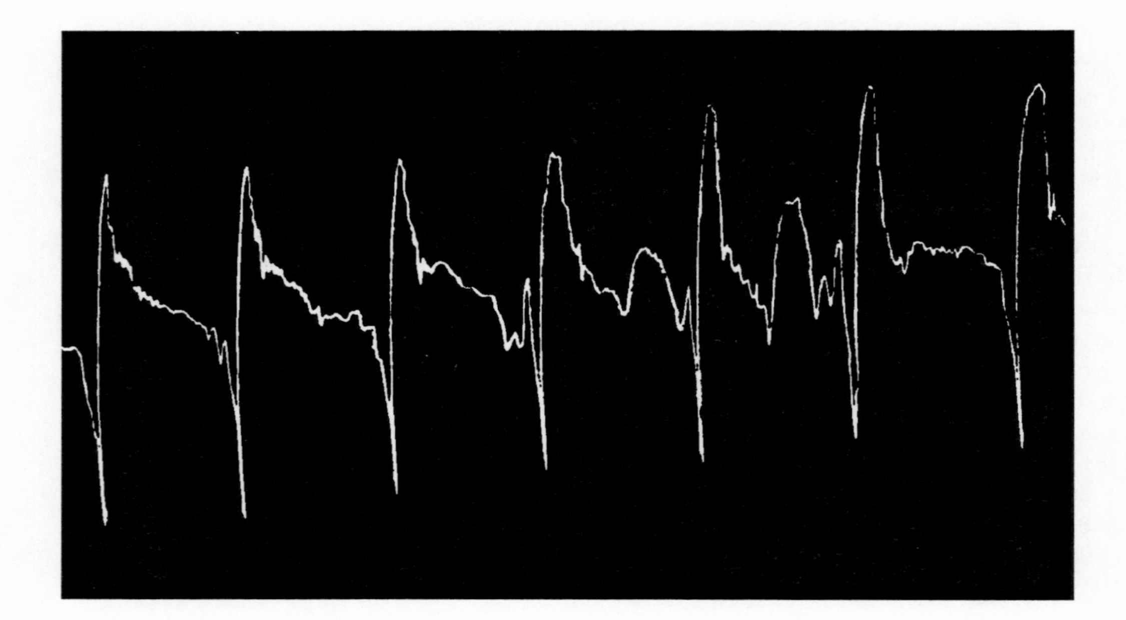

Figure 3-14. Measurement of Time by Lugworm. At intervals of 40 minutes, the worm moved backwards to the sand surface and emptied its intestine. This causes movement which is recorded on the kymograph. Six such periods are shown. Modified from Wells 1959.

(Pittendrigh and Bruce 1957). However, Harker (1958) stated that the concept of the localized center for the circadian rhythm is an over-simplification; she felt that there is a basic circadian rhythm in the cells of all animals, and any group of cells may constitute a clock, so that any animal may have a number of clocks operating in phase or at variance with each other. Later in this chapter, we will refer to this as *desynchronization*.

Could the cell division cycle of mammalian cells act as a pacemaker? Cell divisions do represent the measurement of time. Since these divisions are relatively imprecise, the mechanism is probably a pacemaker that controls cell divisions rather than the reverse (Edmunds and Adams 1981).

Multicellular chronometers are not essential; Bruce and Pittendrigh showed the presence of endogenous rhythms in Protozoans (Fig. 3-15), and several workers have described rhythms of oxygen consumption and CO_2 production of plant tissue samples (Wilkins 1960). An early, well-developed model for the molecular design of the clock was described by Ehret: his hypothesis states that oscillations might be expected to occur in cellular control mechanisms that operate by closed-loop repression pathways involving DNA, RNA protein and metabolites (Ehret and Trucco 1967).

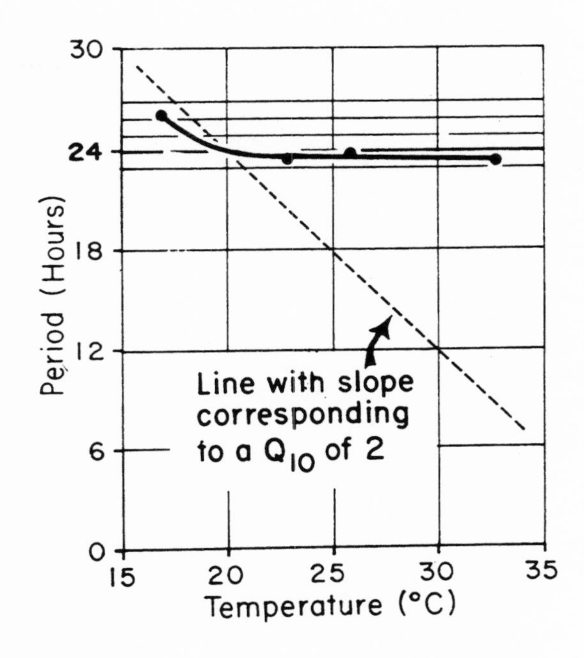

Figure 3-15. Temperature-Independence in Daily Rhythm of *Euglena.* The periods of free-running rhythms of phototaxis in *Euglena gracilis* were recorded at different temperatures. Note that in this experiment, a circadian rhythm was demonstrated in a one-celled organism. From Bruce and Pittendrigh 1956.

Another theory emphasizes the influence of Ca++ ions upon oscillations in activity of the mitochondria (Follet and Follet 1981). The timing mechanism has been further dissected in *Euglena*; here six enzymes, transport systems and calmodulin have been found to act in a cascade as "clock gears" in a self-sustaining circadian oscillating loop (Goto, Laval-Martin, and Edmunds 1985).

Eventually, the experimental question was bound to be asked: If there is no apparent central anatomical control of the clock in the mammal, do individual organs have their own clocks, or, to paraphrase Pittendrigh: *Are* they a clock? A stimulating but simple experiment by Bünning (1964) so excited our curiosity that we spent several years following his lead: he isolated pieces of hamster intestine which then showed circadian rhythms of both tone and amplitude of contraction for three days. We questioned whether isolated rodent organs (adrenal glands, heart, and heart cell networks) would show a circadian rhythm of function when these organs were cultured for three to six days (Fig. 3-16). The positive results are shown in Table 3-1; note that in one case the amplitude was 60% above and below the mean for the adrenal glands (Andrews and Folk 1963; Andrews 1971). Similar evidence for the isolated rodent heart was presented by Tharp (1964), who showed a heartbeat of 200 changing in 12 hours to a heartbeat of 80, and then back to a heartbeat of 160 approximately 24 hours later, followed by a low value in 12 hours of 80 beats again (Fig 17). These results strikingly support Harker's statement that an animal may have a number of clocks all operating at once.

Adrenal glands were also cultured by Andrews from another species, the brown lemming; in this study he measured only the adrenal cortical steroids secreted into the nutrient medium, but the same striking circadian rhythm was evident (Andrews 1971). This secretory rhythm had a period of 23.4 hours.

Chick pineal glands were cultured for 4-day periods in Menaker's laboratory by Kasal (1979). A circadian rhythm of enzyme activity persisted, thus demonstrating, as the authors expressed it, that the pineal gland of the chick contains (or is) a self-sustained circadian oscillator.

Other mammalian tissues have also been maintained for four days, namely the retina of the hamster. The retinal cells produced melatonin on a 24-hour cycle that peaked at night. The test-tube retinas could be reset or "entrained" with light (Tosini and Menaker 1996).

Insect tissues in culture can display the circadian rhythm of cellular activity; Rensing (1971) kept *Drosophila* salivary glands in culture and found a bimodal rhythm of nuclear volume which persisted in several light cycles and matched the rhythm found *in vivo*.

The isolated eyestalk of the crayfish (Fuentes-Pardo, Verdugo-Diaz, and Incian-Rubio 1985) and the cultured eye of the marine mollusk *Bulla* also proved to contain a circadian pacemaker (Block and Wallace 1982).

Other experiments have shown that an individual neuron can not only measure time, but also store information on time due to some past environmental experience (Strumwasser 1967). In the sea hare, *Aplysia californica,* Strumwasser worked with a particular cell, which is referred to as the PB neuron. He found that he could condition these individual neurons in a ganglion isolated from the animal. When he used 12 hours of light followed by 12 hours of darkness, there was subsequently a large peak of impulse rate measured from an impaled microelectrode, the peak coming at projected dawn in most cases. About 10% of his animals did show a large peak of impulse rate near the projected dusk. The peaks of impulses showed a circadian rhythm of about 26.9 hours. When exposing sea hares to constant light for 1 to 2 weeks, the critical experiment showed that the isolated ganglion contained a PB neuron which emitted an impulse rate that still fluctuated with a circadian rhythm, but the form and timing were clearly

Figure 3-16. Circadian Rhythms in Cultured Adrenal Glands.
A. Normalized oxygen comsumption of control glands cultured at 37.5°C; donors were maintained in a solar light cycle (light on 800 to 2000 hours).
B. Normalized oxygen consumption of glands cultured at 37.5°C; donors were maintained in a reversed light cycle (light on 2000 to 800 hours).

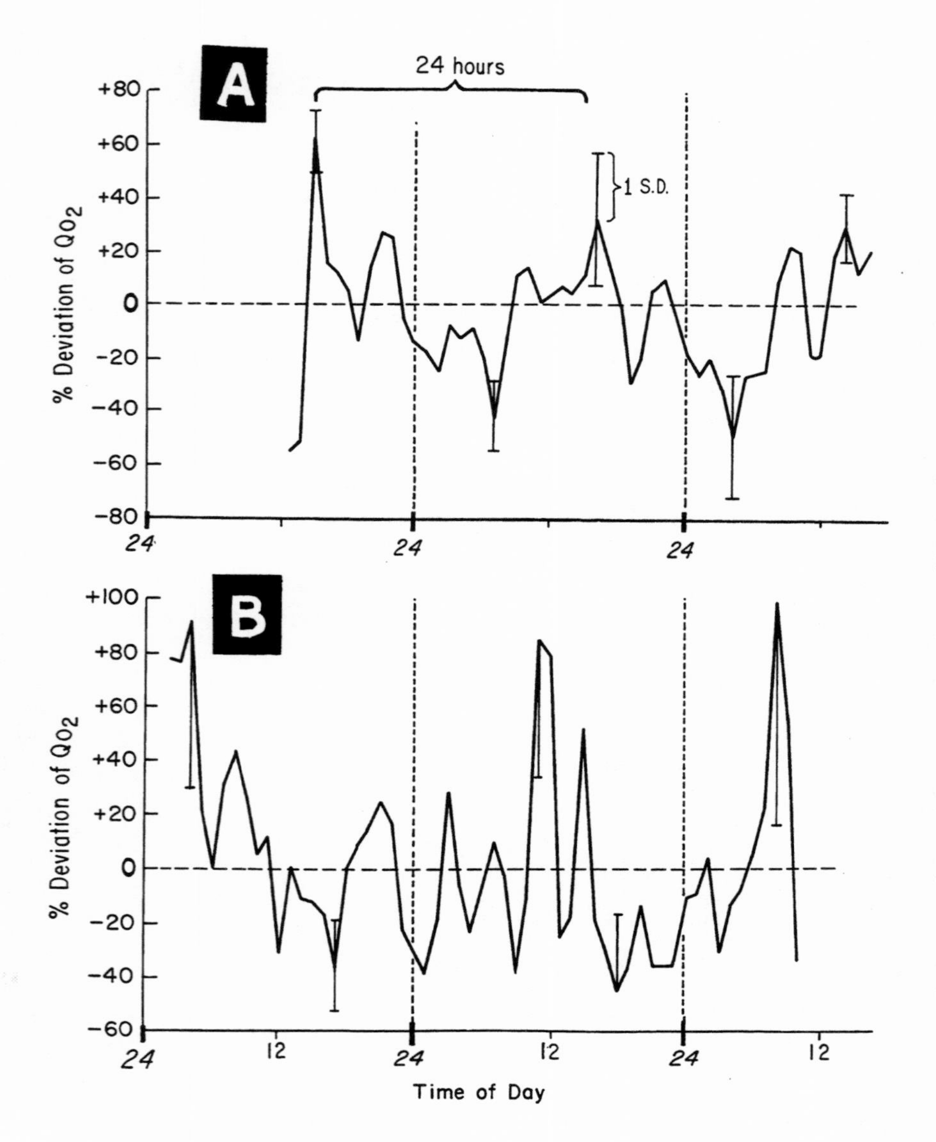

different from that of the light-dark conditioned individuals (Fig. 3-18). Strumwasser went on to postulate that the readout of stored information by a neuron as a pattern of impulses may be understood in terms of messenger RNA release and excitation of the membrane mediated through some newly synthesized product.

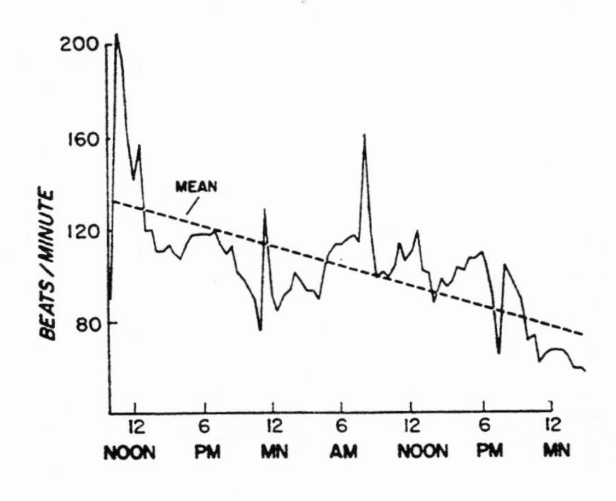

Figure 3-17. Two Day Record of Heartbeat of a Perfused Rat Heart. A typical time series, obtained from an isolated rat heart, illustrating the decrease in rate during the experiment and the oscillations of rate about a linear regression line.

The Circadian Pacemaker in Birds

The search for an anatomically localized master clock or pacemaker in birds has been carried on by Mennaker. The definition of a pacemaker (analogous to the pacemaker of the heart, for example) is that *it is a structure capable of spontaneous circadian rhythmicity in the absence of all periodic inputs, and its rhythm is used to synchronize other biological rhythms.* The preferred term is *pacemaker* rather than *clock*: another comparable term is the *primary oscillator*. In contrast, the secondary rhythm or oscillator has a much less stable periodicity, and while it is spontaneously rhythmic, it does not convey temporal organization to a wide range of other physiological systems the way the primary oscillator does.

The pacemaker in the bird is the pineal gland. This gland appears to play different roles in the circadian systems of birds and mammals. It does act as a primary oscillator in birds, whereas in mammals it merely expresses a driven rhythm. There is, however, a similarity in pineal biochemistry in these vertebrate classes. In both birds and mammals, the gland expresses an overt rhythm that results in the nocturnal synthesis and secretion of the hormone melatonin. This event in indoleamine metabolism takes place by a dramatic (greater than tenfold) increase in the activity of the enzyme N-acetyltransferase (see sidebar, Chapter 2). In mammals, the rhythm of melatonin synthesis originates from the hypothalamus via signals over sympathetic nerve fibers to the pineal gland. In the birds, the avian pineal does not depend on neural connections for rhythmic melatonin production (Reiter 1981). The bird pineal gland contains a circadian pacemaker; the removal of the gland eliminates the free-running rhythm of locomotor activity and body temperature. This rhythmicity can be restored by transplantation of pineal tissue from a donor into the anterior chamber of the eye of an arrhythmic host. The restored rhythm will have the phase of the donor (Follett et al. 1981). When the avian pineal gland is cultured, the rhythm of melatonin release remains robust in light-dark cycles, and with a lower

Table 3-1. Circadian Rhythms in Isolated Organs

Tissue	Measurement	Circadian Rhythm
Hamster adrenal* (isolated)	oxygen consumption	60% above and below mean
Hamster adrenal* (isolated)	steroid secretion	increase in corticosterone
Rat heart** (isolated)	rate of beat	high rate 1500 to 0300
Rat heart cells** (cultured)	rate of beat	high rate 2000 to midnight

*Andrews and Folk 1963 **Tharp and Folk 1964

amplitude can be detected for at least four cycles in constant darkness. Thus we must say that there is direct evidence for a pacemaker role for the pineal gland of the bird, as well as for portions of the nervous system of insects as described earlier.

An Anatomically Localized Pacemaker in Mammals

The evidence for a pacemaker or synchronizer in the hypothalamus of mammals, unlike the transplant experiment of the bird, is indirect, although compelling. The search for the pacemaker in the hypothalamus has a long history. It is significant that Richter began looking for rhythmic control by the hypothalamus in 1922. In 1927, his experiments centered upon the anterior hypothalamus. By 1967 he had made a series of 200 hypothalamic lesions, and he found that in only one location in the ventral hypothalamus did these lesions result in loss of circadian rhythmicity in activity, feeding, and drinking behavior (Moore-Ede, Sulzman, Fuller 1982). As a white-haired man well into his nineties, Curt Richter was still doing similar experiments each day in his laboratory.

Other workers believed that by tracing visual pathways from the retina to the anterior hypothalamus, it should be possible to locate the pacemaker. It was Moore and Lenn (1972) who finally identified a retino-hypothalamic tract that terminated specifically in the *suprachiasmatic nuclei* (SCN) of the hypothalamus. The same laboratory later showed that lesions of the SCN resulted in a loss of the circadian rhythm of adrenal corticosterone; soon other laboratories demonstrated a loss of circadian rhythmicity in drinking and locomotor activity (Stephan et al. 1972; Fuller et al. 1981, 1983; Albers 1984; Rusak et al. 1982). Drinking rhythms were lost due to partial SCN lesions (Fuller et al. 1983), and phase shifts of activity rhythms were caused by stimulation and microinjection into the SCN (Albers et al. 1984; Rusak and Groos 1982). Further evidence that the SCN may be referred to as a pacemaker comes from experiments in which this section of the anterior hypothalamus is isolated by Halasz knife surgery from the rest of the brain (Moore-Ede et al. 1980). Rhythmicity does continue within the hypothalamic island that contains the SCN pacemaker in the form of action potentials of the nerve cell bodies of the SCN (Green and Gillette 1982; Inouye and Kawamura 1979; Moore 1983). It must be these cells that control the sleep rhythm (Enright 1979).

Figure 3-18. Spike Output of Single Neuron of the Sea Hare *(Aplysia californica)*. This experiment was done as evidence that a neuron can store information concerning the time of some past environmental experience. In this case the animal had been living in continuous light. The discharge from the cell (designated in the literature as "PB") had a period of 21.4 hours. Adapted from Strumwasser 1967.

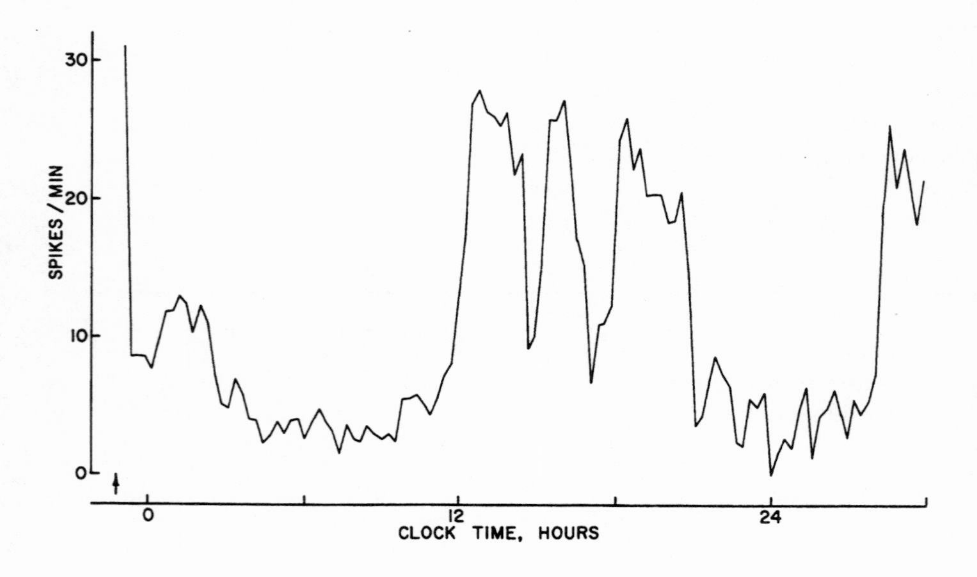

The Pineal Contribution

We must not "lose" the pineal gland within the detail of the retino-hypothalamic tract and the SCN (see Chapter 2). Nor should we forget that the light signal produces and is reinforced by a controlled flood of melatonin (Klein 1985). In the hamster at least, the duration of each light signal, when varied from 12 hours per day down to 4 hours per day, exactly matches the duration of the flood of secreted melatonin (Carter and Goldman 1983). Thus, the shortest day produces the most melatonin. In the monkey, this flood of melatonin is found even in the cerebral spinal fluid (Taylor et al. 1985).

Serotonin is responsible for melatonin but may act independently; added to the medium around the eye of *Aplysia*, it changes the phase of its circadian rhythm (Corrent et al. 1978). Assenmacher (1987) argues that the serotoninergic (5-HT) system is critical for all neuroendocrine rhythms, originating from the midbrain raphe nuclei. It is not yet certain whether this midbrain system is part of the circadian machinery as part of a control loop attached to the SCN, or as a specific circadian component separate from the SCN.

Two Pacemakers

The SCN is probably not the only circadian pacemaker in mammals (Moore-Ede, Sulzman, and Fuller 1982). There is evidence for another pacemaker which coordinates multiple physiological functions on a circadian basis (Johnson and Hastings 1986). The evidence comes from the descriptions of the splitting of normal rhythmic functions into two groups with different frequencies. For example, in Aschoff's experiments in which he studied human subjects isolated in bunkers, some subjects eventually showed a separation of the sleep-wake cycle and the core temperature rhythm so that they could be recorded at totally different frequencies.

The second pacemaker is not located within the SCN. In monkeys, a total bilateral SCN lesion results in circadian arrhythmicity of the locomotor activity cycle; this operation does not eliminate the circadian rhythm of body temperature (Fuller et al. 1981). Furthermore, there is evidence in rodents that this second pacemaker can be entrained by food, even after the SCN has been totally destroyed. Some authors have suggested that this second pacemaker is found in the ventromedial nucleus of the hypothalamus or in the lateral hypothalamic area, although the presence of a pacemaker in either of these locations has yet to be confirmed. Possibly, the second pacemaker can be manipulated (Jacklett 1977).

Many recent papers are based on the assumption that the SCN is a dominant pacemaker for mammalian circadian rhythms. For example, the hypothesis of Rusak and Groos (1982) was that a true pacemaker should regulate both phase and period of distant circadian oscillations. They looked for support for this hypothesis by the manipulation of spontaneous SCN neuroactivity. Electrodes were placed in or near the SCN in both rats and hamsters, and free-running feeding rhythms and activity rhythms were studied. Stimulation delivered through electrodes to free-running rodents produced both delays and advances of the physiological rhythms, depending upon the circadian phase of stimulation. Changes in period often accompanied phase shifts. Stimulation at sites remote from the SCN failed to produce phase shifts. Their results were plotted against the well-known phase-response curve produced by 10-minute light pulses given to hamsters in constant darkness during subjective night. There was agreement between the two phase response curves. The authors concluded that the SCN functions as a pacemaker by generating a circadian rhythm of neuroactivity that is entrained by photic input and which regulates the activity of other oscillators in the circadian system of mammals (Rusak and Groos 1982).

To summarize, there is adequate evidence to describe one or two pacemakers in the brain which are primary and are used to synchronize numerous secondary oscillators. The hypothesis that assumes the existence of at least one secondary oscillator was presented earlier to explain phase delay and phase advance (see phase-response curve). A typical secondary oscillator is exposed when the adrenal glands of hamsters are studied. In an earlier section, it was demonstrated that these glands show a free-running rhythm of oxygen consumption with a peak 60% above the mean and a nadir of 60% below the mean. We will refer later to other secondary oscillators associated with various mammalian functions. In the next section, we will give particular attention to two secondary oscillators that Pittendrigh refers to as "slave oscillators."

INTERACTIONS BETWEEN RHYTHMS AND SECONDARY OSCILLATORS

If the circadian ebb and flow of energy exchange in the neurons of the SCN is the basis of pacemaker activity, what is the basis of activity in secondary oscillators? A secondary clock in the animal body could be a function that depends on several physiological processes; perhaps time measurement depends upon the interaction of several internal rhythms or oscillators. (Although these two terms may at times be interposed, it is convenient to say that an oscillator *has* a rhythm.) Let us suppose that there are two physiological oscillators continuously present in the animal, one with a period of 4 hours and the other with a period of 4.8 hours. (A specific example of the 4-hour physiological rhythm of body temperature in the arctic ground squirrel was presented earlier.) If the first oscillator is superimposed on the second oscillator, their phase relationships can be diagrammed as seen in Fig. 3-19. At first, the two oscillators are out of phase and they become progressively more out of phase as time passes. At 12 hours they are completely out of phase, then they get closer together until they are again in phase (a phenomenon at 24 hours). The cycle will repeat it-

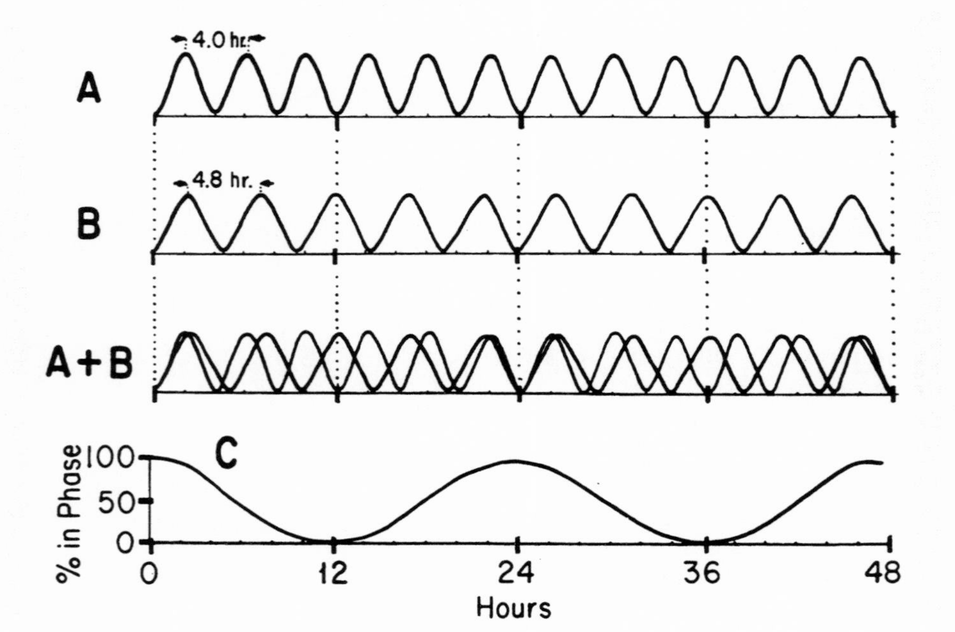

Figure 3-19. Interaction of Internal Rhythms. Rhythm *A* shows a period of 4 hours with a frequency of 6 per day. Rhythm *B* has a period of 4.8 hours and a frequency of 5 per day. If the two are superimposed, the two rhythms come in phase once every 24 hours. The time of coming-into-phase can be plotted as rhythm *C*. This is a 24-hour rhythm that has been created from the phase relationships of two rhythms of higher frequency. Modified from Beck 1963.

self with the oscillators in phase with each other every 24 hours. This relationship can be plotted as a third rhythm, which is circadian (rhythm C in the diagram). If the two oscillators were physiological types involved with body temperature, adrenal activity, heart rate, and kidney function, then the rhythm of their phase relationships (rhythm C) could be very important to the measuring systems.

Secondary Oscillators

Several investigators have suggested that photoperiod comes into this picture if one of the two hypothetical oscillators (Fig. 3-19) is found in the brain and is sensitive to light stimuli. The phases of this particular rhythm might be adjusted by photoperiod, and this would change the phase relationships between this first rhythm and the second. A change in these two would also change the shape of rhythm C. The consensus of several writers is that a biological clock must be made of at least two oscillators; otherwise one could not account for all properties of a clock. It is postulated that one, at least, must be associated with brain function and be sensitive to the photoperiod. Yet our earlier experiment involving the isolated adrenal glands of the hamster indicates that accurate measurement of time in the mammal does not depend upon the presence in the system of a brain. Furthermore, a review in 1983 by Kadle and Folk found several examples of the response of cultured organs or cells to photoperiod; in one case, when the cells in culture were exposed to a reversed photoperiod, the rhythm of the cultured cells entrained to the new light cycle. The dilemma of whether timing depends upon the brain is solved by the well-accepted concept of one or two pacemakers in the brain of the mammal which synchronize secondary oscillators elsewhere in the animal body. These secondary structures are referred to as "slave oscillators." As many as nine slave oscillators have been described in some systems (Pittendrigh et al. 1976).

Blocks of Activity. In order to understand the behavior of secondary oscillators, we must think in terms of blocks of activity (activity time) of nocturnal rodents. The first evidence for the theories concerning secondary oscillators depends upon manipulation and changes in these blocks of activity. In an earlier illustration (Fig. 3-2) the blocks of activity for three species of rodents are presented. Most of our attention will be given to changes in the hamster block of activity. The length of the block of activity varies with the species, and of course, with the photoperiod. For example, in the hamster in

Table 3-2. Variation in Daily Duration of Activity and Period in Free-Running Rodents

Species	*Activity Extent*	*Circadian Period*
M. auratus (Golden Hamster)	10 hr 20'	23 hr 55' - 24 hr 05'
P. leucopus (White-footed Mouse)	9 hr	23 hr 45' - 24 hr 10'
M. musculus (House Mouse)	13 hr 30'	23 hr - 23 hr 45'
P. maniculatus (Deer Mouse)	12 hr	22 hr 30' - 23 hr 30'

constant light of increasing intensity, the period of the activity rhythm gradually lengthens, while the activity time (the block) constantly decreases (Pohl 1985).

Pittendrigh reported the extent of the block of activity and the circadian period of four species of rodents while free-running in darkness. This data is summarized in Table 3-2; note that each species has its own characteristics. Three of these rodents have a single block of activity. Some mammals are biphasic such as the house mouse (*M. musculas*) (Pittendrigh et al. 1976), and the house cat (Randall et al. 1987). The shape of its activity record will have significance in the development of the evidence for a multi-oscillator theory.

Phase Delay and Phase Advances. Blocks of activity while free-running in continuous darkness can be delayed or advanced by means of a single pulse of light. Review again Figure 3-9; the locomotor activity rhythm of the golden hamster (*M. mesocricetus*) free-running in DD and perturbed by pulses of white light is changed systematically. At *A* the pulse falls upon the early part of the activity block; this because of the free-running condition is referred to as "the subjective night." The pulse causes a phase delay which is accomplished almost immediately. At *B*, the pulse falls later on the activity block (late in the subjective night) and causes a phase advance that passes through five or six transient cycles before becoming steady. Later we must relate this phase delay and phase advance to the effect of changes of seasons upon the block of activity of animals in the free environment. Observe in Figure 3-9 that the hamster activity blocks are squeezed between the later arrival of dusk and the earlier arrival of dawn in the springtime. In this case, one peak has received a phase delay and the other has received a phase advance.

Phase Advance and Aschoff's Rule. In an earlier chapter and earlier sections, there has been discussion of Aschoff's rule which states that in nocturnal rodents, the block of activity is delayed by different amounts depending upon the intensity of the continuous light. The significant illustration discussed above whereby the early light pulse introduced a phase delay, and the late night pulse introduced a phase advance appears to be an exception to Aschoff's rule. In other words, if a flash of light at the right time of the activity cycle can produce a phase advance, why would not continuous light tend to do the same? The explanation is simply that if the phase delay depends upon one secondary oscillator and the phase advance depends upon a separate secondary oscillator, the phase delay oscillator is dominant and overrides the other oscillator. However, when we discuss the splitting of rhythms, it will be seen that the phase delay oscillatory may not always be dominant.

Does Aschoff's rule apply to all hamsters? Pittendrigh presented the records of 12 hamsters observed under continuous light (LL). Nine of these showed a typical Aschoff effect with the delay of time of starting activity each day for 30 days. Three of the hamsters, however, showed a modification, a splitting of the block of activity in continuous light. We will discuss this modification below.

Splitting of Rhythms. Under a variety of circumstances, some species modify the block of activity so that it then consists of two components, sometimes with the same period, but at other times with two different periods (Pickard et al. 1984). Consider a hypothetical case of a hamster; in this case, under LL, after a few days one component of the block of activity shows a phase advance while the other component continues with the same period of activity. After a few days, the maverick portion of the block of activity becomes synchronized with the consistent portion of the block of activity, and they then take on, although separated, the same period of circadian behavior (τ). Seven species of vertebrates have been occasionally shown to demonstrate the splitting into two components of their circadian block of activity. There is a variety of means to induce this splitting, as follows: 1) light intensity decrease (LL), 2) light intensity increase (LL), 3) high light intensity (LL), 4) hypothalamic lesions, and 5) testosterone implants. The above details concern locomotor activity. The body temperature rhythm of the hamster can also be split (Fuller et al. 1981). The new rhythms remain concurrent with the split locomotor record, suggesting that the temperature and locomotor rhythms are coupled.

Evening and Dawn Oscillators. We have now discussed three circumstances where a secondary oscillator appears to be responsible for changes in activity blocks: 1) the case of phase advance and phase delay due to a pulse of light, 2) the case of the bimodal distribution of activity in the house mouse, and 3) those cases of the splitting of the activity block.

The theory of this control by these secondary oscillators is so well accepted that they are now named the *E* for evening and the *M* for morning oscillators. In other words, the evening peak of the house mouse and the dawn peaks are under the control of these two oscillators, which usually work in synchrony within a single block of activity. In the case of splitting, the responsibility for the breaking off of part of the block in a phase advance is, of course, attributed to the dawn or morning oscillator. This would be the same oscillator responsible for the phase advance described earlier. Especially within hamsters, there are other patterns produced by the independent action of the *M* oscillator and the *E* oscillators. These oscillators have a correlated endocrine partner. There is a single surge of pituitary luteinizing hormone in concert with the hamster locomotor rhythm; when the hamster block of activity splits, there are two surges of luteinizing hormone (Swann and Turek 1985).

Applications in the Free Environment. The combination of labile blocks of activity that can under some circumstances break into components, and the control of the *M* oscillator and *E* oscillators, help us to understand the flexibility required in rodents in the free environment. These animals' environments are constantly changing, as in England, where dusk comes at 1600 hours for part of the year, and then at 2200 hours for another part of the year. A related point concerning blocks of activity is that they are very apt to be considerably less than 12 hours in duration. In fact, a common procedure in many mammals is to begin a block of activity at approximately 19 hours and to end major activity at 0300. This applies in the case of those nocturnal rodents that have a single block of activity rather than the bimodal distribution found in the house mouse. This cessation of major activity in a 12-hour light-dark cycle permits the animal to maintain somewhat its own behavior pattern when in the free environment the light cycle is increased so that there remains only 6 or 8 hours of darkness. The possible relationships between the *E* oscillator and the *M* oscillator can be pictured quite simply for many nocturnal rodents in that the block of activity is influenced mostly by the changes in time of dusk under the influence of the *E* oscillator. If this oscillator drives the block of activity so that there could be an overshoot in phase delay, this is compensated for by the *M* oscillator which should prevent the overshoot. It is a typical check and balance system which is found so often in the physiological design of the mammalian body.

What about the circumstances when the season produces a long period of darkness, again, such as might occur in England? It would seem that the block of activity which has just been described under summer conditions might now act as in a free-running situation (DD), except that the period of the animal might be longer or shorter than 24 hours. Note that the free-running periods described in Table 3-2 vary on both sides of 24 hours. Depending upon the species, we can picture a block of activity changing systematically in one direction or the other and being kept within a reasonable boundary by the occasional tap on the pendulum of the *M* oscillator and the *E* oscillator. One must not forget that when the SCN is destroyed, the activity rhythms are lost in all animals that have been studied. Now we may think in terms of three components, the SCN, which synchronizes the two secondary oscillators, and the two secondary oscillators responsible for the timing of the overt locomotor rhythm, which can occur as a single block or be spread into different components.

TEMPERATURE INDEPENDENCE OF RHYTHMS

One of the most important characteristics of biological rhythms is temperature independence. Even the mammal may display a wide range of body temperature, so that its hypothetical biochemical clock may be challenged to measure time at low temperatures. Some mammals have this range of body temperature in both a warm and cold environment when they enter hibernation. Furthermore, in this age of hypothermia as a procedure in human surgery, even human subjects undergo large changes in body temperature. More important than these practical applications are the fundamental implications of this biological principle. Let us use as an illustration the classic experiment of Welsh, who observed that crayfish eyes show a pigment migration which turns them light-colored for about 12 hours and dark-colored for about 12 hours in a persistent circadian rhythm. Let us assume that this rhythm is controlled by the subesophageal ganglia. What happens when the crayfish are cooled to 6°C? The cooled clock might run slowly, since ordinarily, biochemical reactions in tissue like crayfish brain obey the law of Arrhenius (that is, the Q_{10}=2.3). The eyes of the crayfish should show some prolonged rhythm, perhaps of 30 hours. Instead, Welsh found that cooled crayfish (DD) showed the original circadian rhythm throughout the 4-month period of cooling.

This control of the crayfish rhythm represents a biological process independent of temperature. Rhythms are altered or lost when experimental animals reach a critical body temperature: usually below 16°C to 6°C depending on the species. Therefore, rhythms are temperature-independent, but they are not temperature-insensitive (Cloudsley-Thompson and Constantinou 1981). When animals in the cooled state warm up, the rhythms they recover may be out of phase with the solar timetable; then zeitgebers will gradually synchronize these rhythms with solar events again.

A survey shows that so many biological rhythms are temperature independent that this must be a specific characteristic of endogenous rhythms. Examples of this temperature independence are found in pure rhythms, interval timers, and plant rhythms. Typical data showing temperature independence are illustrated by Bruce and Pittendrigh (1956) for rhythms in the one-celled animal, *Euglena* (Fig. 3-15). Many observations also show that the continuously-consulted clocks of bees are also temperature independent; otherwise they would be useless. Pittendrigh showed temperature independence in the *Drosophila* clock system (1954), and Wehner (1981) in the clock system of ants.

A clear-cut illustration of a mammalian circadian rhythm independent of temperature is found in body temperature records of hibernating bats recorded by Menaker. A simplified graph of one of his illustrations of the temperature of a single myotis bat is reproduced here (Fig. 3-20). Some mammals were studied in artificial hypothermia to see if their circadian clocks were temperature compensated. The rat, a non-hibernator, was less well compensated (Q_{10}=1.2-1.4) than the hibernator hamsters and bats (Q_{10}=1.01-1.13) (Richter 1975; Gibbs 1980). A hibernator, the hamster, comes into this picture once again. Tosini and Menaker (1996) were able to culture the hamster retina in a cool medium (27°C), yet a circadian rhythm of melatonin production was retained; the normal blood temperature of a sleeping hamster is 36.3°C (Folk 1974, page 75).

Attempts to explain the temperature independence of biological rhythms cannot be separated from the search for the nature and location of the biological clock. Is the characteristic of temperature independence so unusual that our credulity is less tried by supposing a discrete localized clock or pacemaker which has a special mechanism to permit performance at low temperatures? If we assume the alternative (a diffuse series of clocks in each animal), there must be a broader distribution than we have supposed of the rare biological material which can display temperature independence. Presumably in mammals, if the pacemaker is temperature independent, it would then synchronize secondary oscillators, even when their tissue temperatures are as cold as 5°C or 6°C. The clock-identification is complicated as usual by the fact that one-celled animals have temperature-independent endogenous rhythms.

There is little information about the mechanism of temperature

Figure 3-20. Body Temperature of Hibernating Bats. Consecutive records of one hibernating *Myotis lucifugus* show temperature independence of the daily rhythm. The period of the rhythm is slightly over 24 hours, with a body temperature near 10°C. Modified from Menaker 1959.

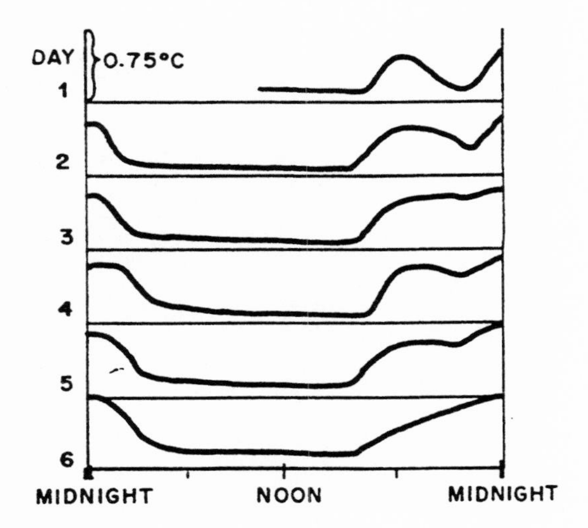

compensation; as an explanation, some workers are searching for enzymes that increase their activity as temperature is lowered. There is variability in enzyme design and there are isoforms of specific eyzymes. One of these isoforms may have a different optimum temperature from another isoform. Another author postulates a chemical mechanism that has no temperature dependence. If part of the mechanism were concerned with supplying a particular substance while a second process destroyed it and both processes were equally temperature dependent, then the substance would accumulate at the same rate irrespective of temperature.

HUMAN CIRCADIAN RHYTHMS

Introduction

Ever since our early human ancestors kept fires blazing during the night, humans have succeeded in providing themselves with approximately 16 hours of daylight and 8 hours of darkness per 24 hours. These artificial conditions damp out the harsher effects of seasons as experienced by most domestic and wild animals. But this downgrading of the seasons has not appeared to make the list of human biological rhythms particularly different from those of other day-active mammals. In the previous sections of this chapter, we have discussed the theory of circadian time-keeping. We will now proceed to apply these principles to the circadian rhythms of human subjects.

Using Human Subjects

There are both advantages and disadvantages to the design of experiments on human subjects compared to those with laboratory animals. It is relatively easy to make measurements on the human animal because the subjects will usually control their activity, and to a certain extent their behavior, upon request. This relative ease of making measurements is counteracted by the variability of each human subject from day to day, and the wide range of results from different subjects (intra- and extra-individual variability). To illustrate the last point, consider the homogeneity of body temperature rhythms obtained from 14 hamsters: every animal showed the same day-night rhythm. A similar experiment of 10 human subjects showed much variability (Timmerman, Folk, and Horvath 1959). Using either body temperature or heart rate as a criterion in experiments under carefully controlled conditions, the series was found to include mostly what Kleitman called "early energy" types of individuals; however, two subjects showed rhythms the exact reverse of the others (high temperature at midnight). Such variability is common in human experiments on circadian rhythms. In order to improve this type of experiment, particular attention has to be given to subsidiary factors such as preconditioning subjects, and obtaining background information (individuals' histories and subjective comments) to aid in interpretation of results.

Selection of Functions

The importance of using human subjects to work out the principles of circadian rhythms cannot be overemphasized; when experiments are done with domestic animals, the influence of the circadian rhythm of activity is not always recognized. In previous discussions, we have found the measurement of running activity or total activity to be useful for understanding the contribution of pacemakers and secondary oscillators in the animal body. However, when it comes to other physiological measurements such as blood sugar, body temperature (see Lefcourt 1982), or oxygen consumption, how are the measurements to be made? If one selects midnight as the peak of locomotor activity of a nocturnal animal, then in between bouts of activity the animal can be removed from the cage and immediately the three physiological measurements can be made. These, however, are only useful indications of the effect of *muscular exercise* on the three functions. The investigator of circadian rhythms would not find these data interesting because he or she wishes to study the basal state of the animal at the time of the habitual large bouts of locomotor activity. This is technically difficult, and usually the investigator has to settle for a resting state that is still influenced by the specific dynamic activity (diet-induced thermogenesis) of food.

Not all biochemical measurements are influenced by exercise. For example, in the classic experiments of Scheving in which plasma corticosterone of rats was compared with the same measurements in human subjects, the peak in both animals occurred before they became active; in the rat, the peak occurred just before its activity period began, and in the case of the human subjects, they were still asleep at the time of the peak. Nevertheless, for many measurements there is a considerable advantage in being able to sample physiological parameters of human subjects while they are in a rested condition.

The Rested Human Subject. In our laboratory we became interested in looking for the relationship between several circadian rhythms in human subjects, with special attention given to the phase relationship of four physiological events, all of which might have separate secondary oscillators, namely the adrenal cortex, the posterior pituitary, the hypothalamus, and the medulla oblongata. We wanted to compare the physiological status of human subjects from 0600 to 1900 with the same person under identical standardized conditions from 1900 to 0100. We've referred earlier to the day-animal and the night-animal, implying a Jekyll-Hyde existence for the person's daily routine. Actually, we wanted to study not two but three states in human subjects: the day-animal, the night- (evening) animal, and the sleeping animal. We set out to eliminate effects on the experiment from the observer, posture, variable meals, exercise, excitement, and sleep.

In one experiment, the investigation with 11 controlled male subjects was carried out in a quiet experimental chamber; each test lasted 33 hours or more. Phase relations of heart rates were compared with rhythms of core temperature, rate of urinary flow, and of electrolyte excretion. Data were collected hourly with the subject reclining for the last one-half hour (the last 15 min. in darkness) before the recordings were made. After the recording, subjects could leave the experimental chamber for 30 minutes if they chose, though most were rarely inclined to do this.

All subjects were given a carefully worded questionnaire to determine that they were in typical health at the time of the experiment. As stated above, we particularly wanted to compare the physiological condition of each subject in his daytime condition from 0600 hrs until 1900 hrs, with his evening condition from 1900 hrs to 0100 hrs; sleep would not interfere with the measurements. Because each subject lay down during the daytime for one-half of each hour, we found that not one of them was ready to sleep before 0100 hrs; this increased the value of the resting awake readings from 1900 hrs to 0100 hrs.

A small radio capsule placed in the pocket of the undershirt of the subject was employed for telemetering of heart rate. One electrode of this capsule was taped above the heart and the other below it. The subject never knew when the heart rate was being recorded. Since the antenna was in the experimental chamber, it was only necessary for the subject to walk into the chamber for recording of heart rate to begin. Subjects were told to drink water in equal amounts at hourly intervals. The most crisp electrolyte rhythm turned out to be that of potassium excretion. The influence of ingestion of food was studied in a variety of ways, including three equal standard meals or else administration of equal, small amounts of a liquid dietary formula at hourly intervals. In all cases, these observations (from 0600 to 0100) were recorded with the subject awake. However, recordings were continued during the sleep period from 0100 hrs to 0600 hrs; the subject was awakened once an hour to obtain a urine sample.

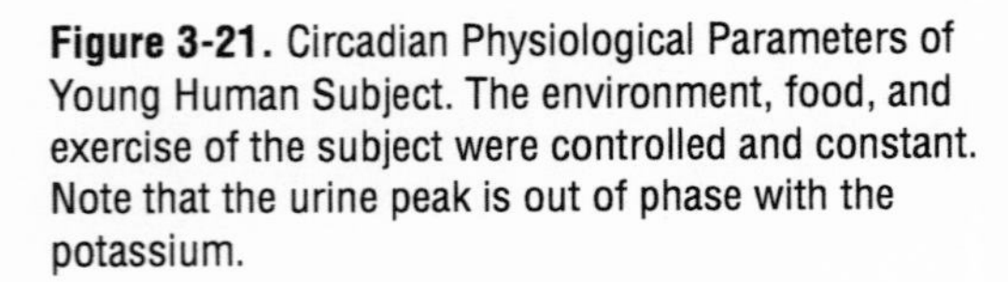

Figure 3-21. Circadian Physiological Parameters of Young Human Subject. The environment, food, and exercise of the subject were controlled and constant. Note that the urine peak is out of phase with the potassium.

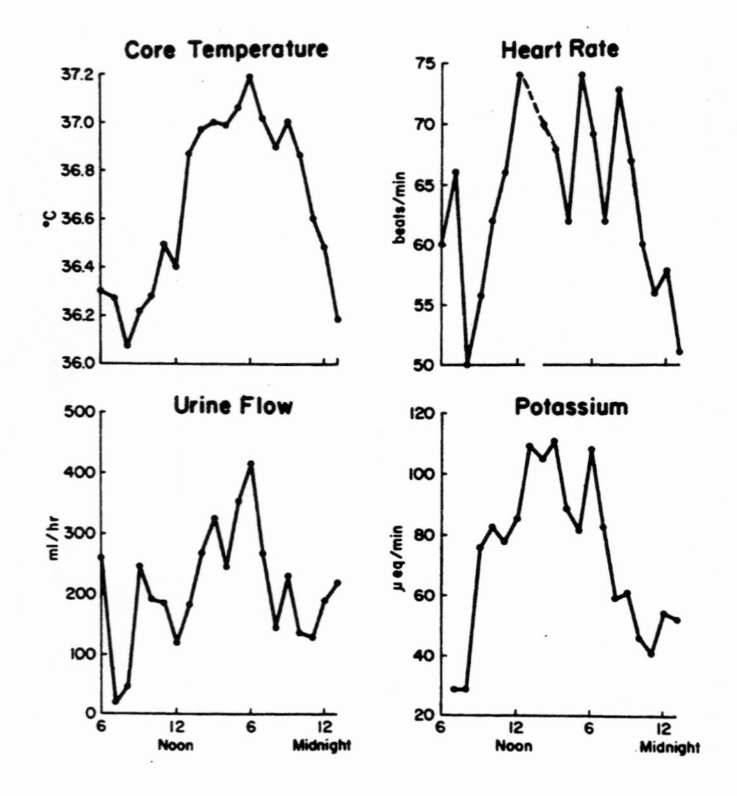

A typical example of the phase relations of the four variants is as follows (Fig. 3-21): the temperature and urine-flow peaks were at 1800, while the heart rate and potassium peaks were near noon. There was a clamping down of urinary flow from 2000 hrs on (still found at 0700); this is, of course, a useful function so that humans do not have to rise in the middle of the sleep period in order to empty their bladders. Note that hydration pressure was constant for each of the experimental hours.

The majority of the 11 normal male subjects showed similar physiological rhythms for the four measurements. Of the four variants considered, body temperature demonstrated the clearest sinusoidal curve. A biphasic rhythm of heart rate was sometimes noted. With some subjects, the peak period for each rhythm was in the morning, although reverse types of subjects were encountered with peaks between 1900 hr and midnight (Folk 1961b).

Experiments in which placebo meals were given indicate that the specific dynamic action (diet-induced thermogenesis) of food elevates heart rate by about 8% over the values characteristic of the non-postprandial period, but does not obscure the fundamental rhythm pattern. There are many other demonstrations of this lack of importance of meals in the literature. With some subjects, the basal resting standardized heart rate in the evening was 25% higher than the same reading in the morning (due only to time of day), and the reverse relationship occurred frequently. These differences between the morning resting values and the evening resting values can be attributed to a variation in a secondary oscillator, perhaps in the medulla oblongata. However, note that in an earlier part of this chapter we called attention to the circadian rhythm of heartbeat to be found in isolated rodent hearts (see Table 3-1).

On several occasions the same subject was studied twice or more at intervals of one day, one week, one month, and in one case the same subject was studied one year later. Individual performance was found to be nearly identical or very similar in the successive periods of testing.

In summary, we felt that these data were collected with all of the important environmental influences either removed or measured so that we were studying pure circadian physiological rhythms.

Ultradian Rhythms. Short physiological cycles between one and two hours in length were also apparent in our data; we referred to them as having a saw-tooth shape. Apparently we had picked up a phenomenon described by Kleitman and studied by Johnson (1972), who demonstrated a 90 min. cycle amounting to as much as 10% from trough to peak and occurring in oxygen consumption, carbon dioxide production, and nitrogen excretion (Fig. 3-22). This 90 min. rhythm also occurs in the activity of the human fetus and during sleep; it should be taken into account in the design of work-rest schedules (Folkard and Monk 1985).

Effect of Exercise on Tenseness. In the experiments just described, we felt that the influence of exercise and activity had been controlled, but to some investigators there is still some doubt. This point had been brought up many years ago in experiments on monkeys by Simpson and Gailbrith. They wanted to study the body temperature of resting and quiet monkeys at noon and at midnight, but considered the possibility that at noon the resting animals looked quiet but were not really resting; perhaps they had tensed major skeletal muscles. They had already measured body temperature difference of approximately 2°C at noon and midnight. They solved the question about the possibility of the tone of muscles by exercising the animals at noon and at midnight to remove any possible tenseness. After the exercise they still found an approximately 2°C difference.

This question came up with our human data. It was asked whether possibly the subjects at noontime, although resting for preconditioning for a half-hour, were in a tense condition compared to subjects at midnight who again rested for a half-hour. We decided to

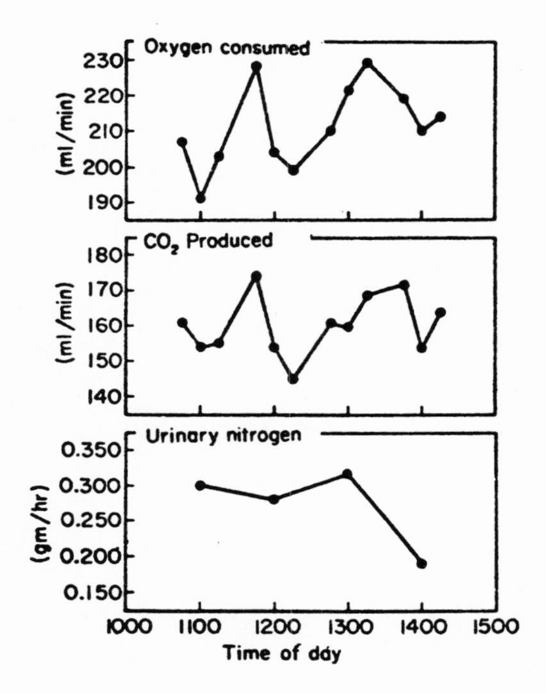

Figure 3-22. Short Physiological Rhythms. Daily rhythms on the order of 1-4 hours have received little attention. The term *ultradian* is applied here. The above rhythms of approximately 90 minutes in oxygen consumption, carbon dioxide production, and nitrogen excretion were recorded after the subject had been in the postabsorptive state for 12 hours. (R. E. Johnson, 1972)

try the experimental procedure used by Simpson and Gailbrith. We continued our usual resting measurements but after they were completed, each subject got on a treadmill and walked for 20 min. at 5 mph. These tests were repeated three days in a row; as usual the eleven subjects were in typical standardized working conditions (at the same time during the work week and well after or well before a vacation period). The measurements of heart rate and body temperature were collected while the subject was still on the treadmill. The results were the same as in the Simpson and Gailbrith experiments. The noon and midnight differences were still found after exercise. This is illustrated in the subject HF in Fig. 3-23; 9 of 11 subjects showed this difference, although sometimes the peaks were reversed (they came in the evening instead of at noontime). (The subject JT will be discussed later because his results were very different.)

Was each subject's large physiological difference between noon and midnight due to the subject's tenseness at one of the times? We believe that the exercise removed any possibility of this influence. There are many implications of these data; they should be of interest to sport physiologists and coaches. If a track meet is held at noon compared to the evening, the team will be in a different physiological condition. With some members of the team, this might be an advantage, and with other members of the team, a disadvantage, depending upon where the physiological peak of heart rate and body temperature are to be found. We have found it useful to apply the term "setting" to the two conditions at noon and midnight.

High-Low Physiological Settings

In discussing circadian rhythms with athletic coaches and with clinicians, we are justified in introducing the word *setting* in the same sense as used with a thermostat for heating a house. The justification of the term can be found by examining the individual daily records of human circadian rhythms of many functions. It is true that many rhythms demonstrate a smooth sine wave and the statistics of rhythm analysis are based upon this fact. However, in our own experiments with monkeys and in those reported by Moore-Ede, the functions often appear to attain a plateau either during the night or during the day. In fact, some functions practically demonstrate a square wave. The human sleep rhythm is an example; another illustration is the body temperature rhythm for 10- to 14-year-old subjects in Fig. 3-4. The plateau of body temperature is even more evident in the supplementary data of Abe et al. (1978). Many rhythms show the opposite extreme, which is a sharp spike perhaps at noon and at midnight. Plasma corticoid in the human is an example of this. Some free-living animals show a circadian plateau or setting in their heart rate recordings, while occasionally a free-living animal such as the fox shows no evidence of this setting whatsoever. Nevertheless, we have found it convenient in explaining to the medical profession that, on the whole, a patient at noon and at midnight is in a totally different physiological condition or setting and therefore medication should be varied accordingly. This concept seems to be grasped more readily than that of a peak and nadir of a rhythm. The concept of a setting assists in the explanation of Kleitman's early energy types *(larks)* compared to the late energy types *(owls);* it is merely necessary to explain that some people have their high physiological settings in the morning and others have their high settings in the evening. With most subjects, one would expect their physiological functions to be in phase at the time of both the high and the low setting. In our experiments, however, we found several conditions or subjects in which this was by no means the case. We have referred to these functions that were out of phase as being dissociated from the other functions of the body in respect to their rhythmical condition. Our term *dissociation* must be distinguished from another term which has a different usage, *desynchronization.*

Figure 3-23. Human Daily Physiological Rhythms After Exercise. The rhythms of resting heart rate and body temperature are illustrated by graphing values obtained at noon and midnight. The day-night differences are still found after exercise, showing the presence of a day-setting and a night-setting of these physiological factors. Note that the two rhythms of subject J.T. are out of phase (ie., dissociated).

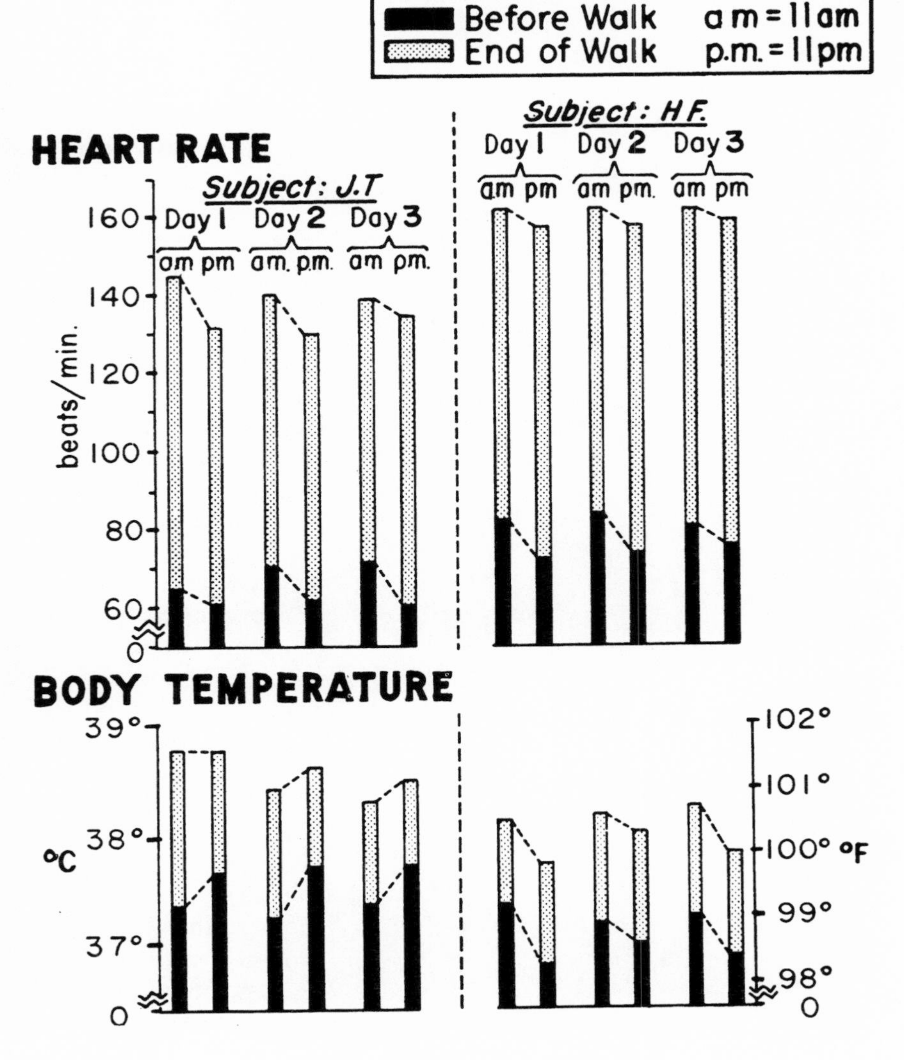

Dissociation and Desynchronization of Function

The circadian rhythms of physiological function which were illustrated earlier were those of core temperature, heart rate, urine flow, and potassium excretion as an example of electrolyte excretion. In most human subjects, the peaks of these functions will occur at the same time of day, probably near noontime; this means they have the same frequencies. On the other hand, a few subjects will have functions with the same circadian frequency, but their peaks occur near midnight (Kerkhof 1985); they are called *late energy subjects,* or owls. The subject HF in Fig. 3-23 is an example of a typical subject, but the reader should examine carefully subject JT in the same figure. Here the resting subject has a peak of heart rate at noon, but a valley (nadir) in body temperature; this means there is a nadir of heart rate at midnight, along with a peak of body temperature. These same relationships held true after exercise. Considering these two functions alone, there are many implications; for example, the heart rate rhythm could not be caused by the peak of body temperature since they occur at opposite times of the day. This is additional evidence of the heart rate being under control of its own secondary oscillator and not slavishly following the body temperature rhythm. However, the situation in this subject is more complicated because the peak of potassium excretion was at noontime while the peak of water excretion was at midnight.

Earlier we noted that there is considerable advantage in the kidney's clamping down on urine flow at the nadir of the circadian rhythm so that human subjects will not have to get up to empty bladders in the middle of their sleep period. JT is not the only subject with this reversed bladder rhythm; Kleitman, an early expert on circadian rhythms, reported in the literature his own reversed rhythm of urine flow. Both of these two healthy subjects continued throughout their lifetimes to get up in the middle of the night to empty their bladders, an unusual necessity for most healthy human subjects.

It is easily shown in such subjects that these four dissociated rhythms, however, *have the same frequency.* The best term to apply to the condition is that of a phase disorder. In a series of 23 subjects studied, we found two subjects with this phase disorder. For reasons to be explained below, the term *desynchronized* cannot be used for this situation. We have employed the term *dissociated* to the phenomenon; this is a term not clearly defined in the past, but used by Kleitman, Halberg, Aschoff, and Pittendrigh.

In the above paragraphs, we have discussed functions that are usually in phase. There are other occasions when these functions are not in phase; in Aschoff's famous bunker experiments, human subjects were isolated from zeitgebers. Perhaps not at the beginning, but eventually, most of these subjects showed circadian rhythms with totally different frequencies. The body temperature rhythm of the same subject could have a frequency of 23 hours, and the potassium one of 26 hours. A similar condition could be found when a person is on the night shift or has recently moved across six time zones and has not yet readjusted. Most authors use the word *desynchronization* when there are two or more rhythms which show *independent frequencies.* Internal desynchronization is the loss of synchronization between two or more rhythms so that they free-run with different frequencies within the same organism (Gander and Moore-Ede 1982).

Morning Types and Evening Types. We stated earlier that most human subjects have their physiological rhythm peaks near noontime; this was the case when several hundred subjects were studied in the laboratory of S. Folkard (1985). The researchers also found that 5% of their subjects were extreme evening types or owls, and 5% were extreme morning types or larks; these larks went to bed at night and arose in the morning about 80 minutes earlier than the owls (see also Kerkhof 1985). However, Horne studied 150 subjects in a similar way and found a much larger percentage of both larks and owls (1976).

The question of a "post-lunch dip" has been studied extensively (Folkard and Monk 1985); this is a period of sleepiness and lowered arousal level after lunch which is found in both the typical (noon-peak) subjects as well as the extreme types described above. Its origin is both exogenous and endogenous.

Other Circadian Rhythms

Although we have emphasized the four circadian rhythms as described in the above paragraphs, there are of course other clear-cut circadian physiological rhythms that have been studied in human subjects. There is an increase in skin blood flow and therefore an increased loss of heat by the body at the time of the body temperature nadir (Lefcourt et al. 1982). Skin temperature of the fingers, however, behaves differently; a typical subject under controlled conditions will have a finger temperature of 26.5° C at 1300 hours, and a finger temperature of 23° C at 1600 hours. There are changes in excretion of 17 keto-steroids with a peak between 0700 and 1200, and a decrease by evening (Krieger 1979); there is a minimum count of eosinophil cells in the blood near 0900, with a gradual increase to a maximum after midnight; there is also a minimum in mitotic activity in the epidermis in the morning with a maximum in the evening. Note that unlike the expected phase relationships mentioned for the four physiological functions discussed above, in these latter cases, the peaks are not in phase. These last three examples are responses that arise from the pituitary body under the control of the hypothalamus. Out of all the physiological rhythms, however, probably the sleep rhythm is the most interesting.

Sleep as a Free Function

The sleep rhythm is a *free function* (terminology suggested by Aschoff) in the daily sequence of events that make up a circadian composite rhythm; this sleep rhythm is often referred to as the *sleep-wake cycle* (Winfree 1982a; 1982b). The pacemaker is probably the SCN (Obál and Benedek 1981). The sleep-wake cycle is most conveniently studied in humans, because in most, sleep is essentially uninterrupted for approximately eight hours. In older subjects, this figure is nearer six hours (Bixler et al. 1985). This cycle is more difficult to study in domestic and free-living animals: for example, the domestic cat has short bouts of sleep and waking throughout the day and night with a very weak expression of the circadian sleep-wake cycle. The white rat falls between these two extremes with most of its sleep occurring during the daylight phase of the 24-hour cycle. Other animals have a wide range of variation in the amount of sleep: the donkey spends only about 13% of its time in sleep, the sloth 70% (Filho, Huggins, and Lines 1983), while the opossum sleeps 80% of the time and still somehow manages to earn its daily living. To complicate the issue even more, note that although some animals are diurnal, others are nocturnal, and still others, such as the laboratory mouse, are called *crepuscular*, which means they have a peak of activity only around dawn and around dusk.

A characteristic of the human sleep rhythm is the presence of profound changes in the depth of sleep. There are bursts of rapid-eye movement (REM) which are linked to the period of dreaming. The important physiological associated information is that REM sleep is accompanied by a loss of muscle tone, because the activity of spinal motor neurons is inhibited. Otherwise, we would probably act out our dreams much more frequently. Young adult humans spend approximately 20% to 25% of sleep time in the condition of REM. The alternation between REM and non-REM has an ultradian rhythm of approximately a 90 to 100 min. period. In human infants, the REM-non-REM cycle has a 50 to 60 min. period, which gradually lengthens in adults.

The circadian sleep rhythm persists in the absence of environmental variations; it has a typical free-running period of about 25 hours, although in most human subjects studied over a long enough time period, the frequency eventually lengthens to 30 to 50 hours (Winfree 1982). Under these conditions, this rhythm becomes dissociated from the ultradian REM-non-REM cycle. Thus, the two rhythms are essentially autonomous. We have some capability to postpone sleep by exerting conscious control; by adding to this control the assistance of other people, a 17-year-old boy in San Diego was kept awake for 11 days. During such times, there is a continuing circadian rhythm in the desire for sleep; if a persons stays up all night, he or she will still be more alert in the morning, but at night the person feels more fatigued at what is ordinarily bedtime. The effect of sleep deprivation is to strain the biogenic amine system; there is an increase in renal excretion of amines, demonstrating increased secretory activity. To be specific, serotonin is increased in urine and decreased in the blood.

What is the relationship between the sleep rhythm and the body temperature rhythm? Although there will be much more to say about this in a discussion of polar, cave, and night-shift rhythms, it is sufficient to say here that when Aschoff (1979b) maintained his subjects in his bunkers, isolated from environmental influences, internal desynchronization occurred in that the rest-activity cycle began to free-run with a frequency that was much longer than that of the body temperature rhythm. Under these conditions, the length of the sleep rhythm is modulated as it passes through the various phase relationships with the body temperature rhythms.

The duration of periods of sleep will also vary according to the stage of other rhythms in the body. In Aschoff's experiments, when subjects went to bed at the nadir of the average temperature cycle, sleep episodes were short. When they went to bed at the peak of the temperature rhythm, sleep was extended. According to Moore-Ede, this indicates that sleep duration is the product of an interaction between the separate pacemakers driving the sleep-wake cycle and the body temperature rhythms.

The student is apt to look upon the sleep rhythm as being an independent or modifier variable. This is by no means the case, since it must be considered the equivalent of the other physiological rhythms. It is actually a dependent or modified variable responding to zeitgebers, or time change, or geographic displacement in the same fashion as other physiological rhythms.

Caves and Polar Conditions

Many of the principles of circadian rhythms described in earlier pages have been defined by novel and ingenious experiments involving living for long periods either in caves (a time-free environment without zeitgebers), or in the continuous light or continuous lack of sunlight in polar conditions. Kleitman (1963) studied the effect of non-24-hour routines on several measurements; Lewis and Lobban (1957) provided their subjects in the continuous light of the Arctic with watches that were altered so that, although all subjects thought they were on a 24-hour schedule, one group was actually on a 21-hour schedule according to the watches, and another group unknowingly followed a 27-hour time-period. These studies were followed by an extensive series of cave experiments by J.N. Mills et al. (1977). Mills described his studies as using subjects living on self-selected schedules but deprived of knowledge of time; one subject descended into the Boulder Cave in Cheddar where he remained alone and without time clues for four months. In about 1964, Aschoff and Wever began their well-known experiments in which they isolated human subjects in bunkers where they were free of time cues. They gave particular attention to introducing external zeitgebers of various modes (Wever 1979). Subsequently, Mary Lobban made a number of expeditions to Point Barrow and to Hudson Bay Inuit villages to study these people under the condition of continuous light in summer, and continuous lack of sunlight in winter.

The most conspicuous result from all these experiments is the finding that groups of overt rhythms when free-running become internally desynchronized. In other cases, when some overt rhythms were controlled by the external zeitgeber, other overt rhythms were free-running. For example, in the experiments of Lewis and Lobban, when all 12 of their subjects were on 21-hour days or 27-hour days, none of them had difficulty in sleeping; this means they changed their sleep patterns to match the artificial periods. However, especially in the first two weeks, most of them had kidneys still on circadian time. In another case, in an experiment with a human subject living in isolation without time cues, Wever showed by periodic analysis that the subject's rectal temperature, urinary potassium, sodium, and water excretion came to a free-running peak at 24.8 hours, whereas the circadian rhythms of activity and urinary calcium excretion came to a free-running peak at 33.5 hours. To generalize, the experiments showed that any one of the following human rhythms can be on some abnormal times, such as 21 hours, when all the rest are on circadian time: sleep, heart rate, body temperature, water excretion, and electrolyte excretion. It also became clear that some human subjects are "true or complete adapters" to artificial times (21-hr days or 27-hr days), while particularly older subjects were more apt to adhere rigidly to circadian time in spite of the unusual zeitgebers to which they were exposed. It must not be forgotten that in our normal routines, our overt rhythms are usually kept in phase by our bedside alarm clocks.

To give another example of the type of information obtained in these isolation experiments, consider the entrainment of a human circadian rhythm. When subjects were exposed in the bunker to an artificial 24-hour light/dark cycle by means of illuminating the whole ceiling of the room during the daytime period, the subjects ignored the zeitgeber as long as they could make use of a small reading lamp. The subjects continued to free-run with a period longer than 24 hours, despite the unpleasantness at times of sleeping with a bright light overhead. We say from this evidence that the cues of selecting one's own schedule by use of the reading lamp is a social cue which is more important than the bright LD cycle.

A group at Stanford University studied five rhythms in a blind man. All of his rhythms had a period of 24.9 hr. which is indistinguishable from the period of the lunar day (Miles, Raynal, and Wilson 1977). Was he free-running, or was there a lunar-linked zeitgeber?

Some comments about sleep habits during the continuous light of the Arctic are appropriate here. We were able to make observations at the village of Barrow where there are 82 days of sun that never drops below the horizon; this village is the largest Inuit village in the world. We frequently visited the village in continuous sunlight during the usual time of darkness in the temperate zone, namely at times such as 2300, 0100, and 0300. These observations extended over many years and the generalization can be made that at all times of the temperate zone "night," there were children playing in the streets of Barrow. Our observation was confirmed by Stefansson in his book, *Hunters of the Great North:* he stated that the native Inuit way of sleeping was completely random in the summertime, sometimes lasting for an hour, sometimes for five hours and sometimes for ten. Some children had random sleeping habits, partly because of the arrival of visitors. One mother in a remote village told him that her 8-year-old daughter had been awake continuously for five days and nights; he himself observed several children who were awake continuously from three to five days. Of course, present-day Inuit children have a regular routine in the wintertime because they all must keep a schedule for being at school and doing homework. Nevertheless, they revert to ancestral sleep habits in the summer.

People from the temperate zone do not take on these informal activity patterns when they come to live in the polar regions. In our experience at Point Barrow, we noted that such persons usually made a considerable effort to provide themselves with dark bedrooms so that they could sleep at the usual hours as practiced in the temperate zone. This was observed as early as 1953 by Kleitman in the Arctic city of Tromsø; the same observations have been made on expedition members living on the Antarctic continent.

Navy Watch-Standing and Space Conditions

We have been considering the polar environment and studies on human subjects living in caves or isolation chambers. Very similar environments are found in Navy watch-standing circumstances, or while doing military guard duty, and in the surroundings of space capsules. The polar environment always includes long periods of continuous light, with the sun going in a circle overhead, and a similar period of continuous darkness, with the sun below the horizon much of the time. Similarly, navy personnel on atomic-powered submarines cruising beneath the surface for several months also must depend upon artificial light for creating a relatively normal environment. A practical question concerns whether the four-hour cycle of duty routine used in some military services, or a similar eight-hour cycle used in other circumstances, is an efficient procedure. A system of eight hours of rest followed by eight hours of travel over the polar ice and snow has been used in the continuous light of the Arctic, and this procedure was also tested for use in space flight.

The environment of space capsules is modified more than in any other circumstances because when the astronauts are in orbit around the Earth, their satellite circles the Earth every 90 minutes so that they are exposed to sunlight for about 50 min. and to relative darkness for about 40 min. However, this is the environment outside the space capsule, and as we discuss the physiological implications, it is clear that it is desirable for the crew to set up an artificial day based on their original local time (Alyakrinskii and Stepanova 1985).

Strughold (1967) evaluated the relation between circadian rhythms and the selection of duty-hours in space. He was the first Professor of Space Medicine, at the Air University in San Antonio, Texas in 1958. In definitive articles each year beginning in 1961, he discussed the solved and unsolved medical problems of the human in space. Year after year, it was necessary for him to reiterate that the circadian cycle in humans can be shortened to only about 18 hours, and extended to 28 hours. He also stressed that the preservation of this physiological pattern within an inherited time frame is a precondition for human health and performance ability. Astronauts, therefore, have to follow the dictate of their physiological clocks by a well-regulated sleep and activity regime. Some interesting questions remain, such as whether the better astronaut is the one with the large amplitude or the small amplitude of circadian physiological rhythms.

Strughold's findings also apply to the futile attempts at maintaining a 16-hour day with 8 hours of rest followed by 8 hours of work. In a similar way, the Navy's 4-hour cycle of duty routine comes into question. With either the 8- or 16-hour cycle of rest and activity, much of the time individuals are attempting to sleep when their physiology is on a daytime setting, and attempting to work when their physiology is on a nighttime setting (Folkard and Monk 1985). We will discuss this further when we consider shift workers.

Jet Lag

We have been comparing free-running rhythms with imposed rhythms, which are somewhat artificial and abnormal. The free-running rhythms have been observed in caves and isolation chambers that simulate the environment of atomic submarines submerged for long periods. We have also been considering imposed rhythms of 21 hours and 28 hours. During some of this isolation, abnormal zeitgebers have been used, including, in some cases, noise produced by gongs. The result of these experiments is considerable information about the driven desynchronization of such physiological rhythms as body temperature and excretory rhythms. Now we must consider the somewhat more normal situations involved in crossing time zones, or being asked to work on the night shift. These can be referred to as more normal (instead of experimental), because the circadian rhythm remains but is simply turned like a dial by as much as 12 hours (i.e., phase shifted).

Let us first consider the North American who travels east to Europe, or west to Hawaii, and remains there for a period of time. The immediate effect upon the individual's physiology is referred to as jet lag. The condition of the person is best understood by introducing the words *physiological time* and *local time*. For those interested in the process of jet lag, it is convenient to carry two watches, one remaining on the original home time (the *physiological watch*) and the other set for the new geographic environment (the *local watch*). For example, a businesswoman flying from the midwest to Europe will pass six time zones. A typical west-to-east flight leaves New York at 1900 hr and arrives (by personal clock time) in Amsterdam at 0100. Our traveler is now ready for a good night's sleep. However, the local time is 0700. She takes a taxi to a hotel, where she is invariably informed that the reserved room will not be ready until 1300 hr. The exhausted traveler must walk along unfamiliar streets, sit in the hotel lobby, or doze in the tourist boat traveling the Amsterdam canals. This is a poor start for a European tour.

The real time-zone effects take place over the next few days. After a few hours, the traveler looks at her physiological watch and finds that it reads 0300; her local watch reads 0900. Over the next few days, she must advance her physiological condition until it is in phase with the local time. The evenings go easily; a tour of the city's night spots may begin at local time 2100, while the visitor's physiology time of 1500 is lagging behind. This afternoon "setting" provides a good physiological condition for enjoying the evening. The next morning, however, when another tour starts at 0800, the individual, who has probably had a poor night's sleep anyway, must join the tour with a physiological setting of 0200. We say that this person now experiences a transitory temporal disorder which usually results in a loss of performance and could have immediate effects upon health. Probably the traveler is in a weakened condition and is more susceptible to diseases carried by the local population. Experiments with rats showed that "jet lag" produced retrograde amnesia in that species (Tapp 1981).

If our traveler is unlucky, in spite of an 0200 physiological setting, she may have to serve as a conference chair, or attend a business meeting. This example applies especially to diplomats traveling from North America to various European cities. When the local time is 1000, the physiological time or setting of the diplomat may be 0400. A curious physiological condition is apt to result: no matter what one is doing, even if important matters of state are at stake, there are short moments of blackout, with sleep passing in and out very rapidly for a few minutes. Yet the overall condition is not that of being sleepy or having to fight sleep. Apparently the State Department attempts to send its envoys on a Friday before a Monday morning conference, so that the person can partially recover over several days from the jet lag condition.

Studies have been done in several countries to define how long it takes for individuals to change their physiological time until it matches local time. Mills et al. (1978) found considerable difference in the ease of adjustment; out of 19 subjects, one was completely adjusted from the second day onwards, and another was still completely unadapted on the fifth day after the time zone change. No difference was seen in the ease of adjustment between the ten male and nine female subjects. In this experiment, adjustment was better in the westward than in the eastward flights, as shown by a change to local time on the fourth day in six out of seven subjects flying westward, and only one out of nine flying eastward.

Further laboratory proof of the chronological insults suffered by subjects crossing several time zones have been summarized by Aschoff (1981) from his own work and the work of Halberg, Cline, and Wever. He found that the various rhythmic functions are reentrained at different rates resulting in the usual transitory temporal disorder. The mean shift rate was greater when all variables were averaged after an eastbound flight, amounting to 80 min per day. When he synthesized the results for westbound flights, the delay shift averaged 57 min per day. When people travel from the Midwest or the East Coast of the United States to Fairbanks, Alaska, or to Hawaii, their physiological time on arrival may be 0300 hrs while the local time may be 2100 hrs. Over the next few days they must delay their peak of physiological function until it fits local time. Note that it is easier and more pleasant in a physiological sense to sleep later in the morning (this is to delay the peak of activity), than it would be in the European set of conditions (i.e., to get up earlier than usual in the morning to advance the peak of physiology function in order to match local time).

There is another difference between eastbound flights and westbound flights; the amplitudes of the rhythms are drastically reduced during physiology advances (eastbound flights) but not during physiological delays (westbound flights) (Moore-Ede 1986).

It is not only tourists and ambassadors who travel through time zones, but athletic teams as well. Of course all the generalizations described above apply to athletes. Coaches today need to be informed about the negative effect of jet-lag and morningness/eveningness (Winget, DeRoshia, Holley 1985); not surprisingly, Hawaii's football team seems to have the jet-lag handicap more than other teams, as its direction of travel is more often eastward. Quite clearly, athletes seldom have several days to adjust to a local time zone because they must return rapidly to their home base for other athletic events. This means that teams and individuals sometimes have to compete when they are at a low physiological setting while their opponents are at a high physiological setting. The effect of physiological settings upon performance will be discussed in more detail under the section on night-shift workers, but it is clear that efficiency cannot be the same at a peak and at a valley of the circadian rhythm.

The situation for tourists, state department workers, and athletic teams is compounded by individual variation in the physiological design of normal circadian rhythms. The reader will recall that there are human owls and larks who have such extreme rhythms that they can be called *phase disorders*. Some of the owls customarily work until 0300 and sleep until mid-afternoon—larks may pursue the reverse. Such individuals may have difficulty adjusting to a new time zone. It is not yet clear how common these disorders are, how much they affect a person who travels across time zones, or how the disorders prevent these individuals from adjusting.

Now we should mention some specific rhythms that must be adjusted to local time regardless of whether the flights are to the east or to the west. In general, the body temperature and heart rate rhythms, and possibly the sleep rhythm, can be reentrained first. The electrolyte excretion rhythms and the rhythms reflected in the endocrine systems appear to be very slow in their reentrainment to local time.

What steps can be taken to speed up the process of adjusting to a new local time? Melatonin supplements have become popular, though there is little to no evidence that melatonin helps combat jet lag. Though less convenient, light therapies used in treating Seasonal Affective Disorder (see Chapter 2) may hold some promise. Rosenthal (1993b) states, "The scope of light treatment is expanding to encompass problems of circadian rhythms, such as jet lag." Although definitive results are not yet available, light-treatment devices are already being marketed. Perhaps the simplist approach is to control the diet: according to Ehret (1993), it seems best to avoid excesses of coffee, tea, or the use of sleeping pills. He has noticed that there is a peak of epinephrine in late afternoon or early evening, which then subsides as the body builds serotonin, a material that prepares the system for relaxation and sleep. He found that these peaks can be supported by diet, and recommends that on a new schedule, people eat most of their protein for breakfast and lunch, because protein stimulates this epinephrine active phase; then he recommends carbohydrates at dinnertime because they enhance the sleep pathways.

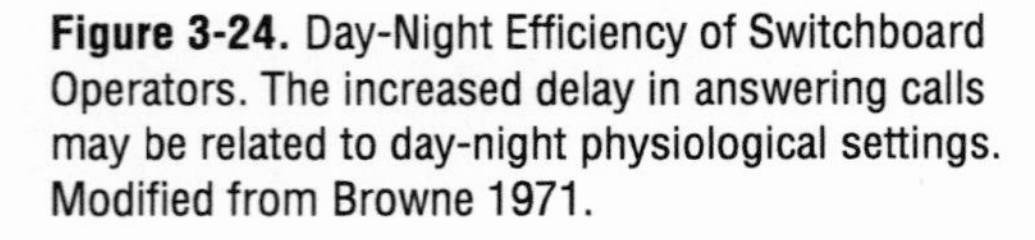

Figure 3-24. Day-Night Efficiency of Switchboard Operators. The increased delay in answering calls may be related to day-night physiological settings. Modified from Browne 1971.

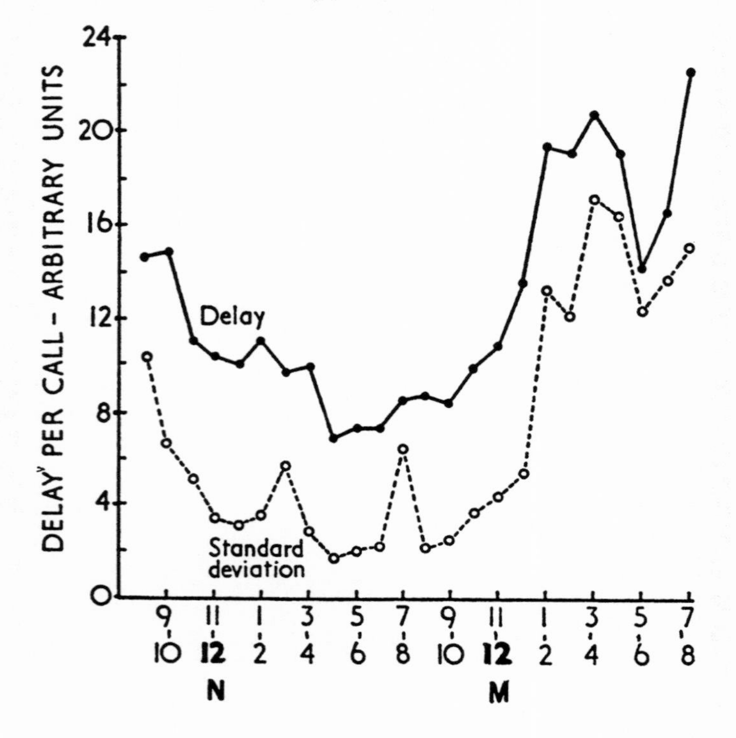

Night-Shift Workers

If a person has customarily been going to work at 0800 and then is asked to begin work at 2000, this new requirement is like passing through 12 time zones. Once again, the body's rhythmic functions are thrown out of synchronization; again we say the subject is in a state of transitory temporal disorder. There are a great many more individuals affected by shift work than by crossing time zones. It is probably an underestimate that 27% of the labor force in the United States has been asked to carry out complex activity at an unusual time of the day. The round-the-clock operations of many industries, public utilities, transportation and communication companies, police and fire departments, hospitals, military service, air traffic controllers, and nuclear plant operations, all necessitate the employment of multiple shifts with a dislocation from the usual circadian rhythm of sleep and wakefulness.

There is little doubt that night workers make errors and have health problems (Fig. 3-24); they often complain of fatigue and report a variety of ailments, particularly digestive disturbances and gastric ulcers. The interpretation of these ailments is difficult. When a person first goes on a night shift, he or she experiences internal desynchronization of rhythms; just as is found in jet lag, the parts of the body are no longer synchronized and the peak of the temperature rhythm may be out of phase by 12 hours with the peak of potassium excretion. Is the problem a lack of health in the shift worker due to the circadian rhythm disorder, or some other factor such as lack of sleep, or a behavioral disorder due to lack of former social contacts? In a number of carefully studied cases, the ailments could be blamed upon lack of sleep (Fuller, Sulzman, Moore-Ede 1981).

Even more serious are the frequently-changing work schedules sometimes demanded by management; for example, some air traffic controllers were asked to be on the night shift every other day. In other industries, the night shift workers rotated between days and nights on alternating weeks. At one time, the Navy simply allowed each atomic submarine commander to set his own watch-standing regimen. As a result these might be 4 hour, 7 hour, 8 hour or even 12 hour shifts. An additional problem is that the shifts selected did not allow the workers to have any of their usual time for sleep. If some workers have a shift from 1600 to midnight, then there must be a "graveyard" shift from midnight to 0800. This latter shift would require working during the time when body temperature is lowest and sleepiness is greatest, and would necessitate daytime sleep. To require a daily change of schedule, or alternating weeks of this graveyard shift, is what those who study biological rhythms would refer to as "chronobiological cruelty."

There are two solutions to this problem: in the first place, when shift-periods are designed, they should last a mandatory 21 days. Second, the best physiological timing, as pointed out by Kleitman years ago, would provide for shift hours of 0400 to noon, noon to 2000 and 2000 to 0400, thus doing away with the graveyard shift and allowing for some of the usual sleep time for each individual. Investigators studying biological rhythms recommend that shift workers retain their sleep schedule over weekends, since it is better not to try to readjust the sleep cycle, not even to get in synchrony with their families; above all, it is not desirable to use sleeping pills to try to overcome the pressure of the biological rhythm. This same caution applies to melatonin: it has become popular to take melatonin to induce sleep at odd daytime hours (melatonin sales were up 20-fold in 1997). Though it does seem to induce sleep, its after-effects include reduced concentration, a depressed mood, lower efficiency, and the development of hostility (Dille 1996).

How long does it take to adjust to the night shift? Work has been done with rodents to test this; the speed of reversal of superficial rhythms of rodents with a 12-hour change has been determined also to be about four days with ordinary laboratory lighting conditions. The feeding pattern of these animals changed simultaneously with locomotor activity. This may provide a clue to how long it may take human subjects to adjust.

The question is often asked whether a person on a night shift is performing more work than he or she would on a normal schedule. The same question applies to athletic performance. Doing physical work at midnight seems easier at the start because some of the workers have a lower oxygen consumption and heart rate at that time than they do at a noontime. However, as the work is continued, it has been found that the increase in energy consumption is much greater than it would be in the daytime, in some cases as much as 44% more. Furthermore, where dexterity is involved, more errors are made, even after a person has been on a night shift for some time. This is because often one of the requirements of night-shift jobs is to manipulate or to study a variety of complex instruments in a monotonous environment. The explanation for the errors seems to lie partly in the body temperature of the individual. When body temperature varied due to the circadian rhythm over 24 hours, or when the body temperature was raised by hot baths or by exercise, under all three circumstances the physical performance and the psychometric performance was improved at the time of the temperature peak. Also related to psychometric performance is the evidence that the shift worker is more prone to traffic accidents.

The effect of the night shift upon heart rates was studied on 20 healthy men by Banaszkiewicz (1979). Heart rates during the first hour of the 8-hour shift were compared with heart rates for the same work during the 8th hour. During the night shift, the heart rate response to the task began at a low value (mean 105 b/m), and then showed a larger increase by the 8th hour of shift work than in the daytime (118 b/m). It should be noted that these averages, as in many other reports in the literature, do not take into account those subjects who are larks and those subjects who are owls. If only one kind of subject were studied, the results would be even more striking.

Have methods been found that will increase the performance of individuals on shift work? Although articles have appeared in the literature offering dogmatic generalizations about the time that it takes to readjust to a new local time as required in shift work, there are many variations within the circadian rhythms of human subjects (Czeisler et al. 1982). For example, some individuals, even after 21 days of being on the night shift, may still have their kidneys on a noonday shift while their body temperature, heart rate, and sleep rhythms are now on a night shift (Conroy et al. 1970a,1970b). Some individuals find it impossible to ever adjust to shift work; these are more apt to be larks than owls. We have discussed Ehret's recommendations for overcoming jet lag through a change in eating habits; he recommends the same strategy for shift workers, advising that protein-rich meals stimulate the body to wake up, and carbohydrate-rich meals produce chemicals that bring on sleep.

Clinical Applications

Beginning with the writings of Nathaniel Kleitman (1963), there have been many books and symposia on the topic of clinical applications of circadian rhythms (Hildebrandt et al. 1987). Examples will be provided related to drug dosing and to epileptic seizures.

Dosages of some drugs must vary with the time of day. Recall that in an earlier part of this chapter we refer to the day person and the night person as different individuals. Drug dosage for anesthesia effectiveness and for cancer chemotherapy may need to be varied depending upon time of injection. An example from animal experimentation will illustrate: *E. coli* endotoxin was injected into two groups of mice at noon and at midnight. The mortality was 83% at noon, but only 13% at midnight. The physiological "setting" of the mice differed by that amount (Moore-Ede et al. 1982). Epileptic seizures were studied for 20 years on the same patients. There were two classes: those who had 91% of their seizures at night (night type), and those who had 66% of their seizures in daytime (day type).

Cardiovascular disorders vary with circadian time. There is a morning increase in the risk of myocardial ischemia and infarction, sudden cardiac death, and stroke (Muller and Tofler 1991).

There is a circadian reactivity of human skin to the following, which were tested over 24 hours: house dust, penicillin, and histamine. The results explain some of the variability in skin test results at the allergist's.

Another entire area of chronomedicine is concerned with the treatment of disorders of sleep, mood, season, and jet-lag (Rensing 1987).

Circadian Rhythms and the Elderly Human

One approach to understanding the change in circadian rhythms in the elderly is to study hours of sleep. Rats sleep mostly in daytime, but this changes with aging. Most elderly rats had lower daily mean body temperatures and smaller daily amplitudes of circadian temperature rhythm. (Li and Satinoff 1995).

Those elderly rats on a reduced caloric intake had a lower metabolism and less circadian rectal temperature variability (Moradian and Shinkar 1993).

Elderly humans developed deterioration of normal sleep and an increase in daytime napping. Elderly men showed a weaker circadian rhythm than did elderly women (Buysse et al. 1993). In another study with cold exposure, control values of skin temperatures and metabolic rate were higher in the young; amplitude of circadian body temperature was larger in the young. Upon cold exposure, core temperatures fell by 0.4°C in the elderly only, and the rise in blood pressure in the elderly exceeded that of the young (Collins et al. 1995).

HOW MANY PACEMAKERS?

To recapitulate, we have been discussing the human circadian rhythms of body temperature, urine volume, electrolyte excretion, heart rate and sleep. The next question involves the search for what oscillators or pacemakers order the daily peaks and troughs of these rhythms. The following evidence leads us to look for a separate control device for each rhythm: namely the observation that internal desynchronization occurs when these rhythms are tested in extreme environments. In about 16% of the subjects studied by Aschoff in his bunkers, these human rhythms shared the same frequency during the early days of isolation from zeitgebers, but eventually showed desynchronization and free-ran at radically different frequencies. This implies different pacemakers; however, during the period when they were in phase, they must have been mutually coupled. Thus, as part of this discussion we must be looking for one or more pacemakers which can control groups of rhythms or functions. First let us reexamine the evidence for the separable autonomous oscillators which control the circadian rhythms listed above. As one example, the autonomous control over the daily rhythm of plasma cortisol is evident in the information from isolated adrenal glands. These glands show a rhythm of secretion into the medium supporting them and also a circadian rhythm of over 60% above and 60% below the mean in oxygen consumption; this continues for day after day while still isolated (Andrews and Folk 1963).

Another autonomous oscillator is related to the heart rate rhythm. That this rhythm does not depend upon the body temperature rhythm was pointed out by Kleitman (1963), Timmerman et al. (1959) and Tharp and Folk (1964). In summary, we point out that some workers refer to the oscillators for these separate rhythms as being weak or perhaps slave-oscillators.

How are these rhythms governed? In rodents and primates, there is ample evidence for a governing pacemaker and possibly a second pacemaker which controls groups of functions. Experimentally in mammals, the SCN is the first pacemaker; when it is removed, it leads to random behavior of intact animals. Specifically, the SCN is known to control the activity cycles, the drinking cycles, and the sleep cycles of experimental animals. However, tampering with the SCN does not affect the body temperature rhythm of the animal.

Moore-Ede proposes that there are two separate pacemakers that drive the body rhythms and that they are autonomous when left to their own devices. Moore-Ede has designated the two pacemakers as *X* and *Y*; pacemaker *Y*, which is presumably the SCN, controls the rhythms of slow-wave sleep, skin temperature, plasma growth hormone, and urine calcium excretion. We would like to suggest that possibly within this group there is to be found the heart rate rhythm carefully described above. Pacemaker *X* is presumed to control REM sleep, core body temperature, plasma cortisol, and at least urine potassium excretion. Presumably the other major electrolytes are also controlled by pacemaker *X*. At the time of the numerous cases of internal dissociation and internal desynchronization, one or perhaps both of the two pacemakers have become uncoupled from the major circadian functional rhythms which they usually control. Because the various circadian rhythms usually remain synchronized even in some free-running conditions, there must be internal coupling mechanisms between the two pacemakers *X* and *Y*. However, the coupling mechanisms between the two pacemakers and between the pacemakers and the various rhythms they control is not strong enough to maintain internal synchronization indefinitely in the absence of zeitgeber inputs from the environment.

Progress has been made at the cellular level. Probably, intracellular calcium plays a role in SCN rhythmicity. An index to calcium homeostasis is the ryanodine receptor, which is a calcium release factor. There was a circadian oscillation in ryanodine activity exclusively in the SCN and not in other brain areas (Diaz-Muñoz 1996).

Clearly, the experimental methods and surgery applied to monkeys and rodents has not been and cannot be applied to human subjects. There is some indirect evidence, however, that these generalizations may apply to these subjects. Moore-Ede, after years of work with monkeys, reviewed medical records back to the turn of the century to find patients whose tumors or brain damage affected the area where the SCN is located. The reports revealed repeated instances of irregular and uncoordinated patterns of sleeping and eating habits in these patients. If more of these patients can be studied in the next few years, then an adequate comparison between the pacemaker control of human circadian rhythms with those of experimental animals can eventually be described.

We have learned that the biological clock mechanism requires the integration of both the entire animal body and of individual cells. We will now proceed to analyze the parts of this integrated system, beginning with the study of temperature regulation.

4. TEMPERATURE REGULATION AND ENERGY METABOLISM

Avenues of Heat Loss in a Neutral Environment

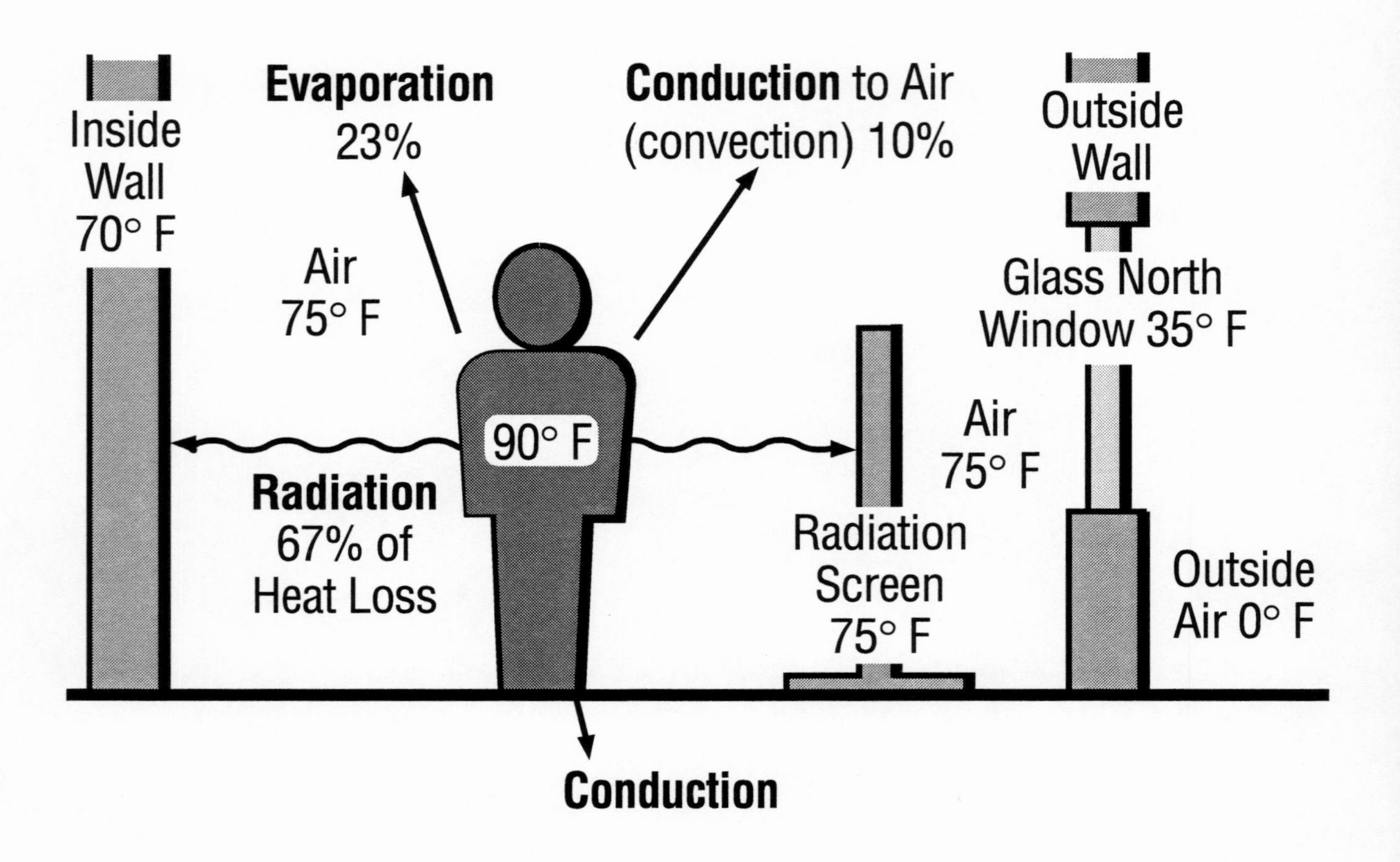

Environmental Diagram: Chapter 4

4. Temperature Regulation and Energy Metabolism

4. TEMPERATURE REGULATION AND ENERGY METABOLISM

PRINCIPLES

Before we consider mammalian responses to extreme heat and cold, we must first review the principles of temperature regulation as they apply to moderate *corrective changes* in the body temperature of the mammal. These changes may be represented either by large oscillations of body temperature, as experienced by some temperature-labile mammals, or they may take the form of rather minute oscillations, as with human beings. Essentially, the small corrective events that keep a person's temperature constant are the responsibility of the hypothalamus, which serves as the control tower of the negative feedback circuit.

The teaching physiologist looks forward to discussing the role of the hypothalamus in regulation because he or she has probably been dismantling the mammalian physiological architecture for many weeks for the benefit of the class, and now must reassemble it into an integrated functional organism. At last, the professor can describe how *all* systems and mechanisms of the mammalian body contribute to temperature regulation.

Temperature regulation is an excellent example of the principle of homeostasis: the maintenance of constancy of the bodily state within narrow limits by a dynamic equilibrium. In the human, this applies not only to body temperature, but also to body water, pH, ionic equilibrium, blood pressure, and body weight. To accomplish its assignment in overall homeostasis, the hypothalamus must act as an astonishingly accurate thermostat. Imagine moving a sensitive laboratory bath and regulating system into extreme environments: will the thermostat maintain at least the interior of the bath within 0.01°C of the desired temperature?

It is often stated that there is a *set point* in the hypothalamus around which body temperature is maintained; a more suitable term would be *variable set point*. We will discuss this concept later in the chapter.

The role of the hypothalamus in regulating body temperature is accomplished by two centers: one, in the anterior hypothalamus governs heat-dissipating events, and the other in the posterior hypothalamus calls on heat-conservation events (Fig. 4-1). The anterior and the posterior hypothalamus not only have conspicuous anatomical characteristics, but physiological ones as well; years ago, McCook et al. (1962) measured a consistent temperature difference between the posterior hypothalamus and the anterior hypothalamus. The temperature of the anterior hypothalamus was always higher (mean difference 0.15°C, SE 0.006°C). These findings help us understand the blood supply to the two portions of this temperature regulatory center.

In an earlier chapter we mentioned the contribution of behavior to temperature regulation. The white rat, for example, carefully adjusts nest thickness to different external temperatures; thus, it would be possible to estimate the temperature of a cold box containing a nesting white rat by measuring the thickness of its nest. Behavioral thermoregulation has been tested further by maintaining rats in the cold with a bar they could press to turn on the heat lamp. Normal rats do not use this capability appreciably (0.02% of the time). Rats with anterior hypothalamic lesions could not keep their body temperatures up physiologically, and lost an average of 3.18°C after 2 hours in the cold. When given access to the heat lamp, however, they kept it on enough (32% of the time) to maintain normal body temperature. The interpretation of this is that the lesion-impaired physiological thermoregulation left some forms of behavioral thermoregulation intact (Satinoff and Rutstein 1970).

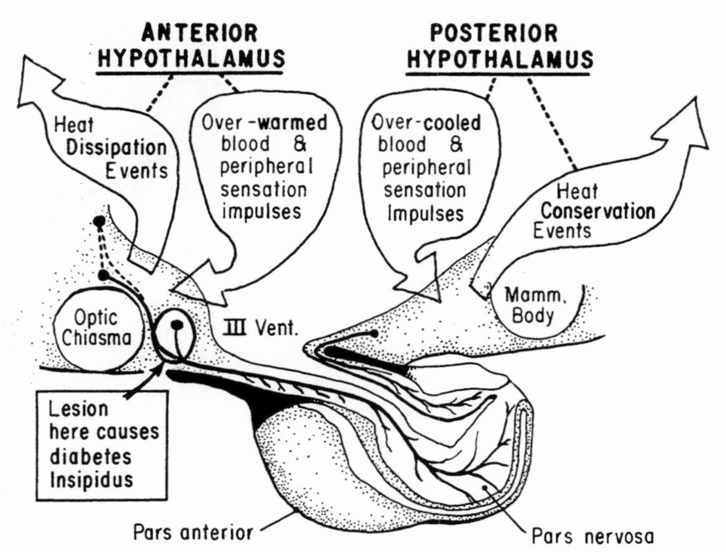

Figure 4-1. Some Regulatory Events in the Hypothalamus. The anatomical landmarks near the hypothalamus are shown, along with the probable physiological processes associated with control of temperature regulation by this structure.

Localized thermoregulation in the vicinity of the hypothalamus has been found in fish, reptiles, and birds as reviewed by Mills and Heath as early as 1970. Their experiments on heating and cooling the preoptic region of the brain in the house sparrow showed that this area must be an important center for thermoregulation.

CONSTANCY OF BODY TEMPERATURES

Most mammals are *homeothermic*, which means that they can maintain a relatively constant body temperature independent of the environmental temperature. For example, the core or interior body temperature of healthy human subjects seldom varies as much as 1°C (1.8°F) over 24 hours.

A convenient term to apply to mammals and birds is endothermic, meaning that they produce and control their own sources of heat. The remaining vertebrates (fish, frogs, salamanders, reptiles) have body temperatures that fluctuate with changes in the external environment. Such vertebrates are called *poikilotherms* or *ectotherms* (needing external sources of body heat).

However, even among mammals there are exceptions to the rule of maintaining a constant body temperature. The marsupials and monotremes are considered to be thermally unstable (heterothermic), although they are not hibernators (Fig. 4-2). Other mammals become dormant at certain times, but awaken when a critical body temperature of about 15°C to 20°C (59°F to 68°F) is reached. Still

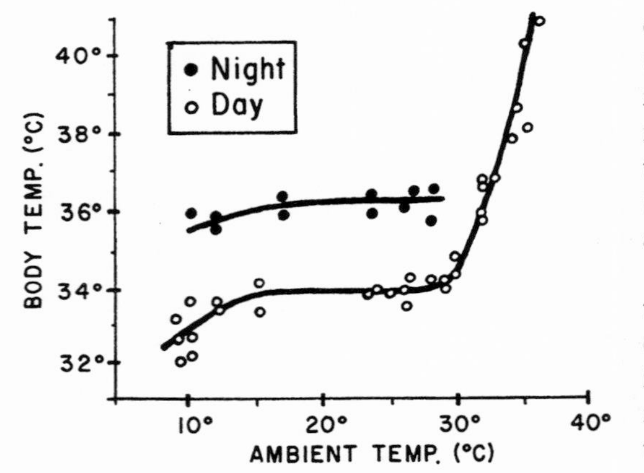

Figure 4-2. Temperature Regulation of a Heterotherm. Body temperatures in variable air temperatures in a nocturnally active Central American opossum, *Metachirus*. From Morrison 1962.

other mammals, such as the hazel mouse, can maintain dormancy (hibernation) until a critical body temperature near 0°C (32°F) is reached; then they rewarm and begin to regulate as homeothermic mammals again. The topic of hibernation has received scientific attention for 100 years, and is still a popular area of research (Wassmer and Wollnik 1997). In our next chapter, we devote special attention to the mammals that become dormant.

Even honey bees demonstrate temperature regulation in the cluster they form in both winter and summer. We may consider a colony or cluster of some fifty bees or more to behave like one intact organism. Temperature regulation is brought about in summer so that the middle of the cluster often remains at about 34°C to 35°C; however, temperatures up to 39°C (102.2°F) in midcluster have been reported. Of course, there is a thermal gradient to the outside of the cluster that may measure 25°C. In cold weather the cluster temperature can remain within a range of 5°C, while the outside temperature varies from -40°C to +40°C. The usual minimum temperature for the center of the cluster is 20°C (68°F). This excellent temperature regulation must be carried out successfully because bees die if they remain a day (d) or two at 8°C (46.4°F). The basic reason for the relatively high temperature in the middle of the cluster is the high metabolic rate of bees. An inactive bee at 35°C consumes as much oxygen per kilogram (kg) as a person doing hard manual work (Shiraki and Yousef 1987).

The overall cluster maintenance is called *eusociality*, defined as having (1) overlapping generations, (2) reproductive divisions of labor, and (3) cooperative care of young. There are only three known cases: social insects, the African mole-rat, and a coral-reef shrimp (Seger and Moran 1996).

The temperature regulation of small mammals has been analyzed for years; only since the sixties have studies been done on large mammals. According to Luck and Wright (1962), the core temperature of the rhinoceros varies within a range of 4°C. Cena (1964) reports the early morning core temperature of the hippopotamus as ca. 34°C, and the afternoon temperature as ca. 39°C. In water, the sublingual temperature dropped from 36°C to 25°C. The camel, as part of its temperature regulation mechanism, normally has a day-night range from 36°C to 39°C (Schmidt-Nielsen 1997). One of our captive bears, recorded in the autumn by implanted radio-capsule, varied over 24 hours from 36.7°C to 38.7°C. Perhaps it is a characteristic of large mammals to have a greater 24-hour body temperature variation than does the human.

The work of McGinnis (1968) supports this concept: he placed body temperature radio-capsules in elephant seals (in the alimentary canal). The capsules recorded a wide range of body temperature in two of the animals (34.6°C to 40.6°C; 30.5°C to 39.0°C). He found that the low temperature occurred with inactivity on land, even in full sunlight. Other data have been obtained on elephant seal temperatures by White and Odell (1971); weanling seals had temperatures that varied from 35.7°C to 39.1°C, while the temperature of bulls varied from approximately 33°C to 36°C. These authors have emphasized the importance of behavioral thermal regulation in this species; the main behaviors observed were throwing sand on the back and leaving the island for short dips in the water.

The extended series of studies by Bligh and Robinson (1965) supported in part the suggestion that large mammals have a large amplitude 24-hour body temperature variation Figs. 4-3, 4-4). Their results (also by radio-telemetry) show a dozen species with fairly wide variation, but the eland and giraffe have the same very fine control over their daily temperature regulation as is found in three breeds of domestic sheep. The relatively constant daily body temperature of domestic sheep was also found in the field in bighorn sheep (38.3°C to 38.9°C) (Franzmann and Hebert 1971). However, once again, in a marine mammal—the Hawaiian monk seal—a fairly wide swing (35.9°C to 37.8°C) of daily temperature occurs (Kridler, Olsen, and Whittow 1971).

According to da Silva and Minomo (1995), the seasonal change in sheep in Brazil was as follows ($P<0.05$):

	Summer	Winter
Minimum	39.55°C	38.87°C
Maximum	40.03°C	39.33°C

Thus, wild and tame sheep have about the same temperature range.

VARIATIONS IN BODY TEMPERATURE IN HUMANS

The precision of the hypothalamic-thermostat has had a long history since homeothermism began some 150 million years ago when mammals first evolved from their reptilian ancestors. Though its accuracy is remarkable, we must also recognize that the mechanisms that the hypothalamus calls upon are not perfect, and the body temperature can change within 20 minutes after exposure to extreme environments. Body temperature can even change when students take examinations (Briese 1995).

Exercise may have a large effect: if a person in poor physical condition begins to do heavy work daily, at first the rectal temperature during exercise may reach 39°C (102°F); after a week or so when good physical condition is achieved, the exercise temperature might be 37.7°C (100°F). In athletes, the usual high body temperatures after training can be even higher; endurance runners have the ability to dissipate large metabolic heat loads and simultaneously maintain cardiovascular stability (Gisolfi 1987). During competition these athletes can sustain body core temperatures be-

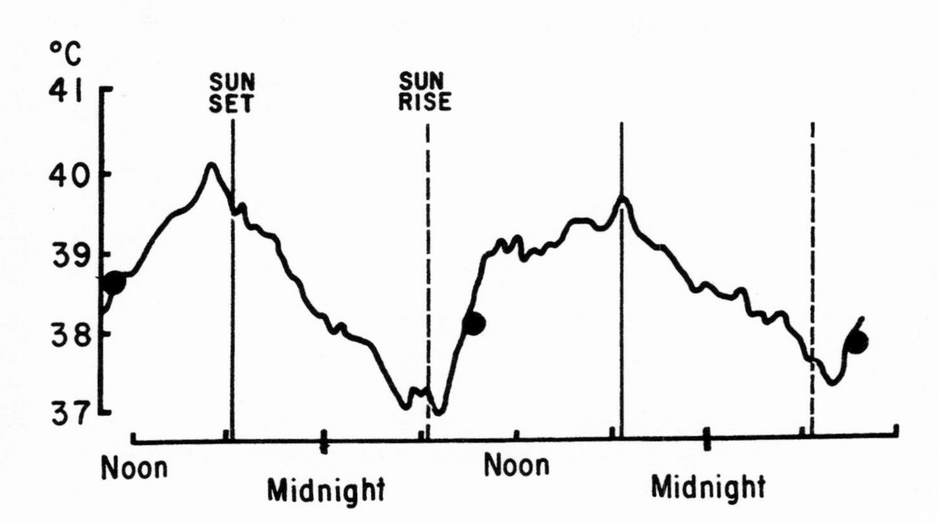

Figure 4-3. The Labile Body Temperature of a Large Mammal. A record of the deep body temperature from the lower neck region of a water buffalo (*Syncerus caffer*) recorded by radio-telemetry. The solid circles indicate spot measurements of rectal temperature. Adapted from Bligh and Robinson 1965.

tween 41°C and 42°C (108°F) for from one to eight hours (Maron et al. 1977; Pugh et al. 1967; Wyndham 1977). This brings up the concept of what Brynum (1978) referred to as the *critical thermal maximum*. For example, artificial fever is used to correct certain diseases (Pettigrew et al. 1974). The top temperature used in fever therapy is considered 41.6°C body temperature (107°F). Thus, these athletes have pushed the critical figure to above 42°C (108°F).

Four diagrams illustrate these variations in body temperature (Fig. 4-5). They also give a preview of what we will study in later chapters, the larger corrective responses that must be made to prevent a thermal runaway in either the cold or the hot direction. Note that the first diagram illustrates two settings of the temperature regulating mechanism, one for the "day-person" and one for the "night-person." Environmentally caused temperature rises are superimposed upon these settings.

Figure 4-4. Deep Body Temperature (°C) of Mammals. These data were collected in East Africa by radiotelemetry. Large mammals tend to have a labile body temperature, but this is not the case with the elephant and the giraffe. Solid circles represent mean temperatures, and lines represent range over 24 hours. Human and dog are also compared. Adapted from Bligh and Robinson 1965.

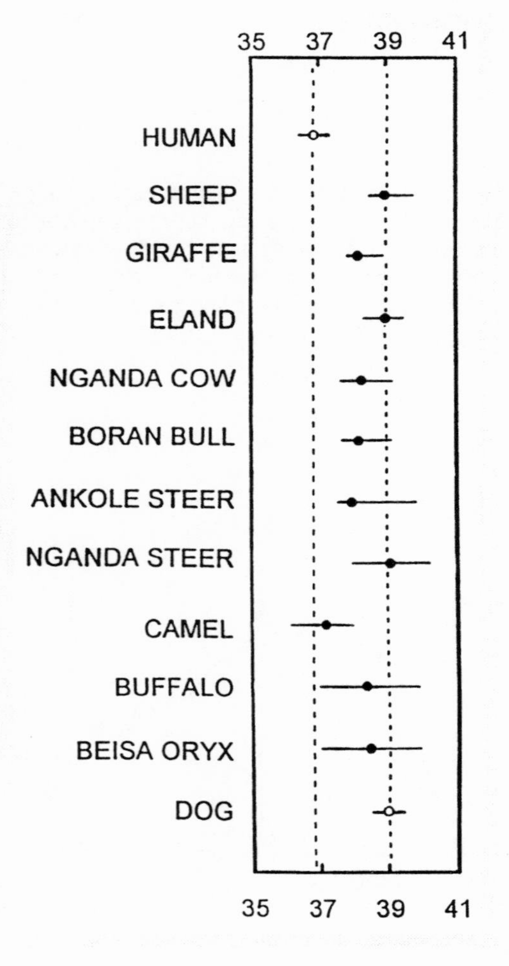

More information on body temperatures of the human fetus near term would prove especially interesting. Temperatures of the amniotic fluid (environmental temperatures) have been measured and were found to be below expected core temperature, with values ranging between 34°C to 36°C. Fetal tissue produces more metabolic heat per gram (g) than does adult tissue, and loses it mainly via the placental circulation. At birth the human neonate undergoes huge heat loss because of its high surface area-to-mass ratio. There is a large fall in body temperature. This fall is reversed by the non-shivering thermogenesis (NST) of brown adipose tissue, by vasoconstriction, and, in some species of mammals, by shivering (Laburn 1996). NST is increased energy cost due to increased permeability of mitochondria in brown adipose tissue. NST will play a large role in further discussions, both in this chapter and chapter 5.

Women experience monthly variations in body temperature, with temperatures being higher in the second half of the menstrual cycle than during the first half. In addition, body temperature peaks at the time of ovulation; this peak is used as an indicator to increase the chance of conception for couples who want children.

As a woman enters menopause, maintenance of a constant body temperature becomes impaired. Menopause may occur naturally between the ages of about 41 to 59, or at any age following the surgical removal of the ovaries. The most obvious event is the complete loss of menstrual cycles, but there may be many other physiological effects, some of which are described below. Ten percent of women will not experience any of these effects, but in 10% to 15%, they may be so severe as to interfere with normal daily activities.

Menopause affects one of the most important mechanisms for temperature regulation, that of vasomotor action; this function takes place in peripheral tissue, expressed either as vasodilation or vasoconstriction. Impairment of this function often causes hot flashes and sudden intense feelings of heat in the chest, neck, and face, accompanied by sweating. These events may last three or four minutes and they may include heart palpitations, nausea, headache, irritability, and anxiety. The vasomotor changes appear to be due to release of adrenaline. At the time of the hot flash, heart rate, skin

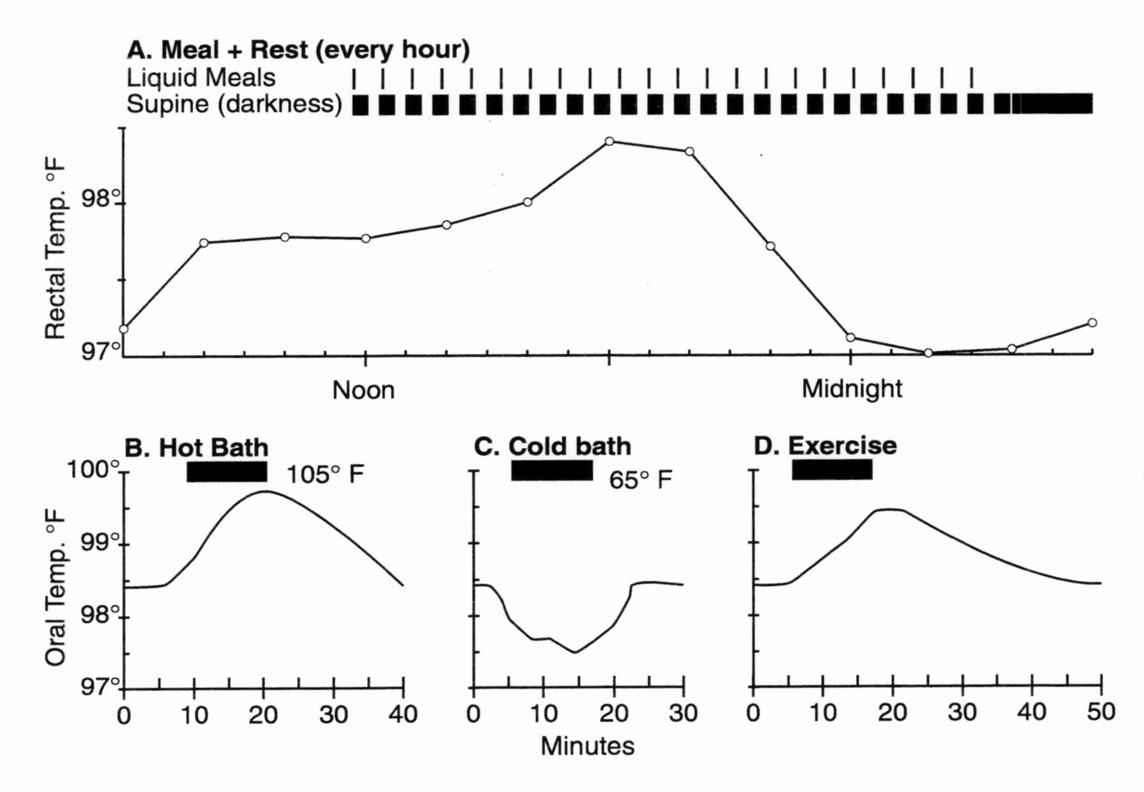

Figure 4-5. Factors Influencing Body Temperature of Humans. The body temperature is nicely regulated, but exact maintenance is altered by many factors such as hot baths, cold water, and exercise. If the influence of exercise or meals is removed, there is a daily resetting of temperature regulation, which persists in a resting individual. If standardized exercise is carried out at noon and midnight, the day-night regulation is still apparent in the exercise body temperatures.

blood flow, and skin temperature all increase because a transient peripheral vasodilation is occurring. This can be accompanied by a release of hormones including leuteinizing hormone (LH), gonadotrophic releasing hormone (GnRH), adrenaline, cortisol, adrenal corticotrophic hormone (ACTH), and neurotensin. The most important correlation is between the hot flashes and increased leuteinizing hormone, which itself results from GnRH release. Hot flashes may be a physiological response to a sudden downward resetting of the hypothalamic thermoregulatory set-point. Although there is a sudden pulsatile release of the hormones mentioned, the overall output is reduced. This has another effect on temperature regulation, because the skin becomes drier with reduced activity in sweat and sebaceous glands; wound healing is slower. The network of capillaries responsible for skin microcirculation declines after the menopause and this results in a thinning of the dermis and epidermis. As we shall see in a later section, lipid metabolism plays a part in temperature regulation. Before menopause, women have increased high-density lipoprotein (HDL) concentrations and decreased low-density lipoprotein (LDL) in their plasma. During the post-menopausal stage this beneficial lipoprotein ratio is impaired. Estrogen replacement therapy reduces the high body temperature of menopause (Brooks et al. 1997; Cagnacci et al. 1992).

The consumption of food influences the measurement of accurate body temperatures. The influence of meals upon basal metabolic rate is well known, as is its effect upon body temperature due to the *specific dynamic activity* (SDA) (Fig. 4-5). Some authors relate SDA

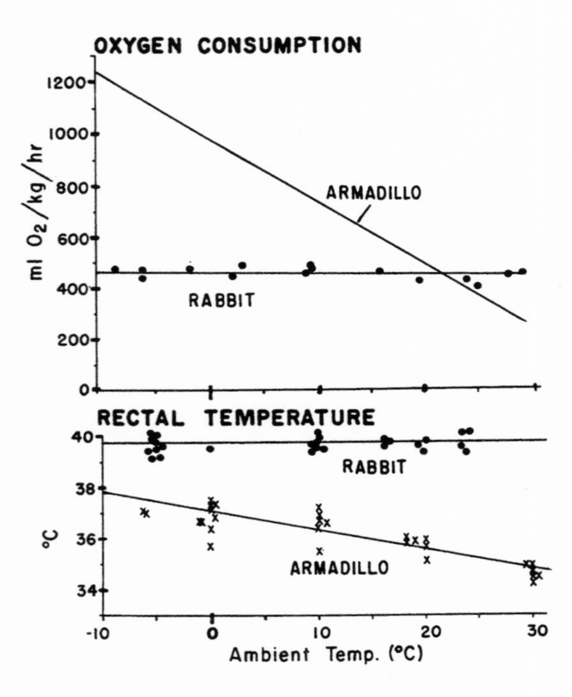

Figure 4-6. Cold Regulation in a Bare-Skinned Mammal. The bare-skinned mammals (armadillos) increased core body temperature in the cold from 34.5°C to 37°C; the rabbits did not. Modified from Johansen 1961.

to the energy cost of absorbing nutrients from the GI tract and uptake of nutrients by cells in body tissues. In an earlier chapter, an illustration is presented of three body temperature peaks due to three regularly spaced meals.

Certain drugs also affect body temperature, with some raising it and others lowering it. The effects on body temperature of a galaxy of types of drugs such as tranquilizers and aspirin have been studied at a constant ambient temperature; when these measurements were repeated at different ambient temperatures, different results were obtained.

VARIATIONS IN BODY TEMPERATURE IN LOWER MAMMALS

Some species raise their body temperature by an increase in metabolism. This increase can have two causes: it may be due to a surge of hormones unaccompanied by shivering, called *non-shivering thermogenesis* (NST), or the animal may resort to conspicuous shivering. This is shown in a comparison of the armadillo and rabbit in Figure 4-6. The armadillo is convenient for certain studies because it is a primitive type of mammal that has a shell instead of fur. Therefore, the vasomotor changes in the skin are more readily understood. Note in this illustration that as air temperature decreased, the animal increased its metabolic rate, vasoconstricted, and raised its body temperature from 34.5°C to 37°C. This does not happen to the rabbit with either measurement.

The domestic cat and dog have a relatively high core temperature; normal for the cat is 38.6°C (101.6°F) and for the dog varies with breed from 38.5°C to 39.2°C (101.5°F to 102.6°F) (Swenson and Reese 1993). Our own measurements for a resting Arctic wolf were lower than the dog at 37.9° (100.2°F).

The dog's core temperature changes by as much as 3°C at the time of parturition. Cooper (1997) recorded from a black Labrador Retriever the following at 6 pm on days leading up to and following parturition:

Day	Reading in °F
1	101.5
2	100.4
3	100.6
4	100.4
5	98.0
6	98.5 (in labor)
7	101.5

Some birds have a remarkably stable body temperature. In a study of Adélie penguins, air temperatures were changed from -10°C to 20°C. The body temperatures remained at a normal 39.3°C. Note how much higher this is than those of mammals (Chappell and Souza 1988).

MEASUREMENT OF BODY TEMPERATURE

We must now turn our attention to accurate measurement of body temperature so that the body mass can be analyzed in terms of physical principles. The most usual measurements are the skin temperature (T_s) and rectal temperature (T_r). The vocabulary used for these locations is based on the concept that the body may be considered to possess an inner core at a temperature near 37°C, and an outer shell of variable temperature.

The standard, or basal, body temperatures of human subjects of various races will fall within the range of 35.6°C to 38.0°C (96°F to 100.5°F). The normal human body temperature is 98.2°F (36.8°C), not 98.6°F as is commonly believed (Eisenlord 1995). Using an accurate thermometer under the tongue, in the axilla, in the groin, or in the rectum it may take up to 5 minutes before the final temperature equilibrium has been reached. The reading in the axilla is about 0.5°F lower than that in the mouth; the reading in the rectum is about 1°F to 3°F higher than that in the mouth (Strydom et al. 1965). Because of these differences, when mothers of young children call their doctor to report a high fever, the first question the pediatrician may ask is, "Is that an oral, underarm, or rectal temperature?" The axilla or groin measurements are used in cases of possible hypothermia in the field, as in auto accidents. The newest body temperature measurements are obtained either by resting a thermistor on the tympanic membrane of the ear, or by swallowing a thermistor—this is called the *esophageal temperature*. The ear temperature is not called a measurement of the core, but is referred to as the temperature of the ear canal.

For the present discussion we will concern ourselves only with resting body temperatures at a standardized time of day. This standardized resting body temperature varies systematically with the individual over a 24-hour period. This variation is not due to sleep, activity, or eating, but should be thought of in terms of a change in setting of the thermostat. A common misconception often found in textbooks is that the maximum setting occurs in the evening, and the minimum setting in the early hours of the morning. This is an oversimplification since there are actually three patterns of settings. The high setting of the hypothalamic-thermostat may come in the morning, it may be diphasic, or it may be in the evening. Of course, the normal routine of activity complicates and sometimes obscures this setting. In resting subjects, the setting may account for a difference of 1°C when measured at 0600 hr compared to 1800 hr.

Very little is known about the change in hypothalamic setting in mammals other than humans; this is because it is difficult to train most mammals to lie down and relax in darkness for 30 minutes before a standardized temperature is taken for each hour of the day. Spontaneous activity of mammals, even when restrained, has made it difficult to study the settings in temperature regulation over a 24-hour period, even by radio-telemetry.

The single temperature measurement in rectum or esophagus is frequently unsatisfactory, especially for calculations of heat loss from the mammalian body. Experimentally, it has been determined that a useful assessment is the *mean body temperature,* which is derived from two sources: 1) numerous skin temperature measurements, and 2) a core measurement usually obtained from the rectum. This relationship is expressed as follows:

Mean body temperature =
0.33 x skin temperature + 0.67 x rectal temperature.

In human subjects, skin temperatures are weighted by the system of either Belding or Ramanathan (1964) (see Appendix C). If there are 10 sensing devices (usually thermistors) on the subject's body, the values from the trunk might be weighted 45%, while those on the fingers 5%. The average skin temperature for a warm subject is frequently about 33°C, for a cold subject 27°C. However, the trunk may be warm while the cold-exposed palm is 5°C. The major difficulty in measuring skin temperature is the conduction of heat from the thermistor down the wire leads. Very thin wire must be used.

To calculate heat loss under various climatic and insulation conditions, it is essential that measurement of core temperature is accurate. For example, if rectal temperatures are not taken correctly, the probe may move to different locations, causing errors in calculations. This was noted half a century ago by Bazett, an environmental physiologist, who detected what he called, "hot and cold areas" along the rectum. Researchers in the 1960's found that accuracy could be improved by using thermistors in the external auditory meatus, especially if changes in core temperature, rather than absolute values, are desired (Cooper, Cranston, and Snell 1964).

Radio-telemetry is a tool that is adaptable to continuous recording of temperature; the telemetry-capsule can be fastened in the axilla of a person, implanted in the abdominal cavity of an experimental animal, or used in a variety of other ways. It is essential for recording temperatures from large, difficult-to-manage, wild animals. For example, continuous body temperatures and heart rates of wolverines (the largest species in the weasel family) were recorded in an outdoor enclosure where the T_A varied down to -45°C (Folk, Folk, and Craighead 1977). These were abdominal temperatures, which probably come closer to being accurate core temperatures than either the mouth or rectal temperatures.

Another factor to consider in measuring temperatures is the relationship between body temperature and the reproductive cycle. As we have noted, body temperature in women rises at the time of ovulation. In men, the main relationship between reproduction and body temperature involves the local temperature of the testes, which must be cooler than the rest of the body. This has been demonstrated in experiments where the testes of rams were heated, thereby making these animals sterile. In tests on a primitive mammal (the marsupial), the requirement of external testis location also held. Some mammals withdraw the testes into the body cavity during the season when they are not in breeding condition. There is still a great deal of comparative anatomy and physiology to be studied in this area. Not only are there wide variations in body temperature among mammals, but there is environmental variation as well; for example, many seals live most of the year in sea water that is at a temperature as low as -2°C.

Core temperature also changes with the seasons; surprisingly, one of the most thorough studies of this topic was done very early by Simpson by taking cloacal (rectal) temperatures on the domestic hen. He found that the core temperature was 0.9°C higher in summer than in winter (42.6°C vs. 41.7°C).

There is a wave of interest in present-day field work in taking stomach and alimentary tract temperatures, encouraged by the ease with which mammals in the free environment can swallow radio capsules. Some mammals, such as porpoises, will pass this capsule through the tract in about an hour, but other mammals may retain them for a period of days. In some laboratories, esophageal temperatures are used as standard core temperature. One must object, however, to temperatures taken by radio-capsule in the stomach; as we mentioned earlier, the influence of food upon body temperature is well known. As early as 1833, Beaumont showed that stomach temperatures were not normal core temperatures. His data came from work with the French-Canadian subject, Alexis A. St. Martin. This man had a permanent opening in the stomach due to a gunshot wound. Beaumont reports that the temperature of the stomach of his subject, after a 17-hour fast, always varied markedly with the weather between 34.4°C (94°F) and 37.7°C (100°F). As soon as food was added, the temperature of the stomach rose to 38.8°C (102°F)

The solution to where to place physiological radio capsules was found: implant them in the body cavity. We implanted thirty species of mammals and birds with capsules and not only was there never any infection, but also the recipients seemed unaware of the capsules in every case (Folk and Folk 1980). The capsules have been gradually improved so that now they can transmit simultaneously blood pressure, heart rate, body temperature, and locomotor activity (Folk et al. 1995).

THERMOGRAPHY

Instead of taking mean skin temperatures with thermocouples or thermistors on many points of the skin, another technique is to take pictures of the skin with an infrared camera. Although some experts say their studio should be darkened and the temperature of the room should be between 20°C (68°F) and 22.2°C (72°F), still others have used it in the field (see next chapter, *Responses to Cold Environments*). This technique was developed by astronomers to take temperature readings of the planets.

Informally, the instrument has been called "a camera with an eye for heat"; the resulting picture reproduces a thermal map of the skin or animal fur. The apparatus scans the surface to register the emitted infrared rays. Where blood concentrates close to the surface in veins, and where there are infections or abnormally rapid growths, the thermal map shows a high temperature. Where there are areas of low metabolism, such as hair, scars, or inactive growths close to the surface, the thermogram registers lower temperatures. Where an abnormal growth is under the skin, the temperature may be indicated as 1.6°C (3°F) higher than in a corresponding area of skin. In some instruments, colors may identify temperatures (Folk et al. 1985) (see Fig. 4-7).

In a modern and unusual application of the infra-red camera, Ek, Nelson, and Ramsay (1997) used thermal images and body measurements to develop an equation to determine body surface area from image profiles of bears. Readings of their mean surface and environmental temperatures were used in heat loss equations to determine energy expenditure of bears in kilocalories per hour. Results compared to Kleiber's formula within ± 4.5% (see Figure 4-12). After measuring skin temperature, metabolism can be calculated.

The instrument can also be used in the field; Figure 4-8 is a thermogram of a wild polar bear. We see that the eye and the nose, which are not insulated by fur, are hot; so too is the fur around the mouth, heated by exhaled air, while the neck is cold.

These thermometry readings can be translated to energy expenditure. This was first applied in free-ranging polar bears as noted above. Later, it was applied to preterm human infants in an incubator by Bell and Ek 1997. They confirmed the data obtained earlier by Karlsson et al. (1995), who used calorimetry, and cooled a series of infants by 3°C to 4°C. The only vasoconstriction occurred on the feet of the babies.

This camera has been especially successful in the study of frostbite; the picture demarks several weeks in advance the parts of fingers or toes that will eventually have to be amputated. Temperatures recorded on the film can be read to 0.1°F (Whipple 1964, Folk et al. 1985).

ENERGY METABOLISM

The analysis of temperature regulation depends upon heat balance, which refers to the temperature equilibrium of the body. The basic unit for temperature equilibrium is the *calorie*, which is a unit of thermal energy. Essentially all the energy expended by the animal body is converted to heat. The chemical energy (food) is made up of heat energy plus work energy plus or minus stored chemical energy. Work energy gives off heat.

Historically, the calorie is defined as the amount of heat required to raise the temperature of *one milliliter* (ml) of water from 14.5°C to 15.5°C at an atmospheric pressure of 760 mm Hg. The specified temperature interval is necessary because the amount of heat required varies slightly with temperature. The most useful unit is the large *Calorie* which is the amount of heat required to raise the temperature of *one liter* (l) of water 1°C. The large Calorie is expressed four ways:

1 Calorie = 1 Cal = 1 kilocalorie = 1 kcal

Two ways to measure the energy produced by the body as heat are

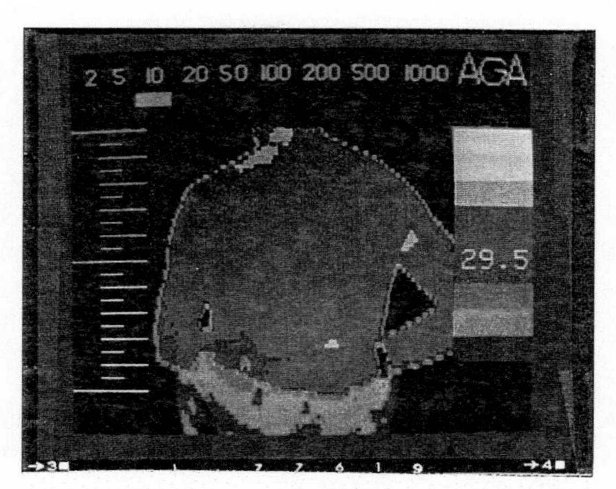

°C	
31.50	YELLOW
31.00	ORANGE
30.50	RED
30.00	VIOLET
29.50	PURPLE
29.00	GREEN
28.50	BLUE

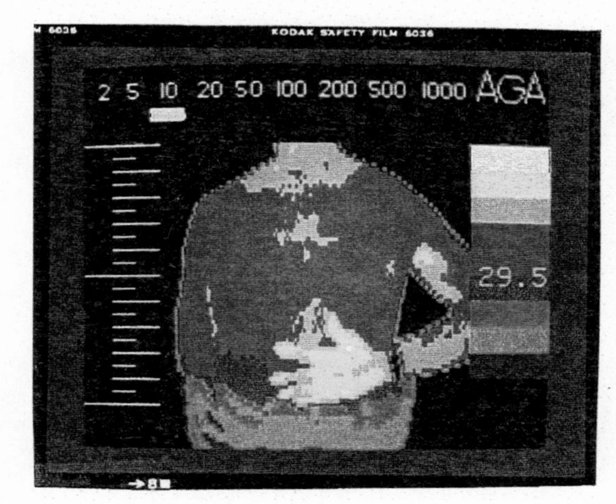

Figure 4-7. Color Thermography. The picture shows the back of the subject with hand behind back, palm toward the reader. After the control (first picture) was taken, the subject drank ethyl alcohol. The second picture was taken twenty minutes later, showing vasodilation of the neck, the brown fat region between the shoulder blades, and the arm and palm. Yellow is 3.5°C warmer than green.

direct and indirect calorimetry. The first involves placing the person or animal in a chamber and measuring directly the heat production, O_2 consumption, and CO_2 production, and determining the energy content of feces and urine. Because the direct calorimetry apparatus is expensive, the most common measurement is by indirect calorimetry. This measurement depends upon the fact that the combustion of food is attended by a fixed requirement for oxygen and by the production of a fixed amount of CO_2. Thus, in indirect calorimetry, the oxygen consumed and the carbon dioxide produced are both measured. The food itself can be measured in a bomb calorimeter with the following results: one gram of carbohydrate yields 4.1 Cal, one gram of fat, 9.3 Cal, one gram of protein 5.6 Cal (in the calorimeter), one gram of protein 4.3 Cal (in the body, since urea has an energy content not available to the body). The supply of O_2 is not a limiting factor. Note the high caloric value of fat; this will be referred to several times in other sections of this textbook.

We stated above that indirect calorimetry depends upon the measurement of oxygen consumption. This is a useful approach for the study of energy metabolism because under some circumstances, such as in space travel, the initial preparation depends upon providing enough tanks of oxygen to serve the needs of encapsulated individuals for periods of time. In other words, it is necessary to know the absolute amounts of oxygen required, rather than the heat given off. Note that measurements of oxygen consumption of fresh samples of tissue weighing one gram or less can be recorded in instruments called respirometers.

When the heat given off by a human subject or other animal is measured, we call this the metabolic rate. Having measured the oxygen, we can convert to heat depending upon what food is being metabolized. We know the correct figures, because food and one liter of oxygen can be put in the bomb calorimeter and ignited. For the usual conversion, we say the caloric equivalent of a liter of oxygen when mixed food is being metabolized, is 4.825 Cal. The supply of O_2 is limited. The equivalents of specific types of food are shown below. The comparison shows that when carbohydrate is metabolized, it provides the most heat.

Food	Caloric Equivalent
Carbohydrate	5.06 Cal
Fat	4.70 Cal
Protein	4.50 Cal

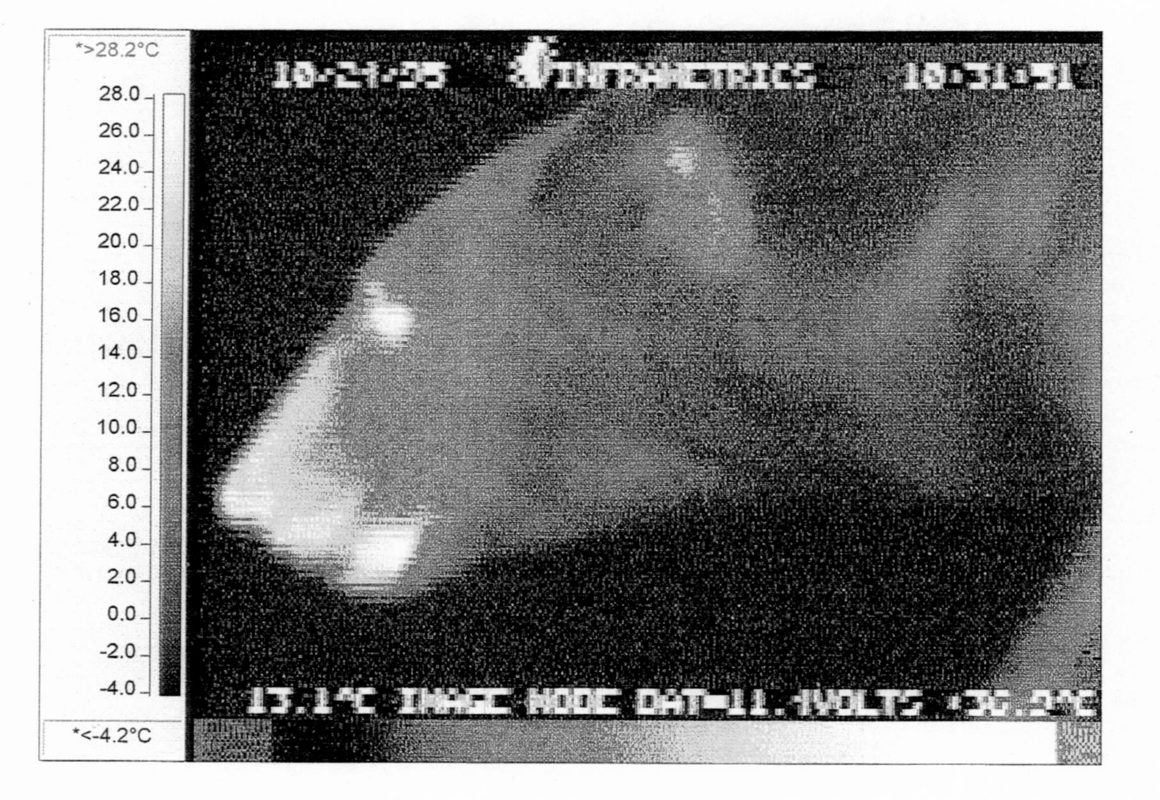

Figure 4-8. Surface Temperature of a Free-Living Polar Bear. In the thermogram, the temperatures can be read on the scale, or read by computer, spot by spot. Areas not covered by fur are hot, but those covered by fur are cold: eye (26°C), nose (22°C), mouth (26°C), neck (1°C). (Photo by Ek, Nelson, and Ramsey 1996)

There are other units to describe energy expenditure. Some modern biologists express their calculations in terms of *joules* (J):

$$1 \text{ cal} = 4.18 \text{ J} \qquad 1 \text{ Cal} = 4180.00 \text{ J}$$

Because biology deals largely with aqueous systems, the Calorie has a more direct and obvious meaning (Kleiber 1972); in addition, the figures used to express joules are much larger than those for Calories. The classic literature is all expressed in Calories; thus, in this textbook you will find illustrations expressed in either Calories or joules.

To further confound the issue, the engineering profession does not even express heat in metric units. They use Btu or B.T.U. which stands for British Thermal Unit. One Btu = 252 calories, the heat to raise one pound of water from 11°C (52°F) to 17°C (63°F).

Furthermore, oxygen consumption is sometimes converted to power in terms of watts (W). A watt is a unit of electric power or of other expressions of power:

$$\text{one W} = \frac{1}{716} \text{ of one horsepower}$$

A watt-hour expresses electrical energy or work in terms of one watt acting for one hour. Examples of usage are found in the sidebar.

In Table 4-1, the caloric cost was expressed in terms of square meters of the subject measured. A small person might have a surface area of 1.5 m^2, a large person 2.0 m^2. This expression might be used on human subjects or domestic cattle or other large animals. This system normalizes the difference between a tall thin human subject and a short stout one; the surface area of such individuals is easy to estimate from common normograms designed by L. J. Henderson. Not all physiologists accept this "surface law" for the BMR. They object because a part of the total body mass does not consume significant amounts of oxygen. This inert infrastructure has the functions of (1) structural support, (2) storage of nutrients, and (3) transport and distribution of these materials. With increasing size, the weight of the metabolic inert infrastructure increases disproportionately (Spaargaren 1994).

Most military and public health physiological handbooks (Consolazio et al. 1963) refer to caloric cost in terms of kcals per kilogram of weight of individual. However, all of these methods of

ALTERNATIVE CALCULATIONS OF ENERGY EXPENDITURES

1. A person may have a surface area of 2 m^2
2. This person's energy expenditure can be expressed as: $100 \text{ kcal h}^{-1} \text{ m}^{-2}$
3. Expenditure of 200 $\text{kcal}\cdot\text{hr}^{-1}$* = 232.6 watts (W)
4. The person may weigh 100 kilograms (k). His or her energy expenditure is 2 $\text{kcal}\cdot\text{hr}^{-1}\cdot\text{kilo}^{-1}$, or it is expressed as 2.33 W kg^{-1}
5. Units: 1 kcal = 0.86 watt (W); 1 watt = 1.16 kcal; 1 kcal = 1 Cal; 1 Cal = 4180 joule (J)

*Not normalized to surface area

Table 4-1. Amount of Heat Produced in Sleeping and Walking

Activity	0_2 Used	In $\text{Cal}^{-1}\cdot\text{m}^{-2}\cdot\text{hr}^{-1}$	Heat Given Off
Sleep	1/4 L/min	40	50 watt bulb
Slow walk	1/2 L/min	80	100 watt bulb
Fast walk (4 mph)	1 L/min	160	200 watt bulb

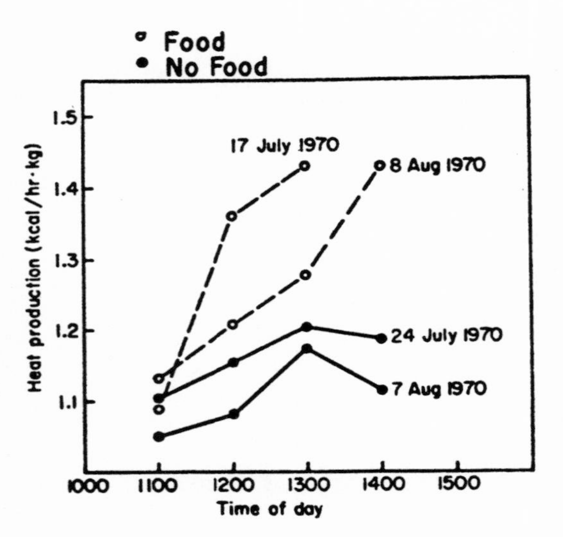

Figure 4-9. Heat Increment of Food Eaten. In designing environmental experiments, meals must be carefully controlled and standardized. The subject in a cold environment benefits from the metabolic stimulus of ingesting a meal. In this experiment, the SDA of a high protein meal is illustrated. The subject was inactive in each experiment, and a high protein meal was consumed at 11 AM on two occasions. No food was consumed during the other experiments. From Johnson 1972.

expression are unsatisfactory for the energy exchange of small mammals and small birds. The convenient expression for them is *ml* 0_2 *per g per hr.*

Human Basal Metabolic Rate (BMR). For medical purposes it is necessary to know the standardized resting metabolic rate of some patients or subjects. This can be done as an estimate by measuring just the oxygen used without the measurement of the carbon dioxide that has been absorbed in the respirometer. More accurately, both the oxygen used and CO_2 given off are measured. Measurement of a person's *basal metabolic rate* (BMR) must always be made before 10 AM. Careful directions have to be given to the person concerning his or her actions the evening before. For example, a large protein meal the night before would influence the BMR measured the next morning (Fig. 4-9). This effect on metabolism is known as the *specific dynamic action (SDA)* of food, and is also referred to as *diet-induced thermogenesis* or *meal-induced thermogenesis.*

Here are some examples of basal metabolic rates expressed in $Cal{\cdot}m^{-2}{\cdot}hr^{-1}$:

Age in Years	Male	Female
1	53.0	53.0
20	38.6	35.3
80	33.0	30.9

Note that the young child has the highest BMR. Also, because of growth, some of the energy taken in by the child will not be expended but will be stored. Women generally have a BMR from 5% to 10% lower than that of men. The BMR decreases as one grows older (see Fig. 4-10).

The Respiratory Quotient. This quotient (RQ) is a figure acquired either from the bomb calorimeter or from measurements of energy metabolism of animals and plants. It is obtained by dividing the carbon dioxide produced by the oxygen used, concisely written as

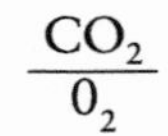

$$\frac{CO_2}{0_2}$$

When fat burns, the value of the quotient is 0.71. When sugar burns, the quotient is 1.00. Protein gives a value of 0.81. Clearly, the advantage of this calculation is information about what foods are being used. One biologist was working with standardized experimental animals that were to be using carbohydrate. To determine that his preparations were correct, he simply put them in a tight jar with a manometer. If the manometer did not change, indicating a respiratory quotient of one, then he knew his preparation was correctly standardized. He did not have to make numerical measurements.

To illustrate, if the respiratory quotient is 1 during heavy work, then this is evidence of carbohydrate metabolism. Later, as the quotient falls, it indicates that now fats are being metabolized. In marathon races, the greatest exhaustion is shown by those contestants whose blood sugar level is the lowest, indicating that they were more successful in mobilizing carbohydrate.

When the RQ is greater than 1.0, or less than 0.7, it is traditionally referred to as the *respiratory exchange ratio.* This departure may occur under the following circumstances:

1. When there is hypoventilation. The RQ may be 0.5 or 0.6. The condition under these circumstances will be uncompensated acidosis.
2. When there is hyperventilation. Here the RQ may be 1.5 to 1.7, a condition of uncompensated alkalosis.
3. When farm animals are fattened. Here there may be a departure in the usual RQ. Fat has little carbon dioxide and is called oxygen rich. Therefore, the farm animal is converting carbohydrate to fat. Extra carbon dioxide then is released; thus, RQ may be 1.4.
4. During mammalian hibernation. Here, fat is converted to carbohydrate. This means carbon dioxide is retained; RQ may be 0.5.

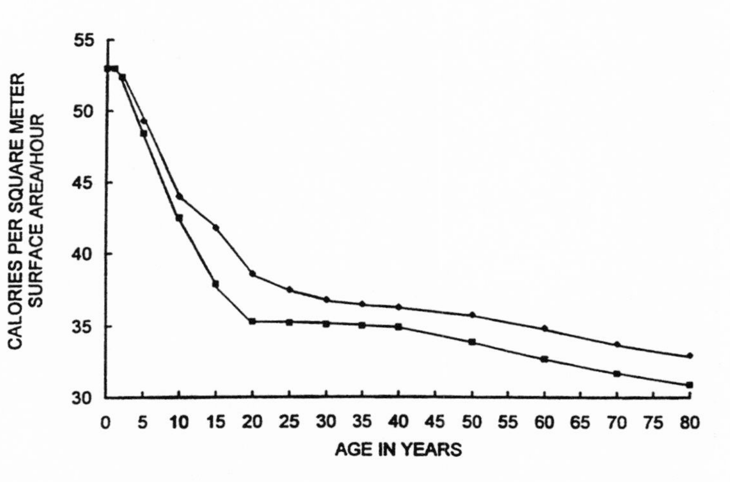

Figure 4-10. Average Values for Basal Metabolic Rate Measured at Different Ages. The upper graph represents data from male subjects; the lower graph is for females. Modified from Fleisch (1951).

Let us consider some daily figures for energy expenditure. In Table 4-1, the figures for resting and walking quietly are expressed in terms of $Cal \cdot m^{-2} \cdot hr^{-1}$. It is useful to think in terms of caloric expenditure over 24 hours for average individuals, or even domestic animals. A typical person resting in bed uses 1900 Calories per day. If the person sits up for a day, 2,000 Cal are used. Some athletes or heavy construction workers use between 6,000 and 8,000 Cal per day. Note that these figures are not expressed in terms of surface area or bodily weight. This commonly accepted figure is regularly exceeded by some male Tarahumara Indians, who habitually expend more than 10,000 kcal per 24 hr. About 50,000 of these Indians live in a rugged area in the Chihuahua, Mexico (Balke and Snow 1965).

A women's expedition skied across the Antarctic Continent to the South Pole, pulling their own food on sleds (see frontispiece). Their leader recently reported that they consumed 5,500 Calories per day, which was much lower than most male expeditions doing similar work. Animal fat made up 35% of their diet. Most climbers on Mt. Everest require 6,000 Cal of food per day until they reach 22,000 feet; then they need 10,000 Cal per day (Finley 1995).

Now let us consider the energy metabolism of small animals. The comparison with human subjects is best made on a per gram basis. The energy expenditure of small animals can be measured with simple equipment consisting of a closed container, a manometer, and a syringe. Of course it is difficult to have the animals in basal condition; in our experience a low dosage of an anesthetic provides the best results to obtain basal figures of rodents. The calculation per gram can be obtained from our knowledge of human energy expenditure, assuming a person of 50 kilograms:

- 1800 $Cal{\cdot}d^{-1}$ divided by 24 hr = 75 $Cal{\cdot}hr^{-1}{\cdot}person^{-1}$.
- 75 Cal divided by 4.825 (the caloric equivalent of a liter of 0_2) = 15.5 liters $0_2{\cdot}hr^{-1}{\cdot}person^{-1}$.
- 15.5 L·50 $kg^{-1}{\cdot}hr^{-1}$
- 300 $ml^{-1}{\cdot}1000\ g^{-1}{\cdot}hr^{-1} = 0.3\ ml{\cdot}g^{-1}{\cdot}hr^{-1}$.

These are the terms in which small animal metabolism is usually expressed. Table 4-2 gives further examples in alternative units of expression.

Now let us compare the figures for hamster metabolism, human metabolism and what is possibly the highest metabolic rate per gram of tissue ever measured: oxygen measurements taken by O. Pearson while a hummingbird flew in a bell-jar. The figures are:

Human	0.3 $ml{\cdot}g^{-1}{\cdot}hr^{-1}$
Hamster	3.0 $ml{\cdot}g^{-1}{\cdot}hr^{-1}$
Hummingbird in flight	119.0 $ml{\cdot}g^{-1}{\cdot}hr^{-1}$

The large difference in metabolism between the human and the hamster can be explained in terms of the surface-mass law, described below. The highest O_2 demand of all is from bacteria (see Table 4-3).

Small Animal Energy Production: The *surface-mass law* states that as body size falls, the mass of heat-producing tissue falls, but the relative area over which heat is lost increases. The smaller the animal, the higher the metabolic rate must be per gram of tissue, because the relatively high surface area drains the body of heat. Engineers learned not to insulate small pipes because the delivery of heat was inadequate; infants' fingers in the cold lose more heat if they wear thin gloves. This is because adding insulation to small cylinders increases the surface area and thus the heat-loss, without adding heat-producing tissue. Horses standing outdoors at -30°C (-22° F) T_A do not shiver; they are protected by their relatively small

Table 4-2. Examples of Small Animal Resting Metabolisms In Alternative Units of Expression

Animal	Weight	Resting Metabolism
Mustela rixosa (male)	70.0 g	19.8 $Cal{\cdot}d^{-1}$
Mustela erminea (male)	231.0 g	52.2 $Cal{\cdot}d^{-1}$
Honey Bee Cluster at 20°C	———	6 $W{\cdot}kg^{-1}$
Honey Bee Cluster at -20°C	———	13 $W{\cdot}kg^{-1}$
Australian Honeyeaters (daytime)	———	2.87 ± 0.37 SD ($kJ{\cdot}g^{-1}$)
Australian Honeyeaters (night hypothermia)	———	1.69 ± 0.17 SD ($kJ{\cdot}g^{-1}$)
Peromyscus maniculatus (at sea level)	21.4 g	3.33 ± 0.15 SD ($ml{\cdot}O_2^{-1}{\cdot}g^{-1}{\cdot}hr^{-1}$)
Peromyscus maniculatus (at 13,000 feet)	25.2 g	2.58 ± 0.13 SD ($ml{\cdot}O_2^{-1}{\cdot}g^{-1}{\cdot}hr^{-1}$)
Peropteryx macrotis (tropical bat)	5.0 g	2.3 ($ml{\cdot}O_2^{-1}{\cdot}g^{-1}{\cdot}hr^{-1}$)
Natalus tumidirostris (tropical bat)	5.0 g	1.6 ($ml{\cdot}O_2^{-1}{\cdot}g^{-1}{\cdot}hr^{-1}$)
Djungarian Hamster (P.s. sungorus)	35.9 g	2.1 ($ml{\cdot}O_2^{-1}{\cdot}g^{-1}{\cdot}hr^{-1}$)
Djungarian Hamster (P.s. campbelli)	31.4 g	1.9 ($ml{\cdot}O_2^{-1}{\cdot}g^{-1}{\cdot}hr^{-1}$) Max. Metabolism = 3.5 x resting metabolism

surface area. An illustration of the surface-mass law is found in Figure 4-11, in the form of two cylinders. The surface of the larger cylinder is relatively small. We must ask why this relationship holds. The initial calculations are easier with spheres; if an animal were spherical, its body surface would be described:

$$SA = 4\pi r^2$$

where SA = surface area and r = radius.

The mass of the animal would be:

$$M = 4/3\ \pi r^3$$

where M = mass and r = radius. Therefore, as size increases in these animals, the mass (or volume) increases in proportion to the cube of the radius, but the surface area increases only in proportion to the square of the radius. Thus, heat-producing tissues would occupy greater mass in proportion to the surface area *in larger animals.* Larger animals would be better adapted to survival in cold climates than smaller animals. This relationship is discussed later in this chapter as Bergman's rule.

The shrew-to-elephant graph (Fig. 4-12) shows an exponential increase in energy per gram of tissue. The metabolism of the 4 gram shrew is and must be 7 times higher than that of a mouse so that the shrew can maintain its constant body temperature. Perhaps its high surface area/volume ratio sets a lower size limit for homeotherms. The overall result seems to be that there is a 2 to 3 gram lower limit for the size (mass) of birds and mammals. A further demand for oxygen may well be beyond the design limits of the vertebrate cardiovascular and respiratory systems. The smallest shrew has a heart rate of about 1,000 per minute, while the respiratory rate is higher at 1,200 respirations per minute.

Table 4-3. Metabolic Rates Showing Demand for O_2 (From Irving)

Animal	Condition	QO_2 liter/kg hr
Part A		
"Bacteria"		100 to 500.000
Tabanus affinis	Flying diptera	210.000
Vanessa	Flying butterfly	100.000
Calypte anna	Flying 4-g hummingbird	68.000
Breslaua	Motile ciliate	5.000
Ameiurus nebulosus	Active bullhead, at 30°C	0.560
Pelomyxa	Ameboid protozoa	0.200
Anguilla	18°C, starved eel (low)	0.026
Part B		
Budgerigar	Flying, Max. Speed	35.000
Sorex cinereus	Shrew in cold air	10.000
Human athlete	Max. For 9 min.	3.700
Nomadic Lapps		3.300
Norwegians		2.700

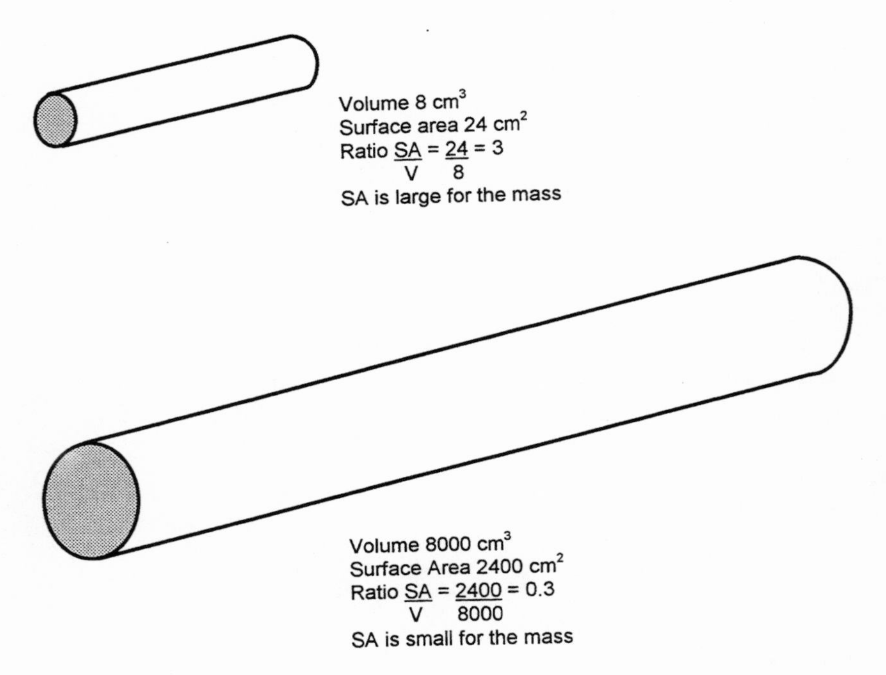

Figure 4-11. The Surface-Mass Law. The relationship between mass (volume) and surface area (SA) (not drawn to scale). The animals are assumed to be cylindrical in shape as are Aplesia and Lumbricus.

Figure 4-12. Semi-Logarithmic Plot to Show the Relationship Between Body Weight and O_2 Consumption: Shrew-to-Elephant Curve. Note the exponential increase in energy expended per kg body weight as body weight falls. This relationship is often referred to as *Kleiber's formula.* Modified from Schmidt-Nielsen (1997).

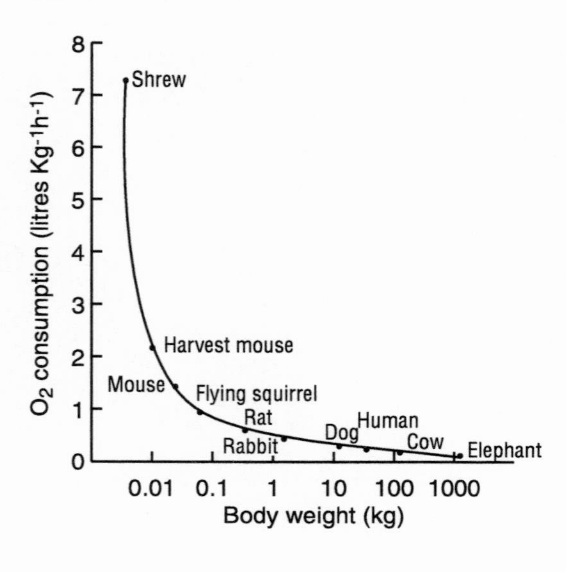

The shrew family is known not only for its high energy consumption, but also for the unexpected habits and adaptations of some of its members: one lives in water much of the time, and another, *Scutisorex,* has eleven vertebrae in the lower spine instead of five; each vertebra is supported by bony buttresses, giving the animal unusual flexibility and strength (Cullinane 1996).

We have been discussing high metabolic rates as recorded in the shrew-to-elephant curve in Fig. 4-12; how are these figures expressed as an equation? There have been many attempts to describe power functions of the form:

$$M = aW^b$$

where M equals metabolic rate and W equals body mass. There have been debates as to the meaning of numerical values of the parameters a and b. For example, metabolic rate might scale with $mass^{0.67}$ (Kleiber 1975), because surface area scales with $mass^{0.67}$. In fact, metabolism scales to $mass^{0.75}$, (Cossins and Bowler 1987), an exponent very different from that of surface area.

Many papers in the comparative physiology literature state that the basal metabolic rate of the mammals they measured is proportional to body mass raised to the 3/4 power. In other words, they showed that a doubling of body mass raised metabolic rate by 68%. However, an equal number of papers state that measurements of their particular species do not fit what is referred to as the mouse-to-elephant equation. Yousef et al. (1989) state that the resting VO_2 of camels sitting quietly on the ground was 21% lower than values predicted from the mouse-to-elephant equation. This expression is used because in the early papers describing the relationship there had been no measurements made on the metabolic rate of shrews.

Hayssen and Lacy (1985) were even more explicit; they state, "No single equation adequately describes the allometric relation between body mass and BMR for mammals." There are several reasons for this lack of conformity: one is simple differences in classification that may or may not identify primitive mammals. Hayssen and Lacy point out that the Edentata have the lowest BMR, but many mammal classifications fall into an intermediate group, and carnivores and even-toed, hoofed mammals (*Artiodactyla*) have the highest BMR. The latter include pigs, deer, and cattle. The two groups of primates do not support the $mass^{3/4}$ rule, but instead tend to separate out by classification. Tree shrews, bush-babies, and the slow loris (*Prosimiae*) tend to have a low BMR, while marmosets and small monkeys (*Simiae*) of about the same weight as the

Prosimiae, have a higher BMR (Muller 1985). The relationship among these three groups beginning with the Edentata does not show a body weight association.

Going beyond the classification argument, Lewin (1982) states that variations in diet decide the BMR. He says, "All mammals will have as high a metabolic rate as their diets will allow, because of the reproductive benefits it confers." Weiner (1989), on the other hand, lists a number of constraints which make conventional regression analysis of size referring to BMR impossible. For example, in some cases the limiting factor is the surface area of the gut.

To summarize, the early workers showed a systematic relationship between mammals of very different body weights and BMR. This relationship was called the mouse-to-elephant curve. However, many experiments since then show points well away from the early classical curve. The preceding section explains in part why these points are not on the curve.

Maximum Metabolism. Having considered the BMR, we must now reflect on maximum or summit metabolism. What is the most perfect "machine" for transporting O_2 from alveoli to blood to muscles? Table 4-4 shows selected examples of maximum work; examination of it seems to give the prize to flying hummingbirds at 119 $ml \cdot g^{-1} \cdot hr^{-1}$. There are a few generalities about summit metabolism. For example, marsupials and desert species have both a low summit metabolism and a low BMR (minimal metabolism) (Hulbert et al. 1985). Is there a correlation between BMR and maximum metabolic rates? To generalize, the rate of oxygen consumption ($\dot{V}O_2$) in mammals increases with exercise intensity up to a point, beyond which further increases in exercise intensity are no longer accompanied by an increase in $\dot{V}O_2$. Instead, the additional demand for energy by the muscles is supplied anaerobically; lactate accumulates, and the duration of the exercise becomes limited by the resulting metabolic acidosis (Taylor and Jones 1987).

For mammals as a whole it has been stated that maximum metabolic rate and BMR scale proportionately to $M^{0.75}$, and remains a constant multiple of 7 to 10 times BMR. Figure 4-13 shows that this relationship is modified by mammals which are more active or athletic. In fact, Koteja (1987), reports on 18 species of wild mammals

Table 4-4. Examples of Maximum (Summit) Metabolism*

Species	Weight	$\dot{V}O_2$ $ml \cdot g^{-1} \cdot hr^{-1}$*	Experimental Condition
Mice	17.30 g	13.80	Simulated cold
Rats	253.00 g	5.20	Simulated cold
Hummingbirds	3.30 g	119.00	Flight
Budgerigars	37.00 g	42.80	Flight
Butterfly	0.30 g	100.00	Flight
Bat (Brown)	8.00 g	8.63	Simulated cold

*Selected from Prothero 1979.

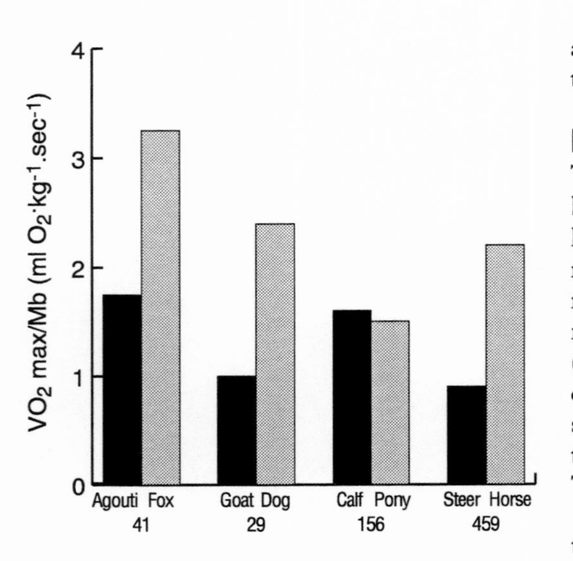

Figure 4-13. Cost of Being Athletic. Maximal oxygen consumption (VO_2 max·Mb-1) in four pairs of mammals with similar body masses. Left bar of each pair represents a mammal with average metabolic scope, right bar is a more athletic species. Masses are means of the two species studied. Trend shows decrease in VO_2 max·Mb^{-1} in both athletic and nonathletic species as Mb increases. Modified from Taylor and Jones 1987.

and finds in them no correlation between BMR and maximum metabolism.

HEAT BALANCE

The body temperature is maintained at a relatively constant level because of the balance that exists between heat production and heat loss. The contributing mechanisms are shown in Table 4-5. Some new vocabulary will be useful in understanding these mechanisms, namely the terms *heat sources* and *heat sinks*. We will be discussing metabolic activities of the organism (a heat source) and evaporation (a heat sink). The other three factors (radiation, convection, and conduction) are usually grouped under the symbol H. H is a heat sink when T_s (temperature of the skin) is higher than T_A (temperature of the air or the ambient surroundings). H is a heat source when T_A is higher than T_S .

These various factors will be discussed individually in the next three chapters. If there were no heat loss, the factors listed above, even in the resting subject, would produce sufficient heat to raise the body temperature by 1°C every hour; a 69-kg human produces 70 calories of heat per hour due to the basal metabolism. The temperature rise is expressed in the equation:

Heat Gained (in kcals) = Mass (in kg) x Specific Heat x Temperature Change (in °C).

The specific heat of the mammalian body (70% water) is often considered as 1.0. To be more accurate, one may use the value of Pembrey: 0.83 (1898).

Heat Gained

Heat is gained by the body not only from internal metabolism but also from the external environment, such as desert sand and rock, if it is at a higher temperature than the body. There will also be a small gain of heat from any hot foods ingested. On the other hand, if the heat production is insufficient to maintain the body temperature, further metabolism is brought about by the involuntary contraction of the skeletal muscles. Haider and Lindsley (1964) reviewed three probable types of involuntary contractions: body microvibrations, physiological tremors, and shivering. The frequency of all three is often between 7 to 13 cycles per second. Shivering is believed to be a natural amplification of the continuously present tremor.

The onset of shivering depends upon the integrity or "health" of the temperature regulating mechanism (probably the hypothalamus), and upon whether the mammal is a hibernator or not. For example, the white rat, when chilled, can rewarm spontaneously by shivering from a colonic temperature of 23°C, and in some cases from 16°C, but not lower; its shivering mechanism will not function beyond a critical temperature (see Chapter 5). Yet a hamster (a hibernator) may begin to rewarm partly by some shivering with a colonic temperature of 3°C to 5°C, although more extensive shivering begins at 15°C (for details, see Chapter 6). Even a monkey can rewarm by shivering in cold air from a rectal temperature of 25°C; this was first demonstrated in a classic paper by Simpson (1912).

Chemical increase of heat production (thermogenesis) is brought about by the body's production of epinephrine, norepinephrine, and thyroxin; experimental mammals exposed to cold develop the capability of producing heat, independent of muscle contraction. Curarized, cold-acclimated rats double their heat production in a cold room at 5°C with maintenance of rectal temperature. Non-acclimatized rats cannot call upon this mechanism. The important link seems to be norepinephrine liberated from sympathetic nerve endings. This change in the calorigenic effect of norepinephrine has been confirmed by Carlson in men who are cold-acclimated (1964), and by Kang on the famous Korean woman divers (1971). The effect of cold and of the injection of norepinephrine produces almost identical increases in heat production (Jansky 1971) (Fig. 4-14).

The interrelationship of the catecholamines (epinephrine and norepinephrine) and of the thyroid hormones was reviewed by Slonim, Heroux, and Davis in the late sixties. Cold increases the rate of thyrotropin secretion by the anterior pituitary. Exposure for several weeks increases the output of thyroxin, sometimes more than 100%, an amount that can increase the basal metabolic rate of the animal 20% to 30%. It is claimed that human subjects in an Arctic region sometimes develop basal metabolic rates 15% to 20% above normal. This endocrine reaction to cold may help to explain the higher incidence of thyrotoxic goiter in cold climates. The effects of cold on thyrotropin secretion are blocked when the pituitary portal system is occluded. One is tempted to demonstrate the influence of the thyroid on cold acclimation by proving hypertrophy of the thyroid and increased fecal excretion of thyroxin. Such experiments must be done with care; according to Heroux, these specific changes are diet dependent. They occurred when rats were fed laboratory cubes, but did not occur when they were fed a thyroxin-free diet containing 14% fat (Heroux 1968).

The SDA of food on the BMR results in a warming effect, and is identified by the extra calories produced. As we mentioned earlier, its effect is so long-lasting in mammals that it creates a technical problem when measuring their basal oxygen consumption. To illustrate the principle, if a basal human subject drinks 100 grams of glucose, his or her heat production begins to rise in a few minutes and reaches a peak in 3 h. The extra heat produced is 25 cal, although the value of the sugar is 370 cal (heat value = 7% of food value). Dietary fat has a relatively slight effect, but protein has a larger effect than carbohydrate (protein heat value = 18% of food calories). Furthermore, the protein effect is longer lasting. It is for this reason that only a low protein meal can be permitted the night before the basal metabolic rate is measured (see again Fig. 4-9).

Brown fat is critical to heat gain due to diet or exposure to cold. The extra heat of the NST is produced in the rat and hamster by increased blood flow to the mitochondria of their brown fat, under the control of increased catecholamines (Glick et al. 1984). The cellular mechanism in the mitochondria depends upon a *specific protein*

Table 4-5. Heat Balance Function

Factors Increasing Heat Production (over BMR)
1. Exercise or shivering
2. Imperceptible tensing of muscles
3. Chemical increase of metabolic rate
4. Specific dynamic action of food
5. Disease (fever)

Factors Decreasing Heat Loss
1. Shift in blood distribution
2. Decrease in tissue conductance
3. Counter-current heat exchange

Factors Enhancing Heat Loss
1. Sweating
2. Panting
3. Cooler environment
4. Increased skin circulation (vasodilation)
5. Decreased clothing or shorter fur insulation
6. Increased insensible water loss
7. Increased radiating surface
8. Increased air movement (convection)

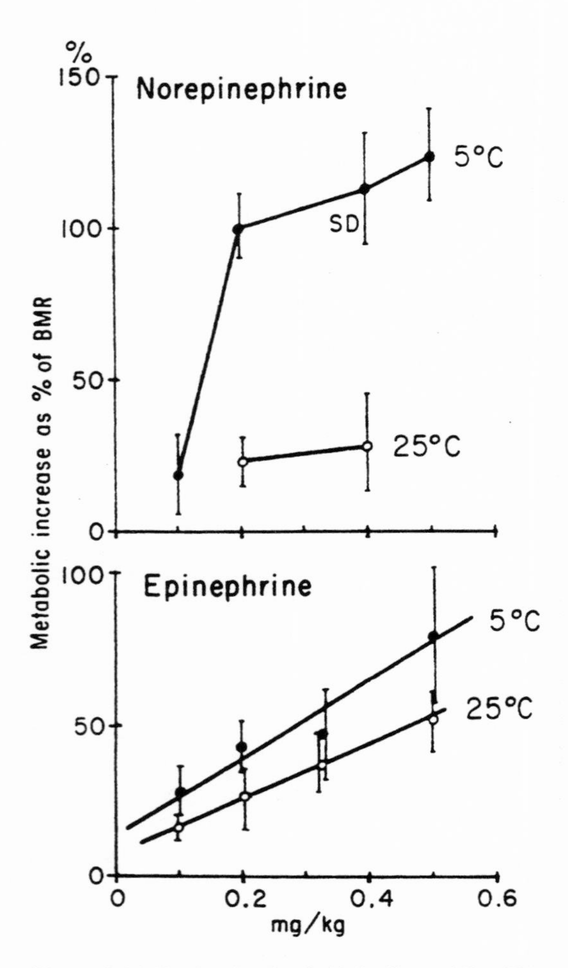

Figure 4-14. Testing For the Cold-Acclimated Rat. The best distinction between a warm-acclimated (25°C) rat and a cold-acclimated (5°C) rat can be made by injecting (IM) about 0.4 mg of norepinephrine. Modified from Bartunkova 1971.

which regulates the proton conductance of the mitochondrial membrane and thus the degree of mitochondrial coupling. With uncoupling, there is increased proton conductance and increased rates of substrate oxidation. Heat is controlled by the amount of functional uncoupling accomplished by the specific protein mentioned above (Horwitz et al. 1985).

Heat balance studies depend upon the measurement of metabolism. These measurements must be made more carefully in the future because of findings concerning the role of nitrogen during the measurements. Traditionally, open circuit methods of measuring oxygen consumption and determining RQ values have depended upon the concept that the nitrogen volumes were equal during inspiration and expiration. In fact, however, there is a significant difference between the volume of nitrogen inspired and the volume expired. Fasting subjects at rest exhibit a retention of nitrogen, while exercising subjects produce nitrogen. Because of this, some of the figures used in the past for heat balance calculations can be in error as much as 12% or more (Cissik, Johnson, and Rokosch 1972; Dudka 1971).

Heat Lost

As can be seen from Table 4-5, mammals lose heat by conduction, convection, radiation, and the evaporation of water. A small loss of heat occurs in expired air that is at body temperature; also, this air is saturated with water vapor. The heat lost by the skin depends partly on the temperature gradient between 1) skin, and 2) air and solid objects. The skin temperature is regulated by the blood flow to the skin and by evaporation. With a low peripheral blood flow (vasoconstriction), the skin temperature is low, reducing heat lost to the environment. With a high peripheral blood flow (vasodilation), the skin temperature approaches the core temperature, maximizing heat lost to the environment. To this relationship, the cooling power of perspiration is added.

The control of blood flow, that is, the blushing of the skin in heat and the whitening of it in the cold, is controlled mostly in the medulla of the brain, although partly locally. We name the areas of control using temperature regulation principles; one area of the medulla through the cardioaccelerator nerve brings about a higher blood pressure when stimulated. Or, stating it differently, one area of the medulla receives cold stimuli; this area calls for constricted vessels, resulting in a tendency for a higher blood pressure. Therefore, this area of integration is called the *pressor area.* Another area of the medulla is referred to as the *depressor area* because it receives stimuli from warmth receptors, and the medulla calls for vasodilation. The result is an object with variable surface temperatures under the control of both external and internal factors. Thus, vasomotor function influences conduction and convection.

Vasomotor changes in skin vessels may depend on arterial-venule (AV) shunts which change blood flow (Fig. 4-15). These shunts in the fingers allow blood flow in the fingers to increase 100 fold. A similar variation applies to the skin of the entire body (Table 6). The term to apply to that area is the shell (Fig. 4-16.)

Conduction

At this point it will be helpful to examine the Environmental Diagram, *Avenues of Heat Loss.* Mammals lose heat by conduction through physical contact with objects and substratum. Loss of heat by conduction is minimized by the insulation of fur and clothing. The effectiveness of fur as insulation permits Eskimos and Lapps to tolerate extreme cold; in northwest Greenland they still dress with one layer of fur facing inward, and another layer facing outward. (See Chapter 5.)

Further details about conductivity can be visualized by placing a glass container full of warm water in a very cold box. This glass container could be placed in a jacket of silver, leather, air, or other insulation material. The loss of heat from the water to the cold box can represent the conductivity. One usually considers that silver has a conductivity of approximately 1. Some materials will have a conductivity of one-fifth of that, such as brass. Actual conductivities, using this method of comparison, are given in Table 4-7.

Conduction is most important where the feet of mammals and birds come in contact with very cold soil or snow. The exposed feet of the wolverine, a member of the *mustelidae*, are an example (Fig. 4-17). The temperature regulation of these large weasels deserves special attention (Folk, Folk, and Craighead 1977). Sexual dimorphism is evident in all weasel species; males may exceed females by 5% to 30% in length and may weigh twice as much (Holmes and Powell 1994).

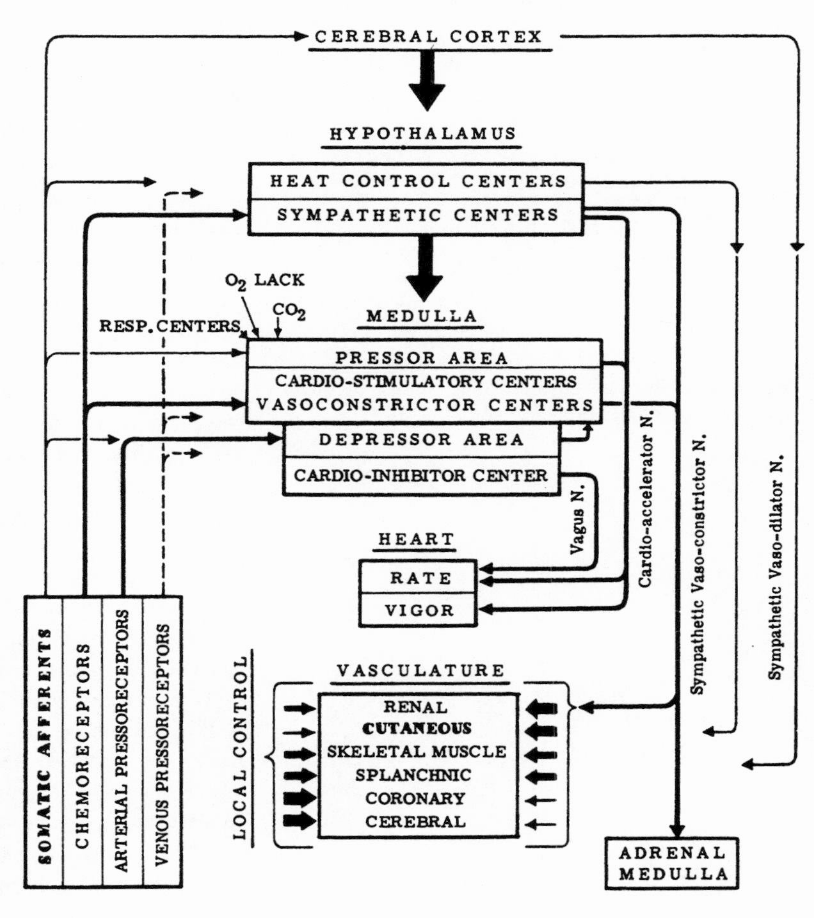

Figure 4-15. Control of Cutaneous Circulation Compared with Other Compartments. The relative importance of local control and central control over the vasculature of various compartments is compared. Our particular interest here is the cutaneous vasculature control. Somatic afferent stimuli affect the cerebral cortex, the hypothalamus, and the medulla. Vasoconstrictor influences originate in the hypothalamus and the medulla. There is some small local control over skin vasomotor function. From Rothe 1971.

Convection

Convection is similar to conduction in that the heat is transmitted from one molecule to another by physical contact, but in convection the heat is transferred to the air, which rises, taking the heat with it. Cooler air comes in to take its place. There are two different forms of convection: 1) *natural convection* (rising of warm air), and 2) increased movement of air (or water) due to the action of an outside source, which is called *forced convection.* Natural convection depends upon the natural buoyancy of the heated material, and the establishment of vertical currents. Forced convection occurs whether the medium (air or water) moves past the stationary object or whether the object moves in the medium. A useful formula for forced convection is:

$$H_c = 2.3\,(T_s - T_A)V$$

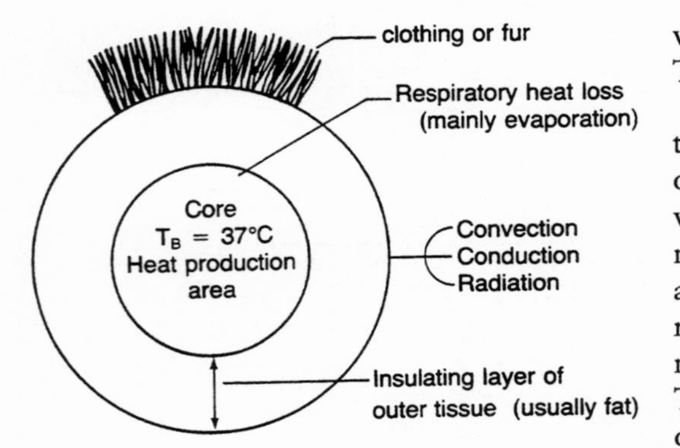

Figure 4-16. Avenues of Heat Balance In the Core and the Shell. Modified from Ferguson (1985).

where H_c is heat loss, V is velocity of the gas (could be air or helium), T_s is temperature of skin, and T_A is temperature of air.

We must now consider the relationship between forced convection and the marginal layer of air next to the skin of humans or of other mammals. This marginal layer of air has insulating properties which vary with the circulation of the air over the skin. The movement of the skin by a person or animal will disturb the layer of air in a variable fashion. Because of this, a large range of values has been reported for heat loss by natural convection from these so-called motionless animals. The range varies from 0.8 to 2.12 $kcal \cdot m^{-2} \cdot {}^{\circ}C^{-1}$. These heat-loss figures can be more than doubled by slight activity of the subjects. The insulation of the air becomes even less if the animal runs, or if the air velocity becomes high (Fig. 4-18). At high air velocities, one must even correct for the heating by friction of the moving air (or water). When there is a reduction in the layer of still air and thus a reduction in insulation, a paradox occurs. If the circumstances include low levels of shivering (or of running) the effect of the increased heat production due to the muscular activity is more than offset by the reduction in the insulation of the air around the skin, which results in a net increased heat loss. It is only at higher levels of shivering (or running) that the increased heat production overcomes the deficit to appear as an effective benefit. The balance between enforced convection and the insulation of air next to the skin or fur can best be expressed in terms of insulation units. These will be discussed in detail in the next chapter, but it can be explained now that one Clo unit represents the insulation of a light business suit on a person who is sitting at room temperature. Although air insulation decreases exponentially as the velocity of the air movement is increased, insulation does not decrease to zero even with extreme air movement. In addition, the changes in insulation values are not uniform.

Radiation

Radiation is the loss of heat by electromagnetic infrared waves. The frequency of infrared is slightly less than visible red, and the radiant energy from the mammalian skin varies in wavelength from 5 to 20 microns. Such radiation does not heat the air through which it passes. There are difficulties in measuring convection and radiation from small mammals, but progress has been made by Heller (1972), who made metal casts of four species of chipmunks and covered them with the pelt of the animal for insulation; the casts were sus-

Table 4-6. Skin Blood Flow In Three Environments

Condition	Blood Flow to Skin	% of Cardiac Output
Cold	30 ml/min.	0.5
Normal T_{air}	300 ml/min.	5.0
Heat	3000 ml/min.	50.0

pended in a vacuum and the effective radiating surface area was calculated. Then the casts were placed in a wind tunnel and the coefficients of convection were calculated.

The amount of heat gained or lost by radiation depends not only on the temperature of an object, but also on its color and texture. Dark rough surfaces radiate maximally, while light shiny surfaces at the same temperature radiate less rapidly. The human skin acts as a black body radiator irrespective of the actual skin color; whether this applies to clothing is controversial. White clothing often has been considered more suitable than black, both in the tropics (less heat gained), and in the Polar regions (less heat lost). Lightweight clothing does protect against heat reflection and from the direct rays of sun. In army tests, the heat gain of clothed subjects seated in the sun was 120 calories an hour; unclothed, they gained 200 calories an hour.

An application of the contribution of radiation to heat loss is evident when a person sits near a large window on a cold winter day. Heat rays leave the person's body and strike the glass, which is a heat sink. If our subject is lightly clad, he will be cold; if he is unfamiliar with heat-loss factors, he will repeatedly look at the room thermostat. This device and the thermometer on it may both read 24°C (75°F), and the air and the walls may be that, but the glass of the window may actually be 2°C (35°F). An actual winter experiment in Iowa, comparing a large uncovered window with a drape-covered window, will illustrate (see Table 4-8). The glass temperature did not change with the weather as much as was expected. The colder the weather, the more the drape (radiation shield) protected the skin. One frequently sees patients covered with a single sheet in winter in the corridor of a hospital near large windows. This window glass is acting as a heat sink (Fig. 4-19). Thus, for the circumstances, this amount of insulation is insufficient and may be placing a thermal stress upon the patient (Lassahn, Folk, and Seberg 1974).

Evaporation

In the mammal group there are three types of skin, each with a different anatomical arrangement for evaporation of water for the dissipation of heat: heavy fur without sweat glands, hairy skin with

Table 4-7. Heat Conductivities (Coefficients of Conductivity) of Various Substances (from Bard)

Substance	Temperature (°C)	Conductivity
Pure Silver	18°	1.006
Brass	Ordinary	0.225
Water	17°	0.00131
Leather (cowhide)	Ordinary	0.0004
Air	0°	0.0000568
Wool	Ordinary	0.000054
Eiderdown	Ordinary	0.000011

Figure 4-17. Feet of Wolverines Act as an Avenue of Conductive Heat Loss. These feet often make contact with unbreakable snow crust as cold as –55°C. This young wolverine is the largest member of the weasel family, with adults weighing as much as 19 kg (42 lb). In winter, the wolverine grows long protective hair that covers the pads; the wolf does not. Photo by D. Gullickson, provided by Naval Arctic Research Laboratory.

Figure 4-18. Air Velocity and Air Insulation. The boundary layer of air covering an exposed surface (skin, fur, or clothing) will always provide some insulation. Increasing air velocity to more than 800 feet per minute still does not reduce the insulation of air below 0.2 Clo. When there is little air movement, the insulation is approximately 1.0 Clo. This figure is important in calculating the insulation of clothing, fur, skin, or subcutaneous fat. Modified from Adams.

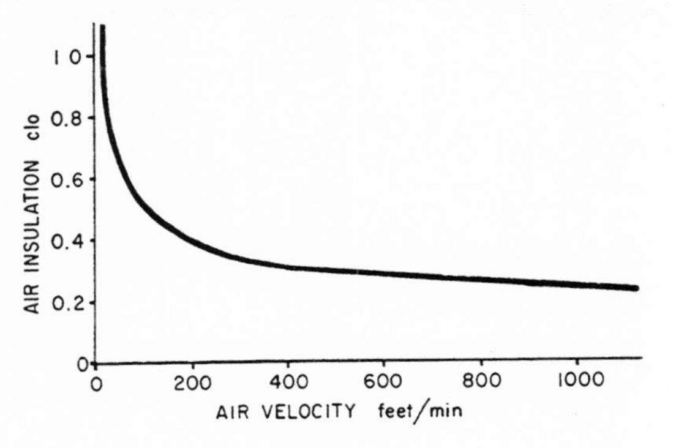

sweat glands, and smooth, relatively hairless skin with sweat glands. The simplest approach is to describe the types of water loss from humans and then to list which of these mechanisms are found in these three classes of mammals. Perspiration from the hairless skin of humans is of three types: 1) *insensible perspiration* (diffusion water); 2) *thermal sweat* (from eccrine glands); 3) *non-thermal sweat* (also called *palmasole* or *emotional* sweat (Folk and Semken 1991).

Insensible perspiration leaves the body at all times unless the ambient humidity is 100% RH. This moisture diffuses through the stratum corneum of the skin, through the pores of the sweat glands, and from the lungs. The existence of this phenomenon of continual evaporation was described by Sanctorius Sanctorius (1561-1636), Professor of Medicine at Padua, Italy. His apparatus consisted of a chair on an iron beam, a fulcrum, and a longer arm of the beam which carried weights. When he measured the weight of a person sitting in the chair, he found that individuals continually decreased in weight. This weight loss of the mammalian body must be partitioned into three factors: 1) the excess weight of the CO_2 expired over the O_2 inspired; 2) the weight lost as the water vapor from the lungs; and 3) the loss by diffusion through the skin. The first two factors are subtracted from the total in order to obtain the contribution of the skin to insensible perspiration. The weight lost from the lungs is calculated with the assumption that the expired air is at 32.70°C (91°F) and saturated at a cost of 0.035 gram of body water for each liter of air expired when the ambient temperature is -6.7°C (20°F) and lower. At 4.4°C (40°F) the cost is considered to be 0.029 gram (Belding et al. 1947). The excess weight in grams of the CO_2 expired is calculated assuming an RQ of 0.88 by multiplying the oxygen consumption in liters per hr STP (standard temperature and pressure) dry by 0.3. Insensible perspiration plays an appreciable part in heat dissipation. This is the only source of evaporative heat loss in subjects that do not have sweat glands. Some of these individuals can tolerate fairly high ambient temperatures depending solely on their insensible perspiration as a means of cooling.

Changes in barometric pressure and vapor pressure alter the rate of loss of diffusion water through the skin and from the lungs. If

Table 4-8. Measurements of Gradients to Cold Glass Windows.

	Jan. 28	Mar. 1	Feb. 24
Outside Air	-17°C	-4°C	-2°C
Glass	11°	13°	17°
Gradient:			
Glass to Outside Air	27°	17°	19°
Skin to Glass	23°	21°	17°
Skin to Drape	15°	17°	14°
Skin to Wall	8°	11°	9°

people move from sea level with its high vapor pressure to a moderate altitude with a low vapor pressure, their insensible moisture loss may increase by 33% (Hale, Westland, and Taylor 1958)(Fig. 4-20).

Thermal sweating occurs from the eccrine glands in the skin. These may produce up to 5 liters per day of sweat, which is a dilute solution of NaCl (often a 0.4% solution). During tests on human subjects in the desert, as much as 4 liters of sweat per hour have been produced, but this production could not be continued for more than an hour or so. The details of the initiation of sweating will be considered in a later chapter.

Non-thermal sweating is of minor importance in the dissipation of heat, but of considerable theoretical interest. The associated sweat glands in humans are *eccrine* on the sole of the foot and the palm of the hand, and *apocrine* in the axilla and pubic regions (Folk and Semken 1991). Non-thermal sweating also occurs on the forehead. When a person is under thermal stress, sweating begins simultaneously over the entire body surface. (For a qualification of this generalization see Hertzman et al. 1952.) However, the non-thermal mechanism (sweating on the palms, soles, axillae, and forehead) is also triggered when the subject is not under heat stress; it is the "cold sweat" of emotional disturbances. Some individuals have chronic excess activity in this sweating mechanism with the result that their palms usually feel cool and moist.

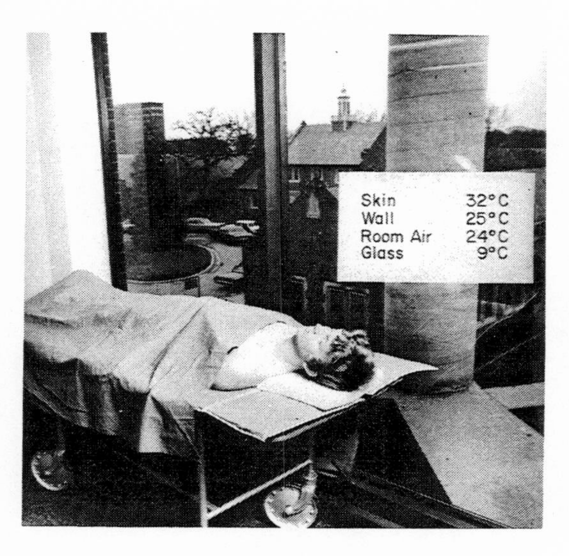

Figure 4-19. Effects of Cold Glass. Some hospital personnel are unaware that a lightly covered patient is not protected against radiative heat loss to a cold window. Likewise, most office workers are unaware that warm air called for by the room thermostat will not protect them from radiative heat loss to a cold window. Courtesy of Larry Beuter.

Paradoxically, some subjects exposed unclothed in a cold room demonstrate the active production of beads of non-thermal sweat. Kuno studied the non-thermal mechanism experimentally and could obtain a sudden increase in this form of sweat by giving his subjects problems in arithmetic (1956). He also described the evolutionary significance of non-thermal sweating. To understand this, one should realize that in many mammals the only sweat glands of any sort are found on the pads of the feet. Probably the pads of most mammals contain non-thermal sweat glands; they even occur on the soft part of the hoof (frog) of the horse. The essential point is that mammals, including humans, produce moisture on the bearing surfaces for grasping and pushing. The importance of this can be demonstrated by the person who attempts to walk up a plank barefooted when the plank surface has been sprinkled with talcum powder; the lack of an adhesive material on the soles makes the task difficult. Kuno points out that another demonstration of the necessity of an adhesive material is seen when a woodsman spits on his hands before grasping an ax; he believes that the expression "spit on your hands" is found in most languages. It would seem that the non-thermal sweat mechanism serves in locomotion and as part of the Emergency Syndrome in mammals. The importance of this mechanism to the temperature regulation of humans is partly a negative one, because the accumulation of moisture in gloves and footgear when exposed to extreme cold is undesirable.

Water (which passes through the skin as sweat) has a high heat of vaporization. It was emphasized by Henderson (1958) in *The Fitness of the Environment* that this property of high heat capacity gives water a peculiar usefulness in the regulation of body tempera-

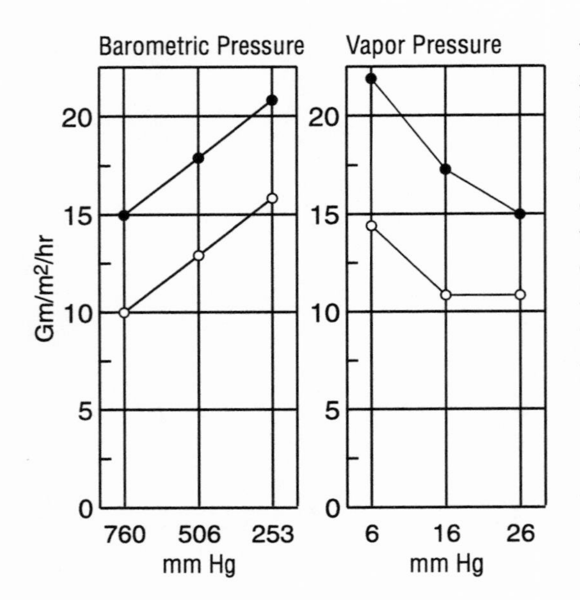

Figure 4-20. Insensible Moisture Loss Corrected for Lung Moisture. The water loss through the skin and the lungs varies with altitude and with the quantity of moisture in the air. Total water loss per hour has been corrected for water lost through the lungs. For example, skin moisture loss at sea level for a person with a surface of 2 m^2 is 20 g/hr, while the same person at altitude would lose 32 g/hr through the skin. Adapted from Hale et al. 1958.

ture. Heat is required to convert water to water vapor; such heat is referred to as *the latent heat of vaporization*. The vaporization of 1 milliliter of water requires 0.58 Calorie; this is the amount of heat lost when 1 milliliter of sweat evaporates from the skin. If sweat appears on the surface of the skin in a copious flood, some of it will drip from the body without removing heat by evaporation. Nonthermal sweat and insensible perspiration leave the body at the rate of 900 milliliters per day in temperate zones. This will result in the loss of 900 x 0.58 = 522 Calories of heat per day. The 400 milliliters of water lost from the lungs each day will dissipate 232 Calories. It is considered helpful today to express the total evaporation from the body in terms of oxygen consumption: mg H_2O/ml O_2 consumed.

The vaporization of body water is the only mechanism available for the reduction of body temperature when the environmental temperature is higher than that of the body. Under these conditions, heat will be gained by radiation and conduction. The heat lost by the evaporation of perspiration will have to include the heat gained from the environment. There is no heat loss by evaporation if the humidity of the surrounding air is too high. The body can withstand very high environmental temperatures if the air is dry, whereas ambient temperatures well below body temperature may be uncomfortable if the humidity is high. These principles will be considered in detail in separate chapters involving temperature adaptations.

EVAPORATIVE COOLING IN LOWER MAMMALS

The avenues of heat loss in a small mammal like the dog, or a large mammal like the elephant, are somewhat different from those in humans. The dog depends largely upon radiation and convection because it does not call upon functional sweat glands except on the pads of the feet. More specifically, the glands are most numerous on the limbs and on the nose. Most researchers think that these glands do not participate in the central thermoregulatory mechanism, but serve to protect the skin from excessive rise in temperature.

Even radiation and convection are not efficient through the heavy winter fur of the sled dog. This reasoning is based on Belding's classic paper on arctic clothing (see Chapter 5); probably the fur of the sled dog blocks 75% of the potential heat loss that would normally occur from radiation and convection. The dog does use evaporative cooling by panting; rapid shallow breathing moves the air quickly over the moist tongue and air passages, and the blood flowing through this area is thus cooled. (We are reminded of this mechanism by watching a sled dog team arrive from a continuous run of 30 miles, traveling at about 15 mph; it is hard to imagine just where the dogs had formerly stored and carried such long, extended tongues. See Fig. 4-21).

According to Dill (1938), this panting does affect acid-base balance: "In some extreme instances the combined carbon dioxide of the blood of the dogs was reduced to less than one-fourth its usual value, and a very alkaline state was established. A human would find such a degree of overventilation very unpleasant—he would be dizzy and might experience tetanic cramps." The normal rate of respiration in a dog is about 15 to 30 breaths per minute, but in panting this rate rises to over 300 per minute. At the same time, the respiration becomes very shallow, such that most of the ventilation involves just the conducting pathway of the respiratory tract. The small amount of lung ventilation reduces the carbon dioxide loss from the lungs and blood.

A dog's requirement for water increases in the heat as its body water is literally panted away. On a hot day, a dog tied in the shade can become dangerously overheated if it does not have a supply of water. When confined in a car, even with a window partly open, the dog's ability to dissipate heat becomes severely compromised. In the summer, a car wall and inside air can easily reach over 43°C (110°F); even the efficient heat dissipation system of the dog cannot cope with this environmental challenge.

How do we apply this same environmental challenge to a human child and a dog of the same weight? The child is in more danger because the gradient is larger, i.e. the difference between a blood temperature of 36.9°C (98.5°F), and T_A of 43°C (110°F). Because the dog has a much higher normal temperature, varying with breed from 38.6°C to 39.2°C (101.6°F to 102.6°F), there will be a lower gradient from dog to hot car than exists in the case of the child. If insulation is comparable in the two cases, then the dog has the advantage. This is assuming the evaporation from the tongue and the

Figure 4-21. Sled Dogs Completing 30-Mile Run. The extended tongues for evaporative cooling are evident in these sled dogs, which have run without a rest for 30 miles in heavy winter coats at temperatures below 0°F.

panting of the dog is the equivalent of the evaporation of sweat by the child.

In the above description, how close would skin temperature be to blood (core) temperature? To illustrate how close these can be, in one experiment in our laboratory, when air temperature was 34°C, the blood temperature of a human adult was 36.6°C, skin of the head was 35.6°C (only one °C different), and average skin temperature was 35.2°C. The subject was nude.

In the dog, the direction of passage of air into the lungs is important; in dogs who are panting due to a heat load, most of the respired air enters through the nose and leaves through the mouth (the opposite of direction of passage of air by a swimmer doing the crawl stroke). In the dog, this unidirectional flow begins over the nasal evaporative surfaces, where the large surfaces of the nasal turbinates come into nearly complete equilibrium with the air in respect to temperature and water vapor. This directional flow is an important mechanism for regulating the amount of heat dissipated in panting (Schmidt-Nielsen 1997). If we look at comparable mechanisms in cattle and sheep, we find that both of these animals also pant; apparently some breeds of cattle and sheep use sweat glands as well. However, neither of these mechanisms in these animals is as effective as the sweating of the human or the panting of the dog.

Like the dog, the elephant does not have sweat glands. The present-day elephant species does not have a thick coat of hair, so radiation and convection are not blocked. However, the extinct Arctic species of elephants did have wool and hair. Heat dissipation from such arctic animals when exercising on warm days must have been difficult because of their relatively small radiating surface and abundant insulation. As for evaporative cooling, consider this example from Kuno (1956): elephants in the Tokyo zoo on a hot day sprinkled water on their backs and sides to cause evaporative cooling. When Kuno removed the water, they collected saliva from their mouths and sprayed this on their backs. This is not a particularly unusual behavior, since a number of species of small mammals have been observed to lick their fur thoroughly when exposed to heat.

In the horse, temperature regulation is accomplished by both sweating and panting. This is successful enough to permit the modern endurance horse, in the hands of an expert trainer and rider, to travel 100 miles non-stop at more than 9 miles per hour, or 50 miles at 10.5 mph with occasional downhill bursts of 20 mph (Robinson 1995).

We can estimate the physiology of sweating during this exercise. Horse sweat is more concentrated than human sweat and has more protein. We say that horses *lather-up*. One would assume that the evaporating power of horse sweat would be less than human sweat, but we have not found any measurements of this. McCutcheon and co-workers (1995) found changes in horse sweat from control values with light exercise; the following increased: osmolality 11%, Na 26%, Cl 9%, K decreased 14%.

These figures have been compared with humans' nearest relatives, the lower primates. The chimpanzee at an air temperature of 24°C loses heat in a similar fashion to a man who is slightly above his critical temperature (that is, at about 31°C or 88°F). Thus, at an exposure temperature of 24°C, the chimpanzee loses about the same amount of heat by the three processes, with the loss by radiation slightly lower than the loss by the other two. Apparently then the hair of the chimpanzee, which is rather coarse and sparse, is a fairly effective barrier to radiation, while it permits some circulation of air to remove appreciable amounts of heat by vaporization and convection (Dale et al. 1970).

TEMPERATURE EQUILIBRIUM QUANTIFICATION

Body temperature is a function of two processes: heat gain and heat loss. As far back as 1947, Nelson and others quantified this relationship, expressed as

$$M \pm S - E \pm C_v \pm C_o \pm R \pm W = 0$$
$$\text{(all terms in kcal·m}^{-2}\text{·hr}^{-1}\text{)}$$

where M = metabolic heat production, S = stored or lost body heat, E = heat lost by evaporation, C_V = heat gained or lost by convection, C_O = heat gained or lost by conduction, R = heat gained or lost by radiation, and W = heat gained or lost by water taken. Any complete study of body temperature regulation must consider all of these parameters. M can be measured accurately, either directly by calorimetry or indirectly by respiratory gas exchange; M is always positive. S cannot be measured easily, nor is there agreement on a formula for its calculation in animals other than humans. It requires measurement of body mass and estimation of mean body temperature and specific heat. E can involve evaporation from body surface or respiratory tract. The surface area, air velocity, and water vapor pressure gradient between moist surface and ambient air determine the extent of evaporation. If we know the weight change due to water evaporation, and the temperature at which evaporation occurred, we can then compute the amount of heat lost by evaporation.

On the other hand, heat gained or lost by convection, conduction, and radiation is difficult to estimate, primarily because the amount of surface available for heat exchange by these avenues is difficult to describe. C_V involves air velocity, thermal gradient between animal surface and air, in addition to surface area. C_O involves thermal gradient between animal surface and surface in contact with the animal, plus the conduction characteristics and area of the soil or other contacting surface. R is determined by the animal's radiating surface area and temperature gradient between the animal and objects in the environment, in addition to the emissivity of the animal. Emissivity, which is the reciprocal of reflectance, can be assumed to have a value of 0.9 within the temperature range most frequently encountered by animals. The radiant heat load of an environment can best be estimated by measuring a black-globe temperature. In a true temperature equilibrium situation, S is zero. Frequently S is not zero; for example, in vigorous exercise, metabolism for a time exceeds the loss of heat by evaporation, convection, conduction, and radiation in which case S becomes positive. On the other hand, during cold exposure, S has a negative value as heat loss is great and the compensatory increase in metabolism is inadequate or delayed.

An error concerning heat balance by voiding is found in some textbooks where it is claimed that heat is lost when the bladder is emptied. Loss of mass is not loss of heat. If two liters of water are at 37°C, and one is removed, the remaining liter is still at 37°C.

Applying the formula under discussion to small mammals has been difficult. There have been a few quantitative studies of conductance of heat, core to surface, in a number of animals exposed to standard conditions (Heller 1972).

SHIVERING AND EXERCISE

One theme throughout the previous sections has been an elucidation of the methods whereby animals combat variable degrees of challenge, such as a cold environment. Often, we have been especially interested in the maximum metabolic rate that the animal is able to obtain. We were interested in witnessing this under the coldest circumstances the animal can tolerate (maximum heat production), and we have given little attention to comparing this maximum heat production with maximum exercise in a temperate environment. Rosenmann and Morrison (1971) devised a technique where maximum heat production can be studied with less damage to the experimental animals; they expose them to the high thermal conductance of an 80% helium and 20% oxygen atmosphere. There are many advantages of this procedure, which is combined with moderate low ambient temperatures; most important is that the risk of cold injury by frostbite is avoided. In three species tested, the ambient temperature to achieve maximum oxygen consumption in helium was about 4°C, while temperatures down to -31°C would have been required to elicit similar metabolic efforts in air. Maximum values obtained were: white rats 4.3 ml $0_2 \cdot g^{-1} \cdot hr^{-1}$, white mice 11.1, and *Microtus oeconomus* 12.5.

As metabolism increases due to cold exposure, one may ask how much of the increased oxygen consumption is due to nonshivering thermogenesis, to tensing of muscles, or to shivering. Tyner and Hemingway (1969), working with cold-exposed sheep, found that prior to shivering, the rectal temperature decreased 0.5°C to 0.8°C and oxygen consumption increased 50% to 100%. When shivering began, oxygen consumption increased further and the rectal temperature was restored.

Figure 4-22 shows the metabolism of bats while exposed to ambient air from 0°C to 40°C. At intermediate exposures (25°C to 35°C) there is a level, or unchanging, zone which is called the *thermal neutral zone;* thus we refer to the entire graph as the *thermal neutral profile*. In this example, the cold exposure extends only to the relatively moderate ambient temperature of 0°C.

Figure 4-22. Upper and Lower Critical Air Temperatures of Bats. The thermal neutral profile of Australian flying foxes (*P. scapulatus*) is shown as determined by oxygen consumption. Each point represents a mean of 3 adults (average weight 362 grams). Modified from Bartholomew 1964.

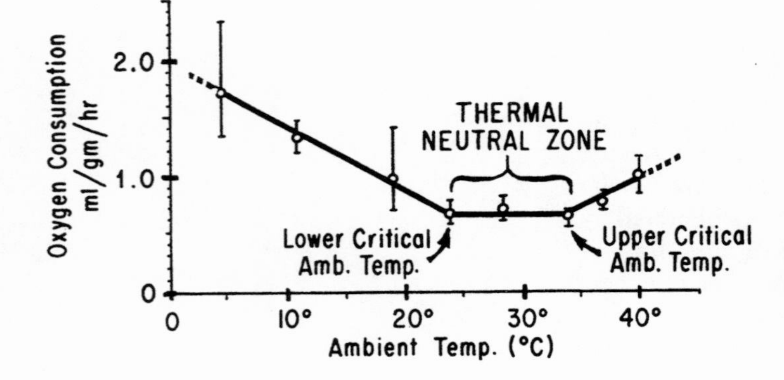

For contrast, Figure 4-23 provides an unusual and extreme case involving the Snowy Owl, an arctic bird (Gessaman 1972). When these Snowy Owls were placed in decreasing ambient temperatures, their metabolism at rest continued to increase without leveling off. In fact, one owl was exposed to -93°C for 5 hours and once again its oxygen consumption increased over its own value at -60°C. Under these circumstances this bird used oxygen at a rate five times greater than at thermal neutral level. This low temperature tolerance was possible because the Snowy Owl has an insulation equivalent to the pelts of the arctic fox and Dall sheep, the best insulated of any arctic mammals.

How does exercise metabolic rate at room temperature compare with maximum shivering metabolism? (Remember, shivering *is* exercising.) In Table 3 we found that a shivering shrew used three times the 0_2 per mile than a human athlete making a maximum effort.

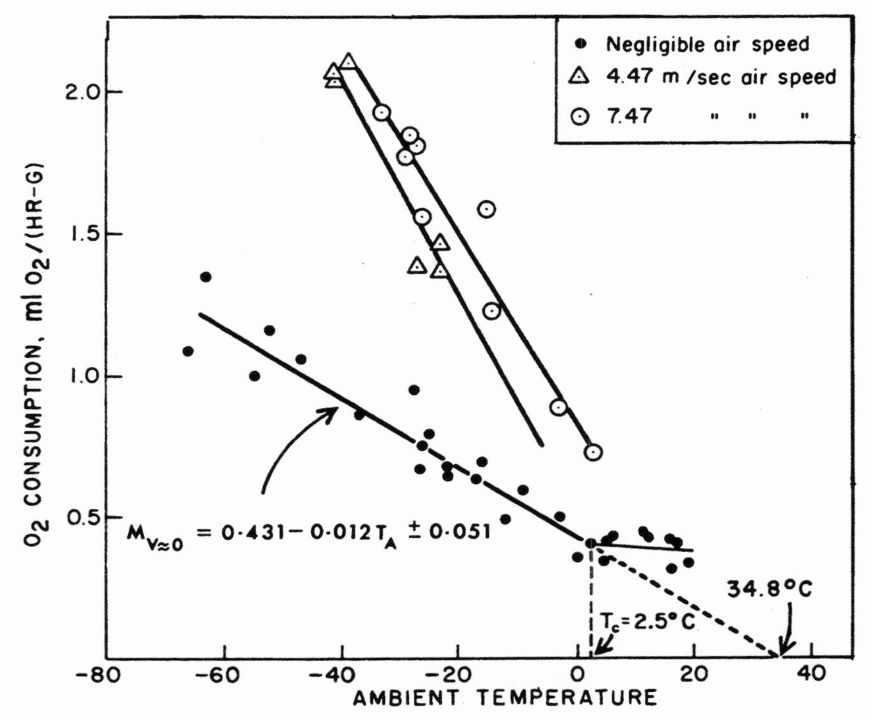

Figure 4-23. Thermal Neutral Profile of Snowy Owls. This graph is presented because few vertebrates have been studied at T_A –60°C. Exposure to wind increased oxygen consumption; however, summit metabolism was not reached. This remarkable ability to tolerate cold is due to the low thermal conductance of the insulation of the snowy owl, which is lower than any published for an avian species, and is equivalent to the pelts of the arctic fox and Dall sheep. From Gessaman 1972.

THERMAL COMFORT

How is temperature equilibrium related to thermal comfort? First, let us look at some measurements. In Fig. 4-24, an unclothed subject is lying motionless in a calorimeter at temperatures that vary in different experiments from 22°C to 34°C. Note that in cold air the feet, unlike the rest of the body, have a temperature close to that of the environment: when air is 23°C to 27°C, the skin of the feet is at 27°C. At 34°C (93°F) ambient, all skin temperatures are similar. These measurements are reported here because it is believed that skin temperature decides thermal comfort. Figure 4-25 describes in subjective terms the relationship between thermal comfort and work level. The environment was adjusted until the subject rated the conditions as comfortable. Thus, the work level of 150 kcal·m^{-2}·hr^{-1} results in a skin temperature of approximately 31°C, which for that work level was deemed comfortable. However, it takes a skin temperature of 34°C to be comfortable at a metabolic rate of 50 kcal·m^{-2}·hr^{-1}.

The contribution of three avenues of heat transfer have been related to comfort (Table 9). When heat loss depends on 23% evaporative cooling, then the condition is comfortable.

When young vs aged males were given a cold exposure test, the skin temperatures of the elderly were lower than those of the young subjects, but the elderly group were not uncomfortable (Taylor et al. 1995).

In the outdoor environment the air temperature (dry bulb, DBT)

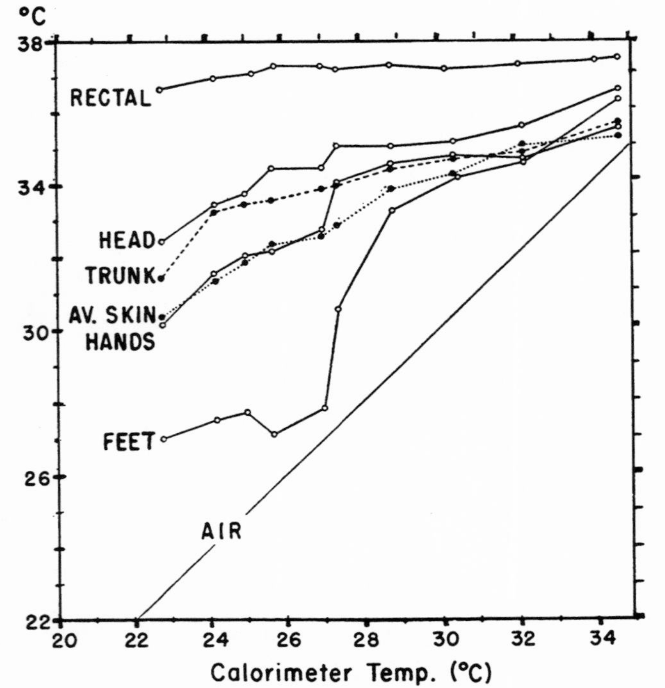

Figure 4-24. Skin Temperatures in Variable Air Temperatures. A human subject lay naked and motionless in a calorimeter for three hours. In cold air the feet have a temperature close to that of the environment. At 34°C (93°F) all skin temperatures are similar. Modified from Hardy 1965.

and the humidity (wet bulb, WBT) must be considered together. Environmental physiologists have devised a Discomfort Index (DI) with the following formula:

$$DI = \frac{DBT(°C)+WBT(°C)}{2}$$

This formula provides one of the methods whereby one can describe hostile heat environments, and possibly predict their effect. For example, in a recent summer, an air temperature of 121°F was recorded in New Delhi; 144 people died as a result. Extreme heat in the summer of 1995 in the midwestern United States affected both people and livestock; in Chicago, several hundred people and thousands of cattle died because of long periods of an air temperature of 90°F along with high humidity. To describe this serious and dangerous condition, a high value of the discomfort index would be 32. An index figure indicating a safe range would be 26.

FEVER

An interest in fever has inspired hundreds of research papers from a variety of biomedical scientists. The Environmental Diagram for the chapter on hot environments details the core temperatures of fever. The anterior hypothalamus/preoptic area is a key locus of fever regulation. Probably prostaglandins are responsible for raising the febrile temperature set-point. Also the thermosensitivity of the anterior hypothalamic/preoptic neurons may be influenced by the neuropeptides arginine, vasopressin, and alpha-melanocyte stimulating hormone to prevent fever or to reset febrile body temperature toward normal values (Cooper 1995; Schmidt and Chan 1992).

ZONE OF THERMAL NEUTRALITY

This *thermal neutral zone* is the *ambient temperature at which warming and cooling of the body are least difficult* (in a physiological sense). An unclothed human at rest in a post-absorptive state is at equilibrium between T_A 25°C to 27°C (77°F to 80°F). The person loses heat to the environment from his or her resting metabolism (BMR), except for normal storage, without calling on reserve heat loss functions. Our subject would not be able, however, to handle the effects of exercise or food without these reserve functions. The thermal equilibrium is handled by regulated skin blood flow (Savage and Brengelmann 1996). Ordinarily, we set a room thermostat at about 20°C (68°F); if the room were left at 26.7°C (80°F), clothing and mild muscular movements would make the factor M larger and the factors C and R smaller.

We usually state that within the zone of thermal neutrality almost all thermoregulation in homeotherms is by physical mechanisms. Outside this zone, the adjustments become exclusively physiological. The physiological mechanisms involve muscular, cardiovascular, and metabolic changes, while the physical mechanisms involve involuntary activation of somatic reflexes and voluntary behavioral adjustments. The behavioral adjustments include postural changes, changes in food intake and water consumption, day active and nocturnal patterns of activity, as well as special parental behavior. The behavioral changes are controlled by the central nervous system and the hypothalamus. Examples of behavioral adjustments for heat loss by the use of posture are seen in bare-skinned mammals such as pigs and water buffalo. They frequent wet spots or dig up soil to get to the cooler subsoil, then turn to expose their moist side to the air. Posture is also used by the raccoon in hot weather; in the central United States they have, even in mid-summer, long hair on the back and sides. They lie on their backs and stretch all four legs, attempting to expose the short hair of the belly. By this means there is some reduction in the insulation of the body. The stretched position increases surface area for evaporation, convection, and radiation. Sleeping children are apt to either curl themselves into a ball or stretch out spread-eagle, depending upon the ambient temperature. It has been calculated that a thinly clad person exposed to cold can reduce the body's surface area by 50% by sitting in a "contained" position (legs together and arms crossed across the chest).

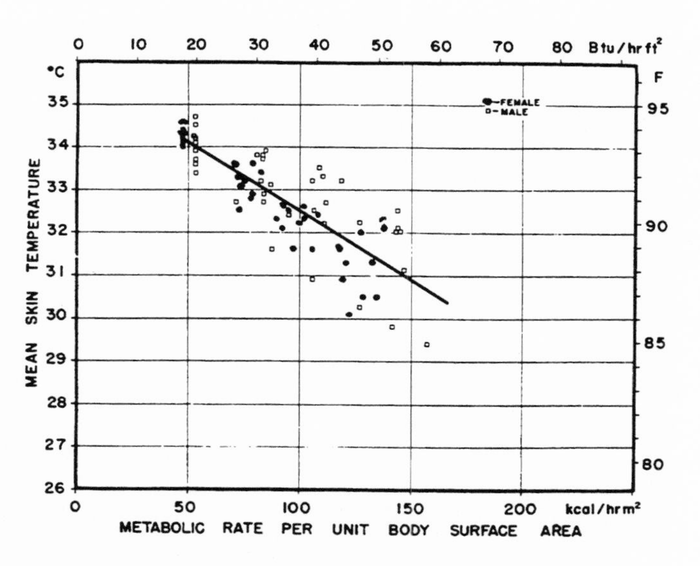

Figure 4-25. A Description of Thermal Comfort. The regression line describes in subjective terms the relationship between skin temperature, the work level of each subject, and thermal comfort. The environment was adjusted until the subject rated the conditions as comfortable. Adapted from Fanger 1967.

Let us look more closely at (1) the physical regulation, and (2) the physiological regulation, in the case of an animal maintained in its zone of thermal neutrality. We can explain physical regulation as follows: Fourier's law indicates that a decrease in the thermal gradient or difference between the temperature of the skin and the surroundings would decrease heat loss from the body. The decrease of gradient is accomplished in two ways: by cutaneous vasoconstriction and by limb blood flow shifts (Fig. 4-26). At a number of environmental temperatures, an animal such as a dog can maintain its body temperature without an increase in metabolic rate by these two methods of physical regulation. If the animal is at a temperature colder than its thermal neutral zone, then physiological (that is, chemical) regulation must be called upon. This is because the animal is maximally vasoconstricted and cannot maintain heat balance by any further vasomotor activity. At first, the necessary increase in metabolism will occur by slight inconspicuous skeletal muscle activ-

Table 4-9. The Partitioning of Actual Heat Loss to the Environment

Room Temperature	Radiation	Convection	Evaporation
Comfortable (25°C)	67%	10%	23%
Warm (30°C)	41%	33%	26%
Hot (35°C)	4%	6%	90%

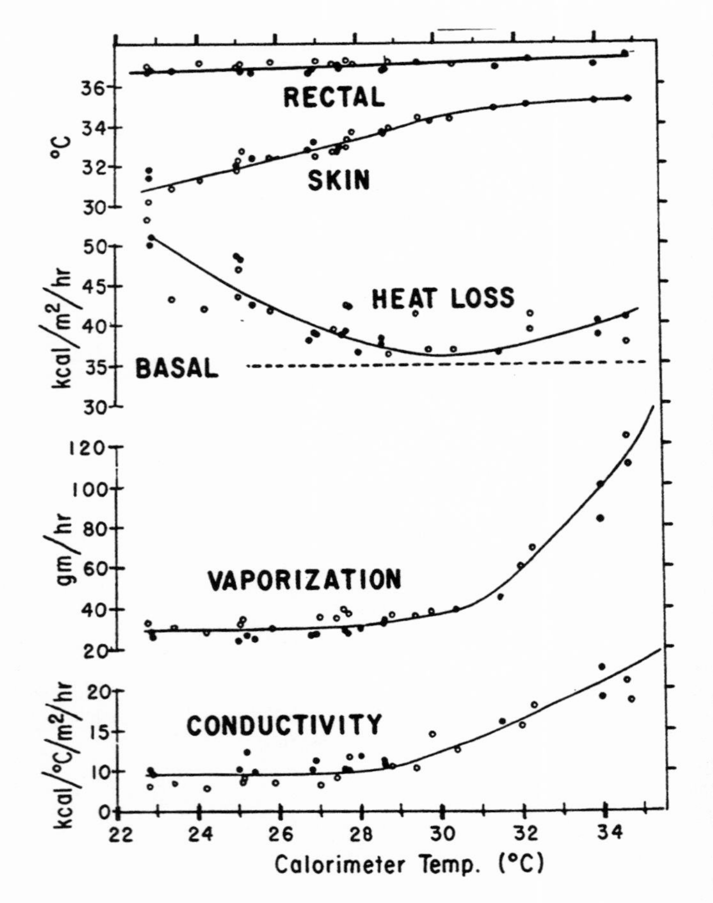

Figure 4-26. Partitioning Heat Loss With Increasing Calorimeter Temperatures. The two male subjects were unclothed and motionless. Heat production ("Basal" line) was uniform over the exposure range; there was no chemical control at lower temperatures. Heat loss and heat production were approximately the same, between 29°C and 31°C (the comfort zone). In this range, skin vessels are somewhat dilated as indicated by a conductivity curve, which measures cutaneous blood flow. Below this zone, the skin vessels constrict. Above it, they dilate progressively, and vaporization suddenly increases because of onset of sweating. Adapted from Hardy 1965.

ity or by shivering. Both of these increase metabolic rate; it is this increase which is referred to as chemical regulation. Both conditions are illustrated in Table 4-10. The thermal neutral data are apt to show an *exponential* rise on the warm side, but it always shows a linear rise on the cold side. We will refer to a graph from such data for an individual animal as its *thermal neutral profile.*

Some environmentalists object to the concept of a zone of thermal neutrality, and prefer to think in terms of *critical air temperature (CAT)*. This is defined as the lowest ambient temperature at which a mammal or bird can maintain its body temperature at the basal metabolic rate. This is 25°C to 27°C in unclothed humans (Erikson and Krog 1956), and in many tropical species of mammals. If we examine the much-used critical temperature graph of Scholander (Fig. 4-27), we find that the CAT can be as low as -40°C in some arctic species. The CAT is a fundamental measure of the overall climatic thermal adaptation.

Note that animals with a low CAT tend to have a lower slope of the metabolism-temperature curve, and for a number of mammals, these curves extrapolate back to body temperature. The slopes for hibernators tend to be steeper than for non-hibernators; also critical temperatures of some mammals change with season. We do not know if this applies to animals living in a constant environment near the equator. Nor have mammals that increase their insulation in winter been compared in winter and summer without their insulation. Water immersion raises the critical temperature. A list of critical temperatures in air is given in Table 4-11; all figures refer to lower critical air temperatures in summer. Some additional and unusual lower critical air temperatures are presented in Table 4-12.

We can make several generalizations from this table. Both of the mammals that respond quickly to cooling (tenrec and gerbil) live in hot climates. The fossorial (burrowing) mole-rat has the lowest

Table 4-10. Heat Production in a Short-Haired Dog at Different Environmental Temperatures (From Bard)

Temperature	Heat Production kcal/kg	
7.6°	86.4	Chemical regulation
15.0"	63.0	
20.00	55.9	
25.00	54.2	Physical regulation
30.00	56.2	
35.00	68.5	

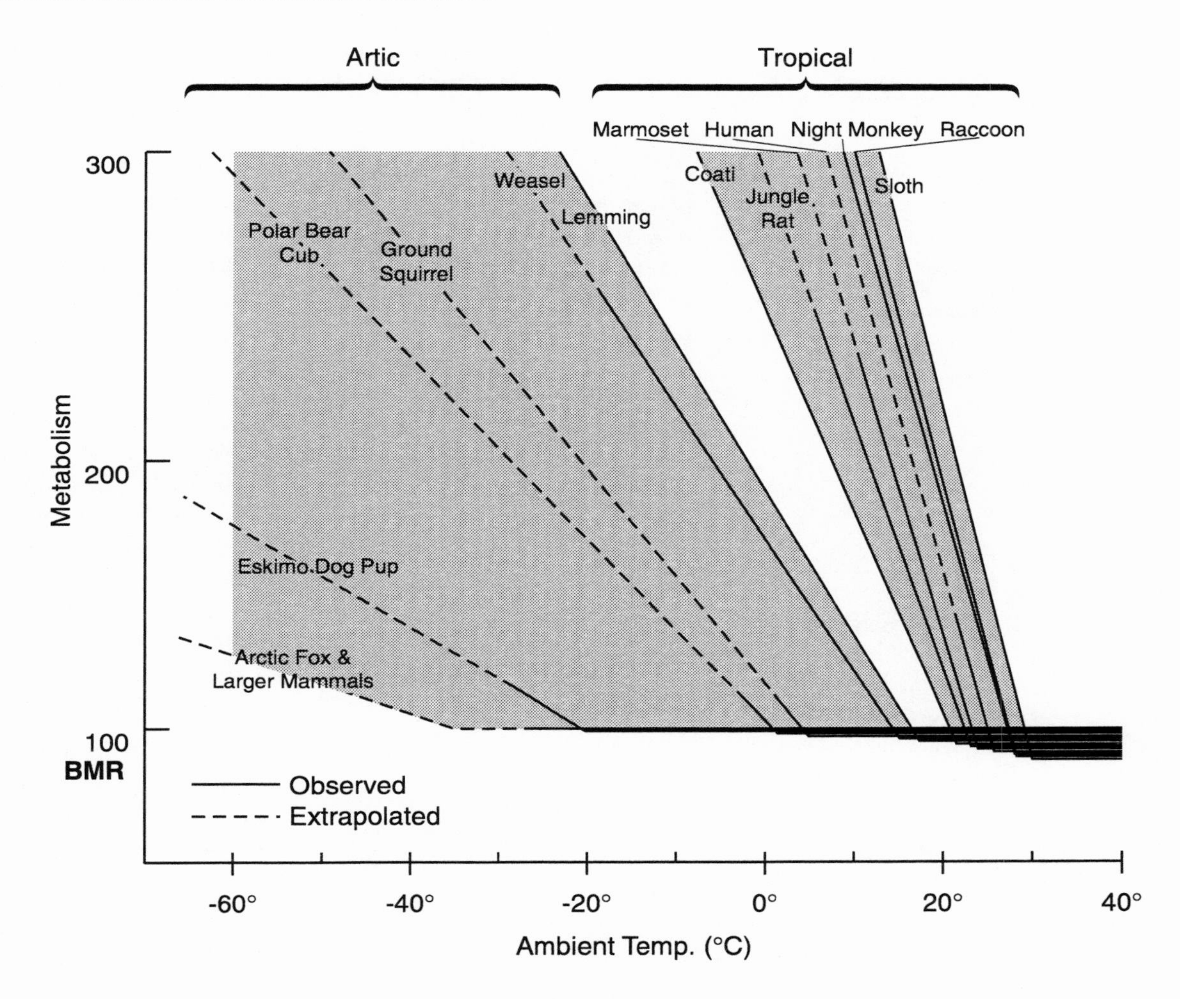

Figure 4-27. Critical Temperatures of Mammals. Resting metabolism of arctic and tropical mammals as a function of T_A. Solid lines represent measurements, broken lines are extrapolations. Critical temperatures are found at points where metabolism graph intercepts standard value. Modified from Scholander et al. 1950.

body temperature and poorest capacity for thermal regulation of any known mammal. Along with other fossorial rodents, it also has a higher rate of evolution of the mitochondrial-DNA cytochrome *b* gene than that found in non-fossorial rodents (Spradling 1996).

Some investigators refer also to the *upper critical temperature;* the range between the upper and the lower is the zone of thermal neutrality. Within this range, compensations for ambient temperature fluctuations are made without increase in heat production. Examination of Bartholomew's curve in Fig. 4-22, shows that we may introduce two new concepts concerned with conductance and insulation: the use of the term *minimal thermal conductance* or its reciprocal, *maximum thermal insulation,* is based upon the experimentally established linear relationship between resting heat production and the temperature difference between the body and the surrounding air (also see Fig. 4-30).

In order to use these concepts, it is necessary to make the following assumptions: 1) minimum thermal conductance or maximum insulation remains constant at low air temperatures; 2) the extrapolated line relating resting metabolism to temperature below

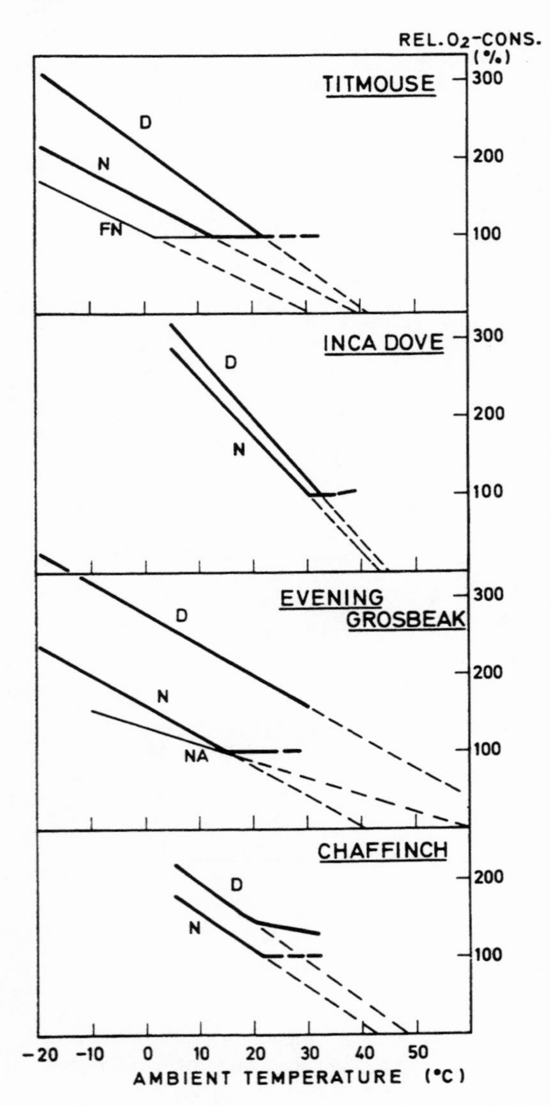

Figure 4-28. Different Avian Thermal Neutral Profiles. The metabolic rate responds differently to cold depending on the physiological condition of the animal. These graphs indicate that the day animal (D) gives a different oxygen consumption in respect to temperature compared to the same animals at night (N), or when freshly caught (FN = first night in captivity). A different slope of a thermal neutral profile is obtained in the evening grosbeak when the bird is in night-migrating condition (NA = night-acclimated birds). Note that only a portion of each thermal neutral profile is shown. Modified from Pohl 1969.

thermoneutrality should usually intersect at a point corresponding to internal body temperature. We will consider conductance again in the next paragraph. These concepts depend upon the thermal neutral profile, but Pohl (1969) introduced some qualifications, stating that the profile model can be documented by the thermoregulatory behavior of about 67% of the birds and mammals that have been tested. Pohl went on to explain that if the following factors or circumstances are present during the time of measurement, the model may not apply: 1) the circadian rhythmicity (the measurements have to be made either during the night or during the day, thus obtaining two different thermal neutral graphs); 2) a variable level of activity; 3) the light condition (which may force different thermal neutral profiles); and 4) differences in the temperature program due to duration of exposure or to sequence of temperatures the animal is exposed to during the test (Fig. 4-28). These qualifications are also illustrated with other data from Pohl based upon six chaffinches (Fig. 4-29). Note that by keeping the birds in 10 lux in one experiment and in 0.1 lux in the next experiment he obtained two different thermal neutral profiles. Possibly this was because a "seasonal" reduction in insulation is associated with a higher light intensity.

The metabolic rate increases as the mammal becomes cold. (Note that the arctic fox does not become "cold" until it is exposed to

Table 4-11. Critical Air Temperatures for Mammals

Order		
Primates:	human	27°C to 28°C
Edentata:	sloth	25° to 27°
Lagomorpha:	rabbit	17°
Rodentia:	ground squirrel	5°
	lemming	20°
	hamster	20°
	red squirrel	20°
	porcupine	7°
	Dipodomys	31°
	Panama mouse	27°
	deer mouse	30°
	rat	25°
Carnivora:	weasel	15°
	harbor seal	<-10°
	red fox	8°
	dog (short hair)	25°
Artiodactyla:	pig	0°
	mountain goat	(-30° in winter)
	steer	7°
	sheep (full fleece 12 cm)	0°

approximately -40°F). A systematic approach to different mammalian metabolic responses to cold is found in the use of "thermal conductance." This is the term applied to the slope of the chemical heat production curve of homeotherms. Thermal conductance units are volumes of respiratory gas, or heat equivalents (absolute, or expressed per unit of body weight) per unit time per unit temperature (cal·g^{-1}·hr^{-1}·°C^{-1}). It includes all forms of heat loss by the animal. It is the rate of heat transfer across the external and internal body surfaces from the animal to the environment, irrespective of means of transfer. Thus, this measurement does not differentiate heat dissipated by evaporation of water, and heat dissipated by radiation, convection, and conduction. For further definitions and examples see Hudson and Brush (1964) and Fig. 4-30.

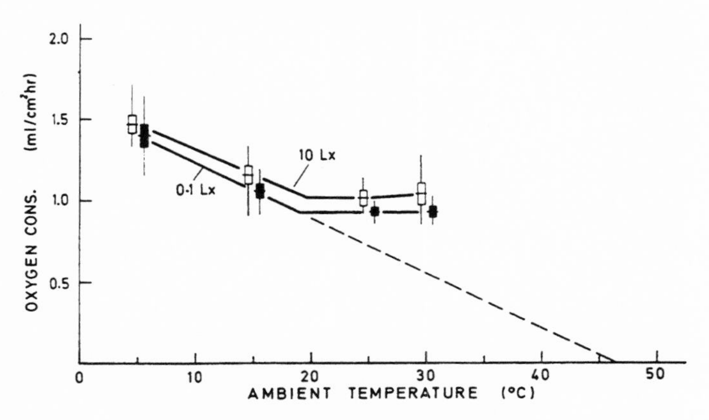

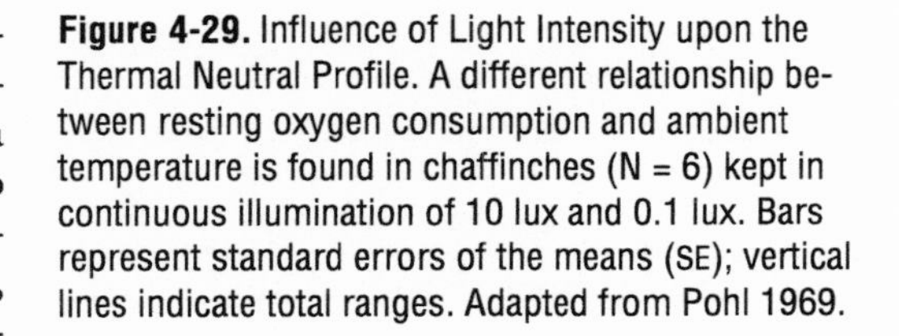
Figure 4-29. Influence of Light Intensity upon the Thermal Neutral Profile. A different relationship between resting oxygen consumption and ambient temperature is found in chaffinches (N = 6) kept in continuous illumination of 10 lux and 0.1 lux. Bars represent standard errors of the means (SE); vertical lines indicate total ranges. Adapted from Pohl 1969.

A study of the chipmunk (Fig. 4-31) further illustrates the usefulness of plotting responses of an animal at different ambient temperatures to expose the thermal neutral profile. Obtaining these data must precede calculation of conductance. It had been debated as to whether marsupials had a much lower metabolic rate and body temperature than eutherian (non-marsupial) mammals. Dawson, Denny, and Hulbert (1969) plotted the thermal balance of a marsupial (Fig. 4-32). Note that both below and above 30°C T_A, heat production tends to rise, and this increases body temperature. The heat production was 30% to 35% below the predicted level for a non-marsupial of similar size. At low air temperatures, body temperature was maintained by a relatively low heat loss and a high critical temperature. At high temperatures, evaporative heat loss was accomplished by panting, and licking of the fur.

Kleiber (1971) plots the same sort of data in percentages so that he can compare animals of widely different metabolic rates (Fig. 4-33). By this means he can compare the metabolic rate of the rat, which decreases as the T_A goes from 15°C to 30°C, and of a frog, which increases over this range. He can also compare the metabolic rate of a frog and a curarized dog. The rate of the frog increases exponentially. The diagram again shows the thermal neutral zone for the rat at between 25°C and 30°C (the exact range depends upon strain, season, and previous thermal history of the animal). Not only is the zone of thermal neutrality useful to describe the condition of

Table 4-12. Unusual Lower Critical Air Temperatures

Japanese Monkey	22°C
Tenrec	30°
Rock Hyrax	25°
Gerbil	30°
Chinchilla	20°
Naked Mole-rat	32° to 22°

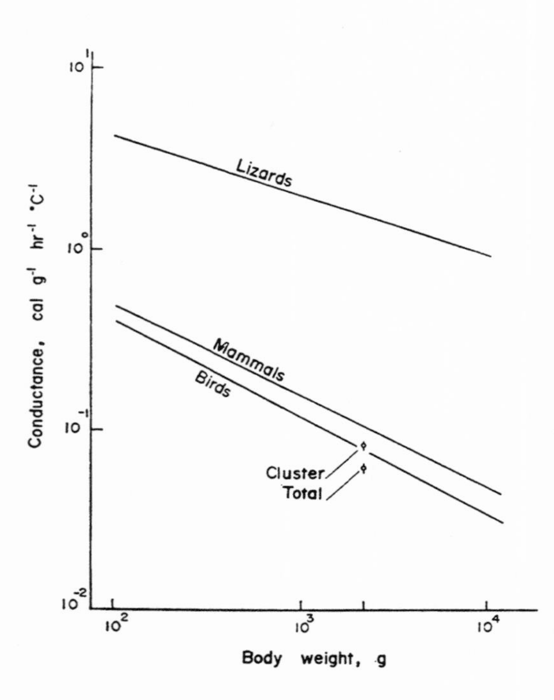

Figure 4-30. Conductance Curves for Vertebrates and a Honey-Bee Cluster. The honey-bee cluster behaves like a complex organism and shows temperature regulation like that of a bird. All conductance values were calculated from the formula:

$$C = \frac{MR}{T_C - T_A}$$

The cluster temperatures were measured directly; for total conductance, the temperature of the center of the cluster was compared with T_A outside the hive. Note that the insulation for birds is consistently better than that for mammals. Adapted from Herreid and Kessel 1967.

minimum metabolism, but Kleiber especially emphasizes that resting heart rates should be measured under such conditions of thermal neutrality. For example, summer resting heart rates of our wolverines were 77 ± 5SD b/m in the thermal neutral zone, but in winter they were 94 ± 7 SD b/m; Coulombe (1970) found that the heart rate of burrowing owls was 192 b/m in the thermal neutral zone; when they began to pant (gular flutter), they raised the rate to 207 b/m.

OTHER VERTEBRATES

At this point we should mention the temperature regulation of vertebrates lower than birds; earlier we discussed how the cluster of honey bees showed temperature regulation. Other resting insects can maintain a constant temperature. The sphinx moth maintains its thoracic temperature at 42 ± 1°C while in free flight at an air temperature from 17°C to 32°C (Heinrich 1970). It is referred to as a *heterothermic insect.* The winter moth, *O. brumata,* flies to the female at an air temperature of 4°C (Roelofs et al. 1982). The elephant beetle (Scarab) initiated a set-point warming when core temperature cooled to 20°C. The T_B (20°C) remained stable for three hours when T_A was 7.0°C (Morgan and Bartholomew 1982).

In heavy varieties of fish, body temperatures are especially well regulated; some species, like the yellow fin tuna, maintain a fixed elevation of body temperature above the water temperature. The blue fin tuna, on the other hand, thermoregulates, and its muscle temperature changes only by about 5°C when it is in water that varies over a 20°C range (Fig. 4-34). The swordfish and other billfishes have a brain heater surrounded by insulating fat. The heater-tissue is associated with one of the eye muscles. A countercurrent circulatory heat exchanger supplies the brain heater with hot blood. Temperatures by radio-telemetry in free-swimming swordfish were 18°F to 25°F warmer than the surrounding water. This fish makes daily vertical excursions into cold water (Carey 1982). The evolutionary dis-

Figure 4-31. Thermal Neutral Profile of Eastern Chipmunks (*T. striatus*). The regression line below 28.5°C is calculated and that above 32°C is estimated. This graph is presented because data on mammals maintained above the thermal neutral zone are rare. From Wang and Hudson 1971.

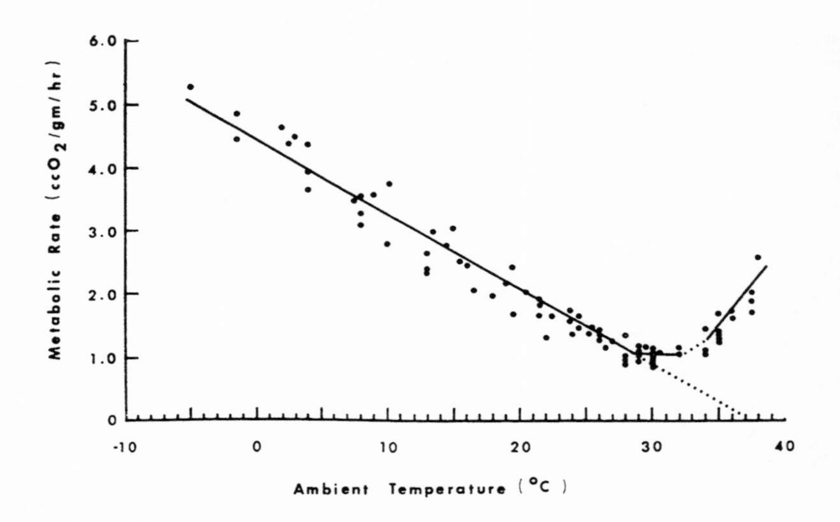

tribution of this endothermy in fish has been studied by Finnerty and Block (1994) by nucleotide sequence data from the mitochondrial-DNA cytochrome *b* gene.

This field of study was begun by Benedict in 1903, when he measured the temperature of an incubating female python curled around her eggs. Her blood temperature was several degrees above air temperature. It is significant that the python had changed its large surface area from that of a long thin cylinder to a mass with a relatively small surface area.

Another example of the unusual reptile that can resist cooling is the gecko. In the following table, a species of gecko that resists cooling (Gecko 1) is compared to one that does not (Gecko 2), as recorded by Zari (1992):

T_A	Gecko 1 ($ml \cdot g^{-1} \cdot hr^{-1}$)	Gecko 2 ($ml \cdot g^{-1} \cdot hr^{-1}$)
20°C	0.245	0.162
25°C	0.226	0.176
30°C	0.385	0.289

The resting metabolism of the first species is significantly larger at the cool temperature (20°C) than at the intermediate temperature (25°C).

Thus we see that the story of temperature regulation in insects and fish has an equally interesting parallel in the reptile group. Other examples of regulation of body temperature by reptiles can be found in a review article by J. E. Heath (1970).

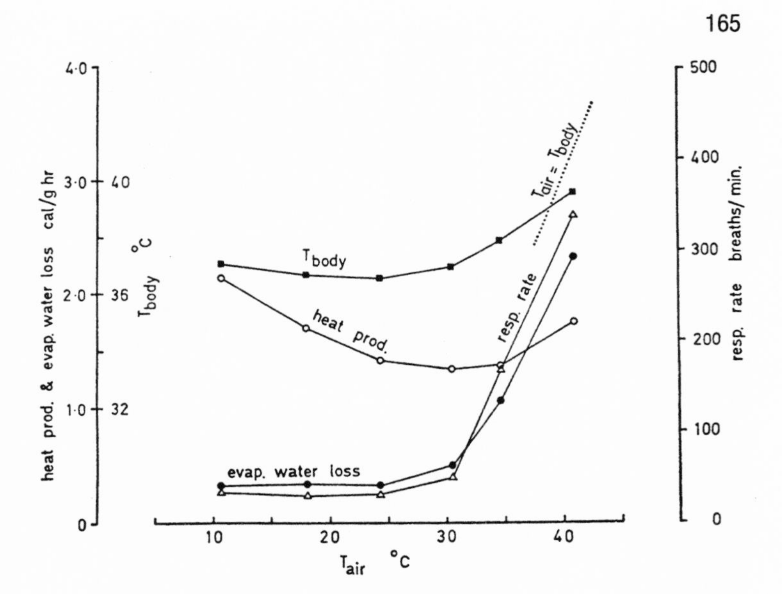

Figure 4-32. Heat Balance Sheet of Wallaby Marsupial at Different Ambient Temperatures. This thermal neutral profile is expressed in calories per gram per hour. The basal portion of the curve shows a level 30% to 35% below that predicted from Kleiber's equation for eutherian mammals. The curve shows the lower critical temperature to be 21°C to 22°C. Below this temperature, there was visible shivering. Above T_A 30°C, the metabolic rate, respiratory rate, evaporative water loss, and body temperature all rose. Below T_A 26°C the percentage of heat production lost by evaporation was less than 22.5%. Adapted from Dawson 1969.

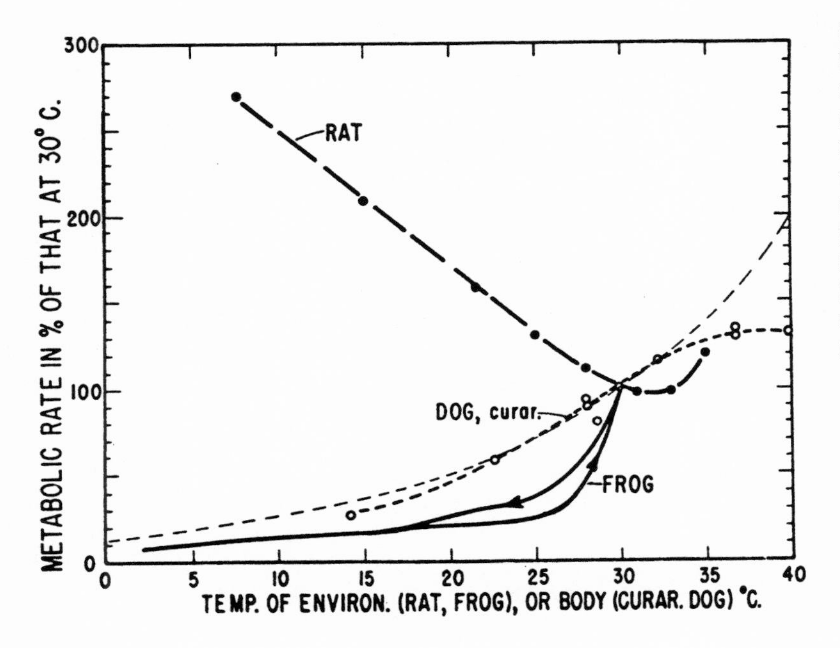

Figure 4-33. Thermal Neutral Profiles Compared in Percentages. By expressing results as percent of metabolic rate of that at 30°C, a normothermic rat, a nonregulating dog, and an ectothermic frog can be compared at different air temperatures. A dog under curare is unable to shiver and combat the challenge of cold air. The rat shows a typical thermal neutral profile. With increasing ambient temperatures, the metabolic rate of a rat decreases, while that of a frog increases, the latter exponentially. Above the thermal neutral zone the rat also increases its metabolic rate exponentially. Adapted from Kleiber 1971.

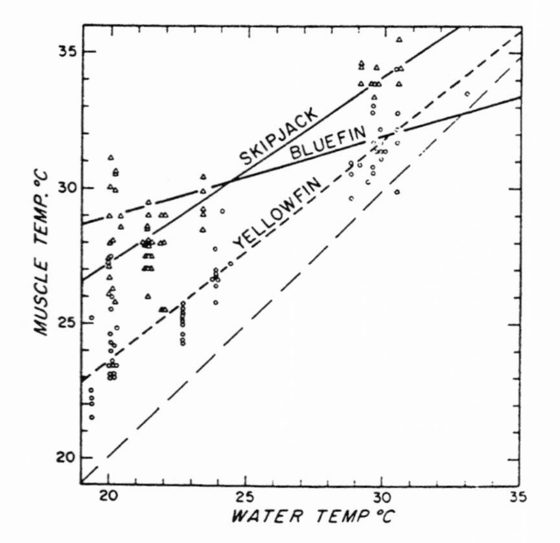

Figure 4-34. Temperature Regulation in Large Fish. It was formerly assumed that active thermoregulation was confined to mammals and birds. It is now known that some large fish thermoregulate (see bluefin tuna); the skipjack and the yellowfin tuna show a fixed elevation above the water temperature. Modified from Carey 1982.

The emphasis in the study of temperature regulation so far has been upon the advantages of measuring tissue temperature and metabolic rate. However, temperature regulation can also be understood by using other measurements, such as weight gain (reflecting depositing of winter fat), hoarding of food, increased food consumption, and the influence of changing daylight. Robert Lynch (1972) paid attention to all of these parameters using the deermouse *Peromyscus* (Fig. 4-35). It is evident from his illustrations that all of these functions are altered by both cold and changing hours of daylight.

ALCOHOL AND TEMPERATURE REGULATION

While attending a football game in cold weather, some people consume alcohol not only for social reasons, but also because the warm flush they experience gives the false impression that they are gaining heat. Actually, at a time when the body needs to conserve heat by vasoconstriction, it will lose heat by vasodilation because alcohol is a powerful vasodilator; as little as 1 oz of 80 proof alcohol will cause the blood vessels to open (See Fig. 4-7). In our own experiments, a subject's cool skin was vasodilated and warmed as much as 3°C by ingested alcohol.

Even more important is the influence of alcohol on the brain's thermostat (the hypothalamus): there is a downward setting which impairs heat production. At moderate intoxication (150 mg per deciliter) the effect will be a rapid fall in core temperature at T_A 39.2°F (4°C), which often causes fatal results. This intoxicated individual may have an increased heart rate at first (by as much as 16 b/m), but no change in blood pressure and stroke volume as the body temperature is dropping (Wittmers et al. 1987). For these reasons alone, the use of alcohol in a cold environment must be avoided.

EFFECTS OF AGING

In a recent study in Japan, Nakamura and colleagues (1997) confirmed that oral temperatures in the elderly are lower in winter than in summer, and also that low levels of daily activity result in poorer thermoregulatory control. Their study of 57 men and women aged 63 and older showed that oral temperatures were as much as 0.25°C higher in summer than in winter. (There were no significant temperature differences between males and females in this study.) Poorly regulated indoor temperatures may be a contributing factor, as also noted by Stout et al. (1991). However, even in well-controlled indoor environments, the subjects who had low daily exercise had the lowest winter oral temperatures. This difference in seasonal temperatures occurred despite a higher basal metabolic rate in winter than in other seasons (McArdle 1991); there was a higher heat loss during the winter.

Age-related changes in temperature regulation are related to oxygen supply, or to the control and use of oxygen in body tissues, especially the nervous system. This concept can be tested when there are large changes in oxygen supply, such as at high altitude or in diving.

(Chapter 8 discusses similarities in the effects of aging and high altitude.)

How do we relate other aspects of aging to Environmental Physiology? It is difficult to generalize about the complex facets of aging, but it is clear that the body's ability to use oxygen is a key factor in age-related physiological changes.

It is also well known that the metabolic rate of humans declines with age (see Fig. 4-10). The use of oxygen by mammalian cells generates potentially deleterious oxygen metabolites (free radicals) and results in oxidative stress; the amount of damage increases with age (Harman 1992). Sohol and Weindruch (1996) have observed that restriction of caloric intake lowers oxidative stress and extends the maximum life-span in mammals. Thus, they hypothesize, senior citizens who eat less are extending their maximum life-span. There are dissenting voices. Visser et al. (1995) showed that in their sample of the elderly, there was not a difference from controls in the influence of diet on longevity.

Another age-related change is that the calcium set point for release of parathyroid hormone is higher in the elderly; thus, the serum concentration of this hormone in older men is twice that in young men (Portale et al. 1997). This reminds us that calcium regulation is important for older adults, often requiring the use of calcium blockers to maintain a healthy cardiovascular system.

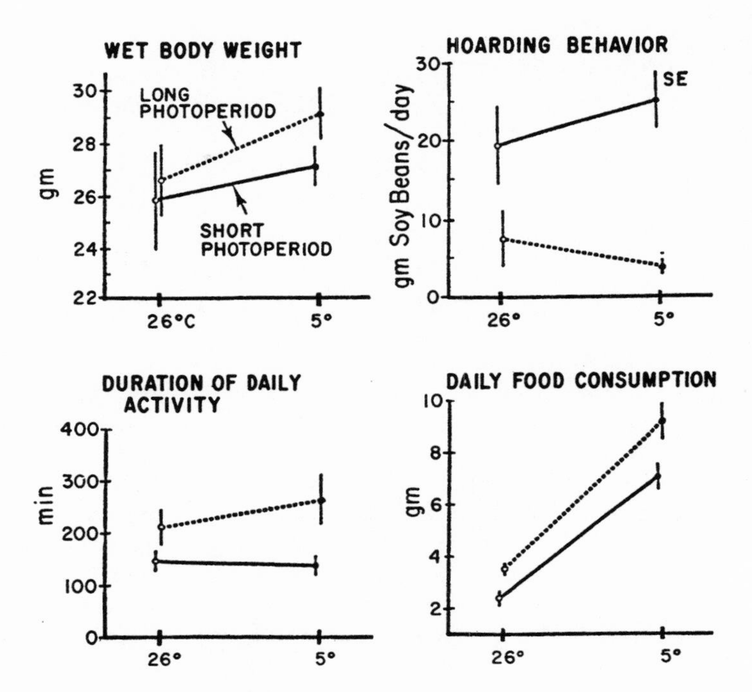

Figure 4-35. Behavior Variables, Thermoregulation, and Photoperiod. Deer mice were exposed to six weeks of summer and fall ambient temperatures and photoperiods. A short photoperiod in a summer environment caused them to begin to hoard food for the winter; this was anticipatory behavior. In a sense, it represents the beginning of cold acclimatization without cold, under the influence of photoperiod. A long photoperiod in the cold appears to affect hoarding behavior as if the springtime season were about to arrive. Modified from Lynch 1972.

MECHANISM OF TEMPERATURE REGULATION

Feedback Loop

Up to now we have discussed some external factors of temperature regulation (evaporation, radiation, etc.) and temperature equilibrium mechanisms. We must now consider the neurological and endocrine details which bring about the exquisite control and fine adjustments of temperature regulation.

The homeostatic mechanisms involved in responses to cold and heat are illustrated in Fig. 4-36. Using the terminology of control systems, we find the *sensor* (the skin), the *controller* (the central regulator, the hypothalamus), and *control processes* indicated as physiological units. The sensory elements are skin temperature (T_s) and internal temperatures (T_1 and T_2). One site for internal temperature sensing is in the hypothalamus, and there may be a second series of sites in the brain stem, in the cerebral cortex, and deep in large veins. The controlled processes are vasoconstriction, the shivering mechanism, and the non-shivering heat production mechanism. The principle controller pathways are the sympathetic nervous system that controls peripheral blood flow, the motor system that controls shivering, and cell metabolism.

Temperature regulation responses are partly neural. These responses are modified and supported by endocrine and behavioral contributions. A model by Smythies relates many of these factors from both the environmental and visceral sources (Fig. 4-37). The response of the regulatory system as a whole depends upon input signals reaching the central integrating system from which output signals effect changes in insulation and heat production.

Thermoreceptors

Body temperature control depends upon the receipt of signals from the skin surface, as well as signals from changes in blood temperature received by the thermosensitive center. Specific cold receptors are located in the skin. The number of receptors varies; the skin of the face and hands has more than the legs or chest. The cold receptors respond to the rate of temperature change and also to absolute temperature. Abrupt temperature changes increase the frequency of impulses in cold receptor fibers. When skin temperature increases, the cold receptors cease sending impulses along their fibers. Individual cold receptors respond to a certain range of temperatures. The nerve fibers that originate in the cold receptors travel in the lateral spinal-thalmic tract where they form synaptic connections that project to the somesthetic area of the brain. Details are illustrated in Fig. 4-37, by Smithies. Some temperature-regulating responses related to blood flow may function through axon and spinal reflexes.

Control in the Hypothalamus

The hypothalamus contains the control systems that regulate feeding and drinking, reproduction, the sleep-wake cycle, and above all, temperature regulation. Yet the human hypothalamus accounts for only 0.3% of the adult brain volume. The volume of this part of the brain is highly correlated with brain size, irrespective of the evolutionary history of the species considered (Hofman and Swaab 1992).

The importance of the hypothalamus in temperature regulation was demonstrated by classic experiments involving lesion procedures, electric stimulation and recording, and changing hypothalamic temperatures by local heating and cooling by thermodes. Such methods revealed functional areas in the hypothalamus; generally, heat loss control operates in the anterior or preoptic hypothalamus, while heat conservation signals originate in the posterior hypothalamic area. The preoptic area has been shown to be temperature-sensitive. As indicated above, some areas in the brain stem, cerebral cortex, spinal cord, and large veins are sensitive to temperature changes.

The classic experiments showing temperature sensitivity of the hypothalamus were done by Hardy in his laboratory at Yale University. He studied the responses of the conscious dog to local heating of the thermosensitive area of the anterior hypothalamus. The hypothalamic temperature was recorded by a thermocouple inside the tip of a needle electrode. When mild heat was applied to the anterior hypothalamus (39°C at the electrodes), the animal became very quiet, laid down, stretched out, and appeared drowsy. When the

COLD AND HOT WEATHER HAZARDS FOR OLDER PEOPLE

Decreased ability to regulate body temperature can put older adults at risk of becoming hypothermic and hyperthermic. Certain age-related illnesses and medications also affect the way the body handles high or low ambient temperatures.

Hypothermia

Health factors that may increase an older person's risk include hypothyroidism, stroke, severe arthritis, Parkinson's disease, poor circulation, and memory disorders. Drugs that can affect body heat include those used to treat anxiety, depression, nausea—even the common cold. Chapter 5 explores the subject of hypothermia in more detail.

Hyperthermia

Heat can be a problem for older adults who have heart, lung, or kidney disease, high blood pressure, poor circulation, inefficient sweat glands, or other changes in skin caused by the normal aging process. Certain drugs can cause an inability to perspire; these include diuretics, sedatives, tranquilizers, and certain heart and blood-pressure drugs. Hyperthermia is discussed further in Chapter 7.

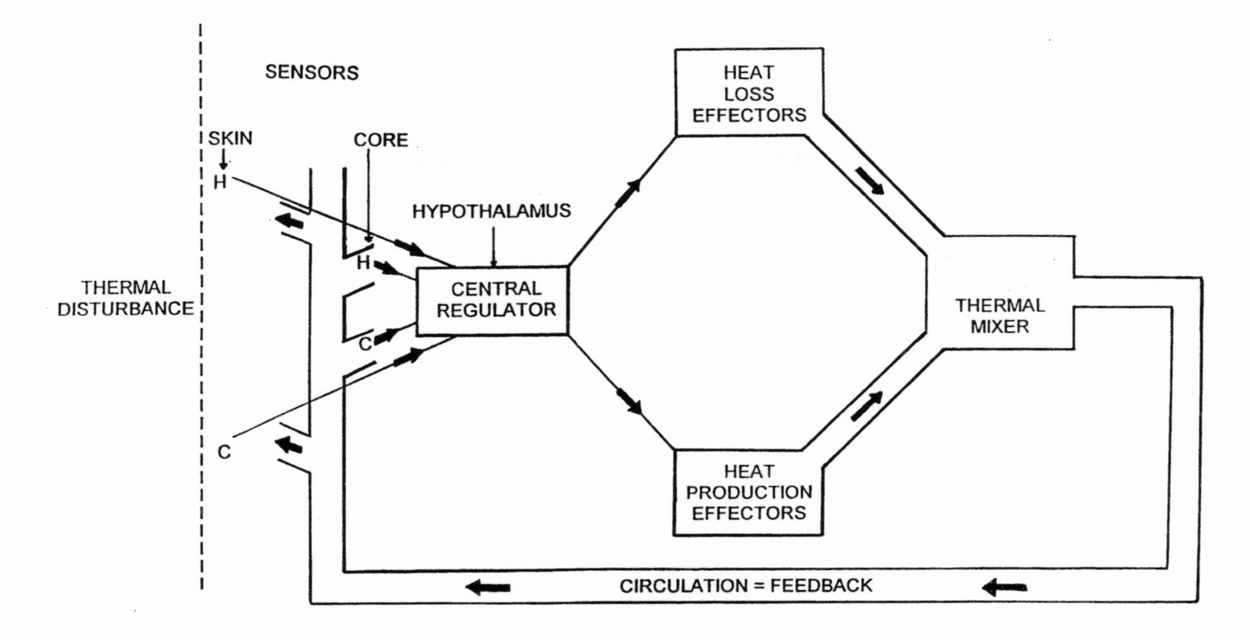

Figure 4-36. The Negative Feedback of Temperature Regulation. Changes of central temperature begin responses which eventually affect (or feedback to) central temperature. This is a negative feedback because deviations of body temperature in one direction induced mechanism to displace body temperature in the opposite (or negative) direction; if temperature starts to go up, mechanisms act to bring temperature back to normal. Temperatures from cold and warm receptors (H and C) are diagrammed. Modified from Ferguson (1985).

heat was increased (41°C) the dog sat up, became alert, salivated, and began to pant; it continued to pant as long as the central heating was maintained. After one hour the heat was turned off and the dog stopped panting. The animal had lowered its core temperature by panting until it was 2°C below the usual body temperature of 38°C, at which time it began to shiver vigorously. When heat was reapplied to the hypothalamus, the shivering became inhibited and the panting began again. This sequence could be repeated many times.

The cellular analysis of thermosensitivity in hypothalamic neurons has been studied by Boulant at Ohio State University. Using horizontal and frontal tissue slices of the rat hypothalamus, he studied the firing rate, input resistance, membrane potential, and the rate of rise in the depolarizing prepotential, with changes in temperature. He could identify warm sensitive neurons, cold sensitive neurons, and temperature insensitive neurons. The warm sensitive neurons displayed the highest spontaneous firing rate. During temperature changes, hypothalamic neurons did not show changes in their membrane potentials that would account for their thermosensitivities. Of the measurements made, the rate of rise of the prepotential was the most strongly temperature dependent, and this must be the primary determinant of neuronal thermosensitivity (Boulant and Griffin 1994; Boulant 1994).

Chemical Control of Thermosensitivity

It is apparent from the above experiments that changes in blood temperature in the hypothalamus bring about temperature-regulating responses. There has been considerable interest in determining the exact mechanism of the responses to stimuli of hypothalamic cells. When nerve impulses arrive or originate in the hypothalamus, the response depends on neurohumors; therefore, many experi-

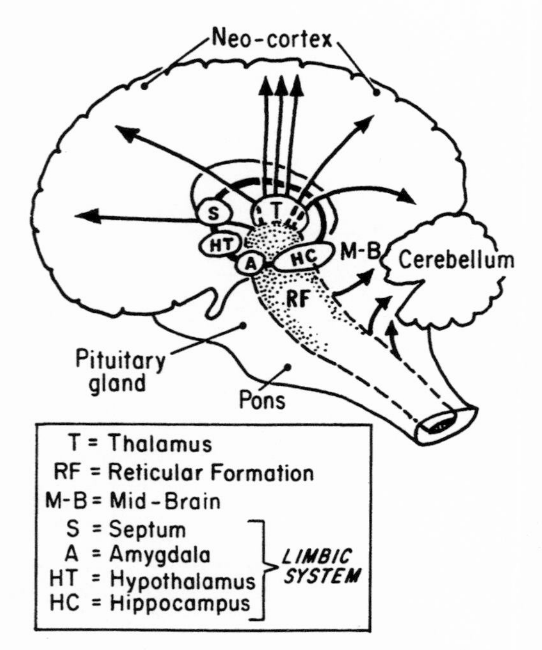

Figure 4-37. Smythies' Model Relating Environmental and Visceral Information. Environmental information is fed to the reticular formation, and then to the thalamus, to the sensory cortex, to the hippocampus, and to the limbic circuit. A reticular formation "alerts" the brain concerning this environmental information, receives and correlates visceral and environmental information, and then informs the motor cortex of the behavior pattern best suited for the situation.

ments attempted to simulate the mechanism of action in hypothalamic cells. Early experiments by Feldberg, Meyers, and Veal led to the monoamine theory of thermoregulation; this plausible theory helps explain fever, and even hibernation. A result of this theory was the concept of a set-point body temperature. Later, Hellstrom and Hammel (1967) developed an additional concept of a *variable* set-point. The monoamine theory depended upon the control of thermoregulation by the influence of two neurohumors upon the cells of the anterior hypothalamus. These two neurohumors are serotonin (5-HT) and norepinephrine (NE). According to this theory, these neurohumors are released as transmitters from the anterior hypothalamus, and function in opposition to one another to control the temperature of the animal around the given set point. Before going into more detail about this theory, we should note that serotonin was only the third neurohumor to be discovered (see sidebar in Chapter 2).

The neurohumor mechanism of temperature control may act in this fashion: one substance in one particular area of the hypothalamus may activate the heat production pathway when the animal is cold, while the other substance may stimulate the heat loss pathway when the animal is warm. This does not explain, however, the mechanism in the central nervous system by which temperature is selected, set, and maintained at some particular constant level, usually near 37°C. One criticism of the monoamine theory is that different experimental animals respond differently to the procedure of perfusing solutions of the test substances either through the third ventricle or directly into the hypothalamus. For example, when serotonin is perfused on the hypothalamus of cat, dog, monkey, and chicken, the body temperature rises. When serotonin is perfused in the same fashion on the hypothalamus of rabbit, hamster, rat, and sheep, the body temperature drops. What we *can* say about this important substance is that it influences temperature regulation in one way or the other.

Another aspect of the fine adjustment by the hypothalamus is the influence of the two ions, Na^+ and Ca^{++}. One theory is that the posterior hypothalamus is the set-point regulator; when the body is too cold, Na^+ ions cause a rise in set point; when it is too hot, Ca^{++} ions lower the set point (Kobayashi and Takahashi 1993).

In summary, we must admit that, while we do understand some of its component parts, the exact design of the physiological thermostat is elusive.

EVOLUTION OF TEMPERATURE REGULATION I

Environmental physiologists have an instinctive interest in the biologic history of the past. One area of interest is called "Evolution Observed," (see Chapter 1). Does the genotype of domestic mice bred for several generations in a coldroom change? There is a suggestion that it does (see Fig. 4-38). At this point we ask—how did temperature regulation develop in the first place? This leads us into paleontology. There is a renewed interest in this discipline today. It has even been suggested to try to take DNA from a frozen mammoth and to introduce this into an elephant ovum. Perhaps by this method one may obtain a living woolly mammoth. (There should be abundant material since the maximum depth of permafrost, for example, near Point Barrow, is 1,330 feet. There are few temperatures recorded, but Brewer (1958) recorded a -10.6°C below the depth of measureable seasonal fluctuation (70 to 100 feet).)

Earlier we stated that the lower vertebrates can be called ectotherms and the birds and mammals, endotherms. Clearly one had to evolve from the other. Most reptiles have a body temperature just above that of the environment unless they have been exercising or basking. An exception is that it has been shown that warm blood can course through tuna in cold water (Fig. 4-34); large reptiles such as the alligator regulate by basking in the sun or cooling off in the shade or water. This is very different from the feedback system of the mammal, which is partially based upon the hypothalamus.

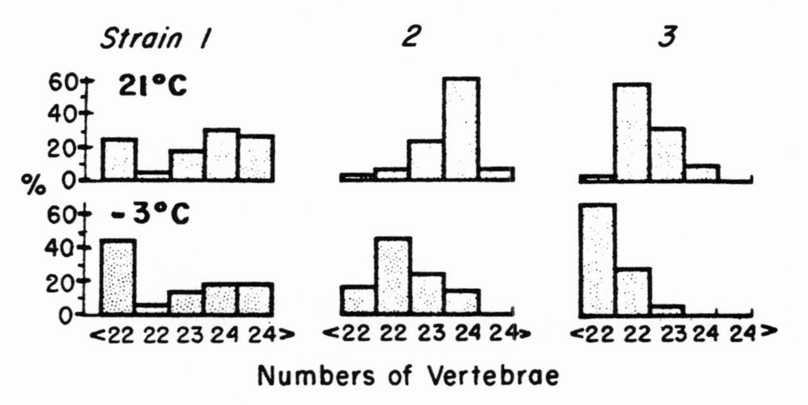

Figure 4-38. Mice Bred in a Cold Environment Have Shorter Tails. The tails of three inbred strains of mice born at -3°C had fewer vertebrae than the controls. For example, in strain 3, at T_A 21°C, 55% of the mice had 22 vertebrae, while at T_A-3°C, 62% had *fewer* than 22 vertebrae. Adapted from Barnett 1965.

Is the energy balance of a reptile as efficient as that of a mammal? In a study of the white-throated monitor lizard in Namibia's National Park, Phillips (1995) realized that large reptiles require far fewer calories to maintain their ectothermic lifestyle, and are therefore better suited to arid environments than are warm-blooded mammals of the same size. He reasoned that as a group, carnivorous reptiles in an impoverished environment should outweigh, in pounds per acre, their mammalian counterparts. He calculated the biomass of adult and sub-adult monitors in the National Park to be 100,000 pounds; this was greater than his calculation of the Park's lions, cheetahs, jackals, hyenas, and leopards. The annual cycle for ectothermic monitors includes four months of a hot, wet season during which the reptiles engorge themselves with hearty meals. This period is followed by eight months of fast. Some mammals also show this remarkable ability to fast for long periods (Folk, Kaduce, and Spector 1991; Folk 1980).

In order for temperature regulation to evolve, some fish-like amphibians had to develop means of walking on land, perhaps at first because of lakes which dried up each year. This probably took place in Devonian times 350 million years ago (see Table 4-13). The family tree (Fig. 4-39) indicates the stages of this large change; a group of fishes within the crossopterygians, namely the coelacanths, developed bones similar to the radius, humerus, ulna, and phalanges of humans and other mammals. This permitted their descendants to develop walking feet. Researchers at the University of Geneva in Switzerland propose that toes and feet were developed before, not after, these animals actually climbed onto land. This is because they discovered that genes associated with the formation of fins in fish are the same ones that orchestrate the development of paws in mice. They suggest that these genes are the well-known homoeotic homeobox genes (or Hox genes for short). About 38 of these genes

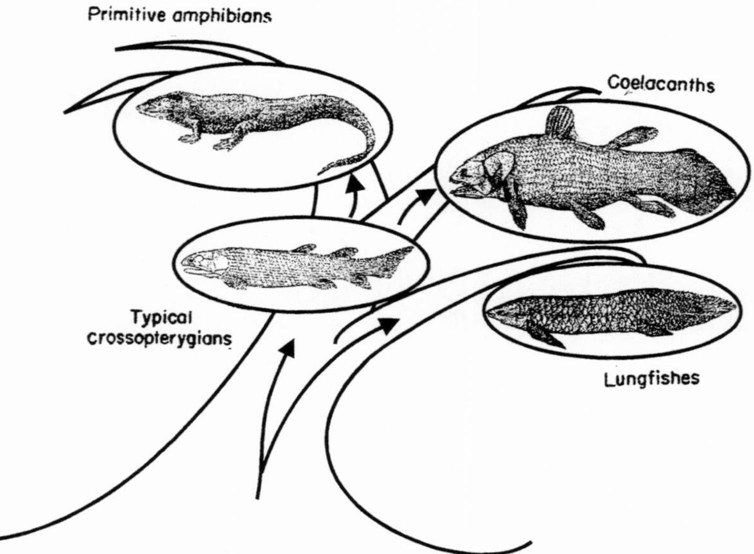

Figure 4-39. The Evolution of Temperature Regulation. Before the temperature-regulating hypothalamus could evolve, the above stages of becoming a land-living animal had to occur. The coelacanth limbs show the beginning of a human-like radius, ulna, and carpals. These fish were presumed to be extinct over 70 million years ago during the Cretaceous period. Recently, however, some living specimens have been found. Adapted from Romer (1977).

seem to be responsible for much of the embryonic development. Possibly these genes were associated with the progression of fish fins to amphibian feet, and then on to higher vertebrates. Duboule of the Geneva team suggests that the Hox genes may switch off early or late. He speculates that, if the genes can be tricked into staying on longer, the fins of their experimental fish might sprout appendages suggestive of primitive feet.

This ancient hand-paw gene is not the only such finding. The "master control gene" for eyes in the mouse can trigger eye formation in flies. After half a billion years of evolution, the eyes of flies and mammals rely on the same master control gene (Culotta 1995a). The principle is that a gene may represent a fundamental and evolutionarily preserved controller; researchers have found another such homeobox gene which identifies the region for the caudal embryonic development, in each case of unrelated animals: the nematode *C. elegans, Drosophila,* and the mouse (Erwin et al. 1997).

This genetic material is not static; the sequence DNA-to-RNA-to-protein is basic. But there are other repetitive elements in the evolutionary process. For example, only about 10% to 20% of human DNA encodes functional sequences. The remaining DNA is called Alu, and is a source of evolutionary genetic variation (Novick 1996). There was unlimited time during the Permian and Triassic

Table 4-13. Geologic Periods

Era	Period	Estimated Duration (in Millions of Years)	Estimated Time Since Beginning (in Millions of Years)
Cenozoic	Quaternary	1	1
	Tertiary	69	70
Mesozoic	Cretaceous	50	120
	Jurassic	35	155
	Triassic	35	190
Paleozoic	Permian	25	215
	Carboniferous	85	300
	Mississippian		
	Pennsylvanian		
	Devonian	50	350
	Silurian	40	390
	Ordovician	90	480
	Cambrian	70	550

periods (see Table 4-13) and unlimited numbers of *trial* vertebrate species as *test* animals for the Alu DNA to convert reptile ancestors to mammalian stock (Ferguson 1983).

If we look in the paleontological record covering the groups from low reptiles to mammals, the fossilized hypothalamus cannot be found. (Actually, some fossilized brains have been described because of findings of well-preserved brain cavities, especially in the Ostracoderms.) However, the overall skeletal design of a mammal as evolved from a reptile has been painstakingly worked out. The following changes must be found (Fig. 4-40):

1. The elbows-out locomotory progression of reptiles is changed to the elbows-in design of most mammals.
2. Reptiles do not have a bony roof just above the tongue in the mouth; this so-called false palate is found in mammals.
3. The skull moves on the first cervical vertebra on a smooth projection of the skull called the occipital condyle. In reptiles this is like a ball and socket joint. There is one condyle. Mammals, however, have two.
4. Reptiles have very similar teeth with occasionally two canine teeth different from the others. However, mammals have a full complement called "polydont" teeth including a variety of shapes and sizes.

Let us first discuss the elbows-out progression which is so easily seen in the large alligator or crocodile. This is not necessarily a handicap, as these reptiles can still run fast enough to catch a human (see sidebar). They can also cover much ground: the Namibian monitor lizards mentioned earlier patrolled a home range of about ten square miles, despite "elbows out."

The false palate seen in the mammal, but not in the reptile, serves to protect the nostrils. On the skull of an alligator or lizard, for example, you can pass a probe directly from the two nostrils into the front of the mouth. These changes from reptile to mammal have all been documented in the Karoo beds of South Africa by Broom: some of the mammal-like reptiles had a partial palate, and others show the complete roof of the mouth protected (Hotton et al. 1987).

The occipital condyles of the reptile, similar to a ball-and-socket joint, can be found intact in some mammal-like reptiles, in other species it is gradually dividing in two, until the mammalian form is achieved.

The evolution of polydont teeth means the gradual acquisition of incisors, canines, pre-molars and molar teeth.

It is important to note that all of these four changes were by no means found within the same species. Essentially, in one part of the Karoo beds, there would be a development of one of these characteristics, and in another part of the beds the development of a different characteristic. It was not that all four characteristics developed slowly, together, and simultaneously. However, in the most recent beds, of course, all four characteristics were found together and this meant that mammals had evolved. It may be that some skulls may be found which at least indicate the presence of the pituitary if not the hypothalamic region.

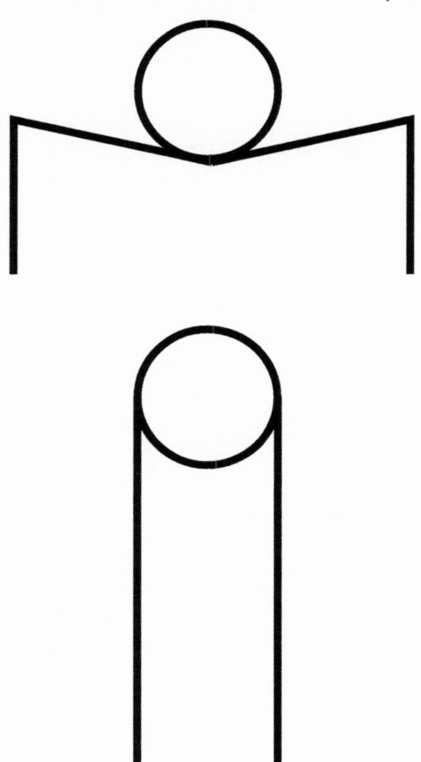

Figure 4-40. Reptilian Leg Articulation Compared with That of Mammals. Reptiles that have legs move with elbows out. Mammals evolved an elbows-in position, close to the body.

CROCODILE NOT SLOWED BY ELBOWS-OUT LOCOMOTION

While traveling in Africa along the Nile, the famous big-game expert, Temple-Perkins, observed the following incident:

A crocodile was chasing a man who was in a small boat, paddling furiously to get to a long sandbar. The man reached the sandbar and ran down its length with the crocodile in close pursuit. The man turned at the end of the sandbar and ran back to his boat, but was captured by the crocodile, dragged into the water, and consumed.

In addition to illustrating that the elbows-out anatomy of the crocodile (and lizards) does not hamper their movement, this episode illustrates something about the lung capacity of a crocodile.

What about reproduction? It is not always realized that many of the reptiles retain the developing fetus within the uterus and give birth to live young. Over this evolutionary path the mammalian reproductive system developed, along with the coat of hair that protects mammals, and the mammary glands that provide milk. Two of the mammals that have the characteristics just mentioned also lay eggs, namely the echidna and duckbill platypus.

The temperature regulation of several groups of mammals (whales, etc.) became adapted to water submergence; they evolved from quadrupedal terrestrial ancestors. Efficiency required giving up paddling (33% efficient) for lift-based swimming (80% efficient) The transition required only slight modification of terrestrial neuromotor patterns (Fish 1995).

The evolution of reptile to bird is equally interesting: birds acquired maternal behavior and care of eggs from reptiles. We know today that alligators build and protect a nest, and crocodiles use their great mouths to gently open each shell for emerging babies. A clue to the origin of this behavior is evidence that the dinosaur, oviraptor, brooded its eggs and wrapped its forelimbs around the nest, 80 million years ago.

The topic, Evolution of Temperature Regulation, is discussed further in Chapter 7, *Responses to Hot Environments.*

ECOGEOGRAPHIC RULES

Three classic, but controversial, rules, attributed to Golger, Bergmann, and Allen, are described below. These generalities should not be thought of as *laws,* which would mean that they were invariably true; they do apply, however, in many of the cases which have been studied. They apply especially to races or subspecies, but also to species in the same genera. These rules are abstracted below:

1. Golger's: in mammals and birds, races in warm regions tend to be dark brown in color, and races in a cold climate tend to be light in color or white.
2. Bergmann's: races of mammals and birds in a cool climate tend to be larger than races of the same species in warmer climates.
3. Allen's: protruding body-parts, such as tails, ears, and bills of mammals and birds of the same species are relatively shorter in the cooler portion of their geographic range than in the warmer portion.

Applications of Golger's rule are presented in Chapter 2. Is the usefulness of white coloration associated only with camouflage in ice and snow? Some workers suggest that in addition, the white fur permits less heat to be radiated to and from the environment. On the other hand, evidence collected by Hammel (1956) and Hock (1965) suggests that the opposite is the case (more heat is radiated through white fur). Do feathers, perhaps, give different results from fur? One author studied white feathers on birds; he found that they prevented a substantial amount of radiative heat gain and therefore must result in body heat retention (Lustick 1969).

Bergmann's rule came under attack by Scholander (1956), especially because the biometrical measurements which support the rule did not appear to him to be useful for appraising the thermal adjustments of the animal. Mayr (1956) took an opposite view, stating: "In a cool climate there should be a selective advantage in the relative reduction of surface resulting from increased size, since the metabolic rate is more nearly proportional to body surface than to body weight." Newman (1966) also reviewed the evidence in favor of this rule, lending weight especially to the classic data on the Puma provided by Young and Goldman (1946). The interested student is urged to read Schreider's (1968) article, "Ecological rules, body-heat regulations and human evolution," and along with it the comments by Bresler, the editor of the collection of papers. In respect to Bergmann's rule, McNab (1971) took a view that most latitudinally widespread mammals in North America simply do not vary in size with a dependence upon the physics of heat exchange. He stated that those who do are usually carnivores, and in their case, change in body size may reflect a change in the size of their prey.

Other single species have resonated with climatic change, some over the past 25,000 years. The woodrat (*Neotoma cinerea*) is a good example. Wood rats adapted to the varying environment by phenotypic plasticity and evolutionary change. Confirming Bergmann's rule, they are smaller in warmer regions (Smith et al. 1995).

Less attention has been given to Allen's rule, which also came under the criticism of Scholander's puckish pen. He was disturbed by the supposed selective value of a species of bird acquiring smaller bills and tails in the cooler parts of the geographic range. However, Barnett (1965; Monro and Barnett 1969) demonstrated experimentally that mice raised in cold rooms have shorter tails; Barnett considered his results as evidence for the support of Allen's rule (Fig. 4-38). Further support for Allen's rule was provided by Allee (1949). There is a systematic incremental decrease in the size of ears of the genus *Lepus* from south to north, specifically in four species: the Arizona jack rabbit, the Oregon jack rabbit, the varying hare, and the Arctic hare. The ears of the first are about twice the length of those of the last.

SPECIAL ADAPTATIONS

Temperature regulatory adaptations are sometimes difficult to recognize and interpret. An illustration is found in the horns of the *Bovidae* (sheep, goats, cattle). The difficulty here is because these structures serve several functions, including defense, possession of territory, and even food-getting. We believe it is justifiable to describe the spectacular horns of an ibex or a caribou as "monuments to calcium metabolism." These massive accumulations of calcium arise from a small button on the skull. (These caribou horns, possibly the most spectacular in the animal kingdom, are actually seldom used against another male caribou, but merely *shown* to it.) In 1966, Taylor added brilliantly to the functions of horns the phenomenon of temperature regulation. We recall helping to lift a buck goat into a truck; the horns felt to the touch as if the temperature were above 38°C (100.5°F), but we did not recognize the significance of this at the time.

The horns of goat-like animals may have the following temperature relationship to the environment:

T_A	At Tip of Horn	At Base of Horn
0°C	1°C	10°C
5°	6°	25°
23°	32°	34°
33°	34°	35°

These horns show broad swings of vasomotor function; they vasodilate in response to heat, exercise, and a blocking of their nerves, and they vasoconstrict when the animal is placed in the cold. A goat usually loses about 2% of its total heat production through the horn (if T_A 22°C). Maximum heat loss from the horn, about 3%, occurs at about 30°C with a low wind, or, with vasodilation as a result of running at 0°C; the loss is 3% while running and 12% after stopping, until horns vasoconstrict. In large desert species there appears to be a trend toward larger horns in both sexes, hence greater heat loss from horns (Finch et al. 1980).

Another special adaptation is found in the arteriovenous anastomoses in the skin of the Weddell seal. These structures are usually infrequent and localized, as in the fingers of humans. One structure for temperature regulation in seals, the sweat glands, are found only on the flipper. In Weddell seals the arteriovenous anastomoses were found just beneath the epidermis, among the hair follicles, and surprisingly, over the entire body surface. This system is clearly a heat dissipating structure when the animal is out of the water (Molyueux and Bryden 1975).

In the next chapter, we will explore some of the coldest regions on earth as we consider acclimatization and adaptations to cold environments.

5. RESPONSES TO COLD ENVIRONMENTS

THE COLD ENVIRONMENT

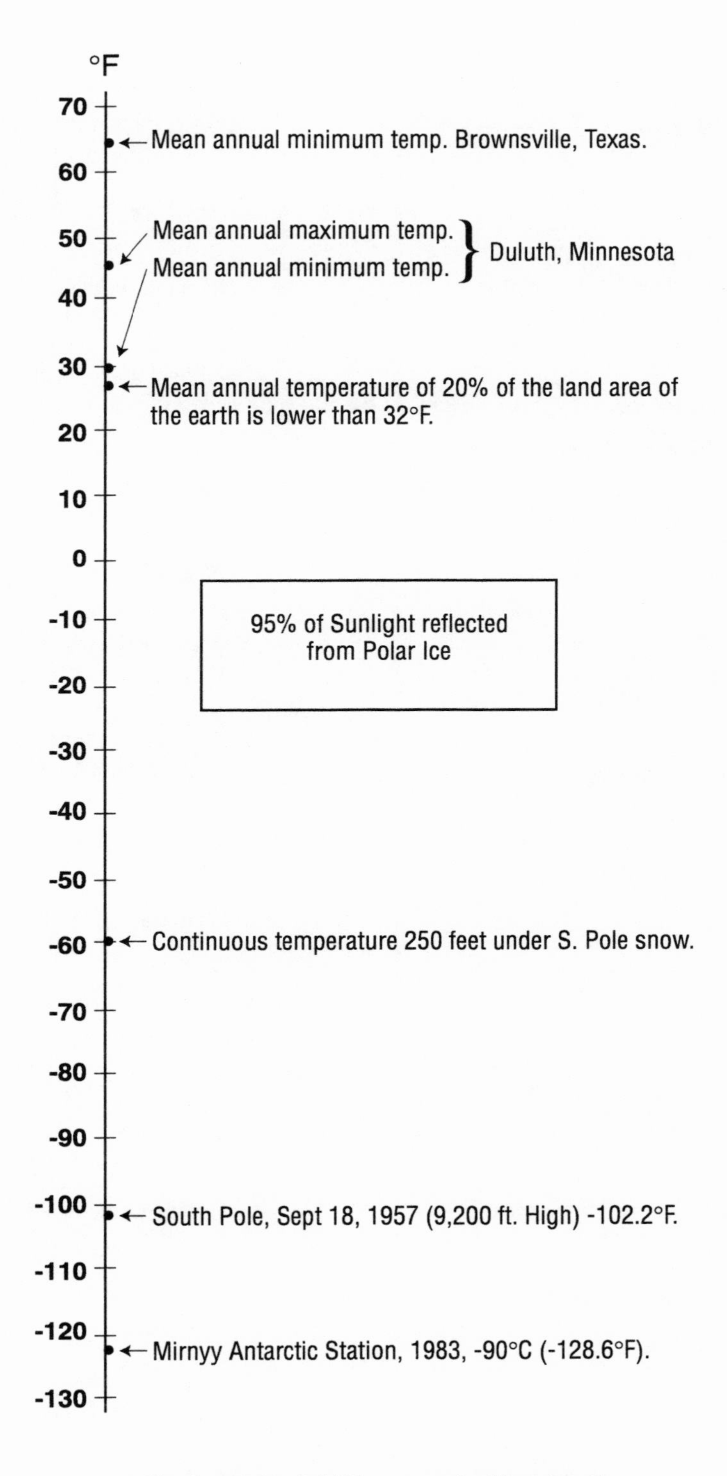

Environmental Diagram A: Chapter 5

COLD INDUCED CORE TEMPERATURES

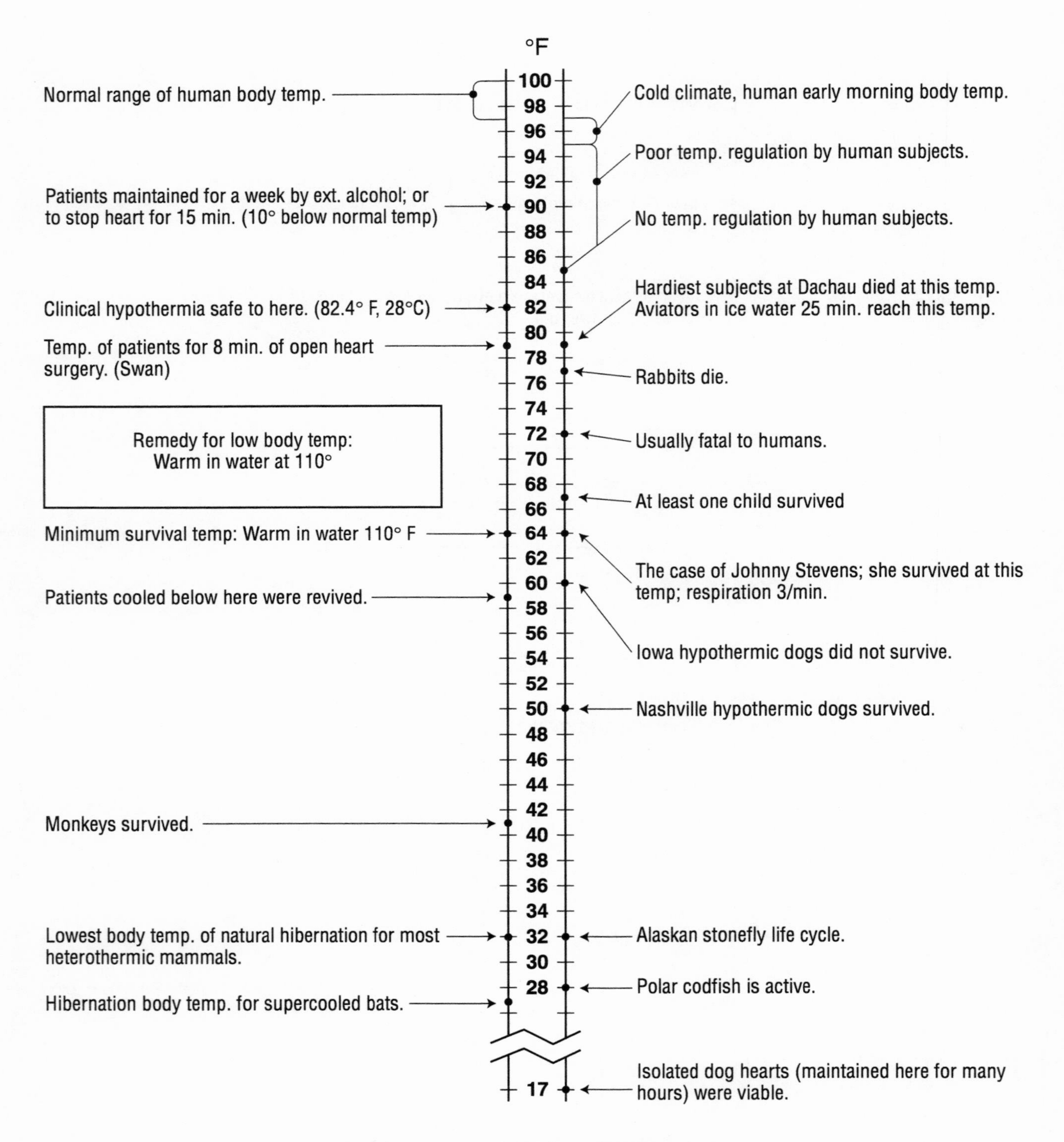

Environmental Diagram B: Chapter 5

5. Responses to Cold Environments

5. RESPONSES TO COLD ENVIRONMENTS

INTRODUCTION: THE IMPACT OF EXTREME COLD

In this day of heated cars and homes, the average person in North America often forgets that his or her total physiological reserves for combating cold may suddenly be called upon. Despite the comfort and convenience-engineering of this age, each year from 600 to 1,000 Americans die of excessive cold with below-normal body temperatures (Doyle 1998). As the following examples show, life-threatening cold exposure can occur almost anywhere, at almost any time. In Wyoming one August, a university professor and his daughter were trapped in their car overnight in a heavy snowstorm; the car, on a main highway, was completely covered by drifts. On the Maine turnpike one winter, over 1,000 people were stranded for 24 hours when a sudden blizzard made it impossible for cars to move; there was a great deal of bitter discomfort, but no lives were lost. A similar situation occurred in 1996 along the East Coast when a series of storms left record snowfalls that brought commerce and government to a halt (Fig. 5-1). One winter in Anchorage, Alaska, 500 cases of severe frostbite were reported by one doctor alone. At the Cleveland General Hospital in Ohio—which can in no sense be considered an extremely cold part of the United States—admissions for cold injury averaged 7 per year. Even in the mild winters of Southern California, unprepared or homeless individuals can find themselves at risk for cold exposure and death. And in Northern Bangladesh in

Figure 5-1. Blizzard Paralyzes East Coast.

January 8, 1996

Blizzard paralyzes East Coast

(AP) A blizzard of historic proportions shut down the East at the start of the workweek Monday, stopping cars, trains, planes and just about anything else that moves. At least 46 deaths were blamed on the weather.

Historically, the city's worst blizzard was the day after Chistmas in 1947, when 26.4 inches fell. But there was little wind, unlike the blizzard of March 12, 1888 when 21 inches got heaped into drifts that reached second-story windows. The 1888 storm killed more than 300 people on the East Coast.

December, 1995, the hostile impact of cold was experienced when fifty people died due to high Himalayan winds and air temperatures of 4.4°C (40°F).

The environment of Fairbanks, Alaska in the winter of 1964-1965 illustrates the barrier that weather can present to the travel of the wage earners of this shopping center for 60,000 people: for 17 consecutive days the official *maximum* temperature was -40°C (-40°F). Temperatures of -51°C (-60°F) were common. A typical case of injury was a woman who stalled her car and walked for 15 minutes with her legs covered only with nylon stockings; as a result she spent a month in the hospital with frostbitten legs. During the winter of 1993, the outstanding dog-team driver, Kate Persons, reported from Nome, Alaska, that there were "many weeks below -40[°F], and many days at -62[°F]."

Cold weather is not a private concession for Alaska; -56°C (-70°F) has been recorded in Montana. Such temperatures are not just a characteristic of old-fashioned winters; recently, a temperature of -61°C (-79°F) was read at Prospect Creek on the Jim River east of Bettles, in Alaska. How do you transport fresh vegetables, not to mention a baby, in weather like that?

Extreme temperatures often must be recorded by human observers. The maintenance of polar winter stations requires that some people be exposed to the most severe circumstances. Extreme temperatures of -75°C (-102°F) were common at the Amundson-Scott South Pole Station, where seventeen times between May and September the South Pole surface temperatures were lower than -70°C (-95°F). A new record was set in 1983 at the Russian Antarctic station, Mirnyy, probably the coldest spot on the globe; they recorded -90°C (-128.6°F). Temperatures as low as this are unknown in the Northern Hemisphere, but large populations nevertheless do find a severe challenge to their survival. The coldest regions are usually almost desert-like (Table 5-1); an exception is Fairbanks, Alaska.

Among lower mammals, a few subarctic animals face higher physiological challenge than most in tolerating prolonged extreme

Table 5-1. Climatic Data at Various Arctic and U.S. Weather Stations

Weather Station	Years of Data	January Means		Minimum Temp. (F)	Frost Free Days
		Temp (F)	Precip. (in)		
Godthaab, Greenland	44	+14°	3.27	-20°	—
Verkhoyanks, USSR	38	-58°	0.16	-90°	—
Barrow, Alaska	25	-17°	0.15	-56°	17
Churchill, Canada	12	-19°	0.48	-57°	52
Fairbanks, Alaska	34	-12°	0.97	-66°	89
Langdon, N. Dakota	35	-1°	0.63	-51°	110
Adak, Aleutians, AK	12	+33°	6.53	+ 8°	165
Chicago, Illinois	40	+25°	1.75	-23°	196

cold. Many mammals (including foxes and wolves) can seek protection under the snow or in dens. This is not possible for some winter residents such as moose, caribou, Canada jays, and chickadees, which do not have enclosed shelter. The temperature regulation of herbivores standing (and sometimes marooned) in deep snow at -51°C (-60°F) requires unusual adaptations; with a core temperature fixed at 100°F, these animals must find enough food to hold this temperature when the environmental temperature is as much as 150°F *lower* than their core temperature. As a sample of snow depth, in a recent year, Riding Mountain National Park, Manitoba, had a maximum winter depth of new snow of 52 cm (20.5 in), and of accumulated snow of 58 cm (23 in). The numbers of moose studied were over 2,000 in some individual years. Around Fairbanks (record low at Tanana -60°C, -76°F), and in the bitter, windy cold of Alaska's Richardson Highway area, it is estimated there are 130,000 caribou and as many moose. Many of the caribou migrate on to Yukon Territory, where -56°C (-70°F) is not uncommon, and where North America's lowest temperature, -63°C (-81°F), was recorded. There may be animals living in even more extreme cold in Siberia, where there were two official temperatures of -68°C (-90°F), and one unofficial one of -79°C (-108°F).

Even more remarkable is the existence of small birds at these temperatures. Both the mammals and the birds have a short daylight period for feeding during the cold season (3 hours and 40 minutes of sunlight on December 21 at Fairbanks), so even the collecting of calories is curtailed. During the severe winter mentioned earlier, Anna Larson, a biochemist at Arctic Aeromedical Laboratories, observed a pair of chickadees coming regularly to a feeding station to peck at ice-hard peanut butter when the temperature was -46°C (-50°F). On a day when the thermometer was near -51°C (-60°F), only one appeared. She noted, "It did not fly well, was very sluggish, and had frost on its feathers." The next day at -46°C (-50°F), the same pair of birds was back. This species is handicapped by a large surface area-to-mass ratio that must require special cold-climate adaptations as yet undescribed. These small birds must seek overhead cover as a radiation screen because the "ceiling" temperature of open winter sky may be as low as -62°C (-80°F).

Bird populations in Alaska experience extremes each year. What is the effect of sudden Alaska-like conditions on birds in a temperate climate? In 1996, an unusual storm covered temperate-zone Iowa, first rain and sleet, followed by two days of snow that totaled 15 to 18 inches, with four days of winds up to 60 mph. The temperature varied for three days, from -36°F recorded at New Mellaray Abbey (1996), to -41°F in Iowa City, to -47°F in other parts of Iowa, tying the record set in the state in 1912. In one city, there were 500 vehicles in ditches or covered by drifting snow. One ornithologist maintained 9 bird feeders, and observed that instead of having more birds needing food, the numbers of birds diminished by about 50%; this figure remained for weeks after the storm. Eleven species were involved, including chickadees. New evidence shows that chickadees can go into dormancy, dropping their body temperature by

10°C (Dawson et al. 1983). It is not known whether the missing birds died, migrated, or went into dormancy.

THE COLDEST REGIONS ON EARTH

A large portion of the earth has a cold climate. The 10°C (50°F) isotherm defines a region where the average monthly mean temperature, even in the warmest season, never rises above 10°C (50°F). (An isotherm is a line drawn on a map connecting points of equal temperature). In winter this delimited region experiences temperatures of -51°C (-60°F). To describe this area differently, it is claimed that 20% of the land area of the earth (50% of Canada and the Soviet Union) is located in zones where the mean annual temperature is below -1°C (30°F). This isotherm, then, includes 1/5 of the globe. In this region, the underlying soil remains perennially frozen (permafrost), while a thin surface layer (active zone) temporarily thaws during the summer.

A definition of the Arctic region in terms of the 10°C (50°F) isotherm, and a definition of the subarctic in similar terms is to be found in Irving's *The Handbook of Physiology—Environment* (1964), which also includes a map describing the southern limits of permafrost.

Now let us compare Arctic and Antarctic latitudes. It is relatively easy to get to the northern edge of Alaska at Point Barrow, where there is a large village and long-standing scientific laboratory. From Point Barrow, the North Pole is about 1100 miles north over the polar ice and water, and the Pole area is routinely crossed by commercial airplanes. Air flights to Barrow in the winter tend to be on a more accurate schedule than those in the summer; summers are apt to be foggy and unpredictable. However, on the Antarctic continent—an area of about 6 million square miles—the opposite situation exists. Although many countries now have permanent stations near there (Fig. 5-2), the weather is unusually stormy all year, and this weather extends to neighboring continents. Sailors call these latitudes "the roaring '40s," "the furious '50s," and "the terrible '60s." Scheduled flights from Capetown, S. Africa to New Zealand, instead of flying straight across Antarctica (7,100 mi), always circle north and around it (12,000 mi). The first flight between Africa and Antarctica was on September 30, 1963, and was an historic event. The pilots, Lt. Commanders Richard G. Dickerson and William B. Kurlack, flew 4,700 mi from Capetown to McMurdo Sound (Reedy 1964).

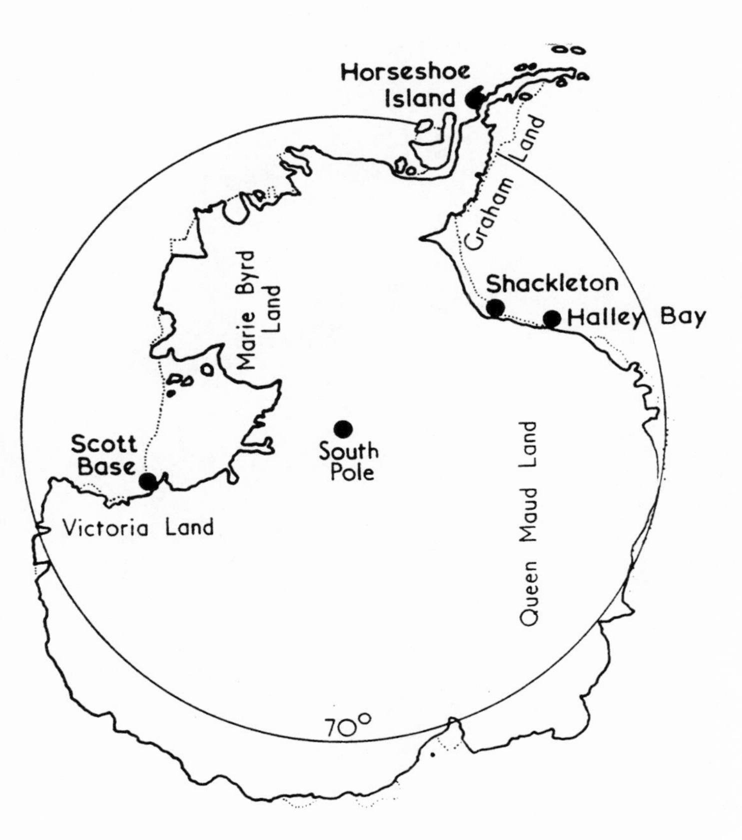

Figure 5-2. The Antarctic Continent. This continent is reached only after a long airplane flight over water. The map indicates some of the many bases where physiological observations have been made on humans and other vertebrates. The faint dotted line represents ice shelves. Adapted from Edholm 1961.

Cold-Weather Native People

Note in Fig. 5-2 that the continental mass of Antarctica is surrounded by S. latitude 70°. Point Barrow, the northernmost point on the North American continent, is on approximately the 70th N. latitude. On the other hand, most of Greenland, the Baffin Bay archipelago, and the town of Thule, Greenland, lie between 70° and 80° N. latitude. Thus, in the north latitudes, there are many indigenous people and other land mammals who show a truly remarkable tolerance to cold (Fig. 5-3). There are about 208,000 cold-weather native people, who subsist on a diet that emphasizes fish, sea mammals, and caribou (see Appendix B; also see sidebar next page).

American field scientists have studied with fascination the means of survival used by the Inuit of Alaska and the Laplanders. Now, in the 1990s, we look for similar knowledge from primitive people in the Russian Far East. For example, photographs taken in 1992 show skin-covered tents still being used by the Chukchi in extreme northeastern Siberia, but in 1960, Folk photographed the last of the skin tents and sod huts in use at Anaktuvuk Pass, Alaska. Thus, we see that the age of plywood and canvas reached the Inuit earlier than it reached the Chukchi.

Studies in 1994 revealed the lifestyle of thousands of Nenet reindeer herders: they eat raw fish, drink reindeer blood, and live in reindeer-skin tents (Specter 1996). These people use the same homemade equipment used over 1,000 years ago, yet in winter they must tolerate -60°F.

The Arctic explorer Stefansson (1964) has said that if you extend a line at the latitude of Edmonton through the Russian Far East, there would be 10,000 people living north of the line in America, while there would be "10 million in Russia." He described the Eastern Russians as an "arctic-minded" and "Arctic-inhabiting" people. We have much to learn about their techniques of tolerating long periods with an air temperature below -40°C.

Figure 5-3. Inuit Wearing Fur Out and Fur In. All are wearing traditional labrets in their lips. The man in the middle front is wearing a parka made from seal intestines. (Photo by E. Nelson, 1881. E.W. Nelson Collection, National Anthropological Archives, Smithsonian Institution.)

Explorers of the Poles

The people who first explored and settled these cold regions were vigorous athletes who endured prolonged discomfort from cold and bitter wind. The tragic story of Scott and his fatal trip to the South Pole is well known. Less well known is the remarkable individual who traveled with him, Edward Wilson, who had previously made a journey from Scott's base to a distant cape where the emperor penguins nested. Wilson's journey to obtain penguin eggs for embryological studies has been called "the worst journey in the world." Typical of many polar explorers, Wilson went on these expeditions as doctor, dog-team driver, scientist, and ornithologist. His art work is well displayed in the book *Edward Wilson's Birds of the Antarctic* (Roberts 1967). A later explorer accomplishing a no less strenuous and dangerous job was Sir Edmund Hillary, who traveled with his team from Scott Base to the South Pole. Also outstanding is Vivian Fuchs who, with a team of ten men, made the 2,200 mile land crossing of Antarctica from Shackleton Base to Scott Base. The latter two expeditions used motorized track vehicles. Another courageous individual is Ann Bancroft, pictured in the frontispiece of this book. In 1993, Bancroft led an all-women's expedition overland to the South Pole; the women skied more than 660 miles, pulling all of their own equipment with no airplane supply.

In the Arctic, the early counterpart to Wilson was Frederick Schwatka, who was a cavalry officer in the American West before distinguishing himself in the Arctic. In 1876, Schwatka obtained his medical degree in New York. As a military surgeon, he served in many parts of Arctic North America; in the bitter winter of 1880, the worst in the memory of the oldest native, he arrived at Hudson Bay with a party of 18 Inuit after searching for the Franklin scientific records in the North. In 1886 he was still making mapping expeditions to Alaska; he mapped Sherman Inlet, Hayes River, and King William Island, covering 3,250 miles by dog team. In later years he wrote at length on topics concerned with environmental physiology (Johnson 1968).

A similar individual, in the sense of being a brilliant leader and a

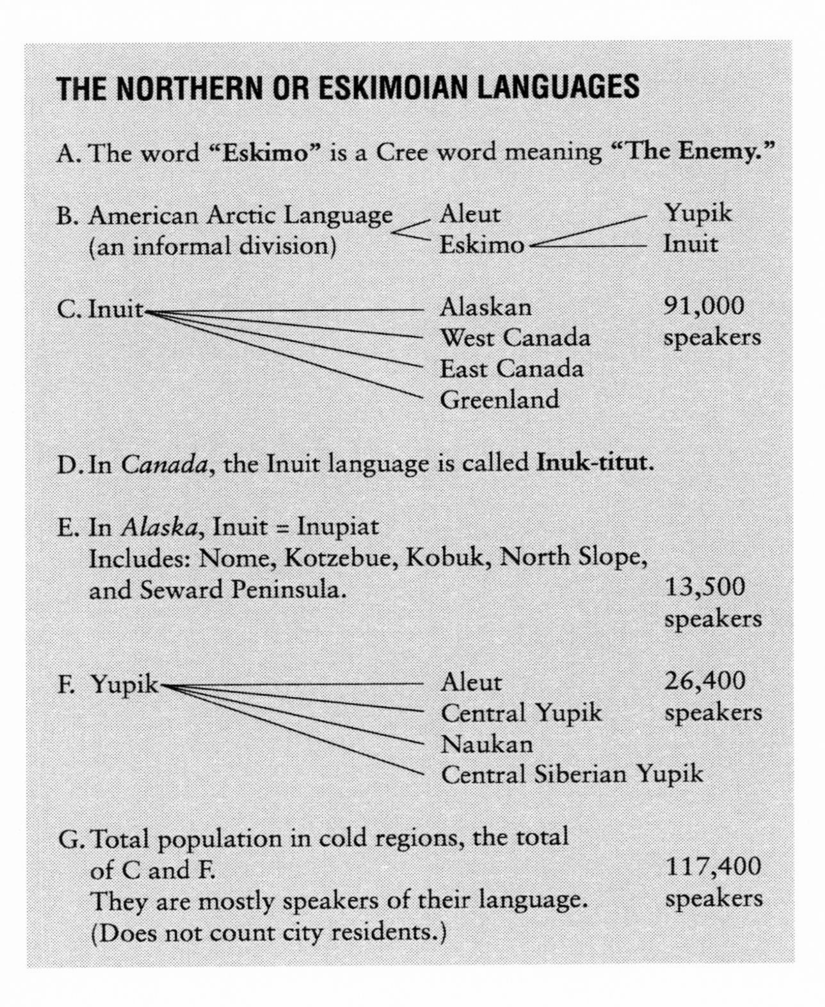

vigorous athlete, was Admiral Robert E. Peary, who was the first to reach the North Pole in 1909 (Weems 1967). His first trip to the North was in 1886, and he returned to the Arctic most years after that. Peary crossed the Greenland Ice Cap several times in preparation for the trip to the Pole. These Greenland expeditions were remarkable accomplishments; he would often travel 40 miles per day by dog team in a 10-hour period. Most people do not realize that the driver does not ride; he or she runs and guides the sled.

A later polar scientist, Vilhjalmur Stefansson (1879-1962), found and named unknown Arctic islands, lived with the Inuit for about 5 years, discovered the essential fatty acids, and wrote 24 books and 400 articles about the far north and its peoples. His widow, Evelyn Stefansson Neff (1996) continues to write and publish extensively about Alaska.

There have been many other overland team expeditions to the South and North Poles between 1980 and 1995. Team expeditions are probably the most valuable because members can be assigned diverse jobs such as recording diets and navigation. For example, one member of Bancroft's South Pole expedition was a dietitian who recorded caloric consumption. Will Steger and his team in 1989 crossed from the North Coast to the South Pole and to Mirnyy, with three sleds and 40 dogs. His meteorological observations were especially valuable.

Then there is another group of athletes, those who, with blistered lips and iron will, want to test their fitness and endurance alone. These solo expeditions have included Hargraves, age 34, mother of two, to the summit of Everest in early 1995 (she died later in August 1995 on K2); Etienne, North Pole (NP) 1986; Arnesen, woman skier, South Pole (SP) 1994; Ousland (NP) 1994; Kaminski (NP) 1995, (SP) 1995; Malakhov (NP) 1995; Mear (SP) 1995. The later six explorers dragged sleds weighing between 265 to 305 pounds.

Figure 5-4. Admiral Robert Edwin Peary. While wearing heavy arctic clothing of this sort, this explorer drove dog teams to reach the North Pole in 1909. This photograph was taken aboard his ship, *The Roosevelt.* Courtesy of Library of Congress and Houghton Mifflin Company.

Note the typical garment worn by Peary and other explorers in the early days (Fig. 5-4). The heavy arctic clothing was a tremendous but necessary encumbrance for the task of covering large distances driving dog teams. In the late 1960's, Wally Herbert (1971) and a team of three other men drove dog teams over the ice pack from Barrow to Spitzbergen (February 1968 to April 1969). Surprisingly, these men wore clothing similar to that of Peary.

As we examine photographs of these explorers—or if we ourselves put on heavy arctic clothing—we are impressed by the weight of this added insulation. Yet, there is much more to the problem than the extra weight; in the Fatigue Laboratory in 1946, we noted that as the bulk of the arctic insulation was increased, there was a caloric expenditure much greater than could be accounted for by the weight of the clothing. We named this effect the "hobbling" effect. This added handicap increases our respect for the athletic achievements of the explorers of the polar regions who ran with their dog teams—or pulled the sleds themselves.

TYPES OF COLD EXPOSURE

You often hear physiologists say that their animals have been "exposed to cold." This expression is inadequate for two reasons: first, there are several types of "cold air"; and second, the cold-exposed animal does not necessarily experience a low temperature. The secondary factors that must be considered are the amount of moisture in the air, the amount of air movement, and the duration of the exposure to cold. A number of authors have stated that wet cold near 0°C (32°F) feels colder (is more unpleasant) than dry cold at more frigid temperatures. It has been difficult to make specific measurements to explain the supposed difference in response to wet cold. Part of the impression by human subjects that wet cold is more unpleasant than dry cold may be based upon sensation alone. Also it may be that the stratum corneum of the skin actually does conduct heat differently or does trap a different layer of air insulation at different degrees of humidity. Horvath studied six unclothed human subjects at 10°C (50°F) at relative humidity 50%, and at relative humidity 93%, in 12-hour exposures. Subjective comments and objective measurements failed to show a difference in reaction to the two humidity conditions. Horvath concluded that previously reported differences in humidity conditions must be a reflection of the influence of clothing in the experiments.

Describing conduction of heat in different wind velocities is more straightforward than explaining wet cold. Physical models allow us to show different amounts of heat transfer with different wind velocities. Figure 5-5 is a wind chill chart in which actual air temperatures can be converted to equivalent wind chill temperatures due to the effects of various wind velocities.

There is some misunderstanding about the use of wind chill figures. Some published tables state incorrectly that they represent a

Figure 5-5. Wind Chill Index. A chart used by the armed forces to make decisions about activities and clothing in very cold weather. Three zones of danger are presented to cover all the equivalent temperatures shown.

EST. WIND SPEED IN MPH	Actual Thermometer Temperature (°F)											
	50	40	30	20	10	0	-10	-20	-30	-40	-50	-60
	Equivalent Temperature (°F)											
Calm	50	40	30	20	10	0	-10	-20	-30	-40	-50	-60
5	48	37	27	16	6	-5	-15	-26	-36	-47	-57	-68
10	40	28	16	4	-9	-21	-33	-46	-58	-70	-83	-95
15	36	22	9	-5	-18	-36	-45	-58-	-72	-85	-99	-112
20	32	18	4	-10	-25	-39	-58	67	-82	-96	-110	-124
25	30	16	0	-15	-29	-44	-59	-74	-88	-104	-118	-133
30	28	13	-2	-18	-33	-48	-63	-79	-94	-109	-125	-140
35	27	11	-4	-20	-35	-49	-67	-82	-98	-113	-129	-145
40	26	10	-6	-21	-37	-53	-69	-85	-100	-116	-132	-148
(Speeds >40 mph have little additional effect)	LITTLE DANGER (for a properly clothed person)				INCREASING DANGER			GREAT DANGER				
					Danger from freezing of exposed flesh.							

person's sensation of cold. Of course, the wind chill figure represents the condition driving actual heat loss. The equivalent temperature is the actual *effective* temperature. For example, if the actual thermometer reading is 20°F (-6.7°C) and the wind is 25 mph, a person loses heat as if the dry-bulb were -15°F (-26.1°C). The chart in Fig. 5-5 also includes the hazardous extremes where exposed flesh freezes. Note that wind speeds of greater than 40 mph have little additional chilling effect.

The empirical formula upon which these wind chills have been based was worked out in 1945 by Paul Siple (the Eagle Scout who accompanied Richard E. Byrd on his second Antarctic expedition in 1933). The formula is

$$H = (\sqrt{100V} + 10.45 - V)(33 - T_A)$$

where H = the wind chill or rate of heat loss in kilocalories per square meter per hour, V= the wind speed in meters per second, and T_A = the air temperature in degrees centigrade.

The formula assumes that the wind chill represents heat loss from skin at 33°C (91.4°F). The calculated figures for wind chill begin with 50 which is considered hot, to 2500, which is intolerably cold. The examples listed in Table 5-2 were compiled by Gates (1972).

Experiments in Antarctica showed that all cases of frostbite occurred at wind chill factors between 1400 and 2100. At one location on the coast, the wind chill factor was over 1400 on more than half the days of the year.

Siple's formula for wind chill equivalent temperature is accepted, effective, and practical, but, of course, it has been criticized (Brauner and Shacham 1995). This is on the grounds that the calculation is for only the bare, damp skin of the face. Most of the heat loss is actually through the clothing. New experimental calculations try to take this into account.

Wind chill extracted a toll on Mt. Everest on May 10, 1996. A sudden storm quickly enveloped the summit; the snow was blown

Table 5-2. Wind Chill

Wind Chill Factor kcal $m^{-2}h^{-1}$	Stages of Relative Human Comfort
600	Very Cool
800	Cold
1000	Very Cold. Unpleasant for skiing on overcast days.
1200	Bitterly Cold. Unpleasant for skiing on sunny days.
1400	Freezing of exposed flesh. Skiing becomes disagreeable.
2000	Exposed flesh will freeze in one minute.
2300	Exposed flesh will freeze in less than one half-minute.

horizontally at 75 mph, there was a white out, and wind chill was -140°F. Eight climbers and guides died.

Communities in more moderate climates occasionally must tolerate hostile environmental extremes ordinarily expected only in the polar climates. In 1965, the city of Edmonton, Alberta experienced for over 21 days a temperature of -40°C (-40°F) with a 30 mile wind. This provided an equivalent temperature of -81°C (-109°F). Very few cities have been known to experience such conditions.

Now let's consider the *degree* or *intensity* of the exposure to cold. A number of qualitative terms have been used, but there is no agreement as to their meaning in terms of time and dry-bulb temperature. A frequent classification follows:

- *acute* (short and severe);
- *chronic* (long-term and usually mild);
- *moderate* duration;
- *long-term or seasonal* duration
- *multi-seasonal* duration

In much of the literature on the physiology of cold, the investigators have not indicated in which category they consider their particular cold exposure to fall. It will not be possible to consider examples in each of these categories, but we will use the terms from time to time.

The final problem concerned with describing cold exposure is that often the animal is in the cold but does not experience it. The early experiments attempting to show acclimatization to cold in humans gave negative results because the subjects had adequate insulation, and the skin cooling on face and extremities was inadequate to produce any lasting physiological effects. There is a tendency today to describe much of the exposure of people to cold in the Arctic and Antarctic as being chronic and moderate. Lewis and Masterton (1963) presented figures that indicate that the skin temperature (the microclimate) under their arctic clothing of men working on some of the expeditions was the same as that of men working in the temperate zones with light insulation. This is only one side of the picture, because it is certainly evident that many on such expeditions have had severe frostbite of the face and hands; Pugh points out that one can ordinarily recognize which people have been on Polar expeditions because of their permanently reddened faces. At any rate, it is constantly necessary to ask the question: Does the animal or human experience the cold in which it is exposed? For example, if the arctic fox (in winter coat) is exposed at -40°C (-40°F), it shows little increase in resting metabolism. Much colder temperatures must be used in order to introduce the experience of cold in this species.

Chronic and Acute Exposure

Are mammals in the free environment subjected to chronic or acute cold exposure? Ordinarily, climatic changes are gradual. Is there a precedent for the environmental physiologist who moves an animal from a warm control temperature directly into cold exposure of 4°C (40°F) or lower? Such experiments are easily justified; for example,

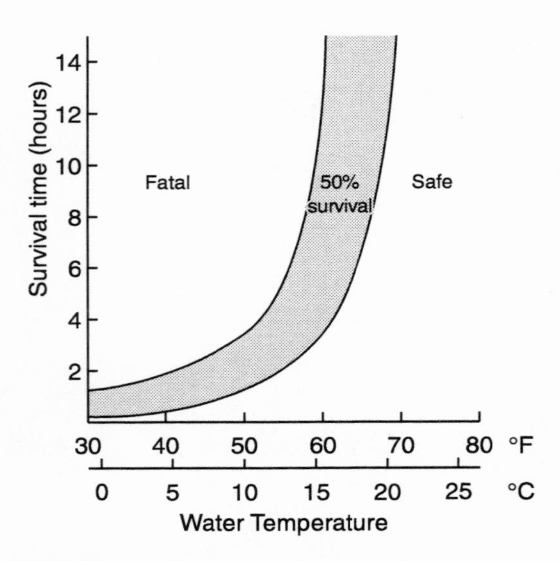

Figure 5-6. Predicted Survival in Cold Water. Adapted from Meurn 1993.

the Kalahari Desert in the winter is as hot as 71°C (160°F) in the daytime, but falls to the freezing point at night (Wyndham and Morrison 1958). In another example, living organisms (including humans) in Browning, Montana experienced a 24-hour temperature change from 7°C (44°F) to -49°C (-56°F). Rapid dry-bulb changes are also common, especially in the spring in the Rockies and in the Alps. On April 13, 1971, in Chicago, temperatures dropped in two hours from 29°C (85°F) to 9°C (50°F). At one point during this event, thermometers showed a 25° change in 2 minutes.

The degree of cold exposure that humans experience is strongly influenced by cultural habits. Certainly, many more British school children are acclimatized to cold in the winter than are children in the United States (Rudge 1996). A common winter bedroom temperature for British children of university-educated parents is 10°C (50°F), while this room temperature in the United States is nearer 20°C (68°F). In the winter in Britian, the office air temperature for civil servants is 10°C to 16°C (60°F).

Immersion in Cold Water

Death in cold water is only too common. During one year after World War II, 717 people died from accidental immersion in the coastal waters of England and Wales (suicides and deaths from injury or disease are not included). At that time, few people realized why these deaths occurred. When the *Titanic* struck an iceberg and sank at 0°C (32°F), 1,489 people were soon floating wearing life jackets in the water. Within 1 hour and 50 minutes of the sinking, the *Carpathia* was picking up people from the water, but everyone was dead. The report by the superintendent of the port of Southampton listed each of these deaths as drowning. Yet, of course, the main threat to life in water of this temperature is the cold itself. An unclothed man of average build will be helpless from hypothermia after approximately 20 to 30 minutes in water of 5°C (41°F), or after 1 1/2 to 2 hours in water at 15°C (60°F) (see Fig. 5-6). With thick, conventional clothing, these times are increased to approximately 40 to 60 minutes in water at 5°C and 4 to 5 hours in water at 15°C. Unusual body builds alter this situation; very fat people may survive almost indefinitely in water near 0°C (32°F) if they are warmly clothed. There is one case of a man who survived for 9 hours in water at -1°C (30°F). Probably these exceptional cases occurred not only because of differences in body build and in clothing, but also because of variation in exercise in the water. This is clearly shown in a table from the book by Keatinge (1969) (Table 5-3). The lowest fall in rectal temperature was 0.29°C when the water was gently stirred and the subject did not exercise and wore clothing. The next lowest drop was again when clothing was worn. When humans are studied without taking several factors into account, the person with the most subcutaneous fat has the best protection (Fig. 5-7). There must have been differences among Keatinge's subjects in heat production and in body size, yet this relationship held very nicely. Some of the scatter is explained by small individual differences in peripheral blood flow which was monitored by finger heat loss. It is apparent, considering all the tissues from core to skin, that this subcutaneous fat must provide the major part of the body's insulation.

It is evident that the cold-water sailor or fisherman should put on warm clothing before taking to the water, if possible adding a waterproof suit over the clothing. Unfortunately the entire picture is not covered by the explanation of hypothermia; there is a series of carefully studied, documented cases called "Sudden Death in Water." In these cases, athletic individuals who tried to swim in cold water died either immediately or within 4 or 5 minutes, much too rapidly for death from hypothermia. Many of these cases remain unexplained; some are recorded as death resulting from a reflex effect of sudden immersion. This matter is illustrated in Figure 5-44, and will be discussed later in more detail. Keatinge's excellent book presents these problems, and includes methods for treatment of people in immersion hypothermia.

Now we should turn our attention to physiological mechanisms for combating severe cold exposure.

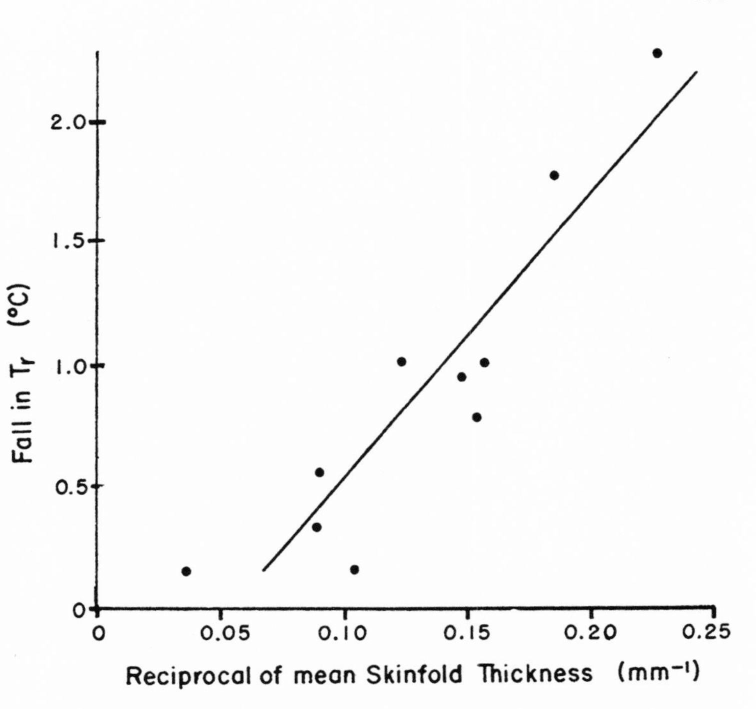

Figure 5-7. Subcutaneous Fat Thickness and Heat Loss in Cold Water. When men were exposed to water at 15°C for 30 minutes, the major factor of survival was the presence of subcutaneous fat which formed an inert layer of insulation; heat production was secondary. Each skin fold thickness point is a mean of four standard body sites. Adapted from Keatinge (1969).

FOURIER'S LAW OF HEAT FLOW

Many of the examples we will present of physiological responses to cold are concerned with acute cold exposure. Since the heat exchange under these circumstances is quite conspicuous, we must first consider the physical interaction between the warm, living mass of protoplasm and the air around it with its deficit of heat.

For years, physiologists have used an incorrect physical description by referring to Newton's law of cooling. This law states: a warm body without internal heating surrounded by a cold environ-

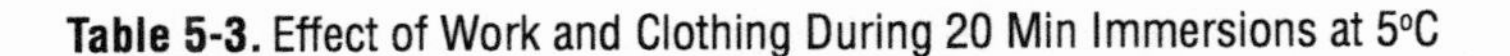

Table 5-3. Effect of Work and Clothing During 20 Min Immersions at 5°C

Mean values from 5 subjects. Falls in rectal temperature in all working experiments, with or without clothes, are significantly greater ($P<0.05$) than in corresponding still immersions. Metabolic rates are the gross values measured.

	Temperature (°C) Skin-Water Junction at End of Experiment	Fall in T_r	Metabolic Rate kcal/min
Still, stirred, unclothed.	5.8°	1.23°	4.55
Working, unclothed	5.4	1.81	6.90
Maximum work, unclothed	5.4	1.61	8.94
Still, stirred, clothed	12.7	0.29	3.39
Working, clothed	8.3	0.61	8.97

ment loses temperature faster, the greater the difference between the body temperature and the ambient temperature. Of course this law does not apply to homeothermic animals, which keep a relatively constant body temperature in a cold environment (we may neglect the difference in the day-night body temperature setting). Kleiber pointed out in 1932, and again in 1972, that Newton's law applies only to *temperature loss*, while Fourier's law applies to *heat exchange*, or *heat flow*. According to Kleiber, "heat is a form of energy, and we no longer believe that it is a material substance; yet we speak of heat flow as if heat were a liquid. This is a very useful fiction applied in our calculations of heat transfer."

Whether the student thinks of heat flow, or heat transfer, the convenient physical description is found in *Fourier's law of heat flow.* This law may be written as a simple proportion in which heat loss per minute is directly proportional to the body surface and the difference between the temperature of the body core and that of the environment, and inversely proportional to the thickness of the body shell; that is, heat flows from the body core to the external environment at a rate which increases with the surface area and the temperature drop between core and exterior, and which decreases with greater thickness of the barrier between core and exterior. This leads us to a discussion of this barrier, which we refer to as insulation.

Figure 5-8. Insulation Values of Winter Fur. The insulation of variable thicknesses of fur is compared with the artificial insulation of a black surface and cotton. Measurements were made with a hot-plate guard ring kept at 37°C (98.6°F) with the outside air at 0°C (32°F). Each point represents measurement of an individual specimen. Modified from Scholander *et al.* (1964).

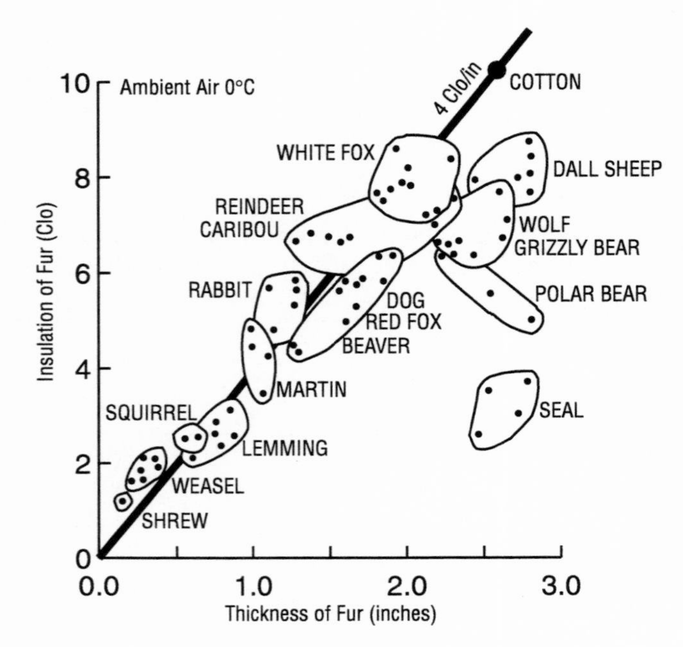

INSULATION

One of the most important factors in the control of heat exchange is the barrier between the core of the animal and the external environment. How is this measured? Two descriptive units have been devised: the *Met* and the *Clo*. These have been defined as follows: a resting person produces about 50 kcal·m^{-2}·hr^{-1}. This unit is called 1 Met. When such a person is sitting at rest in an air temperature of 21°C (70°F), in turbulent air of 10 centimeters per minute, with a relative humidity less than 50%, the insulation can be calculated. This is done by subtracting the evaporative heat loss of 25% amounting to 13 kcal·m^{-2}·hr^{-1}. If the body temperature is to remain constant, the rest of the heat loss, 38 kcal·m^{-2}·hr^{-1}, must be offset by insulation, which in a human amounts to a light business suit. This insulation is called 1 Clo unit and it has been corrected for the insulation of the layer of *still* air next to the clothing surface (0.8 Clo); in a high wind the insulation of the air next to the clothing is 0.2 Clo. Clo units are always measured by physiological means, and they can be calculated for the clothing of humans and the fur of animals. For example, a heavy arctic uniform provides about 5 Clo, and the fur of the arctic fox provides about 8 Clo (insulation of still air not included). Scholander et al. (1950) expressed the relationship between thickness and Clo value graphically with a scale beginning with the fur of the shrew, which consists of a few millimeters in thickness, and ending with the ordinary sheep, which has nearly 6.5 centimeters of thickness. The graph describes a slope of 3.7 Clo per inch of insulation (Fig. 5-8).

Fur: Practical Aspects

The practical aspects of this insulation as used in the fur industry provides interesting data for the biologist. Schakelford of the University of Wisconsin found that the fur of the female mink is more dense than that of the male: it consisted of 24,163 underfur fibers per square centimeter and 380 guard hairs per square centimeter. The number of hairs in the male pelt were about 12% lower than this. Expressing fur in different units, we can say that one twentieth of the body weight of a lemming is fur; this means that a 40 g lemming has 2 g of insulation in the winter. Regardless of one's opinion about the humanitarian aspect of trapping, or of wearing fur, the fur harvest from the various states does give us some clue as to whether the predator-prey relationship in the various states may be endangered (Table 5-4). These figures seem to indicate a decreasing fauna in some states; we cannot be sure as we do not know the total numbers or expertise of the trappers.

Insulation, Surface Area, Exercise

The relation between surface area and mass plays an important part in the effectiveness of insulation. Adding insulation to small objects with a large surface-to-mass ratio, such as the shrew or the fingers of babies, may actually *increase* the heat loss. This effect was learned many years ago by engineers who found that they could not insulate small pipes. Likewise, a thin glove on the fingers of an infant may increase heat loss from the fingers because the source of heat does not balance the large surface area added. (See Figure 4-11 and discussion of surface-mass law in the previous chapter.) This indicates why a much thicker fur would not benefit the smallest of mammals, the shrew. (Also, it would trip on the fur.) The problem of "small cylinders" does not apply to the whole human body. In this case, to provide more than 6 Clo units would fail only because it would impair mobility.

The insulation outside the skin of humans and arctic animals creates a physiological paradox when these mammals must exercise in the cold. We will describe only the problem here; the physiological solutions will be covered under the section entitled "Exercise, Insu-

Table 5-4. Harvest of Predators In a Typical Year

	Bobcat	Gray Fox	Red Fox	Badger
Arizona	43	84	—	—
Iowa	—	503	15,723	446
Louisiana	110	636	—	—
Pennsylvania	—	3,101	4,637	—
Virginia	—	25	43	—
Wisconsin	148	1,484	29,960	—

lation, and Heat Dissipation in the Sled Dog and Fur-Clad Human." If people dressed in a typical arctic protective unit of 3 Clo undertake heavy exercise, they become *tropical beings in arctic clothing.* With this description in mind, it is not difficult to understand why a soldier (a newcomer) was brought to the Fort Wainwright Hospital in a state of collapse during Operation Polar Siege; he had been pulling sleds and chopping timber under conditions where the temperature was -40°C. The diagnosis at the hospital was heat stroke. He had been so alarmed at the thought of working in the cold that he had piled on every item of insulation he could borrow. Under these conditions, some subjects will sweat 350 g/hr, others 725 g/hr. The efficiency of such sweating may amount to only 40%. The answer for humans, of course, is to ventilate the clothing and take off layers of insulation while working. This is not possible for horses and ponies with heavy winter insulation, or for sled dogs. The biological problem evident here will be considered in a later section.

It is possible to test experimentally the relationship of thick insulation to air temperature when sheep are sheared. In some experiments, the length of the regrown fleece varied from 0.1 cm to 12 cm (Table 5-5).

Changes in Insulation

It is a fairly obvious fact that in most animals living in the outdoor environment fur depth varies in winter and in summer. One of the most conspicuous changes is in the red fox, which increases its measured heat transfer value by a factor of 1.5 (Fig. 5-9). The percent change in winter insulation is greater in small animals than in large. It is interesting that we cannot speak of *the* insulation; it varies over the animal. Bianca (1967) has tabulated differences in hair density, thermal conductance due to skin and air insulation, and wind speed on the flank and belly of several species of domestic animals. In some cases with a wind speed of 8 miles per hour, the conductance was twice as high on the belly of the animal as on the flank (see also Hammel 1956).

Now is an appropriate time to consider not only the insulation of fur, but the actual temperature in fur. With thick, efficient insulation, body heat will not reach the ends of the hairs. In many cases the outer (or guard) hairs are at the same temperature as the environ-

Table 5-5. Cold-Exposed Sheep

Fleece Length	Critical Temperature
0.1 cm	28-30°C
2.5 cm	13
4.5 cm	8
10.0 cm	3
12.0 cm	-3

ment, and this will permit snow or frost to accumulate on the hair tips (see Figs. 5-10 and 5-11).

There are two physiological insulative mechanisms which have not yet been considered: these are vasomotor function in the skin and the contribution of subcutaneous adipose tissue. Some animals combine these factors with fur. Seals must have insulation in addition to skin because even in wintertime they come out of the water to rest on ice. Mammals that remain under water, such as whales and porpoises, use only the subcutaneous fat called blubber for insulation. The usefulness of subcutaneous fat to humans is detectable in Channel swimmers in whom fat appears to be preferentially laid down in subcutaneous areas rather than in the deep fat depots (Pugh and Edholm 1955).

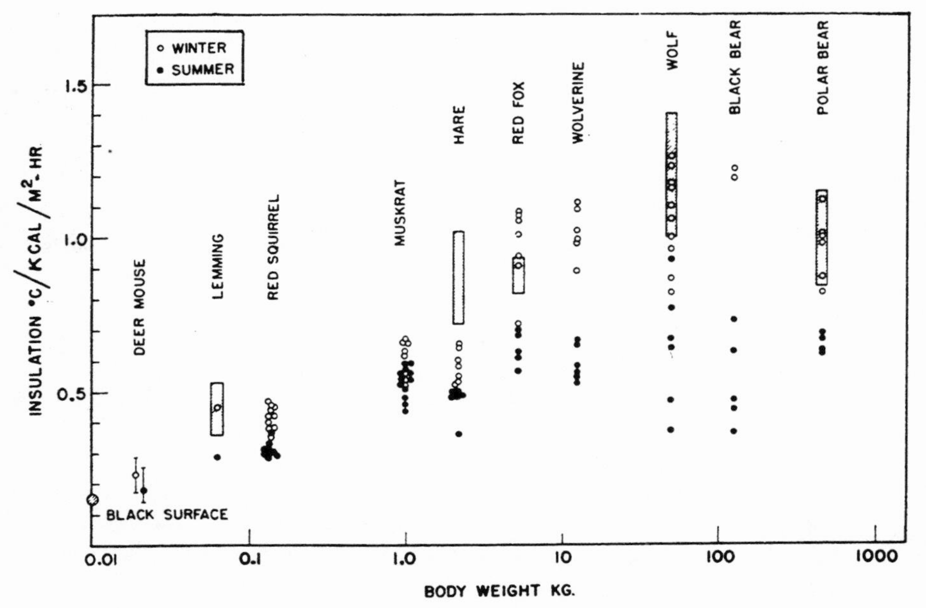

Figure 5-9. Seasonal Changes in Fur Insulation. It is of interest that the polar bear which must swim in ice water has no better insulation that the wolverine or Arctic hare. The rectangles represent winter data from a separate independent study with too many data points to include here. Adapted from Hart (1960).

Changes in vasomotor function have the effect of altering the barrier to heat exchange between the core of the body and the exterior. This vasomotor effect (peripheral vasodilatation and constriction) will be considered in a later section.

Consumption of Fat

There are fashions in medicine related to nutrition: in the 1990's the trend is to persuade the public to eat only from 20% to 30% dietary fat in the hope that it will decrease cancer and cardiac disease. Yet some animals prefer a diet of nearly 100% fat. The polar bear prefers to bite off just the fat layer of a seal, and leave the protein alone (Folk, Cooper, Folk 1994). The white shark near California lives on sea lions, and also chews off the fat layer and discards the rest of the animal. In contrast is the grizzly bear, who starts out as a baby with a high fat intake (the milk has 25% to 30% fat), but ends up being mostly a vegetarian. Plants, which include some fat, make up 75% to 90% of a grizzly diet. How do we relate this to what diet is best for humans? We will discuss the studies that show that it is the type and source of fat, its balance with other dietary factors, and lifestyle, that are important.

Fat Metabolism in Cold

The mammal that prepares itself for the challenge of cold may deposit large quantities of fat in the core or under the skin. The problem and the preparation are the same whether the mammal is a hibernator or a non-hibernator. This preparation may completely anticipate the onset of cold weather, and is often begun under the influence of decreasing daylight; this is part of the picture we refer to as *cold acclimatization without influence of cold.*

In the case of the mammalian hibernator, the success of dormancy depends upon this depositing of fat in the fall. However, it appears that not all animals capable of hibernation or dormancy undergo a period of fall fattening prior to winter. Some rodents we call hibernators hoard food, while others fatten *and* hoard food. Still other hibernators, including bats (Ewing, Studier, and O'Farrell 1970), black bears, and 13-lined ground squirrels, do not hoard food, but increase their weight by fall fattening. Morrison has reported data on one denned bear in which adipose tissue was over 40% of the body weight (1960). Many more measurements should be made through the skin with a depth gage, season by season.

Is fat used during dormancy in the typical mammalian hibernator? The use of stored fat from adipose tissue during cold-exposure or long periods of dormancy is not well understood. There are discrepancies in the amount of winter body weight loss, amount of depot fat mass loss, and above all, the quality of depot fat loss. Fat metabolism has been established as a source of energy during periods of stress produced by starvation and cold-exposure. A respiratory quotient of 0.7 to 0.85 has been reported in bears, bats, marmots and hedgehogs during cold-exposure and dormancy. This figure is indicative of fat metabolism.

Much of the information on this use of stored fat is based upon weighing animals before and after the period of cold challenge. This is unsatisfactory, because the loss of weight could be due mostly to insensible water loss; also, weighing fat masses is unsatisfactory because over the winter, the mass could be partly replaced by water. More observations are needed by biologists because we lack information on where fat stores accumulate. There are numerous patterns of fat distribution in animals preparing for winter: the seal apparently accumulates fat only under the skin, as shown in Figure 5-12; the sea otter apparently does not add any subcutaneous fat at all; some mammals accumulate large inguinal fat pads (hamster), while others do not (opossum).

Figure 5-10. Buffalo With Snow on Fur. The thick fur of this Montana bison demonstrates its insulative value. While the bison's body temperature is approximately 39°C, the ends of its fur are cold enough to allow snow to accumulate, not melt. Photo: Bud Griffith.

The seasonal increase in adipose tissue has been presumed to serve two functions: a) a stored energy source is available, and b) the insulation of the animal is effectively increased. It is known that hibernators such as the arctic ground squirrel lose 30% of their body weight during hibernation, but it is not known what percent of this loss is water or lipid (Hock 1960; Landau 1960). In these animals, this loss could be due to mobilization of stored fat for energy, but there are few figures for *known* loss of just adipose tissue. Beer and Richards (1956) state that their bats lost 70% of their fat in mid-November.

One must ask, however, whether the accumulation of fat deposits is really intended to be used during the winter, or is primarily important for the animal when it emerges in the spring. Hunters have observed that bears do not become very lean and hungry until about 2 to 3 weeks following arousal from their dens. This must indicate that fall fattening serves yet a third purpose: that of protection from starvation following arousal in spring before food is available. This leaves in question the importance that has been attached to fattening for survival during dormancy.

One interesting question concerning fat metabolism of cold-exposed mammals was introduced in papers by Lyman and Chatfield (1956) and by Irving (1964); they asked the same question as to whether deposits of highly saturated fat are altered to a more flexible type (lower-saturation) when the mammal is exposed to cold. Lyman's studies were on fat deposits in the hamster, while those of Irving were on fat in the leg of the caribou. We should also ask whether cold *subcutaneous* fat will have a change in saturation, while at the same time warm depot fat near the core of the body will not. It is reasonable to predict that the cold skin might need to be more flexible and would thus contain a fat which is more like oil. Such studies would need to be done by following quantitatively the utilization of individual fatty acids throughout the year.

Another investigation should determine whether certain adipose tissue deposits are used preferentially for energy sources at the time of cold challenge. With this background on how the mammal uses fat in winter, let us examine the different types of fat depots.

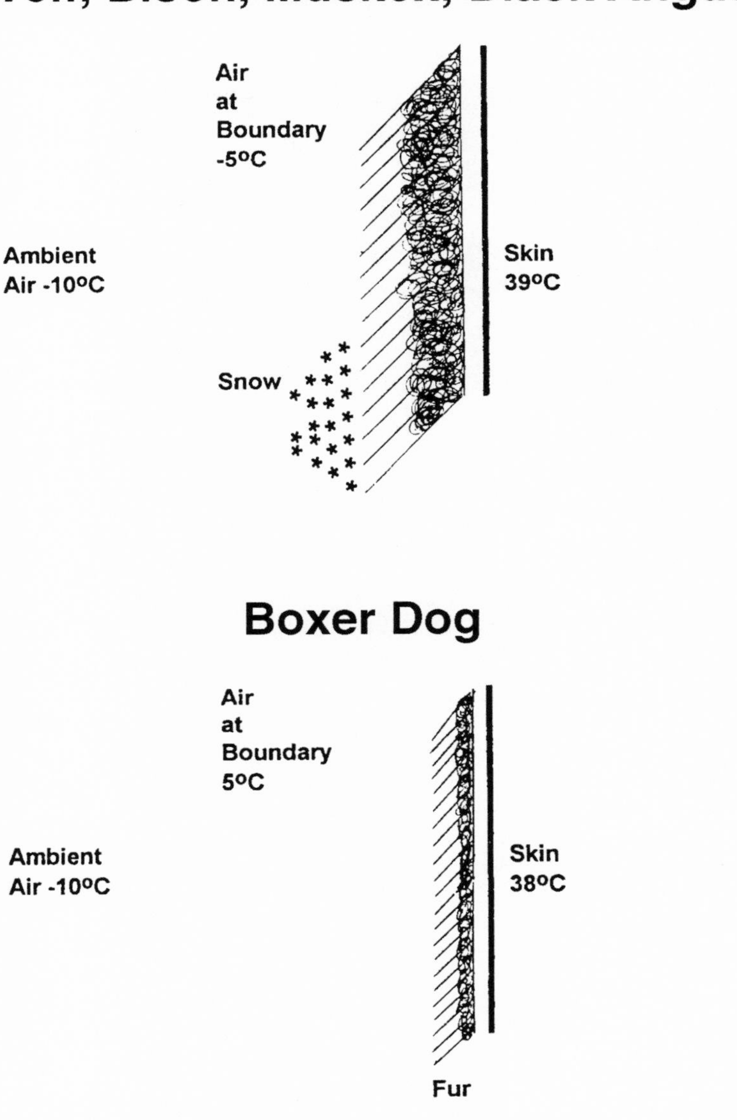

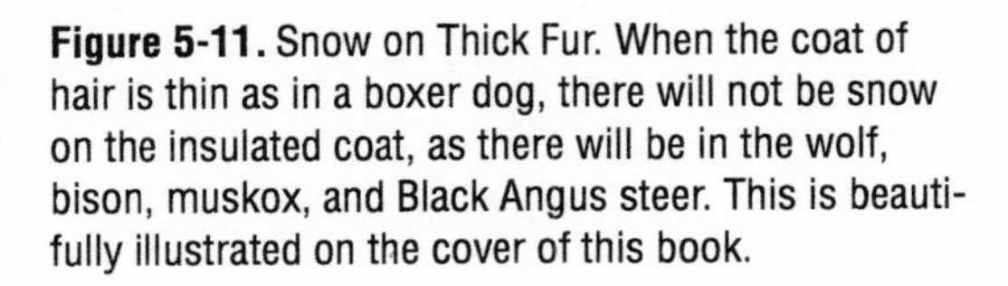
Figure 5-11. Snow on Thick Fur. When the coat of hair is thin as in a boxer dog, there will not be snow on the insulated coat, as there will be in the wolf, bison, muskox, and Black Angus steer. This is beautifully illustrated on the cover of this book.

Types of Fat Masses

The most common fat samples taken by investigators are subcutaneous white fat, inguinal white fat, internal white depot fat, and brown fat. As indicated above, subcutaneous fat serves as an insulating blanket in a variety of cold-exposed mammals, but it may also be used as an energy source. In our own surgical operations on the abdomen of polar bears, we have confirmed the observations of others that we may find either several centimeters of subcutaneous fat, or absolutely no visible subcutaneous fat at all. Polar bears brought in from the field in winter may have used all their subcutaneous fat. There is an abundant blood supply within the subcutaneous white fat that permits its deposit or use. Some environmental physiologists once thought that there is very little circulation in subcutaneous fat. This is by no means the case. In our surgical operations upon seals in a moderately warm room, the incision

Figure 5-12. Section of Ring Seal to Show Blubber. This frozen seal was sectioned at the level of the fore limbs. There is practically no fat in any location except under the skin. The seals and whales have achieved the ultimate by way of combining the functions of storing dietary fat and forming insulation; for further illustrations of subcutaneous fat see chapter on High Pressure. Photo by J. Bruce Ehrenhaft.

through the blubber of the seal is accompanied by excessive bleeding within the blubber mass (contrary to our usual experience of having a "dry field" during the surgical operation to insert radio capsules in the abdominal cavity of other species).

The question of the function of inguinal fat is less clear cut; we presume that this can serve an insulative function as well as an energy reserve function. The brown fat mass down between the scapulae clearly has only the function of a special energy source. It is called a thermogenic tissue. It is well documented both biochemically and physiologically that this brown adipose tissue (BAT) plays a considerable role in non-shivering thermogenesis (see Chapter 4). This thermogenesis is employed under conditions of cold challenge, during arousal from hibernation, and at birth (even in the human species). Triglycerides are stored in BAT under sympathetic nerve stimulation by norepinephrine. The heat production occurs by some type of "waste" mechanism in the numerous mitochondria of the tissue. The breakthrough in demonstrating that BAT shows explosive thermogenesis came with experiments upon bats awakening from hibernation; by direct measurement using thermistors in the different tissues of the body, BAT was shown to be far and above the warmest tissue in these bats while they aroused from deep hibernation. A further discussion of brown adipose tissue will be found in the next chapter, *Hibernation*.

Fats of the Human in Cold

The primary response to cold by the human species consists of a large effort to avoid the experience; humans seek shelter and improve their insulation by wearing extra clothing. However, when humans do receive the stimulus of a cold environment, they will lay down adipose tissue for heat conservation, and of course this extra adipose tissue will serve for a quick energy source. Earlier we used the example of swimmers in the chilly waters of the English Channel who appear to lay down fat subcutaneously rather than in the deep fat depots. This will certainly aid heat conservation, as the heat conductivity of adipose tissue is approximately three times lower than that of water.

A sample of men and women of different body fatness in the cold was studied by Buskirk and others (1963). Subcutaneous fat ranged from 13% to 45% of the body weight. The metabolic response to cold in both males and females was inversely related to the percentage of body fat. Buskirk pointed out that studies such as this are extremely difficult to make. Apparently, the leaner the group of subjects, the more variation is to be expected. Also a "true" metabolic response to cold is difficult to distinguish from certain metabolic artifacts such as the subjects moving frequently when resting or asleep (Buskirk 1966).

If a person is unsuccessful in adding enough insulation, he or she must increase the mode of heat production by shivering. This process liberates heat to the body. The large increase in metabolism during shivering is accomplished by the mobilization of the triglycerides of the adipose tissue (Bernstein 1971).

Ketosis in the Cold

Incomplete utilization of fat by the mammalian body results in the accumulation of several abnormal products called ketones. These are primarily acetone, acetic acid, and ß-hydroxybutyric acid. This abnormal condition is detected in a human subject by the smell of acetone on the breath. It is especially well known in cases of severe uncontrolled diabetes. With human subjects in the cold, ketosis may occur because of the relative carbohydrate deficiency which is also found in starvation, or from being on a high fat diet. It also occurs with abnormally increased metabolic demands such as in pregnancy, lactation, or during periods of rapid growth. If moderate exercise is continued too long, ketosis also can occur. There seems to be an additive effect of cold exposure to the ketosis associated with exercise. Passmore and Johnson (1958) suggest that this ketogenetic effect of cold may be part of a general response to environment and not specifically related to fat metabolism. This explanation is complicated by the fact that people on polar expeditions prefer higher fat in their diet than do people in temperate climates (Table 5-6). Of course, this is true of the Inuit as well, and as was pointed out early in the history of Arctic physiology by Levine (1937; Levine and Wilbur 1949), the Inuit do not show ketosis despite living on a high-fat diet.

Muir (1967) confirmed the diet selection during the British Antarctic Survey, finding that in camp, 47% of the daily calories (3,500 kcals) were derived from fat, while in the field, the daily-diet of 4,300 kcals was 57% fat. To contribute to the ketosis question, urine specimens were frozen and subsequently analyzed for total ketone bodies. Three circumstances were classified: 1) remaining on base and keeping warm; 2) sledging during warm weather, and 3) sledging under very cold conditions. Under these three circumstances, nutrition was approximately the same. Muir concludes from the urine analyses that cold and physical exercise are probably the most important factors in producing ketosis.

The Unusual Case of the Inuit Diet

There are over 117,000 people native to the cold regions who speak Eskimoian languages. They are known to have a diet of 50% fat, much of it from fish and sea mammals. Chronic heart disease is un-

Table 5-6. Environment and Food Intake in Humans

Geographical Location	Mean Temp. (F)	Total Calories (kcal)	Calories Supplied by: Proteins	Carbohydrates	fats
Arctic:					
Churchill, Canada	-17°	5,235	12%	46%	42%
Operation Muskox	+5°	4,400	13%	45%	42%
Tropics:					
Pacific Islands	79°	3,400	13%	54%	33%
38th Div., Luzon	83°	3,200	13%	54%	33%

common in these peoples (Feskens and Kromhout 1993). The protective factor in their diet is the Omega-3 fatty acid. These substances also help to produce an antioxidant balance (Hansen et al. 1994, Wormworth 1995; Mulvad et al. 1996).

Biochemical Background

Above we mentioned that fuel reserves are called upon during shivering; the second mode of extra heat production requiring rapid mobilization is the non-shivering thermogenesis discussed earlier. With either of these modes of increased heat production, there must be an increased oxidation of all foodstuffs; fat in particular is used, not as the preferred fuel, but as the most abundant. The triglycerides of depot fat are the major reserves of energy in the body. They are mobilized under the influence of norepinephrine and are split into free fatty acids (FFA) and glycerol. FFA is further broken down, the end product being the two-carbon fragment called acetyl co-A. This material is used to liberate energy in the tricarboxylic acid cycle. The pathway can be reversed to use excess acetyl co-A to rebuild fatty acids.

Acetyl co-A is also produced during glycolysis from pyruvate. While most glucose is converted to pyruvic acid via the Ebden-Meyorhof pathway, an important fraction of pyruvate is formed via the pentose-phosphate shunt. Johnson, Passmore, and Sargent (1961) emphasize the importance of this shunt in preventing ketosis.

The biochemical story for brown adipose tissue must be quite different because of the explosive thermogenesis it provides. Once again, fatty acids are implicated as a metabolic substrate. There must be high energy compounds available for activation of these fatty acids prior to oxidation. In the resting state, the tissue appears to be in a so-called "coupled" state, that is, substrate oxidation is linked to phosphorylation of ADP to ATP in an energy-conserving reaction. In the active state of the mitochondria, substrate oxidation appears "uncoupled" from phosphorylation, and hence respiration can continue at a high rate without the limiting phosphorylation reactions. The chemical energy thus lost is given out as heat, making BAT, the warmest tissue at that time in the body (see Fig. 5-21). Note in the illustration the rich blood system for dispersing this heat to the cold tissues of the body.

Fat Metabolism During Fasting

There are many examples of long periods of fast (without hibernation) in different species of mammals and birds. To illustrate:

Humpback whale	5 months
Elephant seal	1.5 months
White-tail deer (buck)	1 month
Emperor penguin	4 months
Polar bear	4 months

The most remarkable case is the polar bear of Hudson Bay. About 2,000 of these fast from about July 1 to Oct. 1 along the shore of

Hudson Bay. We refer to this condition as *walking hibernation* because their blood chemistry is similar to that of bears in true hibernation. The fasting polar bear handles glycogen differently from other fasting animals in that their serum glucose and liver glycogen increase significantly (see sidebar on polar bears in Chapter 6). Again, catabolism is supplied from fatty acids, as described in the previous paragraph (Cattet et al. 1997)

All-Fat Diet

Two animals prefer a nearly 100% fat diet: the polar bear and the great white shark. As a result, polar bear blood has a triglyceride level of 199 mg/dl and a cholesterol level of 298 mg/dl; this would be fatal to a dog or rabbit (Folk, Cooper, Folk 1996).

Our discussion of the polar bear's diet has ignored the need for essential amino acids. We had mentioned earlier the three neurotransmitters in the brain. There are others; it is apparent that mammalian neurons in the CNS use the excitatory amino acids, glutamate and aspartate, for transmission, or inhibitory amino acids, aminobutyric acid (GABA) or glycine. Polar bears must eat enough of the seal skin to fulfill their need for essential amino acids.

Genes and Fat Utilization

We must now consider gene control of responses to cold. In review, note that the same gene or genes may reoccur throughout the animal kingdom and even the plant kingdom. There are genes in mitochondria and plant chloroplasts. You can find identical genes in plant chloroplasts, in E. Coli, and in beef mitochondria.

The single gene related to cold-exposure controls fat storage and especially fat usage; it is called the *obese gene,* and it produces a protein, Leptin. Fat mice or normal-weight mice treated with Leptin shed fat (Culotta 1995b).

RESPONSES TO ACUTE COLD EXPOSURE

It will be convenient to consider the situation of an underclothed human or a short-haired dog exposed in a cold room to -28.9°C (-20°F) and a wind velocity of 5 mph. Within a very few moments the following events will take place:

1. There will be cutaneous vasoconstriction. This will permit the temperature of the skin and deeper layers under the skin to cool, and the surface-to-environment heat loss will be lowered; in a sense this means the effective thickness of the body shell is increased, and this decreases conductivity from the interior. Essentially, this is accomplished by a shift of blood from the shell area to the core area. This means there must be an increase of blood in the viscera.
2. There is a paradoxical increase in heart rate, which is evident in experiments with humans (Smith and Hoijer 1962), with cats (Lipp, Knott, and Folk 1960), and with rodents (Lipp and Folk 1960). This response is paradoxical because of the massive vasoconstriction. This increase in heart rate is also found in the cold-pressor test. According to Glaser and Whittow

(1957), the systolic blood pressure will rise approximately 18%, the diastolic blood pressure will rise about 33%, and the heart rate 14%. A typical case will show a rise from 70 beats per minute to 80 beats per minute.

3. There will be an acceleration of pulmonary respiration.
4. Piloerection in the skin will show itself as so-called goose pimples in the human, and by the erection of hair in the dog, a factor tending to increase insulation. This vasoconstriction in humans may reduce heat loss by 1/6 to 1/3.
5. There will be release of norepinephrine at the muscle beds, and of epinephrine from the adrenal medulla. All items so far are the familiar signs of sympathetic nervous discharge.
6. Within minutes this syndrome will be followed by neurohumoral activation of the hypothalamus, bringing about the release of anterior pituitary hormones, especially those stimulating the thyroid and adrenal cortex. All of the above result from stimulation of skin cold-receptors, which brings about reflex responses, all tending to conserve heat.
7. The next event will be an increased electrical activity in skeletal muscle, which will gradually lead to full development of a shivering response. With most subjects, this shivering will begin as soon as the lightly-clad subject is exposed to cold. Some subjects who are especially heavily built and have a subcutaneous fat layer still do not shiver after 20 minutes of exposure without clothing.
8. As a result of shivering, the metabolic rate will increase in both the human and the dog, 3- or 4-fold. The extent of this increase in metabolism should be found in the extreme limits of the thermal neutral profile.
9. There will be occasional bouts of vasodilatation, which will increase the temperature of the skin. This process is referred to as *cold vasodilatation* or "Lewis's hunting reaction" (1930). The phenomenon is most conspicuous on the palm of the hand and on the fingers and toes, and it is also observed in the arteriovenous anastomoses in the ears of mammals (Fox 1961).
10. Both of the mammals in the case history under discussion may assume a position that will reduce their surface area; this is done by tucking the legs and assuming a ball position.
11. As a result of the shivering, the body temperature may increase 0.6°C to 0.8°C during 20 minutes of shivering. The skin temperature on the chest, however, may be as low as 8°C to 10°C (46°F to 50°F).

When the subjects leave the cold room, there may be a loss in core temperature to as low as 35°C (95°F). This lowering of the core temperature will happen with both vasoconstriction or vasodilatation of the skin. It must be due to lack of metabolic heat when cold blood is still being returned to the core from the periphery.

The mechanism of changes in skin-blood flow (vasoconstriction) now deserves special consideration. This change is not necessarily brought about by nerves. It was demonstrated very early that if the sympathetic nerve supply has been cut surgically and has degenerated, measurements of vasomotor action in the fingers under the influence of local cooling still showed reduced blood flow in the fingers. This was confirmed in other experiments on the toes in human patients and on the ears in rabbits. A small amount of this reduction in flow may be due to increased viscosity of the cold blood, but it has been shown that cold has a direct constrictor action on blood vessels. The conclusive experiments consisted of cooling large isolated arteries from pigs to 20°C (68°F), resulting in a temporary cold-induced contraction, while in smaller vessels the contraction persisted for hours, and at times indefinitely at low temperatures. These cold-induced contractions have also been demonstrated in marine mammals. Thus, there are two mechanisms for reducing conductance of the skin, one by reflex activity and the other by a local stimulus. It is small wonder that in such tissues the relief of occasional bouts of vasodilatation is needed to maintain the health of the skin. Such cold vasodilatation waves have been mentioned above, and are shown in Figs. 5-13, 5-14, and 5-15.

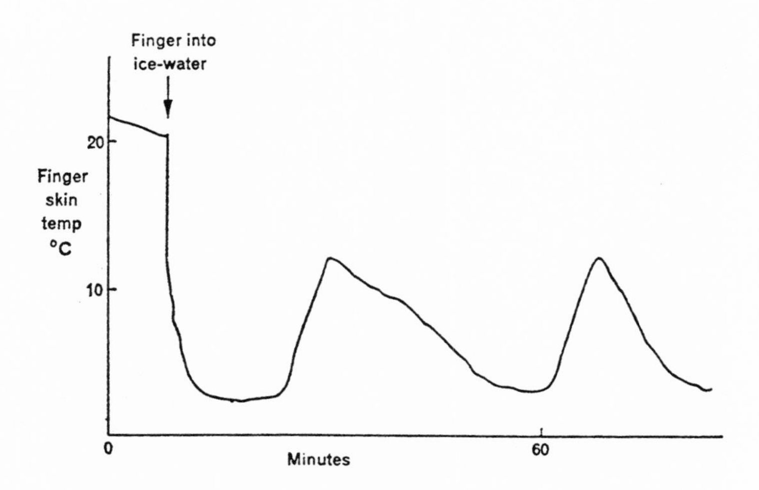

Figure 5-13. Cold Vasodilation in Human Finger. Waves of vasodilation as recorded by Lewis (by thermocouple) in 1930. Adapted from Keatinge 1969.

While Keatinge found that the isolated arteries contracted in response to drug stimuli down to 10°C (50°F), Johansen (1969) found that peripheral arteries from the seal would contract in response to epinephrine at temperatures between 5°C to 1°C (41°F to 32°F). However, central arteries showed no conclusive contractile response below 15°C (58°F). Apparently, Johansen was the first to compare central and peripheral arteries in the same mammal in a cold environment.

The mechanism of cold-induced vasodilatation has been studied using the foot pad of the anesthetized cat (Schwingerhamer and Adams 1969). When the pad is inserted into a water-bath maintained at 0°C (32°F), about every 4 minutes there is a cycle of vasodilatation which warms the foot pad from approximately 0°C to about 3°C (Fig. 5-14). After whole body acclimatization of the cat, this rewarming took the pad up to approximately 11°C (52°F) with the usual quick return to near 0°C. The experiments demonstrated that this reaction did not appear to be dependent upon histamine mediation, nor was it basically involved with neurogenic vasodilatation.

Another one of the rare illustrations of observing the hunting reaction in lower mammals is found in skin temperature in the sheep's ear (Fig. 5-15). In this case, the amplitude of the fluctuations is larger than found in human and cat (Bianca 1967). A special type of cold vasodilatation is found in the human forearm. It probably also occurs in the leg. When blood flow was tested with decreasing ambient water temperatures, the flow fell as expected down to 14°C (58°F). The surprising fact was that it

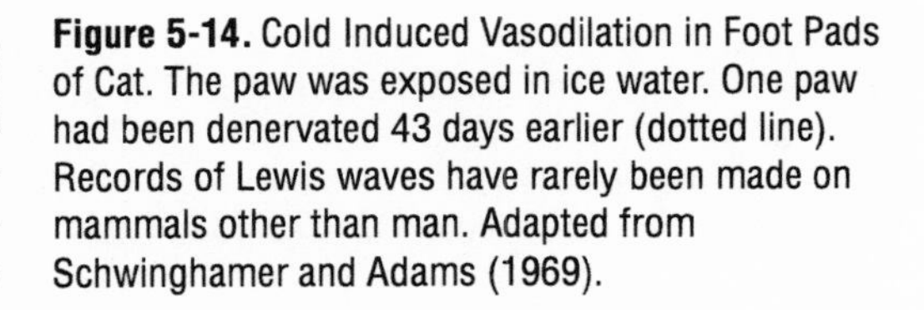
Figure 5-14. Cold Induced Vasodilation in Foot Pads of Cat. The paw was exposed in ice water. One paw had been denervated 43 days earlier (dotted line). Records of Lewis waves have rarely been made on mammals other than man. Adapted from Schwinghamer and Adams (1969).

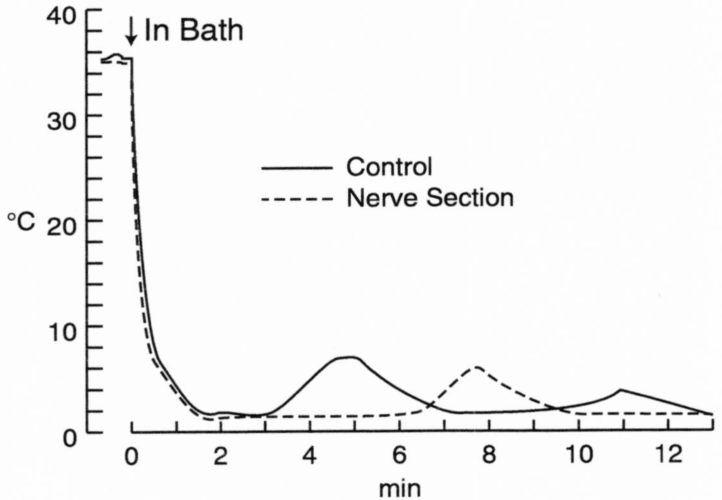

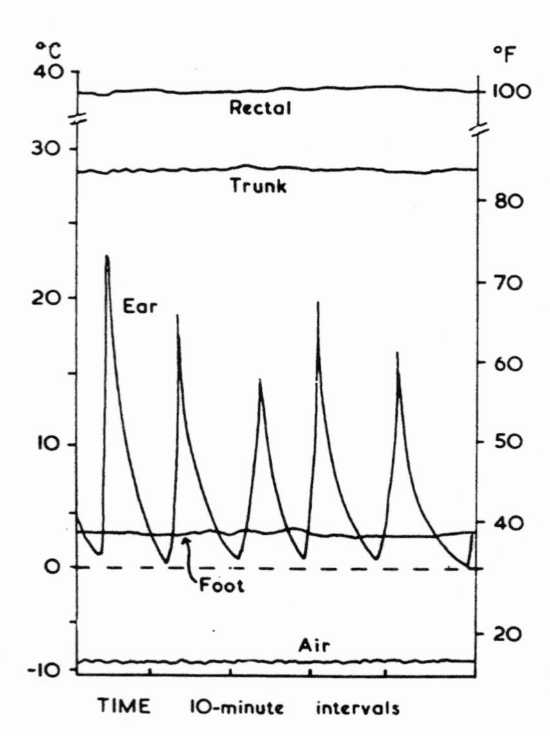

Figure 5-15. Lewis Waves in Sheep's Ear. These rapid fluctuations of skin temperature in the ear were obtained by exposure to air temperatures well below freezing. The foot temperature did not show these fluctuations. Adapted from Bianca (1967).

Figure 5-16. Shivering Threshold of Korean Diving Women. The diving women (Ama) become cold-acclimatized in winter and shiver at a lower temperature than non-divers and men. Adapted from Hong (1989).

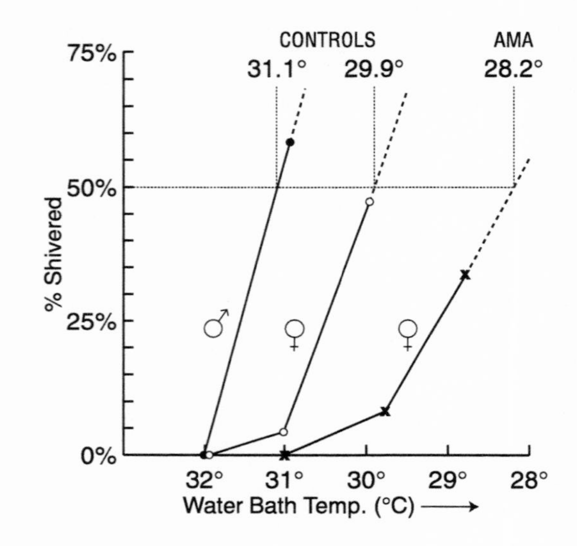

increased again at colder temperatures until at 2°C (36°F) the flow was slightly greater than at a water temperature of 34°C (92°F). This phenomenon is different from the Lewis hunting reaction, since it is predominantly happening within the muscle vessels (Clarke, Hellon, and Lind 1957).

If we may extrapolate from experiments on the dog to experiments on the cat, probably the innervation of the vasodilator system for the Lewis waves is sympathetic. Rolewicz and Zimmerman (1972) have studied the sustained vasodilatation of both the dog's paw and dog's ear. The dog's footpads and probably those of the cat are served primarily with A-V shunts. Earlier investigators realized that *muscles* of the dog have shown sustained vasodilation by sympathetic cholinergic innervation; they demonstrated that it is a different sympathetic vasodilator system that controls the flow of blood to the paw and the ear of the dog.

Vasomotor function of mammals other than the human has been mentioned above and will be discussed in a few instances in this chapter. On the whole, information on this topic is sparse. There is little information as to how much vasomotor activity is taking place under the fur of the various mammals. Some information on domestic animals was compiled by Blaxter (1965); he states that domestic species are capable of altering their insulation by vasomotion in the same fashion as humans.

Shivering

Shivering stimuli arrive from receptors in the skin; the act of shivering depends upon the integrity of the posterior hypothalamus. Also, Hemingway (1957) showed that a region between the posterior hypothalamus and the midbrain in the vicinity of the nucleus of the Field of Forrel must be intact in order for shivering to occur.

Lowering of brain temperature alone does not bring about shivering; this lowering must be accompanied by a reduced skin temperature. This can be demonstrated as follows: the blood can be cooled and the skin can be kept warm. By this means the shivering receptors in the skin can be circumvented. In such an experiment, shivering can be prevented completely. The function of shivering is to add to heat production (it also adds to heat loss); it provides improved protection of core heat by enlarging the thermogenesis to include the muscle mass of the animal. By this means, the temperature of the muscles is raised to approach that of the core. Because the work-function of this muscle contraction is zero, shivering is a very economical thermogenerator (Spurr, Hutt, and Horvath 1957).

The onset of shivering has been used as an effective test of ability to resist cooling. An unusual example is the work of Hong on the diving women of Korea, the Ama (Hong 1963). He used as a criterion the critical water temperature at which 50% of the subjects shivered. This 50% critical water temperature varied from 31.1°C (88°F) in males to 29.9°C (86°F) in non-diving females and 28.2°C (82.8°F) in the divers. These figures illustrate not only the threshold temperature that activates receptors in the skin, but also acclimatization in individuals who lead an outdoor life. Fig. 5-16 shows the water bath temperatures and the number of subjects. We say in this case the shivering threshold is considerably elevated in the divers as compared to the rest of the subjects. The threshold is raised as tolerance increases; if tolerance is lowered the threshold is lowered.

As stated earlier, humans can raise their heat production 3- to 4-fold by shivering, and raise core temperature over 0.5°C. This shivering may be prolonged for a considerable time and even occur during sleep. In experiments at the Lankenau Hospital, Rodahl studied men exposed to moderate cold with little insulation for 9 days and 9 nights; the subjects shivered moderately day and night for the entire period. At times when the subjects slept, they stopped shivering and their body temperatures sometimes dropped as low as 34°C (93.4°F). In spite of this shivering, toe temperature sometimes fell to 8°C (46.6°F) which was the same as the room temperature.

There appears to be a difference in shivering threshold of black subjects compared to both Inuit and whites; according to Adams and Covino, black subjects continue to cool to an average skin temperature of 28°C (82.4°F) before shivering, while the others cooled to only 29.5°C (84.4°F) (Fig. 5-17) (Adams and Covino 1958).

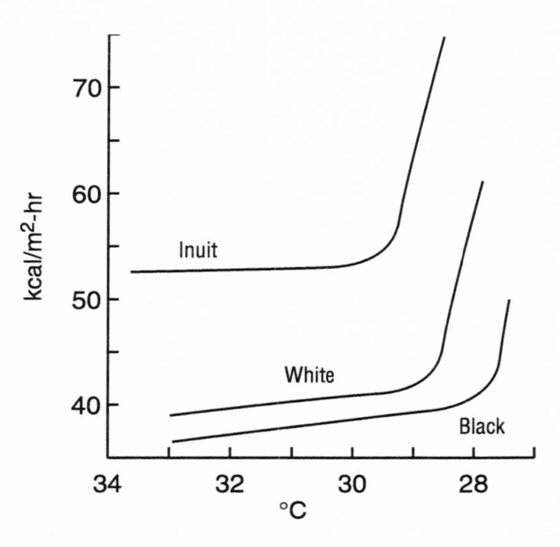

Figure 5-17. Skin Cooling and Metabolic Rate. As skin cools, shivering begins. The high metabolic rate of Inuits may have been due to type of food consumed. Blacks shivered at a lower skin temperature than the other two types of subjects. Adapted from Adams and Covino (1958).

Apparently, all experimental and wild-caught animals shiver as effectively as do humans. In routine experiments, the dog shows a 3- to 4-fold increase in metabolism when shivering. This shivering can be prevented by light administration of ether or other anesthetics; hypothermia cannot be induced in experimental animals until shivering is prevented by these means. Shivering is particularly effective in the wild Norway rat, which has been known to increase its heat production 5- or 6-fold.

RESPONSES TO CHRONIC COLD EXPOSURE

Terms to Describe Responses to Chronic Cold Exposure

The terms we will apply have been discussed earlier in previous chapters. They are

- *acclimation:* responses studied in cold chambers
- *acclimatization:* responses studied in the outdoor environment
- *habituation:* responses explained in terms of the nervous systems, not in terms of physiological thermal differences
- *genetic adaptations:* inherent responses; no conditioning cold-stress is needed to induce these responses.

Comparison Between Acclimation and Acclimatization

Up to now, we have discussed the physiological responses to a sudden and perhaps unexpected exposure to extreme cold. What may be the result if this exposure is repeated daily for two or three weeks, or longer? The animal or person becomes "redesigned"; we call this process *acclimation* when it occurs in a cold box, and *acclimatization* in the outdoor environment.

The technical difference between these two types of long-range responses to cold has been elucidated by Hart (1957)(also see Chapter 1). Some of his findings on laboratory cold-exposure will illustrate acclimation. In the responses of several species of small mammals he found the following: 1) an increase of heat production while in the thermal neutral condition, and 2) better sustained metabolic output of heat in the cold. This acclimation to cold appeared to reach its full expression in white rats after a month, and in white mice after several weeks. In a series of 8 species of mammals and 5 species of birds studied by Hart, the increase in resting rate of metabolism after acclimation was from 0 to 90%. Irving's assessment of this acclimation was: "In spite of the increases in resting metabolism of these animals, their critical temperatures were unchanged except in one case, where it rose. The result of acclimation was an apparent decrease of overall insulation, and life at a greater metabolic expense."

Now, here is an illustration of acclimatization from Hart's laboratory (Heroux, Depocas, and Hart 1959): the investigators exposed white rats outdoors in groups of 10 during summer and winter. In contrast to cold-acclimated rats, the outdoor rats in winter had a lower heat production than rats outdoors in summer. This was true over a considerable range of temperatures. In keeping with the lower heat production, and in contrast to cold-acclimated rats, winter outdoor rats had a greater pelt insulation and lower skin temperatures on the back and tail than did summer outdoor rats when measured at the same air temperature. On the other hand, winter outdoor rats had a higher maximal heat production than summer outdoor rats; the winter rats also had a greater cold resistance when tested at -35°C.

Non-Shivering Thermogenesis

Beginning with the work of Claude Bernard in 1859, studies have shown that the liver and abdominal viscera contribute heat to the cold-exposed animal. It is this mechanism that becomes more efficient in the rat exposed continuously to cold at 5°C (41°F) for 2 to 3 weeks. At the end of this period, shivering disappears so that it cannot be detected by electromyography; however, total metabolism remains elevated by about 80% and body temperature is held near normal. If these animals are moved to a warm environment (30°C)(87°F), most workers report that the metabolism of the rats remains for some time 15% to 20% higher than that of warm-acclimated rats. In another type of experiment, if shivering is blocked by curare and the animals are then exposed to cold, cold-acclimated animals show a rise in oxygen consumption which is sustained, while controls fail to maintain an adequate level of oxygen consumption and become hypothermic. The acclimated animals maintain both rectal and skin temperature near normal (Fig. 5-18).

The locus of the highest percentage of the non-shivering thermogenesis was the special interest of Jansky (1966); he demonstrated that the skeletal muscle of the body is important in thermogenesis that is stimulated by cold or by norepinephrine in cold-acclimated rats. He also demonstrated that these tissues were inadequate to explain all non-shivering thermogenesis, and some must originate from the brown fat. The remaining amount must be from the liver; however, this cannot be stimulated by norepinephrine. It can, however be stimulated by insulin. The experiments on skeletal muscle are illustrated in Fig. 5-18; note that approximately the same effect of non-shivering thermogenesis is produced by exposure to cold and by injection of norepinephrine into cold-acclimated rats.

An examination of the thermal neutral profile of cold-acclimated and control rats shows a difference in the resting metabolic rate at T_A 20°C and 30°C (68°F and 86°F). This graph also shows in a striking fashion the difference between the two groups of rats in their response to low temperatures of -10°C, -20°C, and -30°C (Hart 1960) (Fig. 5-19).

The mechanism of non-shivering thermogenesis has been studied by Brück and co-workers at Marburgh University in Germany. According to their view, non-shivering thermogenesis is inhibited by the anterior hypothalamus and the drive to overcome the inhibition rises from the cold-activated skin receptors on the periphery (Brück et al. 1970).

The development of cold acclimatization in birds is accomplished differently from the strategy used by mammals. Apparently, non-shivering thermogenesis has not yet been demonstrated in birds.

Acclimation in the White Rat

The first factor of cold acclimation is the non-shivering thermogenesis described above. This is followed by a rise in total daily food consumption. Initially, there is a transient decrease in body weight, followed in the young adult by resumption of growth within a week. There is hypertrophy of thyroid and adrenal cortex as well as heart, kidney, liver, and digestive tract. The masses of these organs increase relative to those of the controls and to the respective body weights of the two groups (Smith and Hoijer 1962). These increases occur at the expense of skeletal muscle growth which becomes reduced, relative to controls. The oxygen consumption of each of these tissues is also increased. Thus, the net effect of the product of oxygen consumption per unit weight x relative organ mass is apparently to shift thermogenesis in the direction of visceral regions. With acclimation to cold, heat production by the core region may become more important than that of the surrounding carcass or shell (Sarajas et al. 1967). These changes are not restricted to the rat; Farrand noted the same changes in the cold-acclimated hamster and also described a shift in body water (1959). Added to the above list of organ changes is the marked hypertrophy observed in the brown adipose tissue, also referred to as the interscapular hibernation gland. As stated earlier, this tissue is highly vascular and has a high metabolic rate resulting in considerable heat production. Because of a 2-fold increase in respiration and a 2.5-fold increase in mass as acclimation progresses, this organ (in the 45-day acclimated rat) accounts for an absolute heat evolution of no less than 5 times that achieved in the control. This local supply of heat may be conveyed directly to the posterior regions of the brain. The manner in which this heat is transported is evident in Fig. 5-20, which shows arteries and veins closely juxtaposed in both the cervical brown fat pads and the interscapular pad (Smith and Roberts 1964). It can be seen how brown

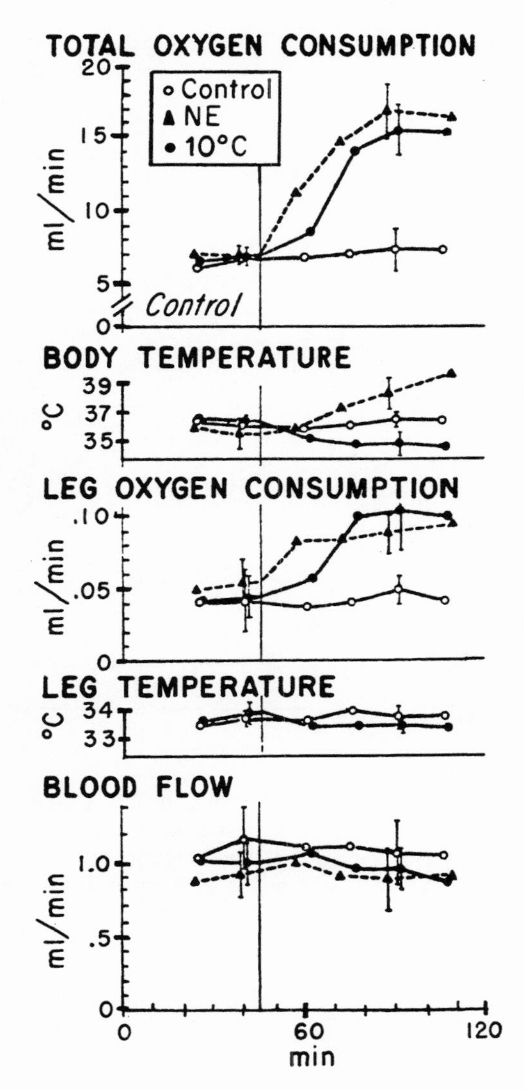

Figure 5-18. Effect of Norepinephrine or Cold Exposure on Cold-Acclimated Rats. Although this experiment was done to determine the site of cold thermogenesis, an important result is the proof that the norepinephrine test gives a normal response; i.e. exposure to 10° C (50°F) gives the same metabolic results as 0.4 mg NE/kg. The site of cold thermogenesis is mostly muscle and fat tissue, but a rise in insulin and blood flow will increase liver metabolism. Adapted from Jansky (1971).

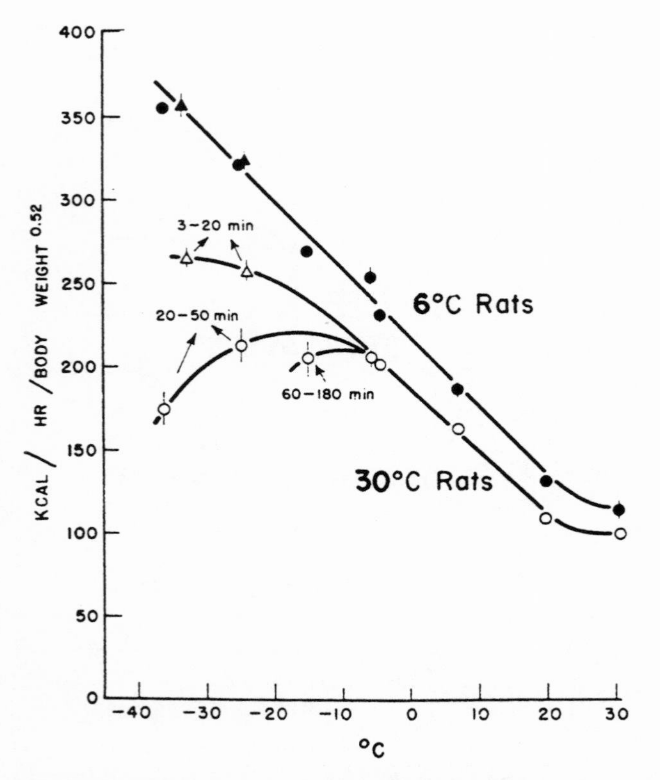

Figure 5-19. Thermal Neutral Profile of 30°C and 6°C Acclimated Rats. The BMR (at T_A 30°C) of cold-acclimated rats is higher than control rats. The maximum driven metabolism of control rats at –40°C can be maintained for only 20 minutes. Adapted from Depocas (1960).

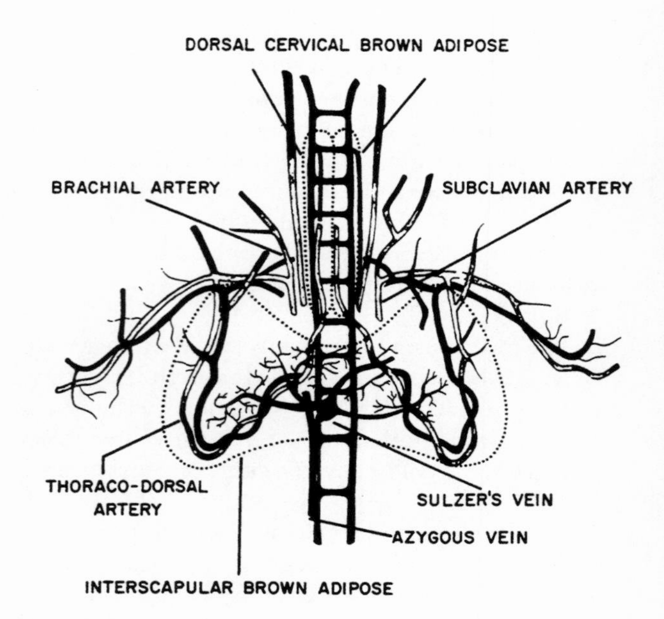

Figure 5-20. Region of Interscapular Brown Adipose Pad. Dorsal view of the bilateral drainage of the interscapular brown adipose region. This circulation permits heating of the cool peripheral blood by passage through "metabolic warming blankets." A consequence is to bathe the thoracic and cervical spinal cord with warmed blood and to supply heat directly to the heart. Adapted from Smith and Roberts (1964).

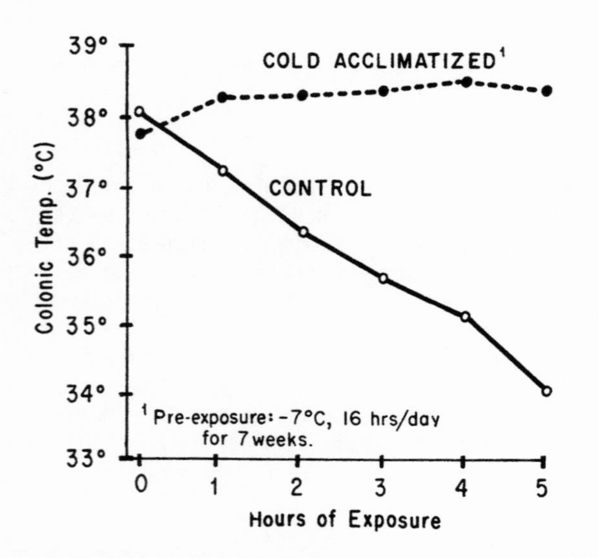

Figure 5-21. Resistance to Cooling of Cold-Acclimated Rats. The mean colonic temperature of 12 rats exposed to –15°C. Note that the experimental rates were pre-acclimated at a warmer temperature than the cold-stress temperature. Adapted from Blair (1961).

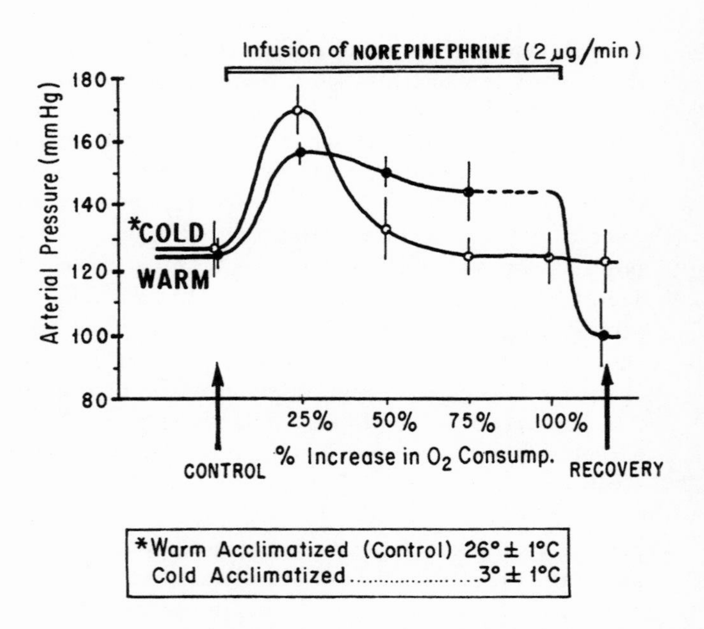

Figure 5-22. Responsiveness of Cold-Acclimatized Rats to Norepinephrine. Cold-stressed rats can be distinguished by a rapid and temporary response to norepinephrine. Controls respond less and return to normal slowly. Adapted from Evonuk and Hannon (1963).

fat returns metabolically warmed blood to the thorax via its venous drainage. One consequence of this is to bathe the thoracic and cervical spinal cord with warmed blood, and to supply heat directly to the heart. The sympathetic chain is also almost covered by brown fat and is therefore subject to the same heating. This principle of heating the cool peripheral blood by passage through this "metabolic warming blanket" serves as thermal protection of the central body core against the peripheral cooling of a cold environment. The significance of the close association of the venous return system and the arteries (while passing through the brown fat) is once again the principle of countercurrent heat exchange, only reversed from what was found in the fin of the whale by Scholander (see later section). It is obvious that the temperature in the paired venous and arterial vessels will be higher in the venous than in the arterial blood, and heat will flow from this venous blood to the afferent arterial stream.

Information from another laboratory added more information on the capability of producing heat independent of muscle contraction. As mentioned above, Carlson (1964) found that curarized, cold-acclimated rats double the heat production when exposed to 5°C while maintaining rectal temperature; in his experiments this capability did not exist in the non-acclimated animal (Fig. 5-21). The important factor associated with this non-shivering thermogenesis seems to be norepinephrine liberated from sympathetic nerve endings (Figs. 5-22, 5-23, 5-24 and 5-25). This additional effect of norepinephrine has been confirmed in men who are cold-acclimated (Kauppinen and Vuori 1988). The enhanced metabolism described here may be just great enough to increase heat content sufficiently to cause shifts in the peripheral circulation so that hands are more comfortable during cold exposure. Thus, it seems justified to refer to norepinephrine as being the hormone of cold acclimation. In rodents, even more than in the dogs described in Fig. 5-23, the effect of norepinephrine may be 4 or 6 times as great in the cold-acclimated animal as in the warm room control. In some small mammals the metabolic rate is increased by a factor of over 6 times by the appropriate injection of norepinephrine. In the white rat there is a fairly rapid increase in the metabolic response to norepinephrine after about 24 hours of cold exposure, which is increased by about 25% in the next few days (Fig. 5-26). Note, however, that the thyroid is also contributing (Woods and Carlson 1956).

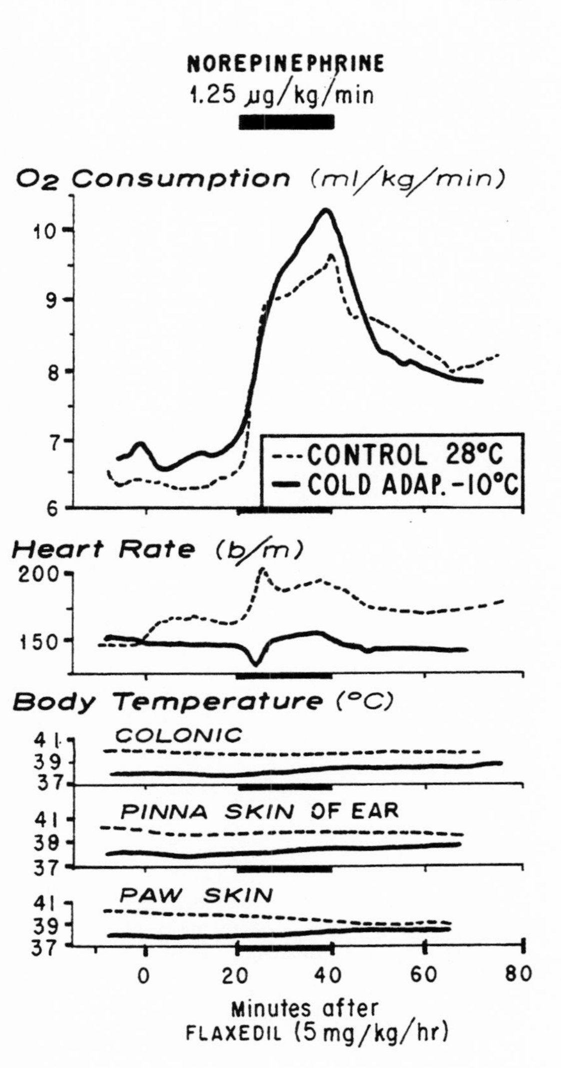

Figure 5-23. Responses of Cold-Acclimated Dogs to Norepinephrine. In most physiological functions the responses to norepinephrine of cold-acclimatized dogs are the opposite to those of control dogs. This was not the case with oxygen consumption; there was a similar response in both groups although a higher amplitude in the cold-exposed dogs. Adapted from Nagasaka and Carlson (1965).

The Origin of Adrenalin and Dopamine

It is time to consider the origin of those amino-acids (the building blocks of proteins) that drive thermogenesis and tolerance to cold. We note in passing that our knowledge of the chemistry of such origins has advanced so rapidly that we now learn that intact proteins can move from the membrane of a blood cell to the membrane of an endothelial cell, and the reverse (Kooyman et al. 1995). Here we may merely comment that the amino-acid tyrosine is transformed in four steps through dopamine to adrenalin (Figs. 5-24, 5-25).

Tolerance to cold is not entirely related to physiognomy. One can observe that one lean person may be regularly tolerant to cold, while

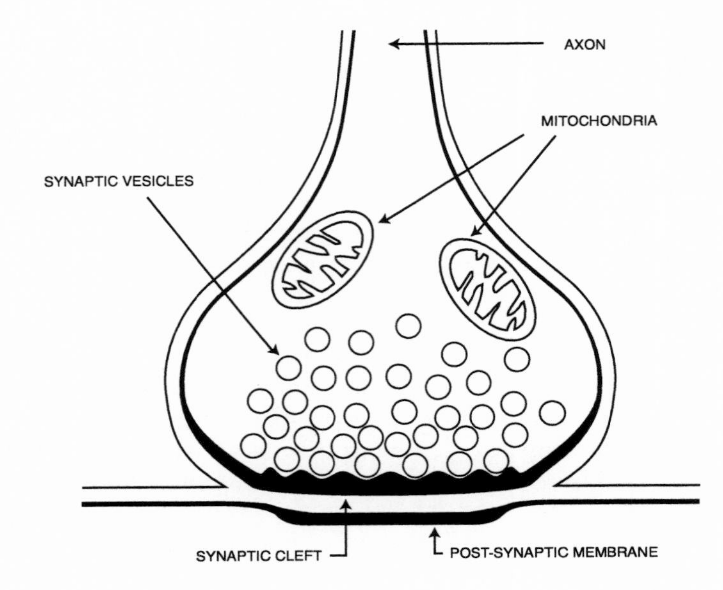

Figure 5-24. The Adrenergic Nerve Terminal. The environmental physiologist has become more interested in amine pools because of their importance in non-shivering thermogenesis. The transmitter vesicles in the nerve terminal normally contain only noradrenaline (norepinephrine) and not the other amines. Apparently this transmitter is manufactured in the vesicles from the amino acid tyrosine which is transformed in three steps to noradrenaline. The two may occur partly outside the vesicles. *(This diagram is also shown in Chapter 2 in a discussion of the amine serotonin.) Adapted from Schmidt-Nielsen (1997).

another may not be. When the human genome is better known, we may be able to relate tolerance to an extra-long DNA sequence on part of a chromosome.

Cold tolerance also may be related to the infamous neural chemical messenger, dopamine (see Fig. 5-25). This is under the influence of a gene called D4DR that controls dopamine receptor formation.

Perhaps a part of tolerance to cold is the presence of stress-relieving chemicals. One of these is a cannabinoid compound found in pig brain (Weesner 1996).

Figure 5-25. Amine Pools in Sympathetic Nerve Terminals. Diagram to show the biosynthesis from the common amino acid tyrosine which is transformed in three steps to noradrenaline, in the granules of nerve terminals. Adapted from Von Euler (1966).

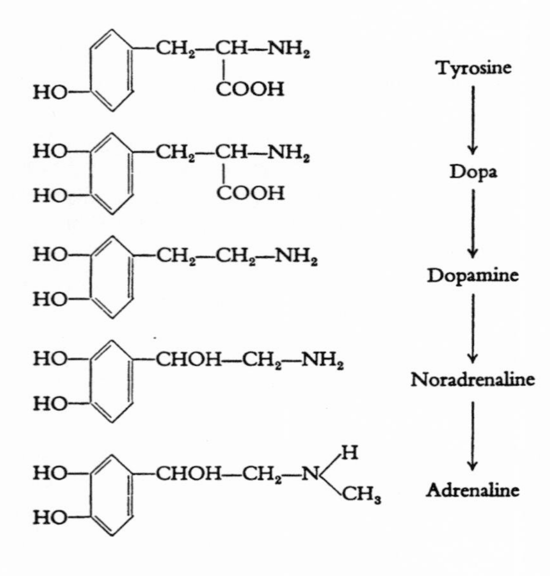

Acclimation in Small and Large Animals

Many cold-exposure experiments have been done with domestic and wild-caught mice; this procedure has the advantage of large samples for statistics. Hart has worked mostly with deer mice and Barnett has worked with domestic mice. Hart demonstrated that mice acclimated to 10°C are well prepared for exposure to -21°C; these data also demonstrate that exposure to 20°C instead of 30°C prepares the animal for exposure to -11°C. In all such experiments, the investigator has the special requirement of determining whether an increased length of hair coat is an important part of the acclimation process. It will be made clear in the next section that often conspicuous acclimation has occurred within 8 to 10 days, and there would not be time enough to increase appreciably the thickness of a coat in that period of time. The time course of rat acclimatization is similar (Fig. 5-26).

Lynch (1972), working on a longer time scale with deer mice, detected two functional syndromes in the development of cold acclimation, the first associated with the basal metabolic rate and liver activity, and the second associated with non-shivering thermogenesis, food consumption, and brown adipose tissue. The first syndrome takes place much more rapidly than the second. In his studies he also went on to compare acclimation and acclimatization and states: "Short-term cold acclimation (3.5 weeks), relative to long-term (12 weeks), better approximates the adjustments in the natural populations."

Another illustration of acclimation of a small mammal is found in the series on arctic lemmings completed by Berberich and Folk (1976). The rationale of the experiments is based on the summer and winter habitat of lemmings. The brown lemming lives in wet tundra which often freezes at night all summer. In winter this species lives under a frozen but protective snow cover eating tundra grass and roots. We reasoned that the lemming might be cold-acclimatized both summer and winter. This did not prove to be the case: both the brown and varying or collared lemming showed marked acclimation to cold in kidney function, blood composition and organ composition (see Fig. 5-27).

Let us compare this response to cold with that of a larger mammal, the rhesus monkey. Although some rhesus species are tropical, one species, the snow monkey, lives in snow and cold all winter. Elizondo (1980) demonstrated two different thermoneutral profiles

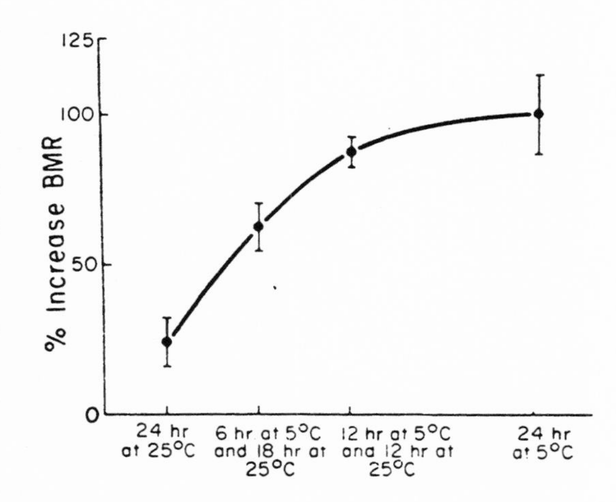

Figure 5-26. Initial Stages of Cold Acclimation. The development of cold-acclimation can be monitored by the cold-norepinephrine test (see text). When a maximum metabolic response is found in a cold-acclimated rat, shivering has been replaced by non-shivering thermogenesis. The second of the 4 experiments above show that the beginning of this process of cold-acclimation can be demonstrated after only six hours of cold exposure in the white rat (dose: 0.2 mg/kg). Although a high response is obtained after 24 hours of cold exposure, it takes about 10 days for the white rat to become completely cold acclimated. Adapted from Jansky (1971).

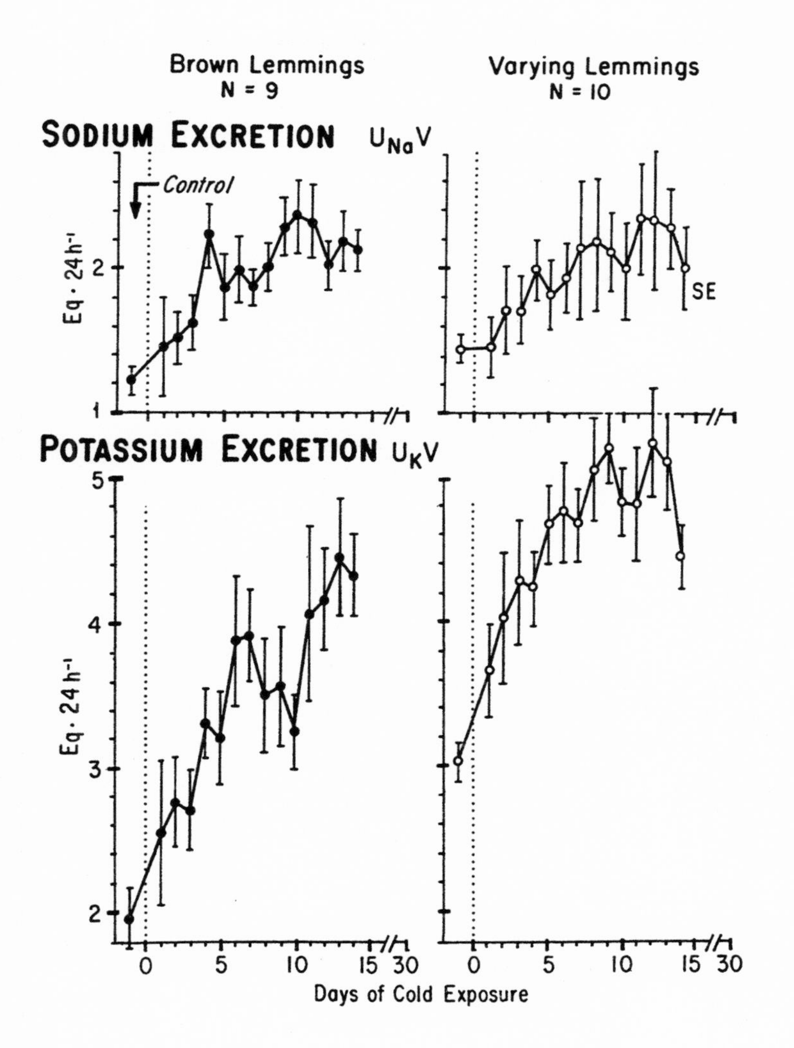

Figure 5-27. Cold Acclimation of Lemmings. Daily excretion for control (18°C) and cold-exposed lemmings (3°C). Excretion expressed as equivalents per day. Berberich and Folk (1976).

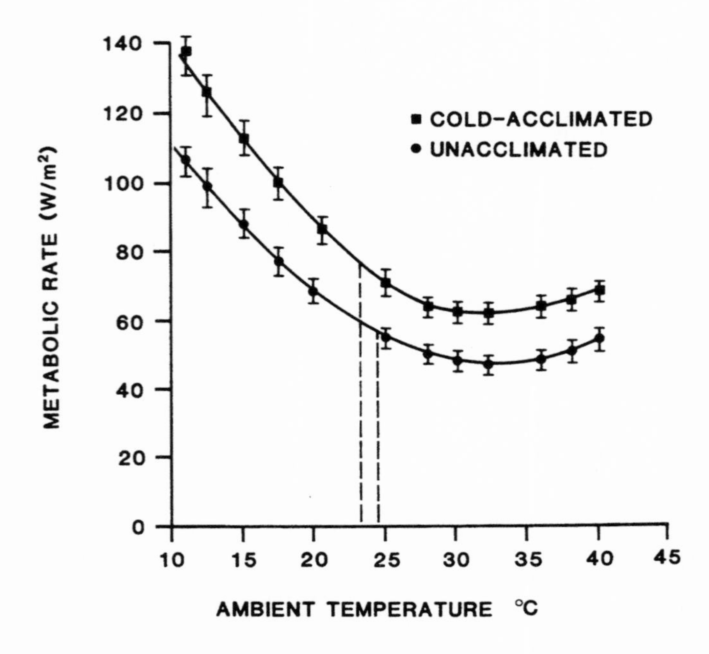

Figure 5-28. Cold Acclimation in Rhesus Monkeys. There have been few experiments of the response to cold of rhesus monkeys. These graphs illustrate the principle of the thermal neutral profile. Adapted from Elizondo and Johnson (1980).

in the rhesus monkey, the higher one representing cold acclimation (see Fig. 5-28). At all ambient temperatures the cold-acclimated animals had a higher metabolic rate.

Time Course of Acclimation

The parameters of cold acclimation which Farrand (1959) measured on the hamster leveled out after 7 days of cold exposure (Fig. 5-29). The timing for rats seems to depend upon the measurement: oxygen consumption leveled off in 2 weeks, but liver metabolism took 4 weeks (Hannon 1960) (Fig. 5-30). Adrenal weight leveled off in 1 week (Fig 5-31), but Heroux (1960) insisted that its secretory activity increased for 2 or 3 weeks. However, Schonbaum (1960) found scarcely any difference in the adrenal secretion between the seventh and the fourteenth day (Fig. 5-32). Hematological and body fluid adjustments in rats were studied extensively by Deb and Hart (1956); ten parameters were measured. On the whole there were large changes in 7 days, but very little change after that. Only in the measurement of hemoconcentration were changes still taking place 3 and 5 weeks later.

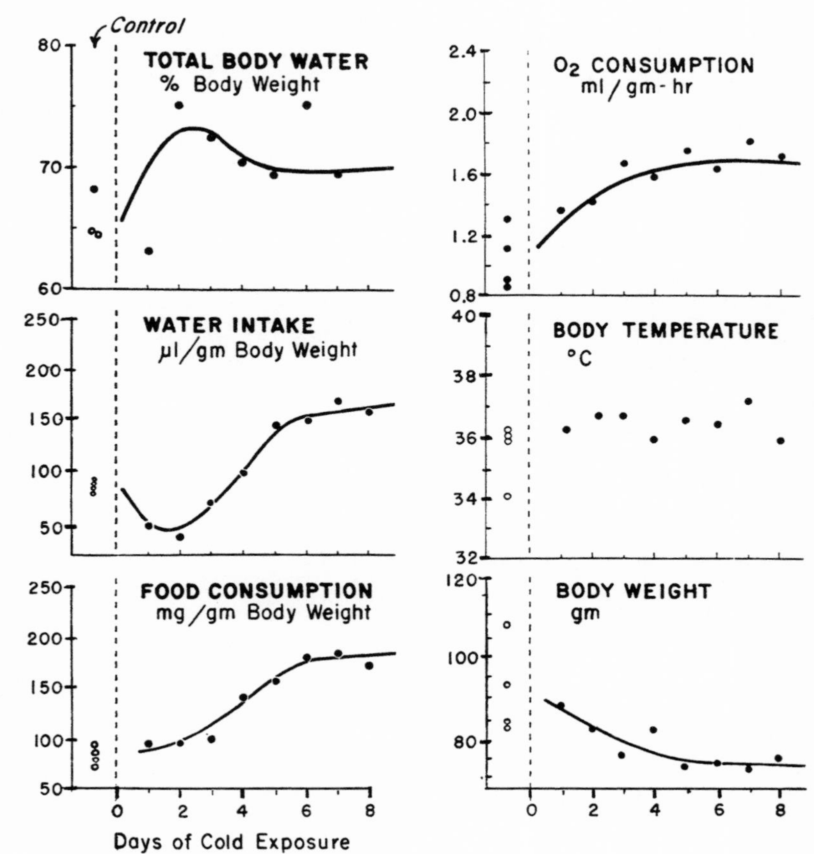

Figure 5-29. Indices of Cold-Acclimation. The conspicuous physiological changes characteristic of cold-acclimation in the hamster have taken place within 8 to 11 days after initial cold exposure (T_A 6°C). Some workers suggest that in some species cold-acclimatization requires as long as two months. Adapted from Farrand (1959).

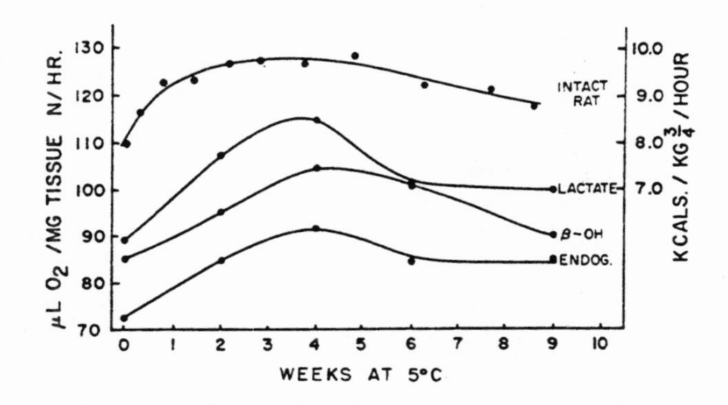

Figure 5-30. Cold Acclimation of the White Rat. The increase in oxygen consumption of the intact rat has leveled off in 2 weeks, but the increase in oxidative metabolism of liver slices, which is due to an increase in oxidative enzyme activity, continues to rise for 4 weeks. Presumably cold acclimatization in metabolic terms is complete at that time. Adapted from Hannon (1960).

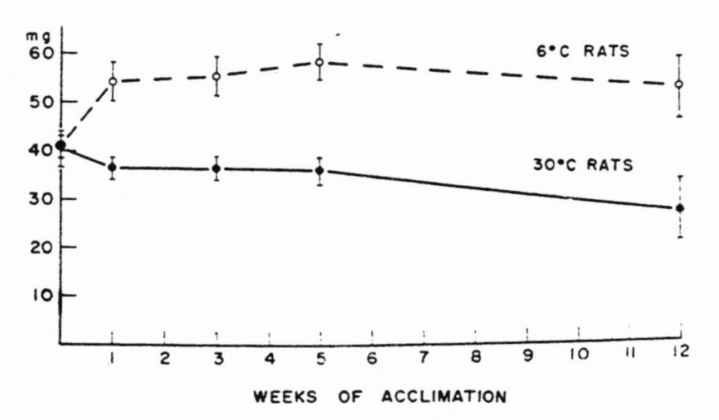

Figure 5-31. Adrenal Hypertrophy and Cold Acclimation in the White Rat. Adrenal weights adjusted to a common body weight of 310 gm. The weight of adrenal glands increased with cold acclimation; there appears to be no significant difference in the increase in weight between the first week and the third week. However, Heroux emphasizes that the adrenal hormonal activity increases for two or three weeks and then decreases to a new level well above normal. Hormonal activity does not parallel hypertrophy in the white rat. Adapted from Heroux (1968).

Another measure of acclimation is heat production by direct calorimetry; this technique revealed that rats at 5°C showed the major part of acclimation by 10 days with minor changes occurring up to 20 days.

The guinea pig appears to be mostly acclimated within 1 week, if controls are taken into account (D'Angelo 1960). The rabbit has adjusted within 1 week in most aspects studied, but changes in body temperature and shivering take 2 weeks to level out (Heroux 1967) (Fig. 5-33). The deer mouse (*P. leucopus*) has shown a maximum response to cold in 8 days in the parameters of increased basal metabolic rate, total body water, and weight of liver, and in 3 weeks in the case of non-shivering thermogenesis, food consumption, and increase in interscapular brown adipose tissue (Lynch 1972).

The final example of the rate of the development of acclimation is seen once again in mean food consumption. Barnett and Mount (1967) presented the results in the adjustment of mice exposed to -3°C showing first a marked drop in mean food eaten and then a steep rise, leveling off at day 9. This new level of food consumption does not at first bring the weight of the mice back to control value.

In summary, we can generalize that for several species the conspicuous expressions of cold acclimation are completed in 7 to 10 days, but some endocrine parameters may take as long as 4 weeks. An earlier impression in the literature that cold acclimation requires as long as 2 months does not seem justified.

Figure 5-32. Initial Time Course of Cold Acclimation in the White Rat. All values are expressed as a percentage of the mean control values. The major endocrine change has been completed by the 17th day of cold exposure: the decline in body weight has leveled off, adrenal weight is no longer increasing, and steroid formation in the adrenal glands is no longer increasing. After two months of cold exposure the animal is still underweight, but the endocrine picture has returned almost to control level. Adapted from Schonbaum (1960).

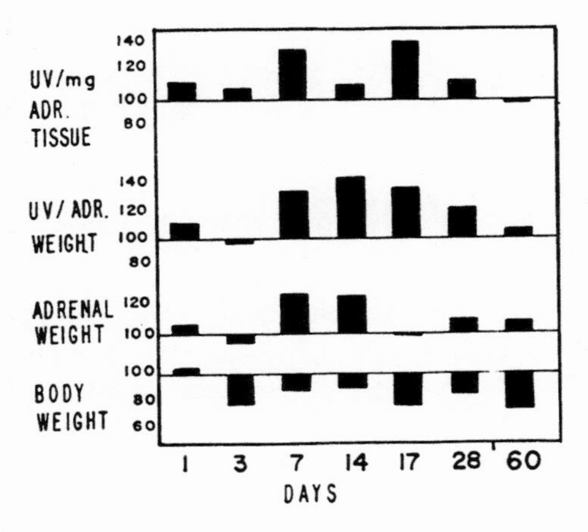

Change in Critical Temperature

It would be reasonable to postulate that as mammals become cold acclimated and increase the thickness of their insulation, there might be a change in lower critical temperature as revealed by the thermal neutral curve. Scholander et al. (1958) did in fact show this with men. These cold-acclimated men, when working in the cold, showed no detectable difference in oxygen consumption, however since

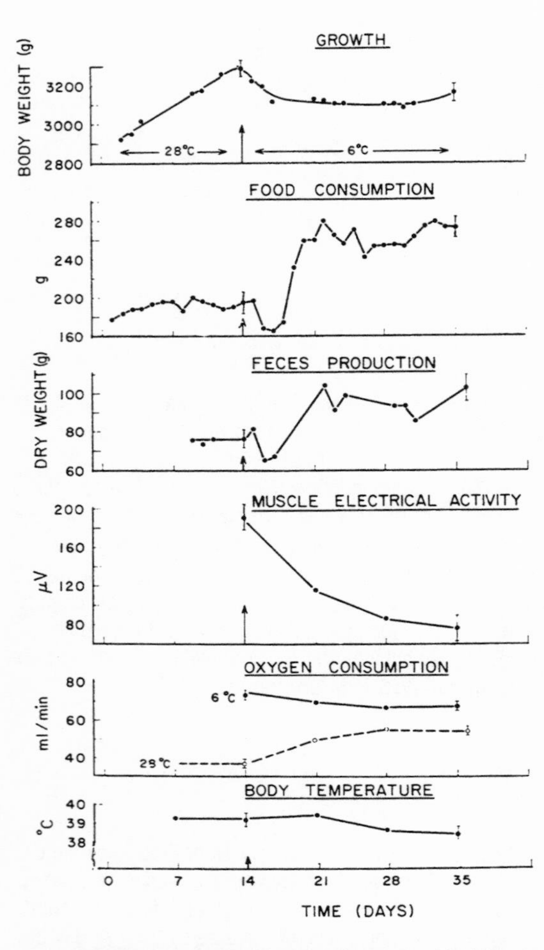

Figure 5-33. Indices for Cold Acclimation of Rabbits. Four shaved rabbits were exposed for 2 weeks at 28°C followed by 3 weeks at 6°C. Most changes leveled off within 7 days, although some required 14 days. Adapted from Heroux (1967).

their basal level without cold was slightly elevated, it follows that their critical temperature had been lowered by about 2°C to 3°C (Fig. 5-34). Hart did not find this to be the case with rats. He surveyed the appropriate literature written about ten species of mammals, and found that in every case as they became acclimated, the critical temperature remained the same. One lower mammal, the seal, seems to be an exception; Lawrence Irving (1970) reports a seasonal change in lower critical temperature of about 4°C.

The change in critical temperature with season is of course amplified in sheep as the fleece grows; this event was discussed earlier (Table 5-5).

Another variation in critical temperature is brought about by latitude. Possibly this change would also be apparent *within* a species, expressed as *change in season.* At any rate, two species of wild rats both showed a higher basal metabolic rate and a higher basal body temperature in the northern latitudes than in more southern latitudes.

LONG-TERM RESPONSES TO COLD IN HUMANS

Next we will discuss the long-term responses to cold in the human during prolonged exposure. These responses differ in various ethnic groups, and the variations will be individually illustrated.

In the following discussion, the term *acclimatization* will be used at first because most of the observations to be considered were made in the outdoor environment. Later, a careful distinction will be drawn between different responses of experimental animals in cold chambers and in the outdoor environment.

The essential question concerned with acclimatization to cold is whether there is improved comfort for the human as the exposure continues over days and weeks. A systematic approach requires that standardized tests be used to determine whether a changed response can be measured. We have already described the *shivering test* used by Hong with the diving women of Korea. A second, more straightforward test is the *immersion of fingers* or hands in ice water (Fig. 5-35). The third test, one developed by Hammel, we call a *cold-bed test;* it consists in having subjects sleep overnight in a cold environment with inadequate insulation (Hammel 1965). Using this method, one of the most striking examples of human cold acclimatization was found in seven nomadic Lapp shepherds; note the distinct separation of curves in Figure 5-36.

The environment of the Lapp shepherds has been carefully described by Milan (1960). He was fortunate to live and travel for 18 days in 1955 with a Lapp *siida*, or herding group, which was still pursuing a nomadic way of life by herding about 400 reindeer. They all lived in a tent while the temperature fell during clear nights to between -25°C and -30°C. A gap of several inches between the tent cover and the ground always allowed cold air to enter as a draft for the fire, which caused considerable discomfort to the scientist. Everyone in the tent slept fully clothed. The under-blanket was a single reindeer hide; a heavy blanket made of skins of long-haired sheep was on top. The fire was always allowed to die down so that the tent took on the temperature of the outside environment.

Variations Among Ethnic Groups

European background: Examples of Acclimatization. When subjects were exposed by Davis (1961) for 8 hours per day at 12.5°C over a 32-day period, the time of initiation of shivering and the total metabolic heat production showed a rapid fall over the first 10 days (Fig. 5-37). A striking example of this reduction in heat production obtained in another laboratory is illustrated in Fig. 5-38 (Carlson 1964). The next illustration of acclimatization was shown by the cold-bed test. As background, note that individuals (especially residents of the United States) working in an indoor environment usually prefer to sleep with a warm body shell, a warm body core, and with relaxed vasoconstriction. Hammel (1963) suggests that all of the skin temperature is above 33°C. What then, is the effect of living outdoors for 6 weeks, sleeping essentially naked in thin sleeping bags in 0°C weather? The unacclimatized person of European stock will be unable to sleep because of constant shivering and a declining skin temperature of the extremities (Fig. 5-39). With acclimatization, the metabolism is elevated by non-shivering thermogenesis, and skin temperatures are maintained during this cold-bed test. Subjects also learn to sleep when shivering is not completely suppressed and when the insulated shell is partially cooled. There are local changes over time in this acclimatization process: gradually the feet are warmer (Savourey et al. 1992).

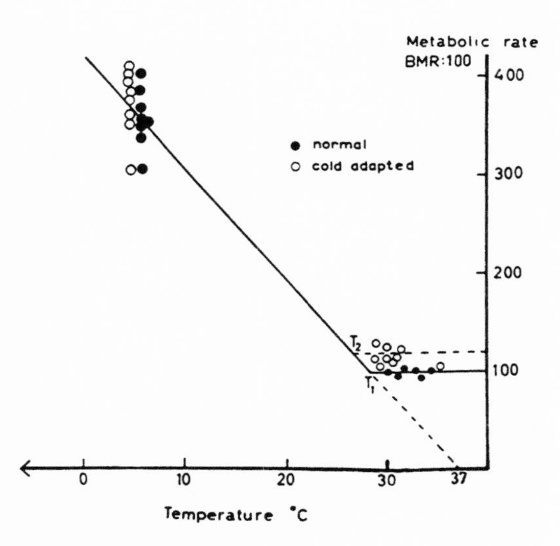

Figure 5-34. Cold Acclimation and the Human Thermal Neutral Profile. This study included a group of subjects before and after cold acclimation. At 5°C the high values were measured when the men were influenced by previous bicycling; the lower values give the basal rates. The solid line at 100 is the DuBois standard; T_1 and T_2 are the critical temperatures before and after cold-acclimation. Adapted from Scholander (1958).

We should now look for evidence of acclimatization through increased body insulation. One example is that of the preferential laying down of subcutaneous fat by Channel swimmers described earlier. Cold acclimatization is affected by the amount of work done in the cold (Virokannas 1996). In working subjects, body insulation is improved by a small increase in body fat and increased vasoconstriction (Jansky et al. 1996). Another remedial factor is increased serum thyrotropin (Reed 1995). There is, however, local acclimatization to cold; northern Norwegian fishermen show clear evidence of this by having warmer finger temperature in ice water (Andersen 1963).

The metabolic type of acclimatization must be discussed with caution and with frequent use of qualifications; the reason is that a comparison of the basal metabolic rate in summer and in winter, or before and after prolonged cold exposure, is apt to include a change in the amount of exercise and in the amount of protein consumed. In only a few instances have these factors been controlled. An example is described in the next section.

Arctic Dwellers: A Possible Case of Genetic Adaptation. Many studies have been done comparing thermoregulation in the Inuit with European (control) subjects. The Inuit begin to shiver at the same skin temperature as the Europeans; they begin to perspire at the same skin temperature and their overall tissue insulation during maximal vasoconstriction is the same. However, the non-shivering metabolism of the Inuit is 30% to 40% greater than that of controls; furthermore, they have a marked ability to withstand hand cooling

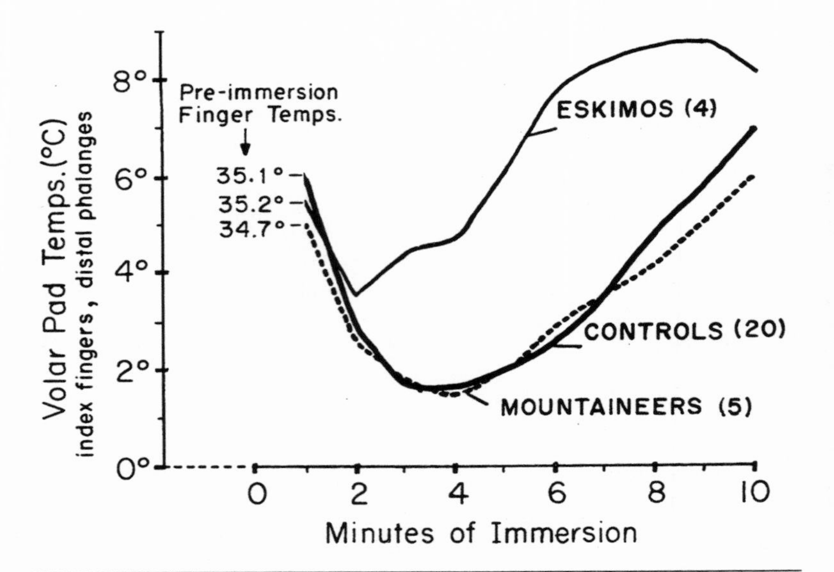

Figure 5-35. Vascular Responses to Ice-Water Immersion. The response to cold of fingers of Eskimos is different from that of other outdoor people. Eskimos' fingers are warmer, which will assist dexterity. Adapted from Eagan (1964).

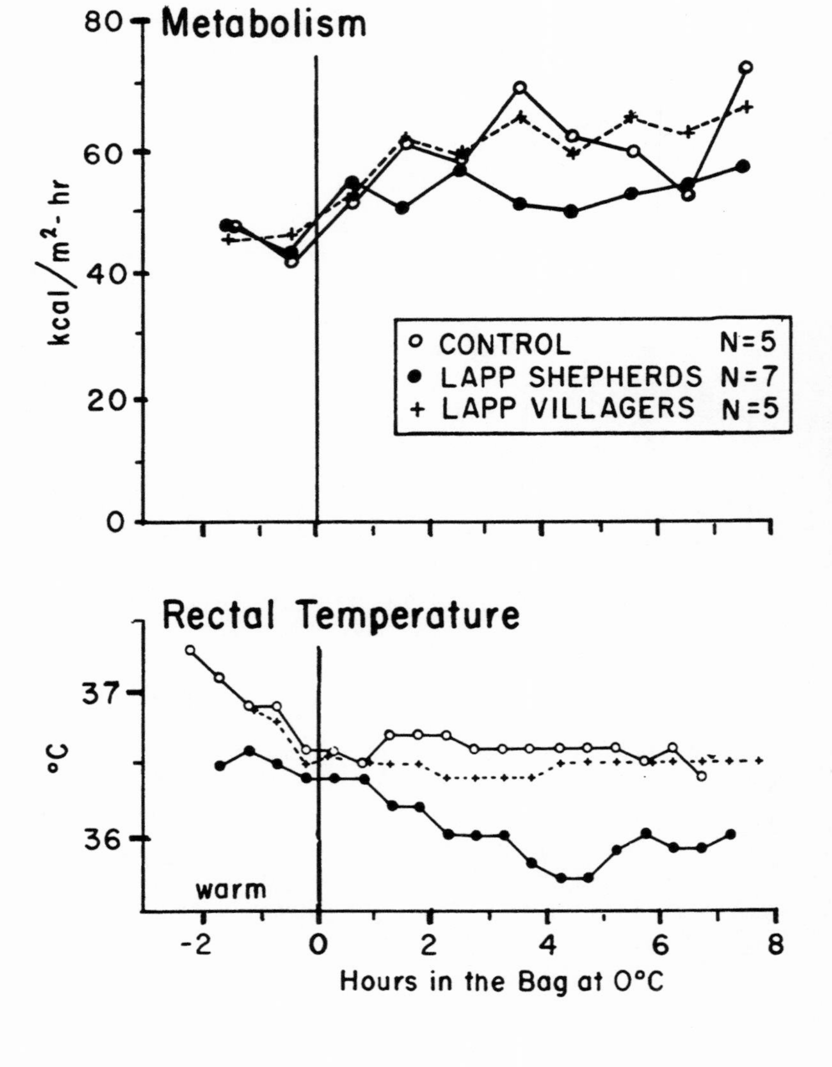

Figure 5-36. Cold-Bed Test of Lapp Natives. Nomadic Lapp shepherds, village Lapps, and control Caucasians were given the Cold-Bed Test to determine whether the Lapps were either cold-acclimatized or cold-adapted. The shepherds slept with a lower metabolic rate, indicating that they experienced less shivering; they also slept with a lower core temperature than the other subjects, indicating a lack of response to, or discomfort from, this noticeable drop in body temperature. Adapted from Hammel (1963).

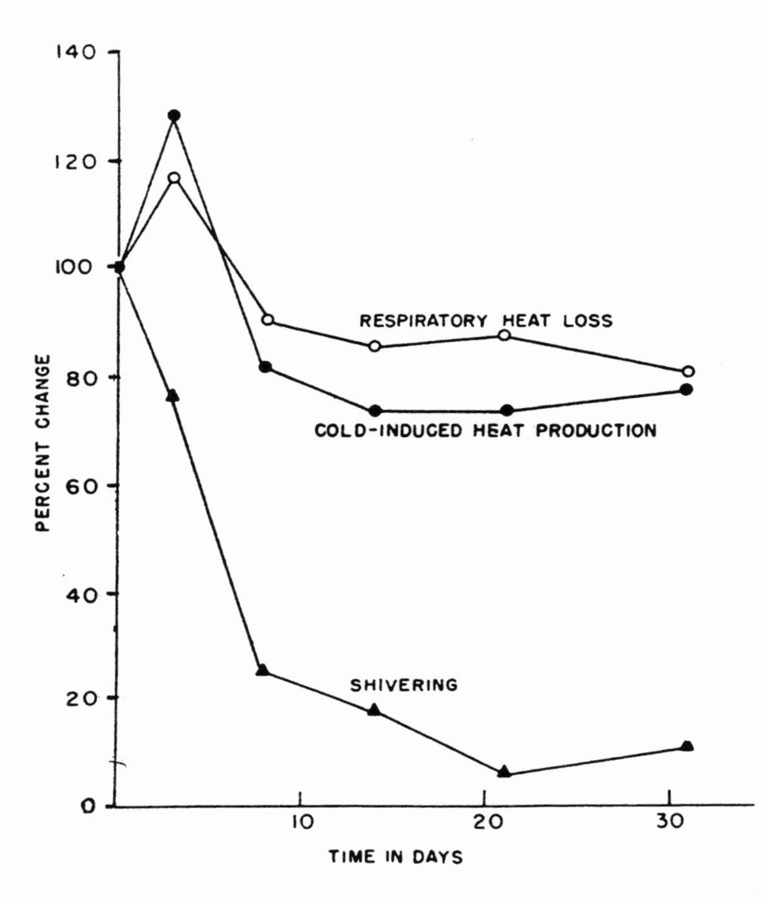

Figure 5-37. Human Cold-Acclimation Within Twenty Days. Unclothed subjects were exposed in a cold chamber (12°C) for 30 days, but they were allowed out of the chamber for the nighttime period. There was a drop in three indices of cold-acclimation within ten days, but especially in shivering. Adapted from Davis (1961).

Figure 5-38. Heat Production of Men Before and After Cold Exposure. Cold-acclimatized subjects had a lower heat production than control in a cold test. The graphs above are remarkable because the results were so similar when obtained independantly in two different laboratories. Adapted from Carlson (1964).

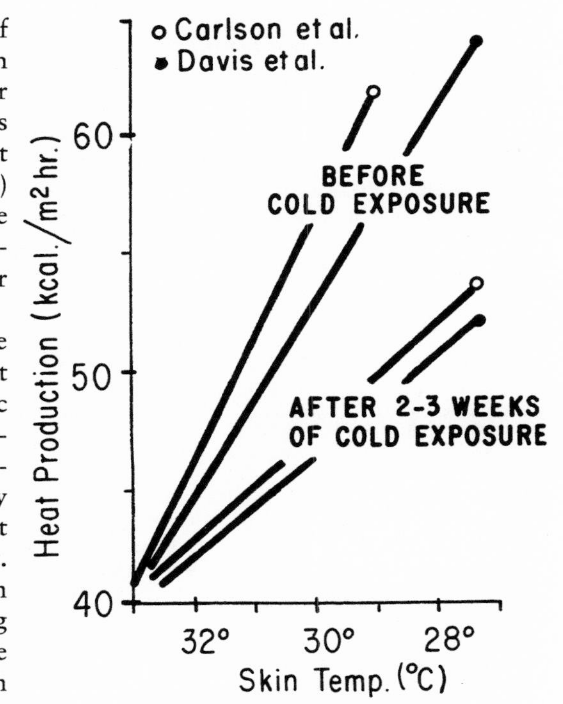

(Fig. 5-35, 5-40). These Arctic dwellers maintained a high rate of blood flow to their fingers during standardized cooling tests in both summer and winter, and after having lived in a temperate climate for several months. In one experiment, the metabolism of the Inuit was higher than that of the controls, and evidence was presented that this was not due to diet (Eagan and Evonuk 1964). Meehan (1955) also observed a cold-induced rise in heat production among the Inuit. It is probable then that the Inuit possess adaptations to combat cold. Milan suggests that they have smaller "cores" and larger "shells."

Australian Aborigines: A Case of Genetic Cold Adaptation. The responses to cold of the Australian aborigines are strikingly different from those of any other human beings. We are indebted to Sir Cedric Stanton Hicks (see Dill 1964) for calling to the attention of physiologists this interesting group of people. The winter night temperatures in central Australia fall to freezing or below, and the night sky radiation temperature is about 20°C lower than air temperature, yet the aborigine customarily sleeps unclothed on the ground at night. Occasionally these people build a scanty shelter or sleep between small fires, but they are capable of sleeping without protection, lying on the ground under extreme conditions. These were, of course, the ideal subjects to test with the cold-bed test. Hicks, later Morrison

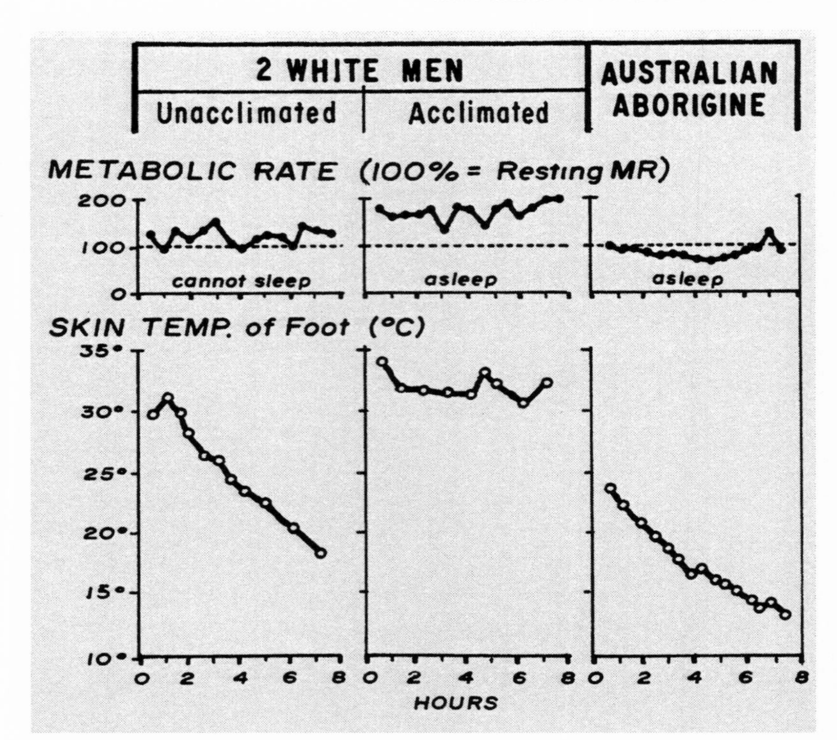

Figure 5-39. Metabolism, Skin Temperature, and Sleep in Cold. These results showed three patterns: the Australian aborigine could sleep with a cold skin and cold core; the acclimatized European maintained a warm skin, could sleep, but lost heat due to the warm shell. The unacclimatized white man could not maintain a warm skin and could not sleep. Adapted from Scholander (1958).

(1957), and then Scholander (1955), found that the metabolism of the natives was not elevated by the cold, and that skin and rectal temperatures were low (Fig. 5-39). Foot temperatures of the natives were 12°C to 15°C. To summarize, the natives increased the insulation of the body shell by vasoconstriction, tolerated moderate hypothermia, and did not elevate the metabolism. These low temperatures are well below the cold pain-threshold of the person of European background (Sinclair 1967).

Further details were obtained when Hammel (see Dill 1964) returned to the native village in the summer to see if the differences observed were inherent or were acclimitizations. By studying the aborigines from two areas in a refrigerated van in summer, he concluded that the Australian aborigines have an inborn ability to tolerate greater body cooling without recourse to metabolic compensation; this tolerance could be increased by prolonged exposure to cold. Therefore, this ethnic group demonstrates both genetic adaptation and acclimatization to cold (Coon 1962).

Native Africans: Responses Similar to White Europeans and White Americans. The first group we will consider are the native inhabitants of the Kalahari Desert. They live on a plateau at altitudes between 3,000 and 5,000 feet. The night climate is sufficiently cold to act as a stress in winter as these people go habitually unclothed. While sleeping on the desert, the inhabitants of the Kalahari avoid cold exposure by wrapping themselves in cloaks and staying near a fire. The cloaks act as a radiation shield between the skin and the night sky (Wyndham and Morrison 1958). During tests conducted by Hammel (1963) and Hildes (1963) the air temperature

Figure 5-40. Finger Temperature During Cold Test. Finger temperatures of a black and a white subject were compared during exposure in a cold room of –12°C (control T_A 33°C). The white subject showed cold vasodilation (Lewis waves). The other subject did not show these and maintained a colder finger skin temperature. Adapted from Rennie and Adams (1957).

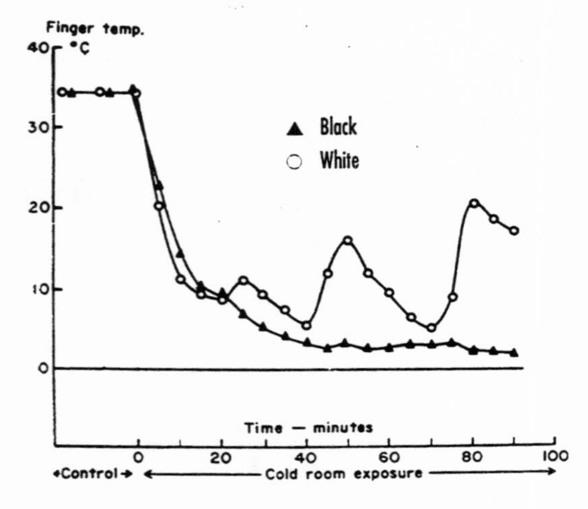

went as low as 0°C; tests of skin, oral temperature, and metabolic responses showed no differences from those of controls. Similar tests, reviewed by Hart (1961) with blacks in other parts of Africa showed comparable results, with the same responses as white controls, or only small differences. For example, there were fewer digital rewarming cycles in the fingers of the black subjects than in white subjects (Fig. 5-40). Although there were no differences in temperature regulation, there were racial differences in blood measurements: black subjects had lower plasma volumes and hemoglobin concentration than did white subjects. These differences would not be expected to reflect on temperature regulation.

Alacaluf Indians of South Chile: A Case of Habituation. A group of Indians of southern Chile, the Alacaluf, have existed as naked people since before the time of Darwin (1845). Darwin saw them at Tierra del Fuego and referred to them as Fuegians. At the time he studied them, there were only about 100, who lived throughout the archipelago of southern Chile (Fig. 5-41). Now, as then, their climate is characterized by cool wet weather, often below freezing. Rainfall is heavy and there is often snow on the ground in winter. These people have little protection from the wind and the rain other than crude huts. They spend much of the day outside gathering food and wood, and they may be soaking wet the entire time. Darwin described his observations of a nude woman in a canoe nursing a baby, also uncovered, while sleet was falling and melting on them both. The Alacaluf now wear enough protection to be described as poorly-clothed, rather than unclothed. When these natives were tested by Elsner (1963), there were slight differences between natives and controls during sleep. It is true that the Alacaluf showed a high metabolism which did not change when they were cooled (Fig. 5-42); this means that in some respects they are like the Australian aborigine, and they allow their bodies to cool without a metabolic response. Here the similarity ends because there was no rectal temperature drop in the Alacaluf as there was with the Australian aborigine. Because of this, one is tempted to decide that in physiologi-

Figure 5-41. Food Sources in Antarctic. These Antarctic limpets were the main source of food for the Alacaluf Indians studied by Darwin. These people were probably the most cold-tolerant on Earth. This mollusk food is tolerant of sub-zero temperatures. The health of mitochondrial membranes serves as an index to cold resistance. The fatty acid composition of mitochondrial membranes in polar bivalve mollusks is totally different from those of temperate zone mitochondrial membranes of bivalves (Gillis and Ballantyne 1997).

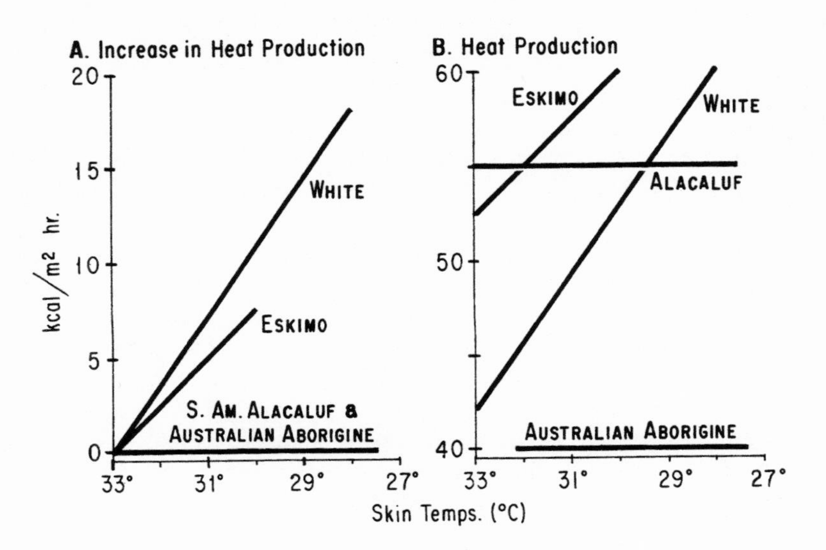

Figure 5-42. Ethnic Summary for Cold Exposure. Differences in response to sleeping in cold air. The Eskimo maintains a higher metabolism than the white person (Caucasian). The native of Tierra del Fuego (Alacaluf) does not raise his/her metabolism when cold. The aborigine sleeps well with a cold skin and low metabolism (Hart el al. 1962).

cal terms the Alacaluf resemble the white controls more than they resemble the aborigines.

In respect to habituation, the situation is very different. Elsner observed these people going about barefooted, occasionally walking in snow, and standing in water that was about 8°C. He decided to study the feet of these subjects with foot calorimetry. After appropriate control periods, the feet of the natives were placed in bath calorimeters at a temperature of 5°C; they sat in this fashion for 30 minutes with amused expressions, indicating that this was a strange procedure and showing no signs of discomfort. When the same experiment was done on the members of the expedition, they were in great agony from the cold water. Nevertheless, there was little difference between the Alacaluf and the controls in the foot heat-loss into the bath water; this indicates that skin temperatures were essentially the same. Elsner considers this experiment a clear case of a very real difference in the abilities of two types of people to tolerate cold exposure by habituation.

Other examples of habituation are increasingly prevalent in the literature. Eagan's (1963) ice water immersion test compares fingers which were immersed 6 times per day with fingers of the opposite hand which were not immersed; after 126 days the temperature of the compared fingers did not differ, but there was a marked difference in pain sensation (see Fig. 5-35). Pain was slight for the test fingers, while it was usually severe for the control fingers. This was considered by the investigator as conclusive evidence for a specific habituation to cold pain. Other habituation experiments by Glaser and Griffin (1962) showed that the pulse rate response of rats to repeated cooling of the tail decreased as the experiment progressed. Hardy (1965) points out a large difference in the characteristics of sensation of cold compared to warmth. He presents experiments to show that there is no significant change in threshold of warmth sensation in spite of changes in room temperature. However, when sensation of cold is studied, there is a marked rise in cold threshold when the skin-temperature is varied and especially when it rises above 34°C. The difference in these responses is functional evidence that sensations of warmth and cold are mediated by entirely different peripheral receptors. Although other workers disagree as to the presence of two different receptors, the rise in threshold for cold has a practical application. This is the basis for the Finnish sauna regimen; after the bathers have a high skin temperature they can then plunge into a very cold swimming pool without any sensation of cold (Kauppinen 1989).

The Diving Women of Korea: A Case of Cold Acclimatization. The work of Hong (1963) on the women divers, or Ama, who harvest plant and animal life from the coastal waters of the Korean Peninsula has attracted the interest of all environmental physiologists. There are some 30,000 of these divers who are initiated into their profession at age 12 and continue to dive to their late 50's. There are similar diving women in Japan. Despite a large seasonal variation in the temperatures of air and sea water, the Ama engage in their diving work throughout the year; the air temperature approaches 0°C and

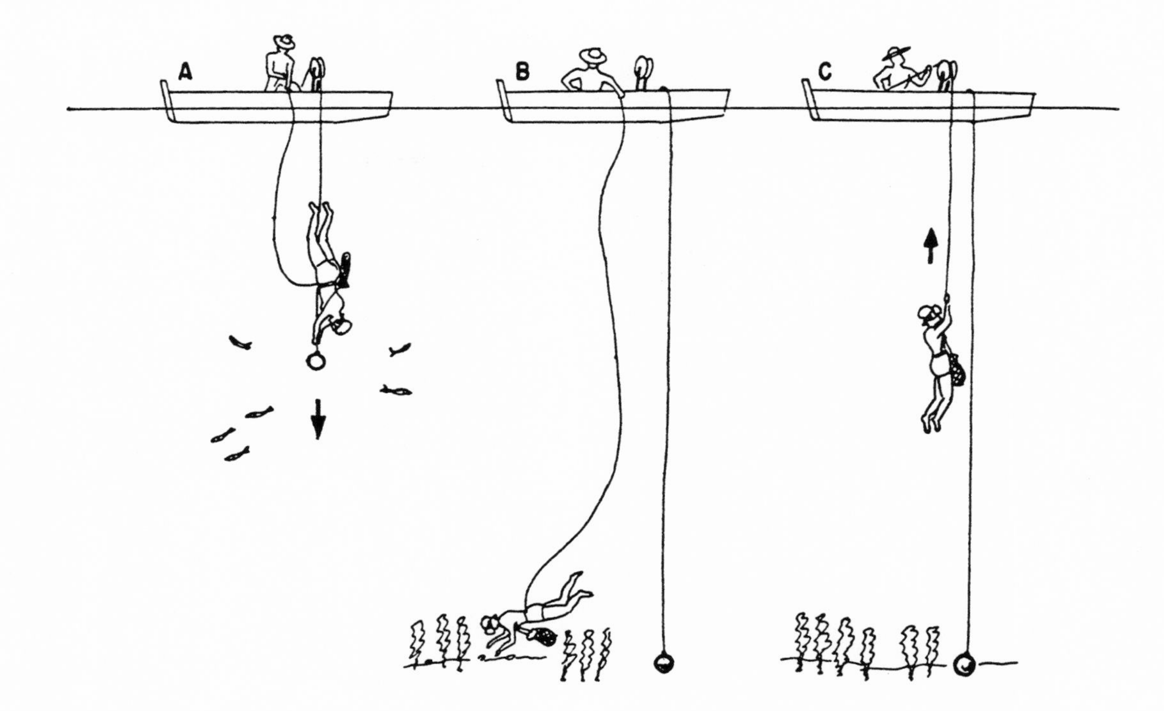

Figure 5-43. Diving Women of Korea and Japan (Ama). The most expert of the Ama are called Funado. They work in the deepest water, descending rapidly by holding onto a weight, and later ascending rapidly with the help of an assistant in the boat who pulls their life rope up over a pulley. Modified from Kita (1965).

the water temperature reaches 10°C (50°F). Hong took oral temperatures during all seasons, and saw the greatest reductions in winter oral temperatures that routinely fell to 33°C (91.4°F) or below. These subjects experience the most severe form of cold stress that human subjects voluntarily tolerate. It is particularly fascinating to physiologists that the sample of individuals undergoing this severe exposure is large enough for safe generalization; the investigator so often has to be satisfied with data gained from 4 or 5 individuals in a cold chamber. The most experienced and skillful divers remain in the cold water the longest period of time; they are referred to as "Funado" (Fig. 5-43). They dive as deep as 20 meters, wearing a ballast belt, carrying a heavy weight (2 kilograms), and holding on to another heavy weight of up to 15 kilograms. The Funado holds on to this weight and sinks quickly with it. As soon as the work is done, the boatman pulls her quickly to the surface.

In this study, evidence of acclimatization was sought in the basal metabolic rates of these subjects. The Ama showed a marked seasonal variation, while the non-diving controls showed a constant BMR throughout the year. The highest metabolic rates of the Ama (35% above normal) were observed in the winter. This increase in metabolic rate cannot be attributed to diet since the analysis of 24-hour urine samples indicated that the excretion of nitrogen was the same in the winter in the Ama and in control subjects. Thus, the elevated BMR of the divers appears to be causally related to the degree of cold stress. According to Hong, these findings represent unequivocal evidence that repeated cold exposure can increase the resting metabolism of human subjects. The finding that the increase in metabolic rate was not due to diet has further support from the evidence that changes in diet in the winter produce small effects upon the basal metabolic rate. Yoshimura (1965) studied this in Japanese subjects and found only a 10% increase in the winter due to diet.

The shivering threshold of the Ama indicates a completely different temperature regulation (see earlier section on shivering). There was no difference in threshold in summer and winter, so this observation cannot be considered a part of the evidence for acclimatization.

Tissue insulation was elevated in the winter in the Ama. This was true despite comparisons with control subjects possessing equivalent subcutaneous fat layers. Therefore, in winter the Ama must have developed vascular acclimatization.

Aging and Tolerance to Cold

Elderly subjects, after sixty years of age, show poor cold tolerance and poor thermoregulatory efficiency (Mathew et al. 1986; Inoue et al. 1992). In a primitive culture, this failure is postponed by being physically fit (Rode and Shephard 1994). One aspect of the thermoregulatory inefficiency seen in elderly subjects was a lack of cold-induced vasodilatation. It is as if many years of cold exposure acts to stress the tissues and organs of the mammalian body.

There is evidence that this supposition is untrue; the longest living small mammals are often exposed to cold for 1/2 of each year. They are hibernators. This was first demonstrated in hamsters (Lyman et al. 1981). The most conclusive study was done by Griffin, whose team banded 16,000 bats in New England. His longest living bats were 30 years old; with a range extending to 20 years (compiled by Jürgens and Prothero 1987). White mice pampered in the laboratory only live about 2 years.

Figure 5-44. Effect of a Cold Face on Heart Rate After 3 Minutes. The exposures were at 20°C, 8°C, and –4°C. The face was exposed to the cold wind; the rest of the body was well insulated. The bradycardia with a 40 mph wind on the face persisted even when the subjects engaged in heavy exercise. Adapted from Le Blance 1975.

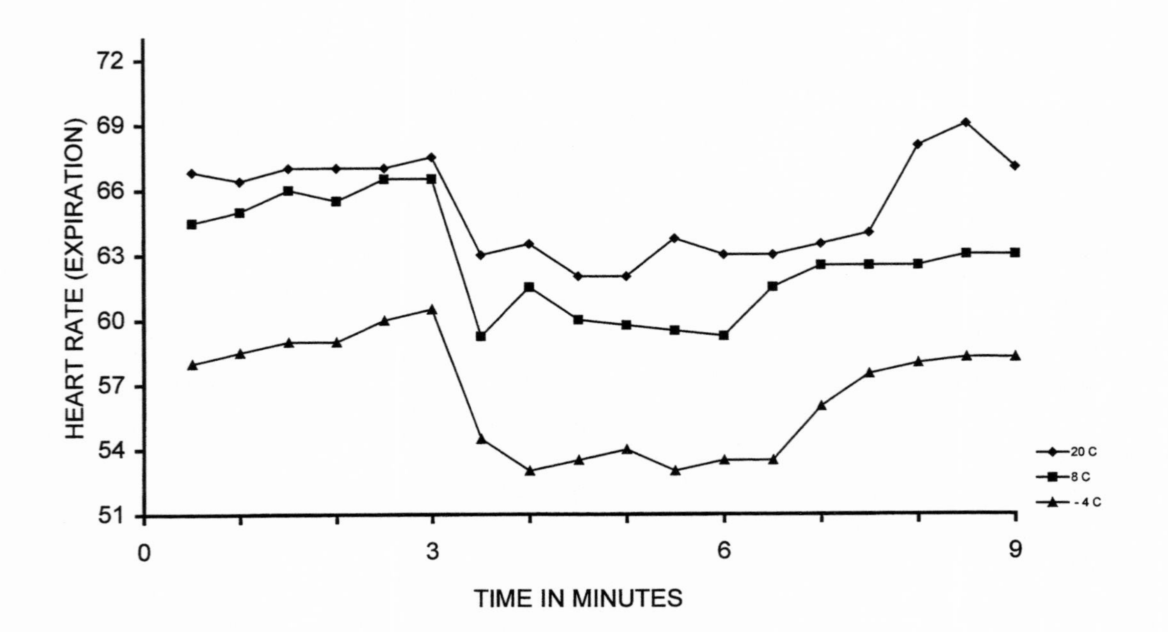

Physical Training and Cold

The addition of exercise to consideration of responses to cold can be introduced by a series of experiments with the variables of cold air, exercise, and a cold wind on the face. Examination of Figure 5-44 shows that when the cold wind is turned on during exercise, the heart rate unexpectedly drops. This could be a health hazard; it will be discussed in a later chapter in its relationship to diving bradycardia.

Investigators have wondered over the past few years whether individuals in good physical condition could adjust to cold and also to altitude more rapidly than unfit individuals (Nair and George 1972; Hellstrom et al. 1970). This is a most enticing question, but unfortunately, the experiments to prove the point are extremely difficult. Probably psychological factors coming into the picture represent the main obstacle to obtaining conclusive data. There is at least one definitive work, however, relating training and cold exposure. Andersen and his collaborators (1966) studied a group of students in Norway who received vigorous physical conditioning and who were then exposed to the cold. The most persuasive information was obtained while the subjects were sleeping. The heart rate of the students was lowered during training approximately 11% (Fig. 5-45). Upon exposure to cold it was apparent that tolerance was increased by the period of training; the trained subjects were able to sleep and rest more comfortably under cold conditions than before training.

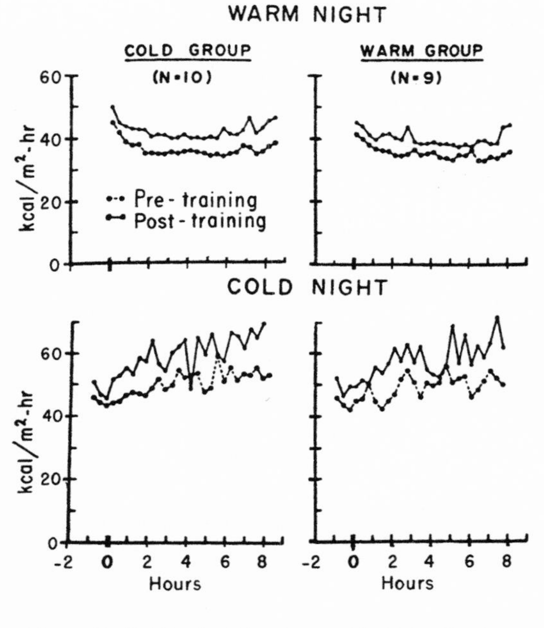

Figure 5-45. Physical Training and Cold Acclimation. The same subjects presented in the previous figure were given the cold-bed test but also in the warm room after being well trained with exercise. However, a non-cold-acclimated group of control subjects gave the same results. One can only conclude that there is an increase in basal metabolic rate due to improved physical condition which must be of advantage when sleeping in the cold. Modified from Andersen (1966).

Now let us compare exercise and shivering. If an individual is becoming quite cold in the outdoor environment, it is not long before he or she is vigorously waving the arms to produce body heat. A recent example proves its effectiveness: on Christmas Eve, 1997, a woman stranded without shelter in the snow-covered mountains of Utah danced to keep warm as overnight temperatures dropped into the single digits (°F). She was rescued unhurt after 18 hours.

The amount of body heat produced in the cold by similar activity was studied systematically by Hart (1967). He compared mice and men exposed to cold and working at varying levels. In the men, oxygen consumption in the cold increased only when they were resting, whereas the oxygen consumption of mice was increased by the cold over and above the exercise at all levels of activity (Fig. 5-46). In the small mammals, adding heat from exercise to the extra heat required in the cold was associated with non-shivering thermogenesis. This, of course, is only present with cold-acclimated animals. The failure to demonstrate the addition of heat production from exercise and from cold exposure in humans may be because the men in the illustration were not acclimated to cold and possessed no appreciable degree of non-shivering thermogenesis. One of the reasons it is easier to produce cold acclimation in small mammals is that along with the cold exposure, these small mammals must drink water which is at the temperature of the cold room (often 2°C). When experiments are done with people to produce cold acclimation, the usual procedure is to bring them hot coffee in the cold room.

A final observation between the effect of exercise and the effect of cold is that any person who has shivered violently and maximally

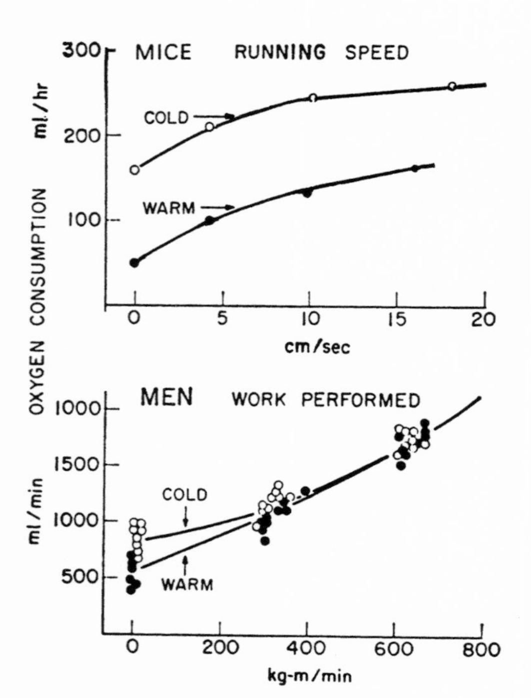

Figure 5-46. Heat Production, Exercise, and Cold Exposure. In this comparison of mice and men, cold-acclimated white mice were tested in a warm environment and a cold environment. In the cold the mice showed non-shivering thermogenesis. Therefore the graph shows the addition of heat production from exercise and cold. In the case of the men, there is a replacement of heat production due to cold with exercise heat. In the latter case, there is no extra heat production in the cold during the exercise. Adapted from Hart (1967).

"for science" is well aware that this is to a certain extent an athletic feat. It is even possible that maximal oxygen consumption is higher under the influence of cold alone than it is with exhausting exercise.

Discussion of Responses to Cold in Humans

In a 1997 symposium, Hammel classified the types of response to cold: metabolic, insulative, and hypothermic. These types of responses have all been illustrated in the above sections which deal with different geographic groups of humans. The metabolic acclimatization is typified by that of the Ama; the insulative acclimatization is found in the cold skin of the Australian aborigines when they sleep at night; the hypothermic responses are also found in the aborigines, who reduce rectal temperature to 35°C; and other metabolic examples are seen in the dexterity-assisting warming of the fingers in ice water by the Inuit. Some of these responses can be interpreted as a lowering of the body thermostat to more economic levels.

There has been some speculation about the advantages of body size and response to cold. Tromp (1963) has reviewed the following data: the average weight of Finns is 154 pounds, of Spaniards 132, of Algerian Berbers 124; in Asia the figures are North Chinese 142; Annamites 112; Andamese 98; African Kalahari natives 89 pounds. These weight differences are due to differences in kind and quantity of food and to metabolic characteristics, although each of these factors is correlated (either directly or indirectly) with the average temperature condition of the region. There is nothing in the data to contradict a rather interesting and practical hypothesis that an efficient adjustment of animals to cold requires a large body mass. Tromp suggests possible additional requirements: short extremities, much fat, deep vein-routing, high basal metabolism, or a combination of these factors (Lasker 1969; McNab 1971).

A common topic of social conversation is the difference in skin temperature between men and women when cold exposed. A husband will often remark: "My wife always has cold hands." There are now occasional papers that put this matter on an objective basis. Toda (1967) reports that women have colder skin (Table 5-7), while a more recent study of Japanese in nursing homes found no significant temperature differences between older men and women (Nakamura et al. 1997). We encourage students to further investigate this topic.

Table 5-7. Cooling Curves of Human Subjects

	Mean Temps. °C	
	Males	Females
Starting skin	33.0°–32.0°	29.1°–29.7°
Cooled skin	9.9°	7.4°
Cooled little finger	4.0°	0.6°

MEETING THE CHALLENGE OF A HEAT-HUNGRY ENVIRONMENT

We must now consider in more detail the technique of survival of large arctic mammals. These mammals and birds have greatly extended the thermal limits within which their survival and activity are possible. The ectotherms (fish, amphibians, reptiles) have not succeeded in extending their survival to temperatures much below freezing. Only among arctic fish do we find that prolonged exposure to cold may lower their lethal limits so that some species may be active when supercooled to levels of -1.7°C. In contrast, arctic mammals may be active indefinitely at temperatures of -40°C and below. The small arctic mammals make use of the snow layer which shows a steep thermogradient. The so-called subnivean layers are relatively warm: Johnson gives figures of -10.5°C (15°F) on the ground under 12 feet of snow when the air temperature was -40°C to -46°C (-40°F to -50°F). Thus, the most interesting Arctic adaptations will be found in the large mammals since the small ones seldom venture out on the surface of the snow at temperatures below -20°C. The same applies at warmer temperatures such as 4° to 6°C (Shaw 1967) as well as below freezing.

The challenge of life in what Scholander calls a "heat-hungry" environment has been met in a variety of ways by the evolutionary process; among the adaptations we will discuss are the cooling of the extremities of large mammals, peripheral nerve conduction at low temperatures, and counter-current vascular heat exchange systems. We will give separate, more in-depth consideration to the adaptations of sled dogs and wolves, and to bare-skinned mammals in the cold.

Arctic Land Mammals. Let us first consider the caribou and moose, herbivores that undoubtedly spend long hours standing in an extreme winter climate that frequently reaches -51°C (-60°F). There are two possibilities for the thermal regulation of the long extremities of the moose: 1) the legs and feet may be warm, requiring a large expenditure of energy and fuel on the part of the animal, or 2) these extremities may remain cold, making special demands on the cellular architecture and function of these extremities. The rule seems to be that arctic animals solve the problem by cooling the extremities. Although the examples to follow relate primarily to mammals, we include data from herring gulls also, because more work has been done on the gulls, and their physiological adaptations appear very similar to those of the mammals.

Some measurements by Irving (1957) will illustrate: on reindeer, when the air was -31°C, the hoof was 9°C and the skin of the smallest diameter of the leg was also 9°C; the rectal temperature was 38°C and the forehead was 36°C. Measurements made on sled dogs at -30°C showed that the pad of the foot was 0°C and the top of the foot was 8°C. As usual with all such animals, the body temperature gradient rises sharply once the limb becomes thickly covered with insulation. On the dog, the skin at the junction of leg and body on the outside was 35°C.

Do the unusual physiological adaptations of moose, caribou, reindeer, and sled dogs apply to only a few scattered specimens? Can these be considered vanishing species? Consider that in one year in Alaska in the early 1970's, more than 7,000 moose were taken by hunters; the biologist in charge felt that the number of licenses issued could have been much higher.

Canada's Barren-Ground caribou are of particular interest because of their long migrations and interesting life cycle. Caribou migrate from Great Slave Lake down to the Trans-Canada Highway in Saskatchewan; some migrations exceed 500 miles. In the Northwest Territories, the Thelon Game Preserve is home to several herds, the largest of which is the Beverly Herd. From this preserve, 400,000 caribou migrate south to their winter feeding ground. Biologists in Canada could sometimes count an accidental damming-up of 100,000 migrating caribou. One estimate for Canada put the total number of caribou at 668,000. There has been a decline in Labrador: one herd was reduced from 2,400 animals to 800 in 5 years. The picture in Alaska is more optimistic: 1997 estimates from the Alaska Department of Fish and Game revealed 32 herds totaling 960,000 caribou (Fig. 5-47). Gordon Haber, the renouned Alaskan authority on wolves, makes an interesting point: these herds should not be considered entities; there is a dynamic flow and exchange of individuals in both directions between herds.

The world's population of reindeer (*Rangifer tarandus*) was estimated in the mid-1980's to be between 3.3 to 3.9 million animals, with approximately 184 herds located primarily in North America, Europe, and Asia. Eighty-three percent of North American, 88% of European, and 68% of Asian herds were stable or increasing at that time.

Figure 5-47. Migrating Caribou in Interior Alaska. Game biologists have estimated as many as 450,000 of these cold-resistant animals in the Western Arctic herd alone, almost half Alaska's total count of 960,000. Photograph by Charlie Ott from National Audubon Society.

Now we will consider the sled dog, a remarkable animal whose footpad measures 0°C when T_A is -30°C. Sled dogs are an example of what the great biochemist Hans Krebs named the *August Krogh Principle*: "For many problems there is an animal on which it can be most conveniently studied." (Recently, Wayne and Staves, 1996, applied the Krogh principle to plants.)

First let us address whether sled dog numbers are declining. In 1960, when Folk spent several months at the largest of all Inuit villages at Point Barrow, Alaska, much of the winter work was still done by sled dog team. Fifteen years later, Barrow had about 500 snowmobiles and very few sled dogs. However, there are numerous camps and groups throughout Alaska and Canada where the inhabitants scorn life in the towns, believing the people who live there to be decadent and unhappy. Most of those Inuit living in remote groups do have snowmobiles, but, being skeptical about their reliability, also continue to maintain working dog teams for survival. Moreover, sled dogs will undoubtedly be maintained and perfected for many more years in Alaska and other northern areas because of the pleasure and occasionally high prize money to be had from racing. Thus, it appears that the physiologist will have a continuing supply of this remarkable animal for measurements to elucidate its adaptations.

Rapid progress has been made in this field by Henshaw with some measurements on sled dogs, but especially with measurements on arctic foxes and arctic wolves (Henshaw, Underwood, and Casey 1972). When these animals are standing on extremely cold snow, or when the foot is tested in a -35°C bath, there is a remarkable mechanism that regulates foot surface temperature at approximately 0°C. This mechanism depends upon an increase in blood-borne heat input to the foot pads to augment tissue metabolism; this blood-borne body heat is served to a cutaneous vascular plexus in the foot pad. This system provides a maximum energetic efficiency because the unit of heat exchange is located on the pad surface, which is in contact with the cold substratum. The blood providing the heat is not distributed throughout the pad. This system should be called "proportional thermoregulation to stabilize the subcutaneous temperature of the foot pad at just above the tissue freezing point." This peripheral plexus system is independent of central arterial regulation. However there are deep arteries supplying the core of the pads which are not independent of central arterial regulation. Apparently, the mechanism of control is a simple, balanced vasodilatation. The well-known Lewis hunting reaction that provides cycles of vasodilatation as found in the cat's paw was not present in the paw of the wolf or fox. The innervation of the pad (not the shunts) has been described for the dog (Brody and Shaffer 1970), and presumably this is active on the pad of the fox and the wolf; the innervation is double consisting of histamine fibers as well as sympathetic cholinergic fibers, both of which produce active reflex vasodilatation.

The physiology of the arctic porcupine has not yet been studied, but the animal provides an excellent example of bare flesh being exposed to extreme cold. They walk on the snow on their large bare fleshy feet at temperatures from -35°C (-30°F) to -40°C (-40°F); their feet are almost as large as the palm of an adult human hand. They show no apparent concern for the snow, so that their extremities must in some way be protected (Irving 1964).

Turning now to arctic gulls, it is remarkable that they can swim with large bare feet in open water and walk on the ice without the webs being frozen white. Irving (1957) made measurements of one gull when the air was -16°C. The web varied in temperature from 0°C to 4.9°C. The middle of the exposed leg was 7.9°C, but the skin under the feathers at the upper thigh varied from 15°C where the feathers began, to 37.8°C a few inches closer to the body. One would suppose that such cooling would reduce the metabolism of these tissues to an ineffectual level, yet in the webs of the feet of these gulls, Irving observed active circulation in small vessels (while a thermocouple in the adjacent tissue registered 0°C).

In another species of gull (*Larus argentatus*) kept outside during the winter in Boston (Chatfield, Lyman, and Irving 1963), the leg tissue temperature was 6°C. Excised tibial nerves from these legs showed responses that were different in the distal and proximal portions; the nerve from the cold bare part of the leg became blocked by cold between 2°C to 5°C, while the central part of the nerve from the warm part of the leg under the feathers failed at 8°C to 13°C. These results are particularly unusual because they show that parts of cells (the neurons are cells) can be acclimated to a relatively cold temperature (2°C to 5°C), while at the same time other parts of the same cells will only function at the warmer temperature of 8°C to 13°C. Of additional interest was the finding that in summer no part of the nerves from gulls would conduct at the cold temperatures (2°C to 5°C). We will call the first type *cold conduction* and the other *normal temperature conduction.* When hens were kept in the cold in outdoor runs in winter, and their leg nerves and temperatures were studied in the same way, it was found that their feet were warmer than those of cold-exposed gulls (even summer gulls). These "winter hens" represent the second possibility we mentioned as a direction the arctic mammals could have taken (having warm legs). Excised nerves from the legs of winter cold-exposed hens were sensitive to cold; they demonstrated "normal temperature conduction." In summary, we can say that the legs of arctic gulls possess several interesting adaptations concerned with the use of cold extremities for conservation of heat; also the physiological behavior of these legs demonstrates acclimatization to cold.

A very similar example of acclimatization to cold of a part of a limb is found in the work of Miller and Irving (1964). They were studying nerve function in several species of Alaskan mammals. The caudal nerves of muskrats were excised; action potentials from these tissues were recorded at temperatures as low as -6°C. Conduction persisted but velocity was reduced 20- to 30-fold. The investigators showed that these nerves were supercooled (by transient increases in temperature at the moment the nerve actually froze). The capability of caudal nerves to conduct in a supercooled state was lost in several muskrats maintained indoors for several months during the winter. Apparently these nerves behave like the tibial nerves of the arctic gull. When sciatic nerves from muskrats were compared with caudal nerves, the sciatic nerves would not function at temperatures much below 4°C. This seems appropriate since these nerves are not subjected to the large temperature changes which take place in the tail (Miller and Irving 1963).

These adaptations and evidences of cold acclimatization illustrate the successful attainment of homeostasis by small and large Arctic animals so that their body temperatures are nearly identical with those found in mammals in the tropics. Irving found the mean body temperature for 20 species of mammals in Arctic Alaska was

38.5°C, and for 60 species of mammals from temperate and tropical zones the mean figure was 38°C (Irving 1957).

Arctic Marine Mammals and Fish. The countercurrent vascular heat-exchange system as an adaptation is found in a number of types of mammals extending from the tropics to the Arctic. In no case is this system better illustrated than in the fin of the porpoise. Scholander asked the question: "What prevents whales in the polar seas from being chilled to death from heat loss through their large, thin fins?" The fins that Scholander and Scheville studied (1955) included the flippers, the caudal fins, and the dorsal fin. In all cases they found that each major artery was located centrally within a multiple venous channel. No matter what else the function of such an arrangement, these bundles must exchange heat between the arteries and the veins. The system is referred to as an arterial-venous counter-current system, serving for heat preservation. In such an arrangement the warm arterial blood is cooled by the venous blood which has been chilled in the fin (Fig. 5-48). The result is a steep temperature drop from the body into the appendage; the heat of the arterial blood does not reach the fin but is short-circuited back into the body in a venous system. Body heat is conserved at the expense of keeping the appendage cold. There are also separate superficial veins associated with this system; their function might be as follows: if the animal needs maximal heat conservation, blood circulation through the fins would be slow and the venous return would preferentially pass through the countercurrent veins; on the other hand, if maximal cooling is needed, as during exercise in warm water, this could be accomplished by a high rate of blood flow through the fins with a venous return through the superficial veins. One author (Tomilin 1950) made observations on a dolphin out of water and found that the fins could vary between 25°C and 33.5°C, while the core temperature varied only 0.5°C. This appears to be an example of adaptation which is not associated with seasonal acclimatization. It is an adaptation for heat exchange.

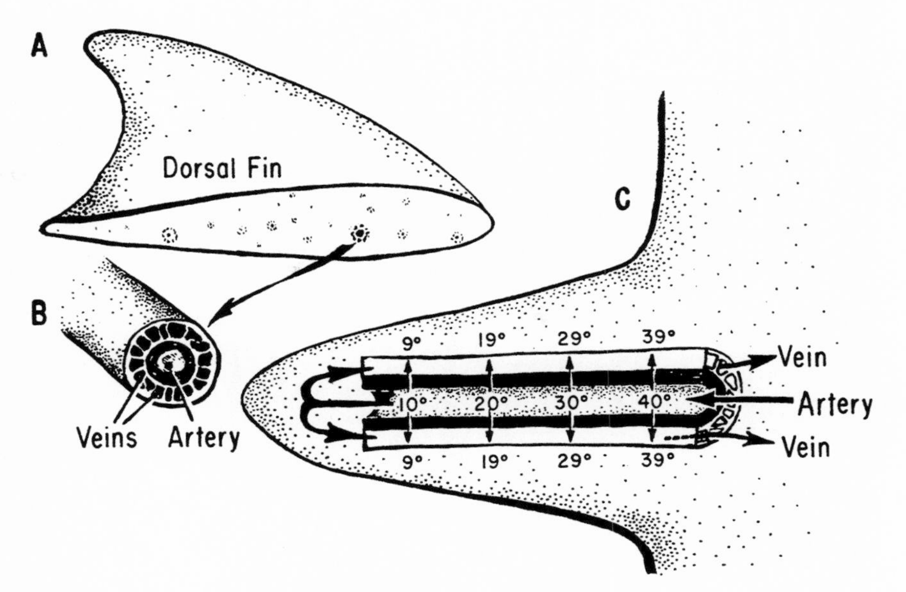

Figure 5-48. A Mechanism for Rewarming of Blood. In a series of mammals from humans to whales, Scholander has accumulated information on the rewarming of cold venous blood by arterial blood. **A** shows a section of the dorsal fin of a bottlenose porpoise where the veins surround each artery. **B** shows a close-up of the artery surrounded by the multiple venous channel. **C** shows a hypothetical temperature gradient in this concentric countercurrent system. Adapted from Scholander and Scheville (1955).

The same adaptation for rewarming cold venous blood is found in both the arm and the leg of humans. This efficient arrangement for heat economy was first pointed out by Bazett (1948). Some of his data are shown in Fig. 5-49. He found that the temperature of arterial blood near the chest was about 37°C, but the temperature of this blood at the wrist could be as low as 25°C. Most importantly, when venous blood nears the chest (leaving the arm or the leg) it is at a temperature of almost 37°C.

The use of fatty acid oxidation is found in special cases of envi-

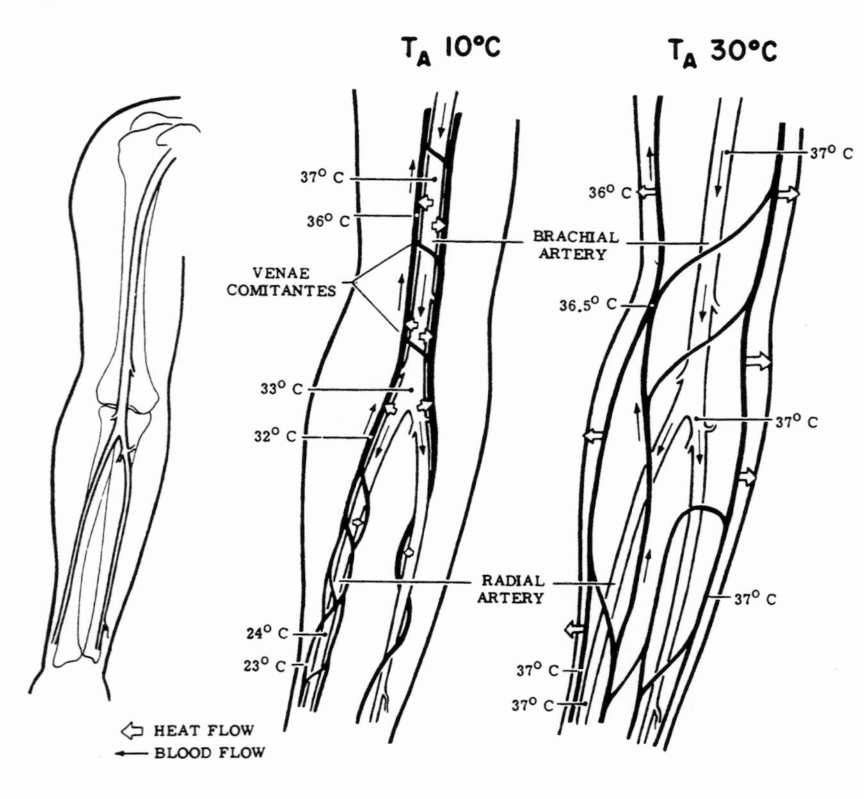

Figure 5-49. Countercurrent Heat Exchange in the Human Arm. When the room is cold, heat is given off from the brachial artery to veins around the artery; as a result, radial arterial blood may be 24°C. When the room is warm, heat is given off from veins near the skin; as a result, radial arterial blood may be 37°C. Adapted from Bazett (1961).

ronmental challenge. Diving mammals have densities of mitochondria 1.7 to 2.1 times the densities of those in terrestrial mammals, and the fuel for these mitochondria are fatty acids, metabolically poised for diving (Kanatous et al. 1997).

Earlier we mentioned the supercooled tissue of the muskrat. Such a protection is common in fish living in the cold regions, and it is achieved by the presence of either glyco-proteins or antifreeze peptides (AFP). The latter have received the most study. AFP's bind to and prevent the growth of minute internalized ice crystals, thereby inhibiting the freezing of extracellular fluids (Sicheri and Yang 1995).

The Antarctic marine environment is characterized by temperatures that approach the freezing point of sea water, -1.9°C. This condition has been present for 40 million years, resulting in the evolution of 120 unique species of fish (order *Notothenidae*). Their plasma osmolality is twice that of temperate and Arctic marine teleosts. For many of these fish, an environmental temperature of 4°C represents a "heat wave," and they cannot tolerate a warmer temperature. In respect to antifreeze peptides, we find parallel evolution in Antarctic, Atlantic, and Arctic fish: within one family, the *Zoarcidae*, the AFP's are found in three widely separated geographic areas (Gonzalez-Cabrera et al. 1995). Within this family, the AFP's are 50% identical in sequence (Osadjan et al. 1996).

Figure 5-50. The Arctic Tundra Wolf in Heavy Winter Coat. These animals chose to sleep outside their shelters on the shores of the Arctic Ocean at the Arctic Research Laboratory, Point Barrow, Alaska—even in the coldest winter weather. This male arctic wolf is carrying abdominal radio transmitters for body temperature, heart rate, and EKG. Temperatures of -54°F and winds up to 25 mph were recorded at his enclosure at Barrow.

Physiology of the Sled Dog and Wolf. If a biologist is fortunate enough to follow closely, by airplane, a black Arctic wolf running in great leaps over the snow of the Arctic slope, he will remember it as an unforgettable experience. Of particular interest are the physiological adaptations of this beautiful animal, as well as those of sled dogs, some of which are genetically part wolf (*Dog Mushers' Magazine* 1965). The ability of both to tolerate cold is excelled only by the Arctic fox. With Max Brewer of the Arctic Research Laboratory, we maintained groups of wolves at Point Barrow, Alaska for several years. During their period of rest, even when the Wind Chill was -60°F, these wolves would be sleeping or resting in the exposed part of the pen instead of using the warm, dry, protected parts of the enclosures (Fig. 5-50). As these animals get up when approached, their customary bed is evident since their heat has melted a bowl-shaped depression of ice replacing the original fluffy snow. If you place the tip of a pencil (for measurement) through the fur of the side of one of these animals, the insulation, irrespective of scattered guard hairs, amounts to 66 mm (Fig. 5-51). This depth of insulation is exceeded only by that of the white Arctic Dall sheep, which is 70 mm. The insulation of sled dogs is similar; it is a common observation of those who chain their dogs to a small kennel that even in the most severe weather these animals curl up in the snow or on top of the kennel rather than inside their shelters (Fig. 5-52). Scholander et al. (1950) studied the critical temperature of arctic dogs, which showed no rise in metabolism at -25°C.

Acclimatization to Cold in Dogs. Undoubtedly we need only the thickness of the insulation of dogs and wolves to explain their extreme tolerance to cold when lying quietly. It is doubtful that one could demonstrate acclimatization to cold in these species. A thick fur in sled dogs is always grown twice a year, even in a very warm environment (Meehan 1957). This is by no means a disadvantage, since a heavy fur coat assists mammals to tolerate the heat. Some of the other breeds of dogs appeared to show acclimatization; McMillan (in Dripps 1956) states that in his laboratory, and in the laboratories of three other investigators mentioned by Swan, they have lost a great many more hypothermic dogs from ventricular fibrillation in the summer than in the winter. This evidence in favor of cold or seasonal acclimatization is supported by work from another laboratory (Covino and Beavers 1957) indicating that dogs submerged in cold water expired with a rectal temperature of 18.6°C, but the ones that had been acclimated to cold survived to a rectal temperature of 14.9°C.

We must now give further consideration to the dog's lower critical temperature, which is made possible by its excellent insulation. As indicated earlier, Scholander (1950) concluded that the critical temperature of dogs and foxes was between -40°C and -50°C, and they calculated that only a 40% increase in their metabolic rate would be needed to maintain body thermal balance at -70°C. Since the winter temperatures in most Arctic areas are almost always well within the thermal neutral zone (+30°C to -40°C) of sled dogs and foxes, no

PULLING POWER OF DOGS

Polar history owes much to the contributions made by these remarkable, tough animals. On average travel days, explorers and their dogs faced harsh conditions full of cutting winds and blinding snow, with the dogs constantly pulling heavy loads. On one journey, Amundsen's six-dog hitch pulled 677 pounds for 62 miles in one day. Sverdrup once traveled with nine sleds, each weighing approximately 670 pounds and pulled by six dogs (Fairley 1959). Sverdrup had expected they would make 20 to 25 miles per day, but their average was 19 owing to rough ice; he described crossing pressure ridges that were 80 to 120 feet high. Nevertheless, on one day he did make 37 miles. Some accounts of polar history have documented the remarkably low calories fed to sled dogs each day in spite of their heavy work. Sverdrup's dogs were relatively well fed, with either 1 pound of pemmican per day or 1.8 pounds of fish.

For a more recent example we turn to Alaska's famous 1,100-mile Iditarod race, and to Libby Riddles, the first woman to win it. In mid-race, Riddles tied her dogs to a dead tree while she was occupied with trail tasks. In their eagerness to run the team broke off the tree and disappeared, dragging it down the trail. Fortunately, another musher caught the team, and Riddles continued on to win (Riddles and Jones 1988). A racer who never stinted on her dogs' rations, Riddles also received an award for the excellent health and condition of her dogs, something closely monitored at mandatory checkpoints throughout the annual race. One last example of the pulling power of dogs: observers witnessed a team of 10 dogs dragging the truck to which they were tied; the vehicle was in gear with the parking brake engaged.

Figure 5-51. The Heavy Coat of the White Wolf. Few people see the white arctic wolf. The color acts as camouflage, but not in summer. Photo on Ellesmere Island by L. David Mech, 1996.

seasonal variations which might represent acclimatization would be predicted.

In an experiment at the Arctic Aeromedical Laboratory (Fairbanks) two investigators decided to test for acclimatization in the sled dog by using daily caloric intake and body weight as criteria of metabolic energy expenditure (Durrer and Hannon 1962). The well-insulated sled dog was compared with the poorly insulated beagle dog. During the period of measurement, the average monthly ambient temperature ranged from 17°C to -22°C. The caloric intake of the huskies rose from a midsummer low of 49 $kcal{\cdot}kg^{-1}{\cdot}day^{-1}$ to a November high of 87 $kcal{\cdot}kg^{-1}{\cdot}day^{-1}$. Mid- and late- winter values averaged about 79 $kcal{\cdot}kg^{-1}{\cdot}day^{-1}$. The beagles studied during the summer had minimum values of approximately 85 $kcal{\cdot}kg^{-1}{\cdot}day^{-1}$. With the onset of winter the beagles increased their intake to a high of 144 $kcal{\cdot}kg^{-1}{\cdot}day^{-1}$ in November, and 131 $kcal{\cdot}kg^{-1}{\cdot}day^{-1}$ in March. The relative magnitude of the seasonal changes was quite similar in both groups of dogs, suggesting a paradoxical and unexplained metabolic acclimatization in the case of the huskies.

Exercise, Insulation, and Heat Dissipation in Sled Dogs and Fur-Clad Humans.

The skin temperature of dogs exposed to -50°C is a consistent 30°C, which is also the skin temperature of arctic foxes and many other Arctic mammals upon exposure to cold. Curiously, when these animals are warmed, they still appear to be comfortable when skin temperature is 37°C. One wonders how these active animals can dissipate heat when their metabolism is increased many-fold in exercise. Do dogs and wolves with heavy insulation carry on sustained exercise, or is their work performed in short bouts?

We do know that dogs with heavy coats still carry on heavy, sustained work for the Inuit and other Native American groups today. In 1964, the well-known Inuit guide, Pete Savolik, estimated that there were 500 working sled dogs in the village of Barrow. If one walked in mid-winter toward the village, the sustained running of these dogs was evident because one could see the teams fanning out in many directions to accomplish the business of the owners. One seldom saw these dogs walking or trotting; they were invariably at a dead run (Fig. 5-53).

Many long-haired dogs also compete in sled dog races between Inuit villages, and in the annual national competitions in Fairbanks and Anchorage. In the national races, these dogs are required to cover 70 miles in 3 days: 20 on day 1, 20 on day 2, and 30 on the final day. The winners cover these distances at about 15 miles per hour; this, after some teams have run from their own village the day before the races (Fig. 5-54).

How can these animals perform this work while carrying an insulation of approximately 66 mm in thickness? One of the adaptations they have consists of suppression of piloerection so that the hair lies very close to the skin. The major heat-dissipating system is the tongue which is protruded many inches in most cases, so that panting may take place over the moist surfaces of the pharynx and the tongue.

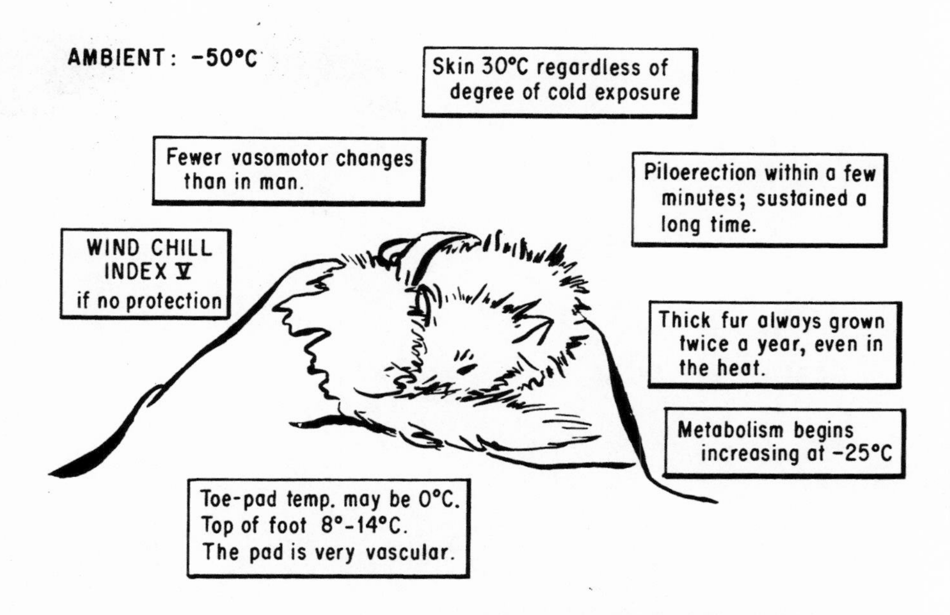

Figure 5-52. Thermal Relations of Sled Dog in Snow Drift. This diagram illustrates an extreme environmental situation which many sled dogs experience each winter. When chained to small kennels they elect to remain in the snow rather than take shelter even at the coldest temperatures. The foot temperatures were obtained (by Irving) from a dog standing in cold air. Heat loss is reduced because the feet are maintained at such a cold tissue temperature.

One wonders if a similar mechanism is necessary with caribou and moose; the caribou and moose do pant, but do not extend the tongue the way the dog does. The problem of dissipating heat is less in these animals because their insulation is only approximately 38 mm thick and because their long slender legs, which have scarcely any insulation on them, must serve as a valuable place to shunt the hot blood from the body. This mechanism is similar in the horse, which has active sweat glands, and, depending upon the breed and environment, grows heavy insulation in the winter.

The adaptations for dissipating heat by the dog and wolf do not appear adequate to explain the efficiency of the working sled dog. This is particularly the case when we learn that there is less vasodilatation in the skin of the dog than in humans (Meehan 1957). We must now consider in finer detail the relationship between insulation and the cold environment. Let us examine and compare humans wearing fur clothing and dogs.

Figure 5-53. Working Dogs at the Tip of North America. These Eskimo Huskies are leaving Barrow for a day's work. On that day the temperature was -50°F and the sun was above the horizon for only 30 minutes. The background is the Arctic Ocean.

The problem of dissipating heat through insulation has been around since the Inuit arrived in Alaska about 10,000 years ago. Among the many problems these cold-weather natives had to solve was how to keep from building up a large quantity of wet or frozen insulation. The fur-clad human does not have the same physiology as the sled dog, and as soon as he or she begins to work there will be

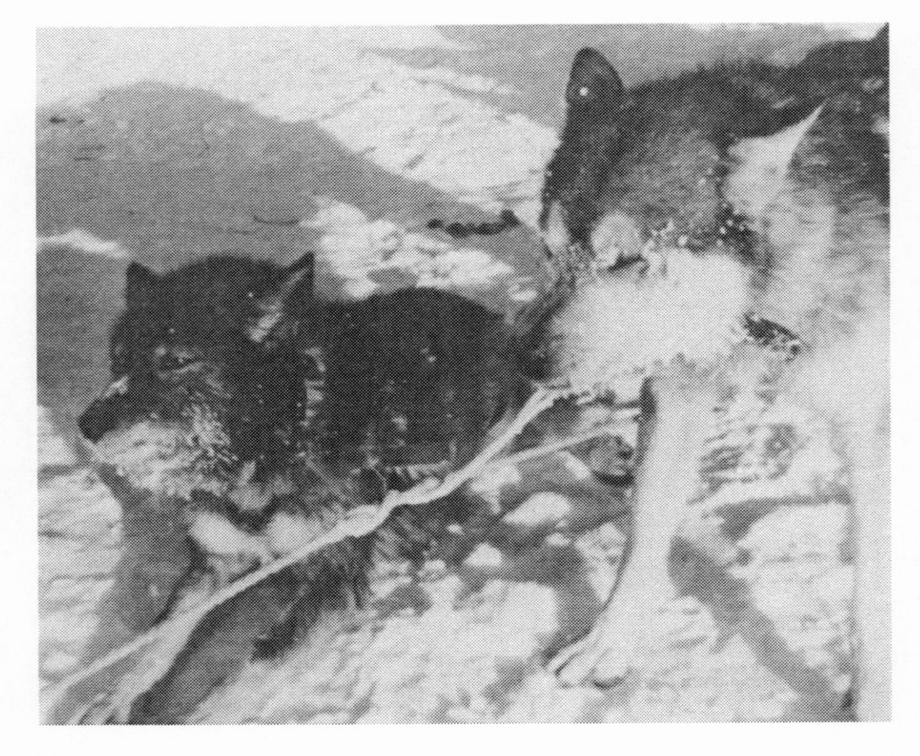

Figure 5-54. Alaskan Huskies After a Race. The run was 30 miles; air temperature was –20° F. Even at this temperature, these hardy animals required no protection after their work. The frost indicates the temperature gradient through their fur. Some physiologist should compare this fur with the wet insulation of Shetland ponies in winter under conditions where they must labor heavily in deep snow.

sweat under the insulation. The person will become, as Stefansson (1960) put it, "a tropical man in Arctic clothing."

The problem is illustrated by a quotation from a member of Scott's Antarctic expedition, Cherry-Garrard (1922), who wrote: ". . . on the most bitter days it seems that we must be sweating; and all of this sweat instead of passing away through the porous wool of our clothing and gradually drying off us, froze and accumulated. It passed just away from our flesh and then became ice . . ."

Does this mean that every time an Inuk runs with his dog team his insulation is coated with frost? The answer is no, because the Inuit have evolved two methods to avoid this. One is to constantly pull off the outer fur parka so that much of the time the skin is quite cool. More importantly, their clothing was cleverly designed with many ventilation areas, gaps, and drawstrings (Fig. 5-55). Particularly in the great barrens of Canada and Northwestern Greenland, those cold-weather natives who are still wearing the time-honored fur parka instead of Sears flannel shirts, invariably show the bare skin of their backs between pants and parkas when they bend over. Air is constantly being pumped in through either the top or the bottom of the parka and out by some other opening. Other venting areas are shown in the figure.

Before leaving the topic of arctic clothing, it is important to point out that the Inuk is not always a tropical person in arctic clothing. According to Richard Perry (1966), some hunters have been described sitting over the breathing holes of seals, or waiting for Eider ducks, for up to 72 hours at a stretch in sub-zero temperatures (Fig. 5-56). It is unlikely that the present generation would be willing to, or would need to, accomplish such a feat.

Before the hunter begins to sit in wait, considerable moisture may have accumulated in the warm fur clothing. For example, if an ordinary subject in arctic military clothing exercises moderately for three quarters of an hour, approximately 630 ml of moisture will have accumulated in the clothing if the work was done at -10°C (-14°F). This of course explains the need for the chimney effect described above.

Unlike the human, the dog and wolf cannot shed their insulation during exercise and, in fact, do accumulate a good deal of frost. It looks as if some remaining mechanism should be present by which we can find more biological adaptability in these species. Part of the answer to this was found in experimental results obtained at the

Harvard Fatigue Laboratory: the Clo value of heavy winter insulation was markedly diminished during movement. This was demonstrated with a heated, sealed cylinder covered with wool and pile, which simulated both a human's leg in Arctic clothing and a dog's leg covered with fur. The cylinder could be swung to simulate walking movements. The insulation was studied at -18°C, first with the cylinder immobilized, then with it moving at 62 cycles per minute, the approximate frequency of leg movement when walking at 3.5 mph. The Clo value of the insulation when the cylinder was not moving was 1.7 Clo, and with the cylinder moving was 0.9 (Belding et al. 1947). This reduction of 47% helps to explain how the dog and wolf can tolerate their heavy insulation when running. One study (Tregear 1965) found that the thermal conductance of pelts was doubled at 20 mph compared with 0 mph.

The effect of wind on the insulation value of fleece on sheep was presented by Bianca (1967), using data from Joyce and Blaxter, who found that a wind of 9.6 mph reduced insulation by approximately 50% compared with 0.6 mph. The effect of various wind speeds with different fleece lengths is presented in Fig. 5-57; it would not be difficult to fit the situation of the sled dog coat into this diagram. Earlier we mentioned a principle, which is evident on the diagram, that at higher wind speeds the percent change in insulation is smaller.

Now that we have considered heat dissipation of sled dogs and fur-clad humans, let us examine further the practical problems of the working sled dog. Prolonged hard work (sled pulling) at mild temperatures (about 0°C) causes an increase of only 1°C to 1.5°C in the rectal temperatures of the average sled dog. At the finish of the Annual North American Sled Dog Championship Race in Fairbanks, Alaska, the average rectal temperature was 39.7°C for 17 dogs in 1963 and 40.1°C for 71 dogs in 1964 (Eagan, Durrer, and Millard 1963). No difference in the averages of teams were found

Figure 5-55. Chimney Effect in Inuit Fur Clothing. When sled dogs exercise, they become tropical animals in an arctic environment. The Inuit prevent this effect by designing their fur clothing so that when exercising in extreme cold, vents can be opened permitting maximum ventilation. Only by this means can the accumulation of large quantities of sweat be prevented. This is called "the chimney effect."

Figure 5-56. Waiting for Eider Ducks. An Inuk at the tip of Point Barrow is dressed in sealskin boots, fur pants, and parka. The heavy clothing is adequate for sitting for long periods at 0°C with a 25 mile per hour wind; this gives an equivalent temperature (wind chill) of -17°C. The challenge here is the wind which causes a climate unlike the dry, still, cold of parts of Siberia, or the wet cold of the Aleutians. The wind can easily halve the protective efficiency of heavy clothing. Photo by W. Dietrich.

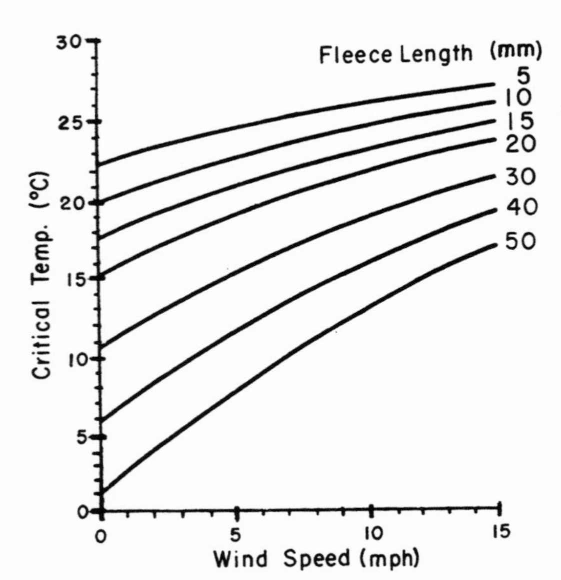

Figure 5-57. Variable Critical Temperatures of Cheviot Sheep. When wind speed and fleece length was varied, the result was a family of curves representing the thermal neutral profile. For example, the metabolism rises above basal rates if the fleece is only 5 mm long, the wind speed is 15 mph, and T_A is about 26°C. On the other hand, if the fleece is 50 mm and the wind speed is 15 mph, the metabolism will not rise above basal rates until the T_A is 16°C. Modified from Joyce and Blaxter (1967).

relating to finishing position in a race. Why is this temperature rise so small? The eating of snow along the sled trail appears to be an important factor in limiting the heat gain of some dog teams. This cannot be the case with all teams because many drivers do not permit dogs to break stride to obtain snow. The advantage of eating snow was studied by Eagan, Evonuk, and Boster (1964) by having dogs swallow temperature-sensitive radio capsules (Iowa transmitters—Essler and Folk 1960) before sled runs; these transmitters (and presumably part of the stomach) cooled to below 15°C in some dogs because of snow consumed.

The Physiology of Bare-Skinned Mammals

In an earlier section we discussed the adaptation of maintaining cold extremities as a response to a cold environment. The cooling of the extremities and the saving of heat by this means has a parallel in another adaptation of arctic animals. This is the maintenance of a cold skin all over the body in mammals that do not have hair. Irving began to study this type of adaptation with domestic swine maintained at Fairbanks; these animals tolerate the extreme cold down to -50°C with apparent ease. The metabolism of two young pigs was studied to look for the rise in metabolism at some critical cold air temperature; this critical air temperature was 0°C. This appears to be a rather low critical temperature for naked animals. When they were exposed to cold, their skin temperature declined steadily until it was as low as 10°C to 8°C.

In large hogs the gradient from cold skin to the warm body interior extended by direct measurement as much as 100 mm, varying in depth with the temperature of the air. The skin showed periodic warming and cooling reminiscent of the Lewis waves of vasodilatation found in human cold fingers. The evidence that the cold skin is an economic biological adaptation is, as mentioned above, that the critical air temperature is 0°C. Irving remarks that as a consequence, the cost of feeding pigs is not noticeably elevated in the cold Alaskan winter.

Irving then turned his attention to the study of seals, which are comparable in many respects to domestic swine. Out of water their thin hair affords them about 1/10 the insulation of the hair on the caribou, but in water it gives them practically no protection whatsoever because the hair is wet and the surrounding water is not separated from the skin by as much as a millimeter. He found that the harp seal, a strictly northern seal, showed no increase in metabolic rate when exposed in ice water. Their skin was practically the temperature of the surrounding water, differing by only 0.05°C. Thus, in both the swine and the seal, the heat loss from skin to environment is greatly reduced by the cooling of the skin. Such a cooling in humans, at least of the extremities, could not be tolerated since dexterity is lost at 20°C.

With harbor seals the distribution of the temperature gradient varied with distribution of blood. The seal controlled this in a different fashion when it was in ice water and when it was in warm air; heat was lost from core to surface over a gradient of 40 to 60 mm when the animal was in ice water, but this gradient was scarcely 20 mm deep in warm air. Such a gradient evidently can be compared with the insulation of fur which is found on other animals; just as is the case with piloerection of fur, the insulation of the seal and swine can be varied in thickness to match the various media to which the animal is exposed. It is, of course, the changing vascular bed that brings about this variability. In the seal the regulatory ability of this vascular bed is severely challenged; this is because the thermal conductivity of water is over 20 times greater than that of air.

The temperature regulation of the elephant, which is also bare-skinned, is entirely suitable to discuss here. Very little information has been obtained. First, one should ask if such a large mammal has a different body temperature from other mammals. Buss and Wallner (1965) provided a figure of 36.3°C (97.5°F). This information was based upon a large series of temperatures of fecal matter taken immediately after defecation of elephants in the field, and of rectal temperatures of recently killed elephants. This mean temperature showed very little variation; this must relate to the functioning of the elephant's testes, which are located internally and do not have the benefit of a thermal regulatory scrotum. This body temperature is not particularly unusual; hamsters, for example, at noontime have a colonic temperature of 36.2°C (97.0°F)(Folk, Schellinger, and Snyder 1961).

Because two species of the elephant group (hairy mammoth and mastodon) lived in the Arctic area in quantity, and because a similar species, the hairy rhinoceros, was also within this area, it is reasonable to raise the question as to whether the elephant with its bare skin can tolerate severe cold exposure as can pigs. This seems to be a controversial subject; on the one hand, keepers in circuses try not to expose their elephants to temperatures below 21°C (70°F), and when it is necessary to move elephants in cold weather they furnish them with blankets. On the other hand, there is some reason to suppose that elephants could resist cold if necessary. The size of the animal alone works in its favor because the surface-mass law works here in reverse. As size goes up, the amount of body heat generated goes up as the cube of the dimensions; but the skin, the surface from which this heat is radiated, is raised by the square. While the elephant's body temperature is around 36.2°C (97°F), its skin temperature according to Benedict lies between 26°C and 31°C (79°F and 88°F). Thus this animal does not appear to show the cool skin found in swine when maintained at warm temperatures. Primarily, especially when the temperature is mild, the elephant's problem is not to prevent heat loss but to enhance it. According to some speculation, the elephant's ears act as radiant devices as do the wings of the bat and the ears of a jackrabbit. There is some evidence that elephants can tolerate cold, since they are found in the wild state almost up to the snow line on the great mountains of East Africa. Furthermore, Jap, the elephant used by Benedict in his classic study of elephant physiology, was kept through a New Jersey winter in a badly heated barn. There she was exposed for several weeks in

Figure 5-58. Time of Hypothermic Collapse for Eight Species of Rodents. These cold tolerance times were obtained during the summer. Times were increased by cold acclimation. The core temperature at which coordination was lost was between 16° and 20°C, recorded between the liver lobes. Ferguson and Folk (1970).

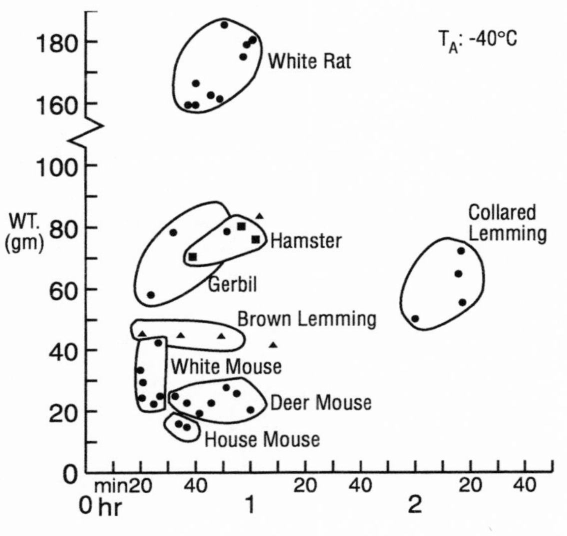

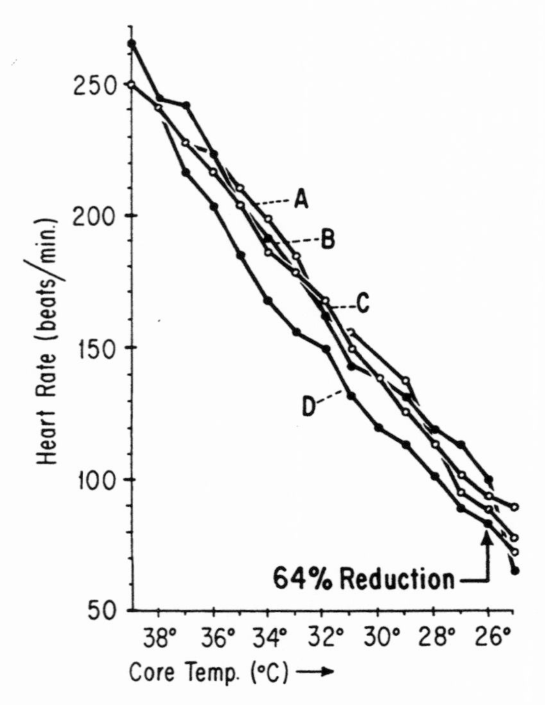

Figure 5-59. Heart Rates of a Cat in Hypothermia. In this experiment the cat was reduced nearly to cardiac arrest four times. There was no evidence of damage or acclimation. She later was retired as a house pet and had at least two litters of normal kittens. Adapted from Lipp (1961).

Table 5-8. EEG Measurements on Hypothermic Cats

Heart Rate (mean) b/m T_c	Heart Rate (N=9)	% Change
Control	211.5 ± 21.3 SD	—
34°C	190.0 ± 19.2	10%
30°C	129.0 ± 15.9	39%
26°C	81.5 ± 15.9	62%
EEG Frequency (mean) cps T_c	**Frequency (N=9)**	**% Change**
Control	32.6 ± 6.8 SD	—
34°C	29.4 ± 6.0	10%
30°C	29.0 ± 2.0	12%
26°C	23.9 ± 3.5	27%

Table 5-9. Index To Cooling Tolerance

Cooling Index = - [Max Survival Time] (in hours) X Temperature (in °C)

Example:

SPECIES	COOLED TO	INDEX
New-born ground squirrels	-3°C (for 11 hrs)	+33
New-born ground squirrels*	-8°C (for 5 hrs)	+40
Rats	5°C (for 1 hr)	-5
Dogs	5°C (for 1 hr)	-5

*Supercooled

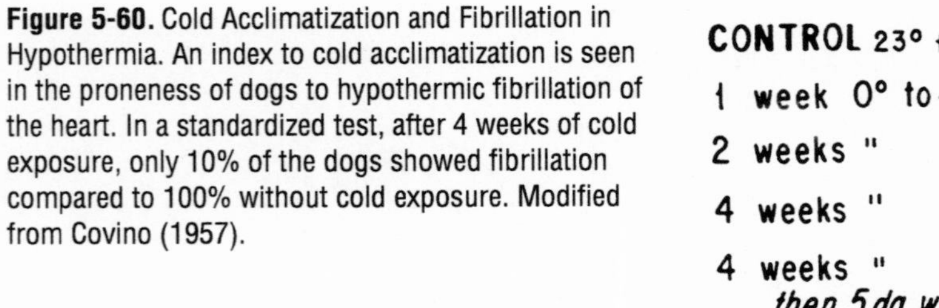

Figure 5-60. Cold Acclimatization and Fibrillation in Hypothermia. An index to cold acclimatization is seen in the proneness of dogs to hypothermic fibrillation of the heart. In a standardized test, after 4 weeks of cold exposure, only 10% of the dogs showed fibrillation compared to 100% without cold exposure. Modified from Covino (1957).

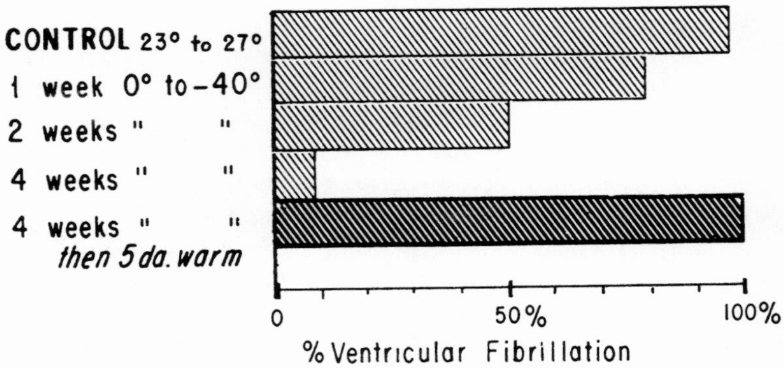

March and April to temperatures down to freezing, always without ill effects.

HYPOTHERMIA

This aspect of physiology has been through three phases: 1) experiments on cooling experimental animals; 2) recovery of humans from accidental hypothermia, and 3) clinical applications.

The work of Andjus, who revived rats after they were mainitained for 40 minutes at -3.3°C (26°F), showed that nonhibernating mammals do not necessarily totter on the knife-edge of homeothermic survival if their body temperature slides downward. Then, eight species of rodents were revived from body temperatures of 16°C to 20°C (60.8°F to 68°F) (Fig. 5-58). Many successful revivals were done with cats, one was cooled to a core temperature of 26°C (78.8°F) four times (Fig. 5-59 and Table 5-8). It was retired as a house pet and had several litters of normal kittens. Monkeys were recovered from a core temperature of 5°C (41°F). Rabbits survived a body temperature of 24.4°C (76°F), and ground squirrels much lower (Table 5-9).

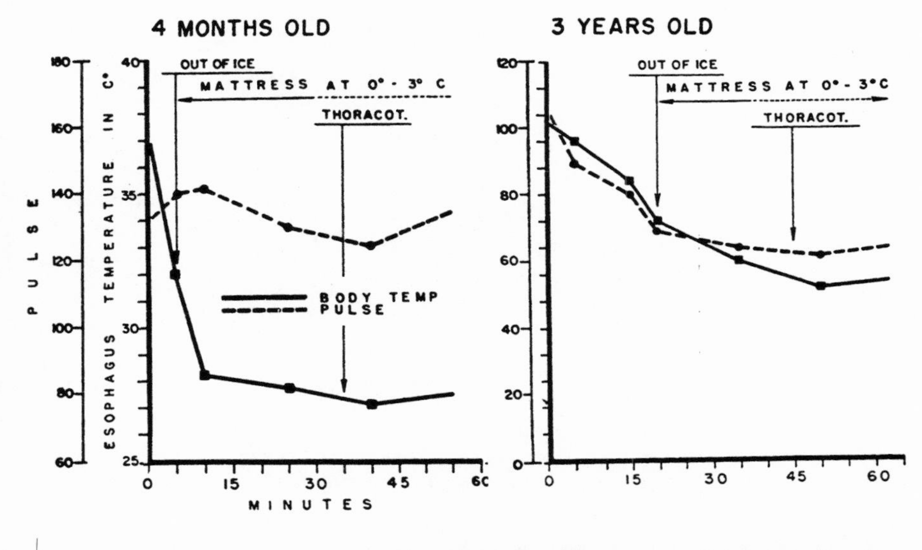

Figure 5-61. Heart-Temperature Relations in Hypothermic Infants. Open chest surgery (thoracotomy) under hypothermia was performed on a four-month-old infant and a three-month-old child. In the four-month-old infant, the heart beat was constant even when body temperature was falling. In the three-year-old, however, pulse depended upon body temperature and dropped from 100 to 50 b/m. Note that body temperature reached 27°C in one case and 32°C in the other. Adapted from Galletti (1960).

The physiology of these mammals in hypothermia includes 1) vasoconstriction of the extremities (but not in the head-neck region), 2) shivering to increase metabolic rate 2 to 6 fold, 3) eventual bradycardia and low blood-pressure, 4) diuresis, especially of sodium and chloride, of tubular origin, 5) augmented urine flow and increased blood viscosity (Granberg 1991).

How common is human accidental hypothermia? In one year, Liverpool Hospital admitted 17 patients whose core temperature was less than 35°C (Gunning et al. 1995). In the United States from 1979 to 1990 there were 9,362 deaths due to environmental hypothermia (Report of Medical Examiner 1993). In New Mexico, Native Americans were 30 times more likely to die of hypothermia than other New Mexico residents (Gallaher et al. 1992). The major element in hypothermic damage is the nervous system (Mouton et al. 1996).

What is the present thinking on re-

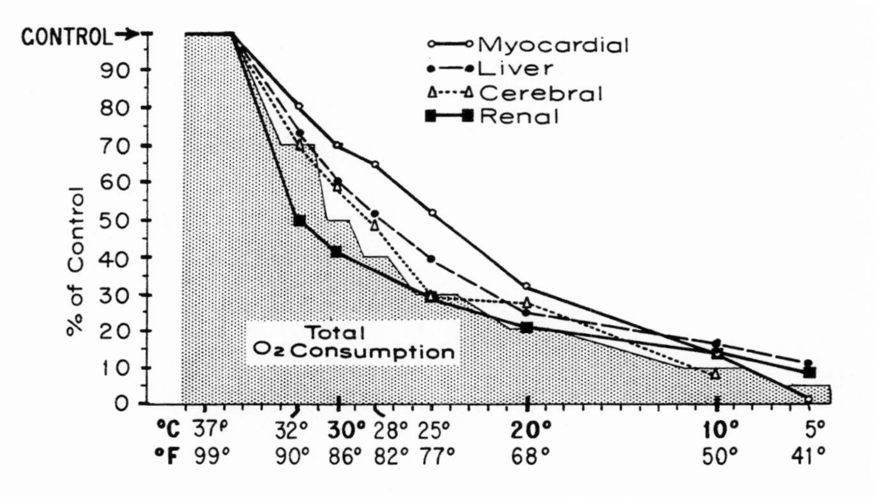

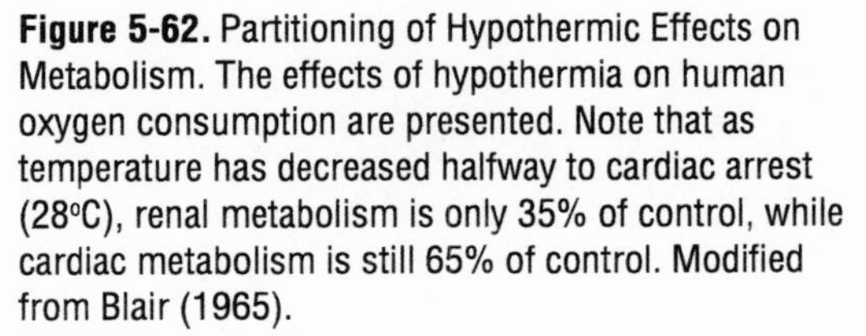

Figure 5-62. Partitioning of Hypothermic Effects on Metabolism. The effects of hypothermia on human oxygen consumption are presented. Note that as temperature has decreased halfway to cardiac arrest (28°C), renal metabolism is only 35% of control, while cardiac metabolism is still 65% of control. Modified from Blair (1965).

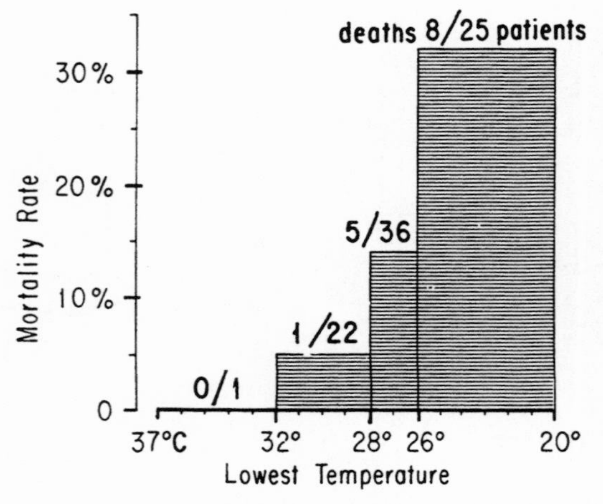

Figure 5-63. Human Mortality in Hypothermia. Results obtained by Swan to determine safety factor for hypothermia during surgical operations. From this series it would appear that body temperatures down to 28°C would be reasonably safe. Modified from Swan (1957).

habilitation from hypothermia? Prediction of survival time is speculative (Tikuisis 1996). The best survival rates have been from airway rewarming and application of warmed fluids (Kornberger and Mair 1996).

At first, when dogs were used as a model of the clinical use of hypothermia, they were found to have cardiac fibrillation after moderate cooling; then it was found that acclimatized dogs did not show this (Fig. 5-60). Later it was shown that dogs were the wrong model—human subjects were able to be cooled to adequate cold blood temperatures without cardiac fibrillation.

When the famous case of Johnny Stevens came to light, the record stated that she was found moribund in an alley with a high blood alcohol content, a body temperature of 17.8°C (64°F), and breathing 3 times per minute. She recovered totally. Her history, coupled with the successes on experimental animals, encouraged the use of light hypothermia for repair of heart defects in children (Fig. 5-61), and in attempts to arrest cancer by cooling patients to 15°C (59°F). There was some clinical success (Fig. 5-62). Out of all this experimentation, it was learned that clinical hypothermia down to 28°C (82.4°F) was safe (Fig. 5-63).

We will continue to discuss responses to extreme cold as we turn our attention in the next chapter to the adaptive mechanism of hibernation.

6. HIBERNATION

A HIBERNATOR IN NORTH AMERICA

The 13-lined ground squirrel is an example that occurs in four climate types.

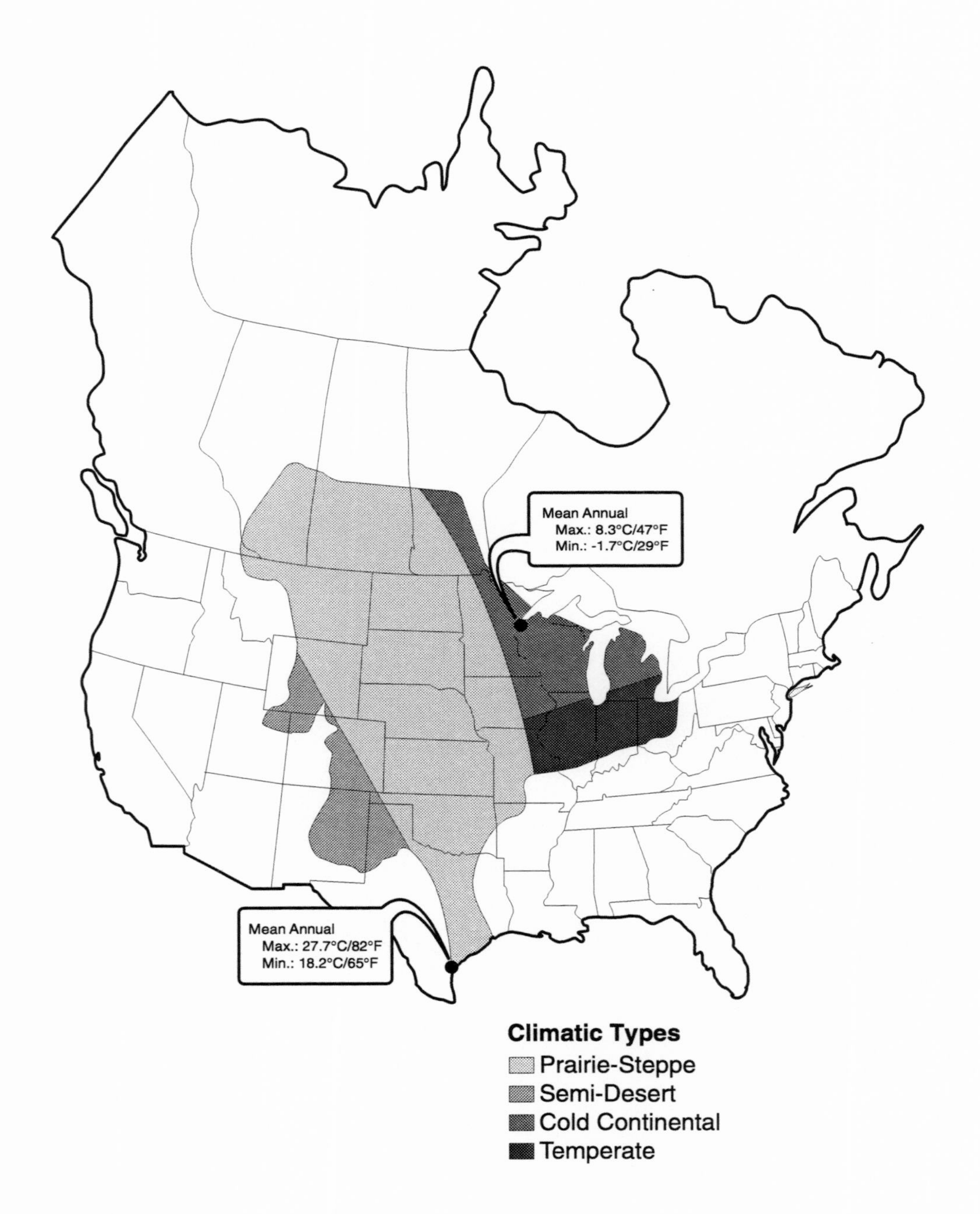

Environmental Diagram: Chapter 6

6. Hibernation

Introduction
Vocabulary of Hibernation
- Endotherms and Ectotherms
- The Definition of Mammalian Hibernation
- Heterothermic Mammals
- Hibernation of Ectotherms

Types of Mammalian Hibernators
- Permissive Hibernators
- Seasonal Hibernators
- Obligate Hibernators
- Daily Torpor versus Hibernation
- The Special Case of the Bats

Ambient Temperatures for Hibernation
- Cold Ambient
- Variable Ambient
- Warm Ambient
- Responses While Dormant

Dormancy of Carnivores
Estivation
Physiology of Hibernation
- Dissociation of Function
- Circulation
- Respiration
- Electrolytes
- Kidney
- Central Nervous System
- Gastrointestinal Tract
- Endocrine Glands
- Brown Adipose Tissue (BAT)

The Process of Hibernation
- Preparing to Hibernate
- Hibernation Trigger
- Entering Hibernation
- Arousing from Hibernation

Theories of Hibernation
Advantages of Hibernation

6. HIBERNATION

INTRODUCTION

The understanding of hibernation has challenged scientists since the time of Aristotle. Field biologists around the world have often become perplexed by this adaptive mechanism, perhaps upon finding tightly curled hamsters under rocks in the winter-gripped mountains of Romania; upon observing dormant pocket mice on a cool midsummer's morning in the Sierra Nevadas; on watching a mountain ground squirrel digging itself out of an 8-month snowpack; upon finding that captive birch mice of Norway become stiff and quiet as death each morning; or perhaps on digging out a dormant winter colony of vipers or lizards from beneath a mass of grass roots above the Arctic Circle in Scandinavia. Many times observers have judged these dormant animals to be "winter-killed," only to see the animal slowly move a leg as it awakens.

In this chapter, we discuss the phenomenon of hibernation, addressing these and other questions: What environmental extremes are associated with this physiological behavior? What is the physiological difference between mammals that hibernate and those that do not? Does a so-called cold-blooded reptile respond to cooling and rewarming like a mammalian hibernator?

VOCABULARY OF HIBERNATION

Endotherms and Ectotherms

The animal kingdom is divided into two groups: (1) the *endotherms* (also known as *homeotherms*), which maintain a relatively constant body temperature (birds and mammals); and (2) *ectotherms* (also known as *poikilotherms*), which take on the temperature of the environment around them (invertebrates, fish, amphibians, and reptiles). We may say that hibernation in some form occurs in all vertebrate orders if we accept for a moment that this phenomenon is *the act of resting in a dormant state for a prolonged period in a protected place*.

The winter season is the usual hibernation period. Most mammalian hibernators do not spend all of this season in hibernation. Rather, they have bouts of hibernation which may last for periods varying from several days to several weeks. Between the bouts of hibernation, the animals tend to maintain a constant body temperature of 37°C (98.6°F), and are referred to as being in a state of *euthermia*. (*Euthermia* refers to body temperature near 37°C, or 98.6°F.) The term *torpidity* usually refers to short periods of dormancy that occur at times other than the hibernation period.

The evidence cited in earlier chapters established the terms *endotherm* and *ectotherm* as being arbitrary. One will find there are variable temperatures occurring in endotherms, and there are uniform temperatures found among ectotherms (this latter partly because the ectotherm may inhabit an environment with a uniform temperature). Although the distinction between them continues to become more vague, these terms are useful as research progresses.

The Definition of Mammalian Hibernation

One of the most useful definitions was reported by Menaker in 1962: *Hibernation is the assumption of a state of greatly reduced core temperature by a mammal (or bird) which has its active body temperature near 37°C, meanwhile retaining the capability of spontaneously rewarming back to the normal homeothermic level without absorbing heat from its environment.* The state differs from hypothermia in non-hibernator mammals, in which case body temperature is reduced (usually artificially), but cannot be raised by autogenic (metabolic) means. Thus, there are two distinct groups of mammals: hibernators and non-hibernators.

Mammalian and avian hibernation is a state of dormancy associated with a reduction in most physiological processes: heart rate, respiration, body temperature, and total metabolism. *Dormancy* usually means *cessation of coordinated locomotor movements*. The body temperature remains about 1°C above the environmental temperature, and the animal usually awakens if cooled to a critical level, which is often at about 1°C (33°F). Two consistent and characteristic changes are found in the blood: an increased production of heparin and a rise in serum magnesium.

An experiment will emphasize the differences between reptiles, mammalian hibernators, and non-hibernators. Using the technique of Andjus and Smith (1955), we can place a 13-lined ground squirrel, a white rat, and a snake in a condition of hypothermia; we bring the body temperature of the ground squirrel down to 5°C (41°F), and that of the rat and the snake to 19°C (66°F). Now let us expose these three animals in a container in which the air and the wall temperatures are maintained at 18°C (65°F). The ground squirrel will spontaneously raise its body temperature and will pass through the air temperature of 18°C until it reaches its usual body temperature (euthermic) of 37°C (98.6°F); the white rat and the snake will retain a body temperature near 19°C (66°F) until eventually they will die. The rat and the snake do not have any ability to regulate body temperature or to counteract the induced cooling. Most varieties of snakes will not eat at this temperature. They will remain in a suspended state with a low metabolic rate unless they can slowly crawl to a source of higher air temperature or radiant heat that will raise their body temperature. According to Benedict (1938), the snake is producing less heat per gram of tissue than the two mammals. He illustrated this point by metabolic measurements on a snake (an ectotherm) and a rabbit (an endothermic mammal), each weighing 2.5 kg (5 1b) and having a body temperature of 37°C (98.6°F):

	kcal/kg 24 hr^{-1}	kcal/m^2 24 hr^{-1}
Rattlesnake	7.7	91
Rabbit	44.8	619

Benedict also stated that mammalian hibernators have a lower metabolism than non-hibernators, but this is incorrect. When comparing hibernators and non-hibernators, it is important to keep in mind that hibernation is much more than a change in temperature regulation and metabolism. In this chapter, we will describe differences at the molecular, cellular, tissue, organ, and organ-system levels of organization.

Heterothermic Mammals

For many years biologists have been aware of occasional mammalian species or groups that have fluctuating body temperature. Prosser and Brown (1961) appear to have been the first to suggest that primitive mammals (Monotremata, Marsupialia, and Edentata) and the higher mammals that hibernate should be called *heterotherms*; many of these show body temperature fluctuations only in certain seasons. Although the temperature-labile mammalian hibernator and the primitive mammal ordinarily maintain body temperature well above that of the environment, changes in air temperature may cause body temperature to fluxuate, especially when the animal is inactive (see Chapter 4, *Temperature Regulation*). During normal activity, heterotherms have lower body temperature than do other mammals. Heterothermy is shown in varying degrees among monotremes, marsupials, and edentates. Unfortunately, the term *heterotherms* is not useful in our present context simply because some heterotherms hibernate and some do not. Thus, we must continue to use the term *mammalian hibernator,* keeping in mind that the hibernation of some birds is very similar to that of the mammalian hibernator.

Hibernation of Ectotherms

We have just described an experiment comparing the thermal activity of a snake, a rat, and a ground squirrel. We should now consider in more detail the thermal behavior and hibernation of amphibians and reptiles. In order to elevate body temperature, these vertebrates must seek out heat and absorb it from the external environment; when a season turns cold, they must always hibernate. Survival depends upon their ingenuity in selecting an area of earth or vegetation that will stay above a critical temperature in winter; most importantly, this area must warm up at the appropriate time and release them from their cold microenvironment. The success of these animals in finding appropriate terrain for hibernation is shown by the distribution of reptiles in Europe as far north as the Arctic Circle.

Because these animals are not able to be active in the cold, we may refer to this type of hibernation as *required hibernation*. Required hibernation means that the *animals in this condition remain inactive all winter until they are warmed directly or indirectly by the sun in the spring*.

As we will see, mammalian hibernation is completely different. During the long winter hibernating period, the mammals, while still in the cold, can awaken and become euthermic, usually for 12 hours or more, and then go spontaneously into another bout of dormancy; these bouts may last from 2 to 30 days, varying among individuals and species.

The extent to which ectotherms undergo physiological changes in preparation for hibernation is a neglected area of research. The land turtle, *Terrapene ornata*, provides us with an example of such physiological change. Turtles taken from the field in winter had different respiration patterns and gas-exchange ratios from those reported for animals during short-term cold-exposure experiments (Glass, Hicks, and Riedesel 1979). Physiologically speaking, the ectotherm can be expected to be a different animal from winter to summer.

TYPES OF MAMMALIAN HIBERNATORS

The diversity among hibernators is evidence that hibernation has evolved many times. Brown adipose tissue (BAT) is the only gross anatomical feature characteristic of hibernators. No doubt there are cellular or molecular features that have not yet been identified, because the classical theory that *form and function are interdependent* must hold true for hibernators. Most of the distinctive features of hibernators are behavioral or physiological. These features provide the basis for R.A. Hoffman's identification of three conspicuous types of hibernators: *permissive*, *seasonal*, and *obligate*. This means that all temperature regulation exists in five forms: (1) endothermy, as in most mammals and birds; (2) permissive hibernation, as in the hamster; (3) seasonal hibernation, as in the arctic ground squirrel, the 13-lined ground squirrel, and the marmot; (4) obligate hibernation, as in the pocket mouse and some bats; and (5) ectothermy, as in lower vertebrates. We will now discuss the three conspicuous types of mammalian hibernators.

Permissive Hibernators

Mammals like the hamster store and use food prior to hibernation and during periods of arousal; hibernation is optional even in the winter. These animals may hibernate, but many individuals never do.

Seasonal Hibernators

The seasonal hibernators each year experience a rhythm of preparation for the coming of extreme environment; the capacity or inclination to hibernate is followed by a season of breeding and active feeding. One of the best examples of seasonal hibernation is provided by the arctic ground squirrel. We have kept these animals under constant conditions of light and temperature for several years in an animal colony, yet despite the lack of environmental clues, these animals have a totally different temperature regulation in the fall, as compared to the rest of the year. At this season, their body temperature drops for part of each day nearly to the ambient temperature (T_A) of the animal room, and they accumulate large quantities of subcutaneous and mesenteric fat. In this season, they can usually be picked up and handled easily, while at other times of the year they struggle to escape.

A second example of a seasonal hibernator is the European ground squirrel. Their behavior and physiological status can be predicted on a *yearly schedule*. This schedule has been aptly named by Pengelley the *circannian* rhythm; he has presented ground squirrel weight curves on an annual basis, including those of blinded specimens, to illustrate this distinct annual sequence of events (Pengelley 1969).

Obligate Hibernators

The hibernation of obligate hibernators must be *triggered*; this was first demonstrated by Bartholomew and Cade (1957) working with the little pocket mouse (*Perognathus longimembris*). This is one of the smallest of North American rodents, weighing 5.5 to 10 g; it is in the same family as kangaroo rats (*Heteromyidae*). When these animals are kept for 2 to 3 weeks at temperatures near 6°C (42°F) with food, they maintain their body temperature; if food is removed, they enter hibernation. The same type of manipulation of body temperature was done by Hudson (1965) with the pygmy mouse (*Baiomys taylori*), which weighs 4 to 8 g. Torpor in the pygmy mouse, induced by removing food and water, occurs only down to an ambient temperature of 20°C (68°F). Apparently, these two species usually go throughout the year without calling upon their ability to hibernate. They do not show a seasonal preparation for hibernation. The differences between the two types of hibernators are listed in Table 6-1.

Daily Torpor versus Hibernation

Earlier we defined day-night torpor as short periods of dormancy that occurs at times other than the hibernation period. Daily torpor

Table 6-1. Differences between Seasonal and Obligate Hibernators

Seasonal Hibernation	Obligate Hibernation
spontaneous	induced
genetically controlled	no preparation
endogenous	exogenous
starts prior to stress	result of stress
imperative	opportunistic
starts slowly	starts quickly
occurs regularly	need not be regular

is usually limited to 22 h or less, whereas hibernation is represented by longer (at least 96 h) periods of torpor during the winter season. Onset of both daily torpor and hibernation are influenced by food deprivation and temperature (Webb and Skinner 1996). The best examples are the hummingbirds: during the day, their body temperature declines while they are relaxed, but quickly rises when they are disturbed (Bucher and Chappell 1997). Nocturnal torpidity is very common among these birds. Under some circumstances, torpor may be costly. *Perognathus* will assume torpidity even at high environmental temperatures; this is a costly procedure to the energy balance of the animal, and they lose weight rapidly. They could not subsist for extended periods in this state. We must consider obligatory hibernation at warm temperatures as useful to the species for only short periods.

Another unique situation is found in the California pocket mouse; if its body temperature is below 15°C (59°F), it cannot arouse from torpor.

After considering published data on body weight, minimum body temperature, minimum oxygen consumption, and duration of torpor in 20 birds and 83 mammals (not including bears), Geiser and Ruf (1996) concluded that body weight and duration of torpor are the major features distinguishing daily torpor from hibernation. The mean body weights were significantly higher in hibernators (2,384 g) than in organisms in daily torpor (253 g). The average maximum duration of torpor was 355.3 h in hibernators and 11.2 h in daily torpor organisms.

The physiological features of these three types of hibernators and organisms entering daily torpor are discussed in a series of symposia that cover more than 35 years in this ever-expanding field: Lyman and Dawe 1960; Suomalainen 1964; Fisher et al. 1967; South et al. 1972; Musacchia and Jansky 1981; Heller, Musacchia, and Wang 1986; Malan and Canguilhem 1989; Carey et al. 1993; Geiser, Hulbert, and Nicol 1996.

The Special Case of the Bats

The temperature regulation of bats has been particularly difficult to study because of the continuously changing physiological state of many species of these animals. The North American cave bats can be described as a "bundle of thermal contradictions." For many years they have been considered as a special group; besides their seasonal hibernation, they show a remarkable lability of body temperature throughout the year. The day-night rhythm in the body temperature of bats was described as early as 1832 by Hall. He proposed the term *diurnation* as opposed to hibernation.

Hock (1951), to the contrary, found no evidence for a difference in the daytime sleep of *Myotis* bats and common hibernation. Hock's statement on this subject has been frequently quoted: "The body temperature and the metabolism of the resting bat varies directly with the environmental temperature." There are a few reservations about this statement, because there are some species of bats to which it does not apply. Furthermore, Stones and Wiebers (1967) state that *Myotis* bats, when provided with sufficient food in a cold environment, will remain active and do not drop their body temperature. Much more work needs to be done in this area using North American cave bats as material.

Hock's statement does, at least, partially apply to the birch mouse. Whenever these mice are inactive, body temperature falls to nearly that of the environment. They are frequently inactive from 0700 hr until 1500 hr. After 1500 hr, they are quite active and maintain the homeothermic state.

The case of the birch mouse is not comparable in some respects to the situation of bats, which fly in the cold and must have considerable difficulty in catching insects for nourishment under these circumstances. A team studying bats in southern Nevada found numerous specimens of *Pipistrellus hesperus* and *Myotis californicus* flying at temperatures between 1°C (34°F) and 5°C (41°F) (O'Farrell, Bradley, and Jones 1967).

Among bats (order *Chiroptera*) are examples of endothermy, seasonal hibernation, and obligate hibernation. The endothermic types are found in the tropics. The common vampire bat of Mexico (*Desmodus*), has an active body temperature above 30°C (86°F) which it maintains in an ambient temperature of 5°C (41°F) by increased food intake and activity. The animals die if they are exposed to this cold environment for an extended period. Other examples of endothermic states (as in fruit bats) are given in the next chapter, *Responses to Hot Environments*.

An extensive series on other tropical bats has been reported by Studier and Wilson (1970) (Table 6-2). Seven other species behaved like *Artibeus jamaicensis;* when cooled to T_A 8°C (47°F), they maintained a constant differential of about 7°C (44°F) above T_A. If the air was below 8°C (47°F), then the body temperature (T_b) was about the same as T_A, and almost none of the bats recovered from this extreme hypothermia. It is evident from these data that about three conditions are represented: (1) some species are poor regulators and simply reflect their environmental temperature; (2) other species, such as *jamaicensis,* when cooled to about 8°C (46°F) maintain a differential of 7°C (44°F) to 11°C (52°F) above the air; and (3) one species, *coibensis,* is an excellent regulator down to 17°C (63°F) air

Table 6-2. Responses of Tropical Bats to Ambient Air Changes (From Studier and Wilson 1970)

Species	T_A	T_b	Recovery if Warmed?
Myotis nigricans	3°C	ca 5°C	yes
Molossus coibensis	17°	33°	(no torpor)
Molossus coibensis	3°	ca 5°	yes
Carollia perspicillata			
one case	6°	38°	(no torpor)
another case	6°	6.3°	yes
Artibeus jamaicensis	8°	ca 15°	(slight torpor)
Artibeus jamaicensis	<8°	8°	no

temperature, maintaining a high body temperature to that point. In 1970, Lyman provided many more examples of types of temperature regulation in bats in his review of 10 species. McManus and Nellis (1972) added to the data provided earlier by Studier and Wilson; the major difference in their results was the finding that some species could maintain body temperature for 6 hours (Fig. 6-1) instead of the 2 hours reported earlier.

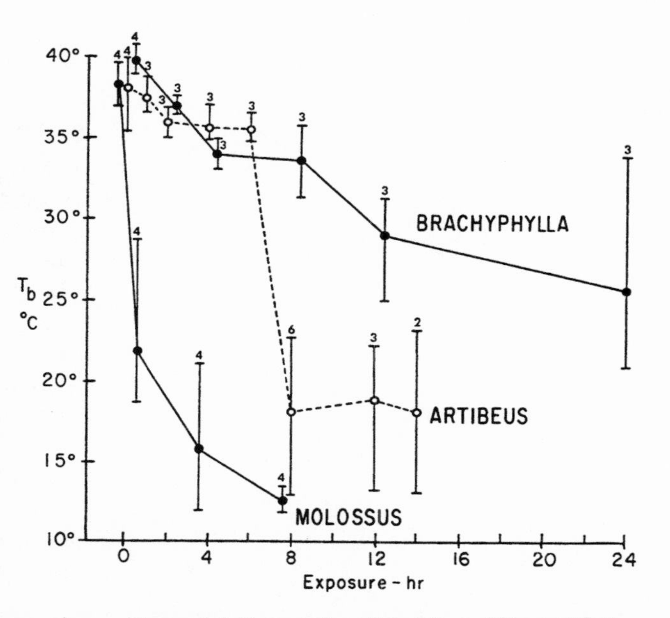

Figure 6-1. Temperature Regulation of Tropical Bats. Three species of bats were exposed to 10°C. One species failed to regulate almost immediately, a second species regulated for eight hours, and the third species did moderately well for 24 hours. Adapted from McManus and Nellis (1972).

Some bats are seasonal hibernators. Prior to experiencing environmental stress, they prepare for a prolonged state of torpor, even though the clue from the environment has not yet been experienced, and although the animals have not hibernated for many months. One of the best examples is the California mastiff bat, which remains endothermic in the summer but goes into daily hibernation during the winter when there is a shortage of food. Another example is the most common of the cave bats and the most extensively studied, *Myotis lucifugus*. This animal has a seasonal pattern of hibernation that it cannot modify by itself. During the summer it lays down fat in preparation for winter hibernation. The winter hibernation period is what would be expected, with frequent arousals during which the individual bat licks water from the cave walls; of course, there is no food to be obtained. When the seasonal sequence of physiological events turns to the summer, we find the animal in a totally different, active condition.

When looking for examples of obligatory hibernators within the bat group, we must look for cases where environmental stresses are quickly and irregularly imposed, and where there is no preparation for hibernation by the bat. An example given by Herreid (1963) is that of the Mexican freetailed bat (*Tadarida*), which maintains an active body temperature day and night in its home cave, but enters torpidity when placed in the confinement of the laboratory respirometer. Some neotropical phyllostomid bats have been described to be facultative hibernators as they adjust metabolic rate to feeding success and their level of fat stores (Audet and Thomas 1997).

We cannot leave the subject of hibernating bats without noting the large metabolic scope of bats, whose metabolism ranges from basal to maximal oxygen consumption. *Myotis velifer*, maintained in an environmental chamber with temperatures ranging from 5°C (41°F) to 35°C (95°F), had its largest metabolic scope of 227-fold at a temperature of 5°C (41°F). Some of the best human athletes have a metabolic scope of 20-fold.

AMBIENT TEMPERATURES FOR HIBERNATION

An interesting detail of hibernation concerns the suitability of variable ambient temperatures for dormancy. These temperatures will affect the amount of activity the animal displays while dormant. Considering first the ambient temperature, we find that some hibernators are limited to a distinctly cold zone, others tolerate a variety of zones, and others require a warm zone.

Cold Ambient

The hedgehog hibernates quite well at -5°C (23°F) with a body temperature of 2.5°C (36.5°F) and a heart rate of 19 beats/min (Kristoffersson and Soivio 1964). However, this is not the usual hibernating ambient temperature for these animals; the investigators of this study concluded that the optimal ambient temperature was near 4°C (40°F). To determine the upper limit of tolerance for hibernation, the ambient temperature was gradually raised. All hedgehogs awoke after 4 or 5 hours at 12°C (54°F); apparently this temperature acts as an arousal stimulus.

Variable Ambient

In our experience, the arctic marmot hibernates well at an ambient temperature of 14°C (57°F) for the entire winter. The same species also hibernated at temperatures considerably below freezing; some individuals nested and hibernated all winter in relatively small quantities of straw in extreme cold (at times as low as -48°C, -54°F). When temperature probes were placed in contact with the marmots, they often recorded -5°C (23°F). This species represents an animal that can adapt to a hibernating environment that varies over a wide range.

A second group that will tolerate a variety of ambient temperatures for hibernation includes the arctic ground squirrel, the 13-lined ground squirrel (see Environmental Diagram), and the Mohave ground squirrel. Both arctic and 13-lined ground squirrels will tolerate hibernation at an ambient temperature near freezing; they will also hibernate well at 15°C (59°F) and 16°C (61°F), and can be in a torpor at 20°C (68°F), which will permit them to be handled.

Bartholomew and Hudson gave an account of the hibernation of the Mohave ground squirrel in which the ambient temperature was raised from 10°C (50°F) to 32°C (87°F). As shown in Table 6-3, when the temperature was raised and the body temperature reached 27°C (81°F), they still appeared completely dormant; however, by the time the body temperature reached 32°C (88°F), they were active and appeared euthermic. When the hibernating ground squirrels were returned to 10°C (50°F) ambient, they re-entered hibernation.

Table 6-3. Reactions of Dormant Mohave Ground Squirrels to Changes in Temperature (From Bartholomew and Hudson 1960)

Body Temperatures	Observations
25°C (77°F)	Squeaks when stimulated; easily aroused
21°C (70°F)	No vocalization; shows coordinated movements
15°C (59°F)	Unable to right themselves when on their backs
10°C (50°F)	Responds to touch by withdrawal

Warm Ambient

The final group of hibernators to be considered are those that restrict hibernation to a warm zone of air temperature. The pygmy mouse will become dormant without food and water even if the ambient temperature is not below 20°C (68°F), and will spontaneously go into and out of dormancy, with arousal usually at night. At times it will maintain the state of hibernation for longer periods; if, under these circumstances, the ambient temperature is dropped below 20°C (68°F), it will arouse. It does not arouse when ambient temperature is increased. When the spotted ground squirrel (*Spermophilus spilosoma*) is placed in a 5°C (41°F) environment, most individuals will not hibernate; however, elevating the temperature to 15°C (59°F) will result in most entering hibernation. Then, if the temperature is subsequently lowered to 5°C (41°F), animals will continue to hibernate. Thus, the optimum temperatures for either initiating or maintaining hibernation may vary among hibernators.

Responses While Dormant

Hibernating mammals show conspicuous variation in their responses while in deep dormancy. Above we described the behavior of the Mohave ground squirrel. At 5°C (41°F) body temperature, arctic ground squirrels and 13-lined ground squirrels show slight flexing movements of the body. Many times we have placed sterile safety pins for electrodes in the skin of such animals, and have made incisions in the skin down to the scapular region for the insertion of cannulae without initiating arousal. The golden hamster is the most inert in hibernation, the only visible movement being 3 or 4 respirations/min that occur after periods of apnea, which last 2 min or more. Many species of ground squirrels and the woodchuck periodically rock from side to side when in hibernation at a body temperature of 5°C (41°F); the woodchuck vocalizes and moves sluggishly when disturbed at this temperature. The California ground squirrel may lift its head, cock its ears, and even vocalize at a body temperature of 7°C (45°F). With a rectal temperature of 4.5°C (40°F), the birch mouse responds to slight disturbance by a piping vocalization; with more intense stimulation, it always starts to awaken.

The activity of any particular species during hibernation seems to be correlated with the amount of electrical activity recordable from the cerebral cortex. At brain temperatures of 5°C (41°F) to 18°C (65°F), no spontaneous electrical activity was recorded from the cortex of the hibernating hamster at any amplification (Chatfield and Lyman 1954); yet a definitive and exhaustive series of experiments with the California ground squirrel showed complex electrical activity, particularly from the motor cortex, at brain temperature of 7°C (45°F) (Lyman 1958). The 13-lined ground squirrel when hibernating has a sciatic nerve that will conduct near 0°C (32°F) (Fig. 6-2). Hibernating poor-wills have been reported to continue hibernation while exposed to wind and hail (French 1993).

Figure 6-2. Conduction of Nerves of Hibernators. Conduction velocity was studied in the 13-lined ground squirrel as a function of temperature and time in hibernation. The nerves of non-hibernating specimens maintained in a warm temperature and in the cold were studied; there was little difference in their conduction. When nerves were tested from animals in hibernation, they were found to function at temperatures that would otherwise block conduction; in addition to that, the conduction velocities had increased (Kehl and Morrison 1960).

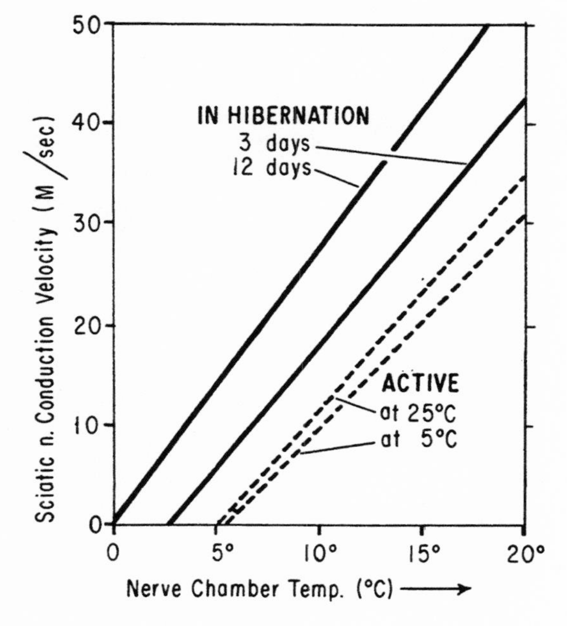

DORMANCY OF CARNIVORES

For many years, the term *hibernator* has been applied to some of the

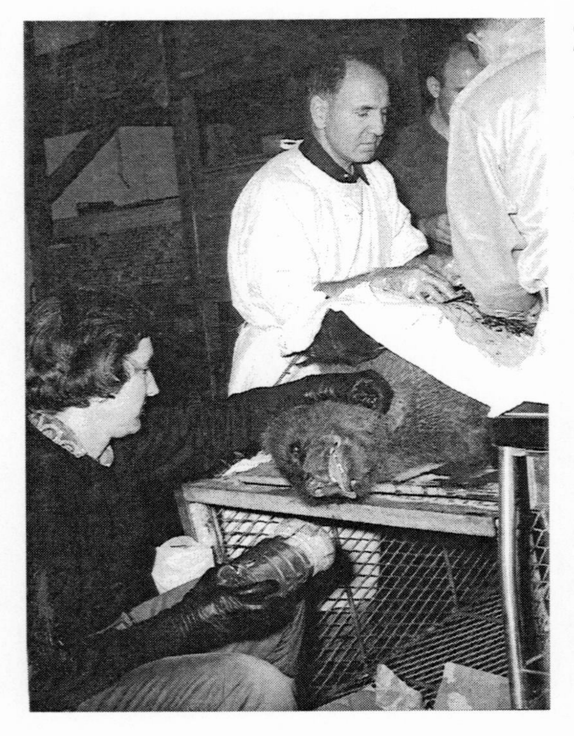

Figure 6-3. Implanting Physiological Radio-Capsules in Bear at Point Barrow: G. Edgar Folk, Jr. (center) assisted by Mary A. Folk. The procedure for implanting physiological radio-capsules in large carnivores such as bears and wolves consists of immobilizing them with tranquilizing drugs and then applying ether anesthesia. Bears weighing more than 800 pounds have been implanted in the field. These operations on bears were done in the bear cages. The radio-capsules transmit from the body cavity the heart rate, EKG pattern, and body temperature. These Iowa radio-capsules are designed for short-range, long-life applications. The same capsules have been used on wolverines, then taken out and used with the same procedure in birds; they have recorded heart rates in the golden eagle varying from 174 beat/min and in the longtailed jaeger a resting rate of 164 beat/min varying to a flying rate of 664 beat/min.

carnivores such as the bear, the badger, the raccoon and the skunk. These species spend long periods in burrows or dens in the winter. In 1960, Hock named this condition *winter lethargy.* This term applies to the raccoon and skunk, but not to the bear and badger. There is a dearth of physiological information on these animals in the winter resting state. Many more measurements need to be made; data should have accumulated rapidly now that radiotelemetry instrumentation is available, but unfortunately this has been far from the case. Only one topic can be taken up here: the condition of bears and badgers in their winter dens.

Hock has obtained rectal temperatures from wild bears and from specimens in captivity. The lowest figure he obtained from a thermocouple, which stayed in place in a dormant bear for three days, was a temperature of 31.3°C (88°F). A similar temperature (33°C) was obtained by Rausch (1961) on a wild black bear immediately after it was shot in its den. This careful measurement was by long glass thermometer which passed beside any possible fecal plug that might have closed the rectum. In two other studies of black bears in dormancy by radio, Craighead et al. (1976) showed a drop of 5.2°C, and Follman et al. (1978) showed a drop of 6.8°C. The usual active temperature of a bear is 38°C (101°F); thus, there is a striking drop in body temperature of at least 6.8°C. This picture of dormancy was supported by measurements on the metabolism of a black bear in its den obtained at -40°C: Hock states that the metabolic rate was 50% of summer rate. Similar data has been reported more recently by Watts et al. (1981).

During seven winters, we were able to obtain heart rate measurements and body temperatures of bears in dormancy by radiotelemetry (Figs. 6-3, 6-4 and 6-5). The summer sleeping heart rates of 4 bears (2 black and 2 grizzly) ranged from 40 to 70 beats/minute (Folk, Hunt, and Folk 1980). When the same bears were recorded in dormancy throughout the winter by radiotelemetry, their lowest daily heart rates dropped steadily each day until in December they reached, in a few instances, 8 beats/minute, with numerous readings of 10 to 12 (Figs. 6-4 and 6-5). In the field, Reynolds et al. (1983) recorded heart rates of an undisturbed female grizzly bear in her den. On March 21, the range was 12 to 6 beats/minute, and on March 23, it was 12 to 22 beats/minute. These data suggest that the dormancy of bears is similar to the hibernation of other mammals.

Further experiments support the concept. When three adult bears were exposed to hypothermia, their hearts stopped beating at a core temperature which varied from 17°C to 21°C (63°F to 70°F), but they did not fibrillate. This is also characteristic of the small hibernators. Johansson (1960) exposed badgers to hypothermia. This species behaved even more like a mammal that can hibernate, for the hearts continued to beat at a core temperature of 13°C (55.4°F) and did not fibrillate.

In an experiment including badgers showing normal behavior in the outdoor environment exposed to cold, Harlow (1979) recorded by radio the heart rate and body temperature of these animals. Two badgers chose to remain below ground for 80 days in winter. The heart rate slowed 50% and the body temperature showed a reduction of 9°C.

The results reported above on bears were first published in the 1970's (Folk, Folk, and Minor 1972); later, we showed the same dormancy in polar bears (Folk and Nelson 1981). Some kidney physiology was done on black bears by Nelson et al. (1978) at the Mayo Clinic and at the University of Illinois; their results confirmed our analyses of the first urine collected from bears when they emerged from dormancy (see Table 6-4). Nelson found a consistent increase in blood ammonia, lysine, lipids, and a decrease in respiratory quotient, none of which offers any explanation for failure of bears to develop uremia. Although fat metabolism increased, a considerable combustion of lean body mass occurred which did not produce uremia. Nelson and colleagues showed that urea is absorbed from the bladder and recirculated (1979, 1983).

To summarize, we note that all three species of bears (black, *Ursus americanus;* grizzly, *Ursus arctos;* polar, *Ursus maritimus*) do not eat, drink, defecate, or urinate for three to five months, and probably as long as seven months in northern Alaska. The small hibernators do just the opposite. The core temperature of bears does not drop in dormancy to the extent observed in small hibernators. This sharp decline would be biologically disadvantageous to bears because several days would be required for the body temperature to be raised at a time of emergency.

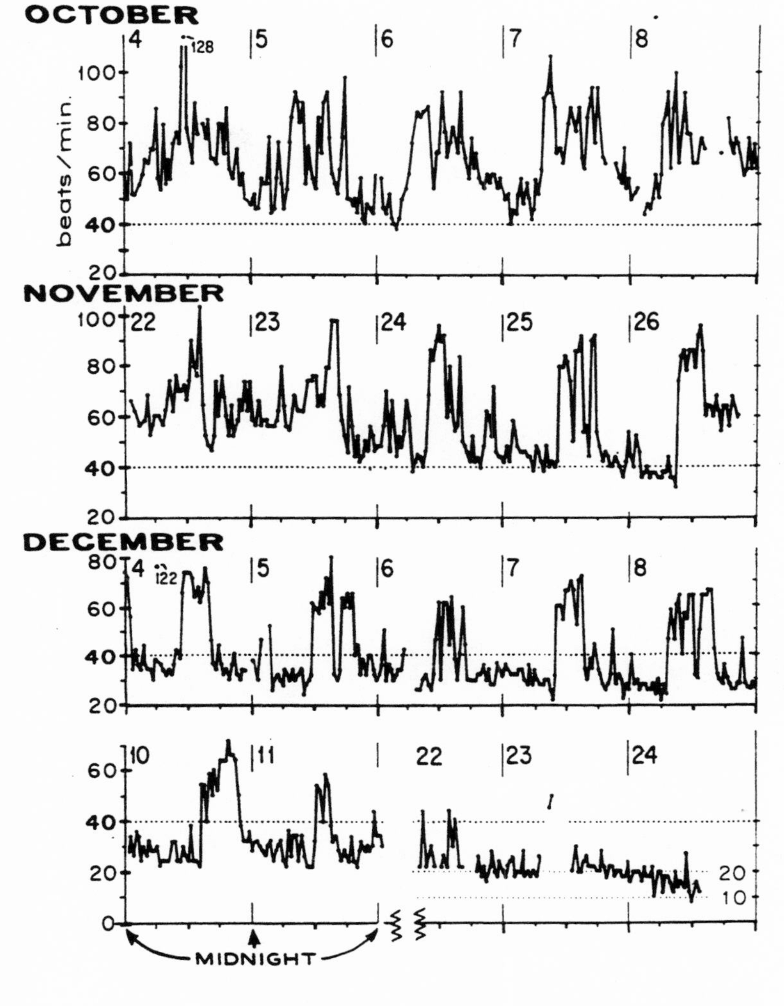

Figure 6-4. Heart Rates of Black Bear in Winter Den. A captive black bear went into winter lethargy for four months at Barrow, Alaska. Its heart rate and body temperature were recorded by radio telemetry over much of this period at half-hour intervals. Note the gradual drop in sleeping heart rates from 40 to as low as 8 beats/min (Folk 1966b).

Now, how do we compare the dormancy of small mammals weighing 25 g with that of large mammals weighing 700 kg? What do they have in common? An adjustable set point for temperature regulation applies to three types of dormancy:

1. that of the small hibernators (ground squirrels, woodchucks, marmots);

Table 6-4. Analysis of Urine Samples from Grizzly Bear.

Condition	Sample Time	Urine, ml	Nitrogen, g
Hibernation	4.5 months	181	1.43
Recovery*	24.0 hours	140	1.80
Normal Summer	24.0 hours	2080	34.00

* Seven to nineteen days; there is no urination during the first half of the recovery period.

HIBERNATION OF POLAR BEARS AND THE OMEGA 3 FATTY ACIDS

The hibernation of the polar bear represents a totally different picture from that of the black and grizzly bears, who hibernate regularly, even (in the case of the black bear) in warm temperate areas such as Florida. We demonstrated in careful experiments that two-year-old male polar bears went into hibernation in the same manner as other bears. However, during the young males' third winter, they were without food in a cold environment, and did not hibernate. This has been confirmed by other naturalists who observed polar bears in the free environment. The situation is different with the female polar bear: throughout her lifetime, she will hibernate every two or three years and bear her young during that period of dormancy.

The reason for the difference in hibernation between polar and other bears is that in the winter, the vegetable food of the omnivorous brown and black bears is unavailable. Polar bears, however, are specialized feeders whose diet consists almost entirely of seals, although occasionally they will feed on a dead whale. This means they are able to hunt for seals all winter long.

There are two rather different populations of polar bears: those living on the polar ice pack, and those that live on the ice in Hudson Bay. The polar bears on the ice pack can hunt seals all year around, while those on the ice in Hudson Bay are limited by the spring thaw that melts the bay ice. In the Hudson Bay there are about 1,600 polar bears living on the approximately one million seals in that area. All goes well during the winter, but in the spring the ice in Hudson Bay begins to melt, and the prevailing winds

2. that of those small mammals that show daily torpor or estivation (no deep hibernation); and
3. the large mammals (three species of bears).

Folk and Hammel (1992) present the following details of the physiological condition in a typical hibernator (Type 1): in hibernation from T_A3°C to 13°C, metabolism was passively determined by tissue temperature and by tissue activation energy, but at T_A2.6°C a critical blood temperature (CBT), a set point, was reached and there was a thermoregulatory response. Similarly, in Types 2 and 3, at first there is passive cooling, but at the CBT, the physiological event again occurs: the regulated response. For Type 2, the CBT is about 15°C; for Type 3 (the bears) the CBT is about 31°C.

Considering all of this background, it now seems reasonable to say that bears hibernate, despite textbooks by uninformed authors who state that they do not.

In one biochemical characteristic—high serum magnesium during dormancy—the black bear, at least, is like typical hibernators. High serum magnesium is not found in non-hibernator mammals when they sleep, and not under any circumstances when in health.

Another study was conducted to compare non-hibernating polar bear plasma lipids and lipoproteins with those in human plasma. The polar bear plasma had much higher lipid concentration than human and other mammals, and the bears had twice the plasma phospholipids of human subjects. Such a level would be fatal to a dog or rabbit. There were larger differences in fatty acid compositions; one fatty acid of the bears contributed 6% of the total, while in human subjects it contributed 13% to 15%. With another fatty acid, the bears varied from 4% to 30%, while this component in humans was never over 1% (Kaduce, Spector, and Folk, 1981). This unusual state of the plasma of non-hibernating polar bears is probably explained by the preference for an almost totally fat diet. When possible, polar bears will eat the blubber from seals or whales, and will leave the muscle protein untouched. Possibly no other mammal tries to obtain such an exclusively fat diet.

An intriguing experiment was done by Øritsland in 1969, using radiotelemetry on polar bears at Spitzbergen. His work concerned the body temperatures of these animals when exercising. It is obvious that they must be unusually well endowed with insulation since, winter and summer, a high percentage of each day is spent swimming in ice water. With such insulation, one would predict a high exercise body temperature when the animal is forced to run. Øritsland reports that the conductance of the winter pelt and the summer pelt is approximately the same. When the animals slept, the body temperature was about 37.4°C (99°F); when they exercised, the temperature rose to a plateau of 39°C to 40°C (104°F). This appears to be a fairly broad spread of temperature, perhaps indicating that the means of heat dissipation are insufficient. Øritsland found that this conclusion agreed with observations on the ice pack; adult polar bears when chased hard by people for 15 minutes tended to dive into the water. It is probable that the bears had a definite need for body cooling; if they could not enter water, they ceased running.

ESTIVATION

Mammals sometimes show daily or prolonged torpor at relatively high ambient temperatures. *Estivation may be characterized as shallow torpor.* The only known physiological difference between estivation and hibernation is the inability of estivators to withstand low body temperatures during dormancy. Above we described the physiological behavior of the pygmy mouse when deprived of food and water; this is a classic example of estivation. These animals will not tolerate a body temperature below 20°C (68°F). This particular temperature is interesting, because many other mammals have a change in physiological behavior below this figure. Fibrillation of the heart, or cardiac arrest, occurs in many mammals near 20°C. Thus, a more specific definition of estivation would be *dormancy at a body temperature of 20°C (68°F) or higher.* There are some species that are dormant from about 1°C (33°F) body temperature to 27°C (81°F); when above 20°C, it is estivation. There is at least one means of distinguishing sleeping animals with a body temperature of 27°C (68°) from animals in estivation at this temperature: the estivating animals have prolonged periods with no visible respiratory movements, while animals that are sleeping have regular respiratory movements without periods of apnea.

The same question that has been asked about hibernation can be asked about estivation: Is the state an accidental one due to poor temperature regulation, or is there an advantage to the species? In the next chapter on hot environments, we find evidence of a measurable advantage when desert species estivate. Apparently, estivation is a physiological tool used by some species to conserve energy and carry the animal over climatically severe times of the year (Riedesel and Folk 1996).

PHYSIOLOGY OF HIBERNATION

Dissociation of Function

The physiology of hibernation illuminates conspicuously some of the principles that are less evident when there are only small changes in body temperature. Let us first consider the statement, "the hibernator shows a dissociation of physiological function." In introducing this line of thinking on this topic, we are reminded of the teaching of George Wislocki, who emphasized not a physiological, but an anatomical dissociation. Instead of emphasizing the close association and integration of anatomical units of the body, Wislocki sometimes stressed the opposite, telling us to think of units such as the brain, heart, kidney, the gastrointestinal tract, and skin as separate compartments or boxes that carry on a quite reasonable existence when isolated from the rest of the body. This approach helps us gain a thorough appreciation of the physiology of each unit. In a similar fashion, we should at times depart from the concept of homeostasis

push the diminishing ice pack to the south and west coasts of Hudson Bay. By July, the ice pack has disappeared, and the polar bears have moved onto the land, where their food supply is unavailable. Consequently, half of them start slowly back up the west coast of Hudson Bay as the other half travel up the east coast. Some of them dig dens down to permafrost to escape insect bites, assuming a condition we call *walking hibernation* because it has some of the characteristics of true bear hibernation. Most of the bears continue walking up the coast and do not feed, although occasionally there is a dead whale for the migrating bears to feast on. A large number of these bears arrive near the city of Churchill, where they wait until the ice forms again on the bay at the end of October. In that vicinity, there may be over 300 bears on the shore one day, and if the ice is firm enough the next day, they will all be gone, out hunting seals on the ice surface.

The condition of these bears before going on the ice is that of fasting. This means that the scientist has the rare opportunity to study a population of well-fed bears that prefer to eat 100% fat (Kaduce et al. 1981), compared to fasted bears, which have probably not eaten for as long as 4 months. We did this experiment, and found that the total cholesterol and triglyceride levels in the fasted bears was 50% to 100% higher than in the control bears that were eating seals. The Omega 3 fatty acids, however, were found to be very low in the fasted bears, and high in the control bears. Thus, it appeared that the Omega 3 fatty acids had protected the control bears from having high cholesterol and high triglycerides (Folk, Cooper, and Folk 1992).

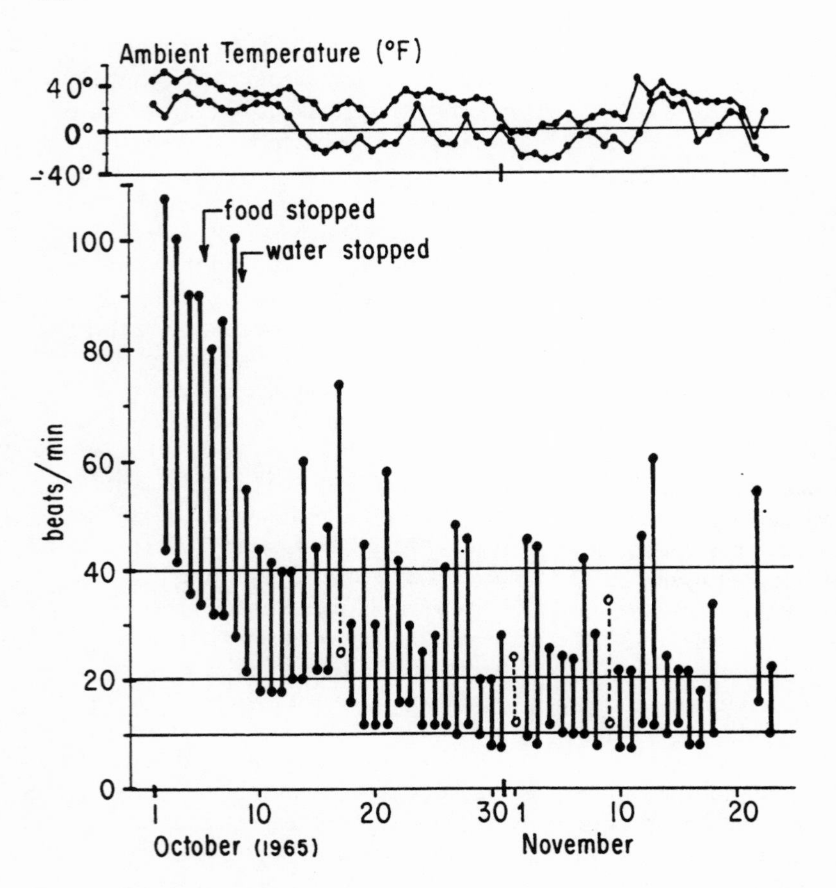

Figure 6-5. Winter Dormancy of Arctic Grizzly Bear. An example of graphing the minimum and maximum heart rate for each day as an index to daily activity. A heart-rate radio-capsule was implanted in the peritoneal cavity. To obtain these data, 48 half-minute records were made each day all winter. Note that the animal went into dormancy in the middle of October (Baker, Folk, and Ashlock 1966).

and its simultaneous integration and coordination of many totally dissimilar systems and mechanisms of the body. For example, students of day-night physiological rhythms realize that a person may have a high body temperature setting but a low heart rate setting at noon, and a low body temperature setting but a high heart rate setting at midnight (see Chapter 3, *Biological Rhythms)*.

Now let us look for dissociation in hibernating mammals. It can be seen as the woodchuck goes into dormancy (Fig. 6-6), a process lasting about 9 hr. During the first 2 hr, the baseline of body temperature, heart rate, and O_2 consumption drops steadily, but there are peaks and valleys superimposed upon this trend. These occasional resumptions of the active state last about 2 hours. During the steady decline of baseline, heart rate drops first, followed by 0_2 consumption, and then by body temperature. When the animal rewarms, the heart rate and O_2 consumption increase before body temperature. In 1959, Popovic provided another example of this physiological dissociation: he was able to give a ground squirrel in hibernation a strong stimulus, and found that O_2 consumption increased 10 to 15 times, but there was no significant increase in body temperature. In another study, the heart rate of a ground squirrel going into hibernation was recorded; the rate changed from 153 to 68 beats/min within 30 min. During this time the brain temperature declined only 0.6°C (33°F). The extensive dissociation must result from marked differences in metabolic rate and blood flow from tissue to tissue. Investigators must take multiple simultaneous readings of tissue temperature, oxygen consumption, and blood flow in hibernating mammals.

Circulation

The homeostasis of the mammal in deep hibernation is not well understood. The animal is regulating at a body temperature of 2°C to 5°C (35°F to 41°F); many hibernators, if the body temperature is lowered, take physiological action by increasing metabolic heat or by awakening. This homeostasis at the usual body temperature of 5°C (41°F) shows some variability. Although the environmental temperature and the body temperature may be steady, there are fluctuations in the heart rate and O_2 consumption. Heart rates of hibernating hamsters, woodchucks, ground squirrels, and hedgehogs may vary as much as 3-fold during one bout of hibernation. In the illustrations of the hibernation of the arctic marmot (Figs 6-7 and 6-8), note that for about 30 hours the animal had a heart rate of about 15 beats/min; twice, however, for at least 1 min, it had a heart rate of 60, and once, for at least 1 min, a heart rate of 118. Some of this departure from a steady state in hibernation is correlated with the day-night change in the physical environment. The cooled biological clock is temperature independent and continues to measure units that are close to 24 hours.

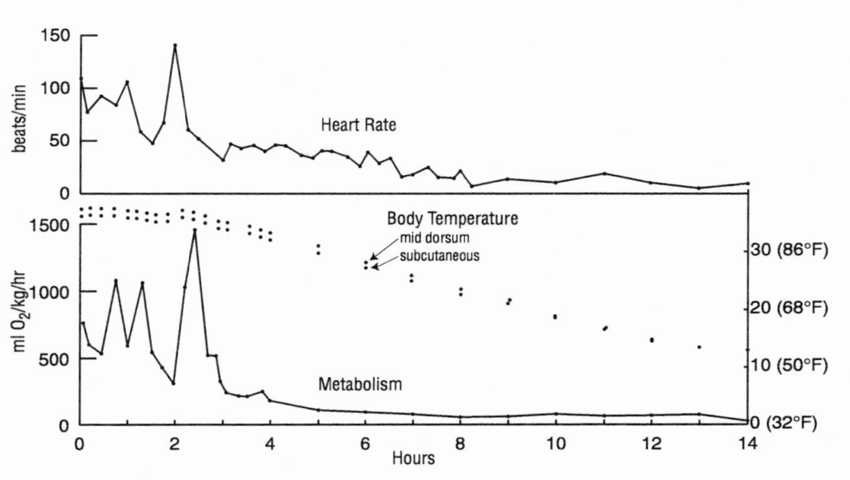

Figure 6-6. Correlative Measurements During Entry into Hibernation. These records of a woodchuck going into hibernation demonstrate a sequence of physiological events. The heart rate drops first, followed by oxygen consumption, followed by body temperature (Adapted from Lyman 1958).

The somewhat unpredictable heart rate in deep hypothermia makes it difficult to assess the extent of depression of the work of the heart. Overall, depression is frequently expressed in terms of metabolic rate: this at 5°C (41°F) is usually about 2% to 5% of the rate at euthermic body temperature. A similar calculation for the heart rate depends more closely upon the particular temperature of hibernation.

As we have said, the hibernating heart rate is often erratic, as illustrated by the following examples:

- *Bats*. The active heart rates of the genus *Myotis* in the bats may frequently be 500 beats/min and sometimes 700 beats/min (Henshaw and Folk 1966). When these hearts were studied in hibernation, the rate varied from 20±5 SD at 5°C (41°F) down to 8 beats/min at -7°C (45°F).
- *Hedgehogs*. A common heart rate for the active hedgehog, is 210±27 SD; sleeping values are 147. Some hedgehogs, however, may have a rate as high as 318 when awaking from hibernation. The variation in the hibernating heart rate of the hedgehog was described in a thorough and detailed study by Kristoffersson and Soivio (1964), as follows:

Ambient Temperature		Body Temperature		Heart Rates
°C	°F	°C	°F	beats/min
4.5	40	5.1	41	8
0.0	32	4.6	40	15
-1.5	29	2.8	37	17
-5.0	23	2.8	37	19

- *Mice*. The figures for the birch mouse also show a large reduction. The range of values for active animals was 550 to 600 beats/min and the reduction at body temperature 4°C (40°F) to 8°C (46.5°F) was to 30 beats/min.

There is no good explanation for the variability of heart rate when animals are hibernating. The abruptness of change in heart rate of an arctic marmot is illustrated in Figure 6-8. The heart rate of this particular animal continued as illustrated throughout a 280-min experiment. The cardioregulatory system is erratic in the hibernating animal. This erratic activity may result from sporadic circulation through various body tissues. The sporadic flushing of a tissue

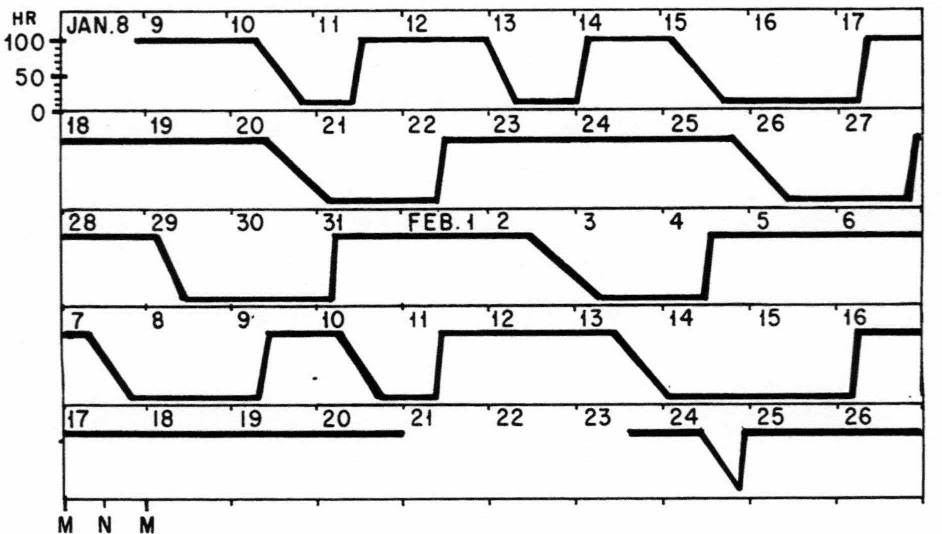

Figure 6-7. Entering and Awaking from Hibernation. An arctic marmot in a controlled environment (50°F) hibernated all winter in a nest of straw. Its heart rate was monitored by implanted Iowa radio-capsule. Ten bouts of dormancy in January and February were recorded. Note the graded decline of heart rate going into hibernation and the abrupt increase in rate upon awakening. All awakenings were spontaneous.

could produce marked changes in blood pH, O_2 and CO_2 content; we may need to have continuous recordings on the blood to identify the causes of the erratic heart beats. The number of days an animal has been in a continuous bout of hibernation may also have an effect on the mechanism controlling heart rate.

One of the large differences between mammalian hibernators and a non-hibernators is the tolerance of the hibernators' hearts for low temperature. The electrocardiographic intervals, P-T, QRS, P-R and R-T, remain uniform throughout hibernation. In contrast, the heart of a non-hibernator often enters fibrillation at temperatures near 20°C (68°F). Ectopic beats occur during hibernation as heartbeats often appear paired; however, the conducting pathways within the heart remain uniform (Steffen and Riedesel 1982). Superior regulation of intracellular calcium ion concentration in the hibernator's heart has been described as accounting for the tolerance of these hearts to temperatures below 20°C (68°F) (Liu, Wang, and Bulke 1993).

The characteristics of the hibernator's heart have also been studied by analyzing cardiac behavior when the animal is entering dormancy and when awakening (Fig. 6-9). The curves of heart rate and

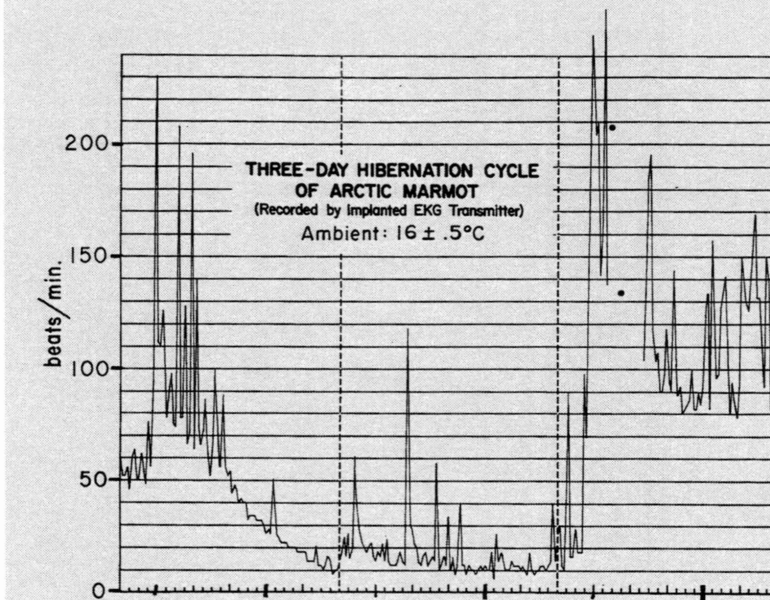

Figure 6-8. Heart Rates When Entering and Awaking from Hibernation. In the previous illustration a continuous three-month record of hibernation bouts of the arctic marmot was presented. In the present illustration, each heart rate is shown every 15 minutes as the animal enters hibernation, remains dormant for about 24 hours and then awakens. Partly because the ambient temperature was relatively warm, this animal showed occasional high heart rates during the period of hibernation.

body temperature do not fit the Arrhenius constant, which describes simple physicochemical reactions. This, in part, may be explained by the heart being under strong inhibition when the animal is going into hibernation, but being driven during arousal from hibernation. When the hibernating heart is isolated, the rate is described by a straight line in an Arrhenius plot.

While a decrease in heart rate is typically seen during hibernation, Figure 6-10 shows an unusual example of a hibernating woodchuck whose heart rate rose as the temperature dropped. By implanted radio, Folk and Folk recorded the hibernating heart rate of "Chuckie," a tame woodchuck nesting outdoors in a large packing case that was filled with excelsior, crumpled newspaper, and cotton. There was a probe in her nest which she bit off; air temperatures were taken on the exterior of her packing crate. The animal regulated her body temperature by increasing metabolism and heart rate.

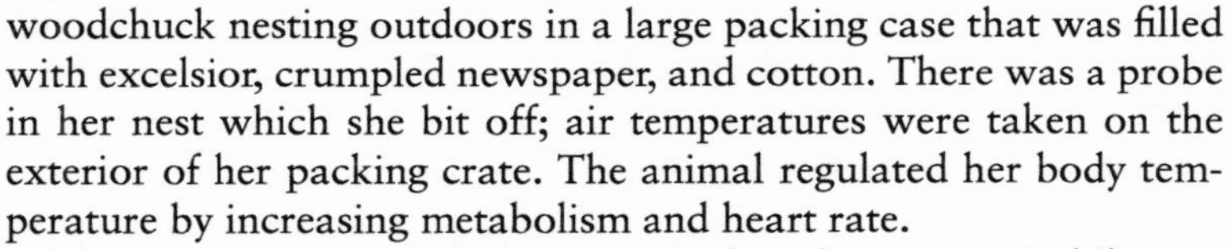

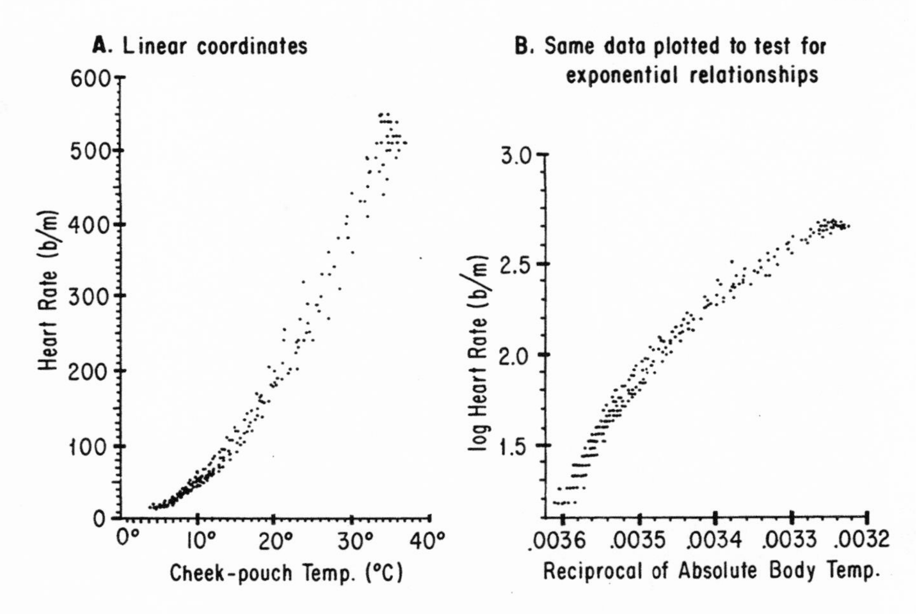

Figure 6-9. Heart Rates of Hamsters During Awakening from Hibernation. These records of heart rates of six hamsters awaking from hibernation contribute to the question as to whether the relationship of heart rate to temperature is a linear function. The body temperature was measured in the cheek pouch. In Graph B, the logarithm of heart rate of waking hamsters was plotted against the reciprocal of the absolute temperature to determine whether the relationship fits the Arrhenius equation. This equation describes simply physicochemical processes. Since a straight line was not obtained, it is evidence of the exponential character of the heart rate-temperature relationship. This is probably due to an increased effectiveness of sympathetic activity (Lyman and Chatfield 1955).

There have been few measurements of cardiac output in hibernation. Popovic (1964) reported a cardiac output in the 13-lined ground squirrel at a hibernating body temperature of 7°C (45°F) of about 1 ml/min, an output about 65 times smaller than in the euthermic state. Bullard (1964) reported changes in regional blood flow and blood volume during arousal from hibernation. He also reported a constant stroke volume in arousing hibernators. The changes in cardiac output and minute ventilation both appear to be dependent upon increased frequency.

During arousal, the anterior half of the hibernator "wakes up first," and blood is moved through this portion of the animal much faster than through the posterior half. Johansen (1961) measured this fractional distribution of blood by use of radioactive indicators. Using the arousing hamster, he demonstrated that the blood flow to the anterior portion of the arousing animal is more than 16 times greater than in the euthermic squirrel. Furthermore, the perfusion rate to the myocardium is twice as large in the arousing animal. To complete this picture, a substantial increase was shown in blood flow to lungs, diaphragm, and brown fat, but the blood flow to the gastrointestinal tract was greatly reduced. Flow to other parts of the body was approximately the same in the awaking animal and the active animal.

In two species of ground squirrels, the blood pressure has been measured during hibernation by chronic intubation; the results from both species indicated that blood pressure varies considerably even when heart rate is constant. According to Lyman, these variations may be due to changes in the degree of vasoconstriction. Popovic

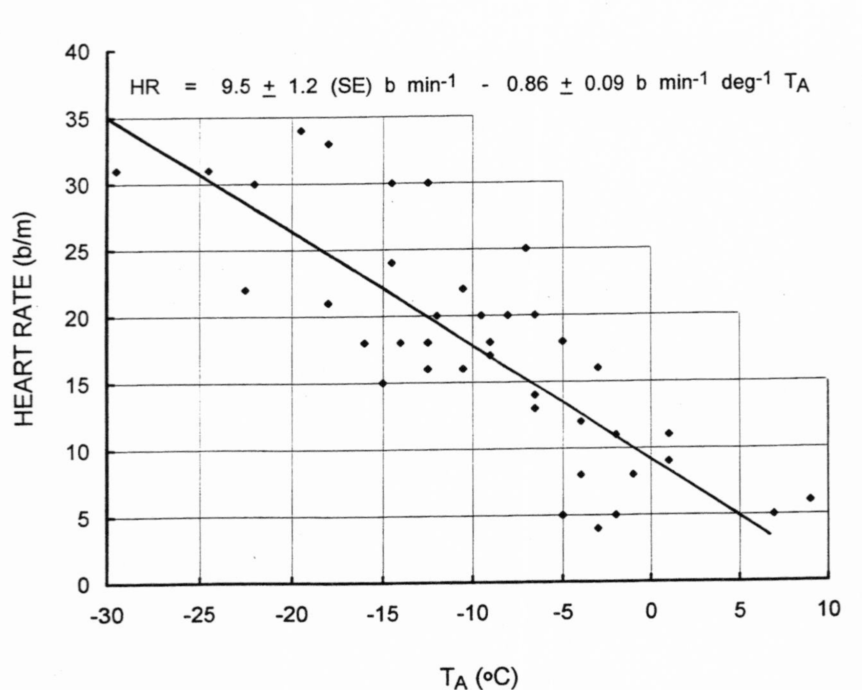

Figure 6-10. Homeostasis by heart muscle contraction in a hibernating woodchuck. In the outdoors, "Chuckie" burrowed into a packing case of insulation. Recordings of heart rate were taken between 7 and 7:30 AM by implanted radio capsule over 40 consecutive days, along with air temperatures. The animal seemed to resonate with air temperature or regulate while in hibernation, with high hibernation heart rate at the colder temperatures.

(1964), in a series of exquisite experiments, used chronic cannulation of the aorta and the right ventricle with plastic tubes in order to measure the arteriovenous difference of O_2 content of the blood of hibernating 13-lined ground squirrels. There was some decrease in the O_2 content of the blood, but the arterio-venous difference was unchanged.

Many hematology measurements on hibernators have been contradictory. The hematocrit reading from mammals in hibernation are very different from species to species. Usually they decrease, as in the 13-lined ground squirrel which has a hematocrit value of 40% in hibernation, compared to the euthermia value of 57%. Clotting time is doubled in hibernation and then drops to one-tenth the hibernation value during arousal. There is a trend for blood sugar to be reduced in hibernation, but this is by no means characteristic. In 1956, Suomalainen and Karppanen reported an increase in plasma protein in hibernation. More recently, molecular biology studies by Kondo and Kondo (1993) support the hypothesis that there are changes in plasma protein during hibernation. These authors identify specific plasma proteins present during hibernation. Whereas most genes are not expressed during hibernation, there are some genes that must be expressed to allow hibernation to continue (Martin et al. 1993). Low concentration of specific proteins may be the cause for periodic arousals from hibernation during the hibernation season.

Respiration

There are many different types of respiratory patterns in the state of hibernation. The dormant hamster frequently has 3 or 4 breaths/min following periods of apnea lasting 2 or more min. The hedgehog, on the other hand, at a body temperature near 4.7°C (40°F) does not breathe at all for as long as 56 min. We have recorded 45-min periods of apnea in the golden-mantled ground squirrel (Steffen and Riedesel 1982). The bursts of lung ventilation appear to be similar to Cheyne-Stokes breathing, which is usually associated with hypoxia. There is little evidence, however, of hypoxia in hibernation. Delivering oxygen and removing carbon dioxide from body tissues is dependent upon adequate function of both the respiratory and cardiovascular systems. Thus, it is not surprising that in many cases there are increases in heart rate at the same time there are bursts of breathing (Fig. 6-11). However, there are many reports of no correlation between respiratory frequency and heart rate during hibernation. The efficiency of the lungs remains unchanged during hibernation as the volume of air ventilated per liter of O_2 taken up is similar for hibernating, hypothermic (5°C/41°F), and euthermic ground squirrels (Milsom et al. 1993). These investigators have suggested that hypoxic ventilatory response in ground squirrels may be more closely related to changes in the O_2 content of the blood than to the partial pressure of O_2. They presented evidence that there was no correlation between the O_2 partial pressure in the carotid and frequency of breathing. Incidentally, the bursts of breathing can vary from 3 to 50 or more breaths in rapid succession (within 2 to 3 min—see Figs. 6-11 and 6-12). The end-tidal O_2 and CO_2 values, presented in Figure 6-12, indicate that breathing, once initiated, continues as though it were controlled by a pneumotaxic center. After the first few breaths, the end-tidal O_2 is high and end-tidal CO_2 is very low. In contrast, there is evidence that the ventilatory response to CO_2 is exaggerated in the hibernating ground squirrel (Milsom et al. 1993). Russian investigators have described adaptation to hypercapnia in exercising laboratory rats (Khitrov et al. 1986). A similar adaptation in the opposite direction (hypocapnia) may occur during hibernation. In searching for the type of respiratory control, perhaps we should consider there may be more than one type. At any given time, it may involve tissue sensitivity to blood gases, and on other occasions breathing rate may be controlled by a center that has an endogenous periodicity.

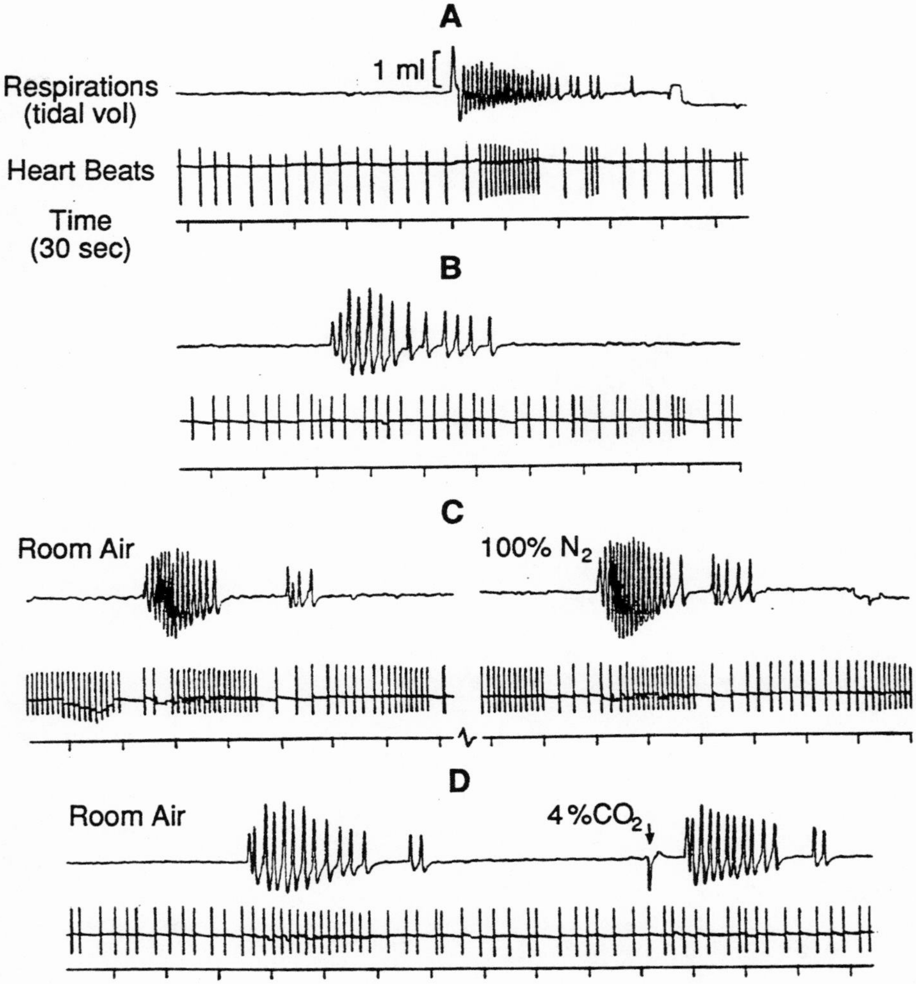

Is the breathing during hibernation similar to the ventilation movements of the fetus? The periodic fetal breathing was first described in the lamb (Dawes et al. 1972) and does have patterns similar to that of hibernators. Fetal breathing movements have been described in many mammals, including the human. Most of the breaths during hibernation involve near maximum vital capacity. This, of course, keeps the lungs and respiratory muscles in a "ready state," which is needed when the animal arouses from hibernation; a similar "ready state" is needed by the fetus prior to being born.

In summary, to date, the mechanisms that control the respiratory and cardiovascular systems are not well defined. It is important to

Figure 6-11. Breathing and Cardiac Patterns, *Spermophilus lateralis.* Recordings were made 30 to 60 min after moving animals to metabolic chambers (Steffen and Riedesel 1982). In A, note increased heart rate while breathing. In B, note no increase in heart rate while breathing. In C and D, note similar breathing and cardiac activity while breathing room air 100% nitrogen and 4% CO_2.

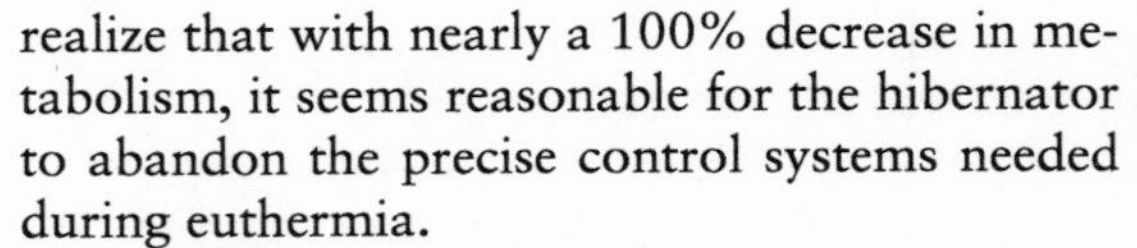

realize that with nearly a 100% decrease in metabolism, it seems reasonable for the hibernator to abandon the precise control systems needed during euthermia.

The respiratory gases in the burrows of hibernators in winter are of particular interest because some authors have suggested that mammals in dormancy are in a condition of hypercapnia and hypoxia. This may differ with species; the arctic marmot (*Marmota broweri*) spends eight months in hibernation, usually as a colony or family group in a single winter den. The entrance to the den is thoroughly closed and cemented with a mixture of soil, stones, fragments of vegetation, feces, and urine; this plug then freezes solid. Some mammalian hibernators do not seal the den as carefully as this. In a family den of these marmots, Williams and Rausch (1971) recorded an atmosphere of 4.5% CO_2 and 15.5% O_2. Studier and Procter (1971) reported even more striking values in burrows of 13-lined ground squirrels: 6.2% CO_2 and 13.7% O_2.

In ectothermic animals, the pH of blood and tissues generally increases as body temperature decreases (Reeves 1977). Nestler (1990) compared pH changes in mammals during hibernation and daily torpor. This study not only described changes as body temperature was lowered, but also noted intracellular pH differences among body tissues.

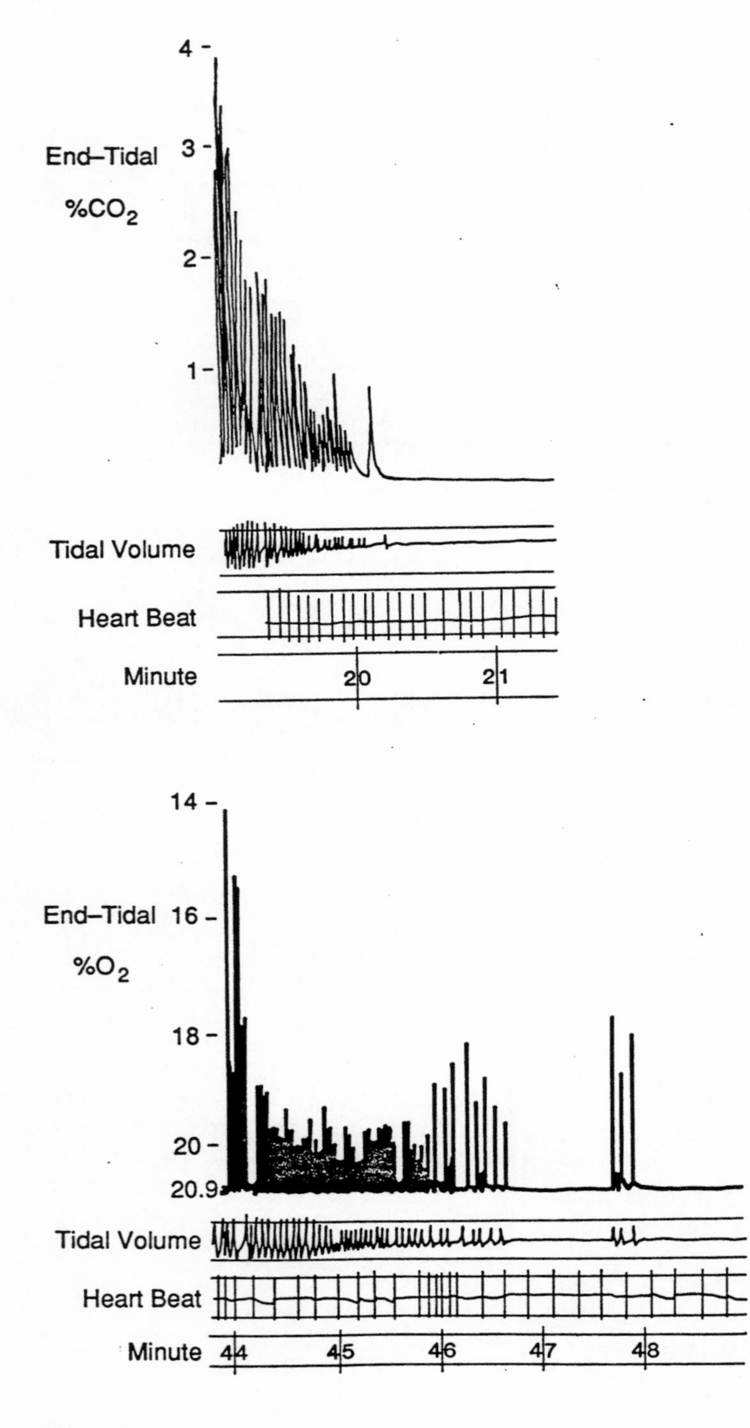

Figure 6-12. End Tidal O_2 and CO_2. The bursts of breathing continue for 1-4 min although end tidal O_2 is high and CO_2 is low. This is evidence for breathing being independent of alveolar air and arterial blood gas concentrations.

Electrolytes

The serum electrolyte that has received the most study in hibernating mammals is magnesium. Temperature regulation can be altered by the ratio of serum magnesium to calcium and by the injection of magnesium. It is also known that increased serum magnesium levels are found in both hypothermic vertebrates and in hibernating mammals. This observation represents an argument against the importance of increased levels of serum magnesium as part of a theory of hibernation. However, the serum magnesium levels in hibernation are higher than those in hypothermia (Table 6-5), and the elevated magnesium levels in the big brown bat are not found until after the core had fallen below 13°C (55°F). Non-hibernators will not tolerate such a low body temperature.

First, we should note that in the cold, the excitable tissues, such as cardiac muscle, have depressed ionic pumping; in non-hibernators, this leads to a lowering of the membrane potential and a failure of conduction. In the tissue of mammalian hibernators, excitability and regulation of water content are relatively normal at low temperatures. Willis (1964) studied the electrolyte gradient of the tissues of hamsters and ground squirrels. Samples of kidney, diaphragm and heart muscle were analyzed to determine if the low body temperature of hibernation might cause a loss of potassium and a gain of sodium. In these tissues, the gradients between tissue and plasma were maintained, or even increased during hibernation.

The mechanism by which different tissues of the hibernator can maintain Na:K gradients may not all be the same. Willis and Marjanovic (1993) compared the Na-K pump and ion (Na^+ and K^+) leakage of guinea pig (cold sensitive) and ground squirrel (cold tolerant) erythrocytes. They presented evidence that erythrocytes of the hibernator have a faster Na-K pump and slower leakage of Na^+ and K^+ ions through the plasma membrane at 5°C (41°F). Future studies at the cellular and molecular level may describe the following differences between hibernator and non-hibernator: amount of free calcium and magnesium ions in cytosol, higher affinity of Na-K pump for ATP in the hibernator, greater efficiency of electrotransport system in the hibernator, differences in intracellular pH, and other physico-chemical features.

Kidney

The question of renal function during hibernation has been of interest because many animals urinate during the periodic arousals from hibernation (Folk 1980). Also, the rates of glomerular filtration and tubular reabsorption of electrolytes must be parallel, or animals would encounter severe dehydration or electrolyte imbalance. Nearly complete cessation of glomerular filtration has been viewed as a major adjustment for water and electrolyte conservation by hibernators. Plasma or serum urea concentrations have been reported to increase, decrease, and remain unchanged (Riedesel and Steffen 1980). Recycling of urea nitrogen has been demonstrated to be important in many monogastric animals (Emmanuel 1981). Gastrointestinal microflora can recycle considerable quantities of nitrogen in hibernating ground squirrels (Steffen et al. 1980). Several investigators have reported that the amount of decrease in lean body mass in a hibernation season is in excess of the amount of nitrogen

Table 6-5. Reports on Serum Magnesium During Hibernation (From Riedesel and Folk 1957)

Common Name	Increase over Controls In Hypothermia
13-lined Ground Squirrel	65%
Woodchuck	63%
Golden Hamster	25%
Little Brown Bat	62%
Big Brown Bat	53%
Hedgehog	92%

excreted. Apparently, urea diffuses into the gastrointestinal tract. Then, microflora convert urea to amino acids. Through transamination, nitrogen released from urea may be incorporated in essential amino acids. Protein metabolism, particularly gluconeogenesis, could be an important regulator of the onset and duration of hibernation bouts. Deprivation of protein has been described to be as effective as food deprivation in inducing hibernation in the garden dormouse.

Central Nervous System

The cardiovascular and respiratory systems are obviously functional throughout hibernation. This strongly suggests that vital centers and the autonomic nervous system are also operational. The role of the central nervous system undoubtedly differs among species. The hibernaculum of bats invariably include water sources. The thirst sensation must be operational as bats have frequently been reported to arouse periodically to drink.

Habituation of hibernating ground squirrels was first demonstrated by Pengelley and Fisher (1967). Their experiment involved picking up the hibernating animals and tossing them in the air, catching them, and placing the animals back into their cages. Within a week of this daily routine, the squirrels did not arouse, suggesting that the animals were very much aware of their peripheral environment. The question of the extent to which hibernators retain memory led McNamara and Riedesel (1973) to train animals in a maize, and subsequently check on performance with and without hibernation. The animals that had hibernated had the better memory retention. These observations support the hypothesis that memory loss is an active process.

Studies by Heller et al. (1993), recording electroencephalograph activity of hibernating golden-mantled ground squirrels, demonstrated that slow wave sleep persists throughout hibernation. These authors also present data indicating that the lack of rapid eye movement (REM) sleep is involved in initiating the periodic arousal from hibernation throughout the winter. There is an apparent need for hibernating animals to awaken periodically and enter into REM sleep.

Numerous investigations have established the point that the temperature-regulation system is operational even in animals in deep hibernation (Lyman and O'Brien 1974; Miller and South 1978). Although most studies place importance on the hypothalamic temperatures, the temperature of the spinal cord and peripheral tissues can also be regulated.

Gastrointestinal Tract

Many hibernators do not eat during the periodic arousals from hibernation, although there are exceptions (Folk 1980). Thus, the gastrointestinal (GI) tract may be in a state of quiescence during hibernation, but must shift to a very active state after springtime arousal. Studies of the GI tract provide us with an opportunity to demonstrate the integration of physiological systems (GI tract and reproduction) and behavior of most hibernators. Keep in mind that shortly after springtime, aroused hibernators begin mating behavior. This is at a time when food availability is still rather scarce. Is the GI tract prepared for intensive activity following hibernation? Mussacchia and Westhoff (1964) gave us a clue that the GI tract is prepared; they demonstrated that the absorption capacity of the hibernating gut exceeds that of active squirrels when both are warmed to 37°C (98.6°F). More recently, Carey (1995) and Carey and Sills (1996) demonstrated the following:

1. The electrochemical gradient in small intestine epithelium is increased during hibernation. This would increase sodium ion coupled nutrient (glucose and amino acid) uptake. We have already noted the importance of the microflora of the intestine in recycling urea nitrogen.
2. The rate of cell division in the gastrointestinal tract is very slow during hibernation, but cells tend to accumulate in the G2 phase of the cell cycle. With onset of arousal, the epithelial cells are ready to divide and apparently ready to provide increased efficiency of nutrient absorption.

The hamster would be an interesting subject for similar studies because these animals will eat during the periodic arousal from hibernation throughout the winter. Does the GI tract of the hamster have increased efficiency during each arousal?

Endocrine Glands

The endocrine system is involved in many facets of mammalian hibernation. The classical point of view has been that all glands involute during hibernation. That the reproductive systems are active within a few days after arousal from hibernation has been a disturbing observation contradicting this point of view. Also, we can expect many hormones to be essential for maintenance of animals during the euthermic state between bouts of hibernation. The contribution of the endocrine glands to the hibernation process represents an extremely complicated picture, partly because of the different response of various species to changes in climate. The pineal gland is an important link between environment and seasonal activity. In particular, the pineal gland modifies the annual reproductive cycle of many mammals, birds, reptiles, and amphibians. Reiter (1981) pointed out that the function of the pineal gland in mammalian hibernators is not well defined. We can consider several reports that confirm this point of view. The pineal is essential to produce gonadal atrophy in the Syrian hamster, and this atrophy must occur before entering hibernation. Harlow, Phillips, and Ralph (1980) reported that ground squirrels can hibernate for one season following pinealectomy, but there can be disruption of hibernation the following year (Ralph, Harlow, and Phillips 1982). Injections of melatonin increased the frequency and duration of hibernation bouts in golden-mantled ground squirrels (Palmer and Riedesel 1976). The pineal may not be the only source of melatonin, and the target tissues for melatonin may include specific organs as well as specific endocrines and areas of the central nervous system. (See Chapter 2 for a discussion of the pineal and melatonin.)

The thyroid gland is essential for hibernation. In a review, Hudson (1981) pointed out how important it is to consider the year-round thermal environment of hibernators. Some rodents appear to need low thyroid activity and low metabolic rate during the hot summer months. Thyroid function has also been described to be important in seasonal changes in membrane lipid transitions.

The relationship between hibernation and the adrenal glands is also complicated. Hibernation does not take place in the absence of adrenal glands, suggesting that a minimal activity of these supposedly involuted glands is necessary for successful hibernation. Musacchia and Deavers (1981) have noted that differences in adrenal cortex glucocorticoid production may explain why hibernating Sciurids have decreased blood glucose, whereas hibernating Cricetids maintain blood glucose. Monthly examination of plasma cortisol led Gustafson and Belt (1981) to conclude that there is a seasonal resetting of feedback control in the brain-pituitary-adrenal axis with respect to ACTH in the little brown bat. These authors also suggested that the periodic spontaneous arousals are essential to permit elevation of glucocorticoid production.

Endocrine function is dependent on synthesis and release of hormones as well as interaction between hormone and receptors on target tissues. To date there are no studies of hormone receptors or the potential effectiveness of receptors to evoke metabolic changes in hibernators.

Brown Adipose Tissue (BAT)

Description. As we discussed in the last chapter, there are two kinds of fat found in many species of mammals: one is the yellowish or white fat occurring mostly in subcutaneous deposits of most mammals, and the other is *brown fat* (brown adipose tissue, or BAT), which is found most abundantly in mammals that hibernate. Some expressions, such as *hibernating gland,* have been coined to describe masses of BAT. This material may have been described as far back as 1551. Small amounts of this same BAT are found in many non-hibernators. In the rat, BAT is found surrounding the aorta as it descends from the thorax into the abdomen, in the back between the two scapulae, and in the axillae. BAT is particularly important in maintaining cardiac activity in the cold-exposed infant rat (Blumberg, Sokoloff, and Kirby 1997). BAT cells have considerably more cytoplasm than the white fat cells, which contain primarily fat droplets. The amount of BAT shows a decrease during hibernation, accompanied by an increase in the brown color. The BAT is most abundant in September and October. The relative size of the interscapular brown adipose tissue in a bat is displayed in Figure 6-13. The animal is depicted during its arousal from hibernation; the major tissue temperatures prevailing at this time are recorded on the diagram (Hayward and Ball 1966).

Function. Brown adipose tissue is currently recognized as being associated with heat production. One of the first studies to associate

BAT with this function was conducted by Robert E. Smith (1964), who traced the distribution of blood vessels in regions of BAT. These vessels are arranged in such a fashion that the use of a countercurrent heat exchange system appeared reasonable. The theory proposed by Smith was that BAT is strongly thermogenic in endothermic animals exposed to cold, and especially prominent in hibernators during arousal from deep hibernation. He believed that BAT is critical for contributing heat to the thorax region, to the cervical and thoracic regions of the spinal cord, and the sympathetic chain. He also presented evidence for control of this thermogenic activity by the sympathetic nervous system. Release of norepinephrine is the primary mechanism for activating BAT. Increased heat production is generated by BAT and other tissues by (1) increased Na^+ pumping across plasma membrane, (2) uncoupling the oxidation-phosphorylation system, and (3) cyclic chemical reactions, which do not produce energy for chemical synthesis or movement. The uncoupling of oxidation-phosphorylation is accomplished by the "uncoupling protein," first identified by Ratner et al. in 1981. Currently, the best method for identifying BAT involves the presence of mRNA involved in formation of the uncoupling protein (Trayburn 1993). The role of BAT in hibernating and non-hibernating animals continues to unfold; in one example, Cummings and co-workers (1996) suggested that activation of a cyclic adenosine monophosphate protein dependent kinase in BAT may be a mechanism for controlling obesity in mice (Cummings et al. 1996).

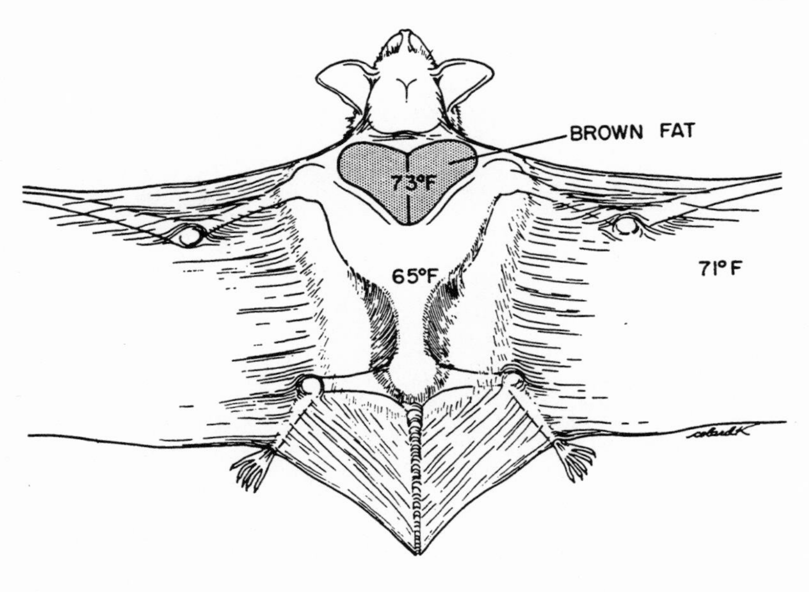

Figure 6-13. Dorsal Surface of Bat Showing Brown Fat. These results were obtained by thermography, showing the location and the temperature of the interscapular brown adipose tissue during the bat's arousal from hibernation. On the actual thermogram, the higher temperature and intensity of infrared radiation from the skin surface over the brown fat caused a brighter image. Modified from Haywood and Ball (1966).

Distribution among Mammals. Brown adipose tissue is found in all mammalian hibernators, but has not been identified in any of the hibernating birds. The group of animals that show carnivore lethargy—the bear, the badger, the raccoon, and the skunk—lack BAT. BAT has been identified in rodents, bats, rabbits, dogs, sheep, cattle and primates, including human beings (Trayburn 1993). In most of these species, cold-acclimated animals have higher amounts of BAT. In many species it is much more prevalent in infants, with the amount tending to decrease with age.

THE PROCESS OF HIBERNATION

Preparing to Hibernate

In some species of hibernators, we find a conspicuous effect of weather and climate in the preparation process. Changing hours of daylight brings about the physiological adjustments and changes in some species before cold weather sets. Temperature, humidity, and change of diet must also be important. The conspicuous response is usually one of fat storage or, as in hamsters, the storing of enormous amounts of food. The amount of conditioning by cold exposure required before hibernation varies with the individual animal and among species. Acclimation to cold does occur with at least some hibernators (Pohl and Hart 1965; Folk et al. 1970).

It is still not possible to generalize about the relationship between physiological preparation and the process of going into deep hibernation. In our colony of 13-lined ground squirrels maintained in an unchanging light cycle for three years (12 h of daylight), we found that some individuals will go into hibernation within 24 hours or fewer after being moved from a warm, constant temperature (24°C ±1°C) (75°F) to a cold room (6°C ±1°C) (33°F). Yet in the hamster, conspicuous acclimation to cold takes at least seven days (Farrand 1959). The apparent lack of dependence upon climatic preparation by some species may be due to the presence of the annual internal physiological rhythm. There is a yearly chain of events consisting of hibernation, mating, raising young, and putting on fat for the winter. Associated with this internal rhythm is a variable dependence (differing among species) upon the action of the physical environment as a synchronizer (timer). For three years, we observed evidence of this internal rhythm in the spring season. Our 13-lined ground squirrels were in hibernation for four months each winter, except for occasional 12-hr periods of euthermic body temperature. Toward the end of each winter, some ground squirrels awakened and refused to go back into the dormant state. They then existed for weeks with an endothermic body temperature in the cold room without going back into hibernation. There was no known clue from the external environment. Undoubtedly, the internal rhythm called for cessation of the lazy period of the cold winter-dormancy.

This laboratory model of ground squirrel hibernation can be matched by similar observations in the field. Signals from the physical environment probably cannot penetrate snow or reach deep into burrows to wake the ground squirrels in March or April; thus, it must be their internal rhythm that causes them to investigate the climatic conditions outside of the burrow.

An experimental variation in photoperiod to test the time of beginning of hibernation in the 13-lined ground squirrel was undertaken by Morris and Morrison (1964). Two groups were studied for 18 months; one group was given a 9-hr day and the other a 19-hr day. The onset of hibernation, however, was the same in both groups. This species represents a physiological type that is independent of photoperiod and can go into hibernation very rapidly at the appropriate time of its internal rhythm. It apparently does not depend upon a build-up of fat as part of its preparation because lean 13-lined ground squirrels will readily hibernate. It should also be noted that at the end of winter hibernation, both 13-lined ground squirrels and arctic ground squirrels may still retain large stores of unused fat.

Another type of hibernator does depend upon photoperiod to prepare for and enter hibernation. An example is the dormouse (*Glis glis*). Morris and Morrison (1964) compared this species with the 13-lined ground squirrel, using the 9-hr and 19-hr photoperiods.

The dormice in the short-day group quickly adjusted their cyclic responses, prepared for hibernation, and became dormant early.

The third type of response to photoperiod is found in the golden-mantled ground squirrel (*Spermophilus lateralis*). These animals were tested with a variety of lighting conditions; the onset of hibernation could be changed only slightly by changing the day lengths. Onset of hibernation did depend upon the occurrence of a specific stage in the annual weight cycle of these animals. However, by manipulation of the length of the exposure to particular temperatures, it was possible to change the weight cycle, causing animals to hibernate in the summer and become active in winter. This type of hibernator seems to depend upon the annual weight cycle more than the 13-lined ground squirrel does.

The importance of the superchiasmic nucleus (SCN) as a circadian pacemaker in mammals was discussed in Chapter 3, *Biological Clocks and Rhythms*. There is now evidence that the SCN is important in determining the hibernation annual rhythm. Removal of the SCN changed both the hibernation duration and frequency among *Spermophilus lateralis* (Ruby et al. 1996).

Hibernation Trigger

The evidence that there are one or more chemicals that initiate hibernation becomes impressive as investigations continue. For a number of years, Dawe and Spurrier (1969) conducted experiments in which hibernation was induced in ground squirrels in the summer by blood transfusion. They called this blood-borne material a *trigger* for natural mammalian hibernation (Dawe and Spurrier 1972). Ryan et al. (1982) induced summer hibernation in the 13-lined ground squirrel through urine injection. The trigger is present in serum and cells of hibernation blood in both ground squirrels and woodchucks. The chemical is apparently bound to albumin. It acts intra- and interspecifically, meaning that the serum of a 13-lined ground squirrel, when transfused, can trigger hibernation in other ground squirrels, and appropriate serum of woodchucks can trigger hibernation in both woodchucks and ground squirrels. The whole blood, serum, and cells retain the hibernation-inducing power for at least six months if kept at freezer temperatures of -15°C (5°F) or below. The trigger has not been found in the blood of either active or aroused animals. Thus, the effective material is not described by the expression *winter blood,* but rather by the expression *hibernation blood.* It is important that other biologists, using the Dawe technique with great care, confirm these experiments on a hibernation trigger.

These experiments are considered to be important, partly because the trigger seems to have some characteristics of a metabolism inhibitor. The trigger has been described to induce hypothermia and hypophagia in the monkey (Myers, Oeltgen, and Spurrier 1981). It appears that a trigger could be a neuropeptide. Bombesin, beta endorphin, somatostatin, neuropeptide Y, and vasoactive intestinal peptide are possible candidates (Boswell and Kenagy 1993; Kalter and Folk 1979). Studies are needed to better identify the environmental and physiological conditions that either permit or prevent initiation of hibernation. Then we will be able to better define the types of control experiments needed to establish a molecule as a trigger. Another approach is to consider chemicals that disrupt hibernation.

Kromer (1980) showed that administering an opiate antagonist, such as naloxone or natrexone, to hibernating hamsters induces premature arousal from hibernation. Opiate antagonists elicit a dose dependent reduction of hibernation bouts in golden-mantled ground squirrels (Beckman 1986). Cui and co-workers (1997) have suggested that changes in the opioid receptor binding sites in the limbic system of the Columbian ground squirrel are involved in regulating preparation for, maintenance of, and arousal from hibernation. Earlier we considered the point that one or more plasma proteins are required for hibernation. Now it appears that there are many essential chemicals required for hibernation (Horton et al. 1996). It should not be surprising that a phenomena as fantastic as hibernation should involve multiple interacting facets. The sequence of action and definition of the specific roles of each chemical awaits further research.

Entering Hibernation

The physiological changes in the animal as it becomes capable of being dormant are among the most interesting aspects of all the phases of hibernation. By understanding entrance into hibernation, we can better appreciate that hibernation involves more than a resetting of the temperature regulation system. Hibernation is a physiological state involving adjustments in many physiological processes at many, if not all, levels of organization. There appear to be four classes of mammalian hibernators, each of which proceeds from the endothermic state to dormancy in its own unique fashion.

The 13-lined ground squirrel and the bat may be called Class I hibernators: these can go into hibernation very quickly. The experiments described in preceding paragraphs were conducted at two different seasons; observers recorded respiratory rate of ground squirrels at intervals. They found some 13-lined ground squirrels went into hibernation within 12 hr. Johnson (1929) also described a rapid and spontaneous change in body temperature of the 13-lined ground squirrel from about 37°C (98.6°F) to 4°C (39°F) within less than 8 hr.

Note that in our own experiments, these animals were maintained at 24° ± 1°C (75°F); some individuals became semi-dormant for short periods at this temperature. Other types of hibernators make progressively deeper short-term body temperature drops called *test drops*. It is very possible that the 13-lined ground squirrel maintained at 24°C (75°F) undergoes its body temperature test drops from 40°C (104°F) to 25°C (75°F) when the appropriate time for hibernation comes in the fall. If test drops occur in the cold room, in many cases they must be very rapid and transient.

The pocket mouse probably belongs with Class I because it readily enters hibernation without needing to prepare for it. How-

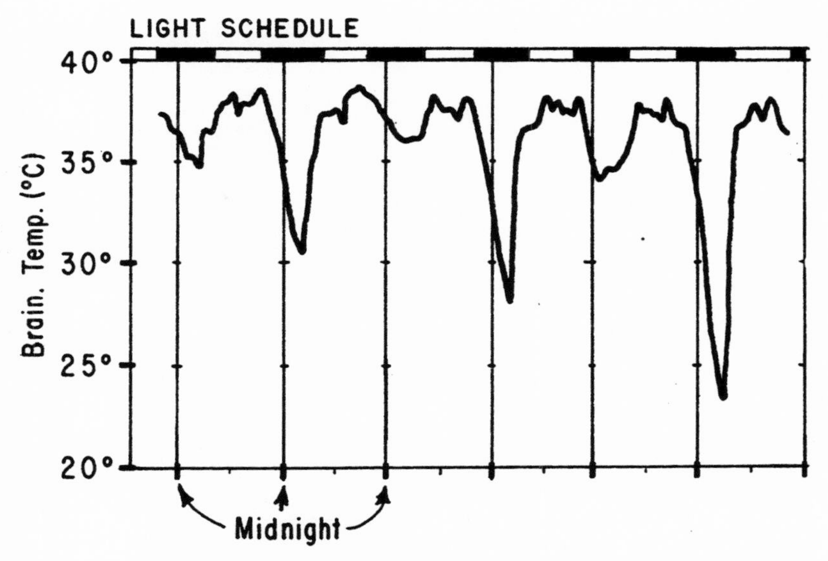

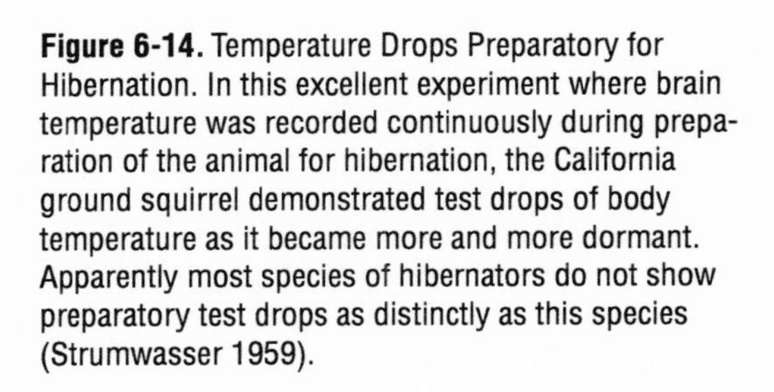
Figure 6-14. Temperature Drops Preparatory for Hibernation. In this excellent experiment where brain temperature was recorded continuously during preparation of the animal for hibernation, the California ground squirrel demonstrated test drops of body temperature as it became more and more dormant. Apparently most species of hibernators do not show preparatory test drops as distinctly as this species (Strumwasser 1959).

ever, in an earlier section, we called the 13-lined ground squirrel a seasonal hibernator and the pocket mouse an obligate hibernator; there are other intrinsic differences between the two species.

The hamster is a Class II hibernator. It seems to be prevented from hibernating until biochemical and physiological preparations are sufficiently advanced for dormancy to commence. It may be pertinent that the body temperature in this species without cold exposure is relatively constant at all seasons and in our opinion, unusually low (Folk, Schellinger, and Snyder 1961).

The mammals that show conspicuous test drops when they enter hibernation may be called Class III (Fig. 6-14). The classic example is that of the California ground squirrel; Strumwasser (1959) pointed out that this species does not enter hibernation in one temperature decline, but undergoes successive preliminary periods of torpor and arousal. It is as if the species enters hibernation while preparations are still under way; it proceeds only to the level of dormancy dictated by the particular state of preparation. We have studied a series of arctic ground squirrels during the first six days of entrance into hibernation. They appear to belong in Class III, because many of them demonstrated semi-hibernation and then re-attained the endothermic state for a few hours. In typical animals, this meant that the attainment of deep hibernation took approximately three to six days. The illustration from Strumwasser shows that the California ground squirrel requires five days to attain deep hibernation.

The woodchuck represents Class IV. Test drops certainly occur in these animals, but only during the first few hours of the initial entrance into hibernation, compared to the five-day sequence of the California ground squirrel. The illustration from Lyman's work on the woodchuck shows that, based on heart rate, the first test drop occurs within 30 min, the next one 1 hr after that, and the last drop 1 hr later (Fig. 6-6). These three test drops occurred over the first three-hour period; after that, any test drops were extremely inconspicuous.

We obtained similar information from four arctic marmots equipped with internal radiotransmitters. One illustration of a three-day hibernation cycle (ambient 14°±1°C, 57°F) shows a very similar picture to that of the woodchuck (Figs. 6-8). The sleeping heart rates of the euthermic animal vary from 80 to 140 beats/min. We took readings every 15 min; as the marmot left the euthermic level, the pattern of change of heart rate resembled the action of a springboard with depressions becoming deeper and deeper. The animal had a low heart rate of 75 beats/min for one reading, then in the next 15-min period the heart rate was 208 beats/min, followed in the next 15-min period by 78. These changes are best described as oscillations. As with the woodchuck, after 8 hours of these oscillations, the marmot settled into a steady drop lasting 12 hr. The first entrance into hibernation of this marmot set a pattern that was repeated closely for each of the next nine entrances into hibernation. The average time for each period of dormancy was 29.8 hr (SD 14.4). The length of time for going into hibernation was about 13.9 hr (SD 3.7). The most consistent physiological behavior observed was in the length of time emerging from hibernation (average time 5.5 hr, SD 0.65). In the illustration, the most striking change in heart rate was the increase from 70 to 250 beats/min within 15 min.

Some individual hibernators refuse to go into dormancy while in captivity. We have noticed this especially with Arctic marmots in winter: one group of three or four will go into deep hibernation in a cold room, while another group of three or four with all conditions being apparently the same will not hibernate. Chaffee (1966) has even described super-hibernating and non-hibernating genetic strains of Syrian hamsters.

Arousing from Hibernation

Hibernation is not a prolonged period of constant torpor. The 10 bouts of hibernation within a two-month period just described for the arctic marmot are typical of most hibernators. The individual periods of hibernation are shorter and more frequent in early and late hibernation season, and longer in mid-hibernation. In the marmot, the longer bouts in mid-winter were characterized by more regular heart beats.

Brown adipose tissue is a major source of heat during the early stages of arousal, and appears to be important for warming vital areas of the body first. Once an animal attains a body temperature of 15°C (59°F), shivering is the dominate source of heat. Shivering may explain the occasional high heart rates that we obtained from the arctic marmot during some bouts of deep hibernation. Mammals in deep hibernation are in one sense hyper-responsive. When stimulated, spinal and muscle action potentials can easily be recorded. Hibernating mammals are remarkably sensitive to vibrations such as digging made by a predator; in the laboratory cold room, muscle action potential may occur each time a cooling compressor starts its cycle.

There is no lack of suggestions as to why an animal arouses from hibernation. As noted earlier, an annual rhythm may be involved; decreases in specific blood plasma proteins, and lack of REM sleep (Strijkstra 1997) are among the multiple types of stimuli that may account for arousal from hibernation.

Awakening is a costly process; it takes as much energy to wake up as it does to stay in hibernation for 10 days. Wang (1978) conducted a comprehensive study of the energy costs of the various phases of hibernation in the Richardson's ground squirrel. Entrance into dormancy accounted for 12.8%, arousal 19%, and the interdormancy euthermia 51.6% of the total energy expenditure during a hibernation season. These various percentages can be expected to vary among species, particularly as a result of differences in body size. The total energy saved by hibernation in the Richardson ground squirrel was 87.7%.

THEORIES OF HIBERNATION

Although several theories of hibernation have been proposed, Morrison's theory, developed in 1960, is proving the most useful. Consider the oversimplified example of white rats being exposed to only one environmental change: cold. The response to cold for this non-hibernator will be an increase in basal metabolism and occasionally an increase in insulation (Hart 1964); the usual body temperature is maintained. Morrison (1960) presented equations that interrelate body weight, fat content, metabolic rate, thermal conductance, ambient, and body temperatures. He theorized that a hibernator may draw upon the mechanism of hibernation by "turning off" the increase in metabolic rate. If the animal is the size of a marmot or smaller, if the air temperature is relatively cool, and if the animal remains resting, then its body temperature will drop steadily. For animals larger than the marmot (i.e., the size of the raccoon or bear), even though the environmental circumstances may be the same, a smaller drop in body temperature will occur. In the small mammals that do drop their body temperature to hibernate, other physiological events precede the decline; some active process appears to depress heart rate and energy metabolism. This compartmentalization of function has been mentioned earlier. Note that in Lyman's illustration (Fig. 6-6), the heart rate of the woodchuck drops first, followed by O_2 consumption and then body temperature. This same sequence occurs in the hamster; as dormancy deepens, O_2 consumption starts to decline prior to temperature reduction, and O_2 consumption reaches minimal levels about 190 min before the temperature stabilizes at the dormancy level.

There have been few studies of breathing rate during the drop into dormancy; in an unusually thorough study, Landau and Dawe (1958) reported that respiration in the 13-lined ground squirrel is the first function to decline, followed by heart rate and then body temperature. We have mentioned that the small hibernator "turns off" the cold-response for combating cold. Hoffman (1964) suggested that when this happens, the thermal regulatory mechanisms are not simply abandoned, but are rather readjusted, probably as a result of depressed respiration, heart rate, and other factors. The key functions in attaining this readjustment are breathing and heart beat. Lyman and O'Brien (1963) suggest that the changes proposed could be mediated only via the autonomic nervous system.

ADVANTAGES OF HIBERNATION

It seems reasonably well established that hibernation and estivation are the same physiological phenomenon. This means that the body temperature of some hibernators will passively follow ambient temperature from 2°C to 32°C (35°F to 88°F) without eliciting arousal. What is the advantage of such behavior to these animals? Some small mammals and birds, because of high metabolic rates, are faced with an acute need for a continually available supply of food and water. An ability to lower their metabolism drastically conserves the food and water supply. Experimental proof was obtained by Bartholomew and Cade (1957), who found that dormant pocket mice at cool temperatures over a few days lost less weight than the non-dormant controls. They also showed in birds that a torpid poor-will could survive for at least 100 days on the energy derived from 10 g of fat (Bartholomew, Howell, and Cade, 1957). The medium-sized hibernators, such as ground squirrels, make use of this ability on a seasonal, not a daily, basis. Some naturalists believe that dormancy may prolong the life of mammalian hibernators; the unusually long life of cave bats of 24 years is cited as evidence. Hibernation extended the longevity of the Turkish hamster (Lyman et al. 1981). Thus, we may conclude that hibernation is not a stress/strain, but rather a diversion from adverse environments.

Can hibernation extend the geographical distribution of a mammal? Using as an example the distribution of the 13-lined ground squirrel, as shown in the Environmental Diagram at the beginning of this chapter, the answer would have to be "yes." These animals *(Spermophilus tridecemlineatus)* inhabit four climatic types: prairie-steppe, semi-desert, cold continental, and temperate, as they range from mid-Canada to the shore of the Gulf of Mexico. The physiology helps to explain how mammals can live where they do, but it rarely explains the exact limits of their distribution. Anatomical features, availability of nutrients, predator-prey relationships, and multiple facets of the ecology contribute to determine geographical distribution.

To summarize the advantages of hibernation, a thought–provoking quotation from Morrison (1962) will help: "Animals with adequate food either in body stores or accessible in the environment need not fast. Fasting (and hibernation) is found only in animals with inadequate external food reserves, whatever the reason. Animals without external reserves must use internal reserves and often hibernate." This illustrates the close association between environmental physiology and ecology.

In our next chapter, we leave the environment of cold and enter the world of heat.

7. RESPONSES TO HOT ENVIRONMENTS

THE ENVIRONMENT OF HEAT

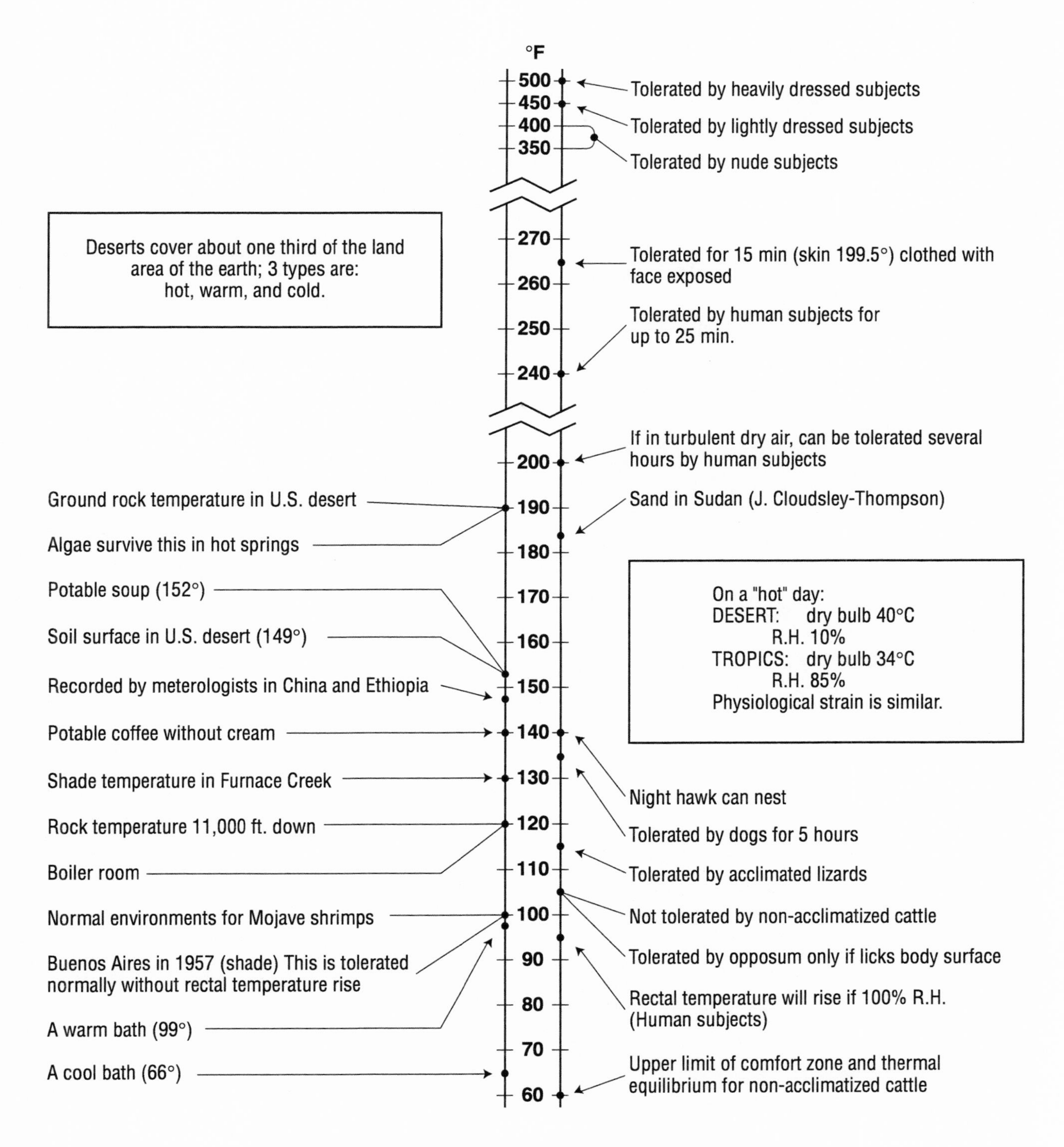

Environmental Diagram A: Chapter 7

UNUSUALLY HIGH BODY TEMPERATURES

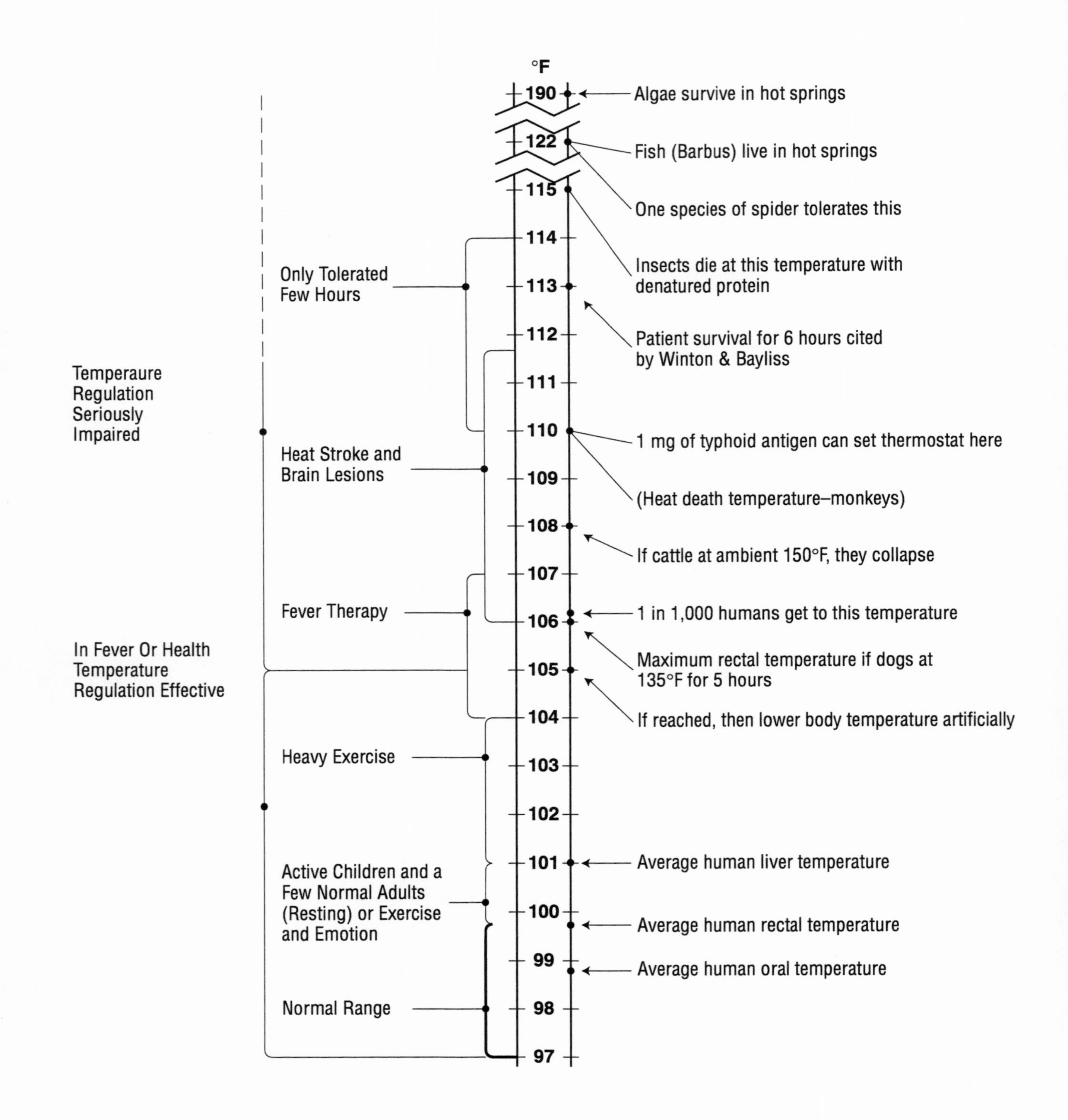

Environmental Diagram B: Chapter 7

7. Responses to Hot Environments

7. RESPONSES TO HOT ENVIRONMENTS

Figure 7-1. Earth's Temperature and Origin of Life. This compilation presents the hypothesis that at the origin of life on this planet, the Earth's temperature was near 70°C (158°F). The possibility that primitive ancestral protoplasm had a high temperature should be taken into consideration as we try to understand the optimal temperatures and lethal heat temperature of mammals today. Modified from Dicke (1962).

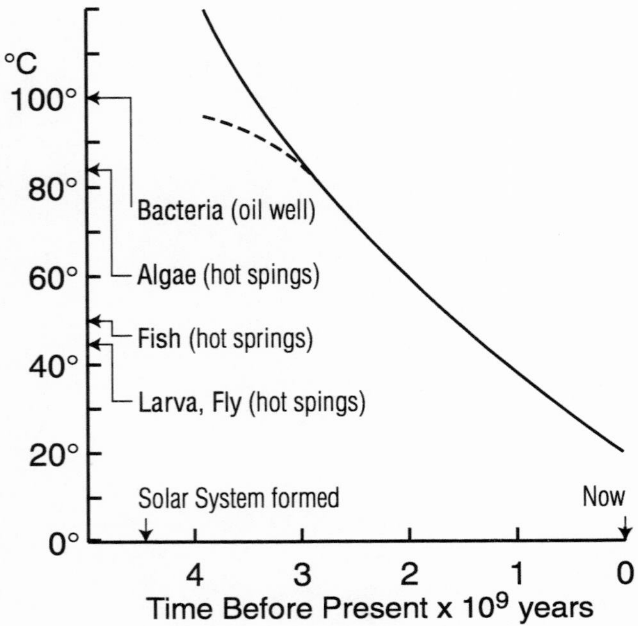

INTRODUCTION

More mammals suffer from heat than from cold, partly because there are more physiological mechanisms for combating cold than for combating heat. Simple behavioral mechanisms are also available to help mammals warm up; these include huddling together to collect insulation. Escape from heat is not so easy to achieve by either behavioral or physiological changes. Essentially, the only physiological devices available for cooling the body are vasodilation, sweating, and panting.

EVOLUTION OF TEMPERATURE REGULATION II

We began our discussion of the evolution of temperature regulation in Chapter 4; here we continue and expand our look at the origins of thermoregulation.

The role of the thermal environment in evolution is particularly interesting because all organisms respond to temperature extremes. The response can vary from subtle behavioral changes to death. Most geologists think that life on Earth originated while this planet was hot (Dicke 1962). If this hypothesis is true, then the early molecular ancestors of modern protoplasm must have had high-temperature tolerance. Dicke presents a hypothetical graph of the earth's surface temperature starting at approximately 100°C (212°F), the calculated temperature for the time one-fifth into the present lifetime of our solar system (Fig. 7-1). The graph drops over time to 17°C (62.8°F), the calculated mean surface temperature of Earth at the present.

Reasoning from the oldest extensive fossil evidence of life provided by the ancient algal reefs, Dicke stated: "Apparently, algae could have lived 3 x 10^9 years ago without violation of any conditions imposed by temperature requirements." Algae, he pointed out, survive at 85°C, and fish live in hot springs at 50°C (122°F). Many bacteria can grow in environments below 90°C, and a few can grow even up to the boiling point of water. The tolerance limits of bacteria for temperature extremes, like every other organism, vary with time. Thus, many bacteria can tolerate steam and freezing temperatures for limited periods. The simple organelles that represented the origin of life probably had biochemical features very different from those present today. A monograph edited by Kushner (1978) provides information regarding current microorganisms in adverse environments.

In analyzing evolution of thermoregulation, there are several basic issues to be considered: (1) Endothermy has evolved numerous times and many of the organisms in which it evolved may now be extinct; (2) The whole ecological system as well as the physiology of the organism must be described before we can recognize the significance of temperature regulation for a given organism; (3) Behavioral, physiological, and morphological changes can be of equivalent significance.

Ideally, tracing the evolution of temperature regulation should involve individual treatment of each aspect—namely, evolution of behavior, fast physiological responses, biochemical changes, and morphological changes. It would be fascinating to examine how these various facets are interwoven. Possibly, in many situations trial and error behavior was replaced by learned behavior. Behavioral responses could subsequently be replaced by development of more reliable physiological responses—vasomotor control, panting, sweating, and shivering—and these can be supplemented by morphological changes such as heat exchangers, or modified protein enzymatic structure and function. Developing our understanding of the evolution of temperature regulation, like many other topics, requires an interdisciplinary approach. In this brief consideration of temperature-regulation evolution, it should become evident that diverse forms of temperature regulation evolved many times.

Endothermy in excess of heat loss is used by many plants. Often, the excess heat production insures flowering and reproduction of the plant. In brooding pythons and other ectotherms, endothermy is associated with reproductive capacity. Some insects need to warm up by muscle contractions prior to initiating various activities (Heath and Heath 1982; Heinrich 1974). The preflight warm-up of insects may be a matter of survival to the species as well as the individual.

The need to examine the ecology of the organisms to evaluate the role of thermoregulation was emphasized in a symposium organized by W. W. Reynolds (1979). There are many aspects of the environment and animal physiology which can change the thermoregulatory needs of the animal. An animal may enter a temperature extreme to seek food: a fish that feeds on minnows in warm surface waters would be an example. Other factors resulting in changes in an animal's behavioral or physiological response to temperature can include season, light intensity, salinity, disease, presence of pollutants or predators, and body size and growth rate. Studies of many fishes emphasize the importance of autonomic temperature regulation. Physiological responses appear to be very important in true tuna and some sharks (Stevens and Dizon 1982; Dizon and Brill 1979). Behavioral responses in fish raise interesting questions about the importance of preferred and optimum temperatures. A fish may prefer one temperature for feeding, yet have a very different temperature for optimum growth and development. Minnows may need to acclimatize to warm, shallow water to avoid predators. In fishes we see primarily a combination of behavioral and biochemical responses to temperatures. Evolution of mammalian temperature regulation has involved interaction of behavioral, vasomotor, and numerous other physiological and biochemical responses. In most instances, behavior, rapid physiological responses, and changes in chemical features preceded morphological adaptations to hot environments.

Temperature regulation as seen in mammals may have originated with amphibians. Amphibians were preadapted as a result of using fluctuations in skin blood flow to regulate respiration. The vasomotor control employed for respiration became useful as animals extended into environments with thermal extremes. It is difficult to evaluate the relative significance of various adaptations because in many environments, amphibians are confronted with the need to regulate water balance, body temperature, and respiration by simple physiological and behavioral methods (Brattstrom 1979).

Among reptiles, we can expect to find many of the features characteristic of mammalian temperature regulation. An important point set forth by Huey and Slatkin (1976) is that, as a group, only a few reptiles demonstrate precise temperature regulation; thus, only reptiles native to extreme thermal environments can be expected to demonstrate temperature regulation. Studies of the marine iguana of the Galapagos Islands (Bartholomew and Lasiewski 1965) and the varanid lizard of the Australian deserts (Bartholomew and Tucker 1964) provide ample evidence of mammalian-like temperature regulation.

The distinctive feature involved in the transition from amphibians to reptiles is the decreased permeability of the skin. Obviously, this feature limits evaporative cooling as a method of heat loss among reptiles. Although this is true, there are exceptions. Many lizards pant; and a land turtle, *Terrapene ornata,* has been described to increase saliva production and evaporative cooling in response to either increased core or surface temperature (Sturbaum and Riedesel 1974). The integration of core and surface temperature apparently is accomplished by the hypothalamus (Morg and Hammel 1975), and establishes the presence of a temperature regulation center in an ancient reptile. These turtles evolved just prior to the dinosaurs.

Basking in direct sunlight is a behavioral trademark of lizards. The efficiency of basking has been improved by vascularization and vasomotor activity because many lizards have heat exchangers on their body surface. One interesting exception is the horny toad, which directs the highly vascularized eyeball toward the morning sun, thereby elevating its body temperature before totally emerging from the sand.

Seasonal acclimatization of reptiles has been a neglected subject. A 1977 paper describes seasonal variation in thermoregulating behavior of diurnal Kalahari lizards (Huey and Pianka 1977). Glass, Hicks, and Riedesel described changes in the respiratory pattern of a tortoise (1979). As discussed in the previous chapter, the adjustments associated with reptilian hibernation may not be as elaborate as those of mammalian hibernation, but certainly the 15°C to 20°C decline in body temperature that persists for months must be accompanied by numerous physiological and biochemical changes.

Temperature regulation implies the presence of a receptor, a regulating center (thermostat), and an effector. The diversity of these components is illustrated by some reptiles that can tolerate a broad range of body temperatures and have developed a remarkable temperature sense. The pit vipers have facial temperature-sense pits, and some boas have labial pits that contain temperature receptors. These pits sense temperature changes, and even shadows passing over the pit are readily detected. From the modulation of the steady discharge of nerve impulses, it has been calculated that the receptors can detect as small a change as 0.003°C in the pit membrane. This means that various objects can be distinguished by their radiation when their surface temperatures differ by only 0.1°C. A 0.4°C temperature rise increases the frequency of nerve impulses from 18 per sec to 68 per sec, an extreme sensitivity to heat used by these snakes to detect and capture prey (Bullock and Diecke 1956).

Why has thermoregulation evolved? Perhaps the answer varies with each species that has developed one or more of the components of thermoregulation. Ecologists may answer by noting that each step in evolution has expanded the ecological niche of the species. An often-cited example is that, for terrestrial animals, the independence from the external environment as a heat source permitted nocturnal existence. Physiologists may answer, "homeostasis regarding temperature resulted in increased reliability of multiple biochemical, physiological, and morphological features of the species." In the future, close examination of each step in evolution is needed because it can yield valuable information about the basic mechanisms involved in detection, control, and response to heat or cold.

THE ENVIRONMENT OF HEAT

We will first consider some extremely hot environments. More people live, work, and play near these extremes of heat than they do to the extremes of cold. The following are typical of extreme hot environments that affect both humans and mammals: a temperature of 127°F was recorded in Queensland, Australia; a temperature of 136°F was recorded in Libya; a temperature of 134°F was recorded in Death Valley, California (1913). These high temperatures are most apt to be in those areas where water is not available for osmotic regulation and for evaporative cooling. For example, in the United States, one area in California had no rain for 767 days, a locale in Chile had no rain for 14 years, and a part of the Sudan had no rain for 19 years. Populations in large cities must sometimes endure extreme dry bulb temperatures along with a shortage of water. In 1965 there were record hot spells in New Delhi, India. In one part of the city the temperature was 121°F and in Delhi itself, even at night, the temperature was frequently 100°F; the Associated Press reported that 144 deaths were attributed to this heat.

One fifth of the earth's land mass is semiarid. The importance of these areas is becoming evident as our population and productivity needs keep increasing. *Desertification*, the process by which arid lands are expanding, and the extent to which overpopulation and other factors contribute to desertification, is of concern to many social and natural scientists (United Nations 1977).

Some individuals choose to earn their living in extremely hot environments. The Chamber of Mines of the Transvall Free State in Africa reports that state-owned mines employ nearly 400,000 tribesmen who, for economic reasons, work for periods of about 1 year. Some of the mines in which these men work are more than 2 miles below the surface. One mine on the Witwatersrand reaches down more than 13,000 feet. Of course, the natural rock temperature increases steadily with depth. In one mine before ventilation was introduced, the natural rock temperature was 105°F at 10,000 feet, and usually 120°F at 11,000 feet. On some occasions, the tonnage of air moved in and out of the mine exceeded the tonnage of ore removed. Over the years, improved cooling and ventilation systems have alleviated the severe environment to a large extent; nevertheless, heat stroke is a daily human problem in these mines (*Mining Survey* 1963).

People also expose themselves to heat in sports. Although racing cyclists and runners frequently experience heat stroke, it appears to be most common in American football players. These individuals suffer a handicap because the protective clothing limits heat loss. Consider the environment for construction workers in deserts and tropical areas, and stokers or train engineers in many parts of Africa or Australia. Modern thermal conditioning of working and living facilities is—for reasons of economic or energy conservation—designed to meet the needs of "average" weather. Periodic extremes in weather can result in exposure of non-acclimatized persons to extreme heat (Folk 1981). Infants, the elderly, and persons in poor health are particularly susceptible (Robertshaw 1981).

Classification of Hot Climates

Hot climates can conveniently be referred to as *hot-dry* or *warm-humid*. The hot-dry climates are usually desert regions, which are widespread over the globe and exist over large areas of the southwestern United States. The warm-humid climates are typically represented by the tropical rain forest areas lying within latitudes of 10° or 20° from the equator.

Hot-dry climates have the following characteristics:

1. High air temperature during the day
2. Low humidity
3. Intense solar radiation
4. Wide day-to-night variations in temperature
5. Scanty precipitation
6. Vegetation is scrubby or non-existent
7. Terrain reflects up to 30% of the incident sunlight
8. Ground absorbs solar energy and heat to temperatures as high as 190°F

9. Ground surface radiates long-wave heat to cooler bodies in the environment (such as humans)
10. Ambient air at a temperature higher than skin and clothing
11. Hot ambient air heats the body by convection instead of cooling it

Warm-humid climates are characterized by the following:

1. Air temperatures not excessive with an upper limit of 90° to 95°F
2. Average relative humidity is 75% or higher
3. High moisture content of atmosphere reduces its transparency to solar radiation, thereby reducing solar heat load on humans; direct solar heat is less a problem than in the desert
4. Little day-to-night or seasonal variation in temperature and dew point
5. Precipitation high, usually varying with season
6. Vegetation abundant, providing ample shade and favorable radiant environment

Now, comparing the two areas in more detail we find that the input of solar energy is high in both areas, but in tropical regions, solar energy is converted into water vapor and thus exists in the atmosphere as high humidity. In the desert, where moisture is lacking, the solar energy directly or indirectly heats surfaces as well as ambient air, and thus exists as radiant heat. Physiologically speaking, in both areas humans experience heat stress.

Vocabulary of Heat Stress

One of the oldest, unresolved controversies in the temperature-regulation field has been: What temperature is regulated? Is it skin temperature, because a subject working in the heat stops sweating when entering a cool or neutral environment? Or is it the temperature of the hypothalamus, as indicated by experiments which monitor hypothalamic temperature (Benzinger 1959)?

In recent years, the most prevalent theory has been that the hypothalamus temperature-regulating center contains four types of cells: (1) cells that increase heat loss, (2) cells that increase heat production and retention, (3) cells sensitive to input from receptors located throughout the body, and (4) cells sensitive to local temperature. The activity of the first two cell types is determined by the activity of the latter two. The overall activity of the temperature-regulating center is proportional to the difference between the *set point* temperature and the actual temperature, very much as a thermostat controls the temperature of a building. The hypothalamus thermostat is unique because the greater the discrepancy between the set point and the actual temperature, the greater is the response—sweat rate, panting, shivering, or whatever.

Another theory has been that it is the temperature gradients within the body that determine the level of activity of the regulating center. This theory has received impetus in the past from reports that changes in panting and shivering rates have resulted from changes in the temperature of various body tissues, for example, liver and spinal cord (Thauer-Simon 1972).

A third theory considers it is the total body heat content that is regulated, rather than the temperature at a specific point or points in the body. The rate that heat dissipates is regulated in proportion to the rate that heat accumulates, and would take into account Nielsen's study, which has troubled environmental physiologists for many years (Nielsen 1938). Nielsen's data indicates that core temperature (T_r) was determined by the subject's work rate independent of the thermal environment. Changes from acclimation or acclimatization complicate responses to heat stress, and must be considered in any theory.

Behavioral responses to heat have been described as the first line of defense against a heat stress; this raises the question of the potential relationship between the centers concerned with consciousness and centers concerned with autonomic responses to heat. Do both of these centers receive input from the same receptors? And what is the extent of interaction between these centers? Motivation is a complicating factor (Satinoff 1981). We often speak of persons who are totally immersed in their work; such a person will not be aware of slight to moderate changes in the thermal environment.

Applied environmental physiologists are continually being challenged by the problem of assessing the severity of a heat stress and reducing thermal environments to within tolerance limits. Belding and Hatch (1955) identified and quantified the various components of the stress—namely, radiation, convection, metabolic heat—and compared these to the evaporative cooling capacity of the environment. The equations for heat exchange are continuously being improved (Shapiro, Pandolph, and Goldman 1982; Goldman 1978; Santee and Gonzales 1988).

Assessments of a thermal environment and subsequent physiological strain vary and are somewhat controversial. Gagge (1981) suggests the use of the "Operative Temperature," a linear average of the ambient air temperature and mean radiant temperature. In some situations, he uses the "Humid Operative Temperature" or "New Effective Temperature." The investigator's decision as to which index to use is often determined by how many measurements of both the environment and subjects can be made, as well as by how accurate the prediction needs to be. Five heat stress indices have been described by Roth (1972), by Leithead and Lind (1964), and by Santee and Gonzales (1988).

HUMAN PHYSIOLOGY IN THE HEAT

The defense against a heat load involves behavior, vasomotor response, heat storage, sweating, and acclimatization. Identifying the interaction and interdependence of each of these responses represents a challenging, complicated situation for environmental physiologists. There are multiple variables in the environment and multiple variables in the physiological systems.

Temperature homeostasis is a predominant role of the cardiovascular system. The system is involved in initial responses to hot environments and is also the system that oftentimes fails—resulting in heat death. The demands of a heat load may result in the cardiovascular system's failing to maintain homeostasis regarding energy, nutrients, and blood gases to exercising skeletal muscle and vital tissues (in particular the brain).

In hot environments, over 90% of the heat produced by exercising subjects is transported by the circulating blood. A nude person is apt to begin sweating at an ambient temperature of 30°C or 31°C (86°F to 88°F), although this threshold varies for different body regions, level of heat acclimatization, sex, and age of the subject. By the time maximum sweating occurs, the blood flow to the skin will have increased the thermal conductance of peripheral tissues 5- to 6-fold.

Mechanisms of Heat Dissipation

During heat exposure, the physiological systems must protect against water loss, salt depletion, and elevation of body temperature. To prevent heat exhaustion or heat stroke, the upper limits of these strains should not be exceeded:

- water loss equivalent to 7.6% of body weight,
- salt loss of 0.75 g/kg body weight, and
- core temperature of 40.6°C (105°F) (Hori 1987).

Cardiovascular Responses to Heat. Measurement of total skin blood flow is a technical problem that has not been solved. Because arm muscle blood flow does not increase with leg exercise or heat exposure, arm venous occlusion plethysmography is used as a measure of skin blood flow. Blood flow to the hands, feet, and ears is not representative of whole-body skin blood flow. Release from sympathetic vasoconstriction results in vasodilation in these acral areas.

Skin blood flow (SBF) is subject to modification by a wide variety of situations: exercise, gravitational changes, posture, arterial blood pressure, central venous blood pressure, mean skin temperature, and local skin temperature, among others. Because there are many situations that can reduce SBF, the full potential of the cutaneous vasodilator system is not realized during heat stress under a wide variety of conditions. The extent to which SBF parallels a given heat stress is difficult to assess. Most change in cutaneous vascular response is mediated by an active vasodilator mechanism that requires an intact sympathetic nervous system (Edholm et al. 1957). Passive vasodilation, or release of tonic vasoconstrictor activity, accounts for 10% or less of the vasomotor response to heat (Shepherd 1963). Whereas norepinephrine is the neurotransmitter involved in passive vasodilation, the transmitter involved in active vasodilation is unknown. The core temperature, measured by monitoring the right atrial blood temperature, has been reported to have a greater effect on SBF than does mean skin temperature by a 20:1 factor (Wyss et al. 1975). Similar values were reported for the baboon (Proppe et al. 1976). At esophageal temperature (T_{es}) above 38°C, Brengelman et al. (1977) reported an attenuation in SBF and a break in the linear relationship with T_{es}. Roberts et al. (1977) examined the effect of physical training as well as heat acclimatization on the SBF/T_{es} slope

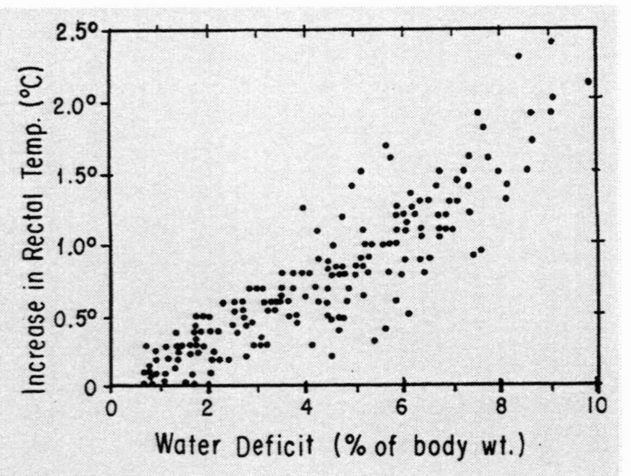

Figure 7-2. Effect of Dehydration on Rectal Temperatures of Humans. When resting men were exposed to desert conditions, rectal temperatures increased linearly with progressing water deficit, reaching a 2°C increase at 10% water deficit. This is probably associated with increased viscosity of the blood and the inefficiency of the heart during heat exposure. This situation has an advantage to humans since the heat flow from the environment to the skin is reduced. These high body temperatures should not necessarily be considered a failure in temperature regulation. Modified from Adolph (1947).

and found no consistent change, but there was a lowering of the threshold for increased SBF.

Increase in blood volume is one of the distinctive changes associated with acclimatization to heat. Just how the increased blood volume changes SBF and improves heat transfer is one of the major challenges and current controversies in the study of heat acclimatization.

Heat Storage. The elevation of body temperature results in an increase in metabolic rate. The physiological cost per degree elevation of body temperature is not well defined. Perhaps this is why many applied physiologists try to avoid elevation of body temperature in industrial situations. The large mass and the high specific heat, 0.83 kcal/kg·$°C^{-1}$, of the body are attributes that can reduce the extent to which costly physiological responses to heat are needed. The mechanism of storing body heat is illustrated in the extreme circumstances of humans being exposed to heat and dehydration at the same time; their body temperatures begin to rise. Some environmental physiologists say about such experimental subjects that they are behaving like large mammals. The significance of this expression will be clear when the comparative physiology of responses to heat is discussed in a later section; it is sufficient to say for the moment that, at least in the case of humans and camels under some circumstances, the body temperature rises in heat exposure.

In experiments with men on the desert, a group from the University of Rochester in New York found that the rectal temperature of humans increased in a linear fashion with progressing water deficit, reaching a 2°C increase at 10% water deficit (Fig. 7-2). Although some investigators consider this increase in body temperature with progressing dehydration to be a failure in heat dissipation, a rise in body temperature reduces the heat load in a hot environment because the difference in temperature between the environment and the cooler body is diminished. The heat flow from the environment is roughly proportional to the temperature difference, and goes down as the difference gets smaller. The rise in body temperature has both advantages and disadvantages, but it is probably better to avoid classifying it as a failure of heat regulation.

This does not resolve the question, "What is the physiology that ultimately determines how much of a heat load is to be stored and how much is to be dissipated by cardiovascular and sweating responses?" We have already noted that Nielsen (1938) observed heat storage is determined primarily by the level of work load (Figure 7-3). However, heat dissipation begins before elevation of body temperature ceases; if there were no heat loss while working at a maximum work load, the body temperature would be elevated 1°C every 5 to 7 min. The extent of body temperature elevation appears to be determined in a conservative manner. Furthermore, the mechanisms available for heat dissipation have to account for most of the heat load encountered in a working day. Once stored, heat is not rapidly dissipated; the heat stored during an 8-hr industrial working day is dissipated during the 16-hr period before the start of the next day. Immersion in cool water can result in rapid lowering of body temperature (Holmer and Bergh 1981), and helps to explain the popularity of a cool swim at the close of a hot day. In hot environments, the skin is usually considered to represent one third of the body mass, and the core, two thirds. Detailed models describing heat transfer within the body have been reviewed by Houdas (1981).

Sweating. The sum of the cooling capacity of the sweat produced (600 kcal/liter evaporated at 35°C [95°F]) plus the total heat stored by body temperature elevation is an excellent indication of the heat load for a given situation. It is fortunate that mammals have a relatively high body temperature. As pointed out by Lyman (1963), if a mammal's body temperature were set at 85°F, it would be physically impossible for it to produce enough water to cool itself by evaporative water loss at an air temperature of 100°F.

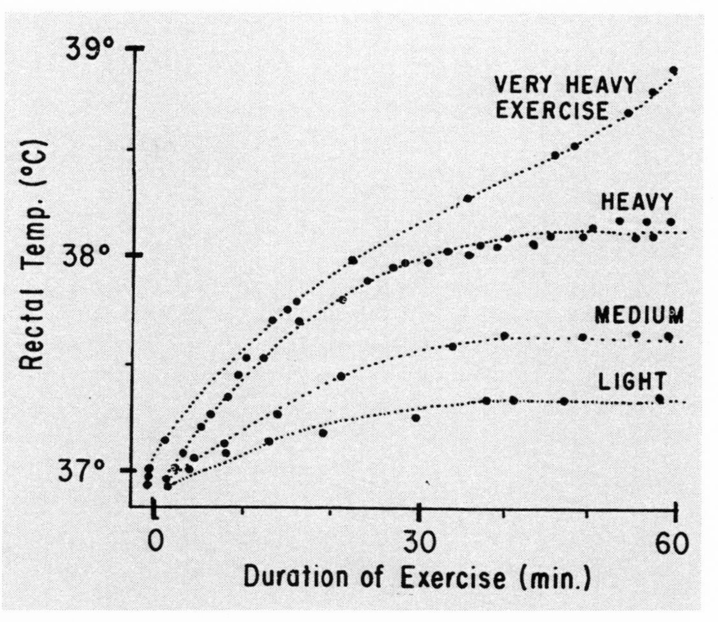

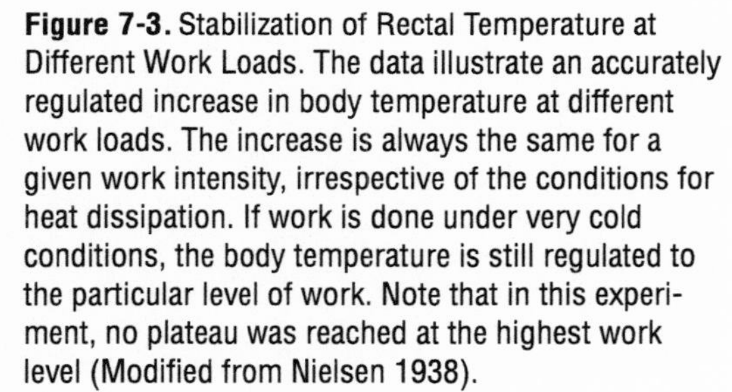
Figure 7-3. Stabilization of Rectal Temperature at Different Work Loads. The data illustrate an accurately regulated increase in body temperature at different work loads. The increase is always the same for a given work intensity, irrespective of the conditions for heat dissipation. If work is done under very cold conditions, the body temperature is still regulated to the particular level of work. Note that in this experiment, no plateau was reached at the highest work level (Modified from Nielsen 1938).

The production of thermal sweat is a conspicuous and costly physiological event that might be predicted to be a factor to upset homeostasis (homeokinesis) completely. Consider for a moment the quantities of water concerned. On a hot day in the desert, most men can produce 12 liters of sweat at a rate of 1 liter/hr. It is true that Ladell (1945) observed some men who apparently could produce only 0.5 liter/hr, while maximum capacities are given by several authors as 2.6 liter/hr (Moss 1924) and 3.5 liter/hr (Eichna et al. 1945). These rates were not sustained of course, but Eichna did observe men in good condition after they had been sweating at rates of 3 liter/hr for 4 hr. This mechanism is obviously a different type of process than those involved in warming the mammalian body in the cold. Shivering is primarily a muscular process not too much different from waving the arms or legs, or running slowly in place. The effects upon homeokinetic mechanisms will be those of exercise. On the other hand, maximum sweating removes a high percentage of water from the body. The hourly rate of 4 liters is of the same magnitude as the total amount of water present in the blood. Although this water is taken from the blood as it passes through the capillaries of the sweat glands, the loss is indirectly replaced from other body compartments. Such a high loss of water cannot be sustained without being replaced in the form of drinking water. Human subjects in moderate to severe heat will not drink water equivalent to the rate of sweating, and the rate of sweating is not altered by moderate dehydration. The rate is still adjusted according to the need for heat dissipation. Furthermore, drinking in excess does not increase the rate of sweating.

What environmental stresses cause the sweat glands to be activated? In a comfortable climate, an unclothed seated person loses metabolic heat partly by evaporation and partly by convection and radiation from the skin. In air temperatures from 28°C to 30°C (82°F to 86°F), temperature regulation results from changes in the amounts of heat lost by convection and radiation from the skin controlled by changes in the vascularity of the skin. As the ambient temperature gets hotter, sweating is evoked in quantities well adjusted to the requirements of thermoregulation. The output of sweat increases 20 g/hr for each 1°C rise in air temperature. Once sweating begins, the skin blood flow must continue to increase as the heat load rises, to transfer more heat from the core to the periphery; here it is dissipated to the environment predominately by the evaporation of sweat. Undoubtedly, the skin blood flow becomes maximal soon after it begins to increase.

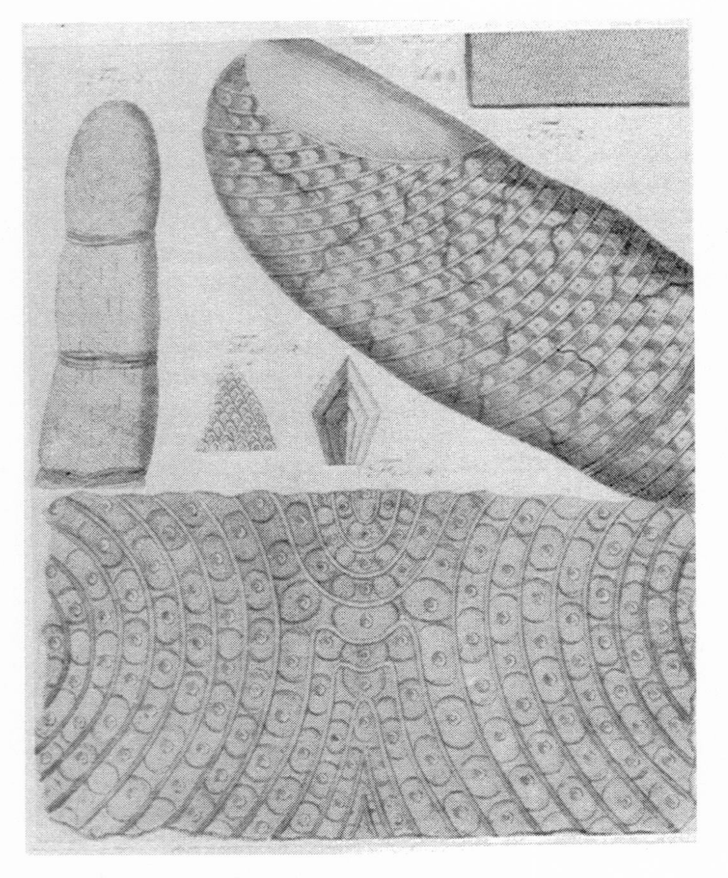

Figure 7-4. Sweat Gland Pores Drawn in 1763. About 100 years after the discovery of the microscope, Martin Ledermüllers recorded the sweat gland openings and drew fingerprints of human subjects.

1. *Sweat glands.* These remarkable glands, which can accumulate such a copious quantity of body water in human subjects, are called *eccrine glands*. They lie deep in the skin and are open to the surface through twisted secretory coils. The precise route for the transfer of water and other solutes through the walls of the glands is unknown. Water and presumably sodium and chloride may be reabsorbed in the sweat gland duct (Bullard 1971).

   Sweat glands in the human skin have been studied since 1763, when Martin Ledermüllers drew and counted the sweat glands (and fingerprints) of human subjects (Fig. 7-4). He did not realize that there are two forms of sweat glands among the mammals; almost all sweat glands in humans are of the *eccrine* type, although we have a few glands of the *apocrine* type in specific locations such as the axilla. Although it is an oversimplification, we may state that the eccrine glands secrete copiously onto a furless skin, while the apocrine glands are associated with hair follicles and produce an oily sweat (Fig. 7-5). Some other mammals have no eccrine glands and are served entirely by apocrine glands; the horse is an example. Apocrine sweat is odorless but it contains sebum which is quickly attacked by bacterial action. According to Kuno (1956), apocrine sweat droplets dry to form a glistening glue-like mass. This material is evoked in humans from our few apocrine glands by pain and fear; these stimuli also produce some eccrine "emotional" sweating. Apocrine sweat is not evoked by heat. Some authors describe the apocrine material they collected as milky; it is secreted by a semiholocrine mechanism. There is fat in the secretion represented by a series of fatty acids including baleric, capoic, and caprylic acids. Kuno states that he doubts the presence of any fat in eccrine secretions.

The composition of sweat formed in the secretory portion of the eccrine gland differs from the composition of sweat emerging at the skin surface (Sato 1977). The upper layers of the epidermis, especially the stratum corneum, are capable of passively absorbing and storing water and electrolytes from sweat. The active ionic reabsorption and secretion by the gland duct, together with the passive absorption and storage by the epidermis, not only markedly alter the composition of sweat appearing at the skin surface, but also establish complex transient and steady-state gradients of water and salt within the skin. These variable water- and salt-storing capabilities of the skin alter its mechanical and thermal characteristics. The water and salt content of the skin undoubtedly account for the differences in sweat production in humid and dry environments. The eccrine sweat gland also maintains a chronic low-level secretion rate without hyperthermic stress, so that even though such sweating is inadequately intense to bring water to the skin surface, it serves to hydrate the outer skin layers by intracutaneous diffusion.

There is evidence that the "resting" eccrine sweat gland functions in human thermoregulation. In resting glands sweat is continuously produced by the secretory portion of the gland. Sub-surface evaporation in the base of the sweat duct, condensation of the vapor at the top of the duct adjacent to the skin surface in the horny layer, and return of the condensate, aided by osmosis and capillary action back to the base of the duct, represents a model of what engineers refer to as a "heat pipe." Reay and Thiele (1977) have calculated that this heat-pipe action of sweat glands can account for the loss of 50 kcal (58 Watts) by the resting human body.

The nerve supply to the sweat glands is sympathetic; eccrine sweat glands respond to both acetylcholine and adrenaline when injected, but it is believed that there is no adrenergic component when sweating is induced naturally by heat. The reflex control of sweating seems to be partly related to the stimulation of peripheral receptors and partly to the temperature of the hypothalamus. One accepted theory is that the reflex act of sweating is initiated by the thermal receptors in the skin, while the degree of activity so induced may be potentiated by a rise in hypothalamic temperature. Another hypothesis is that the sweating is solely affected by receptors lying deep in the skin. Benzinger (1959), however, states that the deviation in temperature from a fixed set point of the thermosensitive area in the hypothalamus alone determines that rate of sweating. An alternative to Benzinger's theory was proposed by Hammel et al. (1963) that the set point of the temperature-regulating mechanism could change; they proposed this changing set point because they found that the hypothalamic temperature of a dog shivering in the cold could be higher than when it was panting in hot environments. The change in set point, they suggest, could be produced by impulses from thermal receptors in the skin, in the core of the body, and from other factors such as state of wakefulness and perhaps also muscular work.

In Chapter 4 we discussed the temperature regulation theory that takes Hammel's variable set point hypothesis into account. This theory by Feldberg and Myers (1964) applies

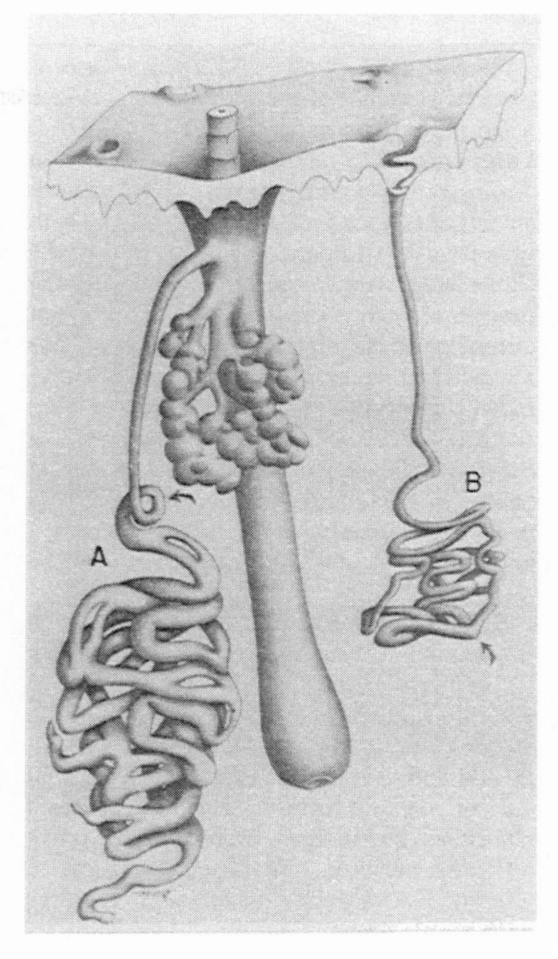

Figure 7-5. Two Kinds of Sweat Glands in Human Skin. The apocrine sweat gland (A) opens into a hair follicle; the eccrine sweat gland (B) opens into a pore (Montagna 1962).

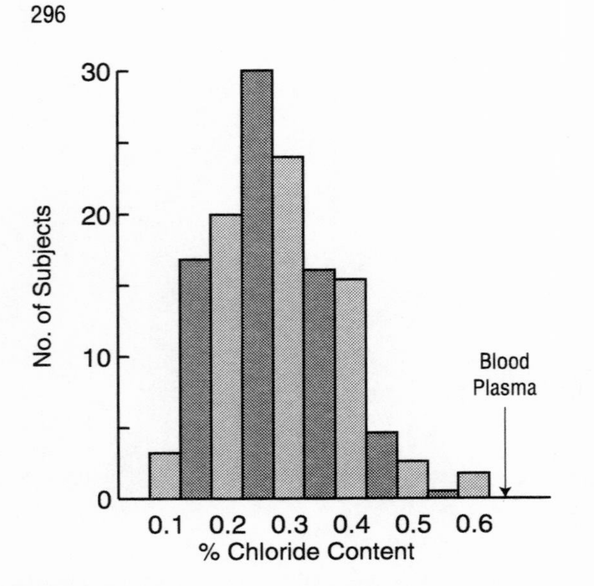

Figure 7-6. Sweat Concentrations Under Desert Conditions. A frequently quoted figure for sweat concentration is 0.2%. These data for concentration of sweat collected from British subjects in southern Iraq show that considerable variation is to be expected. Some of this variation is due to the degree of acclimatization to the heat (Ladell et al. 1944).

not only to sweating but to temperature regulation as a whole; they propose the regulation is mediated through the release of the amines serotonin, epinephrine, and norepinephrine in the hypothalamus. The core temperature may be the outcome of the balance between the release of the three amines. They further suggest that the action of pyrogens and antipyretics may be explained as an interaction with this chemical balance. It was also pointed out earlier that there are interspecific differences in the action of these amines on the hypothalamus, and that there are reverse effects of large and small doses of the same substance (Cooper 1966). Nevertheless, it seems that the normal balance between these particular amines does affect the body temperature level in a large number of animals.

2. *Sweating into water.* The wetness of the skin has a marked effect on sweat-gland activity. In waterbath experiments, the absorption of as little as 30 g of water can reduce sweat rate by 50% (Hertig, Riedesel, and Belding 1961). The mechanism by which soaking the skin reduces sweating is unknown. Possible changes in skin conductivity could be a factor. Reducing sweat production when evaporation is not occurring reduces unnecessary water loss, but the physiology involved is difficult to explain.
3. *Sweat composition.* The salt concentration in sweat is of considerable importance during exposure to extreme heat. Large quantities of salt are lost through the skin during the sweating process. The urinary output of salt may go down virtually to zero (less than 1 mEq/liter). Apparently the skin of humans cannot conserve salt during the process of heavy sweating. (Some mammals conserve chloride more efficiently than do humans; the sweat of the desert donkey contains much less chloride than does the sweat of the human.) Thus, the salt output must be replaced in the diet. This means that water alone is not sufficient to make humans into successful desert dwellers; thus, the price of salt is high in hot countries. Salt has been a main article of trade and taxation; it has caused wars, and at times has been weighed against gold.

Sodium and chloride are always present in lower concentrations in sweat than in the blood plasma. The sweat of most subjects has a chloride content that corresponds to 0.2%, varying to 0.3% (Fig. 7-6), while a few individuals will have a concentration of 0.1% or 0.6%. In other terms, sodium and chloride in sweat range from about 5 to over 100 mEq/liter. There is a tendency for the salt concentration in sweat to decrease with a person's acclimatization to heat; according to some authors, the salt concentration tends to increase with increasing sweat rates. Robinson et al. (1950, 1965) provided evidence that it is a change in skin temperature rather than an increase in sweat rate which brings about the increase in salt concentration. Another variable has been found in controlled experiments where the diet was analyzed along with the salt balance; it was shown that when sweating rates and states of acclimatization are constant, the skin output of salt depends on the dietary intake. In absolute figures, we find that at high sweat rates the total loss may be as extensive as 10 to 30 g of salt per day. Ladell (1945) even reports that when men worked in a room at 33°C with 80% RH, the loss was 25 g of sodium chloride in 162 min. This is a high percentage of the estimated total of salt in the body, which is only 165 g. During sweating, the effects of salt loss are not apparent, but when water is consumed, the body fluids become diluted, and this may lead to *heat cramps*. Heat acclimated subjects rarely experience muscle cramps, as salt balance is maintained by decreased salt in the sweat, increased salt in the diet, and increased salt conservation by the kidneys.

As a matter of historical interest, until approximately 1929 it was assumed that heat cramps were due to the heating of tissue. In that year J. S. Haldane wrote a paper establishing that these cramps depend on the salt loss associated with heavy sweating.

Thirst. In human subjects, sweat rate readily surpasses water intake dictated by thirst. The physiology of thirst is complex. Thirst and drinking are modified by integrating signals from osmoreceptors and volume receptors. Osmoreceptors are activated by cellular dehydration of hypothalamic and other tissues, with probable contributions from osmoreceptors in the oropharyngeal, hepatoportal, and renal regions. Cardiac receptors are important in maintaining vascular volume, primarily by secretion of atrial natriuretic peptides and by modifying the brain antidiuretic hormone secretion (Greenleaf and Morimoto 1996). Nausea or potential nausea is the reason most subjects give for refusing to drink fluids equivalent to water lost in the heat. Many animals replace water loss more readily than do humans. One major control for drinking is in the hypothalamus. Andersson and McCann (1955) demonstrated that in the hypothalamus there are specific osmoreceptors which respond to a reduced osmotic concentration of the blood; when these cells were stimulated in goats, they drank large quantities and overhydrated themselves up to 40% of their body weight.

In some animals under natural conditions, drinking stops before the water has been absorbed from the stomach and could dilute the blood. Apparently the amount of water passing through the esophagus influences the animal to stop drinking. This mechanism in humans is not sufficient or perfect enough to regulate the amount of water that should be consumed to replace evaporated water. By not drinking enough after heavy sweating, a person undergoes a voluntary dehydration which often reaches 2% to 4% of body weight. Thirst is satisfied before water intake equals loss. This strange drinking behavior has not been explained. One handicap that humans must face is a small capacity for drinking water. Schmidt-Nielsen found that while a donkey could easily drink 8 liters in 1 min, he himself had considerable difficulty drinking 1 liter even over several minutes.

Short-term Heat Exposure. We will now consider abrupt physiological changes in humans upon initial exposure to extreme heat (the first 1 or 2 days). Acclimatization, which takes from 4 to 10 days, will be described later. Several physiological reactions will be listed or considered in detail:

1. At first, humans in extreme heat show a transitory increase in plasma volume as fluid shifts from interstitial space to the vascular bed. Several days later, more red blood cells enter the circulation and blood volume may increase from 20% to 30%. This increase does not occur at all when the daily exposure to heat is brief. When water lost by sweat is not replaced, eventually the blood volume decreases.
   Because under conditions of dehydration there is no significant change in volume of circulating cells or plasma proteins, a consequence of the loss of plasma water is an increase in both red cell and protein concentration. This means an increase in viscosity of the blood, which places an additional load on the heart.
2. The oxygen consumption increases if body temperature rises, because of the direct cellular effect of heat and also the increased ventilation.
3. There is an initial hypoglycemia.
4. There is an increase in food consumption. In one study, food intake per person per day was 400 cal higher when they were exposed to heat. There was a body water gain which exceeded the total weight gain, so that the subjects actually lost body tissue during the heat exposure (Consolazio et al. 1961).
5. There is some degree of dehydration, because human thirst does not suffice to replenish body water lost. If an inadequate supply of water is provided, then dehydration continues more rapidly. As mentioned earlier, sweating continues at the appropriate rate regardless of dehydration. If this dehydration continues, it can extend to 8% to 11% weight loss. At 2% weight loss, thirst is violent; at 4%, the mouth and throat feel dry; at 6%, the above symptoms are severe; at 8%, salivary function has stopped and speech is difficult; at 10% weight loss, a person is physically and mentally unable to take care of his/her needs; at 12% water deficit, the subject is unable to swallow and can no longer recover without assistance. According to Adolph (1947), a person dehydrated to this point must be given water either intravenously, intraperitoneally, by stomach tube, or through the rectum. It is significant that formation of sweat did not stop in the subjects in Adolph's group in Rochester who voluntarily tolerated experimental dehydration to 12% weight loss. No information is available about the degree of dehydration at which sweat production fails. Schmidt-Nielsen (1964) suggested that the lethal limit in humans is about 18% or 20% weight loss.
6. As dehydration progresses, human body temperature rises. The Rochester group found that the rectal temperature increases in a linear fashion with progressing water deficit, reaching 2°C increase at 10% deficit.

7. As was shown almost 200 years ago, when humans are exposed to great heat there is an increased pulse rate. There is also a reduction in the maximum heart rate subjects can tolerate. More recent studies (Fortney et al. 1981; Montain and Coyle 1993) have demonstrated that isotonic hypohydration reduces skin blood flow for a given core temperature and therefore increases the potential for dry heat exchange. Fortney et al. (1981) provided a rationale as to why an iso-osmotic hypohydration might reduce skin blood flow and sweating rate. They theorized that hypovolemia might reduce cardiac preload and alter the activity of atrial baroreceptors which have afferent input to the hypothalamus. Therefore, a reduced atrial filling pressure might modify neural information to the hypothalamic thermoregulatory centers which control skin blood flow and sweating. Subsequent research (Gaddis and Elizondo 1984; Moack, Nose, and Nadel 1988) have demonstrated that acute unloading of atrial baroreceptors during exercise with periods of lower body negative pressure impairs heat loss and increases core temperature. Dehydration had been reported (Gonzalez-Alonso et al. 1995) not only to reduce cardiac output, but also to produce increased systemic and cutaneous vascular resistance during exercise.
8. Urine volumes are reduced in a hot climate below the amount found in a moderate climate when water intake is restricted. The urine formed contains excretory products in concentrations as high as physiologically possible. The minimum urine volume can be further modified by changes in diet.

Tolerance to Heat. Still giving attention to the reactions of humans during the first few days of extreme heat exposure, we will now consider unusual cases of tolerance to heat. A dramatic account was given by McGee (1906), who described the recovery of a man lost in the Mexican Desert for 8 days with 1 day's supply of water. The man survived by drinking his own urine, breaking off spines of cactus and chewing the moisture from the ends, and drinking the blood of scorpions.

The detailed study of heat tolerance is over 200 years old. In 1775, Charles Blagden described the following study before the Royal Society in London: after preparing a room at 260°F (127°C), Blagdon entered it with some friends and a dog—the dog was in a basket to keep its feet from being burned. They all tolerated the temperature without ill effects for 3/4 hr. With humor rare in scientific reporting, Blagden described how a steak he took in with him was thoroughly cooked by the heat. The modern "Blagden" is David Gates of the University of Michigan; he took a sauna bath at 230°F (110°C) and recorded the effects in detail in his delightful book *Man and His Environment* (1972). One of his observations follows: "By blowing on the back of my hand, I blew away the boundary layer of air and caused the hot air of the sauna to come into contact with my skin. The hot air burned my skin, leaving a red blister."

There have been experimental studies on men exposed in extremely hot chambers because of the interest in learning the tolerance of humans in friction-heated cockpits. Early experiments were done by Craig Taylor (1951) on himself in a room at 260°F for about 15 min. In the 1960's, 7 men tolerated 400°F in a hot room for periods extending up to 20 min (Murray and Ross 1965). The mean skin temperature rose to 43°C and rectal temperature increased 0.9°C. In Blockley's environmental-chamber experiment (1963) 8 subjects tolerated 160°F for 60 min, 200°F for 40 min, and 235°F for approximately 20 min. Their clothing consisted of standard one-piece underwear and equivalent material on the hands and feet. The capability for a complex psychomotor task was tested; the duration of unimpaired performance was found to be approximately 75% of the total tolerance time.

Totally different experiments were done by Strydom and Wyndham (1963) on groups of gold miners shoveling rock under natural conditions in high heat and humidity. Some groups of men (acclimatized and hyperacclimatized) worked at a wet-bulb temperature of 96°F with a dry-bulb approximately 4°F above this. Ventilation was provided with a velocity of 100 ft/min. At the end of 1 hr both groups of men had oral temperatures of about 102°F. These temperatures did not continue to rise in most cases. Since the risk of heat stroke was high, many subjects were withdrawn from the test after 1 hr.

Considerable attention has been given to the criteria for determining reasonable tolerance time for exposure to heat. Ellis et al. (1960) listed the following criteria of incapacitation which they used to make decisions about terminating the exposure of subjects to high heat:

1. a pulse rate of over 160 beats/min after work
2. a pulse rate of 140 beats/min before work
3. a rectal temperature of 39.2°C or higher
4. a rise in pulse rate during rest period
5. definite signs of physical inability to cope with the task
6. evident cyanosis or circumoral pallor
7. complaints of unpleasant symptoms such as faintness or cramping

As their program of heat-exposure studies developed, these workers carried some experiments close to the criteria listed above. In one experiment the subjects, seated and in shorts, were exposed to dry-bulb temperatures of 37.4°C and wet-bulb 37.2°C; in a second experiment they were exposed to a dry-bulb of 54.5°C and wet-bulb 40.6°C. For 9 subjects, the average tolerance time for the mild exposure was 144 min, and for the extreme exposure, 37 min. In both cases, none of the men actually lost consciousness. Most of them remained in the chamber until they were on the verge of collapse and had to be helped to leave. None was fit to do any responsible work for the rest of the day. The symptoms were of sudden onset, and the men usually felt quite well until 5 to 10 min before they reached the limit of their endurance. Only one man experienced severe muscle cramps. Two men on the verge of losing consciousness were unwilling to leave the chamber and protested that they were all right; their manner was similar to that observed in humans suffering from lack of O_2. Many of the subjects had difficulty in thinking clearly and quickly, and answering questions. Wyon et al. (1981) found that mental performance as well as physical capacity can be changed by heat.

The tolerance limits in various industrial situations have been of concern to many physiologists (Vogtetral 1981; Kuhlemeur 1978) and organizations (World Health 1969).

Exercise and Heat Exposure. Acclimation and acclimatization to heat may not occur with human subjects unless they exercise during the heat exposure. This is surprising because one would suppose that if a person can be acclimated by exposure with exercise at 40°C, then she might become acclimated by sitting and sweating profusely in a room at 50°C day after day. However, simple elevation of body temperature does not result in acclimatization. It appears that all physiological responses to heat must be challenged before acclimatization can occur. The need for exercise during heat exposure to accomplish acclimation emphasizes the key role the circulatory system plays in physiological responses to heat. The dual stress of exercise and heat exposure places extra demands on the cardiovascular system; there is a need to supply the requirements of active muscle, and to transfer extra heat to the body surface. The subject of exercising in the heat has been reviewed by various investigators (Armstrong and Pandolf 1988; Gonzales 1981; Nielsen 1981, 1978; Gisolfi 1979; Nadel et al. 1978; Nadel 1977, 1978).

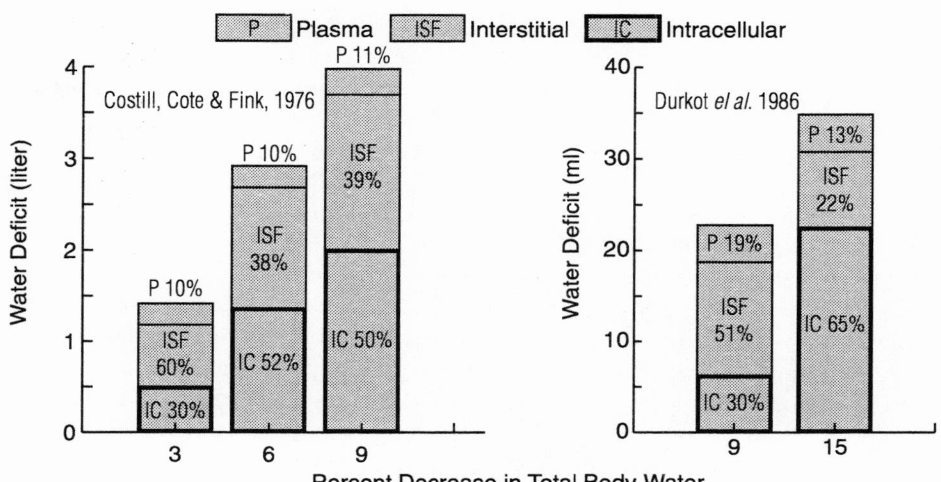

Figure 7-7. Sources of Water Lost during Periods of Dehydration.

Maintaining Body Fluid Balance. Earlier we noted the imbalance between the capacity of the sweat glands to reduce body fluids and the inadequacy of the thirst mechanism to replenish them. In addition, renal blood flow decreases during heat exposure, and as hypohydration develops, renal reabsorption of electrolytes increases. The sources of water lost during periods of dehydration are illustrated in Figure 7-7. Remember, total body water constitutes approximately 60 percent of an average adult's body weight, so a 75-kg person will have a total body water mass of about 45 kg. Therefore, a fluid loss equal to 5 percent of body weight could constitute 8 percent of total body water for this person.

The detrimental effects of hypohydration have been described many times by researchers and include (1) decreased volume of blood returned to the heart; (2) increased heart rate to compensate for decreased stroke volume; (3) elevation of body temperature which increases metabolic rate; and (4) changes in osmotic pressure of the various fluid compartments in the body. The effects of these osmotic changes are not well defined, but muscle cramps are likely

MAINTAINING TOTAL BODY WATER IN THE HEAT

Everyone should keep in mind that maintaining total body water is important. When exercising in a humid environment or a hot and humid environment, the body can lose up to twelve cups of water in the first hour. This amount of fluid cannot be replaced within the hour because the stomach can only empty approximately four cups of fluid per hour. People planning to exercise in such environments should heed the following recommendations:

1. Make sure you are well hydrated prior to exercise; drink at least four cups of fluid an hour or two before exercising.
2. If you don't mind the calories, drink fluids containing carbohydrate and electrolytes; they are retained better than plain water.
3. Drink fluids at regular intervals, for instance, 2/3rd cup every ten minutes during the exercise.

Maintaining body water will reduce the elevation of body temperature and reduce the strain on the cardiovascular system.

the result of osmotic changes. Replacement of 81% of the fluid lost during exercise has been reported to markedly reduce the detrimental effects of fluid imbalance (Montain and Coyle 1992).

Hyperhydration before or during heat exposure has been demonstrated to help maintain fluid balance during heat exposure. Moroff and Bass (1965) examined the influence of excessive fluid ingestion on thermoregulatory responses to exercise in the heat; they reported that hyperhydration decreased core temperature while increasing sweating rates above normal levels. But Greenleaf and Castle (1971) reported that excessive fluid ingestion does not alter core temperature or sweating rate from control levels during exercise in the heat. If hyperhydration did improve performance during exercise-heat stress, these improvements would most likely be mediated by hypervolemia. In fact, some of the thermoregulatory advantages gained through heat acclimation had been attributed to plasma volume expansion (Senay, Mitchell, and Wyndham 1976). The real question is: Can the gastrointestinal tract absorb water, calories, and electrolytes equal to the rate that muscles can burn fuel and sweat glands can produce a hypotonic solution? The answer to this question may very well be "Yes, under some circumstances."

One serious challenge to this response is whether it can be accomplished comfortably. Most athletes will respond with a resounding "No!" A full stomach while exercising is invariably quite uncomfortable. However, drinking while exercising can be advantageous for hypohydrated subjects (Mountain, Kain, and Sawka 1995). The popularity of various sports drinks is evidence that many persons are willing to try to maintain fluid and caloric balance. The type of exercise and variability in individual physiology and preferences make predictions of the optimum solutions difficult. In 1992, Sawka and Greenleaf chaired a symposium concerned with thirst, dehydration, and fluid replacement.

Another approach to improving fluid balance during exercise in the heat has been to hyperhydrate subjects prior to heat exposure by ingestion of glycerol, plus a large volume of water (Riedesel et al. 1987). The theory behind this approach is that glycerol is soluble in aqueous and lipid media, and this causes the glycerol to be rapidly absorbed by the gastrointestinal tract and to be evenly distributed among body fluid compartments. The theory also assumes osmotic action of glycerol causes water to follow the glycerol. A study by Lyons et al. (1990) describes increased sweating and reduced rectal temperature (T_r) of subjects exercising in the heat. Reduced heart rate and increased stroke volume have also been reported in exercising subjects following glycerol-induced hyperhydration (GIH) (Montner et al. 1996). Sawka and others (1993) report that GIH involves proportional increases in all fluid compartments. The discomfort of ingesting fluids just prior to or during exercise has been avoided by drinking selected solutions at regular intervals and thereby maintaining GIH for 49 h (Koenigsberg et al. 1995). Although glycerol is readily metabolized and some is excreted in the urine, once subjects are hyperhydrated, they do not rapidly return to euhydration. This is not surprising; Dill (1938) reported that once men are hypohydrated it takes 12 to 18 hr for them to return to a state of euhydration. Apparently, a similar time period is required to go from hyperhydration to euhydration. Additional research is needed to identify the optimum volumes, timing, and concentrations of glycerol and other fluids (Latzka et al. 1997).

We lack a clear picture of how changes in cell volume are sensed and transduced into regulatory responses. Recent studies in yeast, however, may provide important clues to this problem. Saito and coworkers (Maeda, Takehawa, and Saito 1995; Maeda, Wurgler-Murphy, and Saito 1994) have described a cascade involving a sensor kinase and synthesis of glycerol which results in accumulation of water. These clever genetic experiments may be an indicator of how sensing and regulating cell volume occurs in other organisms.

Acclimation and Acclimatization of Humans to Heat. Acclimatization in response to a hot environment is a process that reduces the amount of heat stored, heart rate, salt content of sweat, and threshold for onset of sweating, while increasing blood volume and sweat rate. In a 1981 review, Horvath provided the following definition: "Acclimatization represents those physiological changes which result in an improvement in performance with successive exposures to or residence in a hot environment." Those who have observed heat-exposed human subjects agree that remarkable improvement in performance and tolerance follows 7 to 10 days of regular heat exposure. Figure 7-8 illustrates the marked increase in sweating, reduction in heart rate, and elevation of core temperature that accompanies acclimatization to heat. Figure 7-9 also illustrates changes in temperature, heart rate, and sweating after heat acclimatization, this time to humid heat.

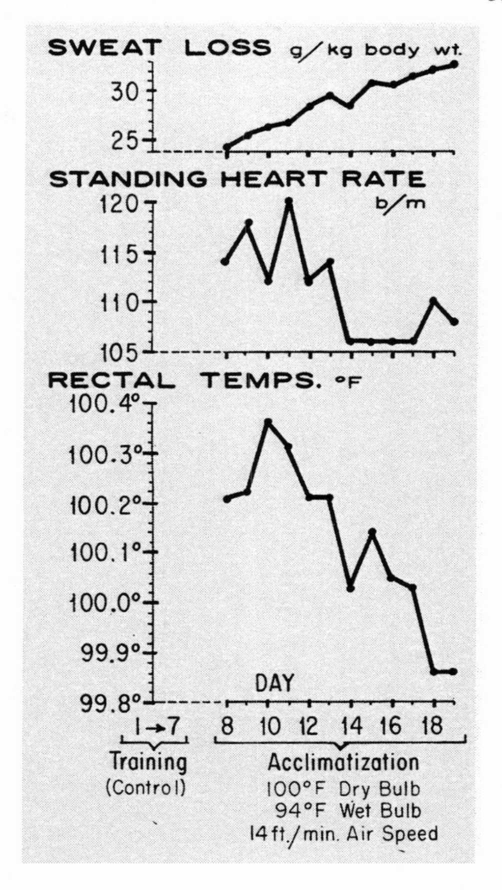

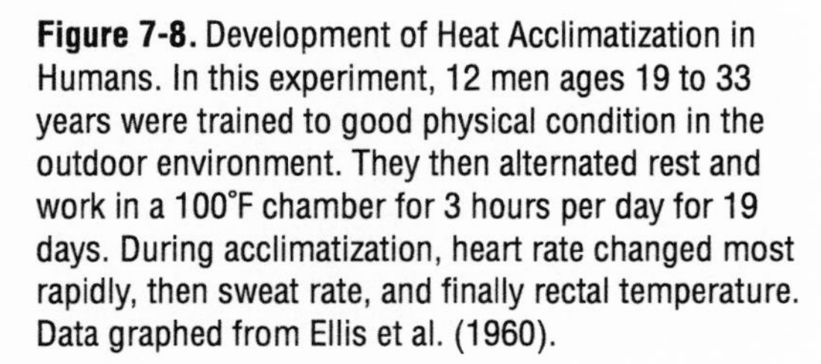
Figure 7-8. Development of Heat Acclimatization in Humans. In this experiment, 12 men ages 19 to 33 years were trained to good physical condition in the outdoor environment. They then alternated rest and work in a 100°F chamber for 3 hours per day for 19 days. During acclimatization, heart rate changed most rapidly, then sweat rate, and finally rectal temperature. Data graphed from Ellis et al. (1960).

The requirement of 7 to 10 days to acquire acclimatization suggests that some adjustments of the endocrine system are involved. Mack and Nadel (1996) found evidence that aldosterone action on sweat glands and kidneys is in part responsible for the decreased NaCl content of sweat and expansion of plasma volume. These authors also note that arginine vasopressin (AVP) secretion increases in response to hypohydration. Changes in the production of renin and aldosterone, or changes in the sensitivity of their respective receptors on target tissues, may also be involved in acclimatization.

Here is an example of a heat acclimatization study: we paid 10 subjects a minimum wage and asked them to walk at 3 mph in a hot environment of 115°F, 50% RH until they were ready to quit for the day. The first day subjects chose to walk 40-60 min; they had heart rates of 150 to 160 beats per min, and rectal temperatures of 101.5°F to 102°F. After a week to 10 days the subjects elected to walk 4 to 6 hr, had final heart rates of 125 to 130, and rectal temperature 100°F to 100.5°F at the end of the exposures.

After having had experience in acclimatizing over 280,000 recruits, Wyndham (1973) gave the following conditions for optimum acclimatization over an 8- to 9-day period:

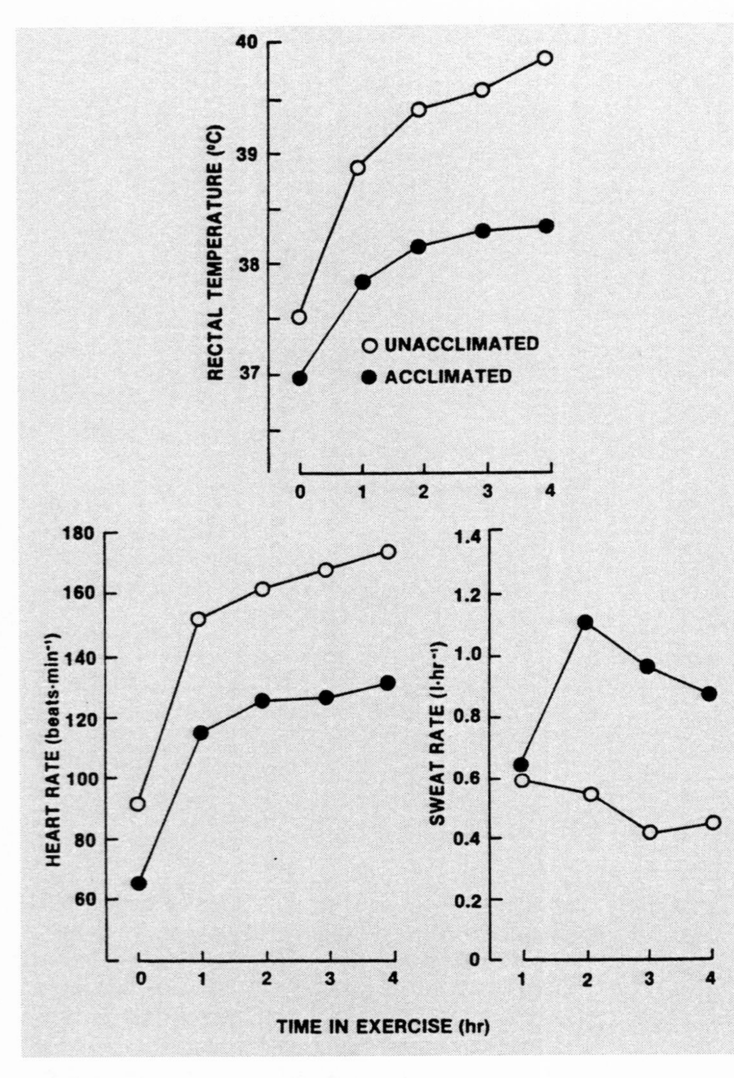

Figure 7-9. Exercise and Acclimatization to Humid Heat. These graphs compare responses of male subjects during 4 hours of exercise in humid heat, before and after a program of acclimatization. Redrawn from Wenger 1988.

Oxygen Consumption, liter/min	Wet-bulb Temperature, °C
0.65	34.0
0.95	32.0
1.45	30.5

The various work loads were selected because the recruits would be assigned different tasks in the humid gold mines.

There is no good comprehensive explanation of why or how acclimatization to heat occurs. It appears to include improved control of the various systems involved in responses to heat. The improved overall performance persists for 3 or 4 weeks after the heat exposures. Some investigators select one aspect of long-term responses to heat. Senay (1979) claimed the "secret" to heat acclimatization is increase in plasma fluid volume. He even went so far as to explain that a group of people who could not acclimatize to the heat also could not expand their blood volume. The expansion of blood volume is reported to result from an increased permeability of peripheral capillaries to interstitial proteins. In contrast, Harrison et al. (1981) reported the changes in plasma volume are incidental to heat acclimatization and vary with the type of exercise conducted during acclimatization.

A safe approach to defining heat acclimatization is to take the position that multiple receptors, control systems, and effectors are involved in the changes. However, this leaves physiologists with an almost uncontrollable urge to resolve the identity of the basic processes involved, and is of little value to applied environmental physiologists in assisting personnel who must experience unanticipated excessive heat and perform assigned tasks.

The extent to which physical training and exercise influence the response to heat stress has been a controversial subject—as has heat acclimatization—for many years. Improved physical fitness usually involves improvement of the cardiovascular system. The dependence of adequate responses to a heat load on the circulatory system's ability to move heat from sources to avenues of heat loss leads to the conclusion that persons in excellent physical condition will perform better in the heat than persons in poor or moderate condition. The elevation of core temperature during exercise also suggests that persons who exercise regularly also regularly experience an extra heat load and will thereby become acclimated to heat. However, intense training in a cool environment cannot serve as a substitute for exercise in the heat if acclimation is desired within a two-week period (Gisolfi and Cohen 1979). To facilitate acclimation to heat, the exercise workouts may need to be for an extended duration. Exercise in the heat has been the topic of several reviews (see earlier section, *Exercise and Heat Exposure*). There is considerable evidence that regular exercise does improve sweat-gland response to heat loads. However, there is no change in the distribution of sweat with repeated heat exposures (Cotter, Patterson, and Taylor 1997). The circulatory adjustments required to meet the needs of active muscles may reduce the extra skin blood flow necessary for heat dissipation, and thereby the changes essential for heat acclimation are not met by exercise conditioning without heat exposure. Also, regular elevation of body temperature, such as in a hot tub, will result in limited heat acclimatization. Hori (1987) reported that persons native to cooler climates do not attain the extent of heat acclimatization as do persons native to the tropics.

We must briefly consider cellular responses to hyperthermic stress and heat shock protein (Hsp). Whereas classical physiological heat acclimatization (lowered heart rate, lowered rectal temperature, higher sweat rate) takes days to manifest, the effects of excess heat on the cell begin within minutes. These cellular responses are measured by increased rates of transmembrane ion flux, inhibition of protein synthesis, induction of Hsp's, and changes in morphology. Recall that there are undoubtedly millions of different proteins in nature. The catalytic and structural functions of proteins are dependent not only upon the correct sequence of amino acids, but also upon the proper folding of proteins. It is the tertiary chemical bonds that are responsible for protein folding, and these bonds are readily disrupted by heat. Breaking of these tertiary bonds makes proteins insoluble in aqueous media, and thereby nonfunctional. Synthesis and repair of proper folding of proteins is accomplished by *chaperone proteins*. Hsp's are a group of small proteins that serve as ATP-dependent molecular chaperones. Whenever protein configuration is disturbed by environmental extremes, Hsp's combine with damaged proteins and provide repair. Disruption of tertiary bonds and folding of proteins occurs in response to numerous physical and chemical stresses including hypoxia, metabolic poisons, drugs, detergents, and heat. These stimuli activate heat shock genes and synthesis of Hsp. These genes are among the most conserved genes in organisms, including plants, bacteria, yeast, and mammals. Data from 140 different species have been tabulated by Nover and Scharf (1991), which indicate that the threshold and temperature range of response (37°C to 42°C for humans) varies with the species and its normothermic state. Whereas the role of Hsp's in vertebrates is not well defined to date, the presence of Hsp in intertidal mussels is reported to make it possible for these organisms to tolerate a 35°C body temperature (Hofmann and Somero 1996). This tolerance enables them to occupy an unusual environmental niche.

Clothing. The convective heat transfer, reflectance, heat-conducting capacity, and water permeability are the physical features determining the effect of clothing in hot environments. Of course, personal and social preferences in fashion can also dictate the type of clothing. Goldman (1981) has described the necessity to conduct both climatic chamber and field studies with a copper manikin to identify the advantages and limitations of various military garments. The copper manikin contains thermocouples to check surface temperatures. Electric heating elements within the manikin can simulate various levels of metabolic heat.

The design of ventilated suits was described by Leithead and Lind (1964), and the use of water-cooled vests and head gear has been examined in several laboratories (Nunneley and Maldonado 1982; Nunneley, Reader, and Maldonado 1982). These various devices provide minimum encumbrance, remove metabolic heat, and permit heat exposures in a variety of industrial and aerospace environments.

The clothing worn in hot climates involves compromises in amount of radiation shielding, insulation value, and ventilation for convective and evaporative heat loss. A quantitative study of the black robes worn by Bedouins in the Sinai desert provides explanation of how the chimney or bellows action of the garments can compensate for the greater absorbency on the surface of black clothing (Shkolnik et al. 1980).

Heat Exhaustion, Heat Stroke, and Fever. For many years, heat exhaustion and heat stroke were considered to be separate entities. Heat exhaustion has been related to circulatory failure, while heat stroke is seen as resulting from neural pathophysiology which leads to anhidrosis. Heat exhaustion occurs in situations where the excessive heat-transfer demands on the circulatory system result in circulatory failure. The resulting circulatory shock causes collapse of the subject. Heat stroke, on the other hand, results from neural pathophysiology; possibly, overheating of nerve cells results in pathological functioning of the temperature-regulation center. Subjects stop sweating, and elevation of temperature results in collapse and unconsciousness. However, in a 1979 review, Hubbard, identified heat exhaustion and heat stroke as representing a continuum. Examination of the rat as a model for heat stroke has led to the conclusion that heat stroke is not exclusively neural and does involve circulatory failure. Exercise contributes to an increased rate of heat stroke mortality and injury at low thermal loads. Hubbard's review demonstrates parallelism in the human heat-stroke cases and the rat model. It is still very difficult to identify the relative roles of high temperature, metabolic acidosis, and anoxia in the viscera due to local reduced blood flow in the development of heat exhaustion and heat stroke.

Comparative behavioral studies have necessitated considerable rethinking about our concepts of temperature regulation and fever. Fever has been described in vertebrates from fishes through mammals in response to infection. Behavioral elevation of body temperature in newborn mammals reduces survival of the infected organism (Kluger 1979). There is no single theory of fever which explains the mechanisms involved in all species. Certainly the classical concept of an elevated thermoregulatory set point has to be reconsidered. In mammals, fever appears to involve the production of one or more endogenous pyrogens, possibly by leukocytes. These endogenous pyrogens subsequently act by changing the rate of neurotransmitter release or modifying the sensitivity of regulatory cells in the hypothalamus to temperature or neural input.

Women in Hot Environments. We cannot readily answer the question of whether women are at an advantage or disadvantage when compared to men working in hot environments. As more studies are conducted, it is becoming obvious that what is needed are experiments in which male and female subjects are matched in terms of maximum oxygen consumption, lean body mass, surface area, and levels of acclimatization to heat. Differences in men's and women's physical features and past life styles account for the rarity of such experiments. The answer to the question will also vary depending upon the characteristics of the environment. In hot, humid environments, Shapiro et al (1980) reported that women had lower rectal temperatures than men, whereas in hot dry environments, men had lower heart rates and rectal temperatures. The advantage of the women in humid heat could be related to their lower and perhaps more efficient sweat production (Nadel et al. 1978). The advantage of men in hot dry environments could result from a better surface-area-to-weight relationship and greater sweating capacity. Frye and Kamon (1981) reported no differences in the sweat rate of acclimated men and women. Avellini et al. (1980) studied men and women with comparable maximal oxygen capacity and observed higher sweat rates per unit surface area for men, yet men had shorter tolerance times and higher body temperatures and heart rates. In a study of 46 men and 22 women recruited from industries, Kuhlemeier et al. (1978) reported males had a substantially greater increase in heart rate per degree elevation of rectal temperature than did females, but had a markedly lower intercept of the regression line relating heart rate to rectal temperatures.

Age and Responses to Heat. Infants have a large surface area-to-weight ratio, and in terms of heat balance this usually causes more problems in cold than in hot environments. Thermoregulating capacity does not equal that of adults until two years of age. The number of sweat glands is constant throughout the life span of humans, although only a few—predominately on the head—are functional at the time of birth and may explain why infants in the first 24 hours of life rank second only to the elderly as the most heat-illness prone group (Ellis 1976). Infants born three weeks prior to term have a defective sweat gland system (Robertshaw 1981). Infants in incubators do make positional changes in response to temperature. The limited tolerance of young children to heat stress has been attributed to an immature cardiovascular system (Drinkwater and Horvath 1979).

The elderly are particularly susceptible to heat stroke, and among 100 cases Austin and Berry reported in 1956, 70 were older than 60 years and 40 of these were between 71 and 90 years. These statistics may reflect an impaired thermoregulatory system that enhances the effects of the reduced capacity of the cardiovascular system in older persons. A significant reduction in the sweating per gland in older males has been reported (Collins et al. 1977; Foster et al. 1976),

THE NATIONAL INSTITUTES OF HEALTH WARNS:

"Heat stroke can be LIFE-THREATENING! Victims of heat stroke almost always die, so immediate medical attention is essential when problems first begin.

A person with heat stroke has a body temperature above 104°F. Other symptoms may include confusion, combativeness, bizarre behavior, faintness, staggering, strong rapid pulse, dry flushed skin, lack of sweating, possible delirium or coma.

It is important to realize that older people are at particular risk of hyperthermia. Many people die of heat stroke each year; most are over 50 years of age."

Source: "Hyperthermia: A Hot Weather Hazard for Older People," National Institute on Aging, NIH Publication No. 89-2763, August 1989

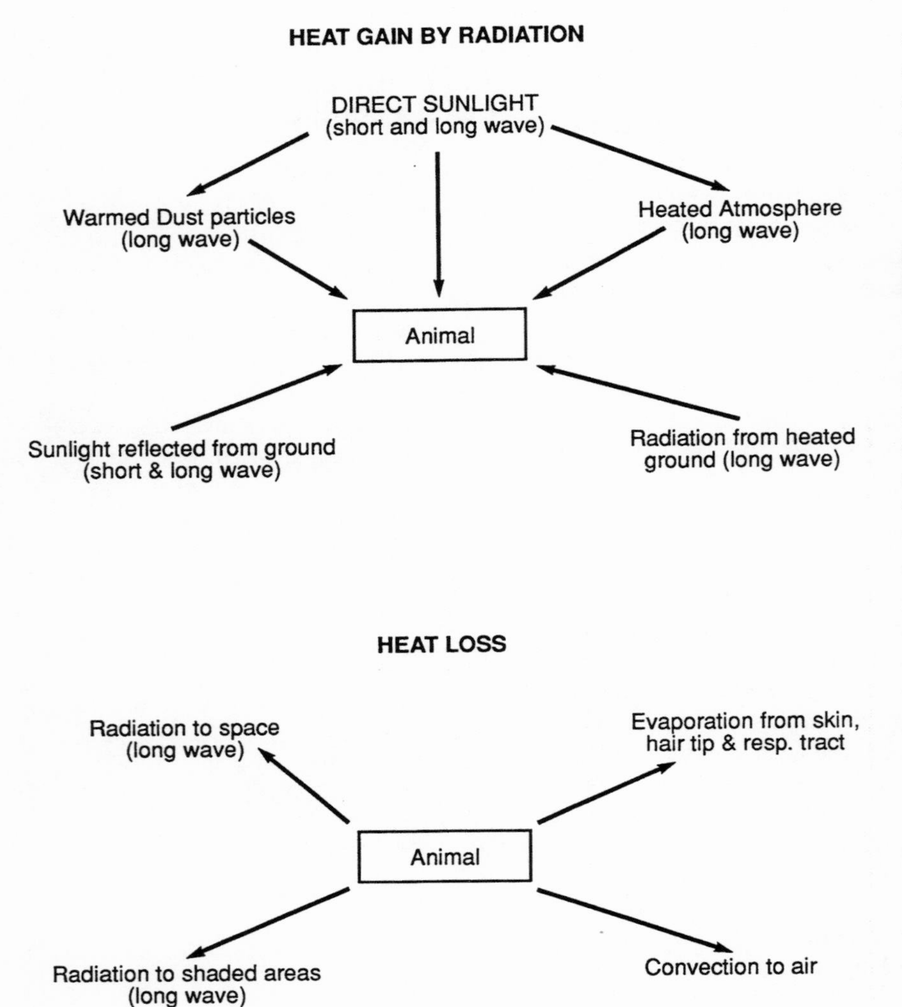

Figure 7-10. Heat Exchange of Animals When Radiant Heat is the Major Component of Heat Load. Surface temperature of the animal is greater than air temperature.

along with an elevation of the threshold for onset of sweating for elderly subjects. Kielblock (1987) reported no change in heat tolerance of South African gold miners over the age range of 18 to 62 years.

COMPARATIVE PHYSIOLOGY OF MAMMALIAN HEAT ADAPTATIONS

Mammals who live on the desert, or those who may suddenly be exposed to heat, have developed a remarkable number of adaptations for survival or for remaining comfortable. The variety and often apparent inconsistency of these adaptations become evident when we see that some non-desert species living in temperate climates are fairly well adapted for exposure to heat and dehydration. In the desert habitat, some mammals have elaborate adaptations, some have very few adaptations, and some have no adaptations that differ from temperate zone mammals. For example, a man has little concentrating power in his kidneys, while his ecological companion, the laboratory rat, has essentially twice the concentrating power. Two rodents living side by side in the desert illustrate the other point: the wood rat of the genus *Neotoma* must be maintained with wet food or free water, while the kangaroo rat can live entirely upon dry seeds.

Heat gained in the desert is primarily radiant heat. However, the thermal balance of desert animals involves multiple factors. Many are illustrated in Figure 7-10. Others to be considered later include wind speed, metabolic rate, reflectivity and thickness of the coat, plus availability of water and capacity to conserve it. Conduction of heat to and from the ground is not considered a major avenue of heat exchange, despite the usually extreme temperatures (120°F to 140°F) of top soil. Moving animals, of course, have a small body surface area in contact. Resting animals usually stay in one spot. They may change posture, but remain in the same area. Body surface temperature will come into equilibrium with the soil. Conductance can be important to burrowing animals as heat may be lost in the burrow. The complexity of heat exchange in a desert rodent has been noted in detail by Walsberg and Wolf (1995).

The adaptations that permit tolerance to heat and a desert environment have been characterized by Kirmiz (1962) as morphological, physiological, ecological, and ethological (behavioral). All of these types of adaptations must be genetic. Our interest must not be restricted to the desert, since acute heat exposure may occur anywhere in the summer, even in the Arctic. For example, according to Bartholomew (1956), fur seal bulls may die due to overheating when driven at an air temperature of 10°C (50°F). Much of the following discussion, however, will be devoted to the more natural circumstances of chronic heat exposure, such as that experienced by desert animals.

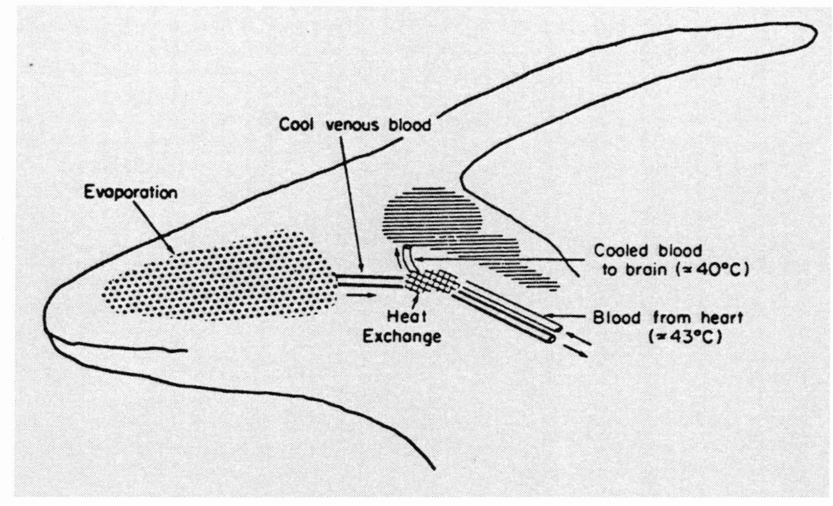

Figure 7-11. Countercurrent Cooling of Cerebral Blood. In some large desert mammals, the brain is spared a rise in tissue temperature; in the donkey, sheep, goat, ox, pig, and onyx, arterial blood in the internal maximal artery flows beside cooled venous blood from the nasal passages before it goes to the brain. (From Taylor, in Lee 1972)

Morphological Adaptations

Heat tolerance probably requires a trend toward a small body mass, attenuated extremities, little fat, and extensive superficial vein routing. The size of the woodrat has increased during cool climates and decreased during warm climates (Smith, Betancourt, and Brown 1995). Although we see the attenuated extremities in the camel, Eisenberg reports a behavioral interpretation of the long appendages in small mammals. He points out that strong evolutionary pressure in four families of rodents has produced independent forms which have bipedal locomotion such as is found in the jerboa and the kangaroo rat. He interprets this as an adaptation to living where there is little vegetation; the bipedal method of jumping is an advantage for escaping predators such as snakes and owls.

One morphological heat adaptation in mammals is designed to prevent some of the rising body temperature in critical organs such as the brain. A provision has been described for doing this in the donkey, the sheep, the goat, the pig, and antelope. This is a countercurrent phenomenon, and is the opposite of the one described in Chapter 5 on the cold environment; this adaptation in animals in the cold is designed to rewarm cold venous blood by means of arterial blood. In the adaptation for the brains of mammals in the heat, we now find cooling of hot arterial blood by cool venous blood (Fig. 7-11).

Physiological Adaptations

Conspicuous Features. Next we will consider some superficial physiological mechanisms to combat acute heat exposure:

1. Evaporative cooling is accomplished in humans and especially in the horse group (*Perissodactyla*) by numerous and efficient sweat glands. To a greater or lesser extent most of the mammals, except the rodents, also have sweat glands, but they are small, not very numerous, and are called upon only under extreme circumstances. Seals have sweat glands on the flippers only. Apparently many of the smaller primates do not have conspicuous sweat glands. In our experiments that involved exposing rhesus macaque monkeys to extreme humid heat, there was no evidence of liquid moisture or droplets of mois-

ture anywhere on the skin (Frankel, Folk, and Craig 1957). Hooten, however, referred to droplets of sweat appearing on the forehead and chest of the large apes (1942). There are other exceptions; the water buffalo is presumed to be without sweat glands. For those animals that do not have sweat glands, excess saliva running out of the mouth and down the chest is an assistance, and many such mammals lick the fur.

Two kinds of sweat glands were described earlier: eccrine and apocrine. Apocrine sweat is a dilute oily material that dries to form a glistening glue-like mass. If you examine a horse that has "lathered-up," the composition of this sweat is evident. In Table 7-1 we present a comparison of the efficiency of the oily type of secretion (apocrine) and the watery type of secretion (eccrine). The human high-sweat values are from six physiol-

Table 7-1. The Secretory Activity of Apocine (A) vs. Eccrine Glands (E)

Animal	Gland Type	Contribution of Panting	Sweat ($g \cdot m^{-2} \cdot hr^{-1}$)	Measured as T_A or T_r
Antelope (6 species)	A	present	30-120 g	air 47°C (116.6°F)
Camel	A	—	240 g	air 57°C** (134.6°F)
Cow	A	1/3 of skin	150 g (maximal)	air 40°C (104°F)
Dog	A	maximum	histological evidence under fur, but no functional value	
Horse	A	does occur	100 g	—
Pig	none functional	none	none	rectal 41°C (105.8°F) collapse occurs
Sheep	A	greater than skin	32 g† (maximal)	panting begins at rectal 41°C
Human (Bantu)	E	none	500 g (seated)	air 34°C (93.4°F)
Human*	E	none	basal 15 g	
Human*	E	none	366 g	air 35°C (95°F)
Human*	E	none	540 g	rectal 37.7°C (100°F) light exercise
Human*	E	none	573 g‡	air 45°C (112°F)
Human*	E	none	884 g	air 34.8°C (92.7°F) humidity wet bulb: 24.9°C (77°F)
Squirrel monkey	A, E	none	6 x basal when terminal	air 39°C (102.4°F)

*These subjects were reported in the literature with no ethnic group specified.
**globe thermometer reading in the sun
†20 kcal/hr by sweating
‡as much as 1000 kcal/hr by sweating

ogy laboratories. Note that humans stand alone in their success at using evaporative cooling by means of a dilute watery material; a few other mammals probably approach the maximum value of the camel of 240 $g{\cdot}m^{-2}{\cdot}hr^{-1}$, but note that an extreme globe thermometer reading was required to produce this sweating in the camel. Since this occurs underneath the woolly coat of the animal, the network of residues from the dried apocrine sweat must be substantial. Most figures in the table for the sweat-rate of eccrine glands of humans are double or triple the maximum capacity of apocrine glands.

It will take many years to resolve the question of the sweat glands of the higher apes. Nakayama (1971) follows the view, as we do, that the sweat glands of monkeys are vestigial. However, the research team at Indiana University considers that they may be able to prime monkey sweat glands by heat acclimation, with the result that sweat can be collected from the skin (Johnson and Elizondo 1972).

2. Some mammals in particular can capitalize on their shape to assist in combating heat: bats and flying rodents can extend their wings or skin-folds and greatly increase their surface area; this advantage is carried further by fanning the wings. All mammals stretch out their limbs in the heat in an attempt to increase surface area for evaporative and convective cooling.
3. Large size has a certain advantage in the heat because quite a bit of heat can be stored with a relatively small rise in body temperature. The camel (and possibly even the human) has made use of this characteristic. It probably explains the large range of body temperature reported for the rhinoceros (34.5°C to 37.5°C). In contrast, invertebrates can overheat or become dehydrated within a few minutes' exposure to desert heat. They are specialized in burrowing, taking advantage of shade, limiting water loss, and adapting to dehydration.
4. A heavy coat of fur, wool, or hair is an anatomical and physiological factor protecting mammals exposed to radiant heat. Blair and Keller (1941) demonstrated that long-haired dogs could withstand a sudden heat stress much better than dogs with short hair, and clipped dogs showed decreased resistance to heat instead of increased resistance. Schmidt-Nielsen (1964) made the same observations on the camel: he found that the temperature of the fur surface on the back of a camel in the sun many be as high as 70°C to 80°C (158°F to 176°F), when the skin temperature underneath was close to 40°C (104°F).

The heavy coat of the camel and the dog, as well as other animals such as the sheep, protects the animal in the heat by certain physical relationships (Schmidt-Nielsen 1964). We must first consider the radiation heat gain from the sun. If we consider merely the visible solar radiation, some of this will be reflected from the coats of the animals depending upon their color. The color will decide whether the reflection is higher or lower than from the skin of humans. The heat gain from the visible spectrum is independent of the skin surface temperature. When the infrared energy is concerned, we find all is absorbed irrespective of coat color and texture. The outer portion of the fibers (hair and wool) will absorb radiant heat and will lose some of the heat to the air by convection, and the shorter underlying fibers serve to insulate the skin from the high temperature attained by the surface fibers.

Conservation of Water. Some mammals tolerate long-term exposures to heat and desiccation, in many cases during the entire lifetime of the species; in these cases a parsimonious saving of water becomes of the utmost importance. Water is conserved by a reduction in the following phenomena:

1. *Pulmonary Water Loss.* This reduction is achieved in large mammals by maintaining a relatively low breathing rate, and in small mammals by entering protective areas where the vapor pressure is high. As another means of conserving water, Schmidt-Nielsen (1964) reported a remarkable reduction in temperature of expired air from small desert rodents. This must represent a countercurrent heat exchange in the respiratory passages. For example, at a room temperature of 25°C (77°F), the temperature of expired air and nasal passage walls of the rodent could be even lower than 25°C, although the body temperature was 38°C (100.4°F). Most heat exchangers in the various biological systems have two streams of fluid moving continuously in opposite directions; in this case the counter-current heat exchange depends upon the cooling of the nasal walls by incoming air and by evaporation. During the flowing expiration, heat flows from the warm air to the cool surface, and water is deposited.
2. *Cutaneous Water Loss.* In small mammals in the desert there is little water loss through the skin because insensible perspiration is reduced by the animal remaining in a burrow with a high vapor pressure. At night they emerge onto the desert when temperature is low and vapor pressure is high. Also, their skin is not interrupted by the water-permeable orifices of sweat glands. If they did have to sweat, they might have to evaporate 9% to 22% of their body weight to cover their relatively large surface area. Large mammals such as the camel have water-conserving sweat glands; their threshold to sweating is an ambient temperature of 35°C (95°F) or above, while in humans it is 30°C to 31°C (86°F to 88°F).
3. *Renal Water Loss.* There are many descriptions of the concentrating power of the kidneys of desert mammals. This water retention is probably not as important as a reduction in cutaneous water loss. It is, however, a characteristic of the most successful of desert mammals that they can produce a very concentrated urine. It is a curious fact that although most desert mammals probably do not ever walk along an ocean beach, they have a remarkable ability to benefit from drinking true sea water (not a mere artificial 3.5% salt solution); this, despite the magnesium salts in sea water that bring about an additional loss of body water by producing diarrhea. The clas-

sic example of the ability to concentrate urine is the desert kangaroo rat studied by the Schmidt-Nielsens (1950a); its maximum concentration of salt and urea in urine is compared with that of humans and the white rat. Concentrated urine compositions are shown in Table 7-2.

Several species of Australian desert mice show higher concentrations of urine. The ability to concentrate urine is associated with the possession of a kidney containing only long nephrons. These nephrons are characterized by an extension of the portion of the kidney tubule, known as Henle's loop. Associated with the long loop of the tubule is a long renal papilla which extends beyond the pelvis of the kidney and into the ureter. Superficially, the length of the loop structure of the nephron is recognizable by the relative thickness of the renal medulla. There is a correlation between the thickness of the renal medulla and the aridity of the habitat in which the mammals live (Schmidt-Nielsen 1964). Rodents, camels, and antelopes of the desert have the greatest relative development of the renal medulla, while animals with an abundant supply of water such as beavers, muskrats, and platypuses have a thin medulla and very short loops. The explanation of the concentrating power of the long Henle's loop depends upon the countercurrent hypothesis for the concentration of urine (Schmidt-Nielsen 1965). In addition to the thickness of the renal medulla, there is considerable variation in the blood vessels of the vasa recta. Some form specific zones in the medulla while in other mammals there are no zones visible; the opossum shows very distinct zones while the beaver has a vasa recta which breaks up into many scattered plexi.

The concentrating power of the kidney is often expressed as a ratio between the concentration of the urine and the plasma of the blood. For example, if the urine is twice as concentrated as the plasma of a particular animal, then the urine/plasma (U/P) ratio is 2. This figure applies to most birds in which the kidney loops are fewer and not as well developed as in mammals. Hudson (1964) has collected data on the concentrating power of a number of rodents that tolerate heat; the first of this series (comparing urine and serum) is represented by the wood rat, which has a ratio of 6. At the other end of this series is the kangaroo rat with a ratio of 17. The data for Hudson's series are presented in Figure 7-12. The sandy inland mouse of

Table 7-2. Urine Compositions of Three Species

Salt	Urea	
Human	0.37 N (2.2%)	1.0 M (6%)
White rat	0.60 N (3.5%)	2.5 M (15%)
Kangaroo rat	1.20 N (7.0%)	3.8 M (23%)

Australia has been found to attain a (U/P) ratio of 14.6 (MacMillen, Baudinette, and Lee 1972). A comparison of the concentrating capacity of many rodent kidneys was compiled by Fyhn (1979).

4. *Fecal Water Loss.* Desert mammals have reduced the water loss from the digestive tract both directly and indirectly. The percentage of water in white rat feces is 68% (calculated on wet weight), in camel feces 43%, in kangaroo rat feces 45%. In addition to having a low water content in its feces, the kangaroo rat utilizes its food more efficiently and eliminates a smaller amount of dry matter. The net result is that white rats use 5 times as much water for formation of feces as kangaroo rats, even when they both eat the same kind and amount of food. The explanation of the more efficient use of the feed in the kangaroo rat may be connected with its habit of eating its own feces (coprophagy). This action is not unusual in rodents and is essential to normal digestion in some rats, rabbits, and pica.
5. *Metabolic Water.* Further conservation of water is attained by increasing metabolic or oxidation water. For an active animal, there is no particular advantage in the formation of oxidation water; usually the evaporation from the lungs exceeds the water formed and the animal must drink. There is little oxygen in fat and the increased ventilation of the lungs to supply oxygen can increase respiratory water loss. It is true that for each gram of food metabolized, 1.07 grams of water per gram of fat is formed. This metabolic water becomes more useful in dormant mammals. The intriguing idea must be considered that the bear in winter lethargy may benefit from the retention of metabolic water from metabolized fat. This may explain why these large animals do not drink during the winter. For a further discussion of this matter, see Chapter 6, *Hibernation.*

Dehydration Tolerance. Conservation of water is related to tolerance of the animal to dehydration. At first one assumes that the better a mammal is adapted for survival during chronic exposure to dry heat, the more tolerant it will be to hypohydration. This is indeed true for most desert species; however, a conspicuous exception is the pack rat or wood rat (*Neotoma*) which has little tolerance to water deprivation. It survives only for about 7 days without water, while white rats under these conditions live for 2 to 3 weeks.

Ordinarily, tolerance to dehydration is a requirement for successful desert existence. For example, Giles compared

Figure 7-12. Maximal Urine Concentration in Rodents. This comparison places one of the desert rodents, the wood rat, on a low position in the rank order of ability to concentrate urine. This animal must eat wet food or drink water. At the other end of the scale, the kangaroo rat shows a high concentration, which probably reaches a ratio of 14. Some kidneys are more powerful than those of the kangaroo rat, especially those of the jerboa, which can produce a 25% urea solution. Modified from Hudson (1964).

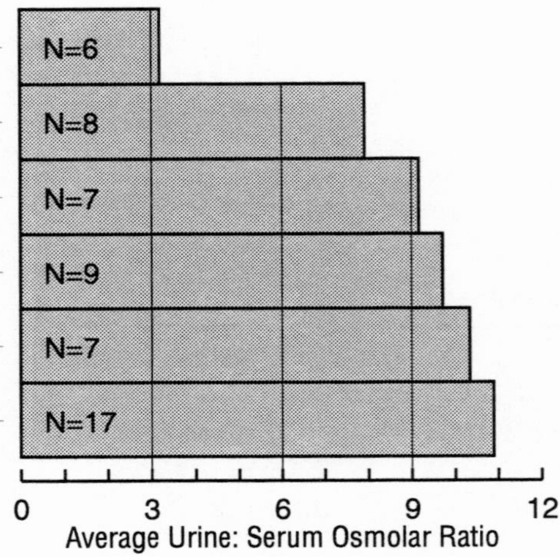

[1]Assuming a serum osmotic pressure of 384 milliosmols
[2]Assuming a serum osmotic pressure of 350 milliosmols

horses and camels during two expeditions across the South Australian desert (1889). On the first expedition in 1874, he found that horses required up to 50 liters of water per day beyond what was contained in their food during the summer. He started out with three horses and two camels, the camels being used to carry the water for the men and horses. In spite of the contribution of the water carried by the camels, all three horses died before the journey was two-thirds completed. The camels were given no water until the end of the 8-day march of 350 km. In 1876, Giles again crossed the desert, this time traveling over 480 km with 22 camels. The camels tolerated this march without water for 17 days.

In more specific terms, we can compare the temperate-zone mammal with the desert mammal: the dog and human tolerate a loss of water corresponding to 14% to 20% of body weight; the desert rabbit tolerates a water loss approaching 50% of its body weight; the camel tolerates a loss of 27% of its body weight. Details of the physiological rearrangements of the body that permit this dehydration will be described later in this chapter.

Even desert animals differ in their ability to tolerate or prevent dehydration. Shkolnik kept a series of rodents from Israel on dry grain for a month at 30°C and 30% relative humidity. Some species tolerated dehydration but were endangered, others prevented dehydration. This was demonstrated by the amount of water consumed after the test (see Table 7-3).

One should expect that survival in extreme environments depends upon very different evolutionary solutions; a sequence of tolerance to dehydration was found by German (1961) in murine rodents of the Steppe Zone (Table 7-4). The rodents were fed only air-dried oats. If judged by this experiment, only the laboratory mouse would be called desert-adapted.

Estivation. A reduced body temperature is a mechanism which may be useful to the animal exposed to extreme heat. This phenomenon may be described as the torpor or dormancy that occurs at relatively high ambient temperatures. Many estivators also hiber-

Table 7-3. Rodents Most and Least Adapted to the Desert

	Water Consumed (% body weight)
Least Adapted to Desert:	
Acomys cahirinus	11.4
A. russatus	9.3
Jaculus jaculus	4.3
Most Adapted to Desert:	
Meriones crassus	none
Gerbillus pyramidum	none
G. dasyurus	none
G. gerbillus	none

From Shkolnik, in Schmidt-Nielsen (1964)

Table 7-4. Tolerance to Dehydration in Some Rodents

	Water Deficiency Tolerated (Days)	Weight Lost (% body weight)
Microtus oeconomus	2-3	25
Microtus arvalis	3-4	35-40
Evotomys glareolus	3-4	35-40
Lagurus lagurus	10-15	50
Laboratory Mouse	over 25	19

nate, although there are some estivators that will not tolerate body temperatures below about 15°C (61°F). Mammals that estivate seem to have a rather high critical air temperature; they also have unusually low basal metabolic rates. These characteristics not only allow dormancy at moderate ambient temperatures, but also are adaptive to hot environments. Hudson and Bartholomew give an account of a series of estivators; the most typical example is the roundtail ground squirrel. Another example is the jerboa, which is reported as undergoing a state of torpidity unlike any other rodent described. Kirmiz (1962) states: "The jerboa can easily tolerate high environmental temperatures up to 45°C (113°F) by entering into a state of lethargy (deep sleep)." This unusual example of estivation requires considerably more study.

Why is a reduced body temperature and its associated low basal metabolic rate an adaptation for exposure to extreme heat? It is an advantage to the heat-exposed mammal because where there is a small difference between the body and air temperatures, there is a minimal amount of metabolic heat to be dissipated. For example, the roundtail ground squirrel has a relatively low metabolism, approximately 60% of that predicted for a species of its size (125 g). Associated with the low metabolism is a relatively inactive thyroid gland. The body temperatures of inactive but non-torpid animals may be as low as 32°C (90°F). The roundtail ground squirrel becomes dormant during the summer so that the body temperature drops to within a degree of ambient air (22°C to 25°C) (71.7°F to 77°F). This temperature drop is also an advantage in temperature regulation. As is usual with a mammal with a labile body temperature, there is a marked day-to-night (circadian) rhythm of metabolism and body temperature, with the higher phase of the cycle prevailing during the day. The result is a lower nocturnal level of pulmonary water loss and a lower level of metabolites to be excreted so that a more dilute urine can be produced at this time. The reduction in metabolism is of the order of 30% to 40% with a 3°C to 5°C drop in body temperature. Hudson (1964) estimates that this slight drop in body temperature will save as much as 1.56 ml of water per day, a significant amount for a 125 g animal. The animal does indeed occupy a territory of high temperature and limited water.

A second example is the case of the jerboa. This animal's central temperature (36.8°C) is lower than the white rat's (37.55°C). The jerboa's basal metabolic rate at the ambient of 30°C is 3.649 kcal/kg·hr^{-1}, compared to 6.156 for the white rat. This desert animal enjoys the same advantages possessed by the roundtail ground squirrel.

Ethological Adaptations

One of the characteristic types of adaptations is ethological (behavioral). Many of the small desert animals live in burrows during the daytime and become active when vapor pressure is higher at night and the heat sink of the cool night sky is apt to be available. We say that the digging of burrows and nocturnal habits represent behavioral adaptations.

A second example is the behavior of the dehydrated camel exposed to radiation from the sun. This animal exposes as small an area of its body surface as possible to the incident radiation. It sits on the ground with its legs under its body, usually facing the sun with the body oriented lengthwise in the direction of the sun's rays. It remains sitting on the same spot, only changing its direction as the sun moves during the day. In this way it shields the ground from the hot sun, thereby reducing heat flow from the ground to the underside of its body. Camels often rest during the hottest part of the day huddled on the ground in small groups, each animal pressed tightly against the next. The clump of camels constitutes a giant social organism that minimizes its exposed surface to reduce heat gain.

Eisenberg (1962) notes that small desert animals have developed a behavioral means to avoid the accumulation of sebaceous material in their fur: organized sand bathing.

Acclimatization To Heat

Associated with adaptations to combat extreme exposure to heat is the phenomenon of acclimatization (or acclimation in environmental chambers). The sequence of changes referred to as acclimatization to heat has been carefully studied in humans. The net effect of this phenomenon is a higher capacity for activity and less discomfort from heat exposure. It is characterized by specific and measurable physiological changes day by day as exposure progresses. There is every reason to assume that when desert-adapted and temperate-zone-adapted mammals (other than humans) are exposed to heat, conspicuous physiological changes in temperature regulation and water balance will take place from the first day to perhaps the tenth day of exposure. There is a dearth of information on this topic in the literature; whether this is due to an absence of acclimatization among mammals, or to lack of study by environmental physiologists is not apparent. Schmidt-Nielsen's book, *Desert Animals: Physiological Problems of Heat and Water* (1964), lacks reference to acclimatization to heat in any mammals other than humans. We have observed kangaroo rats laboratory-housed for several months with access to water become dependent upon drinking water. What happens under field conditions is not known.

There are a few scattered observations that may provide evidence of acclimatization: for example, some domestic animals have a different rectal temperature in different seasons. Dowling (1959) has shown that the heat tolerance of cattle at different seasons can be attributed to changes in the hair coat. Hart takes the view that changes in insulation do represent acclimatization. He points out that photoperiodically induced physiological changes would tend to pre-acclimatize animals to cold in advance of seasonal changes in temperature. He gives many illustrations of what he considers good insulative acclimatizations through seasonal modification of the fur. What happens when the animal is gradually or suddenly exposed to heat while the thick fur developed for cold is still present? The ideal sequence apparently would be for the animal to gradually drop this insulation during the mild periods of spring, and be prepared with a thin coat for the heat of summer. Remember, however, that a thick coat has some advantages during heat exposure. Arctic huskies do not appear to be uncomfortable when exposed to unaccustomed heat in a warm climate (if not exercising). Also, they continue to grow a very heavy coat even in the summer.

One possible case of acclimatization to heat irrespective of any change in insulation is recorded by Hart: "Sheep also undergo a seasonal variation in rectal temperature response when placed at 40.6°C (105°F) and 34 mm Hg water vapor. The heat tolerance indicated by the rectal temperature was highest during the summer and followed the seasonal rhythm of temperature and wool growth. However, statistical analysis of the data showed that there was a seasonal acclimatization in heat tolerance in addition to the changes caused by differences in wool growth."

An interesting description of the responses of heat-acclimated and cold-acclimated rats has been presented by Hale and Mefferd (1963). These authors worked with groups of animals that had lived for months in three air temperatures, 2°C, 24°C, and 35°C (35.6°F, 75.2°F, and 95°F). Rats had a physiological response when moved from one temperature to another. The authors used the expression "heat-acclimated rats"; these rats differed from rats maintained in the other two environments. The heat-acclimated rats displayed their new physiological characteristics the first day they were exposed to heat. The time sequence of the changes in rodent heat acclimatization seems to be lacking in physiological literature. If heat acclimation is acquired in a day, it is hardly comparable to the heat acclimation observed in humans over a 6- to 10-day period.

The work of Hale and Mefferd is cited here partly because of an unusual phenomenon not directly related to acclimatization; they give an account of cold and heat diuresis. A cold diuresis is an unusual situation in comparative physiology; it is common in humans but uncommon in the other mammals. Of comparable interest is the heat diuresis of the rats in 35°C. This is paradoxical because of heat-exposure and the necessity for water conservation; it may be associated with the higher water consumption when the animals were heat exposed. The cold-exposed rats, when placed in the heat, demonstrated heat diuresis but did not show an increased water intake. These data were obtained by abruptly changing the group of rats from one ambient temperature to another; recordings were made over a 24-hr fasting period.

More specific information was provided by Findlay (1963), who stated that during acclimatization to heat, cattle and mice reduce their food intake (by half) and their metabolic heat production. Cattle grow more slowly in the heat at first, and then increase this rate; they also show depressed thyroid activity. Heat-reared mice have highly vascularized ears and longer tails (Harrison 1963).

More detailed information on the hamster was provided by Chaffee and Roberts (1971): the metabolic rate declines in the heat. Associated with this decrease is a change in the oxidative enzyme activity of mitochondria. This decrease in activity levels out between the 2nd and 5th day in the heat, and remains depressed during the whole period of exposure. Rats, on the other hand, showed a significant increase in metabolic rate when first exposed to heat; it lasted from 48 to 72 hr, followed by a decline in metabolic rate until it was lower than control values (Yousef and Johnson 1967). It is of considerable interest that, in these studies, no exercise was used to obtain heat acclimation. In a study on beagle dogs (Folk and White 1970), a deliberate effort was made to use exercise in order to have comparable experiments with those on human subjects. Four dogs in good physical condition and trained for treadmill running were then exposed to treadmill running in the heat. Rectal temperatures on the 9th and 10th days of exposure were approximately 0.7°C lower than rectal temperatures on day one and two; heart rates on the 9th and 10th days were approximately 11% lower than those on the 1st and 2nd days of exposure (Fig. 7-13).

Special Adaptations: Conservation of Water in Birds and Aquatic Animals

Concentrated Milk. The challenge of existence in severe heat is usually solved by judicious spending of water derived from a delicate and precarious conservation system. Part of this system is the control over the quantity of water required for the formation of milk; one possible conservation device would be to produce a very concentrated milk. Perhaps the high fat content in the milk of marine mammals, such as seals and whales, is an adaptation associated with their environment of low availability of fresh water. The fat content of the milk of these aquatic mammals is 30% to 40%, compared to 2% to 5% in most terrestrial mammals. When these high figures were first obtained, the same analysis was done on camel's milk on the supposition that this might also be concentrated. The fat content of camel's milk turned out to be 4%. This is predictable, as the young camel

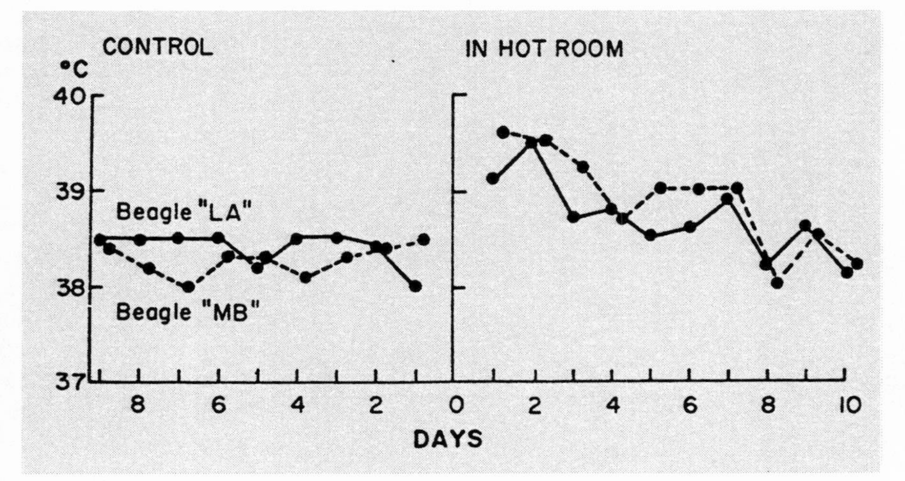

Figure 7-13. Acclimation to Heat of Beagle Dogs. This experiment was designed to be comparable to those on human subjects. Four dogs in good physical condition and trained for treadmill running were exposed to treadmill running in the heat. Based on the criteria of rectal temperature and heart rate, acclimation was accomplished in approximately 8 days.

would be at least as much in need of water as a nursing mother. Some small desert mammals, however, keep their young in a cool burrow and the mother's water supply is almost exclusively oxidation water. It is conceivable that in such a case the nursing young not only depend to a large extent on oxidation water, but in addition are provided with concentrated milk (Schmidt-Nielsen 1964); no analyses have been done.

Salt Glands. As a matter of interest, this list of heat adaptations should include a solution to water economy found in some of the birds and reptiles. Sea water is known to be toxic to humans and most mammals; how then do oceanic birds obtain their supply of body water? Some workers presumed that sea birds could subsist like seals on water obtained from the food. In order to profit from the ingestion of sea water, an animal must excrete salt in a concentration at least as high as that of the water ingested. The bird kidney, however, excretes salts in a concentration only about one-half that found in sea water. Marine birds excrete a major part of their ingested salt through salt glands (nasal glands). These are able to produce a highly concentrated salt solution, making it possible for sea birds to drink sea water. Schmidt-Nielsen and Fange (1958) have demonstrated this excretion in cormorants, pelicans, and Humboldt penguins. The experiments are simple: a 10% sodium chloride solution is injected into the bird and within 1 to 5 min, drops of a clear waterlike liquid appear at the external nares. This secretion continues for 1 to 2 h. The salt concentration in the nasal secretion is a 4% salt solution, while the salt concentration in the urine is 1.5%. More liquid is excreted by the salt gland than by the kidney; thus the amount of salt eliminated by the kidney is about 1/5 of the total amount from the two glands. A similar salt excretion has been demonstrated by the Laysan and Blackfooted Albatrosses (Frings, Anthony, and Schein 1958) and in some lizards. The salt concentration in the liquid from the nasal gland of the albatross was almost twice that in sea water.

CASE HISTORIES

In the above discussion of adaptations to combat heat and desiccation, the mechanisms and physiological devices are rather involved. At times the behavior of the whole animal in its environment is neglected. It thus seems reasonable to consider several "case histories" of mammalian species exposed to heat. The mammals to be discussed separately are the bat, the kangaroo rat, the rabbit, the dog, and the camel.

Bat. There are few descriptions in the literature of response to heat in those bats that hibernate. Because their physiology is totally different from the usual laboratory animal, they have not been studied extensively by environmental physiologists. It has been said that the body temperature and metabolism of the resting hibernator bat vary directly with the environmental temperature (Hock 1951). If this is so, one cannot describe the body temperature of this animal in the terms ordinarily used for laboratory mammals. In 1964, Henshaw at the University of Iowa made an exhaustive study of some specimens of these "bundles of thermal contradiction." He compared two species, *Myotis sodalis* and *Myotis lucifugus* (Henshaw 1965; Henshaw and Folk 1966). He devised a temperature probe 1 mm in diameter for the colon of these animals, which they tolerated during exposure to many combinations of dry-bulb temperatures and humidities. The animals were exposed in metabolism chambers, where Henshaw found that the reputation of these thermal-labile mammals was justified; for example, one of the species (*sodalis*) would tolerate a body temperature of -5°C and still have a heart beat.

The euthermic body temperature, or thermal neutral temperature, for *Myotis lucifugus* is estimated to be close to 39°C. It was not possible to determine a similar figure for *Myotis sodalis,* as it did not appear to maintain a homeothermic-type temperature for long enough to allow a selection of a temperature range. It is possible that this species has a critical air temperature near 15°C (Fig. 7-14). When exposed to air temperatures approaching 45°C, it was obvious that the two species had totally different capacities to resist heat. *Myotis lucifugus* developed large negative temperature differentials between air and body (as much as -6°C); *Myotis sodalis* maintained little or no differential at air temperatures of 38°C to 45°C. The poor temperature regulation in heat of *sodalis* was found in samples collected throughout most of the year including July; body temperatures of 41° to 42°C usually appeared to be stressful to the animal and frequently caused death within the respirometer. During part of the year (the months of March, April, May, and June) *sodalis* became even more sensitive to heat; body temperatures of 34°C to 35°C were usually fatal. During these months when exposed to 35°C (95°F), *sodalis* became excessively active. The heart rate was above 700 beat/min, and this was reached without comparable increases in metabolic rate. When exposed at this season to air of 25°C (76.8°F), the heart rates were those expected at 35°C at other times of the year. At this season, *sodalis* had usually died by the time a body temperature of 38°C to 39°C (110.5°F to 102.4°F) was reached (Henshaw 1965).

The picture of heat tolerance by *Myotis lucifugus* is totally different; this species tolerated body temperatures of 42°C (107.6°F) at all times of the year with little apparent distress. A sample from a summer colony in an attic with an average air temperature during 1 week in July of 54°C showed the greatest capability to withstand high temperature. (The saturation deficit was 50 mm Hg. This measurement is defined as the maximum vapor pressure at existing temperature minus the actual vapor pressure.) One juvenile, 1 to 2 months old, exposed to a chamber temperature of 51°C, was recorded with a body temperature of 48°C (118.4°F) for more than 15 min. There was no evi-

Figure 7-14. Bat Metabolism in Changing Ambient Temperatures. The rate of carbon dioxide production was used as an index to metabolism. Line *A* represents active animals at various temperatures. It may be that the critical temperature for this species is somewhere near 15°C. Line *B* represents hibernating specimens in cold ambient temperatures. Modified from Henshaw (1965).

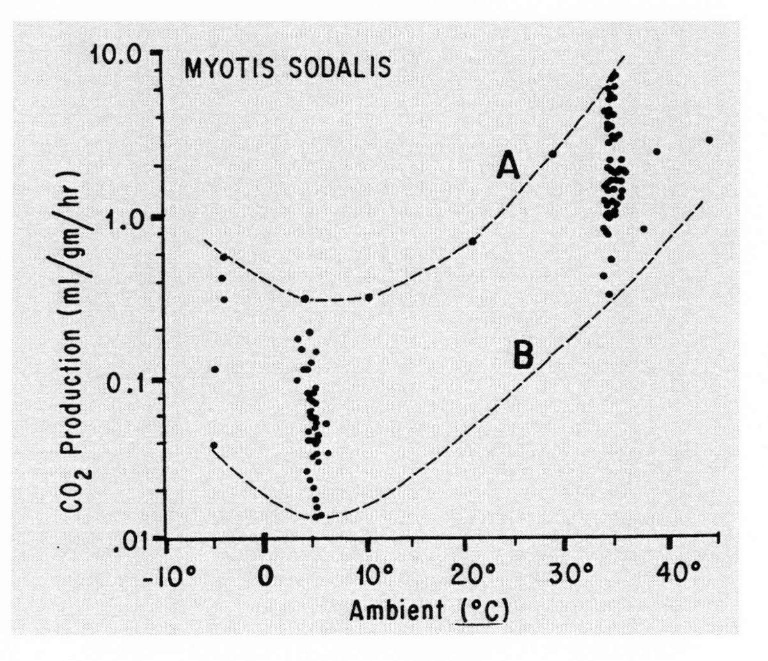

dence of unusual excitement in its behavior or EKG record (Henshaw 1965).

The tolerant species (*lucifugus*) attained its large negative temperature differential of -6°C (21.3°F) (while *sodalis* attained none) by a series of physiological mechanisms. In these experiments, and in others completed by Reeder and Cowles (1951), cooling was accomplished by panting, by a sudden massive engorgement of the blood vessels of the wings, by a gentle wing fanning at a body temperature of 40°C to 42°C (104°F to 107.6°F), and above 42°C by licking the entire body. In the Iowa experiments, panting and vasodilatation were the most conspicuous mechanisms, since there was little room in the respirometer to accomplish wing movements or licking.

We will now compare responses to heat of tropical nonhibernating bats with the North American hibernating types. Bartholomew, Leitner, and Nelson (1964) studied three species of flying foxes in Australia (Table 7-5). These three species represent a delightful series for the environmental physiologist to consider, as each responds so differently. The first maintains a negative gradient by physiological work; the second tolerates hyperthermia that it could prevent (because if air is 41° C or 105.8°F, it calls on the entire "battery" of cooling mechanisms); the third is in distress-hyperthermia (positive gradient of 4°C) even at 35°C (96.8°F). Bartholomew discusses the ecological significance of these measurements in his usual idea-provoking manner.

There is very little data available on the physiology of desert bats. Two species, the pallid bat (*Antrozous pallidus*) and the free-tailed bat (*Tadarida brasiliensis*), inhabit arid treeless areas in New Mexico. G. L. and L. B. Bintz examined them after an 80-hr period of water deprivation, and observed water retention by the liver tissue of the pallid bat and severe dehydration by the free-tailed bat (Riedesel 1977). Once again, here is an example of animals that inhabit similar areas, but apparently have different approaches to solving an environmental stress.

Vampire bats have urine-concentrating capacity similar to the

Table 7-5. Responses to Heat of Three Species of Flying Foxes (in °C)

Species	Common Resting Body Temperatures	Heat Exposure Air Temp	Heat Exposure Body Temp	Cooling Mechanisms
1. *Pteropus poliocephalus*	35.9° ± 0.7 SD	40°	39°	Panting Wing engorging Fanning Licking
2. *Pteropus scapulus*	36.6° to 39.2	40°	41°	Slight panting Wing fanning
3. *Syconycteris australis*	35.3° to 37.6°	35°	39.3°	Stressful panting

kangaroo rat, but the vampires have short loops of Henle (Horst 1969). It is doubtful that vampire bats need to drink water. There is at least one vampire bat which has been reported to live without drinking water. *Rhinopoma hardwickei* does not need free water (Vogel 1969).

Kangaroo Rat. These rodents can live indefinitely on dry seeds and other dry plant material without access to drinking water. They do not depend on body storage of water for their survival. When existing in this fashion, they maintain their body weight and a normal water content in the tissues, and are not gradually consuming water reserves. However, they must be losing some water through the feces, formation of urine, and evaporation from the respiratory tract.

The feces of the kangaroo rat have a low water content (see earlier section of this chapter), and food is well utilized so that water is conserved. For example, white rats use 5 times as much water for the formation of feces as kangaroo rats do when they eat the same amount of food. Furthermore, the kangaroo rat makes extensive use of reingesting its feces (coprophagy). This procedure is common among rodents because during the second passage through the tract, a number of vitamins can be used which had been synthesized by bacterial action in the cecum. The kangaroo rat, however, has a further advantage because there is a reduction in water loss by the second ingestion of the feces.

The kangaroo rat eliminates excretory products in a very small volume of highly concentrated urine. The concentration of urea and salt in the human represents a 6% urea solution and a 2.2% solution of salt; that of the kangaroo rat is a 23% solution of urea and a 7% solution of salt. To test the high concentrating power of the kidney, Schmidt-Nielsen did a very enlightening experiment. These rats do not ordinarily drink. He forced them to do so by feeding soybeans which have a high protein content and therefore yielded large amounts of urea to be excreted. On this feed, kangaroo rats would die without water, so they began to drink. The concentrating power of their kidneys was such that they should have been able to drink sea water, even if the magnesium salts in sea water resulted in diarrhea and an attendant water loss. In this experiment, the kangaroo rats did thrive when given sea water to drink along with soybeans (Schmidt-Nielsen 1950b).

Kangaroo rats appear to lose virtually no water by evaporation from the skin. As in all rodents, there are no sweat glands in the skin. White rats, however, lose water from the skin by diffusion (insensible perspiration); the amount lost is similar to that which evaporates from the lungs of the rats. In the kangaroo rats, evaporation from the lungs was low. The long nose of these rats provides an excellent countercurrent system for conservation of water in the nasal area. The mechanism of conservation depends upon the low temperature of the expired air. These animals not only have a low loss of diffusion water, but also consolidate their gains by being nocturnal; they remain in their burrows during the day and avoid the excessive drain of water to the environment. Their burrows do have a relatively high vapor pressure; this was measured very ingeniously by the Schmidt-Nielsens (1951) by having the rodents pull a small dry-bulb wet-bulb recorder into their burrows. These instruments consisted of watches which turned a small smoked plate upon which the micro-meteorological record was made.

The water resources of the animal must now be considered; these consist of metabolic water and water present in the food. The amount of metabolic water formed varies with the composition of the food, but when dried barley was fed, 54 grams of metabolic water were formed for 100 grams of food. Included in this is a small amount of absorbed moisture in the food which varied with the atmospheric humidity. Having considered the balance sheet of water gain and water loss, Schmidt-Nielsen calculated that they are equal at any atmospheric humidity above approximately 20% RH at 25°C (77°F). It has been assumed that these rodents are able to withstand a greater degree of desiccation than other mammals. As we have said, they can become dehydrated by being forced to eat soybeans which contain about 40% protein. Kangaroo rats tolerate this diet without water for only 2 to 3 weeks. When analyzed at time of death their weight loss amounted to 34% of their body weight, but the average water content of the body was the same as that while living on their normal diet: 67%. Although a considerable amount of water had been lost, the body was not really desiccated. These figures also apply to other small rodents when dying from water restriction; thus, the kangaroo rat is not different from other rodents that lacked the special ability to survive without water (Schmidt-Nielsen 1964).

Jackrabbit. There are two common species in the arid Southwestern United States, and they should properly be called hares rather than rabbits because they remain above ground and have no tunnels for underground escape. It has been difficult to maintain these species in captivity because they throw themselves against the wire cage until they are injured. Some investigators have performed cesarean sections and then raised the young by hand-feeding. When these were studied, it was discovered that jackrabbits lose weight and become unhealthy if they are maintained with just dry plant material from their usual habitat. In the free environment they undoubtedly live upon green food to a large extent since their range is so small they cannot possibly move to open water for drinking. When it is available, they consume approximately 80% green and fresh food. There have also been a number of studies on the rabbit *Oryctolagus cuniculus;* this species was introduced in Australia where it thrives in arid areas. They require green vegetation, but can subsist without water. When the vegetation is dry, they will survive for about 2 months, but lose weight consistently.

The temperature regulation of these three species must depend in part upon their relatively enormous ears. A related hare living on the Sahara Desert (*Lepus capensis*) has ears even larger than the Western jackrabbit of the U.S. Brody's group (1958) studied the temperature tolerance of New Zealand domestic rabbits and found that if the animals were acclimatized, they could tolerate a rectal temperature of 42°C (107.6°F) and an ear temperature of 40°C (104°F). The rectal temperature for a rabbit is probably between 42°C and 43°C (107.6°F to 109.4°F); this seems to be a temperature lethal to most mammals, although it was noted earlier that bats tolerate much higher body temperatures than this.

What mechanisms do rabbits under these circumstances have for preventing a further gain in body temperature? This rodent begins rapid respiration in a fashion quite different from that observed in dogs exposed to heat. The Angora rabbit attains a high respiration rate of perhaps 700/min, but in the New Zealand rabbit, the figures are nearer 200 to 300. This increase is gradual, not sudden and well-defined as in the dog. As the rabbit increases its rate of breathing, the respiratory volume is also increased; it should probably not be called panting because it is not shallow enough. Schmidt-Nielsen calls this type of breathing "hypothermic polypnea" (1964). Some saliva drips from the mouth of the rabbit, but it is not spread over the fur as in cats and bats. Water, however, is evaporated from the skin; this was carefully studied by Lee (1941) who concluded that there is an increase in cutaneous water loss at high temperatures due to increased circulation in the skin. Even with maximum evaporation through the skin and from the lungs, evaporative cooling accounted for only a small fraction of the heat dissipation (Keeton 1924). Brody's group partitioned this evaporative cooling as follows: at moderate ambient temperatures, skin evaporation accounts for 60% and respiratory evaporation for 40% of the water loss; however, at high temperatures the proportion changes to about 80% from the skin and 20% from the lungs. In rabbits, at the high ambient temperature of 40°C, evaporation accounted for only 30% of the heat dissipated. The rabbits could tolerate this high temperature for extended periods (Fig. 7-15); apparently releasing the remaining heat by conduction and radiation. The rabbit maintained a heat flow over a gradient between its rectal temperature of 42°C and the ambient temperature of 40°C. Apparently this small gradient was enough to dissipate two-thirds of the body heat; this must take place mostly through the thin skin of their relatively large ears. Note in Figure 7-16 that at high ambient temperatures, both humans and cattle dissipate heat entirely by evaporative cooling.

The contributions of the various avenues for heat exchange in the rabbit are rather unusual; Schmidt-Nielsen has assembled an elaborate and convincing hypothesis of how the jackrabbit makes use of these avenues when it exists on the hot desert (1964). An important part of the proposed scheme depends upon the micro-climate around the rabbit. Another important factor is the radiation temperature of the sky. When large animals are being considered, radiation temperature is not important because the integrated substratum radiation temperature provides a flow toward the organism. With a small animal like the rabbit, however, the radiation temperature of the sky could become important, because a small animal can escape

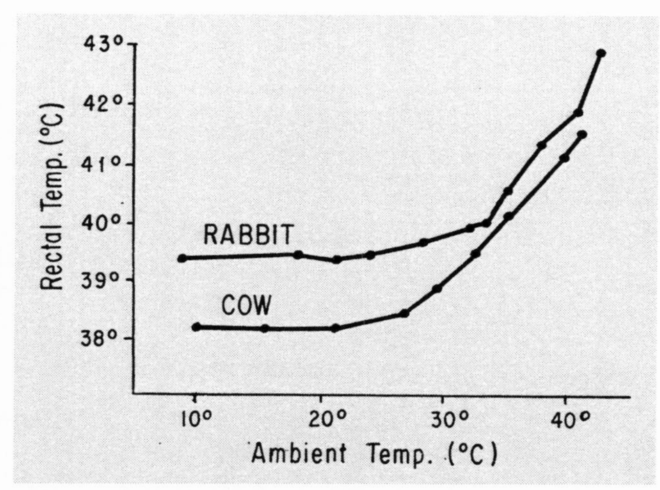

Figure 7-15. Tolerance of Rabbits to Heat. When ambient temperatures are increased gradually, rabbits appear to be more tolerant of high temperatures than are cattle. In this experiment, New Zealand rabbits were exposed to increasing temperatures. Above an ambient temperature of 33°C the rectal temperature rose precipitously. The rectal temperature of cattle began to increase at an ambient temperature of 25°C. Modified from Johnson, Cheng, and Ragsdale (1958).

Figure 7-16. Contribution of Evaporative Cooling in three species of mammals. At high temperatures (40°C), the heat dissipated from a rabbit includes a fraction for evaporation, which is only 30% of the total heat production. Apparently, the remainder of the metabolic heat is dissipated by convection and radiation, especially from the ears. At this same temperature, both the cow and human dissipate the entire heat production by evaporative cooling. Modified from Johnson, Cheng, and Ragsdale (1958).

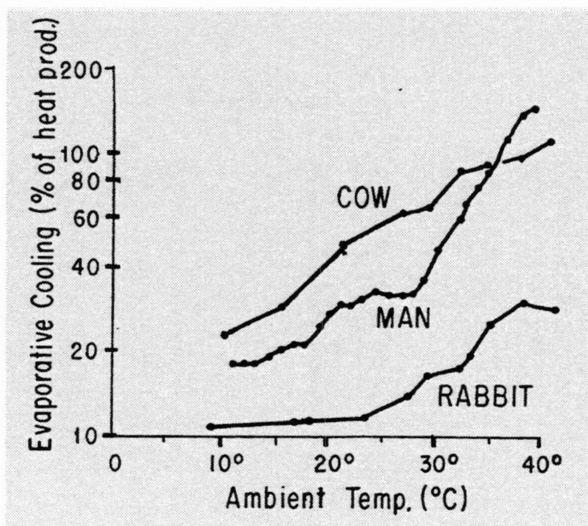

the substratum radiation by finding a shaded depression. Sky radiation temperature on the desert is much below air temperature during a day with a clear blue sky; under these circumstances the entire visible and infrared range are integrated. A thermal radiometer under such conditions records 20°C (68°F) in the daytime and almost freezing at night. On cloudy days, however, the sky radiation temperature increases. Lee (1964) recorded sky temperatures near 13°C (55.4°F) in the desert on clear summer days. The lowest night sky temperature measured by Lee was 7.5°C (45.6°F); with haziness and cloud cover, these radiation temperatures approached air temperature. Now, an animal on the ground, especially on the desert, is in radiation exchange with the visible sky particularly at night. The radiation exchange explains why the ground surface at night is colder than the air. Grass and leaves may reach freezing temperatures when the air is 10°C to 15°C above freezing. Knut Schmidt-Nielsen (1964) gives a delightful and detailed record from the proceedings of The Royal Society in 1775 of the process of making ice in the East Indies when temperatures recorded by air thermometers were always above freezing. By exposing thin dishes of porous clay so that the water is pre-cooled by evaporation, the water at night freezes due to the night sky acting as a heat sink, even when the air feels warm to the human skin. (The history of the use of ice-making in warm climates and modern applications of the principle were reviewed by Hay, 1972).

The contribution of the radiation temperature of the sky is important in the explanation of how the jackrabbit can tolerate the desert environment (Fig. 7-17). On hot days these animals are less active and tend to rest in the shade. If the animal sits in a depression in the earth on the shaded north side of a bush, the wind of the hot desert will blow above the rabbit and have no effect. This wind ordinarily constitutes an increased heat load because of conductive heat flow from air to animal. The shaded ground where the animal is resting is cooler than the sunny surface; radiation from the sun does not reach the shade and re-radiation from the hot ground does not reach the animal in the depression. (The analysis of "hot ground" is not often available for physiologists. One publication reported a temperature of 123°F inside a pebble when the desert air temperature was 100°F). The large ears of the jackrabbit will be in radiation exchange with the bush and with the sky where, the temperature may be assumed to be about 25°C lower than the ears.

In this section we have emphasized the physiological contribution of the ears of the rabbit in a hot environment. Other desert mammals have prominent and even tremendously large ears which are completely out of proportion for the rest of the body. A conspicuous example is the desert Fennec, which is the most common mammal of the Empty Quarter of the Northwestern Sahara Desert. This small fox-like animal seems to live entirely independent of drinking water.

Cattle. The popularity of cattle as a domestic animal has resulted in their being exposed to diverse environmental extremes. The multiplicity of factors involved in cattle and other fiber- or food-producing animals responding to heat is reviewed in other publications

(Olsson et al. 1995; Clark 1981; and Hafez 1968). This discussion of cattle will be limited to the roles of body size, coat characteristics, and sweat because these features are also important to clothed humans and other mammals.

The large size of cattle permits storage of 1% to 2% of the heat load encountered within a day. In contrast, humans can store 0.1% or less of a day's heat load. Seeking shade during the hottest portion of the day combined with night and early morning grazing can be of great benefit to cattle in hot arid regions.

Cattle have morphological changes in their hair between cold and hot seasons. The importance of the coat can best be appreciated after considering the avenues for heat exchange between animals and environment presented in Figure 7-10. Direct sunlight includes long-wave (less than 400 µm) radiation as does the reflected sunlight. The radiation from the ground, dust particles, and atmosphere will be short-wave radiation. Because long-wave radiation makes up most of the radiant heat, black globes are most frequently used to measure the radiant heat load. Color and length of hair coat have marked effects on heat exchange. Light color will reflect more direct sunlight, whereas long-wave radiation will penetrate to the skin. Long hair with dark tips can be an advantage because the dark color absorbs radiation and loses heat by convection to the air before the heat is conducted to the skin surface. The tip of the hair may be 125°F to 150°F, and air temperature may be 110°F. Short hair under the long hair can provide insulation from the extreme surface temperature and maintain skin temperature near 95°F. The combination of long and short hair on the camel has an advantage over the coat of the cow. The distribution and effectiveness of sweat evaporation can be another factor influenced by hair density. Sweat that evaporates on the tip of hair is not effective in cooling the skin. Thus we can appreciate why some animals evaporate body water by panting. Panting provides cooling nearer the heat source, the core of the animal.

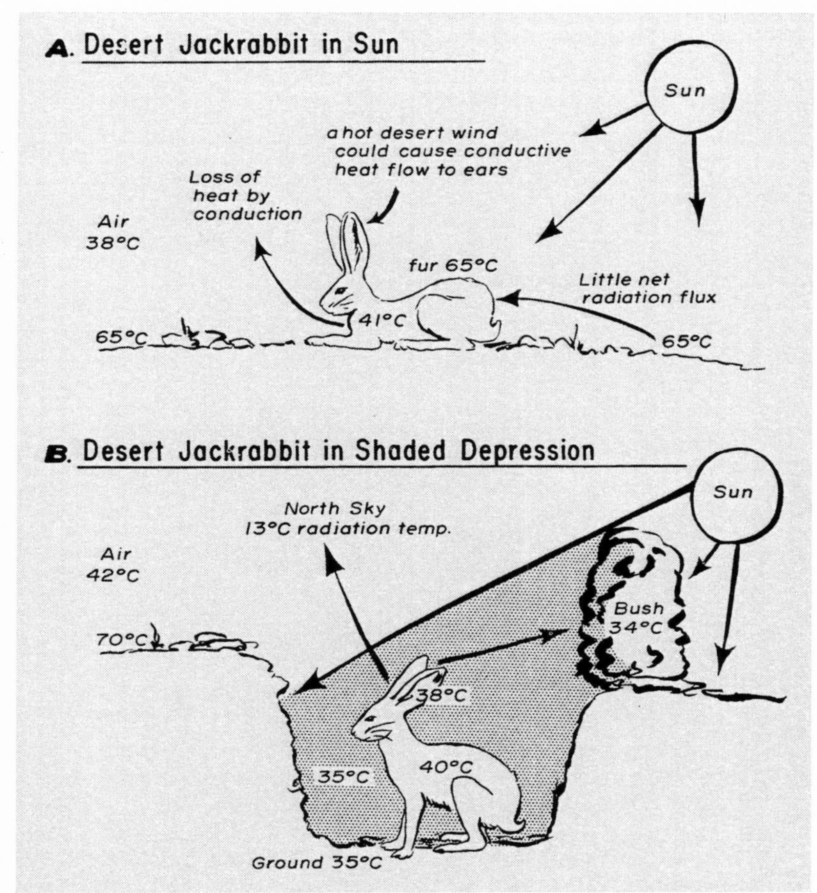

Figure 7-17. Desert Jackrabbit in Two Thermal Situations. These diagrams illustrate a hypothesis of Schmidt-Nielsen. The rabbit exposed to the sun on the desert would gain heat far more rapidly than it could lose it. By remaining in a shaded depression, the animal might achieve thermal balance by radiation to the sky and to cooler objects around it.

Shiny flat hair will reflect more heat than will curly hair. These features explain why investigators must take into account the lipid and protein content of skin secretions when attempting to evaluate the absorbency, reflectance, and insulation value of animal hair.

Dog. As an example of the carnivore group, we will now consider the water balance of the dog. The ecology of carnivores living in hot arid climates has been discussed in detail in the book, *Desert Animals* (Schmidt-Nielsen 1964). This group of animals lives upon prey which have a composition of two-thirds water; this proportion changes little with the seasons. The suggestion has been made that carnivores should be able to live independently of drinking water, at least if no water is needed for heat regulation.

The physiology of the dog depends, of course, upon its breed; we would make very different generalizations about the Chihuahua dog compared to the heavy-coated sled dog. The present discussion will be confined to large dogs with thick coats. These animals were discussed in Chapter 5 because of their remarkable tolerance to cold. In that discussion we also considered their responses to heat because when they exercise, even in extreme cold, they are a "tropical animal" in very thick insulation. Likewise, the present consideration of thick coats in a hot environment will enhance our knowledge of the exercising sled dog in a cold environment.

One of the earliest examples of heat tolerance of the dog was Blagden's experiment mentioned earlier in this chapter. When he took a dog with him into a room heated to 127°C, the animal was able to tolerate this heat in a perfectly satisfactory fashion as long as its feet were kept off the hot floor. Of course, any blistering of the feet would also have occurred with hot sand in the desert; many desert animals have feet covered with stiff hair to prevent this damage.

As the body and surface temperatures of the dog increase, the first and most important mechanism is panting, which causes water to be evaporated from the moist surfaces of the tongue, the mouth, and the upper respiratory tract. A second mechanism which comes with higher body and surface temperatures is sweating from the skin glands. Panting is associated with heat dissipation and not with exercise; a dog swimming in cold water does not pant. The panting response shows itself in a sudden change in frequency from ordinary respiration of 10 to 40 per min to a rate of 300 to 400 respiration/min. Panting is so efficient that dogs could tolerate a hot room of 43°C (109.4°F) with low humidity for 7 hours without any elevation in rectal temperature. At 44°C and 65% RH they can tolerate only 3 to 4 hours and will reach a rectal temperature of almost 42°C. There are some obvious advantages of this process of panting, which is so well developed in the canine species. Human's sweat evaporates from the skin more efficiently if there is a slight breeze; the dog provides its own slight breeze over the cooling surface of the pharynx region. The insulation value of the hair permits the dog to tolerate very high surface temperatures. There seems to be no physiological economy concerned with the distribution of blood in the panting process. Under extreme circumstances of heat exposure, the burden of blood transport to the tongue seems to be as great as it is to the skin in humans. Because of the increased need for blood flow during panting, the flow to the dog's tongue is increased 6-fold when the rectal temperature rises to 42°C. A final advantage of panting is that the dog avoids the salt loss which may cause incapacity in the human. A review of panting in resting and running animals is provided by Taylor (1977).

The disadvantage of the panting process is that the increased respiration gives a greater exchange of air in the lungs, causing the removal of CO_2 from the blood and severe alkalosis. Although there have been statements that panting is so shallow that the gas exchange in the lungs is not increased, studies indicate that this is not the case. In the panting dog, the CO_2 content of the blood is decreased 25%, showing a tremendous overventilation of the lungs. This degree of alkalosis would be intolerable to the human.

A final question to be asked concerns the actual metabolic cost of panting. Does this add to the overall heat load? The design of the panting mechanism as a machine even included a mechanism to reduce the work of the respiratory muscles. The dog minimizes its heat production by using the natural resonant frequency of its respiratory system. In other words, the dog pants at the same frequency with which the respiratory system oscillates naturally; the energy requirement for panting is greatly reduced by this system (Schmidt-Nielsen 1964).

If the body temperature of the dog is extremely high, the skin glands begin secreting fluid. This usually happens in response to a local heating and especially in those areas closest to the source of radiant heat. Two investigators, Aoki and Wada (1951), showed that the threshold skin temperature for this sweating is between 38.4° to 38.7°C (101.3°F). If the outside of a heavy coat of hair is particularly hot, the sweating at the skin level introduces a very efficient system. The insulating dry fur reduces the heat flow from the hot outside air and ground to the cool skin surface. Although the fur may heat up to 70°C (158°F) from solar radiation, there will be a conduction gradient from the fur to the air which might be about 40°C. Dogs were compared with men on the desert; the efficiency of a heavy coat in the dog was apparent; the men evaporated 12% of their body weight during rest periods and the dogs evaporated 15%. Recalculated in terms of surface area, the figure for the dogs was lower than that of the men. Thus, both species evaporate at nearly the same rate when exposed to the same conditions on the desert.

The water balance of the dog is under rather delicate control. When the dog is on the desert, it drinks an appropriate quantity of water to very nearly correct the amount of water evaporated. This brings up the question of the control of thirst. Many of the experiments on thirst have been done on dogs. The urge to drink is controlled from a thirst center in the hypothalamic region of the brain. This cannot be the entire mechanism, since partial experiments indicate that the relief of thirst partly depends upon the distention of the stomach, partly upon the amount of water passing the esophagus, and partly upon the absorption of water from the intestine and dilution of the body fluids. All of these stimuli can send signals to the hypothalamic region of the brain.

With plenty of water, dogs will survive indefinitely in air at 55°C (131°F). If water is removed, they lose weight by evaporation at the rate of about 1.5% body weight per hour. When dehydration reaches 10% to 14% of the body weight, the animal's rectal temperature will begin to rise explosively. A few dogs will tolerate a body temperature rise to 41.7°C (107°F), and occasionally a temperature of 42.0°C (107.6°F). As noted before, this appears to be critical rectal temperature for a number of types of mammals. The bats, which were discussed in an earlier section, are an exception to this observation, and apparently cats also can tolerate rectal temperatures as high as 43°C and 44°C (109.4°F and 111.2°F).

We want to mention again the relationship between the coat of a sled dog and exposure to heat. This thick coat is always grown twice a year, even in a very warm environment (Meehan 1957). A comparison can be made between the effect of the heavy coat of sheep in extreme heat and a dog under the same circumstances. When sheep were tested at 36°C T_a, and compared at different wool lengths, it was found that with the longest length of wool, rectal temperature, respiratory rate, and skin temperature all decreased.

Camel. There are numerous examples of the remarkable ability of the camel as a beast of burden while traversing waterless desert. Many of these accounts have been collected by Monod (1958) of Dakar, West Africa, who traveled across the waterless Empty Quarter of the Sahara on a march of 944 km, or nearly 600 miles. The beast capable of such a feat is reported to be able to store water in its stomach. For several years Schmidt-Nielsen investigated the truth of this statement (1964). The camel does have several compartments that precede the true stomach. The controversy about the water question concerns only the first compartment, called the rumen. The entire contents of this compartment average from 11% to 15% of the body weight; the contents consist of coarsely masticated feed formed into a semi-liquid mass. This mass is 83% water, and the fluid that can be drained from it contains 98% water. This fluid is slightly more dilute than the other body fluids; its function is to moisten the solid food for further digestion. Apparently it comes from the glands in part of the rumen; they may be considered as accessory salivary glands. Although there is a large amount of fluid in the masticated feed in the rumen, it is a mistake to assume that it is water stored for the camel's use on the desert; rather, it is water associated with the process of digestion.

Another misconception is that the hump is used for water storage; the hump consists mostly of fat. It is true that the oxidation of fat results in a corresponding amount of metabolic water being released. The problem is that an increased oxidation of fat for the function of obtaining water would involve increased ventilation of the lung; thus more expired air would be saturated with water vapor and considerable loss of water, instead of gain, would result. There would be more benefit to the camel if the hump contained starch, since more metabolic water is obtained by the oxidation of starch than of fat. If the starch were metabolized, the use of extra oxygen would again result in evaporation from the lungs in excess of the amount of water formed. To summarize the water storage question, it can be said that there are no unusual water storage compartments in the camel. It is true that physiological subcutaneous edema can be produced by giving camels an extra ration of salt. But it is a moot question as to whether this would do the camel any good on a desert march.

We now turn our attention to an entirely different aspect of the temperature regulation of the camel; this is the variable body temperature and the benefit of this situation to the camel exposed to heat. In summer the range in rectal temperatures of camels is large: the temperature is usually quite low in the morning, sometimes 34°C (93.4°F), and high in the evening, usually about 40°C (104°F). In one case, the evening rectal temperature was 40.7°C, the result of a rise of 6.2°C in 11 hr. The pattern of the daily temperature cycle shows an abrupt drop about 0600, undoubtedly due to a sudden vasodilation bringing cold blood from the periphery to the deeper parts. The daily temperature variations were much greater in the animals when they were deprived of drinking water. The advantage of temperature lability is that if the body temperature is permitted to rise, all heat that goes into warming the body can be considered as stored, and this stored heat is dissipated during the cool night by conduction and radiation without expenditure of evaporated water. Furthermore, the rise in the camel's temperature to a level well above 40°C would reduce the heat flow from the environment. We can conclude that the rise in body temperature is not a sign of failure in heat dissipation, but is an actively regulated pattern in water conservation. The capacity to store heat must be considered an important feature of temperature regulation in all large (greater than 30 kg) mammals.

The camel's heavy coat also contributes to its cooling. If the animal is shorn, its water expenditure increases by 50%. Apparently, this is because the sweat from the camel is evaporated at the skin surface without wetting the wool. The skin surface will be the coolest area and heat will flow to it from the core. The heat flowing to the skin from the sun and ground will be blocked by the thick wool. The temperature of the wool surface on the camel's back in the sun may be as high as 70°C (158°F) to 80°C (176°F), while the skin temperature underneath, due to sweating, may be close to 40°C (104°F). The wool surface will lose heat to the air, which is far cooler than the fur. The camel has more control over its initiation of sweating and sweat rate than does the human. Apparently evaporative cooling begins in the camel abruptly when air temperature is about 35°C (95°F).

The threshold for sweating, or at least for a marked increase in evaporation, was measured by a somewhat unusual technique. The objective of the experiment was to compare sweating thresholds in the donkey and the camel. When this experiment was done at different air temperatures, there was not a consistent relationship between the increase in evaporation and high air temperatures. Therefore, measurements were made with the black-bulb thermometer, which consists of a thermometer placed in the center of a black copper sphere (6 inches in diameter). Since the experiment was done in still air, the reading from the black-bulb gave a good measure of total heat load, including the heavy radiation from both the sun and the ground. If there had been a wind, the temperature of the black-bulb would have decreased, while the actual heat load on the camel and donkey would have been higher because heat from the hot air would have been conducted to the cold surface of the animal (Fig. 7-18). This contradiction would probably have applied more strikingly to the donkey than to the camel because of a heavier coat of wool on the camel. The threshold seemed to be about 30°C (86°F) in the donkey, and between 35°C (86°F) and 40°C (95°F) in the camel. The evaporation data measured in both animals in terms of milligrams per square centimeter per minute are included in the book by Schmidt-Nielsen (1964). It should be pointed out that the donkey has sweat glands all over the body surface; it is a nose breather, and it does not pant. (See Figure 7-19.)

The conservation of water by the camel exposed to heat depends upon urine concentration, a reduced urine volume, and fecal water loss. The daily urine volume of camels is relatively low. One camel studied by Schmidt-Nielsen was provided with an abundant supply of water, but produced on average only 0.9 liter of urine per day. Other animals observed during grazing produced from 1 to 4 liter per day. In the laboratory, one camel that weighed 300 kg produced an average of 0.75 liter of urine per day; when it was deprived of water, the urine flow decreased to less than 0.5 liter. This is about the same as the minimum urine volume in humans, although the camel weighed 4 times as much. It is apparent that little water is used for urine formation.

The next question concerns urine concentration. The camel's kidney can produce a urine considerably more concentrated than sea water (perhaps twice as concentrated). The camel urine analyzed in the laboratory by the Schmidt-Nielsens had a total concentration 8 times that in plasma (U/P ratio = 8) (Schmidt-Nielsen et al. 1956). This is twice as concentrated as in the human (U/P ratio = 4). Several investigators have studied the ionic composition of concentrated urine in dehydrated camels, and found that chloride concentration remained constant, potassium about doubled, sodium increased 9-fold, and the sulfate concentration increased more than 16-fold. The camel kidney has an exceptional ability to eliminate sulfate. Because of this concentrated urine, one might reason that this animal produces more than usual amounts of antidiuretic hormone (ADH). To test this, Macfarlane (1960) infused ADH into camels that were dehydrated. Instead of a further reduction in urine volume, he found a several-fold increase in urine output. The cause of this is apparently an increased output of electrolytes caused by the ADH.

Studies on the concentration of urea in camel urine make a particularly interesting story. An earlier worker reported that there was practically no urea in camel urine; (Smith and Silvette 1928) analyzed the urine from a circus camel and found a normal amount of urea which was 60% of the total urinary nitrogen. One young and growing camel was studied by Schmidt-Nielsen. The animal was completely deprived of water and the urea concentration decreased steadily day after day. Toward the end of the experiment the total amount of urea in the urine in 1 day went down to less than 1 g. Schmidt-Nielsen explained the observations of the occasional lack of urea as evidence of a special nitrogen cycle possessed by the camel permitting the re-use of its nitrogen in rebuilding broken-down pro-

Figure 7-18. Tolerance to Heat of Donkeys and Camels. In this experiment, the threshold to sweating was studied as a function of black-bulb temperature. The latter measurement was used to integrate the heat load from the sun, ground, and air. At a breaking-point, evaporation increased in linear fashion with the heat load. This threshold was about 30°C in the donkey but nearer 40°C in the camel (Schmidt–Nielsen et al. 1957).

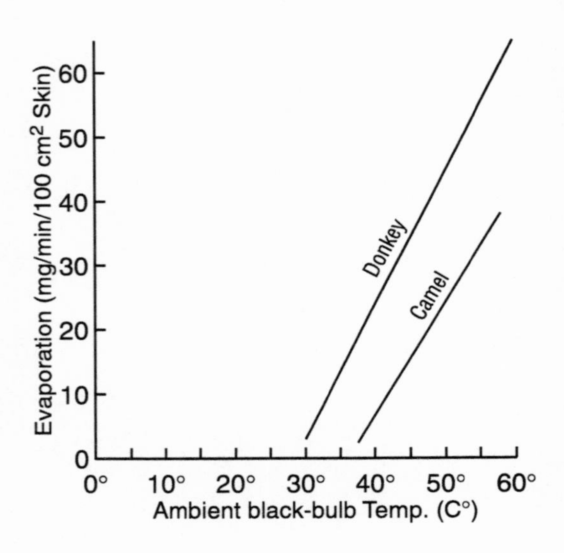

tein. Instead of being excreted by the kidney, apparently urea enters the rumen fluid where bacteria immediately use it in the synthesis of protein. Lower down in the tract this protein is digested. To check upon this special nitrogen cycle, Schmidt-Nielsen gave intravenous injections of urea in amounts up to nearly 30 g; less than 2 g of this was recovered in the urine. The urea had apparently been synthesized into protein. Urea conservation is most apt to be found in pregnant female camels that need to conserve protein for the embryo, and in the young growing animals. Since urea may be either excreted in the usual way or reduced to a very small fraction of the urine when there is need for protein, it is possible that the camel can actively regulate the renal excretion of urea. The concept that urea is filtered and merely diffuses passively back to the blood does not fit this picture.

The final factor in the conservation of water is the amount of water in the feces. When camels in the laboratory were fed on dry dates and hay, they produced about 100 g of dry fecal matter per day. Even when the animal had free access to drinking water, the amount of water eliminated with the feces was only 109 g per 100 g dry matter. In comparable terms the white rat excreted 200 g of water and the grazing cow over 500 g of water. However, when camels have been grazing in the desert, the amount of water lost per 100 g of dried feces was 2 to 3 times higher than on dry feed in the laboratory. It is remarkable that this mammal or any mammal can extract so much water from the intestinal contents.

Figure 7-19. Professor D.B. Dill (see sidebar, Chapter 1) With One of His Experimental Animals at the Desert Research Institute, Nevada Southern University, 1967. Dill's early studies of how sweating permits mammals to tolerate extreme heat attracted large numbers of young people, including high school students, to work in his laboratory. His book, *Life, Heat, and Altitude*, inspired many to seek a vocation in the study of physiology.

All of the camel's means of conserving water have been considered. It is now of interest to consider the maximum use of these water conservation devices; what is the extreme tolerance to dehydration of the camel? Schmidt-Nielsen (1964) maintained camels in the laboratory without water in January for 17 days; apparently the camels could have continued. The experiment was repeated in an outdoor temperature exceeding 40°C (104°F). Two camels were without water 7 days and lost 26.5% and 27.2% of their original body weight. Other species would have had an explosive temperature rise and died at 12% to 14% weight loss.

Attempts were made to study water loss in the separate water compartments of the body. The smallest relative loss of water was in plasma, with less than a 10% reduction. This was verified in other experiments in Australia (Macfarlane et al. 1958). This situation would be an advantage in maintaining adequate circulation. When the animals lost one-fifth of their body weight, corresponding to 30% of the total body water, 20% of this loss was of plasma volume. Sheep would probably have lost one-half of their plasma volume. Of the total water lost by the camel, about 50% might have come from the gut, 30% from intracellular space and 20% from the extracellular space. The blood volume was determined by the Evans Blue technique. Ordinarily when this material is used for blood volume studies, it is gradually lost with a half-life varying from 6 to 12 h; in the camel the half-life of the dye was 2 to 4 weeks. In some way, the plasma proteins which bind the Evans Blue behaved differently in this species.

In the preceding section, we have dehydrated the camel; it is of obvious interest to hydrate the animal once again. In the experimental work described above, Schmidt-Nielsen compared the drinking capacities of men and camels. A man can drink only 2 liters of water in 5 or 10 min, even if this amount does not make up for a water deficit. A thirsty camel, on the other hand, can drink enough to make up its entire dehydration deficit in about 10 min. On several occasions the camels drank amounts equal to 25%, 30%, and 33% of their body weight. During these bouts of drinking, one of the camels consumed 104 liters of water. Follow-up studies on the distribution of this ingested water through the body compartments indicated that even the largest amounts of fluids taken in by a camel were evenly distributed through all compartments in 1 to 2 days.

What are the physiological characteristics of other camel species, most of which live in South America? In one species studied, Rosenmann and Morrison (1963) found that the guanaco (*Lama guanacoe*, not a desert form) has the ability of the desert camel to replace a water loss quickly and to withstand severe dehydration without apparent discomfort. However, the guanaco lacks the ability to become hypothermic and to maintain plasma volume when exposed to high temperatures during the dehydrated state. Here we see partial physiological uniformity within a particular family irrespective of habitat.

Finally, let us briefly compare the South American guanaco with the African antelope. When water is scarce, several East African ungulates can survive in hot conditions without drinking. They develop high body temperatures during the heat of the day; these temperatures aid in reducing the use of water for evaporative cooling. At 45°C (115°F) air temperature, rectal temperature of two desert species exceeded air temperature by between 0.5°C and 2°C, and a rectal temperature of 46.5°C (116°F) could be maintained for 6 hours without observable ill effects (Taylor 1970).

In our next chapter we will venture into the rarefied air found at high altitude.

8. RESPONSES TO ALTITUDE

PHYSIOLOGICAL EFFECTS OF ALTITUDE

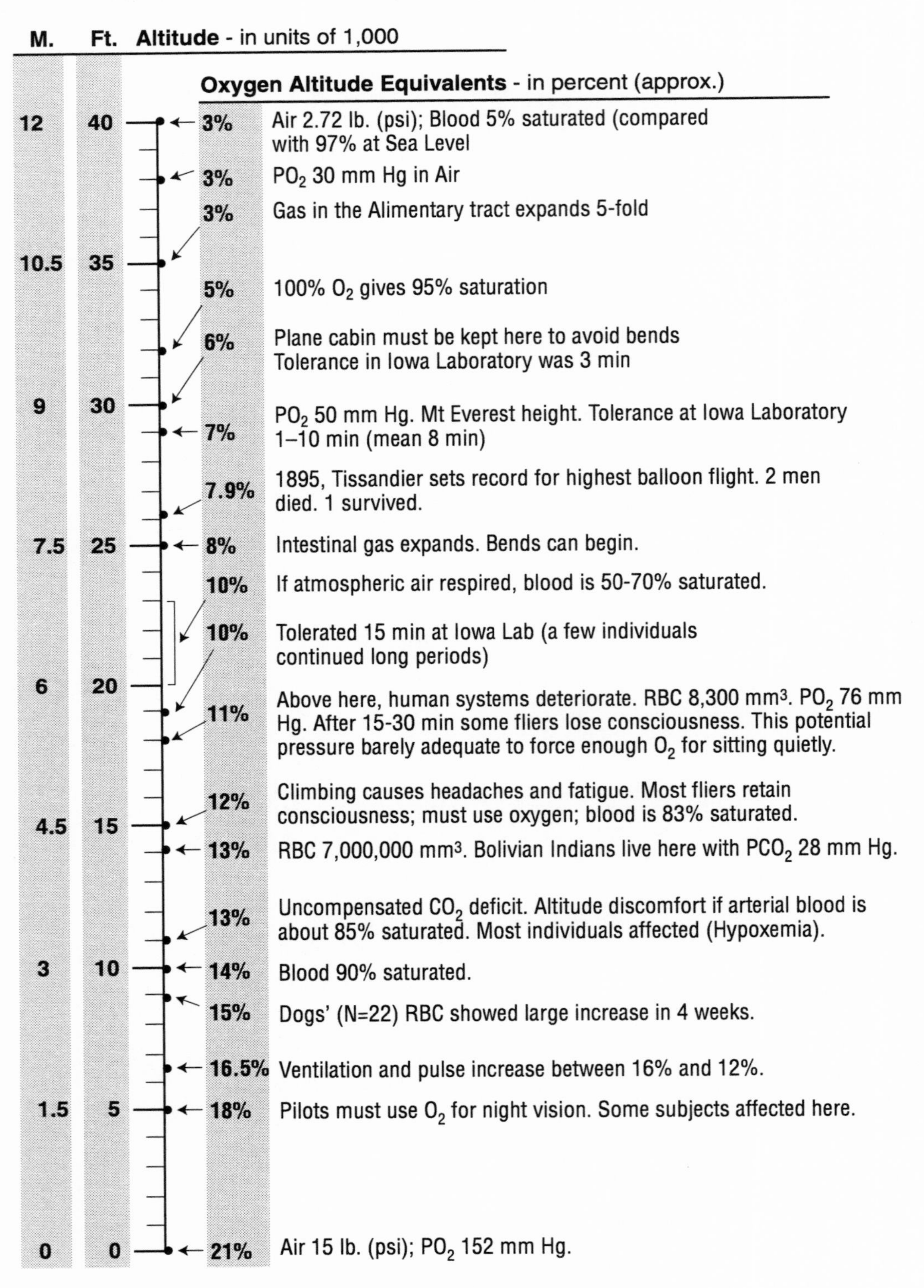

Environmental Diagram: Chapter 8

8. Responses to Altitude

Introduction
Humans at Altitude
  Hypoxia Ventilatory Response (HVR)
  Diffusion Capacity of the Lung
  Transport of Oxygen by Hemoglobin
  Cardiac Response
  Vasomotor Responses
  Problems of Sea-Level Natives Traveling to Altitude
  Acclimatization to High Altitude
  Aging and High Altitude
  High Altitude Natives versus High Altitude Acclimatized Lowlanders
Comparative Physiology of Altitude Effects
  Micro-organisms and Invertebrates
  Fish, Amphibians and Reptiles
  Small Domestic Mammals
  Large Domestic Mammals
  Small Free-Living Mammals
  Large, Free-Living Highland Mammals
  Birds at High Altitude
  Hypoxia-Induced Hypothermia
Comparison of Alpine and Arctic Environments
  What is the Alpine Area?
  What is Tundra?
  Similarities between Arctic and Alpine Areas in the U.S.
  Differences between Arctic and Alpine Areas in the U.S.
  Fauna of the Arctic and Alpine Areas
  Adaptations of Arctic and Alpine Species
Summary of Terrestrial Altitude Effects

8. RESPONSES TO ALTITUDE

INTRODUCTION

Aristotle believed air to be a fundamental element. Not until Toricelli invented the mercury barometer in the 17th century was it possible to establish the relationship between altitude and barometric pressure. Later, the international unit of pressure

$$1 \text{ Torr} = 1/760\text{th atmosphere} = 1 \text{ mm Hg}$$

was chosen in Toricelli's honor. A comparison of oxygen and barometric pressures at various altitudes is illustrated in Figure 8-1.

The French physiologist Paul Bert was the first to use a pressure chamber in his laboratory in the 1870's. This facility permitted him to examine the effects of reduced barometric pressure without the other climatic factors associated with high altitude: cold, wind, low humidity, and high ultraviolet light. Bert (1878) was the first to report that the symptoms associated with high altitude resulted from the low oxygen partial pressure, *hypoxia*. He demonstrated that the symptoms of oxygen deficiency could be prevented by breathing oxygen.

The first field laboratory at high altitude was established in the late 1800s by Angelo Mosso on Monte Rosa, 15,200 ft (4,634 m) (Figure 8-2). The Italian-Swiss border passes through the hut, which looks down on the Matterhorn (4,478 m). The tradition of transporting research equipment to remote mountain laboratories and villages continues today, and has contributed greatly to current concepts of high-altitude physiology. A few of these locations are noted in Figure 8-1.

Figure 8-1. Barometric Pressure and Oxygen Pressure at High Altitude. In this diagram, the pressures of air and oxygen are expressed as percentage of sea level and compared with barometric pressure. On the graph of altitude, the heights of the highest known villages or work areas are indicated.

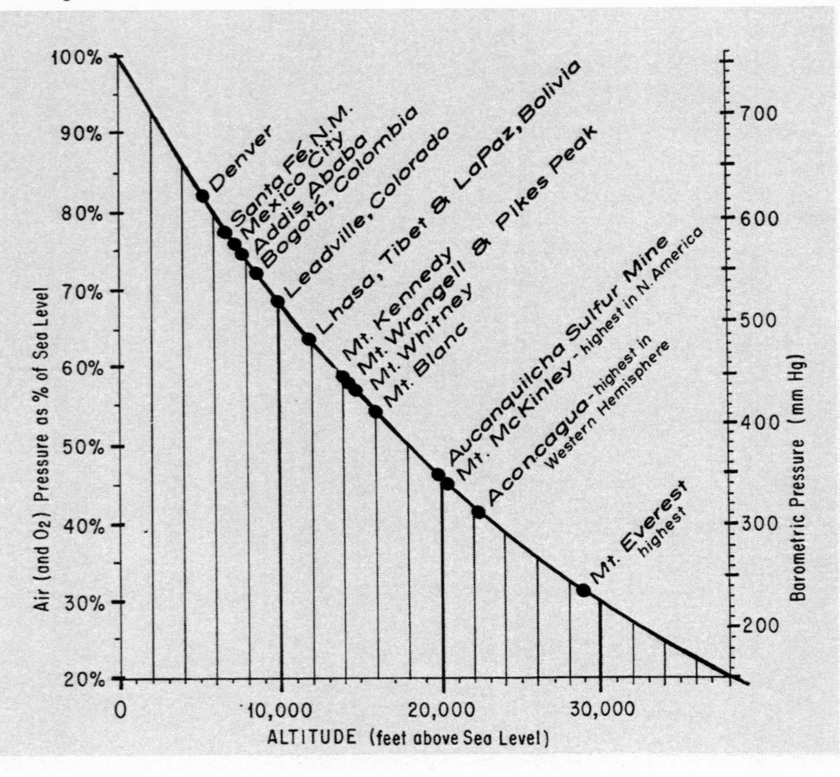

An historical approach to the physiological effects of altitude and why oxygen lack causes the problems it does has been fer-

reted out and delightfully recorded by Houston (1980) to help hikers, climbers, skiers, balloonists, and aviators enjoy high altitude without becoming ill.

Figure 8-2. View of Monte Rosa Near the Matterhorn. On the Monte Rosa there is a series of huts culminating at the summit in a dormitory and research laboratory named for Italy's Queen Marguerita. At this famous summit structure, the initial work on high-altitude physiology was done by Mosso. Picture courtesy Marco Danzi.

HUMANS AT ALTITUDE

Hypoxia, cold, high winds, and low humidity characterize the environment at high altitude. Each of these can be life threatening at extreme elevations. But the ability to adjust to altitude is demonstrated by at least 10 million people living at altitudes over 13,000 ft (4,000 m), and more than 17 million living above 7,500 ft (2,300 m). Adjustment to altitude involves genetic adaptations and physiological acclimatization at all levels of organization—molecular, cellular, tissue, organ, and system. The time required for physiological adjustments varies dramatically; it takes only seconds or minutes for the respiratory system to increase ventilation, but it may be months or even years before the sea-level native can approach the maximum work capacity of persons native to high altitude. Actually, there is considerable evidence that the sea-level native at altitude never attains the maximum adjustments of the high-altitude native at altitude. The maximum adjustments of the high-altitude native seem to develop partly before birth and then continue throughout the individual's growth and development. Differences between high-altitude natives and lowlanders will be discussed further throughout this chapter.

Physiological responses to hypoxia, including acclimatization, begin at elevations of 7,500 ft (2,300 m) and higher. The nature and intensity of the responses to altitude vary with the extent of elevation, previous high-altitude experience of the individual, and duration of exposure. Keep in mind that approximately 90% of the high altitude research has been conducted at elevations of 10,800 ft (3,300 m) to 16,400 ft (5,000 m). The methodology also includes hypoxic exposures in field laboratories at altitude, hypobaric climatic chambers with total pressure less than one atmosphere, and other situations which involve subjects breathing gas mixtures with oxygen at partial pressures below 21% O_2.

An overview of the multiple responses to hypoxia can be gained by considering the oxygen cascade (Figure 8-3). At each stage in the passage of oxygen from ambient air to tissues, the oxygen partial pressure (PO_2) declines. Adjustments to altitude can involve improvement in the PO_2 at one or more of the steps in this cascade (Luft and Finkelstein 1968; Lim and Luft 1965). The object of the

Figure 8-3. Oxygen Cascade. Oxygen partial pressure declines as it travels from the inspired air to the tissue capillaries and mixed venous blood. Modified from Luft and Finkelstein (1968).

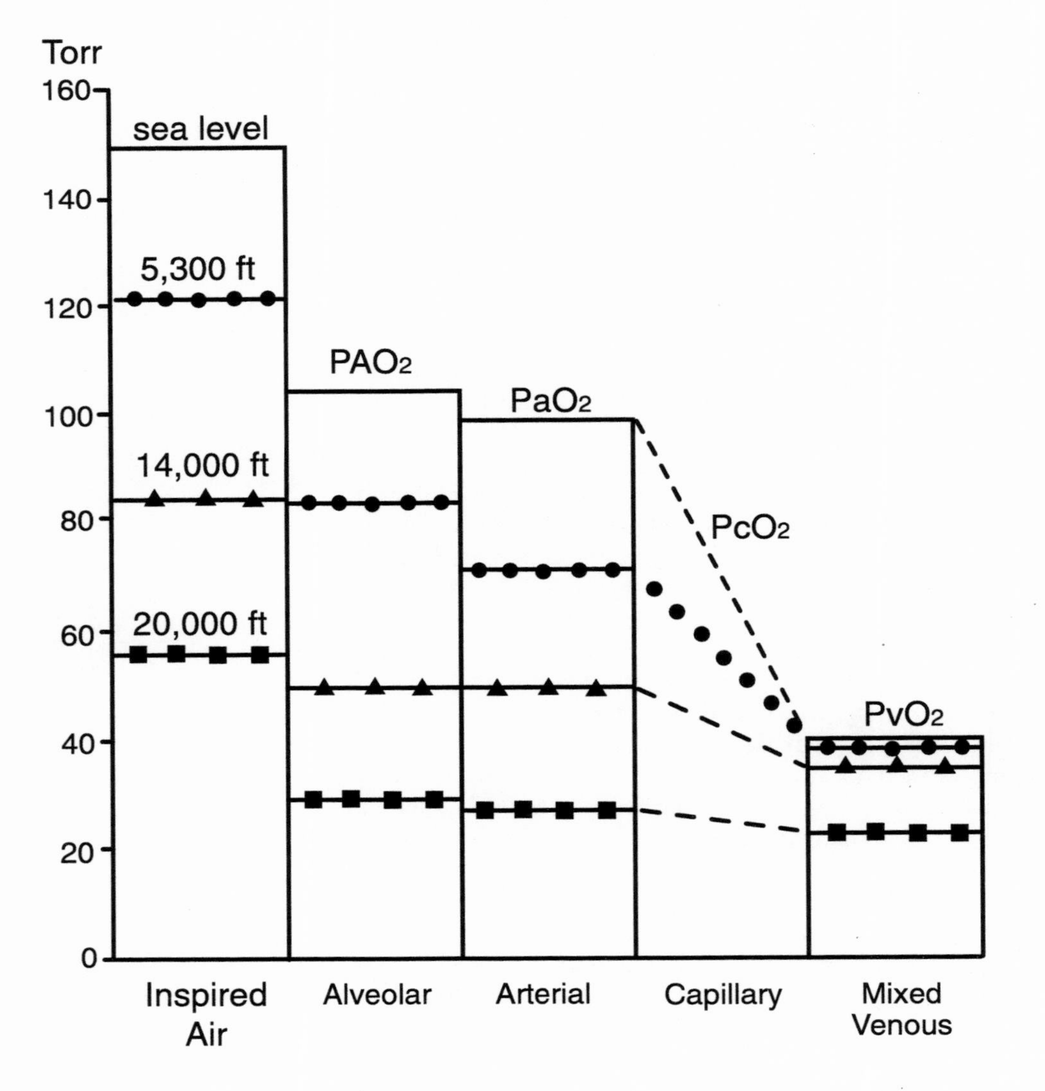

various adjustments to high altitude is to maintain adequate oxygen for aerobic metabolism in body tissues. Anaerobic metabolism occurs in the event of tissue hypoxia and requires twenty times the energy cost of aerobic metabolism. There are multiple lines of defense which prevent or reduce tissue hypoxia. We will look at these now.

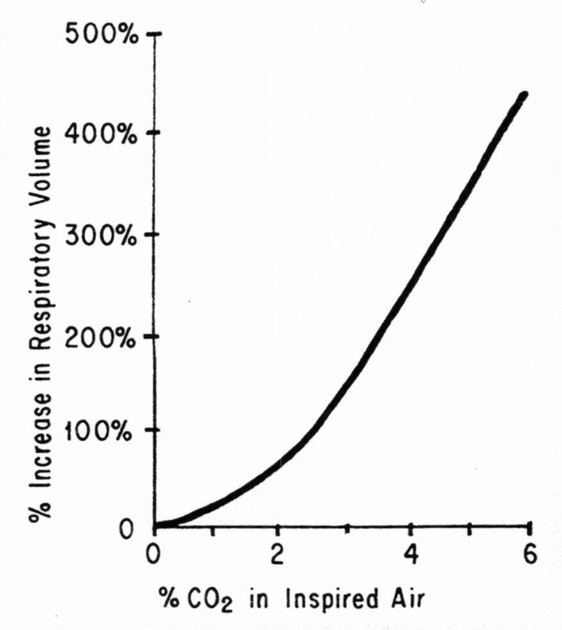

Figure 8-4. Carbon Dioxide and Respiratory Volumes. When the concentration of carbon dioxide in inspired air is raised, the volume of respiration increases to many times the normal level. An increase in respiratory volume does not bring about a corresponding increase in oxygen uptake (Scott 1917).

Hypoxia Ventilatory Response (HVR)

Hyperventilation is the first line of defense to prevent tissue hypoxia. The hyperventilation is a result of carotid body cells responding to reduced PO_2. Action potentials from the carotid body to the respiratory center in the central nervous system result in increases in depth and frequency of breathing. When breathing sea level air, the respiratory system is driven by the PCO_2 in arterial blood entering the respiratory center. The response to breathing CO_2 is illustrated in Figure 8-4. The switch to the oxygen sensors causes a "blow off" of carbon dioxide as a result of HVR. The HVR increases alveolar ventilation 25% to 30%, and increases alveolar oxygen partial pressure (PAO_2). The increased depth of breathing can be evident within a few seconds when acute exposure to hypoxia is in an altitude chamber at 7,000 ft (2,134 m). This initial response to the first few minutes of hypoxic gas breathing is diminished within 20 to 30 min. One factor for this blunting of ventilation is the *braking effect* of reduced partial pressure of carbon dioxide (PCO_2) in arterial blood and cerebral spinal fluid. High altitude natives going to higher altitudes do not demonstrate as much HVR as do sea level natives going to similar increases in altitude. When low levels of HVR are present in sea level natives, it appears to be related to their susceptibility to acute mountain sickness and pulmonary edema, which will be discussed later. Highly trained athletes tend to have a small HVR. HVR is always present during acute exposure to altitude, but the extent of HVR is difficult to predict, as it varies with recent previous experience of the subjects (Roach 1994; Schoene et al. 1990).

Diffusion Capacity of the Lung

Improved diffusion capacity in the lung is the second line of defense against hypoxia. The second stage in the transport of oxygen is the diffusion from alveoli to capillary blood. The extent of diffusion of gases is dependent upon many physical-chemical features including (1) mass and solubility of the gas, (2) distance over which diffusion must occur, (3) the surface area available for diffusion, and (4) the partial pressure gradient over which the gases need to diffuse. These features and the physiological mechanisms which control the distribution of air and blood flow in the lung have been difficult problems for pulmonary physiology researchers to resolve for a number of years. In healthy persons, there is a good match such that the ventilated alveoli also receive an adequate amount of blood flow through the adjacent capillaries. Research in the past 10 years has demonstrated that arteries and arterioles in the lungs constrict in response to hypoxia. This constriction is primarily responsible for reduced blood flow to those alveoli that are underventilated (hypoxic). We

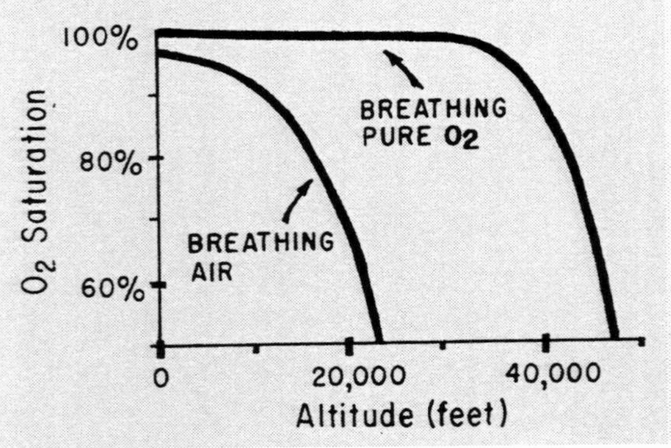

Figure 8-5. Oxygen Saturation of Blood at Different Altitudes. The transport of oxygen is compared when breathing air and when breathing pure oxygen. For example, at 20,000 ft when breathing air, the blood is 69% saturated, but when breathing oxygen, it is 100% saturated. At 40,000 feet when breathing oxygen, the blood is 85% saturated. Adapted from Guyton and Hall (1996).

Figure 8-6. Dissociation Curve for Oxygen and Oxygen Transport. This curve relates partial pressure of oxygen in blood with percentage saturation or with absolute amount of oxygen carried. It can also be called the Association Curve to emphasize what is carried at each partial pressure rather than what is unloaded. The curve demonstrates the Bohr effect by shifting to the right when there is a higher level of carbon dioxide in the blood. The graph at the bottom of the diagram shows the oxygen transported in the water of the blood.

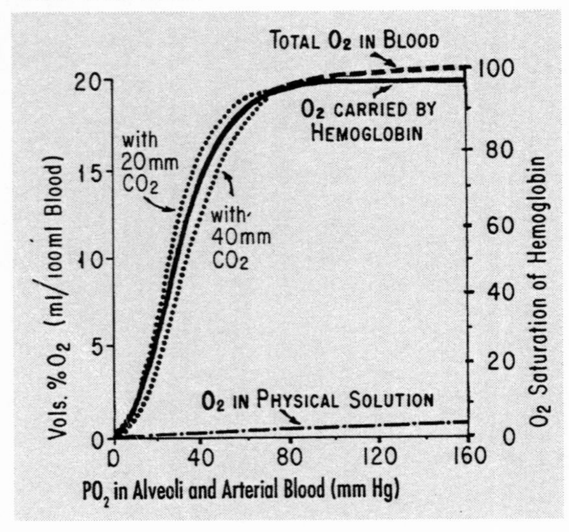

shall return later to consider the problems this mechanism causes during exposure to high altitude when we consider pulmonary blood pressure, high altitude pulmonary edema (HAPE), and O_2 sensitive cells.

Improvement in the diffusion capacity of the lung is recognized as being important in response to exercise training. Whether or not acclimatization to altitude involves an increased area in the lungs for diffusion is a controversial point. However, many investigators report no change in the diffusion capacity of sea level natives becoming acclimated to high altitude unless exercise training is involved. Moore et al. (1994) reported an increased diffusion capacity in all natives to high altitude (above 16,400 ft or 5,000 m) when compared to sea level natives. In contrast, the efficiency of the sojourning lowlander's lung as a gas exchanger is not enhanced by altitude acclimatization.

Transport of Oxygen by Hemoglobin

An increase in the oxygen-carrying capacity of blood is the third line of defense against hypoxia. The hemoglobin concentration increases as the plasma volume decreases during the first few hours of exposure to high altitude. Two factors contribute to this decreased plasma volume: increased urine volume and decreased voluntary fluid intake. A 13-day study at Pikes Peak demonstrated that some of the decrease in plasma volume is due to decreases in plasma proteins (Swaka et al. 1997).

Hypoxic renal tissues synthesize and release erythropoietin hormone which increases erythrocyte production and thereby further increases the hemoglobin concentrations in the blood. The action of erythropoietin on bone marrow accounts for the appearance of reticulocytes (immature erythrocytes) in the blood within one or two days of arriving at altitude. The increased number of erythrocytes increases the O_2 carrying capacity of the blood. Physiologists express this affinity in terms of either "volumes percent" (the number of milliliters of gas which can be extracted by vacuum from a blood sample of 100 milliliters in volume), or as "percent saturation" as in Figure 8-5. Due to the low partial pressure of oxygen at 17,500 ft, the percent of saturation is only 73%, even though the carrying capacity is high (Table 8-1; Figures 8-5 and 8-6). Another way of referring to O_2 concentration and transport is to describe the pressures involved. Oxygen, like any other gas, will only diffuse from areas of higher pressure to areas of lower pressure. Keep in mind that the total atmospheric pressure (760 mm Hg at sea level) is the sum of the partial pressures of the gases in air. Air is 20.9% O_2; thus, the partial pressure of O_2 is 760 x 0.209 = 159 mm Hg. The alveolar air is always saturated with water and at 37.6°C this is 43 mm Hg. The water vapor pressure plus dilution of room air in the upper respiratory tract accounts for the alveolar air having a value of 104 mm Hg (Table 8-1 and Figure 8-3). The advantage of breathing 100% O_2 in terms of increasing the O_2 carrying capacity of blood is illustrated in Figure 8-5.

The shape of the oxygen-hemoglobin dissociation curve (ODC)

can be considered the fourth line of defense against reduction in tissue PO_2 at altitude. The flat portion of the curve insures that hemoglobin will be saturated with oxygen even when the atmospheric PO_2 has declined by 4 to 6 Torr (Figure 8-6). Changes in the PO_2 at which hemoglobin becomes loaded or unloaded with oxygen is another important adjustment to altitude. The reduction in alveolar and plasma PCO_2 (and alkalosis) which occurs as a result of HVR at altitude tends to increase the affinity of hemoglobin for oxygen. The formation of 2,3-diphosphoglycerate in erythrocytes in response to hypoxia and alkalosis facilitates the release of oxygen in systemic body tissues.

Cardiac Response

Elevated heart rate and the accompanying increased cardiac output can be considered the fifth line of defense against tissue hypoxia at altitude. Of course, cardiac output can be increased by either an increase in heart rate or an increase in stroke volume. However, there does not appear to be an increased stroke volume at altitude. It seems that cardiac and respiratory responses to hypoxia are both primarily accomplished by increased frequency of breathing and pulse.

Vasomotor Responses

The sixth line of defense against tissue hypoxia can involve an increased blood flow to vital tissues (brain and heart) and increased capillary density within body tissues. There are also cellular metabolic adjustments that will be considered later.

Problems of Sea-Level Natives Traveling to Altitude

Ascending rapidly (in a matter of minutes) in a climate chamber from sea level to the atmospheric equivalent of the summit of Mt. Everest (29,028 ft or 8,848 m), would result in loss of consciousness in a few minutes and death shortly thereafter. But by ascending gradually over a period of weeks, experienced mountain climbers become acclimatized and often reach Everest's summit without

Table 8-1. Effects of Low Atmospheric Pressures on Alveolar Gas Concentrations and Arterial Oxygen Saturation

Altitude (feet)	Barometric Pressure (mm Hg)	Inspired Air PO_2 in Air (mm Hg)	PCO_2 in Alveoli (mm Hg)	PO_2 in Alveoli (mm Hg)	Arterial Oxygen Saturation (%)
0	760	159	40	104	97
10,000	523	110	36	67	90
20,000	349	73	24	40	70
30,000	226	47	24	21	20
40,000	141	29	24	8	5
50,000	87	18	24	1	1

supplemental oxygen, experiencing only minor symptoms of illness. Modern methods of transportation, including airplanes, cable cars, and automobiles present opportunities for many persons to be exposed to terrestrial high altitude, whether for hiking, climbing, skiing, or sightseeing. Let's consider some of the common inconveniences and pathophysiology associated with rapid ascent (i.e., one day or less) to altitudes above 7,000 ft (2,135 m).

Anorexia is usually experienced. This is thought to be a result of reduced blood flow to the gastrointestinal tract. Loss of body weight may be as much as 1 or 2 lb (1 kg) over a period of a week. Caloric intake may be reduced as much as 1500 kCal per day. Most of the weight loss is due to loss of body fluids resulting from decreased thirst and increased urine volume. Insomnia is frequently reported by lowlander natives at altitude. Lassitude is another feature of the novice at altitude; both the non-athlete and the highly motivated athlete will have a decrease in his/her capacity to exercise. Maximum work capacity decreases roughly 1% for each 300 ft (100 m) above 7,500 ft. Anaerobic metabolism as well as aerobic metabolism is reduced at altitude. This is due to a reduction in the tolerance for elevation of blood lactate while at altitude. There is no good explanation for this reduced lactate tolerance. High altitude natives have lactate tolerances at altitude similar to that of lowlanders at sea level.

Although some athletes may train at high altitude with the idea of increasing the oxygen carrying capacity of their blood, physiologists have difficulty recommending altitude training when competition is to be at sea level. This is partly because of the considerable differences among individuals in level of physical fitness for a given activity. Other differences are due to age, effects of diet changes, weight loss, previous experience, and motivation of the athlete. A careful study of cyclists did describe advantages gained from training at altitude (Terrados et al. 1988). If competition is to be at altitude, training at altitude will most certainly be advantageous.

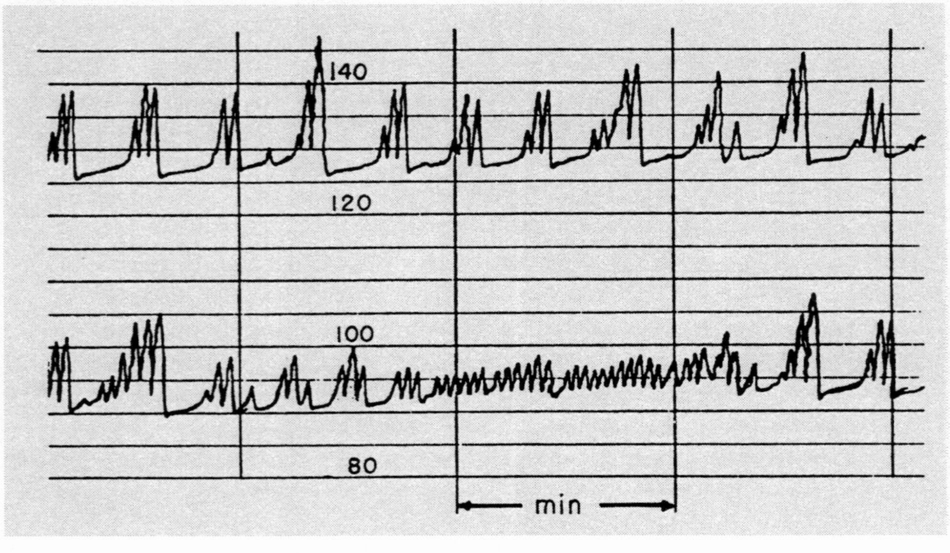

Figure 8-7. Cheyne-Stokes Breathing at Altitude. Lim and Luft (1965) observed that periodic breathing of this sort occurs most frequently in hypoxia combined with cold.

Short term travelers to altitude often experience *periodic*, or intermittent, breathing during sleep. This form of breathing bears the names of the two English physicians who first described this type of intermittent respiration, Cheyne-Stokes (Figure 8-7). Mountain climbers have reported periods of apnea lasting over one minute. This Cheyne-Stokes pattern tends to disappear as persons become acclimatized to altitude. However, periodic breathing may occasionally occur after months at altitude.

Acute exposure to high altitude in a climatic chamber or in the event of decompression of an aircraft at high altitude results in serious mental impairment (Figure 8-8). High altitude airplane pilots must be trained to recognize the symptoms which signal onset of acute hypoxia. Now let's consider some of the pathophysiology which is associated with high altitude.

Acute Mountain Sickness (AMS). The first description of AMS was recorded by Chinese travelers in approximately 30 B.C. when they made references to the "Great Headache Mountain" and the "Little Headache Mountain" (Ward, Milledge, and West, 1991; Krasney, 1994). In addition to severe headache, the symptoms of acute mountain sickness (AMS) include lassitude, irritability, nausea, vomiting, anorexia, indigestion, flatus, constipation, insomnia, and sleep disturbances characterized by periodic breathing. An Environmental Symptoms Questionnaire (ESQ) has been developed by Sampson et al. (1983) in which the scoring of AMS is based primarily on headache, nausea, and insomnia. The onset of symptoms usually begins 6 to 12 h after ascent and peak in intensity in 24 to 48 h. As acclimatization takes place, the symptoms will usually disappear within 3 to 7 days (Rahn and Otis 1949). Lyons et al. (1995) demonstrated that 16 days at Pikes Peak (14,100 ft, 4300 m) followed by 8 days at sea level reduced the incidence of AMS upon return to altitude. The reported frequency of AMS varies widely from

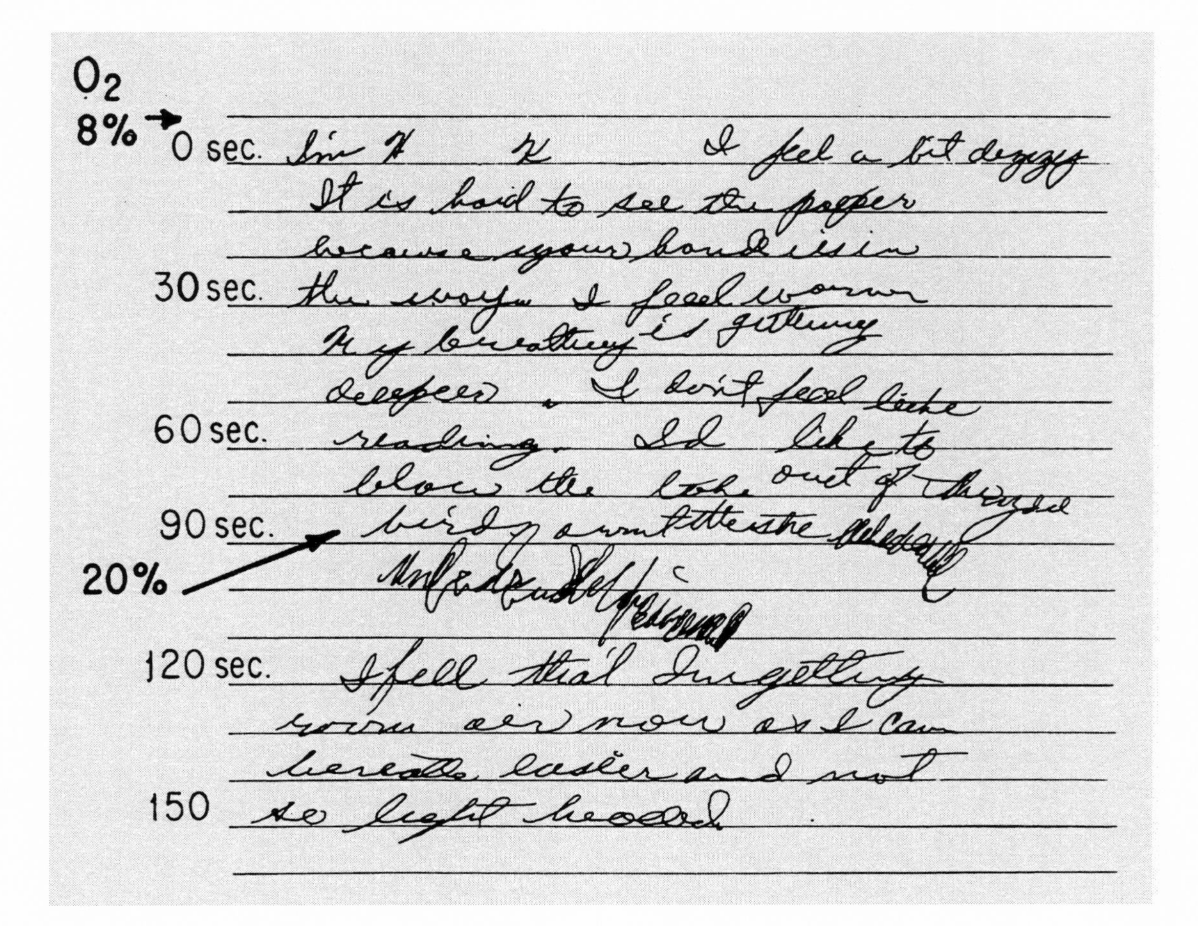

Figure 8-8. Mental Effect of Hypoxia. A common test under conditions of hypoxia is to have a subject write continuously. Such a test is shown here, with a volunteer exposed to severe hypoxia (his name has been removed from the record). The subject was writing on a tablet held above his face during the experiment. He quickly began to make spelling mistakes and duplicate letters. Soon, severe effects set in with complete inability to carry out a mental and mechanical task. When the low oxygen supply was replaced with room air, the subject recovered in about 30 seconds.

ULRICH LUFT (1910-1991) PIONEER IN HIGH ALTITUDE RESEARCH

Ulrich C. Luft, M.D. and Ph.D., is caught in this photograph conducting one of the thousands of gas analyses he made on a Van Slyke apparatus. He enjoyed collecting data with his associates, who included numerous

laboratory technicians, resident physicians, plus 20 graduate students from 12 different countries.

Luft began his professional career in his native Germany. There, he conducted field studies on mountain climbers—first in the Alps, then in the Himalayas, where he was involved in a 1939 expedition on Mt. Everest. His work focused on erythropoiesis and ventilatory changes associated with high altitude. He left his position as associate professor at the University of Berlin in 1947 to become a research physiologist with the USAF, School of Aviation, San Antonio, Texas, where he was recruited as an international aerospace pioneer. He was a leader in

8% to 100% of exposed individuals, and is directly dependent on both the rate of ascent, the extent of physical activity, and final altitude reached (Hackett and Hornbein 1988; Rock 1986). A few individuals experience symptoms at elevations as low as 8,000 ft (2,500 m), but it is much more common over 10,000 ft (3,300 m). Although there is individual variation in susceptibility to AMS, virtually everyone will experience some symptoms of AMS if they go rapidly to elevations of 14,000 ft (4,200 m) or higher (Rock 1986).

The best treatment of AMS is slow ascent. The next best treatment is rapid descent. Supplemental 0_2 (never 100% O_2, which is too toxic) can be helpful in relieving symptoms. It is important to keep in mind that the transition from AMS to serious conditions—such as high altitude pulmonary edema (HAPE) and high altitude cerebral edema (HACE)—is gradual, and increases in AMS symptoms need to be heeded.

High Altitude Pulmonary Edema (HAPE). High altitude pulmonary edema usually manifests itself 12 to 96 h after a rapid ascent to high altitude (Hultgren 1997a; Hackett and Hornbein 1988; Rock 1986). Young active males seem particularly susceptible (Hackett and Hornbein 1988). Onset is often subtle, manifested by fatigue, dyspnea on exertion, nonproductive cough, tachypnea, and tachycardia. Cyanosis becomes apparent. If left untreated, HAPE can run a rapid course progressing to coma and death in less than 12 h (Hackett and Hornbein 1988; Rock 1986). As mentioned earlier, the transition from AMS to HAPE can be subtle.

Recently, the etiology of HAPE has become better identified (Hultgren 1997a). Hypoxia, as the reader may recall, causes constriction in small pulmonary arteries and dilation in systemic arteries. Hypoxic pulmonary vasoconstriction is an important mechanism by which pulmonary blood flow is controlled, such that blood perfusion in the lung is matched to the areas ventilated. As we mentioned earlier, hypoxic pulmonary vasoconstriction ensures that there will be reduced blood flow in areas where alveolar ventilation is small. HAPE seems to result from excessive vasoconstriction in portions of the lung, which elevates pulmonary arterial pressure (Hultgren 1997b). The elevated pressure causes edema in the lung and thus increases the distance over which O_2 must diffuse. In addition to high pulmonary blood pressure and the induced perfusion of fluid from capillaries to interstitial spaces and alveoli, there can also be disruption of the endothelial tissue lining arterioles and capillaries. There is currently considerable research being conducted to identify the cellular and molecular mechanisms involved in detecting hypoxia by cells (Donnelly 1996; Acker and Xue 1995). Cells of the carotid body and smooth muscle cells in pulmonary and systemic blood vessels respond to hypoxia by depolarization or hyperpolarization of smooth muscle in blood vessels. Weir and Archer (1995) provide an excellent review of this subject. In systemic arteries, hypoxia causes an increased current through ATP-dependent potassium channels and vasodilation, whereas in the pulmonary arteries hypoxia inhibits potassium current and causes vasoconstriction.

High Altitude Cerebral Edema (HACE). The early symptoms of HACE resemble those of AMS (Dickinson 1983; Hackett and Hornbein 1988; Hamilton, Cymerman, and Black 1986; Rock 1986). In fact, HACE may be a severe form of AMS. Early symptoms include severe headache, nausea, vomiting, and extreme lassitude. Truncal ataxia, and change of mental status help differentiate early HACE from AMS (Hackett 1980). The incidence of HACE is low, occurring in about 1% of individuals exposed to altitude. Rapid ascent and lack of acclimatization are predisposing factors. It has been reported as low as 9,000 ft (2,700 m), but the majority of cases occur above 12,000 ft (3,600 m). Definitive treatment of HACE is to descend. In general, the greater the descent the better. Supplemental oxygen can be a useful adjunct, but should not be used as a substitute for descent. The close relationship between HACE and AMS would suggest that the measures previously discussed for the prevention of AMS may also be effective for HACE (particularly gradual ascent).

Acclimatization to High Altitude

The rate, extent, and duration of exposure to altitude influence the characteristics of acclimatization. Individual variations in age, physical fitness, and recent high-altitude experience of subjects modify acclimatization. Experienced high altitude climbers often select a base camp at 7,000 to 9,000 ft and spend several weeks at this elevation before proceeding to the next camp at 11,000 or 12,000 ft. They will repeat this process, allowing time for acclimatization to be completed at several elevations. However, once they reach 18,000 ft, they will go on to the peak as quickly as possible, because at higher elevations the human physiological systems tend to deteriorate.

The high financial and labor costs of field studies force most investigators to conduct laboratory experiments. These studies permit separation of responses to reduced O_2 (hypoxia) and reduced barometric pressure (hypobaria). The acclimatization that occurs in the field involves the interaction of acclimation to cold, high winds, low humidity, limitations in diet and fluid selection. To gain an appreciation of the differences between data obtained in the field and in the laboratory, we refer the reader to two recent major studies: (1) a field study, "The American Medical Research Expedition to Mount Everest (AMREE)," led by John West (Boyer and Blume 1984; Schoene, et al. 1986; West 1984; West et al. 1983; Winslow, Samaja, and West 1984; Winslow 1984); and (2) the hypobaric laboratory experiment conducted at the U.S. Army Research Institute of Environmental Medicine, Natick, MA. This study involved six subjects in a barometric chamber for 40 days and nights ascending to an altitude equivalent to Mt. Everest (Andrew, O'Brodovich, and Sutton 1987; Houston, Sutton, and Cymerman 1987; Malconian et al.

developing protocol for high-flying aircraft pilots during and after the second world war. He served as head of the Department of Physiology, Lovelace Medical Foundation, Albuquerque, New Mexico, from 1954 through 1980.

Luft's research interests varied, but always converged with the central theme of hypoxia. He has over 120 publications, monographs, and chapters in books on subjects ranging from techniques of estimating total body water to respiratory pressure-flow relationships in newborn infants. His scientific contributions to respiratory physiology resulted in numerous awards and lectureships. While his scientific accomplishments are impressive, his human qualities of kindness, patience, and perseverance in teaching his research skills and knowledge of respiratory physiology have endeared him to all who have had the privilege of having him as a colleague. Photo courtesy J.A. Loeppky.

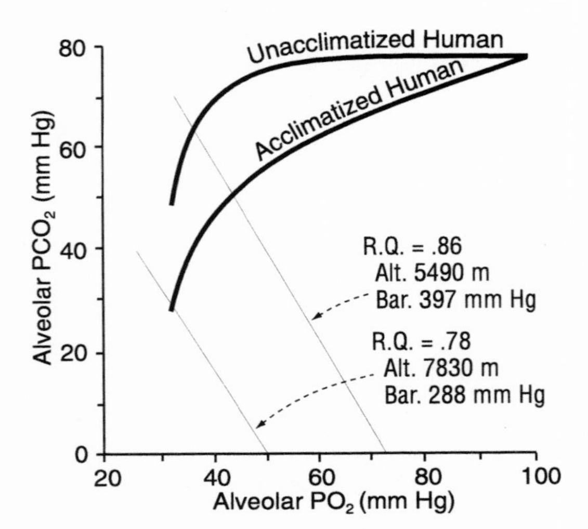

Figure 8-9. Alveolar Gas Tensions Above 18,000 feet. The data are presented on a Rahn-Otis diagram. Plotted points show mean values of end-expiratory Haldane samples taken at various altitudes on two mountains. The upper curve is the line for acutely exposed, unacclimatized subjects. The lower curve is the line for acclimatized subjects. The samples provide a useful index of the degree of hypoxia tolerated by mountaineers after suitable acclimatization. In the unacclimatized subjects, we see that elevation of alveolar pressure of oxygen due to increased ventilation is greatest in the region of 13,000 feet (4,000 meters). Modified from Pugh (1964).

1993; O'Brodovich et al. 1984; Reeves et al. 1987a, 1987b; Suarez, Alexander, and Houston 1987; Sutton et al. 1988).

Respiratory changes with acclimatization. We have already considered the hypoxia ventilatory response (HVR) as being an important initial response to hypoxia. The respiratory response is modified by other factors as hypoxic exposure continues. The HVR produces alkalosis, which acts as a braking mechanism on the central respiratory center and limits a further increase in ventilation (Huang et al. 1984). To compensate for the alkalosis, within 24 to 48 hr of ascent the kidneys excrete bicarbonate, decreasing the pH toward normal; ventilation increases as the negative effect of alkalosis is removed. Ventilation at rest continues to increase slowly, reaching a maximum only after 4 to 7 days at the same altitude. With each successive ascent in altitude, the plasma bicarbonate concentration continues to drop and ventilation continues to increase.

A way to appreciate the importance of the ventilatory changes at increasing altitude is to plot values for alveolar oxygen and carbon dioxide on the Rahn Otis diagram (Figure 8-9). This approach clearly contrasts the effects of acute and chronic hypoxic exposure, and can be used to assess the degree of ventilatory acclimatization (Rahn and Otis 1949). As ventilation increases, the decrease in alveolar CO_2 allows an equivalent increase in alveolar O_2. The blood gases of acclimatized subjects at various altitudes are presented in Table 8-2. It is important to note that as people become acclimatized, the alveolar and plasma PCO_2 become reduced. Acclimatization involves an increased sensitivity of the respiratory center to CO_2. The HVR discussed earlier becomes reduced as acclimatization progresses, and PCO_2 is the driving force for respiration in acclimatized subjects.

Circulatory system. As mentioned earlier, the sea level natives residing at altitude never attain the maximum work capacity (maximum oxygen consumption) or maximum heart rate they have at sea level. Interestingly, myocardial ischemia at high altitude has not been reported in healthy persons. The ratio of heart rate per liter of oxygen consumed remains the same at altitude as at sea level. Increased catecholamine activity on ascent causes an initial mild increase in blood pressure, moderate increase in resting heart rate and cardiac output. Stroke volume is low because of decreased plasma volume, which drops as much as 15% over 1 to 3 days. Resting heart rate returns to near sea level values with acclimatization, except at altitudes above 17,000 ft (5,200 m). As the limits of hypoxic acclimatization are approached, maximum and resting heart rates converge.

Pulmonary vascular resistance increases on ascent to high altitude as a result of hypoxic pulmonary vasoconstriction. This increases pulmonary artery pressure (Lockhart and Saiag 1981) which, as mentioned earlier, can result in pulmonary edema. Echocardiography examination of acclimated subjects describes a smaller than normal left ventricle because of decreased stroke volume, while the right ventricle may become enlarged (Suarez, Alexander, and Houston 1987). We presume this enlargement results from the increased resistance.

Cerebral oxygen delivery is thought to be well maintained with moderate hypoxia resulting from increased extraction of O_2 and increased blood flow (Curran-Everett 1991). Cerebral blood flow increased 24% following abrupt ascent to 12,500 ft (3,810 m), and then returned to control values over 3 to 5 days (Severinghaus et al. 1966).

Blood and body fluid changes during acclimatization. Ever since the observation in 1890 by Viault that the hemoglobin concentration was higher than normal in animals living in the Andes, scientists have regarded the hemopoietic response to increasing altitude as an important component of the acclimatization process (Cerritelli 1976; Winslow et al. 1984). As noted earlier, in response to hypoxemia, erythropoietin is secreted and stimulates bone marrow production of erythrocytes. The hormone is detectable within 2 hours of ascent. Nucleated immature erythrocytes can be found on a peripheral blood smear within 4 to 5 days. The molecular responses to hypoxia have been reviewed by Bunn and Poyton (1996).

As we have described earlier, decreased plasma volume during the first few hours at altitude increases the O_2 carrying capacity of blood. Increased urine volume and decreased fluid intake account for most of the decrease in plasma volume. Mountaineers recognize copious urine volume as a sign of successful acclimatization to altitude. Experimental exposure to acute moderate hypoxia (3,000 to 5,000 m) results in diuresis and natriuresis in humans and many mammals. In contrast, when hypoxia is poorly tolerated or severe (above 5,000 m), renal volume and salt excretion usually decrease.

The physiological mechanisms involved in the changes in fluid and salt balances at altitude are not well defined (Hoyt and Honig 1996). Signals from the arterial chemoreceptors to renal tissue may be modified by hypoxia such that salt and water excretion is reduced. Increased ventilation at altitude also contributes to the decrease in total body water. The alkalosis of the blood and arterial

Table 8-2 Blood Gases in Subjects Acclimated to Altitude

Altitude Feet	Meters	P_B (mm Hg)	PAO_2 (mm Hg)	SaO_2 (%)	$PACO_2$ (mm Hg)
5400	1646	630	73.0 (65.0 -83.0)	95.1 (93.0-97.0)	35.6 (30.7-41.8)
9200	2810	543	60.0 (47.4 - 73.6)	91.0 (86.6-95.2)	33.9 (31.3-36.5)
12020	3660	489	47.6 (42.2 - 53.0)	84.5 (80.5-89.0)	29.5 (23.5-34.3)
15440	4700	429	44.6 (36.4 - 47.5)	78.0 (70.8-85.0)	27.1 (22.9-34.0)
17500	5340	401	43.1 (37.6 - 50.4)	76.2 (65.4-81.6)	25.7 (21.7-29.7)
20140	6140	356	35.0 (26.9 - 40.1)	65.6 (55.5-73.0)	22.0 (19.2-24.8)

Data are mean values and range (in parentheses). All values are for subjects age 20 to 40 years who were acclimatizing well. P_B is barometric pressure; PAO_2 is alveolar partial pressure of O_2; SaO_2 is percent of O_2 saturation in Hb; and $PACO_2$ is alveolar partial pressure of CO_2. Data are from Hackett and Roach 1995.

IS GROWING OLD LIKE GOING TO HIGH ALTITUDE?

Ross McFarland—a former member of the Harvard Fatigue Laboratory and authority on both aviation and aging—compared the mental impairment that sometimes accompanies aging to symptoms of temporary oxygen deprivation in healthy young subjects. In tests of mental function, the loss of insight and judgment experienced by some older people is similar to the mental confusion induced by hypoxia, especially as the oxygen saturation in the arterial blood drops below 70-75%.

Our ability to see and hear is also affected by aging as well as by going to altitude. A study of light sensitivity in subjects from 16 to 89 years showed that the intensity of illumination must be almost doubled for each increase in age of 13 years; the curve was similar in healthy young men who went to high altitude. While vision is influenced early in subjects at high altitude, hearing is influenced last; as with aging, higher frequencies are impaired more than lower ones.

hypoxia may directly inhibit endothelial tissue and lung tissue production of angiotensin-converting enzyme (ACE). Decreased ACE could in turn change the response of the renin-angiotensin-aldosterone system (Honig 1989). Other hormones reported to be involved in responses to acute hypoxia include antidiuretic hormone (ADH) and vasopressin (Walker 1983 and 1986; Claybaugh, Wade, and Cucinell 1989). Clarification of the total body water changes associated with exposure to hypoxia requires additional research (Hoyt and Honig 1996).

Aging and High Altitude

David Bruce Dill, the noted physiologist who made his last expedition to altitude at the age of 87, observed that age does not bar successful adaptation to altitude. Nevertheless, Dill's research with a limited number of older subjects showed differences in the adjustments made by older sojourners to altitude. For example, plasma volume, which decreases in young subjects, either increases or shows no change in older people. Older male subjects also had a considerably delayed increase in hemoglobin and erythrocytes, while in young males these increased early on. In addition, the older men had lower maximum heart rates compared to young men (Horvath 1980).

High Altitude Natives versus High Altitude Acclimatized Lowlanders

When compared to acclimatized sea level natives the high altitude natives have numerous advantages, including greater hypoxic ventilatory response, larger lung volume, smaller alveolar-arterial O_2 diffusion gradients, parasympathetic dominance in the control of heart rate during exercise, greater brain blood flow during exercise, more favorable uterine blood flow distribution during pregnancy, higher birth weight, higher neonatal arterial O_2 saturation, and greater maximal exercise capacity at altitude. Some of these differences have been attributed to being born at altitude. Others may be a result of genetic adaptations (Moore et al. 1994). Interestingly, molecular examination of genes reveals differences between natives of Tibet and high-altitude natives in South America and North America. No one has identified a DNA sequence in genes which is unique to high-altitude natives.

Before leaving the subject of acclimatization, we must note there has never been a permanent population identified above elevations of 18,000 ft. Reproductive capacity is limited at this high elevation because of the difficulty in maintaining adequate oxygenation to support a fetus to term. Experienced mountain climbers will spend as little time as possible above 18,000 ft as their capacity for exercise declines.

COMPARATIVE PHYSIOLOGY OF ALTITUDE EFFECTS

One advantage of studying the comparative physiology of altitude is that the hypoxic environmental stress is continuous, and this makes observation and evaluation of acclimatization and genetic adaptations possible. In the case of thermal stresses, animals in their natural environment often avoid the stress by behavioral methods. Low metabolic rate, anaerobic metabolism, and lethargic behavior may be the only escape for an animal inhabiting areas with low oxygen partial pressure. Note the innumerable similarities in acclimatization processes between human subjects and various animals.

Micro-organisms and Invertebrates

For over a century it has been recognized that microbes adapt to lack of oxygen by anaerobic metabolism. Prevalent in these life forms are various respiratory systems that do not involve oxygen. Our knowledge of the capacity for invertebrates to evolve similar systems has received an impetus from marine biology research. Intertidal pools readily demonstrate the interdependence of microbes, plants, and animals as well as rapid changes in ecology and physiology. The O_2 and CO_2 content of intertidal pools can change tremendously over a 24-hr period, with the O_2 content ranging from zero to 400 to 600 Torr, and the CO_2 dropping to as low as 10^{-4} Torr within a matter of hours (Truchot and Duhamel-Jouve 1980). These extreme environments are often the product of their inhabitants. Metabolic adaptations involving anaerobic pathways are the major adjustments among microbes and invertebrates, which accounts for their reduced need for oxygen. In contrast to metabolic adaptations, there are many organisms (invertebrates and vertebrates) that combine behavioral and metabolic adjustments. This is accomplished by seeking lower body temperature, thereby reducing metabolic rate and reducing the amount of oxygen needed (Wood 1991).

Fish, Amphibians, and Reptiles

A researcher trying to establish whether or not an organism is adapted to living at high altitude may first ask: Does the organism require oxygen? The second question may be: Are the circulatory system and blood adapted to transporting O_2 from a hypoxic environment to body tissues? Some lower vertebrates, fishes, amphibians, and reptiles, facilitate O_2 transport by the presence of multiple hemoglobins. One or more pigments have a high affinity for oxygen and the other(s) a low affinity. Another prevalent adaptation is the presence of ligands which have physical and chemical features that change the pigments' affinity for oxygen. The ligands include CO_2, hydrogen ion, and organic phosphates. Variation in the sensitivity of hemoglobin to ligands can be an important genetic feature of lower vertebrates adapted to hypoxia. A symposium organized by Steve Wood (1980) provides a review of respiratory pigment adaptations. Molecular strategies in the adaptation of vertebrate hemoglobin function has been reviewed by Weber (1992).

Among the lower vertebrates, the greatest amount of study has been on fishes. Of course, the fish most tolerant to hypoxia do not inhabit high altitude, but rather live at great depths in the oceans and in various O_2 deficient aquatic environments (Almeida-Val, Val, and Hochachka 1993). These fish have innumerable physiological and metabolic adaptations. Some of the metabolic adjustments of these deep-dwelling fish are due to hyperbaria, but most of their adaptations are due to hypoxia. At least 10 fish species have been identified to inhabit the bethnic area of Lake Tanganyika, where oxygen content is zero to 0.6 parts per million. The lifestyle of these fish has not been identified, but even if they make migratory trips which permit aerobic metabolism, their capacity for anaerobic metabolism must be extensive. Enzymatic adjustments by fish are not limited to long-term genetic adaptations. Short term, 3 to 5 wk, metabolic responses to hypoxia have been described by Greaney et al. (1980). There are fish in the Amazon that, in response to hypoxia, produce ethanol as a major end product of anaerobic metabolism. This pathway reduces the accumulation of lactate and subsequent lowering of pH (van Waarde, van den Thillart, and Verhagen 1993; Almeida-Val, Val, and Hochachka 1993).

Few amphibians and reptiles inhabit high terrestrial altitudes, although, as noted by Hock (1964a), hypoxia is not a limiting factor. Nevertheless, elevations of 10,000 ft (3,300 m) to 12,000 ft (3,650 m) appear to be the upper limit for amphibians and reptiles. In many ways, this is a curious situation because most of these animals are preadapted to hypoxia. This preadaptation is demonstrated by their ability to readily function with anaerobic metabolism, as evidenced by many amphibians and reptiles spending months or years in ponds or mud with low O_2 tension. These same organisms rely on anaerobic metabolism to sustain exercise at sea level atmospheric conditions.

The use of multiple respiratory organs is one of the most distinguishing features of amphibian physiology. Adult amphibians usually respire with skin and lungs, while larvae and neotenes often use three different organs: skin, lungs, and gills. Breathing through these organs is coordinated by the circulatory system to meet different metabolic and environmental challenges. Respiratory function is determined, in large part, by the circulatory system regulating blood flow to various respiratory organs (Malvin 1993).

The Lake Titicaca frog, native to 11,435 ft (3,812 m), is an exception to the general rule of no amphibians at high altitude. When compared to other amphibians, this frog has considerable genetic adaptations and physiological acclimatization to high altitude. The area for gas exchange by this frog is increased by extensive skin folds. The capacity for transporting oxygen is increased by high hematocrit and erythrocyte concentrations. These frogs also have a low metabolic rate and left-shifted oxygen hemoglobin dissociation curve (Hutchison, Haines, and Engbretson 1976). Most of these distinctive features are genetic adaptations as evidenced by their persistence for one to six months after being at sea level.

Turtles can survive 100% nitrogen for 12 h with no brain damage, and may have the greatest tolerance to hypoxia of any vertebrate (Jackson 1988). The hypoxia preadaptation appears in many forms, including capacity for anaerobic metabolism, low metabolic rate, plus respiratory and circulatory systems that deliver adequate oxygen to tissues from hypoxic environments. The three-chambered heart in non-crocodilian reptiles allows for shunting of blood away from the lungs when ventilation of the lungs is not possible. Hicks

(1993) described the physiological importance of this anatomical feature.

The explanation for the absence of amphibians and reptiles at high altitude is not obvious, but may be concealed in the fact that these are ancient animals. Pough (1980) has emphasized that modern reptiles are specialized animals that reflect selective forces quite different from those that shaped the evolution of mammals. Modern reptiles apparently evolved at a time when there were no pressures forcing them into high elevation areas. The capacity for reptiles to adapt and develop physiological adjustments to adverse environments is represented by these animals inhabiting specialized environments ranging from arid deserts to aquatic streams and oceans.

Small Domestic Mammals

Dogs suffer from mountain sickness and exhibit the same symptoms as their masters. If they are exposed to acute altitude effects in a pressure chamber, they are troubled with sleepiness, vomiting, labored breathing, muscular weakness, and inability to stand on their legs. However, these symptoms are lost rapidly and they acquire excellent acclimatization. In humans exposed in a pressure chamber, after several hours one-fourth of the subjects show no change in the number of red blood cells; the greatest change that can be attained in a pressure chamber is a rise of 5% to 10%. Much larger percentages are obtained from dogs, indicating that they can mobilize red blood cells on short notice.

Dill (1964) observed dogs at 16,400 feet and at various heights up to 20,130 feet. The owners had observed that as the dogs became acclimatized, they lost weight and became more irritable than at lower heights, effects which are commonly observed in people. The dogs' physical ability was not impaired: two of them were observed by Dill at 16,000 feet as they ran behind a truck for about 15 miles at 15 mph without showing fatigue. Another climbed to over 20,000 feet without symptoms of fatigue.

The domestic cat is very susceptible to the action of rarefied air. According to Mosso, cats in South America are never seen in places higher than 11,000 feet; he states that cats transported to an elevation of 13,000 feet, where all other domestic animals thrive, show depressed activity, and after a few days most of them die in convulsions of an epileptic character. Dill, on the other hand, saw healthy cats at 12,000 ft, one at 15,600 ft, but none at 17,500 ft.

Mosso (1898) tested cats in altitude chambers and made the surprising observation that with a resting or sleeping individual, as altitude was attained, the breathing became slower and more superficial. The respiration is ordinarily approximately 30 breaths per minute. A barometric depression to an equivalent of 9,400 feet reduced the rate of breathing by 10 cycles per minute. The explanation of the altitude sensitivity of the cat is not clear.

Many other mammals are essentially like the dog, especially in respect to oxygen capacity (Table 8-3). If blood characteristics were the only criteria, the ability to tolerate altitude would be similar in many mammals, for the oxygen capacity of red cells in a series from humans to sheep is very similar. This is strikingly demonstrated by an artificial test: the measurement is the oxygen capacity of a liter of red cells from each animal. Most types of red cells carry about 45 volumes percent of oxygen (Table 8-3). The cells of the llama and vicuna are an exception, since they could carry 57 to 58 vols % (1/4 more). These unusual members of the camel family will be considered later.

Large Domestic Mammals

The mule seems to have unusual ability to adapt to high altitude. Several were used at the Aucanquilcha mine, a base camp for the International High Altitude Expedition; the mules carried personnel from 17,500 feet up to the 18,800-foot level. Dill suggests that this rugged performance depends upon the ability of the mule to gauge accurately its own capacity for work and to refuse to be pushed beyond a safe limit. The mule stops and pants when it has accumulated an oxygen debt. Horses can acclimatize to 14,000 feet, but are not found higher. The hearts of cattle acclimated to high altitude have an increased number of mitochondria per cell (Ou and Tenney 1990).

Domestic sheep have a low oxygen saturation of arterial blood both at sea level and at altitude. This is in spite of a large increase in hemoglobin at altitude. Sheep studied by Dill at 17,500 feet had an arterial saturation of 56%. In order to exist at this altitude, these animals must unload nearly all the oxygen carried by their blood (venous blood was 6.8% saturated); thus the sheep circulates one-half its hemoglobin without using it in oxygen transport. In spite of this handicap, this animal is able to tolerate 17,500 feet and can reproduce at 14,000 feet.

Small Free-Living Mammals

We may now turn our attention from domestic to free-living animals

Table 8-3. Hemoglobin and Red Cells of Some Vertebrates (Modified from Dill, 1938)

	Red Blood Cells million/mm3	Red Blood Cells ml/liter	Oxygen Capacity of a Liter of Red Cells vols % (ml/liter)*
Human	5.00	460	457
Dog	6.68	517	445
Horse	8.18	380	492
Ox	6.98	416	463
Rabbit	4.55	354	441
Sheep	10.53	353	455
Llama	12.11	275	584
Vicuna	14.90	305	571

*Other values for oxygen capacity of blood at high altitude and sea level are given in Prosser and Brown (1961), page 223.

at altitude. What adaptations do these animals have? Morrison (1962) has devised an interesting test (with a sharp end point) to identify adaptations made by small mammals at high altitude. The mammal is placed in a cold environment so that its metabolism increases; then the oxygen environment is lowered. When the transport of oxygen for this particular species reaches a critical point, hypothermia results. Animals well-adapted to altitude could still maintain their temperature regulation effectively at a pressure about 1/3 that of sea level. Of species collected by Morrison at sea level, the least effective was a rat-sized rodent related to guinea pigs. The critical pressure for this rodent was sometimes reached at a PO_2 of 110 to 120 mm Hg, a reduction of only one-quarter from that at sea level. The other extreme was seen in a species of the high altitude type, a small rodent, which could still be effective at a partial pressure of 50 to 60 mm Hg, or about one-third that of sea level.

The animals from high altitudes were on the whole much more effective than animals from sea level, but this was not always the case. The best species collected at sea level was more effective than several of the species collected at altitude. According to Morrison, the different performance of different species from the same environment appeared to relate to general "fitness." Thus the best performing lowland species was markedly the most vigorous of the lowland species, and its greater metabolic potential would be expected to be effective under the handicap of hypoxia. In a similar manner, wild guinea pigs showed greater performance than their more sedentary domestic relatives at the same altitude. The superior test performance of most rodents from high altitude in the Andes seems to relate to the improvement in transport capacity with which the species has adapted to their hypoxic environment.

What adaptations permit some species at altitude to be effective, and what signs of acclimatization are found there in small mammals? Usually highland species do not have a higher hematocrit (Table 8-3). On the contrary, when lowland rodent species are moved to altitude, they increase all three of the following: erythrocyte number, hematocrit value, and hemoglobin. With deer mice, heart weights increase with altitude (Hock 1964b). When the highland species were moved to sea level (Table 8-4), for the most part their hematocrit level proved to be a genetic adaptation since it did not change at sea level.

The evidence for this genetic adaptation had been reviewed by Steen (1971) in his chapter on respiratory adaptations to life at high altitude. Genetic adaptations to high altitude has more recently been reviewed by Moore et al. (1994). This matter was discussed in detail by Bullard, who makes a marked distinction between any acclimatized sea-level rodent and a rodent native to high altitude (Bullard and Kollias 1966). The effects of forced exercise on sea-level rodents when brought to altitude have been summarized by Altland and Highman (1971).

Large, Free-Living Highland Mammals

The vicuna, a South American camel with highly prized wool, is an

Table 8-4. Hematological Comparisons of Mammals from High and Low Altitude

Animal	Results	Reference
A. Highland Species Studied in Native High Altitude Habitat		
1. Human resident at high altitude	Elevated erythrocyte count and hemoglobin level	Dill 1964
2. Russian mice (mountain race)	Elevated erythrocyte count	Kalabukov 1937
3. Peromyscus (deer mice)	Elevated hemoglobin 45% at 12,500 ft., 90% at 14,300 ft	Hock 1964b
4. 5 species of high altitude rodents (Chile)	Hematocrit not higher than 8 sea level species	Morrison et al. 1963
5. More vigorous species at high altitude (and sea level) (Chile)	20% higher hematocrit than in less vigorous species	ibid
B. Lowland Species Moved to High Altitude		
6. White rat (chamber hypoxia)	Hematocrit of 85%	Highman and Altland 1949
7. Mus (feral)	Hematocrit increase from 47% to 57%	Morrison et al. 1963
8. White rat (field laboratory)	At 13,000 ft a 45% increase in hematocrit	Timiras et al. 1957
C. Highland Species Moved to Sea Level		
9. 3 highland species of rodents in Chile	No reduction in hematocrit after 2 months, nor in offspring	Morrison et al. 1963
10. 4 highland species of rodents in Peru	No reduction in one month, then a reduction of 5%	ibid
11. Peromyscus (deer mice)	A decrease in running performance from that at 13,000 ft.	Hock 1964b

Note: a lowland Phyllotis (in Peru) had a higher hematocrit than any of the highland Phyllotis (three species).

example of a large mammal adapted for life at great heights. Their ability to adapt to high altitude was noted on the International High Altitude Expedition when Dill (1938) observed people in an automobile chasing a band of vicuna at an altitude of 15,000 feet. Dill remarked, "The automobile seemed to be more handicapped than the vicuna at this altitude."

This expedition transported a vicuna by truck to 17,500 feet. Although it is known that in high altitudes the concentration of hemoglobin increases in the blood of humans, this increase was not found in the vicuna or its domesticated relative, the llama. The animals studied had either an 8% increase or a slight decrease. As a basis of comparison, four llamas studied at sea level all had more hemoglobin than the ones studied in high altitude. The blood of these South American camels had a greater affinity for oxygen than that of any other mammal. This physiological advantage is apparent on the dissociation curve (Figure 8-10).

When arterial PO_2 is 40 mm Hg, the vicuna transports 18 vols %, the human transports 14.0, the sheep, 8.0. The vicuna can use 0.7 of the oxygen-carrying capacity of its blood, the human can use 0.6,

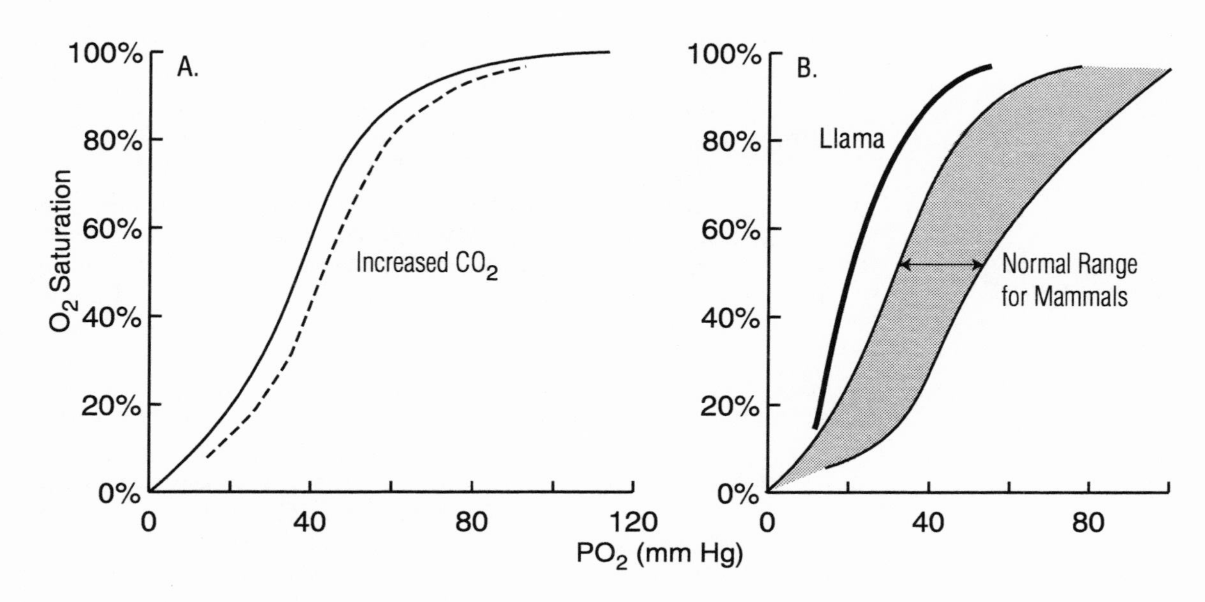

Figure 8-10. Oxygen-binding Capacity of Different Hemoglobins. Graph A demonstrates how the oxygen dissociation curve is moved to the right by the increased presence of carbon dioxide. For a single passage of blood, oxygen loading can be read on the solid curve and unloading on the dotted curve. In graph B, the behavior of blood is demonstrated for the llama, which lives under conditions of low oxygen pressure. This blood binds oxygen more readily than does that of other mammals (Schmidt–Nielsen 1960).

and the sheep can use only 0.4 (under this condition of PO_2, 40 mm Hg). The ability of the vicuna and the llama to deliver oxygen to tissues at high altitude does not depend upon the number of cells per unit of blood; in fact the blood of these camels is relatively dilute (hematocrit 28% and 31%). The explanation of the high oxygen capacity of the two species is the close packing of the protein hemoglobin in the red cells. There are also some unique molecular features of the llama and vicuna hemoglobin which increase the affinity of hemoglobin for oxygen (Perutz 1983; Jurgens et al. 1988).

Human beings and their domestic animals brought from sea level compensate for their disadvantage by increasing the quantity of hemoglobin in the blood. A person could deliver as much oxygen as the vicuna provided he or she had 1/6 more hemoglobin, while the sheep would require nearly double the amount of hemoglobin to accomplish this. The vicuna and the llama do not need an increase in hemoglobin; their tissue oxygen supply is adequate for their needs even at 17,000 feet.

Experiments with mammalian species indigenous to high altitude have revealed longer erythrocyte survival times, particularly in the camel (Cornelius and Kaneko 1962). Approximate erythrocyte survival times (in days) for control and high altitude mammals are:

Control	(# days)	High-altitude Mammals	(# days)
rat	55	Barbary sheep	170
human	120	tahr goats	165
horse	140	wild llama	235

In the high-altitude group, all measurements were made at sea level.

Birds at High Altitude

Birds hold the high-altitude record among all vertebrates. Humboldt (1850) reported condors at 23,000 feet; these were observed when he was breaking a high-altitude record for humans in 1802 by climbing to 19,286 feet. The chough (a crow-like bird) was observed even higher, at 27,000 feet on Everest. Dill observed pheasants at 16,000 feet. However, the record holders are geese, which migrate over the Himalayan mountains. Black and Tenney (1980) compared altitude acclimatization in the bar-headed geese (*Anser indicus*) native to the Tibetan Plateau and a similar-sized Peking duck (*Anas platyrhynchos*) native to sea level. It appears that the bar-headed goose hemoglobin has been "adjusted" structurally to yield the most efficient O_2 delivery between altitudes of 24,930 ft (7,600 m) and 30,176 ft (9,200 m), the altitude range encompassing the summits of the Himalayas. Weber (1992) described the chemical structure and features of the bar-headed goose hemoglobin which contribute to the bird's capacity to fly at such great heights. Swan (1961) points out that the bar-headed goose is an ancient species dating from at least the late Pliocene period, and aside from minor variations caused by glacial advances and retreats, it has apparently used the same wintering and breeding areas, and followed the same migration routes since its appearance. At the time that the species originated, the Tibetan Plateau was a relatively lush, warm area, and the Himalayas were much lower than their present height. During the late Pleistocene, both the Tibetan Plateau and the Himalayas were thrust to their present altitudes over a period of several thousand years, presumably forcing the birds to fly at higher and higher altitudes in order to cross the Himalayas. This may be an example of a behavioral pattern (migration) resulting in a change in molecular structure (hemoglobin).

Thus far we have avoided considering multiple stresses, but the influence of hypoxia on temperature regulation will be considered next.

Hypoxia-Induced Hypothermia

Hypoxia results in a regulated decrease in body temperature in organisms ranging from protozoans to mammals (Gordon and Fogelson 1991; Malvin and Wood 1992, Wood 1991, Wood and Malvin 1991). Mammals decrease body temperature during hypoxia by reducing heat production and increasing heat loss. Ectotherms lower body temperature by selecting an area with lower temperature. Reducing body temperature is an important response to hypoxia because it lowers metabolic rate and amount of O_2 needed when O_2 supply is limited. Lower temperature also increases the binding affinity between O_2 and its carriers (hemoglobin-like compounds), thus facilitating survival. Although this beneficial response is extremely widespread among taxa, little is known regarding the cellular mechanisms that mediate the hypoxia-induced reduction in body temperature. Studies of paramecium suggest that reduction in oxidative metabolism causes membrane depolarization (Malvin, Havlen, and Baldwin 1994).

COMPARISON OF ALPINE AND ARCTIC ENVIRONMENTS

There are characteristics of high altitude other than low oxygen pressure: these are low air temperatures, deep snow, and barren terrain. We will find it profitable to compare two areas that share these characteristics: 1) the alpine or high altitude area, and 2) the Arctic or far north tundra area. This comparison may be justified by the concept of Merriam who pointed out the striking botanical correspondence and climatic similarity between the high altitude portion of mountains and the arctic tundra. Others, however, may question whether there is any more reason to compare alpine and arctic areas than there is to compare a pine forest in Texas and a spruce forest in Canada. The answer, we believe, is yes, and we present our rationale in detail in the next few pages. Perhaps the justification can most simply be illustrated in the case of an ecologist in Texas who has an alpine area within a 6-hour drive from his laboratory. He must be prepared to discuss with his students whether or not this is a relic arctic area. He will also find that some rodents in that alpine area are comparable only to tundra rodents; this naturally turns his attention to the characteristics of other tundra mammals.

What is the Alpine Area?

If we use a physiological index to define the alpine area, our first clue is that altitude effects may begin as low as 6,500 ft (1,980 m); at 10,000 ft (3,050 m) rather conspicuous symptoms may begin. During a research project conducted on Mount Evans, Colorado, some of the investigators attempted to house their families at motels at 10,000 ft. Some individuals were unable to tolerate this altitude and became ill, especially while trying to sleep; thus we see a conspicuous physiological cutoff point for some people at 10,000 ft, although there is no striking clue in the geographical appearance.

Above this level, starting about 10,600 ft (3,230 m), there are apt to be conspicuous changes in botanical, faunistic, and climatic conditions, so that such an area is now referred to as *alpine*. Opinions vary as to the starting altitude for alpine areas, from 10,600 ft to 12,000 ft (3,660 m). This is understandable, since these alpine islands must cap their individual peaks and ridges slightly differently in variable geographical areas.

In the discussion to follow, we will disregard those high mountain tops in Alaska that correspond in structure to alpine areas in the "lower 49" states. The reason for this is that in Alaska, they are permanently snowy, icy, and barren, with very few mammals living there.

What is Tundra?

A good description of tundra is *a wet arctic grassland that is frozen most of the time*. As an example of tundra, we'll examine the Arctic Slope north of the Brooks Range in Northern Alaska. We can define the tundra biome as *a treeless northern plain with no floristic or faunistic source or center of origin*. This thought-provoking description reflects the complicated origin of the tundra. Beginning above latitude 65°, tundra can be found at sea level, but farther south, it will be found only on the sides of mountains.

Similarities between Arctic and Alpine Areas in the U.S.

First, let's consider the biological and physical factors that arctic and alpine areas have in common:

1. A striking botanical correspondence:
   (a) The Presidential Range in New Hampshire has 70 alpine species of plants, all of which are also found in the Arctic.
   (b) In Colorado, there are over 250 species of alpine plants, 40% of which are found in the Arctic.
2. Only a few species of animals and plants live in each area. There is a uniform set of conditions with specific requirements. There are few habitats, thus the few species of mammals. Both places have a simple biologic structure.
3. A master physical factor—lack of heat—rules these lands. Seasonal distortion is found in both areas; the frost-free period is often less than 60 days, but some periods last 80 to 90 days. There is a short growing season and thus, few food types. Insects are present only 2 months. Large herbivores cannot find enough food in any one local area to support themselves. Both areas are nutritionally poor.
4. Both places have high winds, but this has more meaning in the Arctic because there is less protection there. Mount Washington, NH, has a record wind speed of 225 mph.

Differences between Arctic and Alpine Areas in the U.S.

Now let's look at the differences:

1. Photoperiod: 82 days of continuous light and 82 days of continuous darkness at Barrow, Alaska. The growth of mammals is affected.
2. Arctic summers have even temperatures; alpine summers have temperatures that fluctuate.
3. The Arctic has little moisture with thin snow cover. (At Barrow, in the month of greatest snow depth the average snow cover is 15.9 inches.) In some alpine areas there can be as much as 76 inches of snow per day, or 1,000 inches per season.
4. Oxygen pressure is low in the alpine area; breathing air at 14,000 ft (4,270 m) corresponds to breathing 13% oxygen. The Arctic tundra is close to sea level.
5. The Arctic tundra has permafrost. This is several hundred to more than 3,000 ft (915 m) in depth. Each year the upper 6 to 8 inches of soil thaws. This has an influence on the digging of burrows and dens by arctic mammals. Apparently there is no permafrost in alpine areas that can affect mammalian populations.
6. Drainage differences: the alpine is well drained; the Arctic has a great deal of trapped surface water and thus it supports more grass. Near Point Barrow, the lakes and streams cover one-half the surface, but they are frozen most of the time.
7. There is a desert influence on alpine flora; in Colorado, of

more than 250 species of alpine plants, 65 are desert forms.

8. The lichens dominate the Arctic Slope but not the alpine area.
9. There are differences in limitation: the alpine islands are only hundreds of feet in vertical area; the Arctic Slope encompasses 40,000 square miles. In the alpine area, migrations take place down to coniferous forest (Fig. 8-11) there are few species migrations in the Arctic.

Figure 8-11. Bristlecone Pine Zone. A view of the bristlecone zone between 8,500 and 11,500 feet in some areas of the Rocky Mountains. These are among the oldest living trees, with some up to 4,900 years. The zone is of particular interest to students of the arctic–alpine area, because here tundra and forest mammals meet in their distribution. Photo courtesy Gordon Kent.

Fauna of the Arctic and Alpine Areas

We first became interested in comparing the fauna of alpine and arctic areas when considering the origin of some tundra animals. In the Arctic, mountain mammals are still strictly confined to the mountains, and nowhere do they penetrate the tundra. By considering the vertical ranges of the mountain mammals, we may understand more clearly the factors involved in adaptation to altitude. We will also be in a position to select the animals from which we can learn the most about acclimatization and genetic adaptation.

The habitat diagram (Figure 8-12) of the species living in the Inyo-White section of Owens Valley, California (near Death Valley) shows some of the mammals with the largest vertical range, and therefore the ones most likely to show genetic adaptations or an ability to acclimate to altitude. From Owens Valley to nearby foothills, there are 42 mammalian species; actually, in the entire Owens Valley area including Mount Whitney, there are 100 species of mammals. This includes the deer mouse, found at various altitudes. Other intriguing species are the golden-mantled ground squirrel, the bushy-tailed wood rat, the porcupine, and the shrews. In contrast, there appear to be fewer species in the tundra biome: there are 42 mammalian species on the arctic slopes of Alaska, and the tundra of Siberia has 33 species.

Now let us look at the distribution of mammals in two very different areas, both of which extend from sea level to mountain peaks. These areas are Mt. Evans, Colorado and the Arctic Slope north of the Brooks Range. Examples of mammals in the two areas is shown in Table 8-5. Here we have compared the fauna of two extremely rigorous environments: the Arctic tundra and the alpine meadows (which represent an area deficient in oxygen). The similarity in animal life is remarkable considering the distance which separates the alpine and tundra areas.

Adaptations of Arctic and Alpine Species

Now let us consider some adaptations to be found in the two environments; perhaps these adaptations might permit alpine and arctic species to be interchanged. A comparison of adaptations from the two areas is reasonable because there are many characteristics of the alpine area that are not concerned with a limited supply of oxygen.

The most conspicuous adaptation for tundra life appears to be *small body size*. This contrasts with an area like the African Veldt, where there are numerous species of large body size. Seventy percent of the 33 species found in the Siberian tundra are under 1 kilogram, 60% are under 100 grams. A second criterion for adaptation to tun-

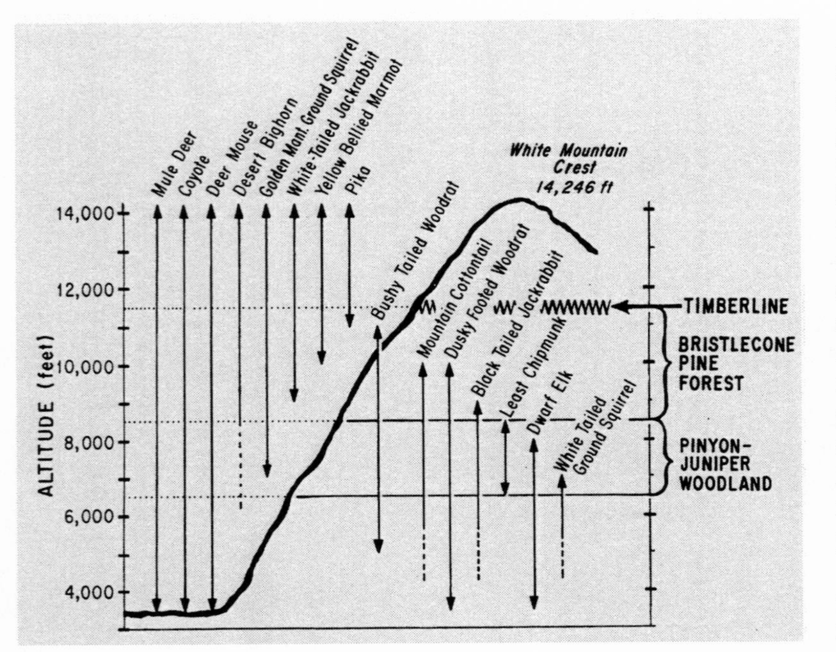

Figure 8-12. Altitude and Mammalian Distribution. A study of the vertical distribution of mammals, as depicted on this typical California peak, stimulates many questions for the student of mammalian environmental physiology.

dra life is *easy acclimatization,* which may not be lost in the summer. A third characteristic is the *ability to thrive on low-grade food.* A fourth is an *increased ability to mobilize bodily reserves* such as glycogen in the liver. The next characteristic applies to *breeding:* tundra mammals frequently produce larger litters of young, although this does not apply to lemmings. Reproduction also occurs at a younger age; for example, a northern subspecies of *Microtus* breeds successfully at 10 to 12 days of age and produces offspring at 30 days. These observations may eventually be found to apply to small mammals in the alpine islands of North America, irrespective of the presence of a low partial pressure of oxygen.

Another adaptation of arctic mammals that we can expect to find in alpine mammals is *independence from photoperiod.* In general, arctic rodents synchronize their seasonal activities without conspicuous dependence on light. This would seem to be a necessary adaptation because of the long periods of near-darkness (except for moonlight and northern lights) and the long periods of continuous light. At Barrow, Alaska, these periods extend for approximately 82 days each. Since this extreme condition of the light environment does not usually apply to the alpine area, especially in the lower 49 states, one would not expect to find an independence from photoperiod in mammals in the high mountains. Nevertheless, there is some evidence that alpine mammals are also independent of photoperiod. For example, the Columbia ground squirrel (*Spermophilus columbianus*) hibernates for 7 months from mid-July to March. Because of the time at which it goes into hibernation, one would have to postulate that it responds to a decreasing photoperiod of about 16.5 hours. But this is an unusual photoperiod for rodents to respond to, and it seems more likely that the animal is relatively independent of the use of photoperiod as a clue for annual activities. As mentioned in the *Hibernation* chapter, the golden-mantled ground squirrel hibernates independently of photoperiod in the laboratory environment.

SUMMARY OF TERRESTRIAL ALTITUDE EFFECTS

In this chapter we have been concerned with the physiological expressions of the stress imposed by the rarefied air at altitude. The resulting physiological strain is conspicuous in the circulatory system, the respiratory system, and the excretory system. Acclimatization to altitude consists of a *partial correction* of the deficits which are so evident in both the person and the domestic animal at high altitude; we use the term *partial correction* because in actuality, the oxygen-carrying capacity of even a native at high altitude is never the same as found in those individuals at sea level. Even acclimatized

Table 8-5. List of Alpine and Arctic Mammals.

Alpine Mammals of Mt. Evans, of Colorado	Arctic Tundra Mammals North the Brooks Range, Alaska
Whitetail Deer	Moose
Mule Deer	Caribou
Bighorn Sheep	Dall Sheep
Red Fox	Red Fox
Coyote	Coyote
Yellow-bellied Marmot	Hoary Marmot
Pika	Pika
Sorex c.	*Sorex c.*
Sorex n.	*Sorex o.*
Sorex p.	*Sorex a.*
Black Bear	Black Bear
Grizzly Bear	Grizzly Bear
Gray Wolf	Gray Wolf
Meadow Vole	Tundra Vole
Mountain Vole	Singing Vole
Boreal Vole	Red-backed Vole
Lynx	Lynx
Short-tailed Weasel	Least Weasel
Cottontail	Tundra Hare
Porcupine	Porcupine
Moutain Lion	Wolverine
Chipmunk	Mink
Long-tailed Weasel	Ermine
Deer Mouse	Brown Lemming
Rocky Mountain Goat	Musk Ox
Marten	Arctic Ground Squirrel
Badger	River Otter
Pocket Gopher	Collared Lemming
Bobcat	Arctic Fox
American Elk	
American Marten	
Long-tailed Vole	
Masked Shrew	
Dwarf Shrew	

athletes at altitude show more distress than do natives of that altitude. Successive acclimatizations are necessary; after physiological adjustments to a moderate altitude, mountain climbers continuing on to more extreme heights will experience acclimatization again at subsequent camps. In people and most domestic animals (with the exception of the domestic cat), the conspicuous evidences of altitude stress gradually disappear between 7,000 and 15,000 ft, and fatigue is no longer apparent. Above 18,000 feet, there is always an underlying process of deterioration.

Some of the characteristics of altitude acclimatization are evident in natives at high altitude. Some are also found in free-living mammals at high altitude, although some high-altitude mammals show no characteristics that differ from their counterparts at sea level.

We also discussed some of the characteristics of the altitude environment that are not concerned with low oxygen pressure. If we disregard this factor (and there is justification for this), we can reasonably compare the alpine area with the arctic tundra. And, because the comparative mammalian physiologist who wishes to work on alpine mammals will find it valuable, we have included some background material on arctic mammals.

At 29,028 ft, Earth's highest point has an environment so extreme that no human can survive there for long. Yet in the past half century we have ventured far beyond Everest's summit into the even greater environmental challenges of space travel, walks on the Moon, and life in space stations. We will examine the challenges of these new frontiers in the next chapter.

9. AEROSPACE PHYSIOLOGY

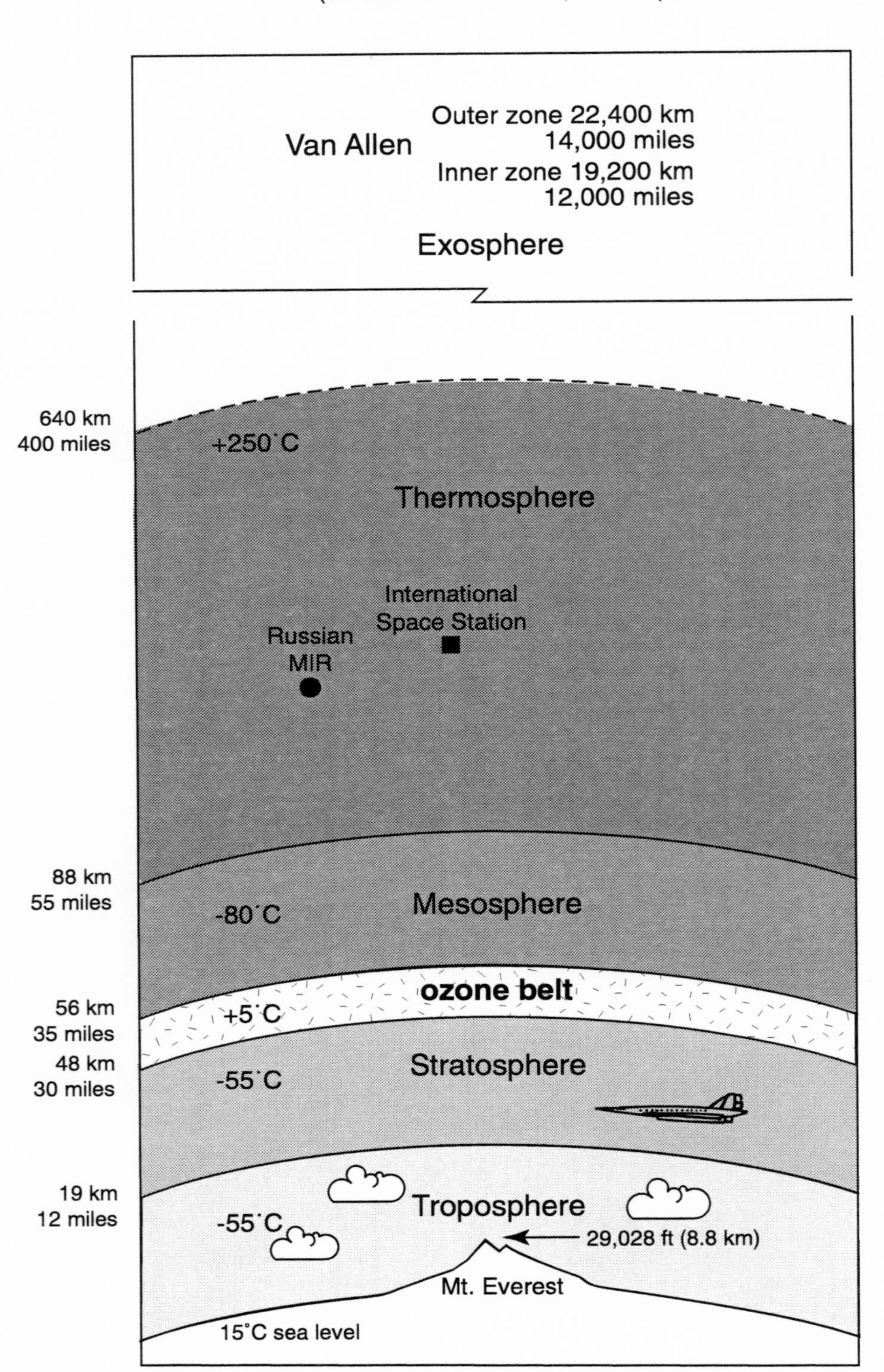

Environmental Diagram: Chapter 9

9. Aerospace Physiology

9. AEROSPACE PHYSIOLOGY

INTRODUCTION

Advances in space travel and aircraft design in recent decades have exposed astronauts, airplane pilots, and multiple organisms to unique environmental conditions, especially changes in gravity, acceleration, vectorial, and rotating forces. Aerospace physiologists, along with occupational and medical physiologists, share an interest in the physiological responses to these unique conditions, as well as to those of ionizing radiation, noise, vibration, and pollution.

GRAVITATIONAL PHYSIOLOGY

All land organisms on Earth have evolved in an environment of one gravitational (1-g) force. Human adjustment to various gravity forces provides environmental physiologists with a unique opportunity to examine modification of physiological systems to a new environmental variable. Can existing receptors and physiological control systems be modified to detect and control responses to changes in gravity? Or is evolution of new receptors and control systems going to be needed? It is beyond the scope of this chapter to consider the responses of all physiological systems to changes in gravitational forces, but we do hope to introduce the reader to an account of some mammalian systems responding to changes in gravitational forces. Furthermore, we will describe some of the experimental procedures used to simulate changes in gravity forces while still in a 1-g environment. For an in-depth coverage of gravitational physiology, we refer the reader to *Space Physiology and Medicine,* by Nicogossian et al. (1994).

Muscle-Skeletal System

Initial studies in this field involved examining organisms in an environment of increased gravitational forces by the use of centrifuges. Because of the relationship between gravitational force and mass, a perspective on the effects of gravity on anatomy and physiology can be gained by noting differences between large and small animals. The greater mass of the larger animal requires a stronger skeleton. More than 300 years ago, Galileo recognized that the breaking strength of a material is a function of its cross-sectional area, whereas the loading force is a function of volume or mass. A larger animal with a three-fold longer bone requires a nine-fold thicker bone to support its weight. A 5 gram shrew has a skeletal mass equivalent to 5% of its total body mass, whereas a 12,000 kg elephant has a skeletal mass equivalent to 35% of its total body mass. Thus, it is not surprising that with increased gravitational force, there are increases in the mass of the skeleton and the muscles involved in support of upright posture (Wunder and Welch 1979). In contrast, microgravity results in atrophy of these antigravity bones and muscles.

In 1953, Hillary and Norgay conquered the summit of Everest, thus attaining Earth's highest point. Sixteen years later, Armstrong and Aldrin walked on the Moon.

Loss of bone mass by weight-bearing bones is certainly of concern to the long-term space traveler. Weightlessness results in marked changes in bone formation and bone resorption. Increased activity of osteoclasts accompanied with no increase in osteoblasts is apparently responsible for the loss of bone mass (osteoporosis). The high serum calcium values that accompany the bone loss have the potential for producing kidney stones during months of space flight. Controlling diet and regular exercise appear to be the best mechanisms to reduce skeletal loss while in microgravity (Zernicke et al. 1990).

Ground-based studies of laboratory rats suspended such that the hind feet do not touch a surface provide a model for the atrophy and metabolic changes involved in muscle disuse (Stump et al. 1990). These types of studies describe changes in muscle metabolism as a result of the disuse, including increased storage of glycogen and decreased aerobic metabolism.

One of the most unexpected results on skeletal muscle in laboratory rats in orbit was the reduced capacity to use fatty acids as a fuel source; carbohydrates were the major source of energy. Rats have been reported to have lost 25% of antigravity muscle mass. After returning to Earth, the rats regained muscle mass slowly: it took twice as long to regain mass as to lose it. Changes in muscle metabolism in astronauts is much more difficult to measure, but all astronauts lose muscle mass during space flight, and may have changes in muscle metabolism.

Antigravity muscles undergo the greatest amount of atrophy in microgravity. The loss of structure and function in both slow- and fast-twitch muscles suggests that a combination of endurance and resistive exercise is required during space flight to effectively protect the integrity of skeletal muscle. Ground-based bedrest studies serve as a good model for changes in skeletal muscle associated with microgravity (Convertino 1996; Baldwin et al. 1996; Convertino, Bloomfield, and Greenleaf 1997).

Cardiovascular System

In an upright human at 1-g force, seventy percent of total blood volume is below the heart, and more than 70% of this volume is in compliant veins. When gravity no longer pulls blood downward toward the legs, body fluids shift from the lower body to the upper torso and head. This cephalic shift of body fluids when astronauts enter microgravity results in congestion of the upper body and formation of bird-like legs as several liters of fluid shift toward the head. The resulting increased volume of the heart and large blood vessels stimulates receptors which, in turn, cause the kidneys to produce urine in excess of fluid intake, which leads to a reduction in blood volume within a day or two. Investigators expected that the shift of fluid to the upper body would increase central venous pressure, but instead, a sudden dramatic decrease has been measured. There is also hyper-perfusion in most upper body tissues as organs are engorged with blood.

The changes in the cardiovascular system in response to microgravity appear to be adequate but are not well defined. The data are difficult to obtain and results have been inconsistent. To determine normative in-flight changes in Shuttle astronauts, Fritsch-Yelle et al. (1996) recorded disturbances in heart rate, arterial pressures, and cardiac rhythm for 24-h periods before, during, and after space flight. During Shuttle flights, the heart rate, diastolic, and systolic pressures, and variability of heart rate tended to be reduced. Premature atrial contractions also tended to be reduced in flight. There is also considerable evidence that the carotid baroreceptor-cardiac reflex is modified (Fritsch et al. 1992). Nevertheless, space travelers are able to conduct strenuous exercise during flights.

In contrast, return to earth results in serious problems; the adjustments made by the cardiovascular system during a tour in space present a challenge to the cardiovascular system to maintain blood flow to the brain and thereby prevent fainting when space travelers try to stand during the first few minutes after landing in a 1-g environment. This is referred to as orthostatic intolerance, which can seriously interfere with the astronaut's ability to egress from the space vehicle.

Orthostatic tolerance is frequently measured by two tests, a Lower Body Negative Pressure (LBNP) test and a standard Stand Test. The LBNP system illustrated in Figure 1 produces shifts in body fluid to the lower body similar to that experienced when the astronaut returns to Earth's gravitational force. A negative pressure of 50 mm Hg is equivalent to standing in 1 g, and -70 mm Hg is equivalent to 1.2 g force, which astronauts encounter during re-entry to Earth. The standard Stand Test involves technicians moving the subject from a supine to a standing position. It is important that the subject not use his/her muscles during the transition from supine

Figure 9-1. Lower Body Negative Pressure Chamber. The subject wears a kayak skirt attached to the chamber and secured at the waist with a belt. A vacuum pump provides negative pressure. A bicycle seat or foot support prevents the subject from being sucked into the box.

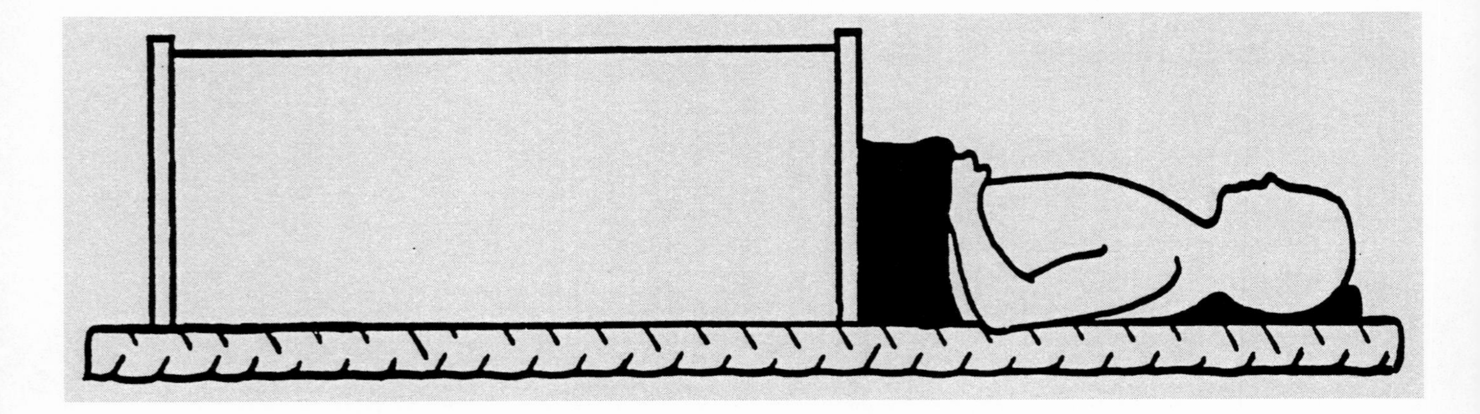

to standing. Most healthy subjects can tolerate the Stand Test for at least 10 minutes without fainting. In contrast, in a study of 14 individuals after space flights of 7 to 10 days, 9 of them could not complete the 10-min Stand Test (Buckey et al. 1996). Ground-based bedrest experiments of 7 to 14 days demonstrate levels of orthostatic intolerance very similar to those experienced by the returning astronauts.

What are the causes of the orthostatic intolerance experienced by return to 1 g? There certainly is no clear-cut answer to this question, except to say that the changes in the cardiovascular system during acclimation to microgravity need to be reversed upon return to 1 g.

Let's consider some of the specific observations noted during the post-flight orthostatic intolerance. There is a 10% to 15% reduction in blood volume, there is excess pooling of blood in legs and viscera (decreased vasoconstriction), there is reduced filling of the heart and decreased stroke volume, the carotid cardiac baroreflex is blunted, and each of these could cause inadequate increases in heart rate. The relative functional significance of these findings has not been established, and an adequate base for physiological countermeasures does not exist.

Orthostatic responses following acute exercise have been reported to be improved by fluid ingestion (Davis and Fortney 1997). Fluid replacement is of benefit to returning astronauts as their total body water (TBW) has been reduced. The reduction of TBW following the cephalic shift of fluid experienced when entering microgravity persists throughout a space flight.

The cerebral autoregulation system may be modified during space flight and return to 1 g. This system involves the cerebral blood vessels automatically responding (dilating or constricting) to changes in metabolic state, pressure, and flow rate such that adequate blood flow to the brain is maintained. Adequate blood flow to the brain is maintained over a wide (50 to 170 mm Hg) range of perfusion pressure (Bondar et al. 1994). The perfusion pressure in cerebral vessels is of course dependent upon the hemodynamics of the systemic circulatory system. Decreases in systemic blood pressure and increased heart rate precede onset of fainting (syncope). However, there is considerable individual variation in these changes and they do not serve as a good indicator of a presyncope situation. Currently, the best method for predicting syncope is to monitor the right medial cerebral artery blood flow velocity with a doppler technique (Bondar et al. 1996). Applying this technique to various regions of the body during future space flights and conducting ground-based studies with LBNP and Stand Test procedures will contribute to the design of countermeasures to ensure the cardiovascular competence of astronauts during and after return from space flights.

Countermeasures developed in the past and currently used include: 1) exercise (dynamic and isometric) for at least one hour twice per day throughout the flight with heart rates of 160 to 180 beats per minute; 2) ingestion of salt tablets plus a liter of water to expand blood volume prior to re-entry; 3) exposure to LBNP several hours prior to re-entry (Fortney, 1991; Hargens et al. 1991); 4) wearing antigravity suit (G suit) during re-entry and post landing. The G suit has somewhat the opposite function of the LBNP, as it increases pressure on the lower body to prevent blood from pooling in the extremities.

Respiratory System

Gas exchange in the lungs appears to be unchanged by microgravity. There is a reduction in vital capacity, presumably due to the cephalic shift of fluid and minor elevation of the diaphragm. During space travel, the relative distribution of blood and air in the lung is similar to that in 1 g. Rebreathing 5% argon demonstrated similar distribution of the gas throughout the lung in microgravity and on earth (Verbanck et al. 1996). The oxygen requirement for a given task is also similar in 1 g and microgravity, but in the space capsules, the potential for contamination is a major concern because of the considerable need to recirculate air and generate oxygen.

Endocrine System

Hormones are important effectors of the body's response to microgravity in the areas of fluid and electrolyte metabolism, erythropoiesis and calcium metabolism. Before 1994, studies of hormones were limited to blood and urine samples collected prior to launch and on landing day. Analyses of these samples is not a good indicator of the endocrine system response to microgravity. Urinary sodium, potassium and chloride generally increase in microgravity, with a concomitant increase in urine osmolality, while serum osmolality and sodium decrease throughout a space flight. Fluid intake is also reduced during the first few days in flight. Anti-diuretic hormone, cortisol, and aldosterone have been considered the most important hormones concerned with fluid and electrolyte metabolism. However, measurements of these hormones in blood and urine specimens collected during exposure to microgravity have indicated that their levels cannot account for the entire measured losses of fluid and electrolytes experienced by the astronauts. Increases in atrial natriuretic peptides (ANP) has been reported early in a flight, but after two days it decreased. These and other changes in fluid and electrolyte metabolism in space flight indicate that weightlessness may affect kidney function. Glomerular filtration rate (GFR) and effective renal plasma flow have not been measured in space. Creatinine clearance studies performed in microgravity indicated that creatinine clearance and therefore GFR increase slightly.

A decrease in red cell mass is one of the most consistent findings in astronauts immediately after flight, and at least two weeks are required for recovery. Because of the role of erythropoietin in erythrocyte production, this was one of the first hormones monitored during space flight. There was a correlation of low erythropoietin and low reticulocyte (newly formed erythrocytes) counts. A recently developed hypothesis states that the reduction in plasma volume during space travel is a key factor causing the drop in circulating red blood cells (rbc). Reticulocytes are large and undergo remodeling, including cell interaction. Under the conditions of reduced plasma volume and low erythropoietin values, there is a resultant catabolism of young rbc, and thereby a low rbc count, among astronauts during and after a flight (Alfrey et al. 1996). With return of plasma volume to control values, the rbc count usually recovers within two weeks.

Measurements of parathyroid hormone and 1,25-dihydroxyvitamin D^3 have varied with the duration of flight. However, in all cases calcium has been lost from the body. The role of the endocrine system in calcium loss during space flight has yet to be identified. There is a need for multiple measurements of hormone production and target cell hormone receptors during the multiple stresses associated with space travel.

Immunological System

Space flight data suggest that some factors involved in immunologic function may be altered in this unique environment. Consistent effects of weightlessness on the immune system include alterations in the relative populations of the types of white blood cells, including neutrophilia, lymphocytopenia, and eosinopenia. In addition, some functional impairment has been suggested by cell-mediated immunity tests and serum protein levels. The increased number of neutrophils, accompanied by decreases in lymphocyte and eosinophil numbers, suggests that microgravity or the multiple stresses associated with launch, entry, and landing contribute to these immune system alterations. Increases in epinephrine and glucocorticoids have been suggested as possible causes for the immune alterations (Tipton, Greenleaf, and Jackson 1996).

Some of the best evidence that gravity is an integral component of cellular function comes from examination of lymphocytes. *In vitro* studies of lymphocytes taken from astronauts on return from space flights indicate that there are changes in the cellular components. The capacity of these cells to produce antibodies is reduced. Schmitt et al. (1996) reported changes in protein kinase C (PKC). PKC is a ubiquitous enzyme that mediates intracellular signal transduction (serves as a messenger between cell membrane receptors and enzymes within cytosol and cell organelles). These authors reported the rate of PKC synthesis increased with exposure to microgravity and decreased in cells centrifuged at 1.4 g. These changes may be related to sensitivity of interleukin 1 Beta synthesis to gravity or responses of the cytoskeleton to gravity.

Despite the changes in the immunological system, astronauts and cosmonauts have remained in good health during and after space flights. On the other hand, the good health may be the result of preflight isolation and other precautions taken prior to, during, and after flights.

Space Sickness

The term, *space sickness,* appeared early in the space program to describe the nausea and dizziness many astronauts experience shortly after entering orbit. Fortunately, the discomfort disappears after a day or so in flight. Unfortunately, there is no good method to identify in advance who is susceptible to space sickness because it is a different phenomenon from sea sickness, car sickness, or any other motion sickness. Generally, if a space traveler does not experience space sickness in his or her first flight, that individual won't experience it in subsequent flights.

The two most prevalent theories on the cause of space motion sickness are the mismatch (sensory conflict) theory and the fluid-shift theory. The *mismatch theory* is based on our knowledge that the image of our body and motion perception is synthesized from input from multiple receptors (vision, touch, pressure, receptors in and around joints, muscle spindles, and otolith receptors) and multiple centers in the central nervous system, as illustrated in Figure 9-2. In our natural environment, this image is derived largely from the forces needed to provide movement. With the possible exception of chewing, these forces are determined by gravity. The forces required for self-motion in microgravity are greatly reduced, and this requires modification of the physical orientation and self-motion mechanism. There are considerable individual differences in the rate at which this modification occurs. Astronauts in flight learn to rely primarily on visual input and reduce reliance on other sensory input. Nausea frequently occurs before these adjustments are made. The sensory conflict theory has recently been supported by studies on fish during a space flight, as their swimming patterns and orientation are modified in microgravity (Mori et al. 1996). The *fluid-shift theory* suggests that the rate of fluid formation and reabsorption in the vestibular organ is modified by the cephalic shift in fluid which accompanies exposure to microgravity. The modification of the organ causes faulty signals to be sent to the central nervous system. This theory is weakened by the point that head-down bedrest causes a shift in body fluids but generally does not result in space sickness.

Figure 9-2. Diagram of the Interaction of Receptors and Neural Centers Involved in Monitoring Self-Orientation in the Physical Environment.

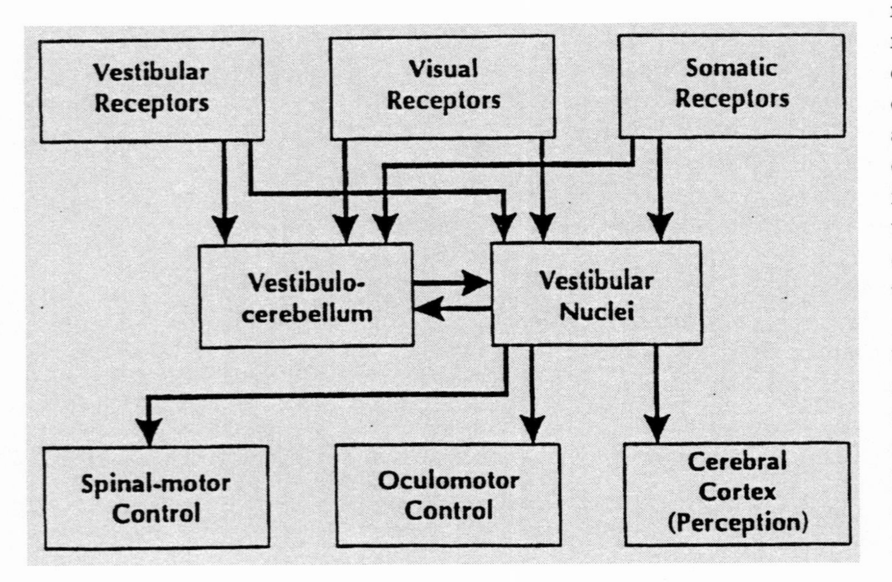

While in orbit, the traveler tends to rely heavily on visual input. Interestingly, reorganization occurs within a 1- or 2-day exposure to microgravity, as evidenced by recovery from space sickness and ability to perform manual tasks with dexterity. In contrast, distinct decrements in gait and jumping are still found several weeks after a flight. The duration of these decrements, in most cases, is related to the length of the flight.

Over 165 flights ranging from 15 min. to over a year involving several hundred astronauts have demonstrated the capacity of the physiological systems to adapt to microgravity. A symposium organized by C.M. Tipton and A. Hargen (1997) describes many of the adaptations and problems associated with space travel.

RADIATION HAZARDS OF SPACEFLIGHT

The severity of radiation exposure is dependent upon whether or not the flight is above, within, or below the Van Allen belts of trapped energetic particles (electrons and protons). These belts, plus the atmosphere, protect the Earth from high-energy particles and solar flares (mostly protons), which can be experienced at unpredictable times. The damage from radiation is dependent upon the type of radiation and the rate at which the radiation particle gives up energy to biological material (linear energy transfer).

The current level established for the general population is 0.1 roentgen equivalents man (rem) per year. An annual dose equivalent of 5.0 rem has been suggested as the upper limit for workers around nuclear reactors. A career limit for these workers may be 250 rem. Flights in low Earth orbit of 240 to 560 kilometers (150 to 350 miles) are below the Van Allen belts, and receive radiation doses similar to those on the Earth's surface.

Outside the Van Allen belts, however, solar flares increase the amount of radiation from a thousandfold to a millionfold. A three-year trip to Mars can be expected to encounter from one to three solar flares with a radiation dose of 100 to 225 rem. It is difficult to predict radiation exposure during such flights beyond the Van Allen belts because in outer space, in addition to solar flares, there are heavy particles (high atomic numbers) with very dense linear energy transfer, particularly iron particles. If one of these particles were to strike the sinoauricular node of the heart or respiratory center in the brain, it could result in a fatality. Accurate estimates of radiation exposure in these flights may not be available until the year 2000 or beyond (Moore 1992). Radiation damage during long-term space flights must be considered an occupational hazard.

NOISE

Noise is defined as an unwanted sound. Because of the great sensitivity of the hearing mechanism, noise pollution is an everyday experience. In the jet aircraft industry, noise is frequently a hazard for personnel; thus protective headgear is essential for pilots and ground crew. In space travel, noise reaches considerable levels during both take off and re-entry to earth. Once in orbit, however, noise within the space capsule drops to a background level comparable to that of a busy office.

The intensity of sound can be measured as dynes per square centimeter. The human threshold for sound is 0.001 dyne$\cdot$cm^{-2}; the sound of a gun firing is 100,000 dynes$\cdot$cm^{-2}. Because of the wide range of sound intensity, this scale is converted to the logarithmic decibel scale, where one decibel is equivalent to 0.002 dyne$\cdot$cm^{-2}. A comparison of common noise levels is presented in Table 9-1.

Auditory Responses to Noise

The auditory system responds to excessive sound by a diminution in the ability to detect weak sounds. If the decrease in sensitivity eventually disappears, it is termed *temporary threshold shift (TTS)*. If the condition does not ameliorate, it is termed a *noise-induced permanent threshold shift (NIPTS)*. Relationships have been identified between the type and level of noise and the resultant TTS or NIPTS (Ward 1973):

- Low-frequency noise will produce less TTS than noise at high frequencies.
- The loss of hearing from a noise of a narrow frequency will be one-half to one octave above the frequency of the noise.
- The TTS increases linearly with noise level: the greater the decibels of noise, the greater is the damage.
- A constant noise is much more likely to produce TTS than an intermittent noise.
- The duration of the noise needed to produce NIPTS varies with the frequency. Frequencies causing the most severe NIPTS are those between 400 Hz and 1,000 Hz, perhaps because these frequencies are transmitted most efficiently through the human ear.

Most recovery from severe auditory injury, such as that caused by an explosion, will occur within two weeks. Little, if anything, can be done to cure NIPTS that persists beyond two weeks; in these cases, the cochlea has been damaged. Common approaches to eliminating auditory damage include preventing exposure to loud noise by using sound barriers such as ear muffs and ear plugs, plus periodic audiograms of persons frequently exposed to noises in the 85 db and louder range (Nielsen 1995; Smith 1995).

Non-auditory Responses to Noise

Noise influences not only the function of the auditory system but the nervous and endocrine systems as well, and thereby potentially influences the function of every organ and tissue in the body. Experiments on laboratory animals and human subjects describe responses

Table 9-1. Some Common Noise Levels.

Noise	Decibels
Near jet plane (pain threshold)	140
Pneumatic riveter	125+
Rock & roll band (peak)	120
Jet aircraft (500 ft overhead)	115
Motorcycle	110
Construction air hammer (10 ft)	110
Subway train (20 ft)	95
Train whistle (500 ft)	90
Heavy truck (25 ft)	90
Average office (multiple occupancy)	50
Average house	40+
Quiet house (at midnight)	30

*(from Lipscomb 1974)

ranging from sudden death in animals to subtle changes in the ability to focus attention. Variation in the mood and concentration of the individual can influence responses to background sound or noise. Our responses to expected and unexpected noise cover a wide range, from indifference to the roar of a jet plane as we approach an airport, to momentary fright at the unexpected slam of a door while we are concentrating, or reading a book.

Multiple neuroendocrine responses to noise have been described (Aruelles et al. 1962). The most frequently cited non-auditory effect of noise is adrenal cortical activation via the hypothalamic-pituitary axis. Rats can have a three-fold increase in corticosterone in response to a loud noise, which can be prevented by destruction of the auditory apparatus. Changes in thyroid gland activity associated with noise involve decreased release of thyroid stimulating hormone (Brown-Grant et al. 1954). Noise can readily activate the adrenal medulla, producing an elevation of the catecholamines epinephrine and norepinephrine in both rats and human subjects. Both oxytocin and vasopressin are released from the neurohypophysis by noise (Jackett 1970; Ogle and Lockett 1966).

Cardiovascular alterations associated with noise range from momentary changes in blood pressure to persistent hypertension. Fluctuations in blood flow to various organs, including the inner ear, have also been reported. Young mice and hamsters are particularly susceptible to noise (Henry and Saleh 1973). In one strain of mice, 16- to 19-day-old animals exposed to a single loud noise of 95 db die from respiratory paralysis or convulsions when exposed to the same noise days or months later. These data indicate why there is concern about the potential effect of noise upon fetuses during pregnancy, and upon the performance of school children when exposed to high noise levels. Good public and personal health practices certainly include the control of noise pollution.

VIBRATION

Just as noise can be defined as unwanted sound, *vibration* can be defined as unwanted movement. Vibration has always been considered a nuisance feature of the transportation industry. Jet plane pilots experience vibration often, as when changing direction at high speed, while astronauts' exposure to vibration may be limited to take off and landing maneuvers.

The effects of vibration are diverse. Severe or long-term vibration can result in trauma, particularly to supporting connective tissue, whereas minor vibration may simply have a sedative or somnolent effect resulting from the monotonous input into the central nervous system. The normal function of structures concerned with movement, including skeletal muscle spindles, respiratory and cardiovascular systems, vestibular apparatus and vision, can all be disturbed by vibration.

Increasing isometric tension reduces the extent of vibration-induced movement in skeletal muscles and can result in increased oxygen consumption. The motor response to sustained vibration of a human limb muscle involves a reflex. The term *tonic vibration reflex* (TAR) has been used to describe a classic segmental stretch reflex initiated from primary muscle spindle endings and resulting in excitation of homologous motor neurons and inhibition of antagonistic alpha motor neurons (Homma et al. 1981). The TAR can be elicited in flexors and extensors, as well as in muscles that elevate or open the jaw.

Receptors in the skin and stretch receptors in the lung have also been reported to be involved in responses to vibration. Healthy subjects experience increases in both tidal volume and frequency of breathing in response to vibration (Hoover and Ashe 1962). In 1957, Balke and colleagues first reported hyperventilation in jet pilots caused by the vibrations accompanying high-speed aircraft. The changes in respiratory patterns vary with the vibration amplitude, frequency, and direction. In contrast, application of vibrators to the chest has been useful in improving the breathing patterns of patients with respiratory insufficiency (Homma et al. 1981).

The most common health problem associated with vibration involves the cardiovascular system. There is a high incidence of vascular diseases among people who operate vibrating tools (chainsaws, jack hammers, etc.) over extended periods of time. The etiology of these diseases is not evident. Modification of neural reflexes, inherent rhythmic contractions of smooth muscles, and mechanical damage to blood vessels are possible causes of various pathological conditions (Hyvarinen, Pyykko, and Sundberg 1993). A pneumatic chisel held in the right hand for a two minute exposure resulted in a bilateral decreased blood flow to fingers and toes (Egan et al. 1996). These exposures to frequencies of 0.4 to 4,000 Hz also increased heart rate. The fatigue, increased heart rate, and increased oxygen uptake associated with whole–body vibration may result from the increased widespread isometric contraction required to maintain posture. Whole–body or limb vibration potentially activates numerous receptors and reflexes.

Bone tissue is responsive to mechanical stress. The mass and geometry of bone changes in response to increased mechanical forces. This response is primarily a result of increased activity of osteoblasts. Thus, it is not too surprising that osteoblasts in tissue culture exposed to simulated launch vibrations had increased mRNA of two growth-related proto-oncogenes (Tjandrawinata, Vincent, and Hughes-Fulford 1997). It is expected that no other body tissues are as sensitive to vibration at the cellular level as are the osteoblasts of bone tissue. The endothelial cells of blood vessels that encounter mechanical stress generated by blood flow may be the exception to this statement.

POLLUTION

It has often been said that *dilution* is the solution to pollution. Unfortunately, within the confines of a space capsule, dilution is not possible, and toxic materials must be removed from the environment as rapidly as they accumulate. Viewing Earth from space has emphasized our realization that Earth is an island unto itself. Thus, even on Earth, the extent to which we can solve pollution problems by dilution is limited.

The nature of pollutants ranges from single atoms of mercury and nitrogen to complex inorganic and organic mixtures of ions and hydrocarbons. The toxic effects can range from neurological damage to minor eye irritations. It is beyond the scope of this text to consider the multiplicities of problems presented by pollution, although we will consider the subject again in our last chapter. For further enlightenment, the reader should consider textbooks on toxicology, earth sciences (including geography and geology), and ecology. Each pollution problem—whether it involves a few individuals in an isolated space capsule or many people in a city, state, country, or continent—is unique in its source, dispersal, damaged target tissue, and social consequences.

In Chapter 10 we return to a world more familiar, yet one with its own environmental challeges, especially for humans.

10. RESPONSES TO HIGH PRESSURE

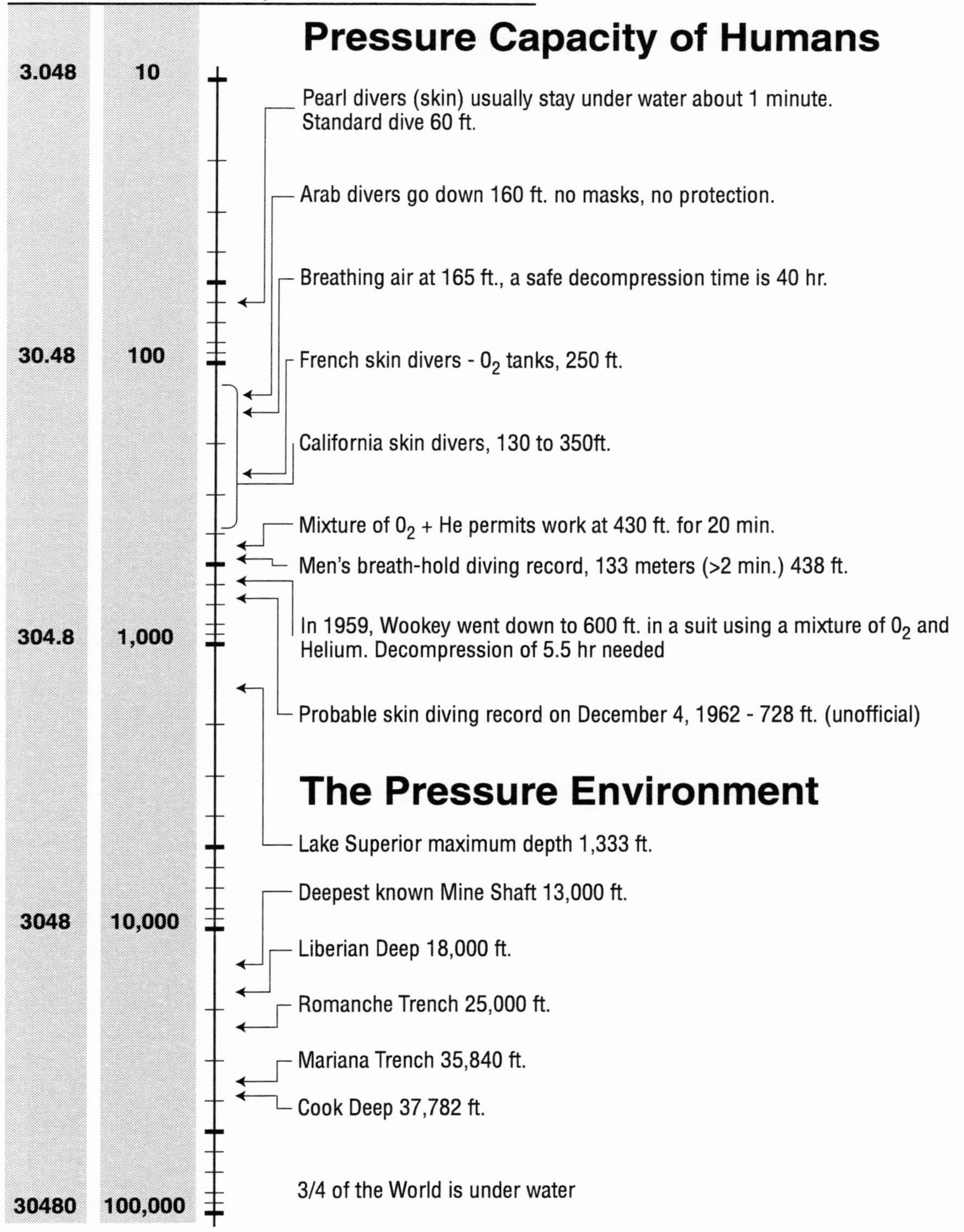

Environmental Diagram: Chapter 10

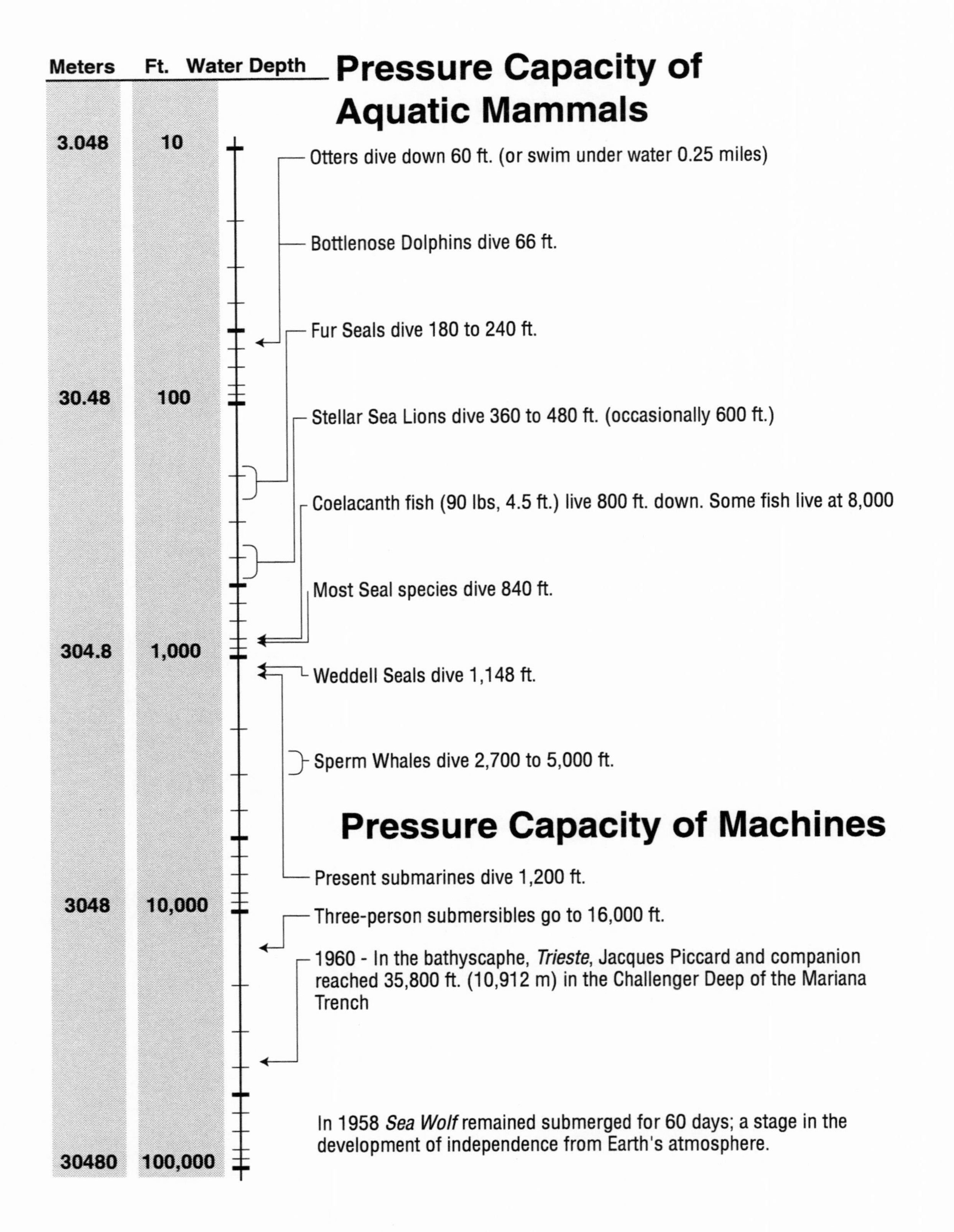

Environmental Diagram: Chapter 10

10. Responses to High Pressure

10. RESPONSES TO HIGH PRESSURE

INTRODUCTION

Mammals experience high pressure when they dive underwater. Humans may also experience it without diving as they go to the bottom of deep mines and into high-pressure caissons to tunnel under large bodies of water. In some hospitals, surgery is performed in chambers maintained at 3 to 6 atmospheres of pressure. In all of these cases, four physiological factors occur that are detrimental to health: 1) the buildup of carbon dioxide; 2) the requirement for atmospheric oxygen; 3) the large pressures, especially on the thorax of the mammal; and 4) at the end of exposure, the effects of bubbles of nitrogen and air in the blood. These four factors may cause discomfort, and even death after an extended exposure. In one recent year in the United States, 93 male scuba divers and 8 female scuba divers died underwater.

The pressure environment that can be tolerated depends upon both the mammal and the origin of the pressure: 1) pearl and sponge and Ama divers hold their breath and dive frequently to 60 feet, thus tolerating 2 atmospheres; 2) construction workers tolerate pressures in caissons of 3 to 4 atmospheres (99 to 132 feet); 3) mice can tolerate pressures in chambers of 166 atmospheres (5,478 feet); and 4) one species of whale (the sperm whale) can dive to a depth of 5,000 feet (152 atmospheres). Once we realize the gigantic capacity some animals have for tolerating high pressure, many obvious questions arise. How common are such pressures, where are they found, and what animal life can survive there? Do mammals that tolerate high pressure of water also tolerate *hypoxia* (severe oxygen deficiency)? Is there acclimatization to pressure and to hypoxia? How do humans fit into this picture? What are the details of physiological Darwinian adaptation to high pressure? We'll consider these questions in this chapter, beginning with a look at appropriate physiological mechanisms.

PHYSIOLOGY OF HUMAN DIVERS

Many individuals still make their living by diving without any mechanical equipment. These groups include the sponge divers in the Arabian area, the pearl divers in the Australian area, and the Japanese and Korean Ama (Rahn 1987; Park et al. 1993). An early record with held breath was set by the Greek sponge diver Stotti Georghios in 1913, when he swam down 200 feet to put a line on a lost anchor. By 1990, the men's record was held by Ferrera, who dove to 112 m (390 ft), while Bandini set the women's record at 107 m (350 ft) (Gamba 1990). More recently, Ferrera reached 133 meters while holding his breath for over two minutes (Kooyman and Ponganis 1997).

Although humans do not remain submerged as long as the other diving mammals (except for polar bears), they do have some of the same adaptations found in lower mammals. Scholander et al. (1962) reported a *bradycardia* (a slowing of the heart beat) during the dives of pearl divers; he also showed that systolic blood pressures remained at or near normal levels, and that there was no release of lactic acid into the blood until after the dive. It was then argued that this delày is also found after surface swimming and sprinting on land (Peterson 1963). The advantage to the animal of bradycardia will be discussed in a later section. We must now consider in detail the major mechanisms which must be adapted for the high pressure of diving. They are all respiratory mechanisms.

Respiratory Physiology

It will be helpful to review some baseline values of respiratory physiology. The atmosphere contains oxygen, carbon dioxide, nitrogen, water vapor, and inert gases, examples of which are helium, neon, and argon (1%). The inert gases are customarily lumped with another abundant inert gas, nitrogen, and together they make up four-fifths of the atmosphere. CO_2 is scarcely measurable in air. The proportion of the three gases is extremely constant and remains the same from sea level to the highest mountains: oxygen 20.95%, carbon dioxide 0.03%, and nitrogen 79.02%.

The pressure of gases is expressed in millimeters of mercury (mm Hg). In a mixture of gases, such as air, the partial pressure of each component depends upon the proportionate number of molecules present. For example, if the normal atmospheric pressure is 760 mm Hg, the partial pressure of oxygen in the alveoli or in oxygenated blood leaving the lungs is 103 mm Hg. Expressed in volumes percent (see chapter on Altitude) it is nearly 20 vols %. Let us now put this vocabulary to use in considering the first respiratory problem to face the diver: the disposal of carbon dioxide.

The Problem of Carbon Dioxide Disposal. The most important single factor in human respiratory regulation is the sensitivity of the respiratory centers to the amount, or *tension,* of carbon dioxide in the blood. If this tension increases only slightly, breathing immediately becomes deeper and faster, permitting more carbon dioxide to leave the blood until the tension has returned to normal. The regulation of CO_2 tension is so exact that the concentration in the lung always remains relatively constant. On the other hand, the respiratory center is at first insensitive to the oxygen concentration in the blood. If there is a serious decrease in oxygen concentration (hypoxia), then the chemoreceptors in the aorta and carotid arteries send impulses to the respiratory center causing an increase in respiration.

To illustrate the influence of carbon dioxide on respiration, we may turn to some of the stories of submarine crews in World War II. At times, the vessel's carbon dioxide "scrubbers" were faulty or became saturated when the submarine was submerged for exceptionally long periods. It was not uncommon to hear former crew members describe how the CO_2 meters, which read to only 4%, were pressed hard off scale (Ebersole 1960; Tonndorf 1974); they described the violent and deep breathing that all members experienced. These men were responding to *hypercapnia*, the presence of excessive amounts of carbon dioxide in the blood. This emphasizes that the regulation of respiration depends primarily upon the accumulation of carbon dioxide and not upon lack of oxygen.

Under such circumstances, including those in caissons and mines, when the carbon dioxide increases to 3% (from 0.03%), the tidal air will increase from approximately 500 to 2,000 ml. The diver will tolerate a higher buildup than this—up to breathing 10% carbon dioxide (Warkander et al. 1990). In compensation, the diver's respiratory volume will increase up to a maximum of about 10-fold (50 to 60 liters per minute). Beyond the 10% level the situation becomes intolerable, and the cellular metabolism of the respiratory center becomes depressed rather than excited; the diver develops lethargy and narcosis, and finally becomes unconscious (Lanphier 1987; Lin 1987b). McDonald and Simonson (1953) tested subjects breathing 30% carbon dioxide; these high levels showed the narcotic effect, and the subjects lost consciousness after 20 or 30 seconds. It is possible that white rats are more tolerant; Giaja states that carbon dioxide had to be raised to 50% before this atmosphere was fatal to these rodents (Giaja and Markovic 1953). He did not report when they showed deep narcosis.

Why does a person take a series of deep breaths before swimming underwater? The increased ventilation of the lungs does not increase the oxygen content of the arterial blood, but it does remove more than the usual amount of carbon dioxide from the blood (Muller 1995). Thus, it takes longer to build up carbon dioxide to the point where the respiratory center overrides the voluntary inhibition of breathing. Most professional divers can stay underwater with held breath for only about 1 minute, although a few train themselves to remain there for 2.5 minutes. If human beings can hold their breath for only that long, then why are there so many mammals that can remain underwater for 15 minutes? In the latter part of this chapter, we review some of the adaptations for long periods of submersion.

Blood gases have been measured in humans during deep breath-hold dives. Analyses revealed that the limiting factor is neither hypoxia nor hypercapnia; the problem seems to be that blood is forced into the thorax under very high pressures (Schaefer et al. 1968, Lin 1990).

When humans swim underwater, there are three ways of handling the problem of CO_2: retain it by breath-holding; use the dangerous closed-circuit oxygen lung which absorbs CO_2; use the open-circuit system which releases CO_2 as bubbles.

The Supply of Oxygen. The pressure of the water around a diver has two important effects: it conditions the air delivered to the diver, and it compresses body air sacks (of all types), including lungs.

To overcome the influence of water pressure, a diver has approximately four choices of equipment: 1) use an open-circuit system which automatically delivers air at the same pressure as the surrounding water and expels CO_2 into the water; 2) breathe from an oxygen-supplying, closed-circuit apparatus—dangerous because it may produce oxygen poisoning; 3) use an ordinary diving suit, thus allowing water pressure to be exerted against all parts of the body including the torso; or 4) use a diving suit that is filled inside with compressed air.

The first diving suit was designed by Leonardo da Vinci (1452-1519), and included webbed gloves and foot flippers. The suit was used in 5 feet of sea water for pearl diving. Today, free divers use equipment called SCUBA, for "self-contained underwater breathing apparatus" (Butler and Thalmann 1986; Warkander et al. 1992).

Pressure Problems. As the diver goes underwater, the pressure will increase by 1 atmosphere for each 33 feet (10 meters) of submersion; this means that the weight of the atmosphere at sea level is essentially the same as a column of water approximately 33 feet high. If a diver then goes to 33 feet underwater, the pressure in her alveoli is 2 atmospheres instead of one; if she goes to 330 feet, then the pressure is 11 atmospheres. These pressures determine the volumes of air that must be pumped to a person in a diving suit. The total quantity of air pumped to the diver must increase in direct proportion to the depth descended. For example, if 1 cubic foot of air is pumped to a diver who is under a pressure of 4 atmospheres, this 1 cubic foot of air will be only $^1/_4$ cubic foot in the diver's helmet. Therefore, a fixed amount of extra air must flow into the helmet each minute in order to wash out the carbon dioxide from the helmet.

On rapid descent, damage can occur if there is an unequal difference of pressure where air is trapped in cavities such as the lungs, the nasal sinuses, and the middle ear. If a person descends without addition of gas to these cavities, a painful reaction called the *squeeze* occurs. The smallest volume the lung can normally achieve is 1.5 liters; when a diver with held breath inspires a maximum breath and descends to 100 feet, his chest will begin to cave in. Severe facial damage can occur if air volume is lost from a helmet or the mask of a free diving apparatus. The first 33 feet of descent causes the greatest squeeze because this is a 2-fold pressure increase; if a person descends from 300 to 333 feet the pressure causes only a 10% reduction in volume.

Rapid ascent if improperly performed can also cause severe damage. The opposite of the squeeze occurs if a diver fails to expel air from the lungs on the way up. Above an alveolar pressure of 80 to 100 mm Hg, air is forced into capillaries. The result is air embolism, not to be confused with the *bends*.

Diver in scuba gear approaching coral reef. Photo courtesy of National Undersea Research Center, University of North Carolina at Wilmington.

Cardiovascular Effects

In the above section there has been no discussion of effects on the cardiovascular system, though many have been described (Hong 1987). This will be analyzed later in this chapter under the section, Comparative Physiology.

Oxygen Poisoning and High Pressures

There are harmful effects of high partial pressures of oxygen (Seals et al. 1991). One of the symptoms is confusion, which is similar to grand mal seizures; this condition ceases as soon as excess oxygen is removed (Bitterman et al. 1987; Butler and Thalmann 1986). If the sea level pressure (100 mm Hg) is increased to 1400 mm Hg, the amount of dissolved oxygen in the water of the blood (now 6.5 vols % instead of 0.2 vols %), plus a little from the hemoglobin, is used by the tissues. This is the situation in about 30 feet of water if a diver wears an oxygen lung delivering 100% oxygen. If the diver goes deeper, the pressure is raised even higher, no oxygen is used from the hemoglobin, the tissues remain saturated, and oxygen convulsions may result (Harabin et al. 1987). Before convulsions occur there can be pulmonary damage (Clark et al. 1991; Eckenhof et al. 1987). To prevent these problems, the rule of thumb for safety is that one should not increase the concentration of oxygen in the alveoli above approximately 7 times normal (Fig. 10-1).

Figure 10-1. Onset of Oxygen Poisoning. Every individual is aware of the absolute necessity for rhythmic inhalation of oxygen many times each minute. Only in the second half of this century has it been realized that this essential ingredient for life can act as a poison as well as a benefit. The safe duration of submersion before pure oxygen acts like a poison is indicated in the diagram. Modified from Duffner (1958).

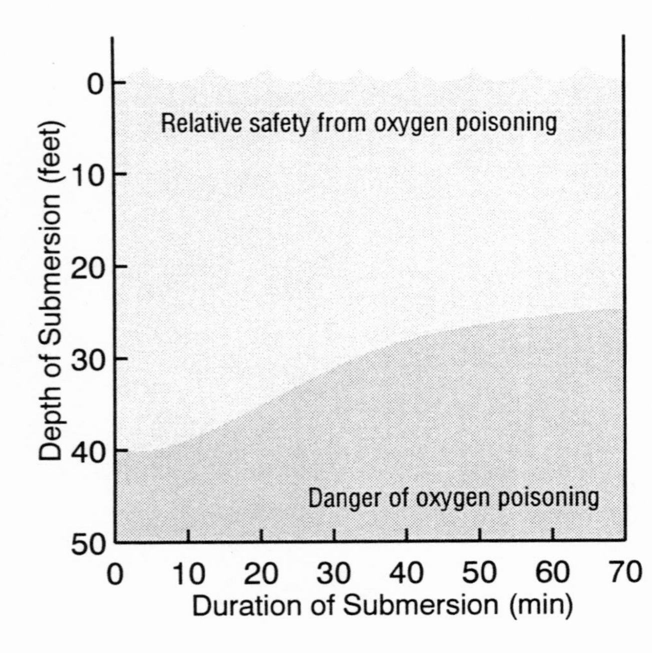

One cause of the convulsions is the large amount of oxyhemoglobin leaving the brain. There is direct tissue damage due to the excess oxygen; also because of the Bohr effect, oxyhemoglobin has less capacity for carbon dioxide transport, and the diver suffers from excess tissue carbon dioxide (Guyton and Hall 1996; Kenmure et al. 1972). Experiments on cats show that another focus of damage is the carotid body (Lahiri et al. 1987; Torbati 1989). Hyperbaric oxygen is toxic even to fish.

The normal metabolism of oxygen throughout the body produces a normal, but toxic, load of free radicals. This load is greatly exaggerated by higher percentages or pressures of oxygen respired in any hyperbaric oxygen situation (Jamieson 1986). Hyperbaric oxygen overwhelms the body's defenses against free radicals; the most important defense is the production of *superoxide dismutase* which mops up the free radicals (Lundgren and Norfleet 1987; Harabin 1990).

If a diver is breathing only compressed air (21% oxygen), he or she can descend to about 200 feet and still be within the safe limit of 7 times normal oxygen in the alveoli. Our diver can descend farther if special mixtures of helium and oxygen are used, as long as toxic amounts of helium are not absorbed.

The imminent physiological damage from high pressure cannot be overemphasized. The highest pressure that has been experienced is 68.7 atmospheres absolute pressure (ATA), equivalent to 686 meters. Divers have performed work while in an environment with a pressure of 66 ATA and a gas density 17 times that of air at sea level. (Lundgren and Norfleet 1987). In recent years, hundreds of new hyperbaric chambers have been added to hospitals for surgical and clinical applications (see later section.) A typical exposure of a patient is at 6.0 ATA while breathing air (Henderson 1987).

The physiological considerations are centered on ventilation and gas transport (Hong 1989). A diver may have to perform work requiring over 2 L/min of oxygen consumption and 60 L/min of ventilation. The accompanying problems are presented in Table 10-1; any of the five environmental parameters on the left may affect any of the six gas exchange functions on the right.

There are numerous physiological effects which result from either (1) exposures in hyperbaric chambers, or (2) free diving (breath-hold or apparatus-assisted) (Claybaugh 1994). One of the most conspicuous is a reduction in expected heart rate when resting and when exercising (bradycardia). This phenomenon will be discussed in detail in a later section of this chapter.

Influence of Nitrogen and Helium

The nitrogen gas from the air is as important to the diver as oxygen and carbon dioxide. If the pressure of nitrogen reaches 7 times atmosphere, it may have an anesthetic function on the central nervous system ganglia, and coma can result. The early stages are called nitrogen narcosis or "rapture of the deep." This state is so much like alcohol intoxication that one diver took off his mouthpiece and offered it to a passing fish. Each diver has his or her own nitrogen narcosis limit; for some it lies between 330 and 350 feet (Bennett 1993a; Fowler et al. 1985). There is a second type of damage from nitrogen: as it bubbles out of the blood during decompression, it causes the bends (see next section).

The breathing of helium is an experimental solution to the damage done by nitrogen at high atmospheric pressures. Experiments with helium have been accelerated because, during both high altitude and space flights, the formation of bubbles by nitrogen escaping during decompression is also a problem. Helium has a small molecule and diffuses through tissue 2.5 times faster than nitrogen;

Table 10-1. Hyperbaric Effects on Gas Exchange

Environmental Parameters:	Gas Exchange Function:
1. Immersion	a. Pulmonary Ventilation
2. Gas Density	b. Intrapulmonary gas diffusion
3. Toxic gas effects	c. Ventilation-Perfusion Matching
4. Pressure per se	d. Pulmonary and Systemic Circulation
5. Compression	e. Tissue Gas Exchange
	f. Gas Transport by Blood

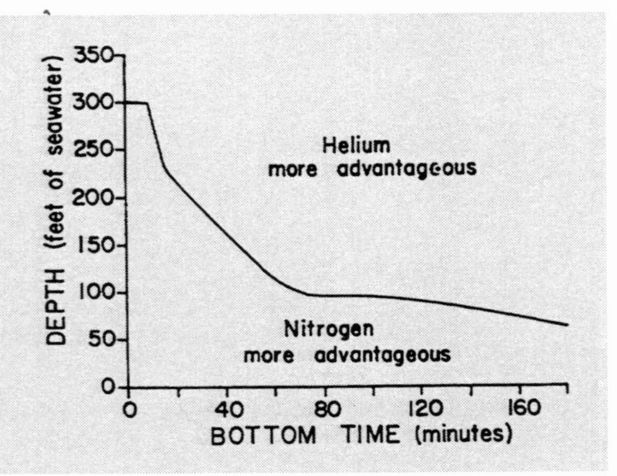

Figure 10-2. Breathing Helium vs. Nitrogen. The depths and times beneath water using SCUBA gear, showing the circumstances when it is more advantageous to use helium compared with using nitrogen. Adapted from Guyton and Hall 1996.

helium is also one-third as soluble in fluids as is nitrogen and has few noxious effects on the body. Because of its diffusability during decompression, helium leaves the body fluids rapidly and easily (Lambertson and Greenbaum 1963; Guyton and Hall 1996).

Some very interesting experiments with helium have been done; on several occasions volunteer subjects have lived in space-cabin simulators for as long as 5 weeks in atmospheric mixtures of 56% helium and 44% oxygen at a simulated altitude of 18,000 feet. Under these circumstances, the major influence on the subjects seemed to be a change in their speech (the Donald Duck effect). Different results were obtained when they breathed 85% helium and 15% oxygen by mask and at a pressure of 15 atmospheres or comparable to 642 feet of sea water. Although the subjects seemed to function and behave quite normally while respiring these mixtures, their performance and behavior deteriorated within 20 to 60 seconds following a return to breathing air.

The relative advantages of helium over nitrogen are illustrated in Fig. 10-2. The breathing of helium will permit extremely deep chamber dives for long periods. An early world record simulated dive was 15,000 feet (4,500 meters). Two volunteers tolerated these conditions for 10 hours. Another group of four divers at the University of Pennsylvania stayed at a simulated depth of 12,000 feet in an atmosphere that had a gas density equal to that of breathing helium at 5,000 feet. This 24-day experiment gave information on the human tolerance to helium, neon, nitrogen, and pressure. It is in these experiments that neon proved to be a respirable gas for humans at high pressures. The pioneer work was done by Lambertson 1971, Morrison and Florio 1971, and Woods and Lythgoe 1971. The two factors that must be kept in mind are the kinds of gases respired, and the effect of high pressure itself upon the central nervous system.

More recently, hydrogen has been the inert gas used in such experiments (Fife 1987; Lemaire 1987), which are called "dry saturation dives" (Claybaugh 1987). Some experiments have been done with xenon (Novotny 1990).

When subjects live at high pressure in an environment containing these rare gases, the kidney and heart suffer specific physiological effects: an increase in daily urine flow (hyperbaric diuresis), caused mostly by suppression of insensible water loss; a lowering of the plasma vasopressin level (Hong and Claybaugh 1989; Miyamoto 1991); and a pronounced lowering of heart rate (hyperbaric bradycardia). An average heart rate reduction is 37% with a peak of 50% (Lin 1996). There are also some orthostatic disturbances (staggering) (Lin et al. 1991). These studies were begun with mice under high pressure (reviewed by Gardette 1989); later studies were on human deep hydrogen dives (Fructus 1987; Gardette 1987).

Decompression Sickness (Dysbarism)

As a diver decompresses, or as the aviator ascends to 30,000 feet, the dissolved nitrogen or air in the blood begins to escape, causing various degrees of damage (Butler and Katz 1988; Brubakk 1989). Small bubbles of this gas may cause pain or blocked nerves, and large bubbles may cause air embolism, pulmonary edema, or brain and spinal cord damage (Francis et al. 1990). Breathing a mixture too high in O_2 can enhance the air embolism form of decompression sickness (Weathersby 1987).

Historically, dysbarism was known as *bends, caisson disease, compressed air disease,* and *divers' paralysis.* Now, we usually refer to the pain in the arms, legs, and head as the *bends;* if the bubbles block capillaries in the lungs, they cause a shortness of breath called the *chokes.* An early monograph from Brooks Air Force Base includes seven symptoms, of which the bends and the chokes are the two most conspicuous (Adler 1964, Lundgren 1989). Another symptom is self-descriptive: *blackout* (Lamphier 1987).

The first experiments involving the formation of bubbles by a change from high pressure to low pressure were done by Robert Boyle in 1670. Boyle exposed a snake to a very low ambient pressure and reported seeing a bubble within the eye. He had picked a very convenient animal for this experiment because snakes do not have eyelids. By 1850, the study of dysbarism became of practical importance to a French engineer, Triger, who at that time built the first caisson for tunnel work. Many of these caissons were used during the last half of the 19th century, resulting in a high death rate and extreme discomfort for the workers on the tunneling projects. In the 1930's, Capt. A. R. Behnke (1955) became a pioneer in understanding and treating decompression sickness.

The principles that control the formation of bubbles caused by a pressure change are as follows: the total nitrogen normally dissolved in all body fluids and in the body fat at one time is approximately 1 liter; if the person is under 2 atmospheres of pressure there will be 2 liters of nitrogen, if under 5 atmospheres then 5 liters of nitrogen. The rate of nitrogen elimination is expressed in the rule that approximately one-half of the excess nitrogen is removed in the first 40 minutes of decompression and half of that remaining is removed in the next 40 minutes. There is 5 times as much nitrogen stored in body fat as in body water, and nitrogen leaves fat tissues very slowly (Van Liew 1991). This provides an obvious advantage to lean divers. The rate of nitrogen liberation from the whole body, from the water of the blood, and from fat is defined in Fig. 10-3.

If the pressure on a diver is 5,000 mm Hg, and the pressure of carbon dioxide and oxygen is in the normal range, then that of nitrogen will be 3,918 mm Hg. If the diver has been at 2 atmospheres, there are 2 liters of nitrogen to be handled; one remains dissolved in the body fluid, while the other forms bubbles. A certain number of these bubbles are tolerated, so that discomfort does not begin unless the total volume of nitrogen available for forming bubbles is more than 1.25 liters (Elliott and Moon 1993).

The rate at which a diver can be brought to the surface depends on two factors: the depth of the dive, and the amount of time spent there. If the person remains at deep levels for only a short period, the body fluids will not become saturated and decompression time can be reduced. Flyers, because of the slow formation of bubbles, can avoid the bends if they do not spend too long over 25,000 feet.

Figure 10-3. Nitrogen Liberation From Body Water and Fat. The rate of nitrogen liberation from the whole body, body water, and fat, when a diver returns to sea level from exposure to compressed air at 33 feet depth. Note that nitrogen leaves body fat very slowly. Adapted from Guyton and Hall 1996.

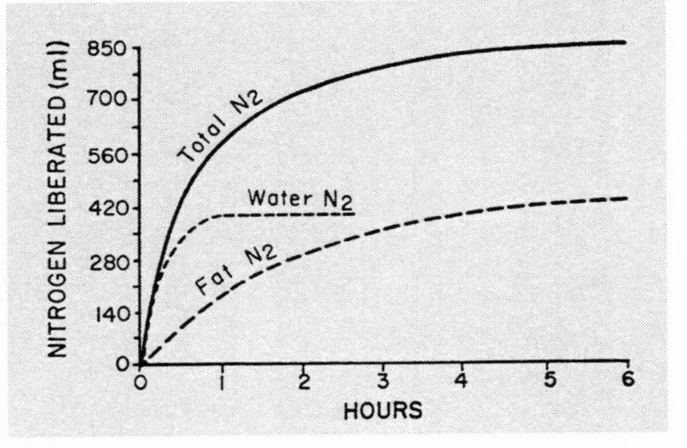

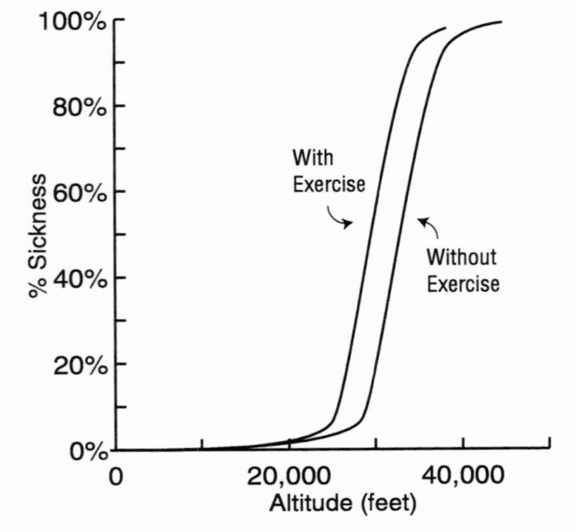

Figure 10-4. Effect of Exercise on the Bends. The range of resistance and susceptibility to the bends is indicated with and without exercise. One might predict that with exercise at the higher altitudes, a much higher percent would become sick compared to the other group. Although this diagram represents the bends at altitude, the principle of the effect of exercise on the bends applies underwater. Adapted from Guyton 1996.

Figure 10-5. Requirements for Decompression After Diving. This diagram emphasizes that a diver must choose between going deep for a period of 25 minutes or remaining near the surface for a much longer period. Adapted from Duffner 1958.

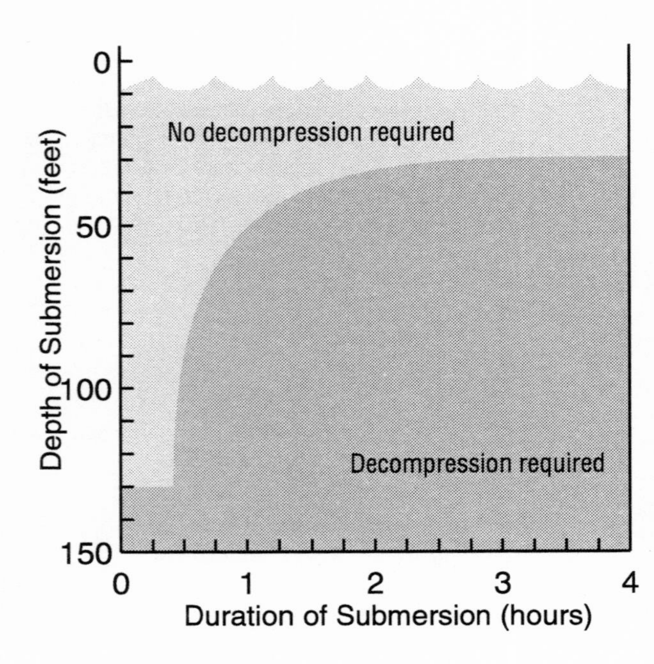

Exercise hastens the formation of bubbles during decompression because of increased motion of tissues (Fig. 10-4).

As examples of decompression time, note that only 20 minutes at a depth of 300 feet requires over 2.5 hours decompression; and 45 minutes at 300 feet requires over 5 hours (Weathersby 1992). In Fig. 10-5 we see that submerging to 30 feet can be tolerated indefinitely without experiencing decompression sickness. However, when one exceeds a depth of 60 feet, the length of time necessary to produce the bends decreases sharply. One can stay at 100 feet only about 30 minutes without absorbing so much nitrogen that slow decompression is essential. The onset of the bends is somewhat unpredictable. One Japanese diver went down to 192 feet 4 times consecutively for 1 hour each without getting decompression sickness, but the fifth time he had a very severe case of convulsions. This symptom is referred to as *high pressure nervous syndrome* (Bennett 1993b). This is probably related to the speed of coming to the surface.

At high altitude the bends usually begin around 25,000 feet; this means that people wanting to ascend mountains similar to Mount Everest could experience the symptoms if they are rapidly transported to altitude.

Is there acclimatization to repeated diving exposures? Asking this question in respect to all hostile environments is a theme of this textbook. Acclimatization to CO_2 does take place during prolonged training, allowing some divers to stay underwater longer. Schaefer discusses some of the unfavorable consequences of this form of acclimatization (1963). Several studies on decompression stress or decompression sickness have shown an improvement after repeated exposures (Hills 1969; Eckenhof and Hughes 1984; Thalmann et al. 1984).

Living Underwater

Naval teams of several nationalities have lived in submerged chambers at 200 feet or more for weeks at a time, where it was learned that divers could move from 600 to 300 feet without decompressing. By 1972 there were 40 undersea habitats in use around the world. Only one of them was lowered into a sub-arctic environment: this particular one was located in 33 feet of water off St. John, New Brunswick where the water is less than 5°C for at least half of the year. As early as 1978, another team lived for weeks under 1600 feet of sea water. These heroic experiments were not without physiological costs, such as debilitation after the exposure (Lin 1987a). Pioneer programs such as the extensive *SeaLab* series contributed much useful physiological data (Hock, Bond, and Mazzone 1966). Such programs continue today (see sidebar).

One reason for occupying this ecological niche is to exploit the food and water resources of the continental shelf. The submerged continents could provide 2 to 3 times the minerals and other resources that have been obtained from dry land.

Aquanauts range out from their shelters in 2-person submarines, or conduct deep-sea research from a bathyscaph, a self-contained vessel capable of reaching depths of 10 kilometers (6.2 miles) or more.

Escape from undersea habitats and submarines poses a special problem: escape is possible from 300 feet without use of special apparatus, but as the person ascends, he or she must consciously exhale or else expanding gases will rupture a vessel in the lungs.

Hospital Applications of Hyperbaric Oxygenation

The applications of hyperbaric oxygen were reviewed by Jennings and Murphy in 1966. The use of a high-pressure oxygen chamber for hospital surgery received its start in 1964, when Boerema demonstrated that he could sustain life in a bloodless pig perfused with a saline solution while the animal breathed oxygen at 3 atmospheres. This use of hyperbaric conditions was hampered at first by lack of information about oxygen free radicals, the agents responsible for oxygen toxicity (Henderson 1987), as discussed earlier in this chapter. It is now believed that the same oxidizing free radicals responsible for oxygen toxicity are also responsible for the therapeutic benefits of hyperbarism (Guyton and Hall 1996).

As interest in this adjunct to surgery grew, technical advances of modern engineering provided a variety of pressure chambers with almost unlimited degrees of sophistication. But along with the advantages of high-pressure came the dangers: like divers, surgical staff and patients face the risk of oxygen toxicity, decompression sickness, and nitrogen narcosis. Pittinger (1966) assembled information concerning attempts to overcome these and other handicaps; for example, some hospitals used 10% oxygen in order to decrease the fire hazard. Patients have been subjected continually for 12 days to a pressure exceeding 6 atmospheres; this pressure permits the oxygen concentration to be lowered to 3.5% (Guyton and Hall 1996).

The problems associated with high pressure can be minimized if the operations are short. The model for this procedure is the landmark in cardiac surgery accomplished by Henry Swan; his work depended upon using hypothermia instead of hyperbaric conditions. Nevertheless, the figures are comparable in that speed is required in both cases. Using direct vision intracardiac manipulations, Swan successfully corrected 35 atrial septal defects, accomplishing the operations in about 8 minutes each. Similar operations are being performed with hyperbaric oxygen.

In the 1990's, hyperbaric chambers are used in a variety of treatments. At The University of Iowa's Hospitals and Clinics, for example, these uses include treating wounds that won't heal, recovery from carbon monoxide poisoning, overcoming the "bends," and retarding growth of bacteria in certain rapidly advancing tissue infections. For CO poisoning, patients breathe oxygen at 3 atmospheres; during each two hour session, patients breathe oxygen for 20 minutes and air for five minutes to avoid oxygen toxicity.

THE AGE OF *AQUARIUS*

After retiring *SeaLab* and *HydroLab*, the National Oceanic and Atmospheric Administration (NOAA) deployed its most advanced underwater habitat, *Aquarius*, in 1993. Bolted to a 200-ton platform sixty-seven feet below sea level at Conch Reef off Key Largo, Florida, *Aquarius* offers scientists a unique opportunity. Here, they can work underwater up to nine hours a day without fear of the bends; this worktime would be reduced to one hour per day if work was conducted from the surface. Scientists can stay in this 81-ton, 43x20x16-foot pressurized, cylindrical chamber up to two weeks, enjoying the homey comforts of 6 bunks, a shower and toilet, a galley, air conditioning, and of course, scientific instruments and computers. A hatch leads to the wetporch, where a pocket of air keeps the water from entering Aquarius through the *moonpool*. This is where aquanauts make excursions to examine sea life.

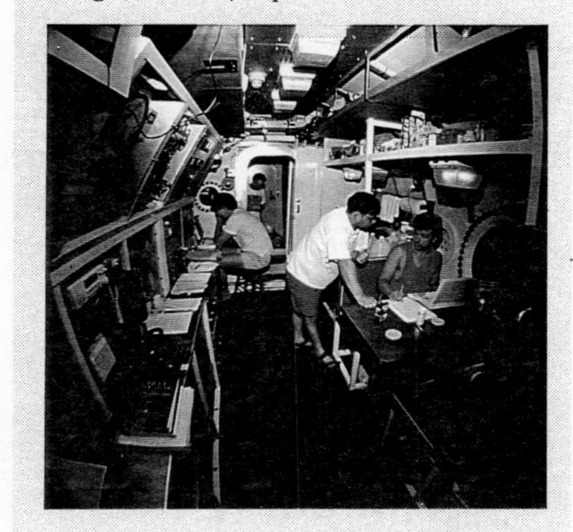

Aquarius is used for saturation diving expeditions, and has hosted more than 17 missions in its first 5 years. Coral reefs provide a fossil record of environmental change, and by examining them, scientists can learn more about the effects of sewage pollution, UV radiation, overfishing, and even climate change. Scientists are also investigating chemicals produced by coral reefs to evaluate their potential for making cancer-fighting drugs.

For more information and an on-line tour of *Aquarius*, visit "the world's first underwater website," at: www2.uncwil.edu/nurc/aquarius

(Photos on this and previous page courtesy of National Undersea Research Center, University of North Carolina at Wilmington.)

COMPARATIVE PHYSIOLOGY OF DIVING MAMMALS

Biology of Marine Mammals

We have been considering the physiological demands on humans as they invade a hostile underwater environment; now let us consider the physiology of other mammals adapted to this environment. What is this physical space? There are limitless areas of the ocean still to be explored; large and undescribed vertebrates inhabit these realms. Five canyons in the Pacific go deeper than 34,000 feet. The deepest, Cook Deep in the Philippine trench, reaches 37,782 feet. These depths interest both physical scientists and biologists: Jacques Piccard, inventor of the bathyscaph, saw shrimp and a foot-long fish at 35,800 feet; it is here that the giant squid live. When Harold Edgerton, inventor of the stroboscope, lowered his deep sea camera to 26,000 feet, his pictures revealed marine invertebrates living on the ocean bottom.

Our interest in life at these depths was sparked by 600 photographs of the ocean bottom taken by Heezen and Holliser (1972). All of this activity has been assisted by the invention of a high-resolution television camera which can view the ocean bottom to a depth of 20,000 feet. As far as we know, mammals have sampled very little of the deep water area. For example, the ecological niche of the sperm whale and the elephant seal extends only to 5,000 feet (1,500 meters) (Stewart 1996).

The ability to tolerate great pressure is no more interesting than the ability to survive and move without atmospheric oxygen; in an informal experiment we placed four fresh water turtles (*Chrysemys*) in jars containing boiled water which would be free of oxygen; these jars were sealed and placed under water at 22°C. For as long as 93 hours, reflex responses were still obtained from the turtles; we took them out and all of them revived (Dodge and Folk 1963). Jackson also studied anoxic turtles, but they were maintained at a cold temperature; he recorded heart rates from them of one per 10 minutes, and increases in blood lactate of one hundred fold (Jackson 1988).

This chapter is concerned primarily with mammals and not reptiles, but it is reasonable to ask the question: If a reptile can survive with such a large oxygen debt, why cannot a mammal? This question is answered in part later as we analyze aerobic and anaerobic metabolism during breath-hold diving of humans and sea mammals.

We can divide the aquatic mammals roughly into two groups: the small freshwater types and the very large marine mammals. The freshwater types vary in size from the water shrew, muskrat, platypus, otter, and nutria, to the beaver. There is, of course, one very large freshwater mammal, the hippopotamus. Then there is the totally aquatic manatee that frequents both salt and fresh water. In the marine group, size varies from the harbor seal to the huge elephant seal, and up to the largest of all mammals, the blue whale, which is over 90 feet long.

New information on the unusual circulation adaptations of the seal group and polar bears has increased interest in their biology and ecology. The physiology of seal pups is especially interesting because a seal pup must be a terrestrial mammal before becoming a marine mammal. Elephant seal pups are weaned, and then lose 31% of their body weight during this process of adjustment (See Fig. 10-6). Many of their blood parameters have been measured. For example, before they enter water, Weddell seal pups have a relatively low hemoglobin. At birth it is about 19 g/dl, while at 17 weeks, after entering the water, it is 27 g/dl (Rea 1995). The Hb of the adult varies with the dive: resting it is 17.4 g/dl, after the dive, 22.7 g/dl (Kooyman et al. 1980). Rea (1995) found that blood urea nitrogen changes from 3 mM to 12 mM, while plasma glucose concentration shows little change.

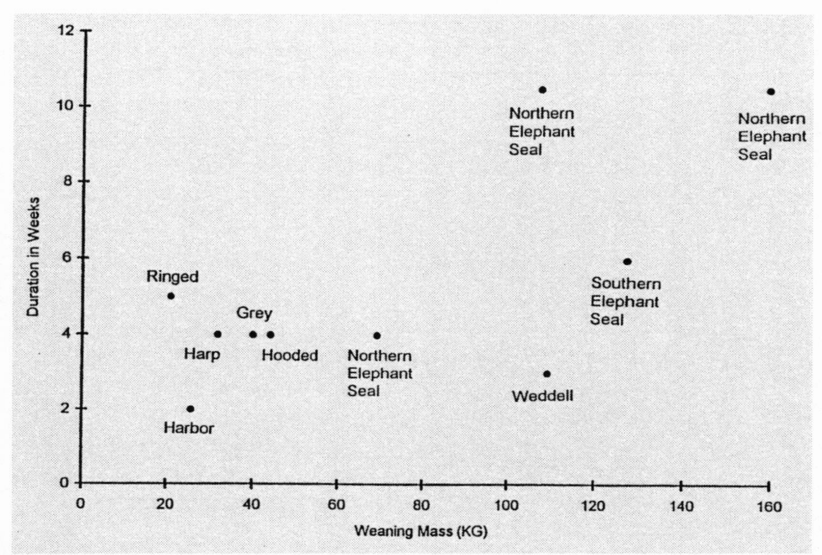

Figure 10-6. Post-weaning fast vs: weaning mass of 8 species of phocid seals. Adapted from Rea (1995).

The environment of marine mammals affects their kidney physiology. We should bear in mind that these animals live in a high salt solution, the sea. The water in the open sea has an average freezing point depression or Δ (delta) of -1.85°C. The ancient seas from which the ancestors of the vertebrate stock arose must have had not more than one-third to one-fourth as much salt, perhaps a Δ of -0.55°C. Most vertebrates have blood Δs not far from this value. Where in this "ocean desert" do sea mammals obtain fresh water? The seal apparently gets its water from the fish it eats (White, Cameron, and Miller 1971), while the whale obtains its water by ingesting sea water (Schmidt-Nielsen 1971); its kidneys take salt out of the circulating blood and throw it back into the ocean.

Populations of Aquatic Mammals

Some aquatic mammals seem to be holding their own against the present-day challenges of civilization: it is not uncommon for a visitor in Africa to observe hippopotamus populations totaling 1,000 individuals. The present-day population of seals, carefully estimated because of their fur value, is considered to be about 26 million. Though herds are culled periodically, some scientists believe that instead of reducing the population, such culling is likely to stimulate breeding and increase population growth.

Pacific walruses, on the other hand, are not doing so well, despite healthy-sounding numbers. Although one estimate reported a population of 100,000, many are very thin instead of full-bodied, and the number of young each year has decreased. The problem is assumed to be lack of mollusk food due to over-dredging by humans. Walrus are hunted by the Inuit for skins, meat, and ivory tusks. Yet a survey of these valuable animals in the North Hudson Bay Islands showed no decline from 1976 to 1990 (Stewart et al. 1993).

Whales are still scarce and some species are still threatened with extinction: in the late 1960's, 30,000 baleen whales a year were being taken from Arctic and Antarctic waters (Fig. 10-7). The largest total number of whales reported killed in one year was 66,000 in

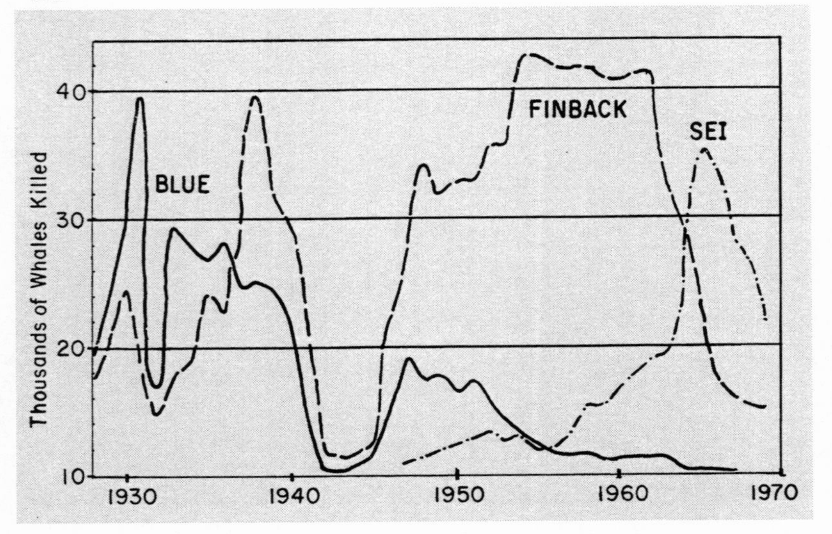

Figure 10-7. Numbers of Baleen Whales Killed in the Mid-Twentieth Century. The graph demonstrates the serious reduction in numbers of three species of whales. For example, in 1930-31, whalers took almost 40,000 blue whales, the largest and most spectacular mammal ever to live on Earth. Today, there are under 5,000 left worldwide. The fin and sei whale populations have recovered, although these whales are still considered vulnerable. Adapted from Animal Welfare Institute 1971.

1961. By 1985, under pressure from the International Whaling Commission, most nations had agreed to a moratorium on the killing of whales, although 3 nations have continued a small harvest, purportedly for "research purposes" (Yablokov 1994; Gambell 1995). (Whale meat is considered a delicacy in Asia, where it sells for up to $140 per pound.) The populations of sei and fin whales have recovered somewhat, although they are still listed as "vulnerable." The blue whale is another matter: still endangered, the world's population is currently under 5,000—compare this to the 40,000 blue whales killed in 1931.

The gray whale is a good example of the results of public protection: once threatened with extinction, by 1996 the gray whale's population has recovered sufficiently that it is off the endangered list and now considered merely "vulnerable". This good news, however, has prompted a few misguided individuals to suggest a renewed harvest of the gray whale.

The moratorium on killing whales does not apply to bowhead whales consumed by the Inuit; the harvest of these whales is an important part of their subsistence living (Fig. 10-8). Records kept at Point Barrow showed bowheads reaching lengths of 61 feet. Of the 5 to 6 whales taken there each year, 4 to 5 were sub-adult (Maher and Wilimovsky 1963).

The Galapagos Islands are home territory for 3,000 sperm whales, a species with individuals that live as long as 70 years, reach a length of 60 feet, and weigh as much as 40 tons (Whitehead 1995). Sperm whale populations seem to be showing signs of recovery, but there is a shortage of males. Recent information on sperm whales diving for food shows an average of 40 minutes between breaths. The deepest recorded dives were to 5,000 feet.

Although sperm whales do not echolocate, the vocalization of other whales has received much attention, as illustrated by the commercial recording, "Songs of the Humpback Whales." The vocalization is seasonal: in the case of the blue and finback whales, this peaks in winter then gradually declines through the spring. Few whales call in May, June, or July.

The polar bear is another of the land and marine mammals that have captured the attention of the public. For years their populations were in decline; in a 30-year period, 8,000 polar bears were killed on Spitzbergen. After international control was introduced, the species recovered. By 1995, the International Union for the Conservation of Nature estimated that the total world population of polar bears had reached 28,370. Polar bears are physiologically unique mammals in that they prefer a diet which is nearly 100% fat (see chapter on *Cold*) (Folk et al. 1995).

Progress on the physiology of diving mammals has been considerably accelerated by the ten annual conferences, with published proceedings, held by Thomas Poulter at Stanford Research Institute. Poulter reported on his success at raising young walruses (Fig. 10-9), and pointed out that the sea lion is capable of listening with its whiskers (Fig. 10-10).

At these conferences, participants plan ways to protect sea mammals. Recently, for example, scientists feared that distemper would spread from sled dogs to seals in the Antarctic. Canine distemper is an easily spread, airborne virus which has infected a wide variety of hosts, including seals and dolphins. To prevent what could be a devastating outbreak, sled dogs were banned in Antarctica, the last ones leaving in February, 1993. (In Africa, unfortunately, the indigenous wild populations were not so lucky; canine distemper killed one-third of the Serengeti's lions, as well as leopards, jackals, foxes, and hyenas [Packer 1996]).

We are interested in the physiological adaptations of these aquatic mammals, whether they are freshwater or marine.

Figure 10-8. A. Whaling at Point Barrow. The top photograph shows Inuit landing a bowhead whale on the edge of the ice floe. Note the *umiak* (skin boat) in the foreground used to capture and bring in the whale. The design of this boat has changed little in 2,000 years, including the waterproof stitches of the sealskin covering. **B.** The bottom picture shows the insulation blubber permitting the whale to live in icy waters. This blubber, considered a delicacy, is sold in the local native stores. Photos courtesy Naval Arctic Research Laboratory.

Breath-Holding Capacities

If one recalls the discomfort experienced when attempting to swim the length of a swimming pool underwater, then one is impressed with the ability of a beaver to stay underwater regularly for 20 minutes, or whales who remain submerged for periods up to 2 hours. Perry reported Antarctic Weddell seals diving deep enough and for long enough to travel 3 miles under solid ice, some of it more than 650 feet thick. It is not known if these seals held their breath the entire way, or if they found pockets of air under the ice.

The data on length of time for breath-holding fall into three groups: 1) mammals that ordinarily do not go under water, 2) those that regularly depend upon specific adaptations to stay down for from 5 to 28 minutes, and 3) those that can remain submerged for as long as 2 hours (Table 10-2).

The interesting generalization about the data presented is that some of the mammals with the most extreme modification or adaptations for aquatic life do not appear to have a particularly marked breath-holding ability; examples are sea otters and porpoises.

The king penguin is reputed to dive deeply, but when studied carefully, 70% of their dives were aerobic, without the presence of lactic acid in the blood (Culik et al. 1996). Most of their dives last only 4 to 6 minutes.

Our knowledge of the breath-holding abilities of marine mammals and the depths of their dives has been advanced by satellite-transmitted measurements and computer-controlled recording instruments (De Long and Stewart 1991). There are large differences between the breath-holding of the smaller seals and elephant seals. Free-ranging Weddell seals only dive for 26 minutes or less (except for 2.7% of dives), while elephant seals dived continuously up to 40 minutes at a time, seldom taking more than three minutes to breathe (De Long and Stewart 1991) and remaining submerged 90% of the time (Hindell et al. 1991). The elephant seals regularly dived to 2,500 feet, and one bull went to 5,150 feet (LeBoeuf et al. 1986). Why do elephant seals and sperm whales dive so deeply? They both live on pelagic, deep-water squid and must dive deeply to find them.

Types of Diving Adaptations

The whales and seals represent a contradiction in physiological terms; they are mammals that depend closely upon atmospheric air and live near the surface of the water. Yet at least some species in the whale and the seal group have a fantastic ability to dive so deeply that it is remarkable that any physiological mechanism can tolerate the pressure and the metabolic strain: the sperm whale dives to 5,000 feet and the Weddell seal to 1,850 feet.

The "challenges" faced by diving mammals are those listed for humans, including the possible build-up of carbon dioxide, the effect of extreme pressure, the oxygen debt, and the possibility of decompression effects. The problem of carbon dioxide can be dispensed with quickly: the Russian biologist, E. Krebs in 1941 demonstrated that the central nervous system of diving animals is singularly unresponsive to carbonic acid (Slijper 1962).

The problem of extreme pressure is exemplified during the dive of a whale to 1,500 feet, where it undergoes a pressure of 50 atmospheres; if it keeps going down to 3,000 feet, it experiences 100 atmospheres. Popular descriptions of diving in deep water emphasize the tremendous crushing force of such an experience; in actual fact, the mammalian body is made up almost entirely of solids and fluids, and the body is no more likely to be crushed than a bucket of water lowered into the depths. Damage can occur if there is an unequal difference of pressure where air is trapped, such as in the lungs, the nasal sinuses, and the middle ear. Sometimes there is gas in the intestine but this gives little trouble because the intestinal walls contract down to equalize the pressure. It is possible that deep-diving mammals control the volume of air in the lungs like the human diver (see earlier section). We have little knowledge of what happens to the volume of air in the lungs of a whale under 50 to 100 atmospheres of pressure and why its ribs are not crushed when the air is compressed. Scholander (1964) states that the anatomy of the lungs must permit a complete collapse in deep diving.

There have been a number of hypotheses over the years

Figure 10-9. Adaptations of the Walrus. The vibrissae of the walrus are adapted for finding shellfish in mud under water. The example below was a 15–day-old walrus weighing 120 pounds from St. Lawrence Island. Photo courtesy Thomas C. Poulter.

on the question of decompression sickness in diving mammals. These theories were complicated and improbable until the simple and obvious solution became generally accepted. In 1935, Hill showed that there is a basic difference when humans dive with held breath or dive with fresh air delivered to a helmet. In the diving helmet and from the tanks of an open-circuit apparatus there is a continuous supply of fresh air and thus of fresh nitrogen with which the blood can become saturated. The diving mammal, however, takes down a small, fixed quantity of air so that despite the high pressure of water, only a small quantity of nitrogen is available for solution in the blood. One important difference between whales and seals is their respiratory behavior immediately before diving. Whales appear to dive on inspiration but seals dive on expiration (Andersen 1969). We will consider the significance of this behavior, and the question of oxygen debt in a separate section.

Anatomy must correlate with function, habitat, and behavior. The bulk of the eye of the whale is an example: in a deep diver like the sperm whale, the wall of the eye would have to be excessively thick and would take up too much room in the skull; thus it must be small. In other words, if the whale had an eye of relative size (for the skull) to which was added many protective layers, the eye would be too large. Figure 10-11 shows the eye of a bowhead whale, a surface feeder. Table 10-3 compares the eye of another surface feeder, the humpback, with the eye of a deep diver.

Figure 10-10. Adaptations of Fur Seals. This fur seal cow and pups particularly illustrate the unusual development of the vibrissae. Experiments at the Biological Sonar Laboratory have indicated that the structures are associated with hearing. They must also be useful when hunting fish under ice and for returning to breathing holes, especially when there is a complete lack of sunlight in winter. Photo courtesy Naval Arctic Research Laboratory.

Circulation and Energy Metabolism

Studies on the circulatory system represent an area where humans and sea mammals can reasonably be compared (Lin 1986; Olszowka and Rahn 1987). Diving mammals are able to sustain themselves underwater under very deep pressure for prolonged periods by reducing their overall metabolism; this process has been variously described as "making themselves into smaller animals" or as "making themselves into a heart-lung-brain unit." Apparently, the blood flow to large muscles and skin areas is markedly reduced

Table 10-2. Breath-Holding Capacities of Mammals (in minutes)

GROUP 1		GROUP 2	
Most humans	1.0	Sea otter	5
Polar bears	1.5	Platypus	10
Pearl divers (humans)	2.5	Muskrat	12
		Hippopotamus	15
GROUP 3		Sea cow	16
Greenland whale	60	Beaver	20
Sperm whale	90	Porpoises	15
Bottlenose whale	120	Seals	15 to 28

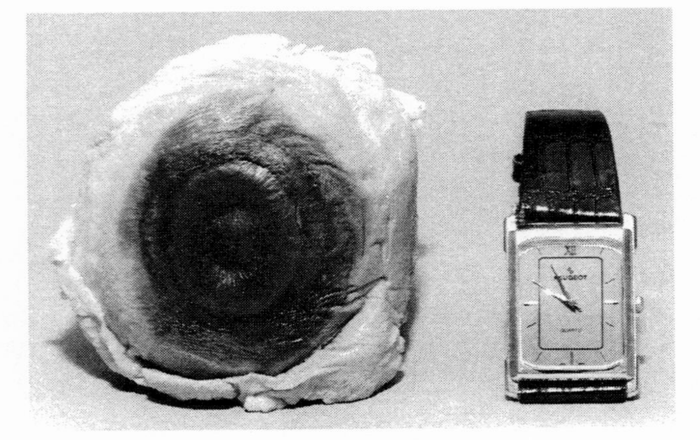

Figure 10-11. The Eye of a Bowhead Whale. It is small for the size of the mammal and protected against water pressure by a thick coat of connective tissue.

(Elsner et al. 1996). The evidence for the redistributed blood flow and attendant metabolic changes (Evans 1984) is as follows:

1. incisions into skin and muscles do not bleed;
2. elevated lactic acid does not appear in the blood until after ascent, suggesting that it is trapped and non-circulating during the dive;
3. the myoglobin oxygen is almost exhausted, while the blood remains 50% saturated with oxygen;
4. direct arterial tracings indicate increased peripheral resistance; part of the evidence is that the rate of fall of pressure during diastole is reduced.

Bradycardia in Humans and Sea-Mammals. Associated with the reduced flow of blood is a marked bradycardia; all diving and non-diving animals when placed underwater demonstrate this (Fig. 10-12) (West and Van Vliet 1986). The development of this effect is gradual in most animals; in the seal, however, the frequency can drop immediately to 1/10 of normal. This drop is abolished by atropine or vagotomy. The electrocardiogram during the dive, showing initially a normal complex, tends to develop a prolongation of the Q-T interval and a peaked T wave which is sometimes inverted (Ferrigno 1991). The bradycardia is maintained while the animal is in its dive, which lasts as long as 15 minutes. In most cases, this bradycardia is of sinus origin, since the electrocardiogram exhibits a normal sinus rhythm. The effect upon central blood pressure is what one would predict with a greatly reduced blood flow: with the heart of the seal beating only 5 to 6 times per minute, it has been shown that the central blood pressure taken in the femoral artery is nevertheless maintained at a normal or even an elevated level (Irving et al. 1942). The extent of the depression in pulse rate may be from 10% to 50% of the surface rate; once developed it persists during physical activity and struggle. The diving reflex in humans is more marked during dynamic exercise than during rest or isometric exercise (Berman et al. 1972; Wittmers et al. 1987).

These studies on human subjects have been amplified in recent experiments (Freund et al. 1991). The focus of interest is the conflicting metabolic demands of sustaining aerobic metabolism vs. exercise (Castellini et al. 1985; Al-Ani et al. 1995; Salzano et al. 1984; and Shiraki and Claybaugh 1995). One at first might expect a normal or accelerated heart beat rather than bradycardia because of the demands of exercise. Some ingenious experiments on exercising

Table 10-3. Whale Eyes Compared

	Body Wt.	Wt. of Eye
Humpback (Surface Feeder)	39.5 tons	980 gm
Sperm whale (Deep Diver, to over 5,000 ft)	40 tons	290 gm

human divers showed that the bradycardia is not due to breath holding, but due to a wet cool face. The bradycardia is caused by cardiac vagal tone through trigeminal cutaneous receptor stimulation; if the subject exercises underwater while breathing air, there is no bradycardia (Fig. 10-13) (Butler and Woakes 1987; Park et al. 1987; Bjertnaes et al. 1984; Smeland et al. 1984).

Associated with the bradycardia is peripheral vasoconstriction. There were early reports of this in human subjects based on reduced blood flow in limbs, but this cannot be assumed unless one makes intra-arterial blood pressure measurements. The reason is because a decrease in blood flow may reflect a fall in cardiac output and not necessarily vasoconstriction. Finally, Heistad and coworkers (1968) simultaneously measured blood flow and arterial pressure in human simulated dives, and proved intense peripheral vasoconstriction.

The bradycardia and apnea observed during diving in seals also often occurs when they are sleeping, whether on land or underwater. These events of low heart rate and low breathing during sleep may last 24 minutes. In some cases, one hemisphere of the brain may be in a sleep state, while the other hemisphere is in a conscious state. Studies on the diving physiology of the elephant seal record that they dive continuously and repetitively for months on end; thus, they apparently do not sleep for 3 months.

These experiments as a whole help in the understanding of Sudden Infant Death Syndrome (SIDS) in humans. As Castellini (1996) suggests, perhaps SIDS can be explained by a wet face due to regurgitation, which results in bradycardia and apnea carried to an extreme.

Figure 10-12. Heart Rate Response to Diving. One of the most consistent of physiological syndromes is the slowing of the heart during submergence, both in non-diving mammals and in aquatic mammals and birds. Curiously, the bradycardia during the dive is more marked in humans, a non-aquatic mammal, than in one that never leaves the water, the manatee. Also, one might predict a similarity in the response of porpoises and seals, but the latter markedly prolongs the bradycardia. Adapted from Irving et al. 1942.

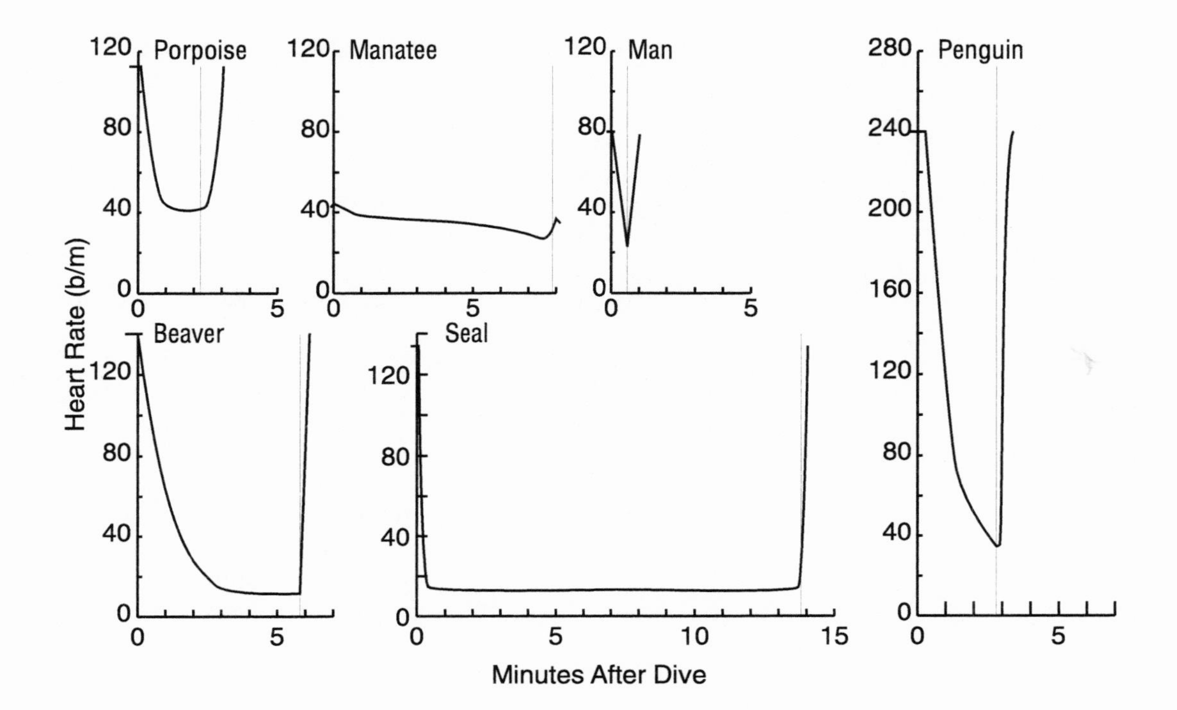

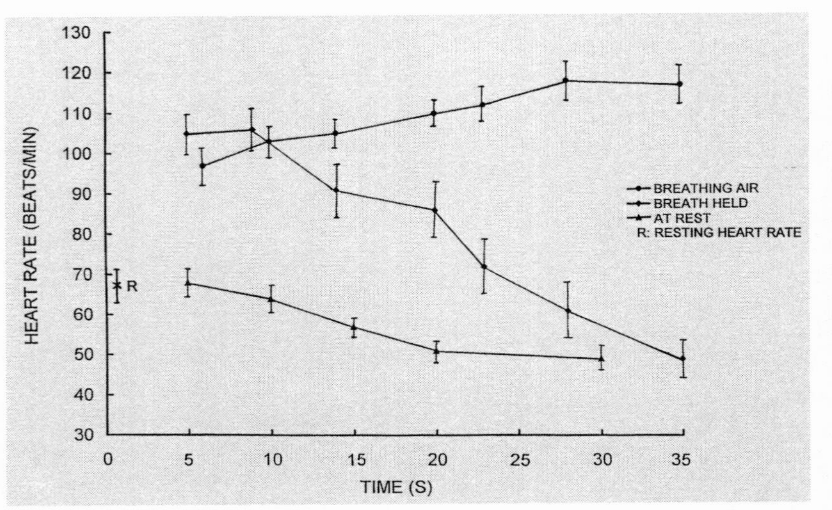

Figure 10-13. This Figure Shows the Definitive Experiment on Exercise and Breath Holding. There is no bradycardia when swimming while breathing air. Adapted from Butler and Woakes (1987).

The comparison of diving bradycardia in humans and seals was possible only through the cooperation of mild-mannered harbor seals who could be studied by radiotelemetry when unrestrained, and trained to submerge the head upon command (Scholander 1963) (Fig. 10-14). Folk, Berberich, and Sanders (1972) used a similar technique to demonstrate bradycardia in polar bears. During the experiment on harbor seals, the natural clamping off of the inferior vena cava during the dive was also demonstrated by x-ray technique (Elsner, Hanafee, and Hammond 1971).

The above early studies on cardiovascular physiology were necessary to demonstrate that a diving sea-mammal must convert itself into a heart-lung-brain unit (Fig. 10-15). The experiments were done on the American harbor seal and the elephant seal. Since then, the emphasis has been on measuring heart rate and oxygen consumption on two new species, the gray seal and the Weddell seal, with the difference being that they were free-ranging while measurements were recorded. At the same time, a debate was settled concerning how many of the bradycardia measurements were influenced by (forced by) the experimenter, and how many were natural (Elsner et al. 1996). It is true that Fedak found that seals in the laboratory reduced heart rate to 5 b/m, while when free at sea to only 40 b/m (Fedak 1986). However, two teams of investigators studying the gray seal (Thompson and Fedak 1993; Reed et al. 1994), and three teams studying the Weddell seal (Kooyman et al. 1980; Guppy et al. 1986; and Ponganis et al. 1993a, 1993b), found a very low bradycardia. The latter investigators were also interested in calculating the mean aerobic dive limit by correcting for the oxygen stores in blood and muscle.

Telemetry of Large Vessels. Early teamwork between electrical engineer and physiologist confirmed Irving's theory (1963) of selective vasoconstriction during diving. Van Citters and coworkers (1964) studied cardiovascular adaptations to diving in the elephant seal in its habitat on Guadeloupe Island, Mexico. They used a telemetry system to measure blood flow in the iliac artery, the carotid artery, and the aorta. On these large animals, resting heart rates averaged 36 beats per minute with marked sinus arrhythmia. (We have also recorded sinus arrhythmia by telemetry in wolves, wolverines, and black bears [Folk and Copping 1978; Folk and Folk 1980]). The blood pressure of the seals was similar to that of humans. Bradycardia occurred with diving, but the immediate response was much less than that commonly observed in smaller seals. Blood flow changes varied with the duration of immersion, but with prolonged immersion iliac flow fell severely, while carotid flow was more adequately maintained. The overall changes in the pressure pulse were consistent with marked peripheral vasoconstriction to combine with the reduced iliac flow. Some of these measurements were repeated and confirmed in the beaver (Ferrante and Frankel 1965).

Another study using radiotelemetry was done by Elsner et al. (1964) at the Scripps Institution of Oceanography. In two species of seals, the radio-capsules were surgically implanted on the main pulmonary artery, the aorta, and the superior mesenteric and renal arteries. Heart rates as low as 10 beats per minute were commonly seen in harbor seals. Cardiac output decreased in proportion to heart rate with little change in stroke volume. Blood flow was reduced in all measured sites falling to virtually zero in mesenteric and renal arteries. Less reduction was noted in the common carotid and coronary arteries.

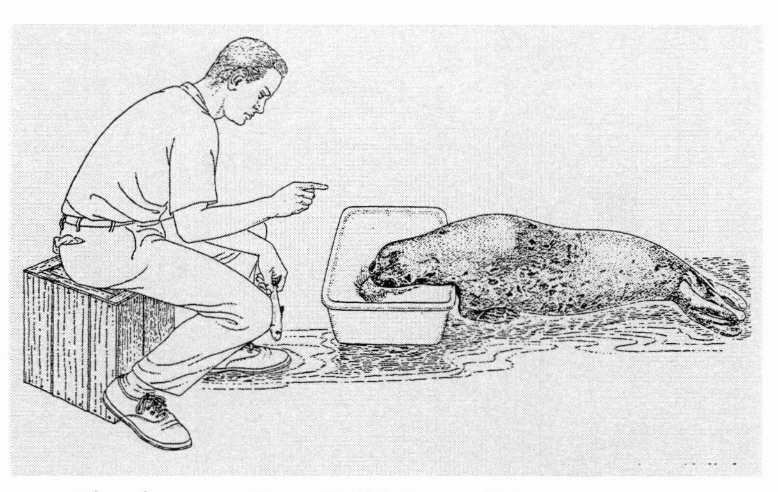

Figure 10-14. Voluntary Diving of Experimental Seal. Heart rate studies were done with harbor seals, which were trained to keep their noses underwater until the experimenter lowered his finger. Adapted from Scholander 1963.

Respiration

Mammals that naturally submerge themselves repeatedly usually have a somewhat larger relative oxygen storage, since their relative blood volumes, myoglobin content, and lung volumes apparently exceed those of humans (Irving 1939; Guyton et al. 1995; Hurford et al. 1996). A partial exception appears to be the lung volumes of the whale; if volumes are referred to body weight, then the lung capacity of whales is one-half that of terrestrial mammals. This apparently relates to the observation that whales dive deeper than any other aquatic mammals. Another surprising adaptation of breathing is found in the sea lion. Even when on dry land, its respiratory cycle consists of rapid breaths followed by a long period of apnea (breath holding) that accounts for 84% of the respiratory cycle. An increase in heart rate occurs a fraction of a second after the onset of inspiration (Lin, Matsuura, and Whittow 1972).

The breathing of whales is of particular interest since it is accomplished through the top of the skull through the blow hole. Whales

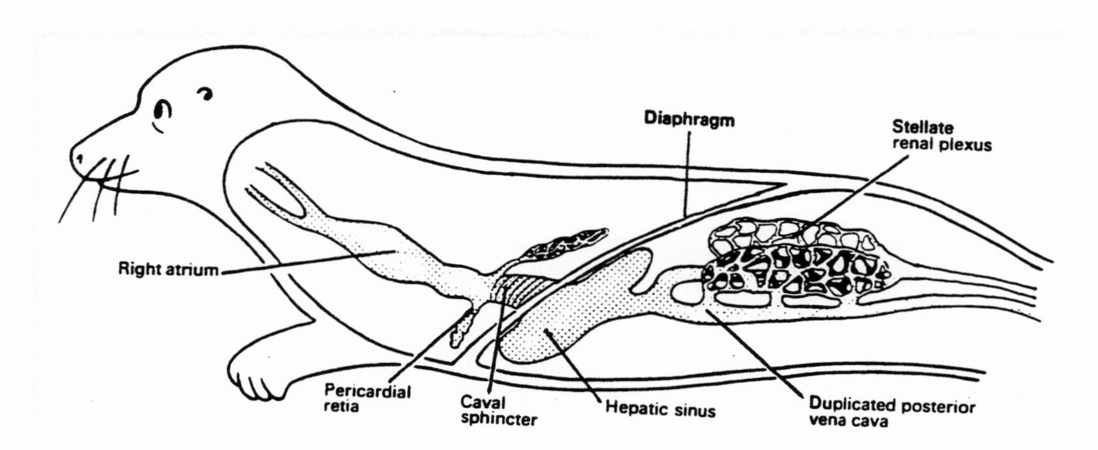

Figure 10-15. Diagram of the Main Venous Pathways, the Hepatic Sinus, and Caval Sphincter in a Seal. When the caval sphincter contracts, it converts the circulatory system so that the seal is supplied with blood as in a "heart-lung-brain unit." Adapted from Harrison and Kooyman 1981.

Figure 10-16. Adaptations of the Leopard Seal. This Antarctic seal was drawn by Wilson in a typical maneuver. This hyperactive seal can be 350 cm long, and can weigh 380 kg. They endangered some of Shackleton's men whom they apparently mistook for king penguins (see sidebar, facing page). From Roberts 1967.

can be identified by species by the pattern and type of spray formed when they spout. Slijper (1962) pointed out that one can determine when a whale is panting by the spouting; panting occurs when whales are chased and are alarmed and moving at top speed. He has observed them change their normal resting respiration from 1 to 30 per minute. Their method of breathing has an important bearing upon their ability to dive; while the tidal air of humans is a small fraction of the total lung volume, it is notable that whales, and porpoises to a certain extent, use a very large part of the total lung volume for their tidal air. The mechanism of breathing deeply with each respiration is attributed to the whales' expanding and contracting their thorax to the maximum; in relative terms this maximum is 10% greater than it is in terrestrial animals.

How does this basic information fit in with diving? Do whales and porpoises breathe deeply while swimming quietly near the surface and also before they dive? A number of investigators have tried to get information on this; it appears that gray seals and sea elephants are known to make a point of exhaling before they dive. Whales do the reverse, and furthermore, Scholander gained the clear impression that they fill their lungs to capacity. It is difficult to explain such a breathing pattern. Because of the forces applied to the thorax during a deep dive, there seem to be advantages in exhaling before diving; in a deep dive, however, the oxygen stored in the lungs is obviously of value. One reasonable explanation for the continuous deep breathing of the whales is that this pattern has survival value because they must at times dive rapidly, and no final gasp is needed before the quick dive.

We must next consider oxygen transport. The hematocrit in whales is much larger than it is in humans. In spite of this, the blood oxygen capacity of whales is not particularly striking. Mammals in general have a blood oxygen-carrying capacity varying from 11 to 24 vols %; that of the porpoise is 19 vols %, while that of the fin whale is only 14 vols %. The answer must be that for prolonged submersion whales depend more upon myoglobin; this is quite possible since some whales have 8 times as much of this material as terrestrial mammals. Seacows and porpoises are also well endowed with myoglobin (Blessing 1972).

All of this shows that there are a variety of ways for the diving mammal to obtain a larger relative oxygen storage; for example, in the seal, the oxygen-carrying capacity of its blood alone is 29 vols %, in spite of its increased myoglobin. This is partly because it has a high hemoglobin concentration (Qvist et al. 1986). The blood volume of the seal is also twice that of a human of the same weight.

The distribution of oxygen in the various stores, i.e., lungs, blood, and muscle, varies among species. People have a total oxygen capacity of 20 ml per kilogram, and Weddell seals have a total capacity of

87 ml of O_2 per kilogram. Penguins are in between, with a total oxygen capacity of 55 ml of O_2 per kilogram. The seal's lungs do not provide O_2 in a deep dive, because they are collapsed, but the O_2 in the bird is provided from an extensive system of air sacs, which do not collapse (Kooyman and Ponganis 1997).

The oxygen supply, no matter how carried, also seems to be conserved by temporary metabolic changes which occur during the process of diving. To consider this we must recall the two phases of the chemical processes that supply the energy for muscle contraction. In the anaerobic phase, the energy supply, which consists of glycogen, is broken down in the muscles into lactic acid without the intervention of oxygen. In the second phase, the aerobic part, some of this lactic acid is oxidized while another part is resynthesized into glycogen. Irving's theory (1963) is that during deep diving, the mammal carrying out this difficult process is undergoing a metabolic phase in which the anaerobic part predominates. If this is accomplished, it obviously saves oxygen. Irving and Scholander based their hypothesis on experiments with rats, seals, and ducks, which showed that during diving, the lactic acid increases considerably in the muscles but does not accumulate in the blood until shortly after resurfacing. Thus, it appears that during diving, the muscles require very little oxygen and the blood is shunted primarily to the heart and brain.

What sort of blood can respond to the osmotic challenge of the ocean's 3.5% salt solution, as well as carry oxygen at a capacity of 29 vols %? The blood chemistry of the bottlenose dolphin has been compared to that of dogs and horses. The components are very different from most mammals and have some characteristics of shark blood. Plasma sodium, plasma chloride, and serum albumin levels are higher, while globulin levels are lower than those found in dogs and horses (Danielsen et al. 1995). Blood urea seems to be considerably higher than that found in the land mammals (Medway and Geraci 1965).

Small mammals (albino mice) have been tested for their ability to tolerate pressure-equivalent depths down to 914 meters (3,016 ft) of sea water. Because of their small size, they have a rapid uptake and later elimination of inert gases. Membery and Link (1964) estimated that mice attain gaseous equilibrium within 1 hour in a hyperbaric environment, while humans require 12 hours. In a 0.65% oxygen mixture with helium, mice tolerated 13 hours at 150 atmospheres of the gas, equivalent to 914 meters of sea water.

Thermal Environment and Food Supply of Humans and Marine Mammals

We find that the problem of carbon dioxide accumulation, oxygen debt, tremendous water pressure, and decompression effects have all been solved quite nicely by adaptations in the mammals that can accomplish deep diving. Many questions remain, however, about the physiology of these fascinating animals. The challenges imposed upon them by their extreme environment are exaggerated by cold. For example, the water about the Antarctic Continent near the Weddell Sea varies from -1.95°C to +5°C all the year around.

CARNIVOROUS SEAL MISTAKES MAN FOR KING PENGUIN

This incident was recorded by Lansing (1986):

"Returning from a hunting trip, one of Shackleton's men, Orde-Lees, traveling on skis across thin ice, had just about reached camp when an evil, knoblike head burst through the ice just in front of him. He turned and fled, shouting for Wild to bring his rifle. The animal, a leopard seal, sprang out of the hole and bounded across the ice after him. The seal had almost caught him when it plunged into the water. Orde-Lees started to cross to safe ice when the leopard seal head again exploded through the ice directly ahead of him. The animal had tracked his shadow across the ice. It made a savage lunge for Orde-Lees with its mouth open, revealing an enormous array of sawlike teeth. Orde-Lees turned and raced away. The animal leaped again in pursuit; Wild arrived with his rifle; the leopard seal rushed at him and was killed by Wild only 30 feet away. Two dog teams were required to drag the animal into camp; it weighed 1,100 pounds."

There are human divers in Korea and Japan who continue their occupation in winter at water temperatures varying from 9°C to 14°C. The story of these human divers must be a delight to anthropologists: four thousand years ago, American Indians were casting nets for fish; today our fishing fleets are still *hunting* fish to net, although fish farming has gained some popularity. How about diving for submarine plants and animals fixed to reefs? Nine thousand years ago, the Alacaluf natives of Tierra del Fuego did this as *hunter-gatherers*. Today, this hunter-gatherer occupation is still carried out by more than 16,000 breath-holding Ama divers—both male and female—in Japan, with another 3,000 in Korea (Elsner et al. 1996; Takeuchi and Mohri 1987). The Ama continue to dive in all four seasons at depths that may exceed 20 meters. Because this includes winter with the ocean temperatures mentioned above, Hong observes that humans *can* acclimatize to cold (Hong et al. 1987) This is covered more fully in the chapter on *Cold*.

What is the effect of these cold ocean temperatures on the body temperatures of sea-mammals? We have said that during the dive, the sea-mammal must turn itself into a heart-lung-brain unit. The cold water must cool blood temperature in spite of the exercise. Ponganis found that the normal core temperature of seals is 37.3°C. During the dive, the aortic temperature (in the heart-lung-brain unit) was 35°C, but swimming muscle temperature was 37°C (Ponganis et al. 1993b).

Now we must consider whales. We realize that a high percentage of them live in Arctic and Antarctic waters where they can consume great quantities of the biomass of plankton abundant there. In the Arctic, the whales avoid the extreme rigors of the winter by migration (Lockyer 1981). Recall that although we have stressed in this chapter the capacity for deep and prolonged diving for some of the whales, there are many of them which seldom submerge for more than 30 minutes. The porpoises usually dive for only 4 to 15 minutes. Therefore sea ice can be an extremely serious barrier to whales and porpoises. The latter species often die in large numbers in the Baltic Ocean when the sea freezes over before they can migrate from it.

Apparently there are more cases of failing to solve this problem in the Antarctic than in the Arctic; observers near Grahamland and Ross Island in the Antarctic observed three species of whales which were obviously cut off from the open water by frozen sea, especially the little piked whales. These fairly numerous animals had breathing holes in the ice which eventually froze over. This situation is even more striking in those Arctic and Antarctic seals which do not migrate in the wintertime. Seals can seldom remain submerged longer than 15 minutes, although the Weddell seal tolerates 43 minutes. This means they must keep breathing holes open all winter which is no small task when the temperature above the ice is -50°C to -70°C (-58°F to -94°F)(Ray and Lavallee 1964). The problem is not simply one of requirements for oxygen for whales and seals under the ice; recall that they are also living in continuous darkness during the winter season. It is a mystery how they select and obtain their food, and orient themselves during the food search so that they can return to breathing holes. We do not know of echolocation by seals, although it is highly developed in whales.

Although some Antarctic seals do migrate as much as 2,000 miles (elephant seals and sea lions), the most interesting one, the leopard seal, remains at the edge of the pack ice in winter. The leopard seal (Fig. 10-16) and the killer whale are the top carnivores in the Antarctic; in the Arctic, the polar bear has this role. Leopard seals feed primarily on penguins, but as our sidebar shows, they are quite willing to expand their diets when the opportunity arises.

The food supply of marine mammals in the polar regions is partly composed of fish that can tolerate an environment below freezing. There are a variety of cryoprotective adaptations these fish may possess. One of them is the production of antifreeze peptides; this adaptation is found sporadically in species in the far-North and in the far-South (see Table 10-4) (Osadjan et al. 1996).

We have discussed responses of mammals and other animals to extreme and challenging environments. In our next and final chapter we will address humankind's impact on Earth's resources, ecosystems, and climate. An obvious question arises: How do our actions rebound to affect us? As we will see, the question may be obvious, but the answer is not.

Table 10-4. The Distribution Of Antifreeze Peptides (AFP).*

Polar Fish with AFP	Atlantic Fish with AFP
Zoarcoid Pachycara (Antarctic)	Zoarcoid Zoarces (North Atlantic)
Zoarcoid Lycodichthys (Antarctic)	Zoarcoid Anarhichas (North Atlantic)
Zoarcoid Lycodes (Arctic)	Flounders
Sculpins	
Herring	
Sea Raven	

*These are by no means limited to fish in the Polar regions. They are an example of parallel evolution.

11. HUMAN EVOLUTION, RESOURCES, AND POLLUTION

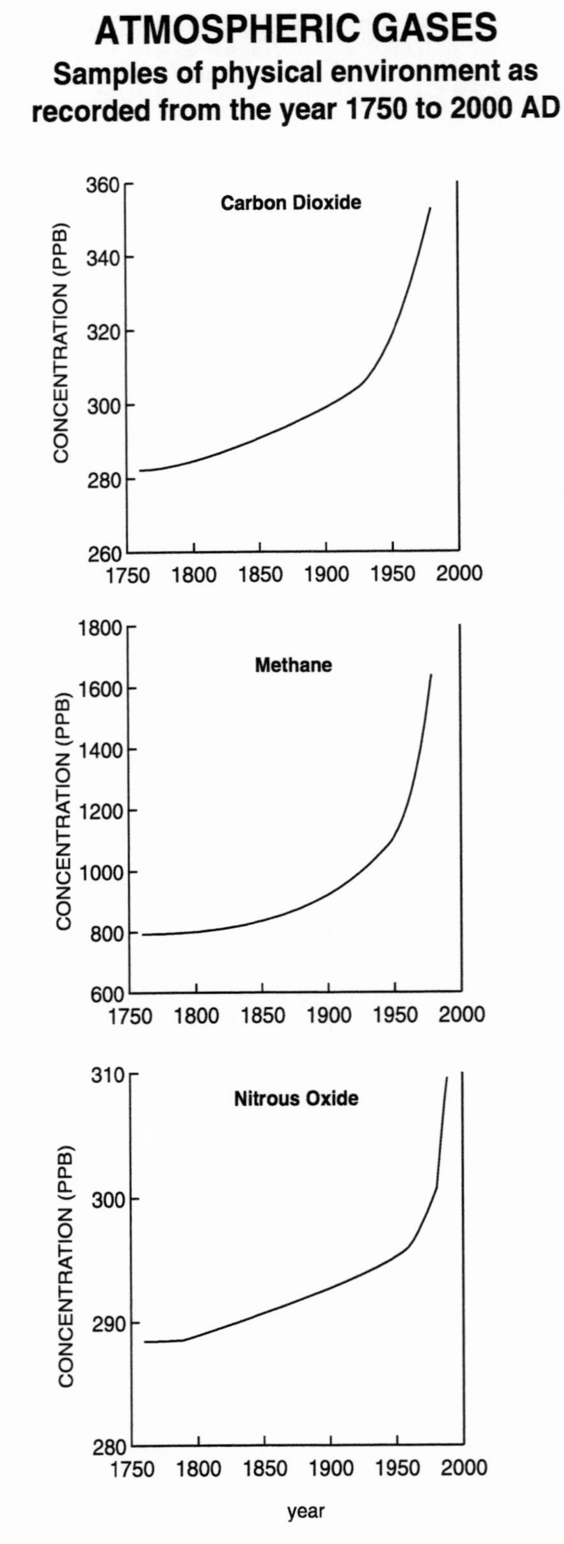

Environmental Diagram: Chapter 11

11. Human Evolution, Resources, and Pollution

11. HUMAN EVOLUTION, RESOURCES, AND POLLUTION

INTRODUCTION

If we were writing a book on *Environmental Problems Facing Humankind Today,* the essential theme would be the "Evolution of Human Ecology." The book would proceed through the topic, "The Fitness of the Human Environment from *Sinanthropus* to Sears." There would be considerable discussion of a series of ecological questions in that book, which, in this final chapter of our current book we can propose, but not fully answer:

1. Primitive humans had to adapt to their environment, while today today, we modify it. Can humans once again adapt to their surroundings? Can we, for example, acclimatize to pollution?
2. Today we are constantly concerned with *resources*. Did primitive humans have difficulty finding adequate resources at certain times of the year? Were there ever periods where their resources were wiped out by unusual climatic events? How do we face the possibility that we may *deplete* irreplaceable resources?
3. We should examine closely those once-great cultures or civilizations that disappeared: will they serve as models to understand our present civilization?
4. Will some cultures or races disappear in the next thousand years? A typical example of an endangered culture is that of the present-day Inuit. We, as a supposedly advanced culture, look with great interest at the Alacaluf Indians at the tip of South America, and at the Ainu in Japan, whose cultures are also endangered. Turning back the time scale to when classic Neanderthal lived, is it possible that some higher or more advanced race exchanged opinions on the last of the Neanderthals—people who were so different from themselves?
5. Which was the greater threat, overpopulation or climate change? Were some groups of primitive people so large in relation to their local environments that they exhausted the local resources?
6. What should be said about changing climate? There is conflicting evidence about the importance of this, and paradoxes are occasionally presented. For example, Dart implied that the climate and type of vegetation in Central South Africa

did not change much in 20,000,000 years. Such a generalization could not possibly apply to the climates of North America, Europe, and Asia, where people had to adjust and adapt to the repeated invasions of ice sheets.

7. Most importantly in our hypothetical book on Human Ecology—as well as in this book—one question should be asked repeatedly: Is the particular culture or nation under discussion *in equilibrium with its environment?*

We will be emphasizing this last question concerning equilibrium with the environment, and will compare prehistoric humans with modern ones. We will be thinking in terms of millions of years for our Proconsul ancestors to evolve into *Australopithecus*. This time scale reminds us of two principles of evolution, which should be borne in mind as we analyze the early history of humans. The first is that there was almost limitless *time* for the evolutionary process to work its magic; the second is that there were almost limitless *numbers and kinds* of animals for the "process" to use for its experiments in evolution.

The concept of unlimited numbers of animals needs amplification. For example, there were about 298 genera of dinosaurs alone. Let us assume two species of dinosaurs per genera; this is reasonable since one genus, *Iguanodon*, includes seven species. Thus, an estimate of 596 species of dinosaurs alone supports the concept of unlimited species for evolutionary experimentation. A final example: dinosaurs were once so numerous that 75 million years ago, one massive volcanic eruption killed and buried an entire herd of 10,000 duckbill dinosaurs.

And now, perhaps a look at the large dinosaurs will also help us to understand the "limitless" time scale. Two hundred and twenty million years ago there was only one big land mass called Pangea, and it was inhabited by large carnivorous dinosaurs. After an incredible one hundred and twenty million years, the land mass became divided into Laurasia and Gondwana. There were still three species of very similar, enormous, carnivorous dinosaurs, all like the famous *Tyrannosaurus rex*, which was 15 meters long. These dinosaurs were living in what was to become the three continents of Africa, North America, and South America. Probably, there was a bridge between Laurasia and Gondwana and genes were being exchanged. By 90 million years ago, the continents were distinct, as were several new dinosaur species (Currie 1996; Sereno et al. 1996). An example is *Carcharodontosaurus saharicus* (Fig. 11-1). Since it is difficult for the human mind to grasp this interval of 100 million years, we can simply say that there was limitless time for the evolutionary process.

Now, armed with a deeper feeling for

Figure 11-1. The Dominant Species Past and Present. Comparison of the size of the skull of the dinosaur, *Carcharodontosaurus saharicus*, with a human skull. The dominant species 90 million years ago was the dinosaur, as compared with the dominant species today, *Homo sapiens*. From Sereno et al. 1996.

the process of evolution, let us proceed to compare prehistoric humans with modern ones, while introducing some ecological principles along the way.

STONE AGE HUMANS

Thousands and millions of years ago there existed the following key populations of primitive people:

Primitive Humans	Years Ago
Proconsul	22,000,000
Australopithecus	1,000,000
Peking	450,000
Swanscombe	250,000
Neanderthal	90,000
Cro-Magnon	Late Ice Ages

The stem-populations that gave rise in Europe to *Homo sapiens* and its races, Neanderthal and Cro-Magnon, were composed of *Homo erectus* (see Fig. 11-2). However, *Homo erectus* is not mentioned in the above table, since the history of this species is changing too rapidly for generalization. The evidence now suggests that *erectus* first migrated out of Africa 800,000 years sooner than the 1.8 million years previously assumed (Larick and Ciochon 1996). Numerous skulls of this ancestor-species have recently been found in Java, where it appears that *erectus* may have existed in parallel with the species to which we belong, *Homo sapiens*. However, *sapiens* and *erectus* were not contemporaries in Africa; the latter became extinct there about a million years ago (Swisher et al. 1996).

What sort of timetable can we apply to ourselves? It helps to realize that Neanderthal lasted for about 25,000 years, and our own type (*Homo sapiens*) has been around approximately double that period. The time scale will be more meaningful when we trace the inhabiting of two vast continents, North and South America, by the highest primate. To do this we must find the causative population reservoir, then we must consider the major geologic events that acted like a gate to so-called migrations into North America, and finally, we must ask, how recent is this inhabiting process? In answer to the last question, consider as an example a central location in the United States (Iowa, Wisconsin, Minnesota): we find that the last of the ice sheets left this area and was replaced by evergreens about 13,000 years ago; these evergreens were replaced by deciduous forests about 10,500 years ago; this deciduous forest was replaced by oak and prairie grass about 8,500 years ago. At first the evergreen forest was devoid of humans; later the same area became heavily populated. What was the *origin* of the new people? Let us discuss this place of origin, the main source of the higher primates who spread out so successfully that they now flood the earth. This source is South Africa.

Figure 11-2. The Peak of Human Evolution Was Not Only in Africa. More than 80 *Homo erectus* specimens like this one were found recently on Java. The age is more than 1 million years. Adapted from Larick and Ciochon 1996.

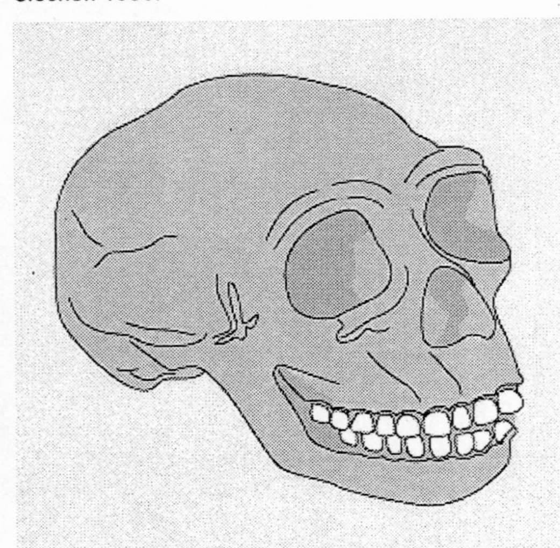

Early African Ancestors

We may picture the habitat of this early human ancestor as being along the border between the dry African veldt and the beginnings of forest. Probably there were two species living there. *Australopithecus robustus* was a somewhat chimpanzee-like animal who probably lived within and on the edge of the forest (Fig. 11-3). If this were the case, these early humans probably needed a heavy coat of hair to protect against thorns and brush and to act as a raincoat (present-day chimpanzees experience about 6 months of rain each year).

The second early ancestor (*A. africanus*) was probably more advanced, possibly had primitive tools, probably lived on the open plain, and assumed an upright position to facilitate running to overtake game in the fashion of the Kalahari-Bushmen of today. This progress depended upon bipedalism, which emerged 4.4 million years ago (Leakey 1995). An argument can be devised to suggest that this early human had copious sweat glands on a hairless, smooth skin, a situation which would not be needed by *robustus* (Folk and Semken 1991).

The discovery of these early humans is the work of Louis and Mary Leakey at Olduvai Gorge, South Africa. Decades of research on primate behavior by Jane Goodall, Dian Fossey, and others provide interesting avenues for speculation on the behavior and society of these earliest human ancestors.

Migration from Africa

It would be an oversimplification to suggest that the descendants of these primitive humans spread out from Africa into Europe and Asia, to Australia, and eventually to North and South America. Such a concept has been called the "Garden-of-Eden" viewpoint (Simons 1963). Probably other human types were developing simultaneously elsewhere, even in Europe. In fact, recent findings, coupled with new dating techniques, have lead some scientists to suggest that the cradle of civilization may lie in Asia, not Africa. Nevertheless, some descendants of *Australopithecus* did arrive in Europe, eventually forming numerous populations of two types of Neanderthal people, designated as the *classic* and the *advanced*.

The Neanderthal culture marks a critical point in human evolution, a time, it can be said, when our early ancestors clearly transcended the boundary between humans and lower animals. Anthropologists have learned much about Neanderthal culture: these primitive people buried their dead, had ceremonies, took care of their old, had language, and were developing artistic ability. Here we see a conspicuous event or model, which will help us understand the condition and predicament of humankind today: Neanderthal, the climax of a gigantic spreading of a species, demonstrates the adaptive nature of the human being. By means of its rapid and continu-

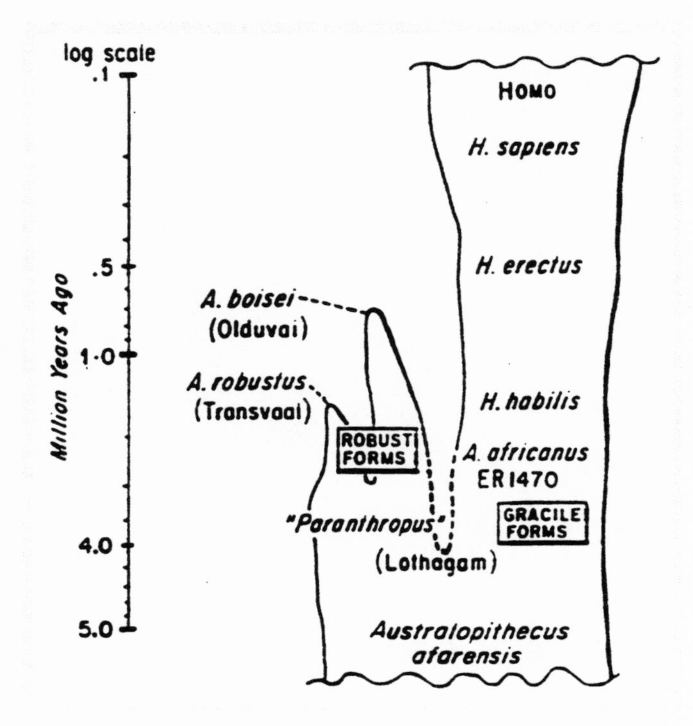

Figure 11-3. Proposed Human Evolution over 5 Million Years. This model of late Pliocene/Quaternary human relationships, one of a number that have been proposed, would allow the robust austrolopiths to maintain a heavy fur coat and permit the gracile forms to have reduced body hair. Alternatively, both could have been involved in fur reduction. Adapted from Clark (1973).

ous process of change, the human race had proven its biological success. From that point forward, culture, rather than biology, would become the primary force directing human survival.

Accompanying the humanoid migration from Africa were the colobus monkeys. At that time, there was species richness in the hominids, but not in these leaf-eating colobus; today this is reversed. So rich is the radiation of species of colobus that one, the langur, has developed resistance to strychnine. It feeds on the leaves of the tree that produces strychnine and tolerates a heavy injected dose, a dose that would kill a macaque (Davies and Oates 1995).

How fast were human populations increasing? In Table 11-1 we find some estimates presented by Ehrlich (1970) of the early changes in populations. He points out that at the time of *Australopithecus*, the world's population of the higher primates might have been 2,500,000. By taking some liberties with his data, we have been able to indicate convenient doubling times for world populations. It took 1,000,000 years to double the population of the South Africa-type humans; it took only 200 to double the 1650 world population.

Let us now picture a large population of classic Neanderthal people living to the west of the Alps, and the later advanced types living to the south and east of this range. These people lived at equilibrium with their environment for a long period, developing individual artistic geniuses who left exquisite drawings of their game animals on cave walls. Hunters were replicating horses, reindeer, mammoths, rhinoceroses, and lions in the French Chauvet Cave 30,000 years ago (Chauvet, Descamps, and Hillaire 1996). These paintings were considered hunting magic: "If you draw it, game will come." Perhaps artists competed to draw the game better and more accurately. About 10,000 years later, animal figures carved from mammoth ivory were found, possibly created by people described as a Cro-Magnon invasion (Mellars 1996).

The next major event in the time scale was the ice age; the known large populations of the classic Neanderthal became trapped by part of the ice sheet in the European area, and then disappeared from the fossil record. The advanced Neanderthal went on to leave descendants in Asia Minor and presumably in Siberia. There is much

Table 11-1. Population Growth

Year	Population*	Doubling Every:
1,000,000 BC	2,500,000	
8,000	5,000,000	1,000,000 years
1650 AD	500,000,000	1,000 years
1850	1,000,000,000	200 years
1930	2,000,000,000	80 years
1971	3,700,000,000	41 years
2006	7,400,000,000	35 years

*Ehrlich 1970; Arizpe and Velazquez 1991

speculation as to why the classic Neanderthal, who had been so numerous, disappeared. Most writers suggest that climatic change was the cause. We will make the point that there are 230 indigenous viruses in wild populations of chimpanzees (Bourne 1970), but that these have not killed off the chimpanzees. Perhaps there existed approximately the same number of viruses within the population of the classic Neanderthal; perhaps they were unable to cope with the presence of an ice sheet and viruses at the same time.

The Neanderthals were physically unique, especially in their skull characteristics. When the classical Neanderthal were at their peak, about 90,000 years ago, the natives of Africa and of what is today Israel, were far more like modern people (or Cro-Magnon) in appearance than were the Neanderthals (Mellars 1996).

Adaptation Required

The next question is why primitive humans had expanded into the northern areas where life must have been more difficult; the mild climate of the Mediterranean region should have been more desirable. The oversimplified answer of course, was overpopulation. Even if these people were learning to be farmers, the population was too large to be at equilibrium with the environment. At this point, competitive exclusion of a race would begin. Thus, early humans were forced to expand into the less desirable climates, and their children would have remained there. Does this sound like the migration of people from our present rural areas into the urban areas where the smog is an unpleasant factor?

One of the races "forced" into the north was a distinct type of Stone Age people comprising the Magdalenian reindeer culture. These are classified as late Old Stone Age people; they had cave art and tools of polished stone and bone. Most importantly, they made clothing by sewing the skins of animals. We can assume that life was relatively difficult for these people if we can compare them with the present-day caribou Inuit culture of northern Canada. The Magdalenians migrated over long distances with the reindeer, and maintained a specific and limited equilibrium with their environment. They were specialized predators, apparently depending about 98% upon reindeer for food. It was probably a culture of this sort that spread across Siberia and eventually to North America.

Once again, one asks why this expansion took place. Was it because of climatic change, loss of game, the threat of another war-like culture, or did the grass look greener on the other side of the hills? No matter what the cause, the pressure was there, and there was in North America a vast area where there would be no intraspecies competition—at least not at first.

Beringia: The Land Corridor

The human species expanded into North America over Beringia, a vast area of land which connected Siberia and Alaska. Two unfortunate terms have been used about this event. One occasionally reads that people "migrated" over a land "bridge." The word "bridge" is totally unsatisfactory because Beringia was a great area extending about 1000 miles from north to south. The word "migration" is unsuitable as well, because the passage was on such an extended time scale; probably people "lived their way" across the corridor numerous times (in the intervals, this land mass was submerged as it is now). As populations outgrew the resources available, new villages expanded endlessly to the horizon.

There are numerous theories as to the approximate time when the first people appeared in Alaska. As physiologists, we are particularly curious about what type of individuals were living on Beringia: were they naked like the Alacaluf Indians in South America, or were they advanced, well-organized hunters with warm, sewn-skin clothing?

Fortunately, the bio-geographer and climato-paleontologist can give us some answers as to the appearance of some of the Beringia corridors. There were at least three discrete corridors widely separated in time; probably each of them supported a totally different type of vegetation. Most workers agree that the last corridor was an area of wide plains with low hills and sparse woodlands; they also suppose that humans were represented by a primitive Magdalenian-type of culture that was "living" its way across. A popular theory states that this happened approximately 24,000 years ago during the simultaneous existence of Beringia and an ice-free corridor through Alaska toward the south, permitting continuing expansion down into lower North America. This time scale probably allows for dispersal over two continents (North and South America) and across 10,000 miles.

The archeological record of these events is very scanty. We do know that the famous Clovis points were widely distributed over middle North America about 11,000 years ago; these points were industriously made pressure-flaked arrowheads and knives. Apparently, ancestors of the Inuit, with a culture much like the Magdalenian (including art and fine tools), lived their way here about 10,000 years ago.

Again we ask, what were the pressures that caused these enormous expansions of the paleo-Indians? With all of the vast Asiatic continent to live in, there must have been tremendous population pressures to bring about such a relatively rapid and long expansion. A comparison with cultures in east Asia at this time (10,000 years BC) tells us that Japan, for example, was indeed heavily populated.

Populating North America

We next ask the question, were the primitive people who spread from Canada down onto the plains of the United States after the last Ice Age in equilibrium with their environment? The time span between 11,000 years ago up to 5,000 years ago is especially interesting; at the earliest date, Clovis people were scattered over much of northern North America. Available game for food seemed to be abundant as well as extremely large: by 10,000 years ago there were still giant sloths, mastodons, mammoths and many smaller species. We can thus picture populations of humanoids spreading down toward and through South America, apparently with abundant game available, because the practice of agriculture had not begun. In one sense, an equilibrium was not present because rather suddenly, about 9,000 years ago, the large mammals disappeared. One of the strong theories concerning their disappearance is that human hunters were responsible for their extinction (Krantz 1970).

Meanwhile, the world's population had increased from 4,000,000 individuals to perhaps 60,000,000. The natives of North America were being threatened with a loss of food resources. Entire nations of Indians flourished over North America. One estimate puts the Huron nation at 60,000, inhabiting the broad valley northeast of what we now call Lake Huron, and northwest of the great Iroquois nation. Agriculture came late to these people; it was not until 500 or 1,000 years ago that corn, beans and squash were cultivated. Life was in a somewhat delicate balance, since there is evidence that these Eastern nations greatly depleted the white-tail deer and caused the woodland bison of eastern North America to become extinct. However, their fish resource was by no means exhausted. They had certainly perfected the technique for obtaining this protein; we know that nets, sinkers, and good cordage existed 3,000 years ago in North America.

Population pressure among these Eastern Native Americans might have been relieved by a westward migration. The Western Indians were very much in equilibrium with their vast buffalo herds, which did not disappear the way the other giant mammals did. On the great plains, at about the time of the disappearance of the other large mammals, there were probably over 60,000,000 of these giant cattle; a mature bull stood almost 6 feet tall and could weigh 2,000 pounds. The Indian in relation to the buffalo was a dominant carnivore like the wolf. Let us now review some simple principles of ecology and physiology to apply to these Plains Indians.

Primary Ecological Principles

The growth of plants represents primary production. Plants are the first organisms to tie up energy from the sun; they are the first and original producers. The plants are basic to the *community;* added to their network are the animals, the soil, water, and many other factors. The ecosystem is made up of communities. Primary consumers, such as grasshoppers and bison, eat the living tissue of plants. Secondary consumers such as predatory insects eat the grasshoppers but not the plants. Other secondary consumers are wasps, tiger beetles, spiders, toads, lizards, and song birds. These are all less abundant species than plant eaters. Although we have called the bison a primary consumer, this is an oversimplification because the abundant one-cell organisms in the bison's digestive tract break down the plants. At any rate, this complex of organisms (the bison) is preyed on by a number of predators; such predators are typically large and speedy, and their population must be smaller than that of their prey. Predators are subject to starvation; plant feeders control their predators better than the carnivores control the plant feeders. Some mammals are generalized predators; others, like the wolf and the early Plains Indian, were specialists.

All these relationships can be visualized as a pyramid. This pyra-

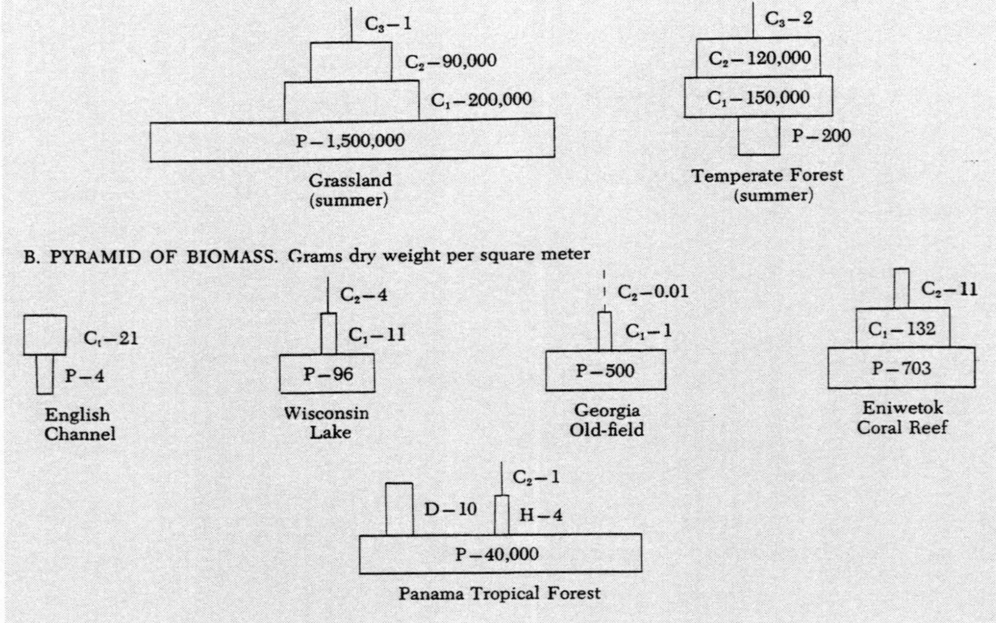

Figure 11-4. Examples of Ecological Pyramids. P = producers; C_1 = primary consumers; C_2 = secondary consumers; C_3 = tertiary consumers (top carnivores); D = decomposers. Adapted from Odum 1971.

mid may be one of numbers, of mass, or of the total amount of the sun's energy (Fig. 11-4). Each level or step in the pyramid is a trophic level. At the pyramid base we find primary consumers. Moving from the base to the next higher level represents *a loss of energy*. The mass of animal life declines in successive layers of the pyramid. On top is the species which is not preyed upon by anything else (under natural circumstances); in the midwest grasslands of the past it was the wolf. In the Arctic, it is the bear.

Now let us consider the above statement concerning *a loss of energy*. Take as an example the Inuk who is living on seal meat. The essence of the energy distribution is that only a certain fraction of the sun's energy is passed on to the herbivores by the process of photosynthesis at the base of the pyramid. An often-quoted example of the total energy budget begins at the apex of the pyramid and can be expressed as follows;

1. each pound of seal consumed by the Inuk must have consumed 10 pounds of herring;
2. this 10 pounds of herring must have consumed 100 pounds of small fish;
3. this 100 pounds of small fish must have consumed 1,000 pounds of algae.

The changes in this energy budget are referred to as *loss of energy transfer*. Some of this same loss of energy transfer applies to a steak which is cut from a bison, elk, caribou, or Black Angus steer. Obviously, the harvest of available energy is much more efficient if an omnivore eats plant food.

Equilibrium with the Environment

Now let us ask about the "equilibrium with the environment" of the Plains Indians. If they had been left alone there might still be good equilibrium, with human beings as the dominant carnivore. However, the human was not a perfect carnivore compared to the wolf because, for the most part, the wolf culls the herd of its prey, taking the over-age and vulnerable animal. The human preferred the young, fat, productive cow.

We should now consider the means of transport of these Plains Indians; for as long as 2,000 years, they used dogs for transport, employing both travois and pack. Then the horse was introduced by the Apaches. These animals brought an "industrial revolution" to the grassland. Life was much easier with horses; they could be ridden in the hunt and used as pack animals. Even with this great change in their way of life, the Plains Indians probably would have lived without damaging the environment or depleting their great food resource; through wars with other tribes, they were limiting their own population. Some Indians did begin to waste resources—harvesting buffalo by driving herds over cliffs—but undoubtedly the great buffalo herds would have tolerated this form of "industrialization."

The resulting situation is called a food chain, with each trophic level being a link in the chain. Four examples of food chains will assist in understanding this:

1. plant → grasshopper → meadowlark → falcon
2. seeds → deer mouse → weasel → coyote
3. grass → vole → skunk → golden eagle
4. grass → bison → Plains Indian

Now let us turn the clock forward to about the year 1800. The world's population was approaching one billion people. Europeans were not living in equilibrium with their environment but were moving in great migrations to North America. About that time Thomas Jefferson gave $15,000,000 for the Louisiana Purchase. By 1900, this region contained one-third of the population of the United States; in the previous hundred years, the population of Louisiana had increased from 5,000,000 to 100,000,000.

What happened to the 60,000,000 bison in the plains? About 1850, the Sharps rifle arrived in the hands of the professional bison-hunter. This rifle was effective at a range of 100 yards, and allowed one hunter to kill between 60 and 100 bison at one stand before the herd fled. By the year 1889, hunters were no longer permitted to kill bison because there were only 100 left. The species might have become extinct if it had experienced a bad epidemic or severe winter. At any rate, the food chain was destroyed for the Plains Indian. Why were these great animals destroyed? It was partly for economic reasons and partly as an actual planned program to break the food chain in order to allow settlement for farming by westward-moving pioneers. This may be considered another example of cultural or social influence on evolution.

ECOLOGICAL PRINCIPLES OF MODERN HUMANS

And so the wave of Europeans swept over North America, bringing more industry and indulging in a great spending spree of resources. Let us look closely at this great spending spree and summon some further ecological principles.

A Model of Overpopulation and Decline: The Lemming Cycle

Picture a vast population of brown lemmings on the 40,000 square miles representing the Arctic Slope. The brown lemming is a grazer. The numerous Arctic predators live in equilibrium with them, and there is by no means a danger of the prey diminishing because of the predators.

J. Forrester (1971) has applied certain rules to this system that

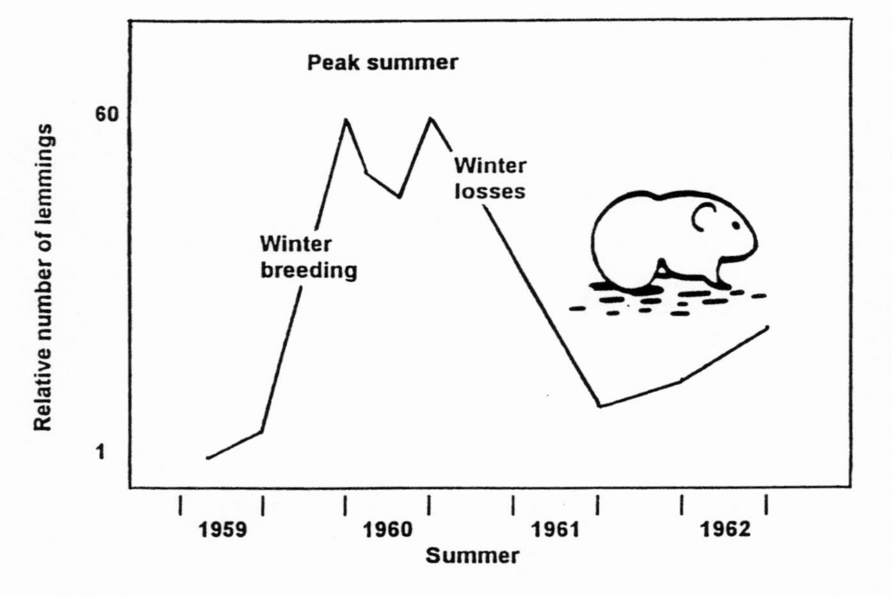

Figure 11-5. Peak and Crash of Lemming Numbers at Point Barrow, Alaska. Occasionally, the cycle occurs every four years. Adapted from Chitty 1996.

include the return to the soil of many nutrients excreted by the lemmings or by the carnivores who have eaten them. The rules are as follows: 1) after the contribution of energy by the sun there is a non-cyclic degradation of this energy; 2) there is recycling of matter; 3) diversity in an ecosystem favors stability; 4) any ecological unit has a limited carrying capacity; 5) rising populations create a pressure on food production; 6) natural resources become exhausted; 7) pollution dissipation capacity of the unit becomes overloaded, this puts a natural ecosystem under stress; 8) overpopulation may result in a homogeneous landscape with far less diversity of species, particularly in respect to vegetation. This is due to feeding on a short supply of plants, as illustrated with an oversupply of lemmings or deer (Diamond 1992).

The above principles are illustrated by a walk on the tundra south of the Point Barrow Laboratory when there has been an overpopulation of lemmings. The grass in many areas looks somewhat like the putting green on a golf course. If one looks closely at the neatly cropped grass one can see conspicuous accumulations of lemming droppings. It is apparent that the grass has passed the stage of its carrying capacity. Temporarily, the conspicuous piles of feces represent an overloaded capacity to dissipate pollution.

What numbers are associated with this overpopulation? D. Mullen designed a transect of traps near Barrow, and on it captured (and released) 4,000 lemmings. The next summer he increased the transect by 25%; in the same period of time, and using the same methods he caught only 20 lemmings. It is apparent then, that associated with this picture of lemming overpopulation is a "crash" in numbers. Frequently there is a cycle of about four years of a peak and crash (Figs. 11-5 and 11-6). Let us see what is known about the cycle, which sometimes is accompanied by a migration.

The history of long distance migrations and cycles of lemmings begins in the 1920's with the work of Elton in Europe. At the same time, Russian scientists found cycles and migrations of *Microtus brandti* in Mongolia. There is no consensus as to causes, but the hypothesis emphasizes intrinsic regulation (stress and disease), or extrinsic regulation (plant productivity, plant chemistry, and predation) (Chitty 1996). Large research sums have been

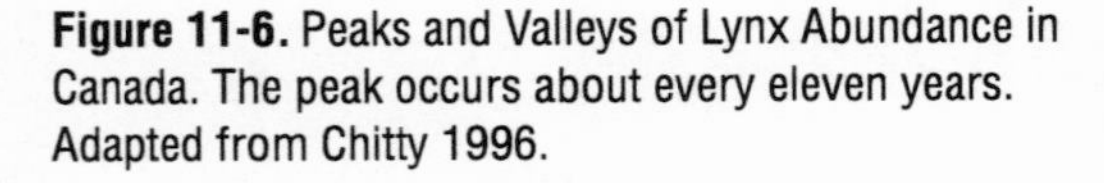
Figure 11-6. Peaks and Valleys of Lynx Abundance in Canada. The peak occurs about every eleven years. Adapted from Chitty 1996.

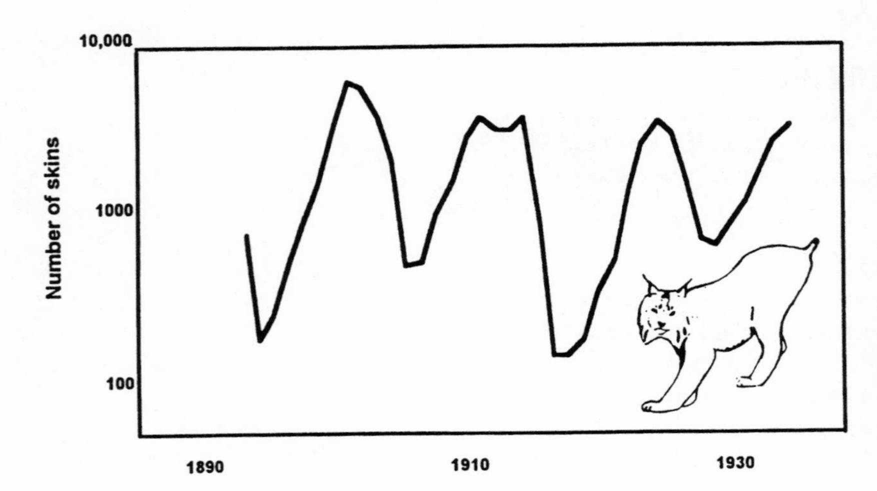

spent searching for the mechanism, as in the International Biological Programme, without success. Two Polish biologists, W. Jedrzejewski and B. Jedrzejewska (1996), have at least found a pattern: rodents in tundra, taiga, steppe, and winter crop farms do cycle; those in temperate forests and desert do not.

The principle illustrated here is *resources limit populations*. The model from Pt. Barrow is found also on the plains of Mongolia, where people graze their livestock. There, rodents eat 25% of the grass in a non-peak year. Today the villages have to move in such a peak rodent year because the livestock cannot live; an ancient Mongolian book describes the same phenomenon.

Although the lemming cycle is not understood, let us use this lemming model to enhance our understanding of our present ecological predicament. In the next few paragraphs, we will consider the causes of the precarious position of humankind, the consequences of this present situation, and solutions which must be considered.

Overpopulation

The human population on the Earth is now over 6 billion. It is doubling about every 30-40 years, but the doubling time has decreased over the centuries (See Table 11-1).

We have to predict then that with the present growth pattern, there will be between 6 and 7 billion people by the year 2000 (Fig. 11-7). Many scientists have attempted to calculate the Earth's carrying capacity because of the rapid increase in populations. They have to make assumptions, of course, in respect to food intake levels, crop productivity, degrees of industrialization, tolerable levels of environmental contamination, and above all, human inventiveness in synthesizing food. Because there are so many assumptions, these evaluations have limited significance (Cohen 1995). Such as they are, the estimates of Earth's final population carrying capacity have varied from 10 billion to more than 50 billion, depending upon the standard of living considered acceptable by various cultures around the world (Henshaw 1971; Moffat 1996).

We could, of course, extrapolate some alarming figures for when the planet's carrying capacity will be reached; before then, let us discuss a factor upon which carrying capacity depends, namely resources.

Figure 11-7. World Population Growth. Estimates vary, but one U.N. committee has estimated there will be 6.5 billion human beings on Earth by the year 2007. This graph demonstrates vividly why the rapid increase in humankind since the year 1,000 is referred to as "The Population Explosion."

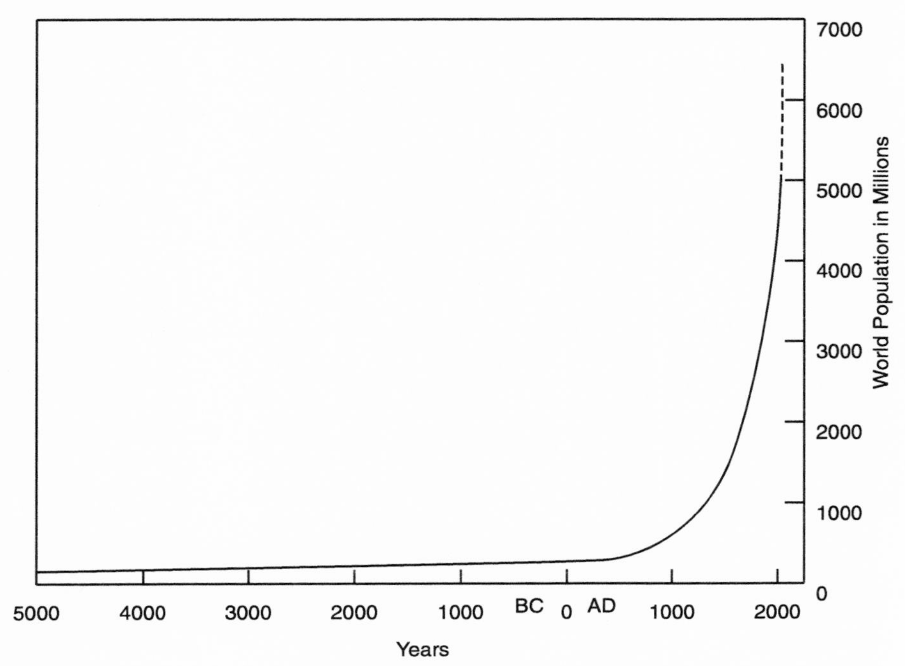

Resources

Though we have seen that there is no consensus in the estimates of Earth's carrying capacity, it seems clear that humankind's choices in how we consume our natural resources will affect the number of people the Earth can support. The point of view we must take toward resources can be illustrated by the problems of energy consumption and food production.

Energy Consumption. The world's consumption of energy is continuing to grow exponentially. In the United States, the use of gasoline is growing at a fast rate, ironically because of emission controls that make automobiles less efficient. Many scientists are concerned about damage to the environment in places where there is a conflict between oil production and environmental protection. The controversy over the Alaska National Wildlife Refuge is a typical example. Here, an attempt has been made to drill in order to avoid dependence on foreign oil. The controversy involves a sanctuary of 9,000-foot snow-clad peaks, glacial valleys and rolling tundra meadows sloping down to the sea ice. The drilling would affect at least 150,000 caribou, whose newborn young are ordinarily protected from mosquitoes on the shore of the Refuge (Kenworthy 1995). The exsanguinated bodies of calves unlucky enough to have been born outside of this refuge have been found. Although the refuge is of great importance to Americans, it is still a minor concern in the world view.

The Office of Technology Assessment compared recent oil usage with projected figures in millions of barrels per day:

	1990	2000
Total Consumption	15.4	15.7
Domestic Production	9.5	8.0

This marked drop in the domestic contribution must be made up for by larger imports or alternative energy sources. The United States appears destined to continue its reliance on foreign oil, which makes it the world's largest international debtor, and keeps it vulnerable to deleterious political and environmental developments (Abelson 1997).

One result, of course, is considerable attention to the question of how we should conserve energy. This can only be achieved through voluntary public cooperation; such cooperation in the past has only been sustained when there was a war or severe power outages.

Americans have been most successful conserving oil in places where fields of wind-powered generators have been built. The state of Iowa is a leader in the development of wind energy. Formerly, this state imported 98% of the energy it used. Soon, a 40 turbine wind-facility will be operational 16 miles west of Fort Dodge. Developed by the Zond Iowa Development Corporation, this 30 megawatt wind farm will provide enough electricity to power a town of nearly 20,000 residents (Bean 1996).

Food Production. Let us next examine an even more serious aspect of the resource problem, the food shortage. Globally, there is already a vast desert of starvation and malnutrition, and only a few small islands of affluence where people are privileged to include abundant meat protein in their diets. However, because of the ecological pyramid, the major reliance for providing an adequate diet must be placed on cereal grain. The reason for this depends upon the point made earlier, that the energy transfer from one trophic level to another is not efficient. To raise livestock for meat requires vast amounts of grain, and the resulting protein yield per acre is very low. However, animal flesh contains ten essential amino acids, while plant proteins are missing one or two of these essential compounds. Plants have a greater efficiency for the production of these remaining eight essential amino acids if calculated on an average pound-to-acre ratio: crops such as soybeans, peas and beans yield 13 pounds of these amino acids per acre, whereas only 1.6 pounds per acre can be found in poultry, beef, lamb, and pork (Majumder 1972). The new push then is to find plants which contain these deficient amino acids; this has been done experimentally with corn in respect to lysine and tryptophan. It might eventually be possible for cereal grains to provide the necessary protein and carbohydrate for almost two-thirds of the world's people. We are woefully behind in developing and using this type of research.

According to Majumder, some frightening statistics indicate that two-thirds of all the children in the world—the future citizens of our planet—now suffer from malnutrition due to protein deficiency. Meanwhile, fish and meat proteins are out of their economic reach, despite the extraction of fish from the seas. Several countries have attempted to reverse the protein shortage by continuing to harvest endangered whales—contrary to the efforts of the International Whaling Commission to stop all whaling. (There has been recovery in one species, the gray whale, but a continued depletion of several others, especially of the Atlantic right whale.)

The world's catch of seafood has been accelerated by technology, and now 50% of the world's food-fish stocks are in a crisis. Lester Brown, editor of the annual Worldwatch Institute Reports, tells us that from 1950 to 1995, factory ships have extracted five times as much seafood as ever before. For example, boats from Vigo, Spain, now must roam from Africa to Greenland, and their freezer trawler may swallow 60 tons of fish per haul. More than one million vessels sift the oceans for seafood—twice as many as in 1970. The biggest ship can process 600 tons of pollock a day. In shrimp boats, 4 pounds of the catch is discarded for every pound of shrimp retained (Parfit 1995). Clearly, this is part of a serious and dangerous trend affecting the quantity of global food resources.

Not only are vertebrate fish depleted, but some invertebrates are in short supply. In 1994, 20 million green sea urchins were shipped to Japan from the coast of Maine, USA, because the supply on the Pacific rim is exhausted. The Atlantic shore cannot sustain this drain (D. Brown 1996).

Is the growing food shortage completely irrevocable and incurable? The leaders of science have for years been divided into two camps, some optimistic and some pessimistic. Historically, these two extremes were represented by Borlaug (1972) for the optimists, and D.H. Meadows (1972) for the pessimists. A sample of Borlaug's reasoning is that we must put up with the damage from pesticides for the greater good they do in combating disease (such as malaria) and increasing our food and fiber production. Today, he still believes that humans can continue to feed themselves by using improved plants and agricultural methods. Meadows, on the other hand, had predicted the exhaustion of resources, such as fertilizer, after the year 2000. From our vantage point in 1998, this seems unlikely. But it has been remarked, "Meadows predicted hell on earth in fifty years, but hell is already present in places such as Calcutta." Currently, Peter Raven (1991) supports the pessimism of Meadows, but predicts exhaustion of resources at a later date.

We must now look at a final resource which may be the crucial link in the chain that holds the anchor of our survival in the present stormy sea: this resource is *phosphate*. The writers of the *Workshop Report* from the Institute of Ecology (Hasler et al. 1972) felt very strongly about the importance of this material: "When the practice of industrial agriculture is interpreted in the light of current knowledge of ecosystems, a picture emerges which suggests that the future dependability of such agriculture is in grave doubt." In respect to natural resource limitation we learn that, "... known potential supplies of phosphorus, a nonrenewable resource essential to life, will be exhausted before the end of the 21st century. Without phosphate fertilizers the planet can support between one and two billion people." This statement placed these writers in the camp of the pessimists. Who will be right?

Figure 11-8. Birds Act As Litmus Paper for the Damage Done to the Physical Environment by Humans. The largest flying bird is the wandering albatross, with a wingspan of eleven feet, occasionally twelve (see below). The longline fishing trawlers catch and kill so many wandering albatrosses that now they are endangered. The lines can be 40 miles long.

Species Diversity and Resources

One aspect of the need for diversity is illustrated by plant survival. Eisner (1992) studied Lake Wales Ridge in Florida, a region which had received an environmental insult so that many species had disappeared. He noticed an endangered mint plant that had a powerful odor and was virtually free of insect injury. The plant contained a new compound, trans-pulego, a potent insect repellent. The plant was found in a few hundred acres of protected habitat.

There is an individualized food chain in all parts or regions of the globe, whether in a city or on a rain forest island. The bird population is part of the chain in many ecosystems—unfortunately it is often a vulnerable link in the chain. As humans upset the food chain by funneling fish to their own hungry populations, bird and other animal populations may suffer. And humans hunting one species often harm another; for example, the wandering albatross, which has a wingspan of 10 to 12 feet (see Fig. 11-8), is threatened by Japanese "longliners."

Longliners are so named for their method of fishing with lines up to forty miles long—lines that carry thousands of baited hooks. Japanese longlines kill 44,000 albatrosses a year, and the global death score could be double that figure. The fisheries are implicated in the decline of twelve albatross species (Slater 1996). On a hopeful note, however, a United States-sponsored resolution to reduce unintentional seabird mortality in longline fisheries was recently adopted by the International Union for the Conservation of Nature. Though not legally binding, it is encouraging that seventy-five nations, excluding Japan and Panama, supported the resolution, and that it was also approved by a fishing industry organization, the North Pacific Longline Association.

Biologic diversity is supported and encouraged by the American Endangered Species Act of 1973 (ESA). There is little criticism of the ESA mandates, but there is a need to improve the regulations that guide its implementation (Easter-Pilcher 1996).

Mineral Rights and Ecosystems

How will animal populations fare against mineral "rights," and the needs of certain groups to exploit mineral resources? As an example, the vast tundra of Canada's Barren Lands is about to be invaded because diamonds and other valuable deposits have been found in the eskers that run through that region. These eskers are high ridges of sand and gravel formed only 6,000 years ago as the ice sheet retreated. They are abundant in parts of the Barren Lands and in some of the United States. In the Barren Lands, lakes and marshes between the eskers make the area an ideal home territory for large populations of caribou, grizzly bear, musk-ox, and wolves. But these animal populations are now being invaded by human prospectors who represent the BHP Diamond Company; they use the eskers as gravel highways to find and develop the diamonds, gold, copper, and zinc buried in them (Kajick 1996).

Learning from History: The Vanishing Rain Forest

The rain forests of South America and the Northwest United States have been called "the lungs of the planet." Unfortunately, their size is rapidly diminishing. This can have serious consequences for our civilization, as history will show. For example, the disappearance of the Mayans, who existed from 1800 BC to 1530 AD, can likely be explained by their unwise use of natural resources. It appears that they exploited the rain forests around them but found the jungle land unsuitable for farming; later, floods released by sterile land completed the damage to their agriculture. Similarly, from about 700 AD to 1500 AD, Cahokia, a city of over 15,000 people, thrived along the Missouri river in what is now Illinois. These early Native Americans built an impressive city with vast plazas, sun calendars, massive earthen mounds, and a city wall of 20,000 logs that had to be rebuilt three times. Their abandonment of the city in 1500 AD was considered a mystery until recently, when scientists found evidence that the Cahokians' need for timber led to deforestation, leaving no plant cover to stem flooding of their fields of crops (Holden 1996).

Today, large tracts of rain forest are being burned so the land can be used for agriculture. The effect of biomass burning in Brazil was studied by an international team; on some days the smoke of distant fires attenuated as much as 80% of the solar ultra-violet B and more than 30% of visible light. Part of the famous Pantanal, a wet region of abundant wildlife in Brazil, was under thick smoke for days at a time (Mims 1995).

The loss of forest has other ramifications: when forest is lost, animal life is lost. In Costa Rica alone, nine species of frogs have been lost. In recent experiments, Blaustein and others (1994) showed that if frog eggs are not covered by some tree leaf protection, our diminished ozone layer (the ozone hole) allows ultraviolet light to kill many of the eggs. However, Licht (1996) suggests that these experiments can explain only in part the reduction in frog populations. The overall causes of decline are reviewed by Sarkar (1996).

Here are medical facts about two other species of frogs at risk as more and more rain forest is destroyed:

Phyllomedusa bicolor: A team of researchers led by John Daly of the National Institutes of Health has found that the skin of the large green frog of Peru produces a peptide, 33 amino acids long, that may manipulate cellular receptors for the common biomolecule adenosine. This discovery may aid in developing treatments for depression, stroke, seizures and Alzheimer's (Amato 1992).

Epipedobates tricolor: The poison arrow frog of Ecuador excretes an alkaloid 200 times more powerful than morphine as an analgesic; only recently have chemists discovered how to synthesize this chemical to aid in their search for a powerful nonsedating, nonopioid painkiller (Bradley 1993).

There are similar stories about the loss of unique plants in the rain forest. Fortunately, in some sections of Central America, villages have ceased "slash and burn" clearing of forests for agriculture, and are collecting allspice, exotic berries, and botanicals for dyes; in the process they are earning three times the nation's minimum wage. We will consider this further under "Sustainable Development."

Species Extinction

One physiological event caused by climate change and habitat loss is species-death, i.e. extinction. For example, bats are a favored mammal because they help in mosquito control and their guano is valuable. There are 40 bat species in the U.S., but of these, over 18 are listed as endangered.

We can learn quite a bit about climate and climate changes of the past from readings of ice-core samples and fossil strata, especially those of marine origin. In a recent example, evidence retrieved from 300 feet below the ocean floor suggests that a huge asteroid hit Earth 65 million years ago, leaving the world nearly dead for about 5,000 years. These findings lend credence to the theory that dinosaurs were killed off in a cataclysmic event.

What is the rate of extinction? Some species of dinosaurs became extinct at a rate of one species per 10,000 years. Today we are losing a species per day.

Climatic Change

When we ask how much modern humans are changing their environment, part of this question must be: Are we actually changing the *climate*? About 200 years ago, we began burning enormous quantities of fossil fuels, releasing the byproducts—water vapor, carbon dioxide, carbon monoxide, and oxides of sulfur—into the atmosphere. It is now believed by some scientists that these materials are assisting natural causes in bringing about a change of climate. There are views to the contrary.

Over the past century, the CO_2 concentration of the atmosphere has increased by about 30% (Fig 11-9). This suggests that any warming of the atmosphere is the result of the *greenhouse effect;* this term is used in the context that carbon dioxide and other gases in the atmosphere are transparent to incoming sunlight, but somewhat opaque to long-wave infrared radiation from the earth. Some have predicted that this increase in CO_2 could be responsible for an increase in global temperature of 1°C to 3.5°C by the year 2100. In 1990, the United Nation's Intergovernmental Panel on Climate Change (IPCC) issued its *First Assessment Report,* describing the damages this could cause.

The essence of the greenhouse problem is the analysis of the *carbon cycle.* In the great Arctic Slope of Alaska, which is a 70,000 square mile marsh, the annual CO_2 cycle is amplified because the ice and snow lock up the plants in winter, causing a rise in CO_2 (see "Barrow" in Fig. 11-9). But in the continuous light of summer, the

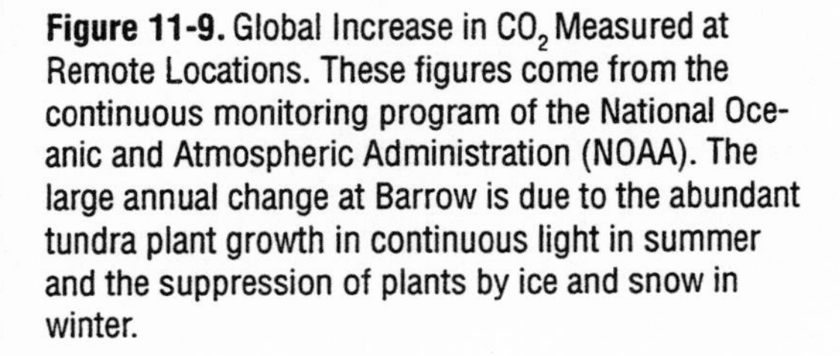

Figure 11-9. Global Increase in CO_2 Measured at Remote Locations. These figures come from the continuous monitoring program of the National Oceanic and Atmospheric Administration (NOAA). The large annual change at Barrow is due to the abundant tundra plant growth in continuous light in summer and the suppression of plants by ice and snow in winter.

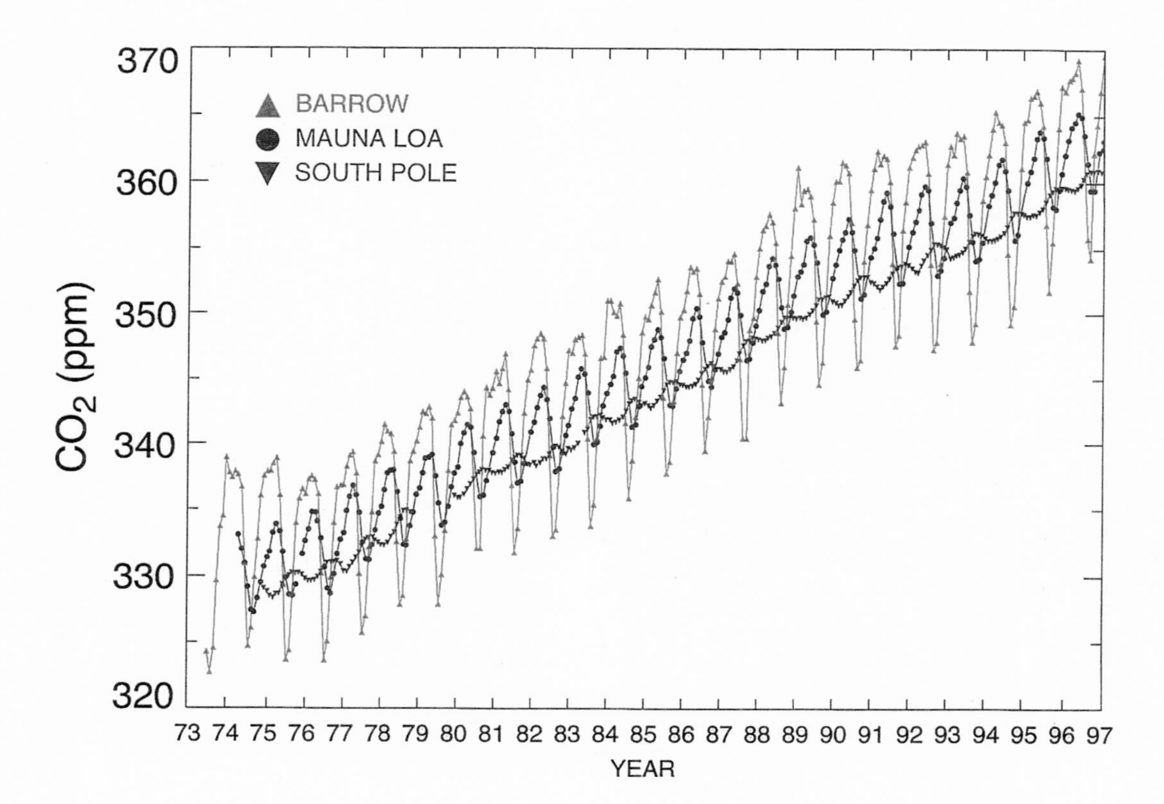

flourishing plants pull "extra" carbon dioxide from the atmosphere. In parts of the great oceans, there is probably near equilibrium, with equal amounts of carbon being *used* by plants and *given off* by decomposition. But when there is a phytoplankton bloom, the plants store carbon, and the balance shifts in favor of utilization. To encourage this, oceanographic scientists have added iron to sections of the ocean to "fertilize" the phytoplankton. In one experiment the result was to pass 2,500 tons of extra carbon into the phytoplankton (Trefil 1996).

However, the unusual warm years from 1987 onward may not be due to an increase in carbon dioxide. There are two other explanations: changes in the temperature of the equatorial Pacific Ocean, or changes in transparency of the atmosphere.

There is much discussion, debate, and disagreement among scientists today about whether there is, or will be, any change in climate. At least 2,500 scientists contributed to the IPCC's *Second Assessment Report,* approved in late 1995 (Kerr 1995). Their release states that the global surface temperature has risen by 0.5°F to 1.1°F, and that there has been a 4 to 10 inch rise in sea level. At a later meeting, this group cited evidence from Scripps Institution of Oceanography and the University of California, San Diego, that the growing season in many countries is a week longer than it was 20 years ago.

Other groups, including the Max Plank Institute in Germany, the UK Meteorological Office, and the National Center for Atmospheric Research, independently reached similar conclusions: that there are changes in the climate that indicate global warming, and that these changes are *anthropogenic* in nature (i.e., resulting from human activity), and not driven by natural causes such as solar activity or changes in Earth's orbit (Colony 1996).

A recent report by British scientists stated that the mean global temperature had risen from 57°F in 1880 to 58.5°F in 1995. They project global temperatures will increase by 0.4°F to 0.5°F each decade.

These changes in temperature and sea level are correlated not only with the 30% increase in CO_2, but also with the human contribution of an increase of N_2O by 15%, and an increase in methane by 145%. Although these climate changes and their correlates have been carefully documented, fossil fuel interests have launched an illogical but aggressive disinformation campaign, one which must be overcome by thoughtful explanations (Goetze 1996).

Not all scientists agree, however. Richard Lindzen of MIT, for example, rejects predictions of climate change and global warming, charging that computer models of the climate used in making these predictions are flawed. Lindzen further asserts that the physics of the atmosphere would permit only a "minor and untroubling warming," despite the buildup of CO_2. Others who believe the case for climate change has not been made include Dale (1995), who criticizes figures used to show rising global temperatures. According to Dale, these figures were recorded in areas that have become urban heat islands; records from a *rural* weather station from 1888 to 1993 showed no warming.

Nevertheless, marine ecologists have detected a warming of the North Atlantic, which has affected mollusk and fish breeding. Farther north, Per Gloersen reported that the pack ice of the Arctic Ocean receded over 9 years, but that a 30-year study was needed to relate this to climate; Regalado (1995) reports a 0.5°C warming of the Arctic Ocean, though El Niño could be the cause; Johannessen (1996) found a significant shrinking of sea ice by 5.8% over 16 years.

An increased global temperature would suggest a greater melting of the ice caps and a rise in sea level. There has, in fact, been a consistent rise in sea level of around 1 mm a year for one hundred years (Untersteiner 1984; Warrick et al. 1996). With global warming, sea level would increase even more due to expansion of ocean water. Snow and permafrost boundaries would probably recede. But, considering just the Antarctic ice sheet—which would contribute nearly 8 times as much water as the Greenland ice cap—Untersteiner suggests there could be a balance between melting caused by a warmer ocean, and increased precipitation back onto the ice cap caused by a warmer atmosphere, all of which would result in no sea level rise *due to melting*. We must not neglect the ice caps of the tropics and sub-tropics in this analysis; Ellen Mosley-Thompson (1996) of Ohio State University presents evidence that tropical and sub-tropical ice caps and glaciers are disappearing worldwide.

Scientists are investigating the role of plants in the climate system. Questions have been raised about the contribution of aerosols released by plants. Fritz Went, the physiologist who discovered plant auxin, showed that the persistent blue haze in the atmosphere, which is especially noticeable in autumn, originates from the volatile organic substances released by plants into the air. Some have suggested that these materials are more important than the aerosols, dust, and chemicals released by industry and volcanoes.

Growth rings in plants provide additional evidence of global warming. Australian and U. S. researchers studied growth rings in Tasmanian pine trees isolated by elevation and geography from other environmental factors. The team found a rapid growth in the pines since 1965; they believe that temperatures have risen faster in the past 25 years than at any time since 900 AD. Such evidence from plants is analyzed in minute detail by Gates (1972 and 1995).

Plants are also the focus in an experiment in Southern California's backcountry in which their carbon-storage capacity is being measured (Graham 1996). In experiments in the greenhouse and in the field, scientists have shown that excess CO_2 acts like a fertilizer to plants, causing them to grow more rapidly. Now, some wonder whether Earth's plants have the ability to "soak up" excess CO_2; if so, they speculate, plants may be able to mitigate global warming. But a weakness in this theory is revealed by examining the top graph in the Environmental Diagram at the front of this chapter, one of many that documents a steady rise in atmospheric CO_2. The steady increase shown in the graph indicates that even if plants do soak up some excess CO_2, existing plants have not been able to keep up with

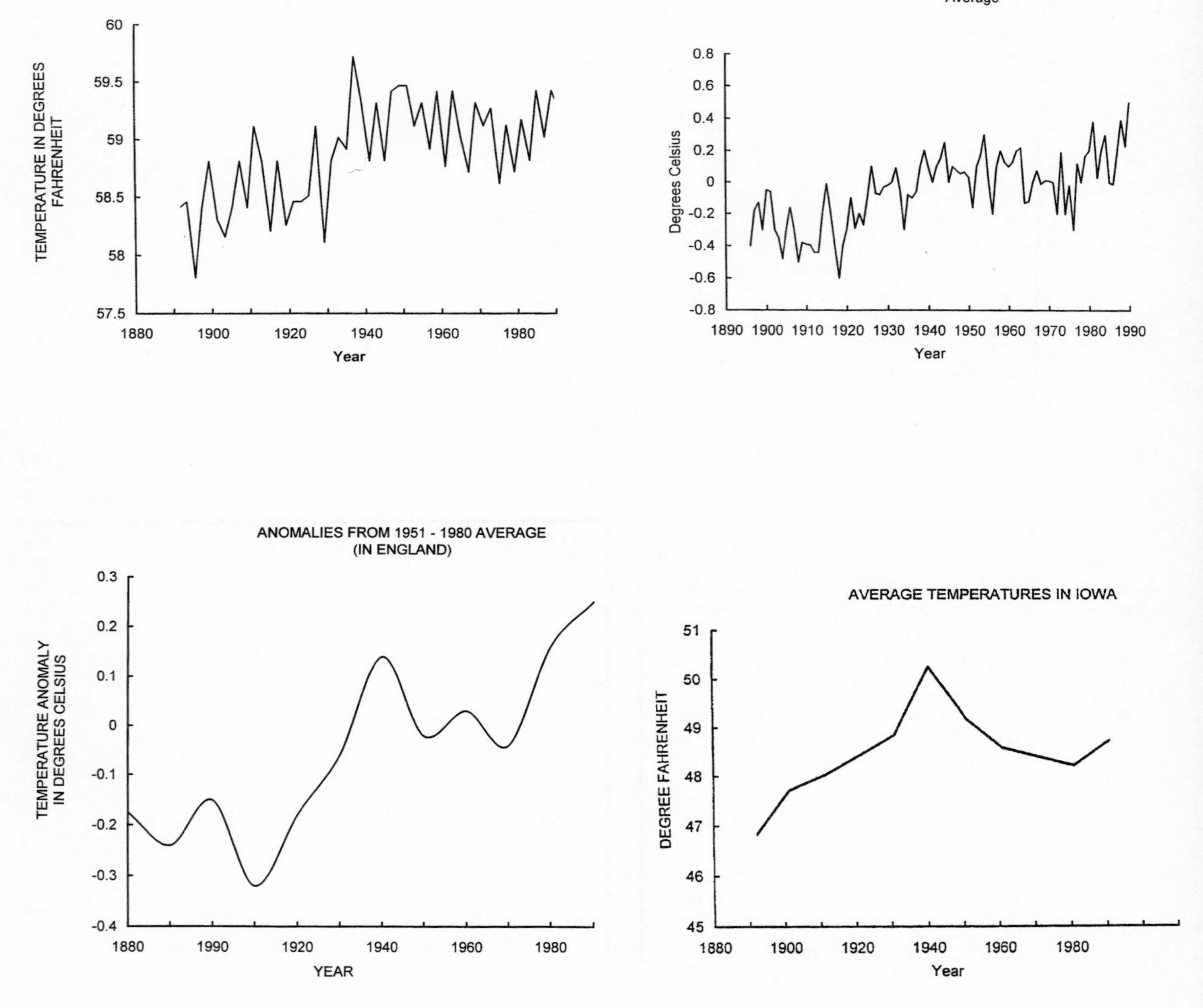

Figure 11-10. What Are the Causes of Global Warming? The warming of the globe may be due to either the greenhouse effect or to natural cycles. The above records of the cooling period from 1940 to 1980 may represent natural cycles. Adapted from Triemstra 1992.

the levels of CO_2 released into the atmosphere over the past several decades.

We notice in Fig. 11-10 that an abrupt cooling set in about 1940 (Kerr 1996a; 1996b). This was worldwide and lasted until 1980. It did not seem to be affected by the steady rise in atmospheric CO_2. This has received little comment and discussion. It is possible that the increased pollution from industry, automobiles, and jet contrails temporarily changed the transparency of the atmosphere; this could have caused more sunlight to be reflected back into space and allowed less radiation to reach the ground, resulting in a cooler ground surface and atmosphere. Or, this may be an example of a *natural* cycle.

Finally, we must also consider the instances of global warming that have occurred in the past. During the last 10,000 years (Holocene Epoch) there have been five oscillations which peaked roughly every 2,400 years. It is possible that these surges are not driven from within the climate system, but by solar activity. These natural oscillations, as opposed to the CO_2 greenhouse effect, must be taken into account as we consider causes of global warming.

Those oscillations have included periods of cooling as well as warming. For 750 years the globe was very cold; this period lasted from 1100 to 1850 AD, and was called the "little ice age." During that time, the canals of Holland and Boston Harbor usually froze over each winter; this has rarely happened since 1850. People had to adapt their clothing and their habits to this period of cold exposure.

What are the time sequences, and what are the methods used to detect these oscillations? The time sequences of these past changes have been measured by reading oxygen and nitrogen isotopes in ice taken from deep-ice bore holes. Apparently, Greenland and much of northern North America were 14±3°C colder than today; Greenland warmed up to its present temperature within 50 years (Kerr 1996c).

Some scientists believe that the climate is not only changing, but that it is doing so in a compressed time frame. In June, 1996, scientists, including George Woodwell, founder of Woods Hole Research Center in Massachusetts, and Ellen Mosley-Thompson of Ohio State University, held a news briefing in Washington, D.C. to highlight evidence of climatic change associated with human "destabilization" of the climate. As evidence, they cited the following: samples from the 12,000-year ice core record show that the last 50 years have been the warmest; several of the earth's major ice sheets are in retreat; and plants, seeking the cooler temperatures they evolved in, are migrating to higher elevations in mountains in Alaska, the Sierra Nevadas, the Andes, the Swiss Alps, and Kilimanjaro.

If there is global warming, it could have an effect on human epidemics. One disease traced to a changed climate pattern was the hantavirus, which leaped easily from deer mice to humans. Other diseases that resonate with climate changes are malaria, dengue fever, yellow fever, filariasis, encephalitis, schistosomiasis, and cholera (K. S. Brown 1996).

Let us return now to the question posed at the beginning of this

DIMINISHED OZONE, CORAL REEFS, AND TOO LITTLE SUNBLOCK

The damage to the ozone layer's protective shield not only affects Earth's terrestrial inhabitants, but also reaches deep into our seas. Like phytoplankton (see text), the organisms in coral reefs are sensitive to UV radiation. Corals produce chemicals, similar to our sunblock, that protect them against the sun's rays. But many corals don't produce enough sunblock, and when UV light penetrates through clear, calm waters, it can damage corals. UV radiation appears to be on the increase because of the growing holes in the ozone layer. Scientists aboard the *Aquarius*, NOAA's underwater habitat located off the Florida coast, have documented the damage to coral reefs caused by increased UV radiation and other factors. (See sidebar, Chapter 10.)

section: Are we, indeed, changing the climate? While there is much evidence to support the concern that we are, there is also much we have yet to learn about how the climate works. Interactions between the land and atmosphere, and between the oceans and atmosphere are not fully understood. For example, does anyone know the ocean's capacity to absorb CO_2? A definitive answer to our question is difficult at this time. Perhaps direct observations of the earth's climate over the next decade will provide us with answers—and even solutions.

The Ozone Hole

In Chapter Two, we discussed the size and causes of the ozone hole, as well as its possible contribution to the "epidemic" of skin cancer seen in the past several decades. Diminishment of the ozone layer affects more than terrestrial plant and animal life: increased ultraviolet radiation due to this ozone hole has caused a decrease in phytoplankton production by up to 25% in Antarctic waters. Low quantities of phytoplankton could cause a complete collapse of life in the oceans since it is the base of the food chain (Thornton 1992).

Urban Growth and Resource Depletion

In 1800, only 3% of the world's population was living in cities; in 1991 almost 50% are city dwellers (Raven 1991). Large ecosystems are endangered by urban growth and resource depletion. Is it important that we protect the survival of the grizzly, wolverine, and wolf? If so, they require hundreds of square miles per individual, and a population of one to two thousand to survive. If their numbers become too small they can be wiped out, especially by inbreeding, disease, or natural disasters. But North American parks are too small; these areas have helped in the battle for the majesty of the big trees, but cannot maintain the integrity of entire biological systems (Friedman 1995).

Diver at work on coral. Photo courtesy of National Undersea Research Center, UNC at Wilmington

Urban demands for energy often affect wildlife and food sources. For example, western salmon migrate upstream to spawn, producing 4,000 eggs per pair. It is estimated that only 800 are hatched. Of those, only 200 survive the downstream journey through a damaging array of spillways and turbine intakes.

Pollution

Urban Effects. We have asked whether certain groups of people have been living in equilibrium with the environment and whether they have adapted to this environment. We should now apply these questions to people living in cities. A characteristic of the urban environment is pollution. Even the ancient Romans described and complained of "the heavy air of Rome" and "the stink of the smoky chimneys thereof." Perhaps we should define this condition in such a way that it applies to city streets, city parks, and to arable land as well. Pollution is the contamination of the environment by one or more of the following:

1. by the introduction of unnatural substances,
2. by the excessive accumulation of natural organic materials,
3. by the buildup of by-products to such a degree that an ecosystem is disrupted.

This definition leaves out unpleasant circumstances such as sonic booms. (See Chapter 9 for a discussion of noise pollution.)

We should ask whether pollution in cities is merely a disagreeable condition or a dangerous one. A dangerous situation or condition is usually expressed in "excess deaths over normal." London's Black Fog of December 1952 caused 4,000 "excess" deaths in 4 days and another 8,000 in the next few weeks. This mortality was similar to London's influenza epidemic of 1918-1919, and their cholera epidemic of 1854.

The U.S. Public Health Service (USPHS) states that a rise in level of air pollution causes the following reactions: 1) a rise in absenteeism because of bad colds, 2) exacerbation of bronchitis and asthma, and 3) increased death rates in older people with heart ailments. The USPHS also reports that there are twice as many deaths from lung cancer in metropolitan areas as in rural areas, even after full allowance is made for difference in smoking habits. The same is true of emphysema. In another example, smog coupled with a sudden temperature inversion descended on Donora, Pennsylvania causing half of the townspeople to become sick, and 26 to die. Yes, metropolitan pollution is more than a disagreeable condition.

The word *smog* means different things in different places. *London Fog* used to mean a special local brew of soft coal smoke plus fog. Smog along the east coast of the United States means a combination of stagnant, sooty, sulfur-loaded air with cool temperatures. Smog in Los Angeles means a local mixture of which the major ingredient is carbon monoxide. This mixture is a manufactured condition called *photochemical smog;* sunshine acts on the products from automobile exhausts, cooks the old familiar poisons, and converts them into new, complicated, and still mysterious substances. This last type of smog occurs mostly in heavily motorized cities where the air is stagnant and the sun is strong. All types of smog contain, in different proportions, the following dangerous materials: arsenic, hydrogen sulfide, carbon monoxide, benzypyrene, nitrogen dioxide, and lead.

The unfortunate oil fires in Kuwait after the Persian Gulf War acted as an unwanted experiment to measure the effect of smoke on a city. Fifty fires reduced solar radiation in one Arabian city by 25%, while 600 more fires reduced solar radiation in that area by 55% to 65% from March to August (Riley et al. 1992).

These oil fires also add to the problem of acid-rain, a phenomenon caused primarily by the burning of industrial fuels. There are now controls intended to reduce those emissions, but it is difficult to detect any resultant improvement in lakes and streams because there is little information about other acidification produced by soil weathering of the banks (Frink 1996).

The study of acid-rain is complex and full of surprises. India seemed a natural candidate for an acid-rain problem. However, the soil there turned out to be alkaline, which provided a protective layer against sulfur emissions (Abate 1996).

Throughout this discouraging picture, is there any voice of optimism? Eisenbud and Ehrlich reported in 1972 that the carbon monoxide concentrations in New York City streets appeared to be far lower that year than they were in 1966.

England took a constructive approach with results that have impressed environmentalists around the world; London is now essentially smog-free because it is a smoke-control area in which only coke, smokeless coal, gas, or electric heat is allowed. Apparently, the required changeover in London homes was accomplished amid loud cries of "invasion of privacy" and "loss of personal freedom." The point of view in England has progressed far enough so that 33 leading scientists, led by Sir Julian Huxley and Sir Fraser Darling, recommend that to avoid a national catastrophe Britain must tax the use of power and raw material, and aim toward reducing its population by one-half.

In the United States, since 65% of smog is provided by automobiles, trucks, and buses, our main hope lies in traffic control around cities, an efficient mass transportation system, and the development of the electric car. In addition, a "war on———" type campaign should be mounted to engineer factories on our deserts to collect energy from the sun. By a combination of such changes, perhaps modern city dwellers can live in equilibrium with their environment. Prompt and effective action must be taken on these matters or else we will face the prediction of Dr. Morris Neiburger, Professor of Meteorology at U.C.L.A., who said, "The atmosphere will grow progressively more polluted until, a century from now, it is too toxic to permit human life."

Agricultural Effects. The rural areas are contributing their share to pollution. The excretory products from feedlots are causing eutrophication in our rivers. Pesticides from rural and urban areas, when carried out to sea, are depressing oxygen production there by affecting the tiny planktonic diatomes that carry out photosynthesis. Researchers from the Louisiana Universities Marine Research Center warn that large amounts of nitrogen fertilizer from farms along the Mississippi River's tributaries are creating a vast "dead zone" in the Gulf of Mexico, which in 1995 stretched 30 to 60 miles from shore. Shrimp and fish cannot survive in these oxygen-starved waters; this *ecosystem stress* reverberates up the food chain, and has economic consequences for those who fish these waters.

In the past, 70% of all photosynthesis has taken place in the oceans. Unless we take great care, the combined pollution from urban and rural areas will start this planet toward an invisible decline.

There are at least two solutions to the problem of toxic runoff from farms. One is organic farming, which is gaining acceptance, but still not widely practiced. The other is the planting of poplar trees at the edges of farm fields next to rivers. The roots of poplar trees draw in and hold toxic chemicals, thus preventing damaging runoff. Poplars also offer economic opportunities to the farmer who plants them: they grow rapidly and can be harvested for wood or paper pulp, while a "new" tree will sprout and grow from the old stump (Licht and Schnoor 1993, Schnoor et al. 1995).

Effects on Wildlife and Humans. Wildlife is an important indicator of pollution damage. For years the ornithologist, Roger Torey Peterson, called wild birds "the litmus paper of the environment." Damage to the eggs of birds first alerted us to the danger of environmental contaminants. In central Alaska, at least, this danger seems to have passed. Environmental contaminants were measured in the eggs of up to 725 pairs of bald eagles nesting in interior Alaska; the contaminants did not reach even sublethal levels (Ritchie and Ambrose 1996). Small rodents in the forest serve equally well as indicators of environmental contamination (Sawica-Kapusta 1994).

Sea mammals have been proven to be contaminated, and contaminants are moving up the food chain to humans; in one study, the milk of Inuit mothers showed ten times the organochlorines and cadmium when compared to the milk of Southern Quebec women. These high values were also found in polar bears and beluga whales (Dewailly 1993). In fact, the few surviving beluga whales in the St. Lawrence river contain so many pollutants in their tissues that the whales themselves could be considered toxic waste. Surprisingly, these pollutants are reported to originate from subtropical and tropical regions (Sanderson 1995).

The contaminants found in beluga whales are ubiquitously present in the environment, and in foods. Researchers have studied the mechanism of damage of these so-called *xenobiotic* compounds: organochlorines (OC), polychlorinated biphenyl (PCB), polychlorinated dibenzo-p-dioxin (PCDD), and polychlorinated dibenzofuran (PCDF). They act on seals by impairing immune responses, which is displayed by suppressed natural killer-cell activity and specific T-cell responses (de Swart et al. 1996). In other studies on aquatic vertebrates, functional and morphological changes have been demonstrated in the thyroid gland and adrenal cortex.

The teeth are also targets of contamination; two species of rodents (cotton rats and prairie voles) living on an abandoned oil refinery developed dental lesions due to fluorosis.

Public health officials have been concerned about the effects of environmental mercury on human and ecological health since the 1960's and 1970's. Mercury has appeared in the bodies of ocean fish and sea mammals since that time, and has caused deformities in humans who have eaten contaminated fish. Now mercury is appearing in the flesh of lake fish in remote regions due to long-distance atmospheric transport. It bioaccumulates so that game fish may contain 225,000 times the mercury level found in water. Some state health departments in the upper midwestern United States have advised anglers to limit consumption of fish caught in lakes there. One of the sources responsible for mercury in the environment is gold mining; in 1989, Brazil released 168 metric tons of mercury into the environment (Murdock 1996). One solution to the problem of contaminated lake fish is to stock lakes with the species that shows the least contamination (Stow et al. 1996).

Mercury found in the hair of Laplanders in Northern Finland is four times greater than that found in the hair of people in Southern Finland. This is explained by the release into the air of huge amounts of waste by heavy industry in the Murmansk region (Mussalo-Rauhamaa et al. 1996).

The world's farmers have higher than average exposures to herbicides, insecticides, solvents, engine exhausts and fuels, some of which are known carcinogens. Farmers also tend to experience higher rates than the general population for several cancers, including leukemia, non-Hodgkin's lymphoma, Hodgkin's disease, multiple myeloma, soft-tissue sarcoma, and cancers of the skin, lip, prostate, testis, brain, and stomach (Blair and Zahm 1991). Investigators find this particularly striking because farmers have a lower than average risk for other fatal diseases, including smoking-related cancers, which is consistent with their lower smoking rate. In epidemiological investigations around the world, associations have been found between many of these cancers and several classes of herbicides and insecticides (Blair and Zahm 1995). Some of these same cancers, including non-Hodgkin's lymphoma, multiple myeloma, skin, brain, and prostate cancer, also appear to be on the increase in the general population of many developed countries.

Sustainable Development

While it is possible to have technological development without a large cost in pollution or resource depletion, this means *future* events or successes, even in rich societies. But for poor societies this progress must be put off even further because there, most individuals are trying to survive a day at a time. Cooper (1995) argues that what is sustainable and adheres to high environmental values may not be acceptable from a sociotechnical perspective. He suggests that the very concept is flabby and elusive, because to apply *our* environmental approach to poor, developing countries would threaten their social and political stability. Our own progress is slow, and must involve commitment over years and decades, as well as a readiness to tolerate glacier-like progress.

The situation in Surinam—an economically poor country in South America—illustrates the problem of attaining sustainable development. Covering an area the size of New England, Surinam is 90% unbroken rain forest, with only one city. It is currently being courted by Malaysian and Indonesian companies who wish to tear down one quarter of the forest. But the neighboring Costa Rican President, Oscar Arias, hopes to persuade Surinam to use his alternative ideas: tourism, carbon offset agreements (maintaining forests to offset carbon emissions), and bioprospecting. Using the latter, "royalties from one new plant-derived drug would dwarf revenues of all foreign concessions combined, without taking down a single tree" (Mittermeier and Bowles 1995).

A Closed Ecological System

To put the above considerations in more specific terms, Ehrlich (1994) has devised an equation that reflects the limits to growth:

$$I = PAT$$

where, the impact (I) of any group or nation on the environment can be viewed as the product of its population size (P) multiplied by per-capita affluence (A) as measured by consumption, in turn multiplied by a measure of the damage done by technologies (T) employed in supplying each unit of that consumption.

If the equation operates in a linear fashion, then impact should increase as population, affluence, or technology increase, and it can only be reduced as population, affluence, or technology are reduced.

Other analysts who attempt to grasp the deepest nature of our present-day ecological predicament are trying to combine economic factors and the other problems we have been discussing into one pattern of thinking. They also try to consider the effects of capital investment, natural resources, geographical space, population, pollution, and food production. The views of Jay W. Forrester are both stimulating and polemic (1971). His concept of earth is that of a closed system. It is true that we all share the oxygen of the atmosphere and are all involved in returning carbon dioxide. The *forest fires of pollution,* however, and the disastrous overpopulation are, in fact, local and regional problems throughout the world. Nevertheless, it is helpful to take a world view and to describe a closed ecological system within the *Space Ship Earth*. Forrester's system is based upon five basic factors: population, resources, industrialization, space and pollution. His analysis includes a circular feedback system in which population creates pressure for increasing industrialization and food production, which increases pressure of pollution and industrialization, which further increases population. The setting within which these components react has certain limits: one is the exhaustion of land and natural resources, and the other is the pollution dissipation capacity of the region or of the earth itself. This feedback concept is similar to the pattern that we have referred to in the natural ecosystem, such as the lemming population when it is no longer living in equilibrium with its environment.

Another factor which Forrester includes is *Quality of Life:* in our circular feedback system, we are in a golden age of quality of life which is higher than ever before in history. This high quality of life cannot continue into the future. The reason? In the past, the produce of industry, (food and material goods) has been able to exceed the rising population. Exhaustion of space and resources will not permit this feedback system to continue (Kendeigh 1965).

And now begins the paradox; if the United States were to introduce population control, this might be self-defeating, since it might increase the absolute amount of food and income, and create pressure for more population growth. Forrester believes it is absolutely essential to stabilize the population or it will be taken out of our hands by what he refers to as ecological principles: a) agriculture

Figure 11-11. Pollution Dissipation Capacity of Earth. This circular feedback system feeds on itself. The artificial ecosystem of civilization develops a pollution dissipation system that becomes overloaded; the result is the same as found in natural ecosystems under stress. In the feedback system, where should the loop be broken? Diagram: M.A. Folk.

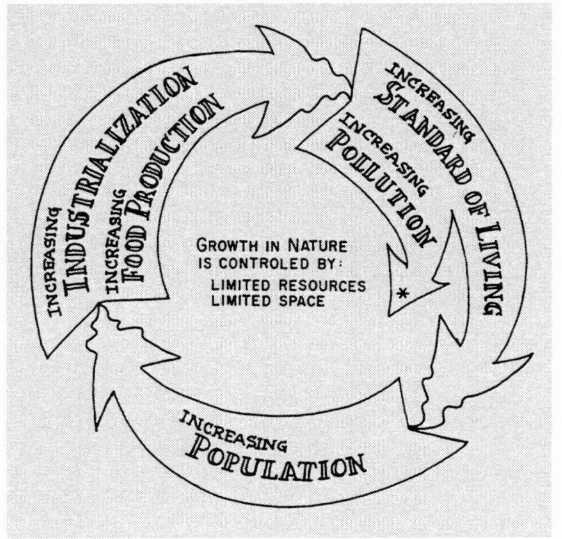

will reach a space limit, industry will reach a resource limit, and both will reach a pollution limit; b) a culture with a high level of industrialization may be self-extinguishing; c) growth and expansion must give way to an equilibrium society. Forrester doubts it is wise to industrialize underdeveloped countries since they may be closer to equilibrium with the environment before industrialization than after. In Fig. 11-11 we have presented his concept of a *Circular Feedback System* in diagrammatic form in order to ask the major question: Where should the feedback loop be broken? Clearly, Forrester believes that our future must include a stabilized population with a declining quality of life.

In this chapter on human ecology, perhaps the collected ideas are too speculative and too negative. If you are moved to say: "They said the whole city is on fire; why, only one-half the city is on fire!" we shall have accomplished our purpose. Environmentalists are asking for that same massive effort which we put into World War II, except it is hoped that there need not be another Pearl Harbor before the massive effort will begin. Will the first great disaster occur in a city of the Orient, or South America, or the United States? The ghosts of *Australopithecus* and classic Neanderthal may be watching us with curiosity as we vacillate about taking steps to save ourselves from ecological disaster.

APPENDICES

APPENDIX A

CLASSIFICATION OF MAMMALIA

Orders	
Monotremata	egg-laying mammals
Marsupialia	pouched mammals
Insectivora	hedgehogs, moles, shrews
Dermoptera	flying lemurs
Chiroptera	bats
Primates	monkey-like animals, man
Edentata	anteaters, sloths, armadillos
Pholidota	scaly anteaters, pangolins
Lagomorpha	hares, rabbits
Rodentia	rodents
Cetacea	whales, dolphins
Carnivora	dogs, cats, weasel-like mammals, seals
Tubulidentata	aardvarks
Proboscidea	elephants
Hyracoidea	conies
Sirenia	manatees
Perissodactyla	horse-type mammals, tapirs, rhinoceros
Artiodactyla	pigs, goats, deer, cattle, giraffes, hippopotami, camels

HIERARCHY FOR THE CLASSIFICATION OF A MAMMAL

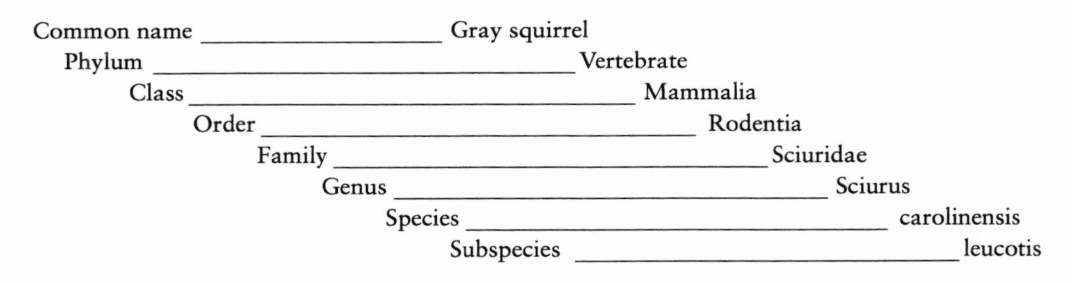

Thus *Sciurus carolinensis leucotis*, The Northern Gray Squirrel

APPENDIX B

POPULATIONS OF COLD-REGION PEOPLE

The population figures in Table 1 and Table 2 below, totaling over 208,000 cold-region people, were first obtained from the older literature starting with Freuchen (1958). More recently, Krauss (1997), a linguist anthropologist, studied and listed all those natives who speak a cold-region or Eskimo-like language (see Sidebar, Chapter 5). An exact census of cold-region natives is difficult, if not impossible. For example, Krauss' language-based census lists 117,000 people, while the earlier data shown in Table 1 totals 208,000; probably this is because some natives today speak only English, not their native language. Table 1 also includes the Lapps and other Eurasian groups, which Krauss did not study. (We have a high confidence level in the figures in Table 1; they also compare favorably with those published in various encyclopedias.)

In general, the cold-region people live above 60° North and follow a diet that emphasizes fish and sea mammals, or they are descendants of parents who used this diet.

Table 1. Estimates of Cold-Region Native People

Location	Population
Eight Eurasian Groups: Lapps, Samoyeds, Chukchi, Nenets, etc.	86,100
Asian Inuit (Siberia)	1,500
Greenland Inuit	45,000
Canadian Inuit	27,290
Alaska Inuit	44,400
Aleuts	4,000
Total	208,290

Table 2. Localized population figures.

The Baffin Inuit live in the territory called Nunavut which extends from Ellesmere Island to the upper third of Hudson Bay. Within this area there are 8,000 Inuit.

In the Disco Bay region of West Greenland near Davis Strait there are 1,029 Inuit living a hunting/fishing existence.

In Alaska, there are 20,000 Yupik Eskimos living on the delta of the Yukon and Kuskokwin Delta. These cold-weather natives live in 53 villages, living entirely a hunting/fishing existence.

There are 27,100 Inuit in Quebec, Labrador, Newfoundland, and the Northwest Territories (1986 Census).

APPENDIX C

PLACEMENT OF TEMPERATURE PROBES ON HUMAN SUBJECTS

Placement of Thermocouples or Thermistors on the Belding Points

1. Inside and end (medial) of great toe on side
2. Lateral side of calf
3. Lateral mid-thigh
4. Medial mid-thigh
5. Below scapula at midpoint of lateral to midline
6. On thorax below nipple on right side
7. Insertion deltoid or lateral side upper arm
8. Lateral lower arm
9. Dorsum of hand on proximal end of 2nd metacarpal. Not a large vein
10. Center of forehead
11. Medial, center, lateral-edge of sole

Obtaining Mean Skin Values

It would be inaccurate to use an average of the above points to obtain the mean skin values. The temperatures for various body segments should be WEIGHTED as follows:

Feet	7%
Lower Leg	13%
Upper Leg	19%
Trunk	35%
Arms	14%
Hands	5%
Head	7%

Locations of Thermocouples for Foot Temperature

If the temperature of the foot is to be recorded at five points, standardized placement is:

1. Tip of great toe.
2. Base of great toe on dorsum.
3. Tip of small toe.
4. Ball of foot.
5. Center on instep, not over vein or artery

APPENDIX D

STANDARDIZATION OF SYMBOLS AND UNITS FOR ENVIRONMENTAL RESEARCH

In a meeting called by the National Aeronautics and Space Administration (NASA) and held at NASA Manned Spacecraft Center, Houston, Texas, on 10 March, 1966, representatives of the various research centers of NASA and the biological laboratories of the Air Force Systems Command agreed on the desirability of standardizing symbols and units commonly used in environmental research. In line with the system agreed upon many years ago by respiratory physiologists the following set of symbols is recommended.

Three major symbols will be used.

T = temperature
H = rate of heat transfer or production
Q = heat quantity

Upper case subscripts will define *physical* variables

A =	air	C =	convective
W =	wall	D =	conductive
R =	radiant	DB =	dry bulb
E =	evaporative	WB =	wet bulb

Lower case subscripts will define *physiological* variables

r =	rectal	s =	skin
e =	esophageal	b =	body
t =	tympanic	m =	metabolism
o =	oral		

A dash above any symbol will indicate a mean value and Δ will indicate a change in a variable. Should it be desired, a dot above any symbol will indicate a time derivative

8 = mean skin
b = mean body
ΔT_r = change in rectal temperature

For metabolic or atmospheric designations those suggested by the Federation of American Societies for Experimental Biology should be used.

VO_2 = rate of oxygen consumption
P_B = barometric pressure

A second subscript will be used, if necessary, to indicate the time of the observation.

i = initial
f = final

Examples:

Q =	$H_R + H_C + H_D + H_E + H_m$	T_A =	air pressure
Q =	quantity of heat	T_W =	*wall* temperature
H_R =	*radiative* heat transfer	T_{WB} =	*wet bulb* temperature
H_C =	*convective* heat transfer	T_r =	*rectal* temperature
H_D =	*conductive* heat transfer	T_e =	*esophageal* temperature
H_E =	*evaporative* heat transfer	T_t =	*tympanic* temperature
H_m =	*metabolic* heat production	T_o =	*oral* temperature
Q_b =	quantity of *body* heat storage	T_{Si} =	*initial mean skin* temperature
		T_{bf} =	*final mean body* temperature

Less universally accepted terms such as "effective" or "operative" temperature, "index of strain," "heart rate," etc., should be defined in publication as necessary. Symbols used should be abbreviations or, in any event, clearly distinguished from those such as T_e (esophageal temperature) or T_o (oral temperature).

Units used in publication should always be in the metric system. In the interest of standardization and uniformity, the Système Interanational d' Unités as agreed upon in the 1975 U.S. Metric Conversion Act is recommended. The SI system is reproduced in reference books such as *The New York Public Library Desk Reference*, 1993.

APPENDIX E

SYMBOLS AND CONVERSIONS

PART I: Conversion Factors*

1 mm H_2O	=	1.36cm H_2O
1 centimeter	=	0.39370 inch
1 meter	=	39.37 inches or 3.28 feet
1 kilometer	=	0.62137 mile
1 milliliter	=	0.03381 fluid ounce
1 liter	=	1.0567 U.S. quarts
1 cubic meter	=	1.3080 cubic yards
1 gram	=	15.4324 grains
1 kilogram	=	2.2046 pounds
1 foot candle	=	10 lux
1 fathom	=	6 feet

*From the United States Department of Commerce, Miscellaneous Publication 233.

PART II: Symbols for Test Chamber Environments which are controlled or "constant" or non-periodic

L	=	artificial daylight
D	=	darkness
CT	=	Controlled Temperature
CS	=	Controlled Sound
CH	=	Controlled Humidity

PART III: Convenient estimates of the forces of wind using the official U.S. National Weather Service designations:

	miles per hour
calm	less than 1
light air	1 to 3
light breeze	4 to 7
gentle breeze	8 to 12
moderate breeze	13 to 18
fresh breeze	19 to 24
strong breeze	25 to 31
near gale	32 to 38
gale	39 to 46
strong gale	47 to 54
storm	55 to 63
violent storm	64 to 73
hurricane	74 and over

REFERENCES CITED

A

Abate, T. 1996. Acid-rain research in developing nations. *BioScience* 45:738–740.

Abe, K., H. Sasaki, K. Takebayashi, S. Fukui, and H. Nambu. 1978. The development of circadian rhythm of human body temperature. *J. Interdiscipl. Cycle Res.* 9:211–216.

Abelson, P.H. 1997. Improved fossil energy technology. *Science* 276:511.

Acker, H., and D. Xue. 1995. Mechanisms of O_2 sensing in the carotid body in comparison with other O_2-sensing cells. *News in Physiological Sciences* 10:211–228.

Adams, T., and B.G. Covino. 1958. Racial variations to a standardized cold stress. *J. Appl. Physiol.* 12:9–12.

Adler, H.F. 1964. *Dysbarism.* U.S.A.F. School of Aerospace Medicine, Brooks A.F.B., Texas. 165 p.

Adolph, E.F. 1947. *Physiology of Man In The Desert.* New York: Interscience. 357 p.

Al-Ani, M., et al. 1995. Exercise and diving, two conflicting stimuli influencing cadiac vagal tone in man. *J. Physiology.* 489(Pt 2):603–612.

Albers, H.E., C.F. Ferris, S.E. Leeman, and B.D. Goldman. 1984. Avian pancreatic polypeptide phase shifts hamster circadian rhythms when microinjected into the suprachiasmatic region. *Science* 223:833–834.

Alder, E.R. 1971. Evaluation of some controller inputs to behavioral temperature regulation. *Int. J. Biometeorol.* 15:121–128.

Alfrey, C.P., M.M. Udden, C.L. Huntoon, and T. Driscoll. 1996. Destruction of newly released red blood cells in space flight. *Med. Sci. Sports Exerc.* 28:542–544.

Allee, W.C. 1949. *Principles of Animal Ecology.* New York: Saunders. 574 p.

Almeida-Val, V.M.F., A.L. Val, and P.W. Hochachka. 1993. Hypoxia tolerance in Amazon fishes: status of an under-explored biological goldmine, p. 435–445. In *Surviving Hypoxia: mechanisms of control and adaptation*, ed. P.W. Hochachka, et. al. Ann Arbor: CRC Press.

Altland, P.D., and B. Highman. 1971. Effects of polycythemia and altitude hypoxia on rat heart and exercise tolerance. *Amer. J. Physiol.* 221:388–393.

Alyakrinskii, B.S., and S.I. Stepanova. 1985. Po Zakony Ritma (*Following Rhythm's Law*). Series ed. O.G. Gazenko, Moscow, USSR: Nauka, 176 p.

Amato, I. 1992. From 'Hunter Magic,' a pharmacopeia? *Science* 258:1306.

American Cancer Society. 1996. *Facts and Figures—1996.* Atlanta: American Cancer Society, Inc. 33 p.

Andersen, H.T., ed. 1969. *The Biology of Marine Mammals.* New York: Academic Press. 511 p.

Andersen, K.L. 1963. Comparison of Lapps, fishermen, and Indians. *Fed. Proc.* 22:834–840.

Andersen, K.L., et al. 1966. Metabolic and circulatory aspects of tolerance to cold as affected by physical training. *Fed. Proc.* 25:1351–1356.

Andersson, B., and S.M. McCann. 1955. A further study of polydipsia evoked by hypothalamic stimulatin in the goat. *Acta Physiol. Scand.* 33:333–346.

Andjus, R.K., and A.U. Smith. 1955. Reanimation of adult rats from body temperatures between 0° and 2°C. *J. Physiol.* 128:446–472.

Andrew, M., H. O'Brodovich, and J.R. Sutton. 1987. Operation Everest II: coagulation system during prolonged decompression to 282 Torr. *J. Appl. Physiol.* 63:1262–1267.

Andrews, R.V. 1970. Circadian variations in adrenal secretion of lemmings, voles and mice. *Acta Endocrinologica* 65:645–649.

Andrews, R.V. 1971. Circadian rhythms in adrenal organ cultures. *Gegenbaurs Morph. Jahrb. Leipzig* 117:89–98).

Andrews, R.V., and G.E. Folk, Jr. 1963. Circadian metabolic patterns in cultured hamster adrenals. *J. Comp. Biochem. Physiol.* 11:393–409.

Andrews, R.V. and R. Strohbehn. 1971. Endocrine adjustments in a wild lemming population during the 1969 summer season. *Comp. Biochem. Physiol.* 38A:183–201.

Aoki, T. and M. Wada. 1951. Functional activity of the sweat glands in the hairy skin of the dog. *Science* 114:123–124.

Ar, A. 1987. Physiological Adaptations to Underground Life in Mammals, p. 208. In *Comparative Physiology of Environmental Adaptations,* ed. P. Dejours. New York: Karger. 224 p.

Arguelles, A.E., D. Ibeas, J.P. Ottone, and M. Chekherdemian. 1962. Pituitary adrenal stimulation by sound of different frequencies. *J. Clinic. Endocrinology* 22: 846–852.

Arizpe, L., and M. Velazquez. 1991. *Population and Societies, Forum Proceedings*. Sigma Xi Research Society, p. 31–64.

Armatich, K.B. 1965. Vernal behaviour of the yellow-bellied marmot (*Marmota flaviventris*). *Anim. Behavior* 13:59–68.

Aschoff, J. 1953. Tierische Periodik unter dem einfluss von zeitgebern. *Z. Vergleich. Physiol.* 35:159–166.

Aschoff, J. 1955. Exogone and endogone Komponente de 24-Studen-Periodik bei tier und Mensch. *Naturwissenschaften* 42:569.

Aschoff, J. 1963. Comparative Physiology: Diurnal rhythms. *Annu. Rev. Physiol.* 25:581–600.

Aschoff, J. 1979a. Circadian rhythms: General features, p. 1–61. In *Endocrine Rhythms*, ed. D.T. Krieger. New York: Raven Press, 321 p.

Aschoff, J. 1979b. Comparative physiology: Diurnal rhythms. *Annu. Rev. Physiol.* 25:581–600.

Aschoff, J. 1981. Circadian system properties, p. 1–17. In *Advances in Physiological Sciences,* Vol. 18. Environmental Physiology, ed. F. Obal and G. Benedek. New York: Pergamon Press, 371 p.

Assenmacher, I., et al. 1987. CNS structures controlling circadian rhythms, p. 56–70. In: *Comparative Physiology of Adaptations,* ed. P. Pévet. New York: Karger.

Audet, D., and D.W. Thomas. 1997. Facultative hypothermia as a thermoregulatory strategy in the phyllostomid bats, *Carollia perspicillata* and *Sturnira lilium*. *J. Comp. Physiol.* B 167:146–152.

Austin, M.G., and J.W. Berry. 1956. Observations on 100 cases of heat stroke. *J.A.M.A.* 161:1525–1529.

Avellini, B.A., E. Kamon, and J.T. Krajewski. 1980. Physiological responses of physically fit men and women to acclimation to humid heat. *J. Appl. Physiol.* 49:254–261.

Avery, D.H., M.A. Bolte, S.R. Dager, et al. 1993. Dawn simulation treatment of winter depression: a controlled study. *Am. J. Psychiatry* 150:113–117.

B

Baker, M., G.E. Folk, Jr., and W. Ashlock. 1966. A recording cardiotachometer for radio-capsules implanted in mammals, with data from hibernating mammals. *Iowa Acad. Sci.* 73:293–302.

Baldwin, K.M., T.P. White, S.F. Arnaud, V.R. Edgerton, W.J. Kraemer, R. Kram, D. Raab-Cullen, and C.M. Snow. 1996. Musculoskeletal adaptations to weightlessness and development of effective countermeasures. *Med. Sci. Sports Exerc.* 10:1247–1253.

Balke, B., and C. Snow. 1965. Anthropological and physiological observations on Tarahumara endurance runners. *Am. J. Phys. Anthrop.* 23:293–302.

Balke, B., J.G. Wells, and R.T. Clark, Jr. 1957. In-flight hyperventilation during jet pilot training. *J. Aviat. Med.* 28:241–248.

Banaszkiewicz, A.C. 1979. Personal communication. Warsaw, Poland.

Barinaga, M. 1997. New clues found to circadian clocks. *Science* 276:1030–1031.

Barnett, S.A. 1963. *The Rat: A Study in Behavior.* Chicago: Aldine, 288 p.

Barnett, S.A. 1965. Genotype and environment in tail length in mice. *Quart. J. Exp. Physiol.* 50:417–429.

Barnett, S.A., and L.E. Mount. 1967. Resistance to cold in mammals, p. 412–477. In *Thermobiology,* ed. A.H. Rose. London: Academic Press. 653 p.

Bartholomew, G.A., Jr. 1949. The effect of light intensity and day length on reproduction in the English sparrow. *Bull. Mus. Comp. Zool.* 101 (3):433–476.

Bartholomew, G.A., P. Leitner, and J.E. Nelson. 1964. Body temperature, O_2 consumption and heart rate in 3 species of Australian flying fox. *Physiol. Zool.* 37:170–198.

Bartholomew, G.A., and T.J. Cade. 1957. Temperature regulation, hibernation and aestivation in the little pocket mouse. *J. Mammal.* 38:60–72.

Bartholomew, G.A., T.R. Howell, and T.J. Cade. 1957. Torpidity in the white-throated swift, anna hummingbird and poor-will. *The Condor.* 59:145–155.

Bartholomew, G.A., and J.W. Hudson. 1960. Aestivation in the Mohave ground squirrel, *Citellus mohavensis*. Proc. 1st Int. Symp. Mammalian Hibernation. *Bull. Mus. Comp. Zool.* 124:193–209.

Bartholomew, G.A., and R.C. Lasiewski. 1965. Heating and cooling rates, heart rate and simulated diving in the Galapagos marine iguana. *Comp. Biochem. Physiol.* 16:573–582.

Bartholomew, G.A., and V.A. Tucker. 1964. Size, body temperature, thermal conductance, oxygen consumption, and heart rate in Australian varanid lizards. *Physiol. Zool.* 37:341–354.

Batkin, S., D.L. Guernsey, and F.L. Tabrah. 1978. Weak A.C. magnetic field effects: Changes in cell sodium pump activity following whole animal exposure. *Research Communications in Chemical Pathology and Pharmacology* 22:613–616.

Bay, F.A. 1978. Light control of the circadian activity rhythm in mouse-eared bats (*Myotis myotis*). *J. Interdiscipl. Cycle Res.* 3:195–209.

Bazett, H.C., et al. 1948. Temperature changes in arteries and veins in man. *J. Appl. Physiol.* 1: 3–6.

Bean, L. 1996. Renewable Energy. *Iowa Conservationist* 55: 52–53.

Beck, S.D. 1963. *Animal Photoperiodism*. New York: Holt, Rienhart and Winston, Inc., 124 p.

Beckman, A.L. 1986. Functional aspects of brain opioid peptide systems in hibernation, p. 225–234. In *Living in the Cold,* ed. H.C. Heller, X.J. Musacchia, and L.C.H. Wang. New York: Elsevier Scientific.

Bee, J.W., and E.R. Hall. 1956. *Mammals of Northern Alaska.*

Univ. Kansas Mus. Nat. History, Misc. Pub. No. 8. Lawrence, Kansas: Allen Press. 309 p.

Beer, J.R., and A.G. Richards. 1956. Hibernation of big brown bat. *J. Mammal.* 37:31–41.

Behnke, A.R. 1955. Decompression sickness. *Military Med.* 117:257–262.

Bekele, A. 1996. Population ecology of the white-footed rat. *Proc. Amer. Soc. Mammalogists.* 76:16.

Belding, H.S., et al. 1947. Thermal responses and efficiency of sweating when men are dressed in arctic clothing and exposed to extreme cold. *Amer. J. Physiol.* 149:204–222.

Belding, H.S., H.D. Russell, R.C. Darling, and G.E. Folk. 1947. Analysis of factors concerned in maintaining energy balance for dressed men in extreme cold. *Amer. J. Physiol.* 149:223–239.

Belding, H.S., and T.F. Hatch. 1955. Index for evaluating heat stress in terms of resulting physiological strains. *Heat, Pip. Air Condit.* 27:129–136.

Bell, E.F., and J.R. Ek. 1997. Infrared thermographic calorimetry applied to preterm infants. *Proc. Amer. Pediatric Society* (In Press).

Benedict, F.G. 1938. *Vital Energetics in Comparative Metabolism.* Washington, DC: Carnegie Inst. Publ. No. 503.

Bennett, M.F., and J. Hugenin. 1969. Geomagnetic effects on a circadian difference in reaction times in earthworms. *Z. Vergl. Physiologie* 63:440–445.

Bennett, P.B. 1993a. Inert gas narcosis. In *The Physiology and Medicine of Diving* (4th ed.), ed. P. B. Bennett and D. H. Elliott. London: Saunders.

Bennett, P.B. 1993b. The high pressure nervous syndrome, p. 194–237. In *The Physiology of Diving and Compressed Air Work,* ed. P. B. Bennett and D. H. Elliott. London: Saunders.

Benzinger, T.H. 1959. On physical heat regulation and the sense of temperature in man. *Proc. Nat. Aca. Sci.* (Washington) 45:645–659.

Berberich, J.J., and G.E. Folk, Jr. 1976. Cold acclimation in Arctic lemmings. *Comp. Biochem. Physiol.* 54: 175–178.

Bergman, C. 1847. Uber die verhaltnisse der warmeokonomie der theiere zu ihrer grosse. *Gottinger Studien* 1:595–708.

Bergman, S.A., Jr., J.K. Campbell, and K. Wildenthal. 1972. "Diving reflex" in man: Its relation to isometric and dynamic exercise. *J. Appl. Physiol.* 33:27–31.

Bernard, C. 1876. *Leçons sur la Chaleur Animale.* London: Bailliere, Tindall & Cox. 125 p.

Bernstein, I.A., ed. 1971. *Biochemical Responses to Environmental Stress.* New York: Plenum Press. 153 p.

Bert, P. 1878. *La Pression Barometrique.* Paris: Masson. 1168 p.

Biali, S., et al. 1995. Influence of perpetual-daylight on periodicity in cholesterol synthesis. *Arctic Med. Res.* 54:134–144.

Bianca, W. 1967. The impact of the atmospheric environment on the integument of domestic animals, p. 83–102. In *Biometeorology, Vol. 3,* ed. S.W. Tromp and W.H. Weihe. Amsterdam: Swets & Zeitlinger. 374 p.

Birau, N., and W. Schloot, eds. 1980. *Advances in the Biosciences 29: Melatonin.* New York: Pergamon Press. 410 p.

Bitterman, N., et al. 1987. CNS O_2 toxicity in O_2-inert gas mixtures. *Undersea Biomed. Res.* 14: 477–483.

Bixler, E.O., A.Kales, R.J. Cadieux, A. Vela-Bueno, J.A. Jacoby, and C.R. Soldatos. 1985. Sleep apneic activity in older healthy subjects. *J. Appl. Physiol.* 58:1597–1601.

Bjertnaes, L., et al. 1984. Cardivasular responses to face immersion and apnea during steady state muscle exercise. *Acta Physiol. Scand.* 120:605–612.

Black, C.P., and S.M. Tenney. 1980. Oxygen transport during progressive hypoxia in high-altitude and sea-level waterfowl. *Respiration Physiol.* 39:217–239.

Black, H. 1995. Female pecking order and fertility. *BioScience.* October:583–585.

Blagden, C. 1775. Further experiments and observations in a heated room. *Phil. Trans. Roy. Soc. Lond.* 65:484–494.

Blair, A., and Zahm, S.H. 1991. Cancer Among Farmers, p. 335–354. In *Occupational Medicine: State of the Art Reviews* Vol 6. No. 3, July-Sept. Philadelpia: Hanley & Belfus, Inc.

Blair, A., and Zahm, S.H. 1995. Agricultural exposures and cancer. *Environ. Health Perspect.* 103(Suppl 8):205–203.

Blair, J.R., and A. D. Keller. 1941. Calibration studies of the regulation of body temperature in normal dogs. *Amer. J. Physiol.* 133:215–216.

Blaustein, A.R., et al. 1994. UV repair and resistance to solar UV-B in amphibian eggs. *Proc. Nat. Acad. of Sci. of the U.S.* 91:1791–1795.

Blaxter, K.L. 1965. Climatic factors and the productivity of different breeds of livestock, p. 157–168. In *The Biological Significance of Climatic Changes in Britain,* ed. C.G. Johnson and L.P. Smith. New York: Academic Press. 222 p.

Blessing, M.H. 1972. Studies on the concentration of myoglobin in the sea-cow and porpoise. *Comp. Biochem. Physiol.* 41A:475–480.

Bligh, J., and S.G. Robinson. 1965. Radio-telemetry in a veterinary research project. *Med. & Biol. Illustr.* 15:94–99. London: British Medical Assoc.

Block, G.D., and S.F. Wallace. 1982. Localization of a circadian pacemaker in the eye of a mollusc, *Bulla. Science* 217:155–157.

Blockley, W.V. 1963. Heat storage rate and tolerance time. *Fed. Prod.* 22:887–890.

Blum, H.F. 1967. Effects of ultraviolet light on man, p 109–120. In *Seminar on human Biometerorology,* ed. J. Dicke. Pub Helth Serv. Pub No. 999–AP–25. 183 p.

Blum, H.F. 1969. Is sunlight a factor in geographical distribution of human skin color. *Geogr. Rev.* 59:557–581.

Blumberg, M.S., G. Sokoloff, and R.F. Kirby. 1997. Brown fat thermogenesis and cardiac rate regulation during cold challenge in infant rats. *Am. J. Physiol.* 272:R1308–R1313.

Boerema, L., ed. 1964. *Clinical applications of Hyperbaric Oxygen.* New York: Elsevier Scientific. 427 p.

Bondar, R.L, M.S. Kassam, F. Stein, P.T. Dunphy, and M.L.

Riedesel. 1994. Simultaneous transcranial doppler and arterial blood pressure response to lower body negative pressure. *J. Clinical Pharmacology* 34:584–589.

Bondar, R.L, M.S. Kassam, F. Stein, P.T. Dunphy, S. Fortney, and M.L. Riedesel. 1996. Simultaneous cerebrovascular and cardiovascular responses during presysncope. *Stroke* 26:1794–1800.

Borlaug, N.E. 1972. Mankind and civilization at another crossroad: In balance with nature—a biological myth. *BioScience* 22:41–43.

Boswell, T., and G.J. Kenagy. 1993. Hypothalamic Neuropeptides, Metabolic Hormones and Seasonal Fattening in Golden-Mantled Ground Squirrels, p. 493–500. In *Life in the Cold: Ecological, Physiological and Molecular Mechanisms,* ed. C. Carey, G.L. Florant, B.A. Wunder and B. Horwitz. Boulder, CO: Westview Press.

Boulant, J.A. 1994. Neurophysiology of thermoregulation. In *Temperature Regulation,* p. 93–101. Boston: A.S. Milton, 376 p.

Boulant, J.A., and J.D. Griffin. 1994. Determinants of thermosensitivity in hypothalamic neurons, p. 9. In *Proc. Thermophysiology Commission of the Int. Union of Physiolog. Sci.* Vol. 8.

Bourne, G.H., ed. 1970. *The Chimpanzee. Part 2, Physiology, Behavior, Serology and Diseases.* Baltimore: University park Press. 417 p.

Bowen, M.F., and S.D. Skopik. 1976. Insect photoperiodism: The "T Experiment" as evidence for an hourglass mechanism. *Science* 192:59–60.

Boyd, C.A.R., and D. Noble. 1993. *The Logic of Life: The Challenge of Integrative Physiology.* London: Oxford University Press. 226 p.

Boyer, S.J., and F.D. Blume l984. Weight loss and changes in body composition at high altitude. *J. Appl. Physiol.* 57:1580–1585.

Bradley, D. 1993. Frog venom cocktail yeilds a one-handed painkiller. *Science* 261:1117.

Brattstrom, B.H. 1979. Amphibian temperature regulation studies in the field and laboratory. *Amer. Zool.* 19:345–356.

Brauner, N., and M. Shacham. 1995. Meaningful wind chill indicators derived from heat transfer principles. *Int. J. Biometeorol.* 39:46–52.

Brengelman, G.L., J.M. Johnson, L. Hermansen, and L.B. Rowell. 1977. Altered control of skin blood flow during exercise at high internal temperatures. *J. Appl. Physiol.* 43:790–794.

Brewer, M.C. 1958. Some results of geothermal investigations of permafrost in northern Alaska. Transactions, American Geophysical Union 39:19–26.

Briese, E. 1995. Emotional hyperthermia and performance in humans. *Physiology & Behavior* 58(3):615–618.

Broadway, J., and J. Arendt. 1990. Human melatonin rhythm responses to season and bright light in Antartica. *Proc. Congress on Circumpolar Health.* 8:11.

Brody, M.J., and R.A. Shaffer. 1970. Distribution of vasodilator nerves in the canine hindlimb. *Amer. J. Physiol.* 218:470–472.

Broghammer, A.M., and C.L. Thurman. 1997. Circadian rhythm of fiddler crabs in Texas. *Proc. Iowa Physiolog. Soc. and Iowa Acad.* 109:20.

Brooks, E.M., A.L. Morgan, J.M. Pierzga, S.L. Wladkowski, J.T. O'Gorman, J.A. Derr, and W.L. Kenney. 1997. Chronic hormone replacement therapy alters thermoregulatory and vasomotor function in postmenopausal women. *J. Appl. Physiol.* 83(2):477–484.

Brown D. 1996. Urchins in the storm. *J. Mass. Audubon Soc.* 35:11.

Brown, F.A., Jr. 1959. Biological chronometry. *Amer. Naturalist* 91:129–133.

Brown, F.A., Jr. 1965. Propensity for lunar periodicity in hamsters and its significance for biological clock theories. Proc. Soc. Exp. Biol. Med. 120:792–797.

Brown, F.A., Jr. 1966. Effects and after-effects on planarians of reversals of the horizontal magnetic vector. *Nature* 209:533–535.

Brown, F.A., Jr. 1969. A hypothesis for extrinsic timing of circadian rhythms. *Canad. J. Bot.* 47:287–298.

Brown, F.A., Jr. 1970. Hypothesis of environmental timing of the clock, p. 13–59. In *The Biological Clock: Two Views,* ed. J.D. Palmer. New York: Academic Press, 97 p.

Brown, F.A., Jr., and H.M. Webb. 1968. Some temporal and geographic relations of snail response to very weak gamma radiation. *Physiol. Zool.* 41:385–400.

Brown, K.S. 1996. Do disease cycles follow changes in weather? *BioScience* 46:479–481.

Brown, W.F., and F. Sargent. 1965. Hydromeiosis. *Arch. Environ. Health* 11:442–453.

Brown-Grant, K., G.W. Harris, and S. Reichlin. 1954. The effect of emotional and physical stress on thyroid activity in the rabbit. *J. Physiol.* 126:29–40.

Brubakk, A.O., et al., eds. 1989. *Supersaturation and Bubble Formation in Fluids and Organisms.* Trondheim, Norway: Tapir.

Bruce, V.G., and C.S. Pittendrigh. 1956. Temperature independence in a unicellular "clock." *Proc. Natl. Acad. Sci.* 42:676–682.

Bruck, K. ,et al. 1970. Nonshivering thermogenesis and its integration in the thermoregulatory system, p. 28–39. In *Physiological Ecology* (a Symposium held in Novosibirsk, Siberia), ed. A.D. Slonipim. 5 Vols. (Paper in Russian.).

Bruguerolle, B., and G. Jadot. 1983. Influence of the hour of administration of lidocaine on its intraerythrocytic passage in the rat. *Chronobiologia* 10:295–297.

Brynum, G.D., K.B. Pandolf, W. Schuette, R.F. Goldman, D.E. Lees, J. Whang-Peng, and J.M. Bull. 1978. Induced hyperthermia in sedated humans and the concept of critical thermal maximum. *Am. J. Physiol.* 235:R228–R236.

Bucher, T.L., and M.A. Chappell. 1997. Respiratory exchange

and ventilation during nocturnal torpor in hummingbirds. *Physiol. Zoology* 70:45–52.

Buckey, J.C, L.D. Lane, B.D. Levine, D.E. Watenpaugh, S.J. Wright, W.E. Moore, F.A. Gaffney, and C.G. Blomqvist. 1996. Orthostatic intolerance after spaceflight. *J. Appl. Physiol.* 81:7–18.

Bullard, R. 1964. Changes in regional blood flow and blood volume during arousal. *Ann. Acad. Sci. Fenn. Ser.* A4 71, 65–76.

Bullard, R.W. 1971. Studies on human sweat gland duct filling and skin hydration. *J. Physiol.* 63:218–221.

Bullard, R.W. 1972. Vertebrates at altitudes, p. 209–225. In *Physiological Adaptations: Desert and Mountain,* ed. Yousef, Horvath, and Bullard. Academic Press: New York. 258 p.

Bullard, R.W., and J. Kollias. 1966. Functional characteristics of two high-altitude mammals. *Fed. Proc.* 25:1288–1292.

Bullock, T.H., and F.P. Diecke. 1956. Anatomy and physiology of infrared sense organs in facial pit of pit vipers. *J. Physiol.* 137:47–87.

Bunn, H.F., and R.O. Poyton. 1996. Oxygen sensing and molecular adaptation to hypoxia. *Physiol. Rev.* 76:839–885.

Bünning, E. 1964. *The Physiological Clock*. London: Academic Press, 145 p.

Bünning, E., and I. Moser. 1969. Interference of moonlight with photoperiodic measurement of time by plants, and their adaptive reaction. *Proc. Nat. Acad. Sciences* 62:1018–1022.

Buskirk, E.R. 1966. Variation in heat production during acute exposures of men and women to cold air or water. *Ann. N.Y. Acad.* Sci. 134:733–742.

Buskirk, E.R., R.H. Thompson, and G.D. Whedon. 1963. Metabolic response to cold air in men and women in relation to total body fat content. *J. Appl. Physiology.* 18:603–612.

Buss, I.O., and A. Wallner. 1965. Body temperature of the African elephant. *J. Mammal.* 46:104–107.

Butler, B.D., and J. Katz. 1988. Vascular pressures and passage of gas emboli through the pulmonary circulation. *Undersea Biomed. Res.* 15:203–209.

Butler, F.K., and E.D. Thalman. 1986. CNS O_2 toxicity in closed circuit scuba divers II. *Undersea Biomed. Res.* 13: 193–223

Butler, P.J., and A.J. Woakes. 1987. Heart rate in humans during underwater swimming with and without breath-hold. *Respir. Physiol.* 69:387–399.

Buysse, D.J., T.H. Monk, C.F. Reynolds 3d, D. Mesiano, P.R. Houck, and D.J. Kupfer. 1993. Patterns of sleep episodes in young and elderly adults during a 36-hour constant routine. *Sleep.* 16(7):632–7.

C

Cagnacci, A., G.B. Melis, R. Soldani, A.M. Paoletti, and P. Fioretti. 1992. Effect of sex steroids on body temperature in postmenopausal women. Role of endogenous opioids. *Life Sciences* 50(7):515–521.

Cajlachjan, M.C. 1936. On the mechanism of the photoperiodic reaction (in Russian). *Compt. Rend. (Doklady) Acad. Sci. USSR* 1:89–93.

Canivenc, R., ed. 1968. Cycles génitaux saisonniers de mammiferes sauvages. Entretiens de Chize, Masson, Paris, Serie Physiologique No. 1. 168 p.

Caradente, F., M.A. De Matteis, R. Mellizzi, and G. Pitari. 1982. Circadian rhythm of rectal temperature in young and adult rats under different conditions of food access. *Chronobiologia* 9:223–227.

Carey, C., G.L. Florant, B.A. Wunder and B. Horwitz. 1993. *Life in the Cold: Ecological, Physiological and Molecular Mechanisms.* Boulder, Colo.: Westview Press, 575 p.

Carey, F.G. 1982. A brain heater in swordfish. *Science* 216:1327–1329.

Carey, H.V. 1995. Gut feelings about hibernation. *News in Physiol. Sciences* 10:55–61.

Carey, H.V., and N.S. Sills. 1996. Hibernation enhances d-glucose uptake by intestinal brush border membrane vesicle in ground squirrels. *J. Comp. Physiol.* B 166:254–261.

Carlson, L.D. 1954. The adequate stimulus for shivering. *Proc. Soc. Exp. Biol.* Med. 85:303–30S.

Carlson, L.D. 1964. Physiology of exposure to cold. *Physiology For Physicians* 2:1–7.

Carter, D.S., and B.D. Goldman. 1983. Antigonadal effects of timed melatonin infusion in pinealectomized male Djungarian hamsters (*Phodopus sungorus sungorus*): Duration is the critical parameter. *Endocrinology* 113:1261–1267.

Caso, A. et al. 1997. Ecological patterns of sympatric ocelot and jaguarundi populations in Mexico. *Proc. Theriolog. Cong.* 7: 70.

Castellini, M.A. 1996. Sleep apnea in seals. *Physiological Sci.* 11:208–213.

Castellini, M.A., et al. 1985. Potentially conflicting metabolic demands of diving and exercise in seals. *J. Appl. Physiol.* 58:392–399.

Cattet, et al. 1997. Intermedian metabolism in wild fasting polar bears. *The FASEB Journal* 11:29.

Cena, K. 1964. Thermoregulation in the hippopotamus. *Int. J. Biometeorol.* 8:57–60.

Cerretelli, P. 1976. Limiting factors to oxygen transport on Mount Everest. *J. Appl. Physiol.* 40:658.

Chaffee, R.R.J., and J. C. Roberts. 1971. Temperature acclimation in birds and mammals. *Annu. Rev. Physiol.* 33:155–202.

Chaffee, R.R.J. 1966. On experimental selection for super-hibernating and nonhibernating lines of Syrian hamsters. *J. Theoret. Biol.* 12:151–154.

Chance, B., A.K. Ghosh, E.K. Pye, and B. Hess, eds. 1973. *Biological and Biochemical Oscillators.* New York: Academic Press, 534 p.

Chappell, M.A., and S.L. Souza. 1988. Thermoregulation, gas

exchange, and ventilation in Adelie penguins (*Pygoscelis adeliae*). *J. Compar. Physiol.* 157(6):783–790.

Chatfield, P.O., and C.P. Lyman. 1954. Subcortical electrical activity in the golden hamster during arousal from hibernation. *EEG Clin. Neurophysiol.* 6:403–409.

Chatfield, P.O., C.P. Lyman, and L. Irving. 1963. Physiological adaptation to cold of peripheral nerve of herring gull. *Amer. J. Physiol.* 172:639–644.

Chauvet, J., E.B. Descamps, and C. Hillaire. 1996. *Dawn of Art.* Abrams.

Cherry-Garrard, A.G.B. 1922. *The Worst Journey in the World.* London: Constable Press. 584 pp.

Chitty, D. 1996. *Do Lemmings Commit Suicide?* London: Oxford University Press. 268 p.

Chizhenkova, R.A. 1966. Changes in the EEG of the rabbit during the action of a constant magnetic field. *Byulletin' eksperimental' noy biologii i meditsiny* 61:11–15. (In Russian).

Cissik, J.H., R.E. Johnson, and D.K. Rokosch. 1972. Production of gaseous nitrogen in human steady-state conditions. *J. Appl. Physiol.* 32:155–159.

Clark, J. A., ed. 1981. *Environmental aspects of housing for animal production.* Proc. 31st Easter School Agri. Sci., Nottingham. Boston: Butterworths. 511p.

Clark, J.M., et al. 1991. Pulmonary function in men after O_2 breathing at 3 ATA for 3.5 h. *J. Appl. Physiol.* 71:878–885.

Clarke, R.S.J., R.F. Hellon, and A.R. Lind. 1957. Cold vasodilatation in the human forearm. *J. Physiol.* 137:84–85.

Claybaugh, J.R. 1994. Renal and endocrine responses to the high pressure environment, p. 295–311. In *Basic and Applied High Pressure Biology*, ed. P. B. Bennett and R. E. Marquis. Rochester: University of Rochester Press.

Claybaugh, J.R., C.E.Wade, and S.A. Cucinell, 1989. Fluid and electrolyte balance and hormonal response to the hypoxic environment. In *Hormonal Regulation of Fluid and Electrolytes,* ed. J.R. Claybaugh and C.E. Wade. Plenum Publ. Corp.

Claybaugh, J.R., et al. 1987. Seadragon VI. A 7-day dry saturation dive at 31 ATA. III. Alterations in basal and circadian endocrinology. *Undersea Biomed. Res.* 14: 401–411.

Clayton, R.K. 1970. *Light and Living Matter. Vol. 1: The Physical Part.* New York: McGraw-Hill. 160 p.

Clayton, R.K. 1971. *Light and Living Matter. Vol. II: The Biological Part.* New York: McGraw-Hill. 256 p.

Clopton, J.R. 1985. Circadian rhythmicity in the flight activity of the mosquito *Culiseta incidens. Comp. Biochem. Physiol.* 80A:469–475.

Cloudsley-Thompson, J.L., and M.J. Chadwick. 1964. *Life in Deserts.* Philadelphia: Dufour Editions. 218p.

Cloudsley-Thompson, J.L. and C. Constantinou. 1981. Effects of light and temperature on the circadian rhythm of locomotory activity of Pimelia grandis Klug (*Coleoptera: Tenebrionidae*) from *Sudan. J. Arid Environments* 4:131–136.

Cohen, J.E. 1995. *How many People Can the Earth Suppport?* New York: Norton. 532 p.

Collins, K.J. 1963. Endocrine control of salt and water in hot conditions. *Fed. Proc.* 22:716–720.

Collins, K.J., C. Dore, A.M. Exton-Smith, R.H. Fox, I.E. MacDonald, and P.M. Woodward. 1977. Accidental hypothermia and impaired temperature homeostasis in the elderly. *Brit. Med. J.* 1:353–356.

Collins, K.J., T.A. Abdel-Rahman, J. Goodwin, and L. McTiffin. 1995. Circadian body temperatures and the effects of a cold stress in elderly and young subjects. *Age & Aging.* 24(6):485–9.

Colony, Roger, 1996. World Climate Research Program, Arctic Climate System Study, Oslo, Norway. Personal communication.

Comperatore, C.A., H.R. Lieberman, A.W. Kirby, B. Adams, and J.S. Crowley. 1996. Melatonin Efficacy in Aviation Missions Requiring Rapid Deployment and Night Operations. *Aviation, Space, and Environmental Medicine* 67(6):520–524.

Conroy, R.T.W.L., A.L. Elliott, and J.N. Mills. 1970a. Circadian excretory rhythms in night workers. *Brit. J. Industr. Med.* 27:356–363.

Conroy, R.T.W.L., A.L. Elliott, and J.N. Mills. 1970b. Circadian rhythms in plasma concentration of 11-hydroxycorticosteroids in men working on night shift and in permanent night workers. *Brit. J. Industr. Med.* 27:170–174.

Consolazio, C. F., et al. 1963. *Physiological Measurements of Metabolic Functions in Man.* New York: McGraw-Hill. 505 p.

Consolazio, C.F., R. Shapiro, J.E. Masterson, and P.S.L. McKinzie. 1961. Energy requirements of men in extreme heat. *J. Nutrition* 73:126–134.

Convertino, V.A., S.A. Bloomfield, and J.E. Greenleaf. 1997. An overview of the issues: physiological effects of bed rest and restricted physical activity. *Med. Sci. Sports Exerc.* 29:187–190.

Convertino, V.A. 1996. Exercise as a countermeasure for physiological adaptation to prolonged spaceflight. *Med. Sci. Sports Exerc.* 28:999–1014.

Cook, J.S. 1986. "Spark" vs. "soup": A scoop for soup. *News in Physiol. Sci.* (Dec.) Vol. 1:206–207.

Coon, C.S. 1962. *The Origin of Races.* New York: Alfred A. Knopf. 468 p.

Cooper, C.L. 1995. *Sustainable Development.* Cosmos 5:73–78.

Cooper, K.E. 1966. Temperature regulation and the hypothalamus. *Brit. Med. Bull.* 22:238–242.

Cooper, K.E. 1995. *Fever and Antipyresis.* New York: Cambridge Univ. Press. 182 p.

Cooper, K.E., W.I. Cranston, and E.S. Snell. 1964. Temperature in the external auditory meatus. *J. Appl. Physiol.* 19:1032–1035.

Cooper, P. 1997. University of Iowa. Personal communication.

Cornelissen, G., et al, 1989. Chronobiology: a frontier in biology and medicine. *Chronobiologia* 16: 383–408.

Cornelius, C.E., and J.J. Kaneko. 1962. Erythrocyte life span in the Guanaco. *Science* 137:673–674.

Corrent, G., D.J. McAdoo, and A. Eskin. 1978. Serotonin shifts the phase of the circadian rhythm from the Aplysia eye. *Science* 202:977–979.

Cossins, A.R., and K. Bowler. 1987. *Temperature Biology of Animals*. London/New York: Chapman and Hall. 339 p.

Costill, D.L., R. Cote, and W. Fink. 1976. Muscle water and electrolytes following varied levels of dehydration in man. *J. Appl. Physiol.* 40:6–11.

Cotter, J.D., M.J. Patterson, and N.A.S. Taylor. 1997. Sweat distribution before and after repeated heat exposure. *Eur. J. Appl. Physiol.* 176:181–186.

Coulombe, H.N. 1970. Physiological and physical aspects of temperature regulation in the burrowing owl *Speotyto Comp. Biochem. Physiol.* 35:307–337.

Covino, B.G., and W.R. Beavers. 1957. Cardiovascular response to hypothermia. *Amer. J. Physiol.* 191:153–156.

Craighead, J.J., et al. 1976. Telemetry experiments with a hibernating black bear, p. 357–371.In *Bears—Their Biology and Management*, ed. M.R. Pelton, J.W. Lentfer, and G.E. Folk, Jr. IUCN Publ. Series No. 40. Switzerland: Morges.

Crawford, C.S. 1981. *Biology of Desert Invertebrates.* New York: Springer-Verlag. 314p.

Cui, Y., T.F. Lee, J. Westly, L.C.H. Wang. 1997. Autoradiographic determination of changes in opioid receptor binding in the limbic system of the Columbian ground squirrel at different hibernation states. *Brain Res.* 747:189–194.

Culik, B.M., et al. 1996. Diving energetics in king penguins (*Aptenodytes patagonicus*). *Jour. Exp. Biol.* 199(4):973–983.

Cullinane, D.M. 1995. The biomechanics of the spine of the hero shrew. *Amer. Zool.* 35:61

Culotta, E. 1995a. Eyes Everywhere. *Science* 270:1903.

Culotta, E. 1995b. Skinny gene. *Science* 270:1905.

Cummings, D.E., E.P. Brandon, J.V. Polanas, K. Motamed, R.L. Idzerda, and G.S. McKnight. 1996. Genetically lean mice result from targeted disruption of RIIb subunit of protein kinase A. *Nature* 382:622–626.

Curran-Everett, D.C., et al. 1991. Intracranial pressures and O_2 extraction in conscious sheep during 72 h of hypoxia. *Am. J. Physiol.* 261:H103.

Currie, P.J. 1996. Out of Africa: Meat-eating dinosaurs that challenge *Tyrannosaurus rex*. *Science* 272:971.

Czeisler, C.A., M.C. Moore-Ede, and R.M. Coleman. 1982. Rotating shift work schedules that disrupt sleep are improved by applying circadian principles. *Science* 217:460–462.

D

D'Angelo, S.A. 1960. Adenohypophysial function in the guinea pig at low environmental temperature. *Fed. Proc.* 19(Suppl. 5):51–56.

da Silva, R.G., and F.R. Minomo. 1995. Circadian variation of body temperature of sheep. *Int. J. Biometeorology* 39:69–73.

Dale, H.E., et al. 1970. Energy metabolism of the chimpanzee (Ch. 5) In *The Chimpanzee: Vol. 2: Physiology, Behavior, Serology and Diseases,* ed. G.H. Bourne. 417 p. Baltimore: Univ. Park Press.

Dale, R.F. 1995. Is our Indiana climate warming? In *Preparing for Global Change: A Midwestern Perspective. Progress in Biometeorology,* ed. G.R. Carmichael, G.E. Folk, Jr., and J.L. Schnoor, p. 91–99. Vol. 9. 298 p. Amsterdam: SPB Academic Publishing.

Danielsen, K. et al. 1995. Effect of diet on plasma variables in endurance horses. *Equine Vet. J. Suppl.* 18:372–377.

Darlington, P.J. 1957. *Zoogeography: the Geographical Distribution of Animals.* New York: John Wiley. 675 p.

Darwin, C.R. 1845. *Voyage of the Beagle.* London: John Murray. 525 p.

Davey, K. 1969. *Australian Desert Life.* Melbourne: Periwinkle Books, Landsdown Press.

Davies, A.G., and J.F. Oates, eds. 1995. *Colobine Monkeys.* New York: Cambridge Univ. Press. 415 p.

Davis, J.E., and S.M. Fortney. 1997. Effect of fluid ingestion on orthostatic responses following acute exercise. *Int. J. Sports Med.* 18:174–178.

Davis, T.R.A. 1961. Chamber cold acclimatization in man. *J. Appl. Physiol.* 16:1011–1015.

Davis, T.R.A. 1962. Effect of heat acclimatization on artificial and natural cold acclimatization in man. *J. Appl. Physiol.* 50:605–612.

Dawe, A.R., and W.A. Spurrier. 1969. Hibernation induced in ground squirrels by blood transfusion. *Science* 163:298–299.

Dawe, A.R., and W.A. Spurrier. 1972. The blood-borne "trigger" for natural mammalian hibernation in the 13-lined ground squirrel and the woodchuck. *Cryobiology* 9:163–172.

Dawes, G.S., H.E. Fox, B.M. Leduc, G.C. Ligginsand R.T. Richards. 1972. Respiratory movements and rapid eye movement sleep in the foetal lamb. *J. Physiol.* 220:119–143.

Dawson, T.J., M.J.S. Denny, and A.J. Hulbert. 1969. Thermal balance of the macropodid marsupial *Macropus eurenii* Desmarest. *Comp. Biochem. Physiol.* 31:645–653.

Dawson, W., et al. 1983. Metabolic ajustments of small passerine birds for migration and cold. Amer. J. Physiol. 245:R755–R767.

De Coursey, P. 1960. Phase control of activity in a rodent. In *Biological Clocks.* Cold Spring Harbor Symposia on Quantitative Biology 25:49–56.

de Swart, et al. 1996. Impaired immunity in seals exposed to biocontaminants. *Proc. Amer. Soc. Mammalogists* 76:5.

Deb, C., and J.S. Hart. 1956. Hematological and body fluid adjustments during acclimation to a cold environment. *Canad. J. Biochem. Physiol.* 34:959–966.

Deleanu, M., and C. Margineanu. 1986. Air ionization and circannual fluctuation of anaphylactic sensitivity. *Int. J. Biometeor.* 30:65–67.

DeLong, R.L., and B.S. Stewart. 1991. Diving patterns of northern elephant seal bulls. *Mar. Mammal Sci.* 7:369–384.

Deviche, P., and B.M. Barnes. 1995. Introduction to the symposium: Endocrinology of arctic birds and mammals. *Amer. Zool.* 35:189–190.

Devlin, T.M., ed. 1992. *Textbook of Biochemistry,* 3rd Edition. New York: Wiley-Liss. 1185 p.

Dewailly, E., et al. 1993. Inuit exposure to organochlorines. Internat. Congress on Circumpolar Health 9:9.

Dewan, E.M. 1967. On the possibility of a perfect rhythm method of birth control by periodic light stimulation. *J. Ob. Gyn.* 99:1016–1019.

Dews, P.B., and T.N. Wiesel. 1970. Consequences of monocular deprivation on visual behavior in kittens. *J. Physiol.* 206:437–455.

Diamond, G. 1994. Regulation, Integration, Adaptation. *The Physiologist* 37:1–98.

Diamond, J. 1992. Must we shoot deer to save nature? *Natural History* 8:2–8.

Diaz-Muñoz, et al. 1996. Ryanodine receptor is a circadian oscillator in the suprachiasmatic nucleus of the rat. *The FASEB Journal* 10: 117.

Dicke, R.H. 1962. The earth and cosmology. *Science* 138:653–664.

Dickinson, J.G. 1983. High altitude cerebral edema: Cerebral acute mountain sickness. *Sem. Resp. Med.* 5:151–158.

Dietz, R.S., and R.F. Dill. 1969. Down into the sea in ships. *Sea Frontiers* 15:2–9.

Dill, D.B., Sec. ed. 1964. *Handbook of Physiology: Sec. 4, Adaptation to the Environment.* Amer. Physiol. Soc. (Publishers) 1056 p.

Dill, D.B. 1938. *Life, Heat, and Attitude.* Cambridge, MA: Harvard Univ. Press. 211 p.

Dill, D.B. 1960. Fatigue and Physical Fitness, p. 384–402. In *Science and Medicine of Exercise and Sports*, ed. W. R. Johnson. New York: Harper, 740 p.

Dille, J.R. 1996. Melatonin: a wonder drug. *Aviation, Space, and Environmental Medicine* 67:792.

Dizon A.E., and R.W. Brill. 1979. Thermoregulation in tunas. *Amer. Zool.* 19:249–265.

Dodge, C.H., and G.E. Folk, Jr. 1963. Notes on comparative tolerance of some Iowa turtles to oxygen deficiency (hypoxia). *Iowa Acad. Sci.* 70:438–441.

Dodge, C.H., and Z.R. Glaser. 1977. Trends in nonionizing electromagnet radiation bioeffects research and related occupational health aspects. *J. Microwave Power* 12:319-334.

Dog Mushers' Magazine. 1965. Alaska Dog Mushers' Association. Box 1212, Fairbanks, Alaska. 37 p.

Donnelly, D.F. 1996. How does the carotid body sense hypoxia? *The Physiologist* 39:181.

Doucet, G.J., and J.R. Bider. 1969. Activity of *Microtus pennsylvanicus* related to moon phase and moonlight revealed by the sand transect technique. *Canadian J. Zool.* 47:1183–1186.

Doyle, R. 1998. Deaths from excessive cold. *Sci. Amer.* 278:26.

Drinkwater, B.L., and S.M. Horvath. 1979. Heat tolerance and aging. *Med. Sci. Sports* 11:49–55.

Dripps, R.D., ed. 1956. *The Physiology of Induced Hypothermia.* (A Symposium). Nat. Acad. Sci. and Nat. Res. Council: Washington, DC Publication 451. 446 p.

Dudka, L.T., et al. 1971. Inequality of inspired and expired gaseous nitrogen in man. *Nature* 232:264–268.

Duffner, G.J. 1958. Medical problems involved in underwater compression and decompression. *Ciba Clin. Symp.* 10:99–117.

Durkot, M.J., O. Martinez, D. McQuade and R. Francesconi. 1986. Simultaneous determination of fluid shifts during thermal stress in a small-animal model. *J. Appl. Physiol.* 61:1031–1034.

Durrer, J.L., and J.P. Hannon. 1962. Seasonal variations in caloric intake of dogs living in an arctic environment. *Amer. J. Physiol.* 202:375–378.

Dyer, F.C., and J.L. Gould. 1981. Honey bee orientation: A backup system for cloudy days. *Science* 214:1041–1042.

E

Eagan, C.J. 1963. Introduction and terminology: Habituation and peripheral tissue adaptations. *Fed. Proc.* 22:930–933.

Eagan, C.J., and E. Evonuk. 1964. Retention of resistance to cooling by Alaskan natives in a temperate climate. *Fed. Proc.* 23:367.

Eagan, C.J., E. Evonuk, and R.A. Boster. 1964. Behavioral temperature regulation in the working sled dog. Proc. 15th Alaska Science Conf. *AAAS.* 15:12.

Eagan, C.J., J.L. Durrer, and W.M. Millard. 1963. *Rectal temperature of the working sled dog.* Technical Report 63–40, Arctic Aeromedical Laboratory, Fort Wainwright, Alaska. 8 p.

Easter-Pilcher, A. 1996. Implementing the Endangered Species Act. *BioScience* 45:355–363.

Eayers, J.L., and K.F. Irelend. 1950. The effect of total darkness on the growth of the newborn albino rat. *J. Endocrinol.* 6:386–392.

Ebersole, J.H. 1960. The new dimensions of submarine medicine. *New England J. Med.* 262:599–610.

Eckenhof, R.G., and J.S. Hughes. 1984. Acclimitization to decompression stress, p. 93–100. In *Underwater Physiology VIII, Proc. Eighth Symp. Underwater Physiol.,* ed. A. J. Bachrach and M. M. Matzen. Bethesda, MD: Undersea Med. Soc.

Eckenhof, R.G., et al. 1987. Progression of and recovery from pulmonary O_2 toxicity in humans exposed to 5 ATA air. *Aviat. Space Environ. Med.* 58:658–667.

Edholm, O.G., H.H. Fox, and R.K. MacPherson. 1957. Vasomotor control of the cutaneous blood vessels in the human forearm. *J. Physiol.* 139:455–465.

Edmunds, L.N., Jr., and K.J. Adams. 1981. Clocked cell cycle clocks. Science 211:1002–1004.

Egan, C.E., B.H. Espie, S. McGrann, K.M. McKenna, J.A. Allen. 1996. Acute effects of vitration on peripheral blood flow in healthy subjects. *Occup. Environ. Med.* 53:663–669.

Ehret, C.F. 1993. Personal communication.

Ehret, C.F., and E. Trucco. 1967. Molecular models for the circadian clock. *J. Theor. Biol.* 15:240–262.

Ehrlich, P.R. 1970. Too many people, p. 219–232. In *The Environmental Handbook,* ed. G. DeBell. New York: Ballantine Books. 367 p. (Reprinted from *The Population Bomb* by P.R. Ehrlich.)

Ehrlich, P.R. 1994. Energy use and biodiversity loss. Philos. Trans. R. Soc. Lond. *Biol. Sci.* 344:99–104.

Eichna, L. W., W.F. Ashe, W.B. Bean,and W.B. Shelley. 1945. The upper limits of environmental heat and humidity tolerated by acclimatized men working in hot environment. *J. Indust. Hygiene & Toxicol.* 27:59–84.

Eisenberg, J.F. 1962. Behavioral evolution in the rodentia. *Amer. Zool.* 2:520–521.

Eisenbud, M., and L.R. Ehrlich. 1972. Carbon Monoxide concentration trends in urban atmospheres. *Science* 176:193–194.

Eisenlord, G. 1995. Normal human body temperature. *Natural History* 104:4.

Eisner, T. 1992. The hidden value of species diversity. *BioScience* 42:578.

Ek, J.R., R.A. Nelson, and M.A. Ramsay. 1997. Energy expenditure of free ranging polar bears by infrared thermography. *FASEB J.* 11:373.

Elizondo, E.S. and G.S. Johnson. 1980. Peripheral effector mechanisms of temperature regulation, p. 397–408. In: *Advances in Physiological Sciences, Vol. 32, Contributions to Thermal Physiology,* ed. Z. Szelenyi and M. Szekely. Pergamon Press,

Elles, F.P, W. Feldberg, and R.D. Myers. 1964. Effects on temperature of amines injected into the cerebral ventricles. A new concept of temperature regulation. *J. Physiol.* (Lond.) 173:226–237.

Elliott, D.H., and R.E. Moon. 1993. Manifestations of the Decompression Disorders, p. 481–505. In *The Physiology and Medicine of Diving* (4th ed.), ed. P. B. Bennett and D. H. Elliott. London: Saunders.

Ellis, F.P. 1976. Mortality and morbidity associated with heat exposure. *Int. J. Biometeorol.* 20(Suppl 2): 36–40.

Ellis, F.P., H.M. Ferres, A.R. Lind, and P.S.B. Newling. 1960. The upper limits of tolerance of environmental stress, p.158–179. In *Physiological Responses to Hot Environments.* Spec. Report, Series No. 298. London: Med. Res. Council. 323 p.

Elsner, R., et al. 1996. Seals and Ama, animal and human diving experts, p. 127–135. In *Physiological Basis of Occupational Health: Stressful Environments,* ed. K. Shiraki, S. Sagawa, and M.K. Yousef. Amsterdam: SPB Academic Publishing.

Elsner, R.W. 1963. Comparison of Australian Aborigines, Alacaluf Indians, and Andean Indians. *Fed. Proc.* 22:840–843.

Elsner, R.W., D. Franklin, R. Van Citters, and N. Watson. 1964. Circulatory adaptations in diving mammals. 15th Alaskan Sci. Conf. Alaska Div. *AAAS.* Abstract No. 30.

Elsner, R.W., W.N. Hanafee, and D.D. Hammond. 1971. Angiography of the inferior vena cava of the harbor seal during simulated diving. *Amer. J. Physiol.* 220:1155–1157.

Emlen, J.T. 1969. Bird migration: Influence of physiological state upon celestial orientation. *Science* 165: 716–718.

Emlen, J.T., and R.L. Penney. 1966. The navigation of penguins. *Sci. American.* 215:104–113.

Emmanuel, B. 1981. Autoregulation of urea cycle by urea in mammalian species. *Comp. Biochem. Physiol.* 70A:79–81.

Engelberg, J. 1995. Integrative Physiology. *Am. J. Physiol.* (Adv. Physiol. Educ.) 269:S55–S70.

Enright, J.T. 1979. *The Timing of Sleep and Wakefulness.* New York: Springer-Verlag, 263 p.

Erikson, H., and J. Krog. 1956. Critical temperature in naked men. *Acta Physiol. Scand.* 37:35–39.

Erwin, D.E., J. Valentine, D. Jablonski. 1997. The origin of animal body plans. *Amer. Scientist* 85:126–138.

Eskin, A., and J.S. Takahashi. 1983. Adenylate cyclase activation shifts the phase of a circadian pacemaker. *Science* 220:82–84.

Essler, W.O., and G.E. Folk, Jr. 1960. The determination of 24-hour physiological rhythms of unrestrained animals by radio telemetry. *Nature* 190:90–91.

Evans, D.E. 1984. Physiology of diving, cardiovascular effects, p. 99–109. In *The Physician's Guide to Diving Medicine,* ed. C. W. Shilling, C. B. Carlston, and R. A. Mathias. New York: Plenum Press.

Evered, D., and S. Clark, eds. 1985. *Photoperiodism, Melatonin and the Pineal.* Ciba Foundation Symposium No. 117. 323 p.

Ewing, W.G., E.H. Studier and J.O. O'Farrell. 1970 Autumn fat deposition and gross body composition in three species of *Myotis. Comp. Biochem. Physiol.* 36:119–129.

F

Fairley, T.C. *Sverdrup's Arctic Adventures.* 1959. London: Longmans. 398 p.

Fanger, P.O. 1967. Calculations of thermal comfort. *A.S.H.R. and A.C.E. trans.* 73:1–20

Farner, D.S. 1957. An essential compound in bird photoperiod. *Physiologist* 1:26.

Farner, D.S. 1961. Photoperiodicity. *Annu. Rev. Physiol.* 23:71–96.

Farrand, R.L. 1959. Cold acclimatization in the golden hamster. *Univ. Iowa Studies in Natur. Hist.* 22:3–29.

Fass, S. 1993. Molecular clocks and biological rhythms. *Forum* 5:1–6.

Fedak, M.A. 1986. Diving and exercise in seals: interactions of behavior and physiology. In *Behavioural ecology of underwater organisms*. Report of the 19th symposium of the Underwater Association, March, 1985, *London. Prog. Underwater Sci.* 11:155–169.

Feldberg, W., and R.D. Myers. 1964. Effects on temperature of amines injected into the cerebral ventricles. A new concept of temperature regulation. *J. Physiol.* (Lond.) 173:226–237.

Fenton, M.B., and J.H. Fullard. 1981. Moth hearing and the feeding strategies of bats. *Amer. Scientist* 69:266–275

Ferguson, J.H. 1979. Effect of photoperiod and cold acclimation on survival of mice in cold. *Cryobiology* 16:468–472.

Ferguson, J.H. 1985. *Mammalian Physiology.* Columbus: Merrill Co. 517 p.

Ferguson, M.W.J., ed. 1983. Zoological Soc. of London Symposia Series, Vol. 52: *The Structure, Development and Evolution of Reptiles.* 720 p.

Ferrante, F.L., and H.N. Frankel. 1965. Cardiovascular responses during protracted apnea in a non-diving mammal. *Fed. Proc.* 24:704.

Ferrigno, M., et al. 1991. Electrocardiogram during deep breath-hold dives by elite divers. *Undersea Biomed. Res.* 18:81–91.

Feskens, E.J., and D. Kromhout. 1993. Epidemiologic studies on Eskimos and fish intake. *Annals N.Y. Acad. Sci.* 683:9–15.

Fife, W.P. 1987. The toxic effects of hydrogen-oxygen breathing mixture, p. 13–23. In *Hydrogen as a Diving Gas,* ed. R. W. Brauer. Bethesda, MD: Undersea Hyperbaric Med. Soc.

Filho, A.G., De M., S.E. Huggins, and S.G. Lines. 1983. Sleep and waking in the three-toed sloth, Bradypus tridactylus. *Comp. Biochem. Physiol.* 76A:345–355.

Finch, V.A., et al. 1980. Why black goats in hot deserts? Effects of coat color on heat exchanges of wild and domestic goats. *The Univ. of Chicago,* p. 19–25. Chicago: Univ. of Chicago.

Findlay, J.D. 1963. Acclimatization to heat in sheep and cattle. *Fed. Proc.* 22:688–692.

Finley, J.W. 1995. Effect of hydrogenated soybean oil. *BioScience* 45:511.

Finnerty, J.R., and B.A. Block. 1994. Endothermy in fishes. *Science* 164:1249–1251.

Fish, F.E. 1995. Transitions from drag-based to lift-based propulsion in mammalian aquatic-swimming. *American Zoologist* 35:4A

Fisher, K.C., A.R. Dawe, C.P. Lyman, E. Schonbaum, and F.E. South, Jr., ed. 1967. *Mammalian Hibernation III: Proc. Third Int. Symp. Natur. Mammal. Hiber.* Edinburgh: Oliver and Boyd.

Fiske, V.M. 1965. Serotonin rhythm in the pineal organ: Control by the sympathetic nervous system. *Science* 146:253–254.

Fiske, V.M., and L.C. Huppert. 1968. Melatonin action on pineal varies with photoperiod. *Science* 162:279.

Folk, G.E., Jr. 1954. The effect of restricted feeding time on the diurnal rhythm of running activity in the white rat. *Anatomical Record* 120:179.

Folk, G.E., Jr. 1955. Modification by light and feeding of the 24-hour rhythm of activity in rodents. Proc. 5th Conf. on Biological Rhythms, Stockholm.

Folk, G.E., Jr. 1959. Modification by light of 24-hour activity of white rats. *Iowa Acad. Sci.* 66:399–406.

Folk, G.E., Jr. 1961a. Observations on the daily rhythms of body temperature labile mammals. *Ann. N.Y. Acad. Sci.* 98: 954–969.

Folk, G.E., Jr. 1961b. Circadian aspects of the circulation. In *Circadian Systems.* Ross Conf. on Pediatric Research 39:86–88.

Folk, G.E., Jr. 1964. Daily physiological rhythms of carnivores exposed to extreme changes in Arctic light. *Fed. Proc.* 23:1221–1228.

Folk, G.E., Jr. 1966a. Introduction to Environmental Physiology. Philadelphia: Lea & Febiger. 308 p.

Folk, G.E., Jr. 1966b. Physiological observations on bears under winter den conditions. In *Proc. 3rd Symp. Mammal. Hibernation.* Edinburgh: Oliver and Boyd. 530 p.

Folk, G.E., Jr. 1969. The roots of environmental physiology, In: *Physiological Systems in Semiarid Environments,* ed. C.C. Hoff and M.L. Riedesel. Albuquerque: Univ. New Mexico Press.

Folk, G.E., Jr. 1974. *Textbook of Environmental Physiology.* 2nd Ed. Philadelphia: Lea & Febiger. 456 p.

Folk, G.E., Jr. 1980. Protein and fat metabolism in mammalian hypophagia and hibernation. *Fed. Proc.* 39:2953–2954.

Folk, G.E., Jr. 1981. Climatic change and acclimatization, p. 157–168. In *Bioengineering, Thermal Physiology and Comfort,* ed. K. Cena and J.A. Clark. New York: Elsevier Scientific.

Folk, G.E., et al. 1995. Daily rhythms of body temperature, heart rate, blood pressure, and activity in the rat. Proc. *Internat. Symp. Biotelemetry* 13: 246–250.

Folk, G.E., Jr., et al. 1985. Cold exposure, ethanol, brown fat, and human vasomotor function. In *Proc. 7th Ann. Conf. on Biometeorol.* Amer. Meteor. Soc., Phoenix 7:357–359.

Folk, G.E., Jr., J.J. Berberich, and D.K. Sanders. 1972. Bradycardia of the polar bear during head immersion and feeding. *Arctic* 26:78–79.

Folk, G.E., Jr., P.S. Cooper, and M.A. Folk. 1996. Polar bears eating nearly 100% fat compared with fasted bears. *Proc. Int. Conf. Bear Res. and Manage.* 9:459–460.

Folk, G.E., Jr., and J.R. Copping. 1978. Telemetry in animal biometoerology. *Inter. J. Biometeorology* 16:153–170.

Folk, G.E., Jr., M.A. Folk, and D. Craighead. 1977. A comparison of body temperatures of least weasels and wolverines. *Comp. Biochem. Physiol.* 58A:229–234.

Folk, G.E., Jr., M.A. Folk, and F. Kreuzer. 1970. Initial stages of hibernation: Is cold acclimatization necessary? *Acta Theriologica* 24:373–380.

Folk, G.E., Jr., M.A. Folk, and J.G. Minor. 1972. Physiological condition of three species of bears in winter dens, p.107–125. In *Bears—Their Biology and Management,* ed. S. Herrero. Internat. Union for Conserv. Nature, Pub. No. 23.

Folk, G.E., Jr., and M.A. Folk. 1980. Physiology of large mammals by implanted radio-capsules, p. 33–44, In *A Handbook on Biotelemetry,* ed. C.J. Amlaner, Jr., and D.W. Macdonald. Oxford: Pergamon Press. 804 p.

Folk, G.E., Jr., K.A. Hagelstein, and P.J. Ringens. 1977. Cold acclimatization and the pineal gland of lemmings. *Fed. Proc.* 36:419.

Folk, G.E. Jr., and H.T. Hammel. 1992. Types of mammalian dormancy. *FASEB J.* 6:954.

Folk, G.E., Jr, J.M. Hunt, and M.A. Folk. 1980. Further evidence for hibernation of bears. *Proc. 4th Int. Conf. Bear Res. and Manage.* U.S. Govt. Printing Office, No. GP 0781–014.

Folk, G.E., Jr., T.L. Kaduce, and A.A. Spector. 1991. Plasma lipid of fed vs. fasted polar bears. *FASEB J.* 5:394.

Folk, G.E., Jr., and J.P. Long. 1988. Serotonin as a neurotransmitter: A review. *Comp. Biochem. Physiol.* Vol. 91C, No. 1, p. 251–257.

Folk, G.E., Jr., and R.A. Nelson. 1981. The hibernation of polar bears: A model for the study of human starvation. *Arctic Med. Res. Report* 33:617–619.

Folk, G.E., Jr., W.L. Randall, C.V. Gisolfi, A.J. Ryan, and R.D. Matthes. 1995. Daily rhythms of temperature, heart rate, blood pressure, and activity in the rat, p 246–250, In *Biotelemetry XIII,* eds C. Cristalli, C.J. Amlaner, Jr., M.R. Neuman. Williamsburg, VA: Inter. Soc. Biot. 436 p.

Folk, G.E., Jr., R.R. Schellinger, and D. Snyder. 1961. Day-night changes after exercise in body temperatures and heart rate of hamsters. *Proc. Iowa Acad. Sci.* 68:594–602.

Folk, G.E., Jr., and H.A. Semken, Jr. 1991. The evolution of sweat glands. *Int. J. Biometeorol.* 35:180–186.

Folk, G.E., Jr., and J.G. White. 1970. Acclimation to heat of the beagle dog. *Int. J. Biometeor.* 14:95–101.

Folkard, S., and T.H. Monk, eds. 1985. *Hours of Work: Temporal Factors in Work-Scheduling.* New York: John Wiley & Sons, 327 p.

Follett, B.K., and D.E. Follett, eds.. 1981. *Biological Clocks in Seasonal Reproductive Cycles.* Bristol: Scientechnica, 292 p.

Forrester, J.W. 1971. *World Dynamics.* Cambridge, MA: Wright-Allen. 142 p.

Fortney, S.M. 1991. Development of lower body negative pressure as a countermeasure for orthostatic intolerance. *J. Clin. Pharmacol.* 31:888–892.

Fortney, S.M, E.R. Nadel, C.B. Wenger, and J.R. Bove. 1981. Effect of blood volume on sweating rate and body fluids in exercising humans. *J. Appl. Physiol.* 51:1594–1600.

Foster, K.G., E.P. Ellis, C. Dore, A.N. Exton-Smith, and J.S. Weiner. 1976. Sweat responses in the aged. *Age and Aging* 5:41–101.

Fowler, A., et al. 1985. Effects of inert gas narcosis on behavior—a critical review. *Undersea Biomed. Res.* 12:369–402.

Fox, R.H. 1961. Local cooling in man. *Brit. Med. Bull.* 17:14–18.

Francis, T.J.R., et al. 1990. Bubble-induced dysfunction in acute spinal cord decompression sickness. *J. Appl. Physiol.* 68:1368–1375.

Frank, K.D., and W.F. Zimmerman. 1969. Action spectra for phase shifts of a circadian rhythm in Drosophila. *Science* 163:688–689.

Frankel, H.M., G.E. Folk, Jr., F.N. Craig. 1957. Effects of type of restraint upoon heat tolerance in monkeys. *Proc. Exp. Biol. Med.* 97:339–341.

Franzmann, A.W., and D.M. Hebert. 1971. Variation of rectal temperature in bighorn sheep. *J. Wildl. Mgmt.* 35:488–494.

Fraser, A.F. 1968. *Reproductive Behavior in Ungulates.* Academic Press: New York. 222 p.

Fregly, M.J., and C.M. Blatteis, eds. 1995. *Handbook of Physiology: Section 4, Environmental Physiology, Vol. 1.* New York: Oxford University Press. 783 p.

French, A.R. 1993. Hibernation in Birds: Comparisons with mammals. In *Life in the Cold, Ecological Physiological and Molecular Mechanisms,* ed. C. Carey, G.L. Florant, B.A. Wunder and B. Horwitz. Boulder, Colo.: Westview Press.

Freund, B.J., et al. 1991. Hormonal, electrolyte, and renal responses to exercise are intensity dependent. *J. Appl. Physiol.* 70:900–906.

Friedman, M. 1995. Conservation of Large Ecosystems. *Environmental Review* 2:1–6.

Frings, H., A. Anthony, and M. W. Schein. 1958. Salt excretion by nasal gland of Laysan and Blackfooted Albatrosses. *Science* 128:1572.

Frink, C.R. 1996. Acid rain revisited. *Science* 273:292–294.

Fritsch, J.M., J.B. Charles, B.S. Bennett, M.M. Jones, and D.L. Eckberg. 1992. Short-duration spaceflight impairs human carotid baroreceptor-cardiac reflex responses. *J. Appl. Physiol.* 73:664–671.

Fritsch-Yelle, J.M., J.B. Charles, M.M. Jones, and M.L. Wood. 1996. Microgravity decreases heart rate and arterial pressure in humans. *J. Appl. Physiol.* 80:910–914.

Fructus, X. 1987. Hydrogen pressure and HPNS, p. 125–140. In *Hydrogen as a Diving Gas,* ed. R.W. Brauer. Bethesda, MD: Undersea Hyperbarics Med. Soc.

Frye, A.J., and E. Kamon. 1981. Responses to dry heat of men and women with similar aerobic capacities. *J. Appl. Physiol.* 50:65–70.

Fuentes-Pardo, B., L. Verdugo-Diaz, and F. Inclan-Rubio. 1985. Effect of external level of calcium on ERG circadian rhythm in isolated eyestalk of crayfish. *Comp. Biochem. Physiol.* 82A:385–395.

Fuller, C.A., F.M. Sulzman, and M.C. Moore-Ede. 1981. Shift-work and the jet-lag syndrome: Conflicts between environmental and body time, p. 305–320. In *The Twenty-Four Hour Workday.* Proc. of Symposium on Variations in Work-sleep Schedules, ed. L.C. Johnson. Cincinnati: U.S. Dept. of Health and Human Services. Publ. 81–127.

Fuller, C.A., R. Lydic, F.M. Sulzman, H.E. Albers, B. Tepper, and M.C. Moore-Ede. 1981. Circadian rhythm of body temperature persists after suprachiasmatic lesions in the squirrel monkey. *Amer. J. Physiol.* 241:R385–R391.

Fuller, C.A., R. Lydic, F.M. Sulzman, H.E. Albers, B. Tepper, and M.C. Moore-Ede. 1983. Auditory entrainment of primate drinking rhythms following partial suprachiasmatic nuclei lesions. *Physiol. Behav.* 31:573–576.

Fuller, W.A., and L. Stebbins. 1969. Overwintering of small animals. *Arctic* 22:34–37.

Fyhn, H. J. 1979. Rodents, p. 95–144. In *Comparative Physiology of Osmoregulation, vol. 2,* ed. G.M.O. Maloiy. New York: Academic Press. 246P.

G

Gaddis, G.M., and Elizondo. 1984. Effect of central blood volume decrease upon thermoregulation responses to exercise in the heat. *Fed. Proc.* 43:627.

Gagge, A.P. 1981. Rational temperature indices of thermal comfort. In *Bioengineering, Thermal Physiology and Comfort,* ed. K. Cena and J.A. Clark. p. 79–98. New York: Elsevier Scientific.

Gallagher, R.P., B. Ma, D.I. McLean, C.P. Yang, V. Ho, J.A. Carruthers, L.M. Warshawski. 1990. Trends in basal cell carcinoma, squamous cell carcinoma, and melanoma of the skin from 1973 through 1987. *J. Am. Acad. Dermatol.* 23:413–421.

Gallaher, M.M., D.W. Fleming, L.R. Berger, and C.M. Sewell. 1992. Pedestrian and hypothermia deaths among Native Americans in New Mexico. *JAMA* 267(10): 1345–1348.

Gamba, R. 1990. New world records of breath-hold diving depth: Francisco Ferrera, 112 m; and Engela Bandini, 107 m (in Japanese). *Marine Diving* Jan (suppl.):6–9.

Gambell, R. 1995. Irish Meeting of the International Whaling Commission. *The Pilot* 12:2–4.

Gander, P.H., and M.C. Moore-Ede. 1982. Forced internal desynchronization between rest-activity and temperature rhythms in squirrel monkeys. *Fed. Proc.* 41:1696.

Gardette, B. 1987. Human deep hydrogen dives 1983–1985, p. 109–118. In *Hydrogen as a Diving Gas,* ed. R. W. Brauer. Bethesda, MD: Undersea Hyperbaric Med. Soc.

Gardette, B. 1989. Compression procedures for mice and human hydrogen deep diving, COMEX HYDRA program, p.217–231. In *High Pressure Nervous Syndrome 20 Years Later,* ed. J.C. Rostain, E. Martinez, and C. Lemaire. Marseille: ARAS-SNHP.

Gardner, W.W., and H.A. Allard. 1920. Effect of the relative length of day and night and other factors of the environment in growth and reproduction in plants. *Agr. Res.* 18:553–605.

Gaston, S., and M. Menaker. 1968. Pineal function: The biological clock in the sparrow? *Science* 160:1125–1127.

Gates, D.M. 1972. *Man and His Environment: Climate.* New York: Harper and Row. 175 p.

Gates, D.M. 1995. Climate Change on Plants and Animals, p 143–154. In *Preparing for Global Change*, ed. G.R. Carmichael, G.E. Folk, Jr., J.L. Schnoor. 298 p. New York: SPB Pub.

Geiser, F., A.J. Hulbert, and S.C. Nicol. 1996. *Adaptations to the Cold: Tenth International Hibernation Symposium.* Armidale, Australia: University of New England Press.

Geiser, F., and T. Ruf. 1996. Hibernation versus daily torpor in mammals and birds: physiological variables and classificiation of torpor patterns. *Physiol. Zool.* 68:935–966.

German, A. L. 1961. The degree of resistance to thirst of some murine rodents of the steppe zone. *Zool. Zhur.* (in Russian) 40:914–921.

Gessaman, J.A. 1972. Bioenergetics of the snowy owl (*Nyctea scandiaca*). *Arctic and Alpine Research* 4:223–238.

Gessaman, J.A. 1973. *Ecological Energetics of Homeotherms.* Monograph Series, Vol. 20. Logan: Utah State University Press. 155 p.

Geyer, L.A., and R.J. Barfield. 1979. Ultrasonic communication in rodents. *Amer. Zool.* 19:411–531.

Giaja, J., and L. Markovic. 1953. L'hypothermie et la toxicité du gaz carbonique. *Compt. rend. Acad. sc.* 236:2437–2440.

Gibbs, F.P. 1980. Temperature dependence of rat circadian pacemaker. *Am. J. Physiol.* 241:R17–R20.

Giesbrecht, G.G., and G.K. Bristow. 1995. Influence of body composition on rewarming from immersion hypothermia. *Aviation Space & Enviro. Med.* 66(12):1144–1150.

Giese, A.C. 1979. The relevance of photobiology. *BioScience* 29:353–357.

Giles, E. 1889. *Australia Twice Traversed.* 2 vols. London: Low, Maston, Searle, and Rivington.

Gillis, T.E., and J.S. Ballantyne. 1997. Influences of sub-zero temperature on marine bivalves. *The FASEB J.* 11:23.

Gisolfi, C.V. 1987. Influence of acclimatization and training on heat tolerance and physical endurance. In: *Heat Stress: Physical Exertion and Environment,* ed. J.R. Hales and D.A. Richards. New York: Elsevier Scientific.

Gisolfi, C.V., and J.C. Cohen. 1979. Relationships among training, heat acclimation, and heat tolerance in men and women: the controversy revisited. *Med. Sci. Sports* 11:56–59.

Glaser, E.M., and G.C. Whittow. 1957. The cold pressor response. *J. Physiol.* 136:98–102.

Glaser, E.M., and J.P. Griffin. 1962. Habituation of rats to repeated cooling. *J. Physiol.* 160:429–432.

Glaser, E.M., and R.J. Shepard. 1963. Simultaneous experimental acclimatization to heat and cold in man. *J. Physiol.* 169:592–605.

Glaser, E.M., M.S. Hall, and G.C. Whittow. 1959. Habituation and adaptation. *J. Physiol.* 146:152.

Glass, G.G., and R.N. Hoover. 1989. The emerging epidemic of melanoma and squamous cell skin cancer. *JAMA* 262:2097. (See also p. 2138, same journal.)

Glass, M.L., J.W. Hicks, and M.L. Riedesel. 1979. Respiratory responses to long-term temperature exposure in the box turtle, *Terrapene ornata. J. Comp. Physiol.* 131:353–359.

Gleason, J.F., et al. 1993. Record low ozone in 1992. *Science* 260:523–526.

Glick, Z., S.J. Wickler, J.S. Stern, and B.A. Horwitz. 1984. Regional blood flow in rats after a single low-protein, high-carbohydrate test meal. *Am. J. Physiol.* 247 (*Regulatory Integrative Comp. Physiol.* 16):R160–R166

Goetze, D. 1996. Global climate change has begun. *Nucleus* 18:1

Goldman. In press

Goldman, B.D. 1980. Seasonal cycles in testis function in two hamster species: Relation to photoperiod and hibernation. p. 401–409. In *Testicular Development, Structure, and Function,* ed. A. Steinberger and E. Steinberger, ed.. Raven Press: New York. 416 p.

Goldman, B.D., and J.M. Darrow. 1983. The pineal gland and mammalian photoperiodism. *Neuroendocrinology* 37:386–396.

Goldman, R.F. 1978. Prediction of human heat tolerance, p. 53–69. In *Environmental Stress: Individual Human Adaptations,* ed. L.J. Folinsbee, J.A. Wagner, J.F. Borgia, B.L. Drinkwater, J.A. Gliner and J.F. Bedi. New York: Academic Press.

Goldman, R.F. 1981. Evaluating the effects of clothing on the wearer, p. 41–55. In *Bioengineering, Thermal Physiology and Comfort,* ed. K. Cena and J.A. Clark. New York: Elsevier Scientific. 289 p.

Gonzales, R.R. 1981. Exercise physiology and sensory responses, p 123–156. In *Bioengineering, Thermal Physiology and Comfort,* ed. K. Cena and J.A. Clark. New York: Elsevier Scientific, 289 p.

Gonzales-Alonso, J.R. Mora-Rodriguez, P.R. Below, and E.F. Coyle. 1995. Dehydration reduces cardiac output and increases systemic and cutaneous vascular resistance during exercise. *J. Appl. Physiol.* 79:1487–1496.

Gonzalez-Cabrera, P.J., F. Dowd, V.K. Pedibhotla, R. Rosario, D. Stanley-Samuelson, and D Petzel. 1995. Enhanced hypo-osmoregulation induced by warm-acclimation in Antarctic fish is mediated by increased gill and Kidney Na+/K+-ATPase activities. *J. Exp. Biol.* 198:2279–2291.

Gordon, C.J., and L. Fogelson. 1991. Comparative effects of hypoxia on behavioral thermoregulation in rats, hamster and mice. *Am. J. Physiol* 260:R120–125.

Goto, K., D.L. Laval-Martin, and L.N. Edmunds, Jr. 1985. Biochemical modeling of an autonomously oscillatory circadian clock in Eugenia. *Science* 228:1284–1288.

Gould, E. 1970. Echolocation and communication in bats, p. 144–161. In *About Bats,* ed. Slaughter and Watson, Dallas: Southern Methodist Univ. Press

Gower, B.A., et al. 1996. Reproductive refractoriness to long day length in collared lemmings. *FASEB* 10:117.

Graham, D. 1996. Plants: A secret weapon against global warming? *Technology Review* (July) 99:16–17.

Granberg, P.O. 1991. Human physiology under cold exposure. *Arctic Medical Res.* 50(Suppl 6):23–27.

Greaney, G.S., A.R. Place, R.E. Cashon, G. Smith, and D.A. Powers. 1980. Time course of changes in enzyme activities and blood respiratory properties of killfish during long-term acclimation to hypoxia. *Physiol. Zool.* 53:136–144.

Green, D.J., and R. Gillette. 1982. Circadian rhythm of firing rate recorded from single cells in the rat suprachiasmatic brain slice. *Brain Res.* 245:198–200.

Greenleaf, J.E., and B.L. Castle. 1971. Exercise temperature regulation in man during hypohydration and hyperhydration. *J. Appl. Physiol.* 30:847–853.

Greenleaf, J.E., and T. Morimoto. 1996. Mechanisms controlling fluid ingestion: thirst and drinking, p 3–18. In *Body Fluid Balance Exercise and Sport,* ed. E.R. Buskirk and S.M. Puhl. New York: CRC Press.

Griffin, Donald R. 1959. *Echos of Bats and Men.* New York: Doubleday and Co. 156 p.

Guillenimault, C., D. Leger, R. Pelayo, S. Gould, B. Hayes, and L. Miles. 1996. Development of circadian rhythmicity of temperature in full-term normal infants. *Neurophysiologie Clinique* 26(1):21–9.

Gunning, K.A., M. Sugrue, D. Sloane, and S.A. Deane. 1995. Hypothermia and severe trauma. *Australian & New Zealand J. of Surgery* 65(2):80–82.

Guppy, M., et al. 1986. Microcomputer-assisted metabolic studies of voluntary diving of Weddell seals. *Am. J. Physiol.* 250:R175–R187.

Gustafson, A.W., and W.D. Belt. 1981. The adrenal cortex during activity and hibernation in the male little brown bat, *Myotis lucifugus lucifugus:* Annual rhythm of plasma cortisol levels. *Gen. Comp. Endocrinol.* 44:269–278.

Guyton, A.C. 1966. *Textbook of Medical Physiology*, 3rd edition, p 624. Philadelphia and London: W.B. Saunders Co. 1260 p.

Guyton, A.C., and J.E. Hall. 1996. *Textbook of Medical Physiology,* 9th ed. Philadelphia: Saunders. 1148 p.

Guyton, G. P., et al. 1995. Myoglobin saturation in free-diving Weddell seals. In *J. Appl. Physiol.* 79(4):1148–1155.

H

Hackett, P.H. 1980. *Mountain Sickness: Prevention, Recognition and Treatment.* New York: American Alpine Club.

Hackett, P.H., and Hornbein. 1988. Disorders of high altitude. In *Textbook of Respiratory Medicine*, ed. J.F. Murray and J.A. Nadine. Philadelphia: Saunders.

Hackett, P.H., and R.C. Roach. 1995. High-Altitude Medicine, p. 3. In *Wilderness Medicine*, 3rd edition, ed. P.S. Auerbach. St. Louis: Mosby.

Hadley, N.F. 1980. Surface waxes and integumentary permeability. *Am. Sci.* 68:546–553.

Hafez, E.S.E., ed. 1968. *Adaptation of Domestic Animals.* Philadelphia: Lea and Febiger. 415p.

Haider, M., and D.B. Lindsley. 1964. Microvibrations in man and dolphin. *Science* 146:1181–1183.

Haim, A., et al. 1987. Urine analysis of moles. *Comp. Biochem. Physiol.* 88A:179–181.

Halberg, F., and R.B. Howard. 1958. 24-hour periodicity and experimental medicine. *Postgraduate Medicine* 24:349–358. (Univ. Minnesota).

Haldar, C. 1996. Impact of humidity on annual pineal testicular cycles of a tropical rodent. *Proc. Internat. Congress Biometeorology* 14:125.

Hale, F.C., R.A. Westland, and C.L. Taylor. 1958. Barometric and vapor pressure influences on insensible weight loss. *J. Appl. Physiol.* 12:20–28.

Hale, H.B., and R.B. Mefferd. 1963. Thermal spectrum analysis of thyroid-independent phases of nitrogen and mineral metabolism. *Fed. Proc.* 22:766–771.

Hamilton, A.J., A. Cymerman, and P. Black. 1986. High altitude cerebral edema. *Neurosurgery* 19:841–949.

Hammel, H.T. 1956. Infrared emissivities of some Arctic fauna. *J. Mammal.* 37:375–378.

Hammel, H.T. 1963. Effect of race on response to cold. *Fed. Proc.* 22:795–800.

Hammel, H.T. 1963. Summary of comparative thermal patterns in man. *Fed. Proc.* 22:846–847.

Hammel, H.T. 1965. *One Method For Assessing Cold Tolerance.* Tech. Report-AF 41 (609), 1970. Fort Wainwright, Alaska: Arctic Aeromedical Laboratory.

Hammel, H.T., D.C. Jackson, J.A.J. Stolwijk, J.D. Hardy, and S.B. Stromme. 1963. Temperature regulation by hypothalamic proportional control with an adjustable set point. *J. Appl. Physiol.* 18:1146–1154.

Hamner, K.C. 1940. Interrelation of light and darkness in photoperiodic induction. *Botanical Gaz.* 101:658–687.

Hamner, K.C., J.C. Finn, G.S. Sirohi, T. Hoshizaki, and B.H. Carpenter. 1962. The biological clock at the South Pole. *Nature* 195:476–480.

Hannon, J.P. 1960. Tissue energy metabolism in the cold-acclimatized rat. *Fed. Proc.* 19(Suppl.5):139–144.

Hansen, J.C., H.S. Pedersen, and G. Mulvad. 1994. Fatty acids and antioxidants in the Inuit diet. *Arctic Med. Res.* 53(1):4–17.

Harabin, A.L., et al. 1987. An analysis of decrements in vital capacity as an index of pulmonary O_2 toxicity. *J. Appl. Physiol.* 63:1130–1135.

Harabin, A.L., et al. 1990. Response of antioxidant enzymes to intermittent and continuous hyperbaric O_2. *J. Appl. Physiol.* 69:328–335.

Hardeland, R. 1997. New actions of melatonin. *Int. J. Biomet.* 41:47–57.

Hardy, J.D. 1965. Thermal radiation, pain and injury, p. 170–195. In *Therapeutic Heat and Cold,* ed. S. Licht. New Haven: Elizabeth Licht Pub. 593 p.

Hargens, A.R., R.T. Whalen, D.E. Watenpaugh, D.F. Schwandt, and L.P. Krock. 1991. Lower body negative pressure to provide load bearing in Space. *Aviat. Space Environ. Med.* 62:934–937.

Harker, J.E. 1958. Diurnal rhythms in the animal kingdom. *Biol. Rev.* 33:1–52.

Harker, J.E. 1964. The Physiology of Diurnal Rhythms, p. 64. In *Cambridge Monographs in Experimental Biology,* No. 13, ed. T.A. Bennet-Clark, G. Salt, V.B. Wigglesworth. London: Spottiswoode, Ballantyne & Co., Ltd. 114 p.

Harlow, H. 1979. Ph.D. Thesis. The University of Iowa.

Harlow, H.J., J.A. Phillips, and C.L. Ralph. 1980. The effect of pinealectomy on hibernation in two species of seasonal hibernators, *Citellus lateralis* and *C. richardsonii. J. Experimental Zoology.* 213:301–303.

Harlow, H.J., J.A. Phillips, and C.L. Ralph. 1981. Day-Night rhythm in plasma melatonin in a mammal lacking a distinct pineal gland, the nine-banded armadillo. *General and Comparative Endocrinology* 45:212–218.

Harmen, D. 1992. Free radical theory of aging. *Mutat. Res.* 275:257–266.

Harrison, G.A. 1963. Temperature adaptation as evidenced by growth of mice. *Fed. Proc.* 22:691–697.

Harrison, M.H., R.J. Edwards, M.F. Graveney, L.A. Cochrane, and J.A. Davies. 1981. Blood volume and plasma protein responses to heat acclimatization in humans. *J. Appl. Physiol.* 50:597–604.

Hart, J.S. 1957. Climatic and temperature induced changes in the energetics of homeotherms. *Rev. Canad. Biol.* 16:133–141.

Hart, J.S. 1960. Energy metabolism during exposure to cold. *Fed. Proc.* 19(Suppl. 5):15–19.

Hart, J.S. 1961. Physiological effects of continued cold on animals and man. *Brit. Med. Bull.* 17:19–23.

Hart, J.S. 1964. Geography and season: mammals and birds, p. 295–322. In *Handbook of Physiology,* Section 4, *Adaptation to the Environment,* Sec.ed. D. B. Dill. Washington, DC: Am. Physiol. Soc.

Hart, J.S. 1967. Commentary: Some differences in responses of small mammals and man to combined exercise and cold stress. *Canad. Med. Assoc. J.* 96:803–804.

Hart, J.S. et al. 1962. Primitive man in the cold. *J. Appl. Physiol.* 17:953–960.

Harvey, H.E., and W.V. Macfarlane. 1958. The effects of day length upon the coat shedding cycles, body weight, and reproduction in the ferret. *Australian J. Biol. Sci.* 11:187.

Hasler, A.D., et al., eds.. 1972. *Man in the Living Environment.* The Institute of Ecology Report of the Workshop on Global Ecological Problems. 267 p.

Hay, H. R. 1972. Solar radiation: Some implications and adaptations. *Mechanical Engineering* 94:24–29.

Hayden, P., and R.G. Lindberg. 1969. Circadian rhythm in mammalian body temperature entrained by cyclic pressure changes. *Science* 164:1288–1289.

Hayes, J.D. 1972. Magnetic reversals and life. *Sea Frontiers.* 4th series. 16:12.

Hayssen, V. 1995. Milk. *Natural History* 104:36–37.

Hayssen, V., and R.C. Lacy. 1985. Basal metabolic rates in mammals: taxonomic differences. *Comp. Biochem. Physiol.* 81A:741–745.

Haywood, J.S., and E.G. Ball. 1966. Quantitative aspects of brown adipose tissue thermogenesis during arousal from hibernation. *Biol. Bull.* 131:94–103.

Heath, J.E. 1970. Behavioral regulation of body temperature in poikilotherms. *The Physiologist* 13:399–410.

Heath, J.E., and M.S. Heath. 1982. Energetics of locomotion in endothermic insects. *Rev. Physiol.* 44:133–143.

Heezen, B.C., and C.D. Hollister. 1972. *The Face of the Deep.* New York: Oxford Univ. Press. 659 p.

Heinrich, B. 1970. Thoracic temperature stabilization by blood circulation in a free-flying moth (*Manduca sexta*). *Science* 168:580–582.

Heinrich, B. 1974. Thermoregulation in endothermic insects. *Science* 185:747–756.

Heistad, D.D., et al. 1968. Vasoconstrictor response to simulated diving in man. *J. Appl. Physiol.* 25:542–549.

Heller, H.C. 1972. Measurements of convective and radiative heat transfer in small mammals. *J. Mammal.* 53:289–295.

Heller, H.C., D.A. Grahn, L. Trachsel, and J.E. Larkin. 1993. What is a bout of hibernation? In *Life in the Cold, Ecological Physiological and Molecular Mechanisms,* ed. C. Carey, G.L. Florant, B.A. Wunder, and B. Horwitz. Boulder, CO: Westview Press.

Heller, H.C., X.J. Musacchia, and L.C.H. Wang, eds.. 1986. *Living in the Cold, Physiological and Biochemical Adaptations.* New York: Elsevier Scientific. 587 p.

Hellstrom, B., and T.H. Hammel. 1967. The variable set point. *Am. J. of Physiol.* 213:547–549.

Hellstrom, B., et al. 1970. Human peripheral rewarming during exercise in the cold. *J. Appl. Physiol.* 29:191–193.

Hemingway, A. 1957. *Nervous control of shivering.* Tech. Report: TN-40. Arctic Aeromedical Laboratory, Fort Wainwright, Alaska. 11 p.

Henderson, L.J. l958. *The Fitness of the Environment.* Beacon Paperback, BP68. Boston: Beacon Press. 317 p.

Henderson, R.A. 1987. Hyperbaric physiology, 5–20. In *Man in Stressful Environments,* ed. K. Shiraki and M.K. Yousef. Springfield, IL: C.C. Thomas. 266 p.

Henry, K.R., and M. Saleh. 1973. Recruitment Deafness: Functional effect of priming-induced audiogenic seizures in mice. *J. Comp. and Physiol. Psychol.* 84:430–435.

Henshaw, P.S. 1971. *This Side of Yesterday: Extinction or Utopia.* New York: John Wiley & Sons. 186 p.

Henshaw, R.E. 1965. *Physiology of Hibernation and Acclimatization in Two Species of Bats (Myotis lucifugus and Myotis sodalis).* Ph.D. Thesis. Univ. of Iowa, Iowa City, Iowa. 143 p.

Henshaw, R.E., and G.E. Folk, Jr. 1966. Thermoregulation and microclimate selection in bats. *Physiol. Zool.* 38:223–236.

Henshaw, R.E., L.S. Underwood, and T.M. Casey. 1972. Peripheral thermoregulation: Foot temperature in two Arctic canines. *Science* 175:988–990.

Herbert, W. 1971. *Across the Top of the World.* New York: Putnam, 347 p.

Heroux, 0. 1960. Adjustments of the adrenal cortex and thyroid during cold acclimation. *Fed. Proc.* 19(Suppl.5):82–85.

Heroux, 0. 1967. Metabolic adjustments to low temperatures in New Zealand white rabbits. National Research Council of Canada. *Canad. J. Physiol. and Pharmac.* 45:451–461.

Heroux, 0. 1968. Specific and non-specific adjustments for metabolic adaptation to cold. *19th Alaska Science Conf.* 19:34.

Heroux, O., F. Depocas, and J.S. Hart. 1959. Comparison between seasonal and thermal acclimation in white rats. *Canad. J. Biochem. Physiol.* 37:473–479.

Herreid, C.F., Jr. 1963. Temperature regulation and metabolism in Mexican free-tail bats. *Science* 142:1573–1574.

Hertig, B.A., M. L. Riedesel, and H.S. Belding. 1961. Sweating in hot baths. *J. Appl. Physiol.* 16:647–651.

Hertzman, A.B. et al. 1952. Regional rates of evaporation from the skin at various environmental temperatures. *J. Appl. Physiol.* 5:153–156.

Hickman, G.C. 1984. Effects of temperature and light on the locomotory activity of captive pocket gophers. *Acta Theriologica* 29:259–271.

Hicks, J.W. 1993. Regulation of intracardiac shunting in reptiles: anatomic vs. effective shunting, p 252–264. In *The Vertebrate gas transport cascade: Adaptations to Environment and Mode of Life,* ed. J.E.P.W. Bicudo. London: CRC Press.

Highman, B., and P.D. Altland. 1949. Acclimatization responses and pathologic changes in rats at an altitude of 25,000 feet. *Arch. Path.* 48:503–515.

Hildebrand, M. 1985. Digging of Quadrupeds, p 107–109. In *Functional Vertebrate Morphology,* Hildebrand, M. et al, ed. Cambridge:Harvard Univ. Press. 460 p.

Hildebrandt, G., et al. 1987. *Chronobiology and Chronomedicine.* Paris: Peter Lang, Inc., 454 p.

Hildes, J.A. 1963. Comparison of coastal Eskimos and Kalahari Bushmen. *Fed. Proc.* 22:843–845.

Hills, B.A. 1969. Acclimatization to decompression sickness: a study of passive relaxation in several tissues. *Clin. Sci.* 37:109–124.

Hindell, M.A., et al. 1991. The diving behavior of the adult male and female southern elephant seals. *Mirounga leonina. Aust. J. Zool.* 39:595–619.

Hinsull, S.M., and E.L. Head. 1986. The effect of positive air ions on reproduction and growth in laboratory rats. *Int. J. Biometero.* 30:69–75.

Hock, R.J. 1951. The metabolic rates and body temperatures of bats. *Biol. Bull.* 101:289–299.

Hock, R.J. 1960. Seasonal variations in physiologic functions of arctic ground squirrels and black bears. Proc. 1st Int. Symp. Nat. Hibernation. *Bull. Mus. Comp. Zool.* 124:155–173

Hock, R.J. 1962. Mammals and Birds, p. 149–159. In *Deepest Valley,* ed. G. Schumacher. Sierra Club, San Francisco. Vail-Ballou Press, Inc. 208 p.

Hock, R.J. 1964a. Animals in high altitudes: Reptiles and amphibians. p. 841–842. In *Handbook of Physiology, Section 4 Adaptation to the Environment*, Section 4, ed. D.B. Dill. Washington, DC: Am. Physiol. Soc.

Hock, R.J. 1964b. Physiological responses of deer mice to various native altitudes, p. 59–73. In *The Physiological Effects of High Altitude,* ed. W.H. Weihe. New York: Pergamon Press. 351 p.

Hock, R.J. 1965. An analysis of Gloger's Rule. *Hvalradets Skrifter* 48:214–226.

Hock, R.J. 1969. Biology of *Peromyscus*. A review. *Lab. Anim. Care* 19:668–673.

Hock, R.J. 1970. The physiology of high altitude. *Scientific American* 222:52–62.

Hock, R.J., C.F. Bond, and W.F. Mazzone. 1966. *Physiological evaluation of SeaLab II: Effects of two weeks exposure to an undersea 7-atmosphere helium-oxygen environment.* U.S. Navy Deep Submergence Systems Project. Anaheim: Nortronics Pub. 116 p.

Hoffman, K. 1968. Temperaturcylclen als zeitgeber der cirdadianen periodik. *Verh. dtsch. zool. Ges.* (Innsbruck) 62:265–274.

Hoffman, K. 1969. Experimental manipulation of the orientational clock in birds. In *Biological Clocks*. Cold Spring Harbor Symp. 25:379–389.

Hoffman, K., et al. 1981. Effect of photoperiod and of one-minute light at night-time on the pineal rhythm of N-acetyltransferase activity in the Djungarian hamster *Phodopus sungorus. Biol. Reprod.* 24:551–556.

Hoffman, R.A. 1964. Terrestrial animals in cold: hibernation. p. 379–403. In: *Handbook of Physiology, Sec. 4. Adaption to the Environment,* Sec. ed. D.B. Dill Wash. DC: Amer. Physiol. Soc. (Publisher)

Hoffman, R.A., and R.J. Reiter. 1965. Pineal gland: Influence on gonads of male hamsters. *Science* 148:1609–1610.

Hofman, M.A., and D.F. Swaab. 1992. The human hypothalamus: comparative morphometry and photoperiodic influences. *Progress in Brain Res.* 93:133–147.

Hofmann, G.E., and G.N. Somero, 1995. Evidence for protein damage at environmental temperature: Seasonal changes in level of ubiquitin conjugates and hsp70 in the intertidal mussel *Mytilus trossulus. J. Exp. Biology* 198:1509–1518.

Holden, C., ed. 1996. Random Samples: The last of the Cahokians. *Science* 272:351.

Holmer, I., and U. Bergh. 1981. Thermal physiology of man in the aquatic environment. In *Bioengineering, Thermal Physiology and Comfort,* ed. K. Cena and J.A. Clark. p. 145–156. New York: Elsevier Scientific.

Holmes, T., and R.A. Powell. 1994. Morphology, ecology, and the evolution of sexual dimorphism in North American *Martens,* p. 72–84. In *Martens, Sables, and Fishers: Biology and Conservation,* ed. S.W. Buskirk, A.S. Harestad, M.G. Raphael, and R.A. Powell. Ithaca: Cornell University Press. 484 p.

Homma, I., T. Nagai, T. Sakai, M. Ohashi, M. Beppu, and K. Yonemoto. l981. Effect of chest wall vibration on ventilation in patients with spinal cord lesion. *J. Appl. Physio.: Resp. Environ. Exer. Physiol.* 50:107–111.

Hong, S.K. 1987. Breath-hold bradycardia in man: an overview, p. 158–173. In *The Physiology of Breath-Hold Diving,* ed. C.E.G. Lundgren and M. Ferrigno. Bethesda, MD: Undersea Hyperbaric Med. Soc.

Hong, S.K. 1989. Diving physiology in man, p. 787–802. In *Comparative Pulmonary Physiology—Current Concepts,* ed. S. C. Wood. New York: Marcel Dekker, 802 p.

Hong, S.K., and J.R. Claybaugh. 1989. Hormonal and renal responses to hyperbaria, p. 117–146. In *Hormonal Regulation of Fluid and Electrolytes: Environmental Effects,* ed. J.R. Claybaugh and C.E. Wade. New York: Plenum.

Hong, S.K., et al. 1987. Humans can acclimatize to cold: a lesson from Korean women divers. *NIPS* 2:79–82.

Hong, Suk Ki. 1963. Comparison of diving and nondiving women of Korea. Int. Symp. Temperature Acclimation. *Fed. Proc.* 22:831–833.

Honig, A. 1989. Peripheral arterial chemoreceptos and reflex control of sodium and water homeostasis. *Am. J. Physiol.* 257:R1282–R1302.

Honma, K. 1995. Ordinary room light suppresses nocturnal melatonin. *Int. J. Biometeorology* 39:57.

Hooten, E. 1942. *Man's Poor Relations.* New York: Doubleday, Doran. 440 p.

Hoover, G.N., and W.F. Ashe. 1962. Respiratory response to whole body vertical vibration. *Aerospace Med.* 33:980–984.

Hori, S. 1987. Natural acclimatization in hot environments. In *Man in Stressful Environments: Thermal and Work Physiology,* ed. K. Shiraki and M.K. Yousef. Springfield, IL: C.C. Thomas. 301 p.

Horne, J.A., and O. Ostberg. 1976. A self-assessment questionnaire to determine morningness-eveningness in human circadian rhythms. *Int. J. Chronobiology* 4:97–110.

Horowitz, B.A., J.S. Hamilton, and K.S. Kott. 1985. GDP binding to hamster brown fat mitochondria is reduced during hibernation. *Am. J. Physiol.* 249 (*Regulatory Integrative Comp. Physiol.* 18): R689–R693.

Horst, R. 1969. Observations on the structure and function of the kidney of the vampire bat (Desmodus rotundus murinus). p. 73–83. In *Physiological Systems in Semiarid Environments,* ed. C. C. Hoff and M. L. Riedesel. Albuquerque: Univ of New Mexico Press. 293 p.

Horton, N.D., P.R. Oeltgen, D.J. Kaftani, T.P. Su, D.S. Bruce, A.S. Krober, and J.F. Jones. 1996. Biochemical characterization of a hibernation-specific 88 kDa protein derived from the plasma of deeply-hibernating woodchucks, p. 333–339. In *Adaptations to the Cold, Tenth International Hibernation Symposium,* ed. F. Geiser, A.J. Hulbert and S.C. Nicol. Armidale, Australia: University of New England Press.

Horvath, S.M. 1980. Aging and adaptation to stressors. In *Environmental Physiology: Aging, Heat and Altitude,* ed. S.M. Horvath and M.K. Yousef. 1981. New York: Elsevier/North-Holland. 468 p.

Horvath, S.M. 1981. Historical perspectives of adaptation to heat, p 11–25. In *Environmental Physiology: Aging, Heat and Altitude,* ed. S.M. Horvath and M.K. Yousef. New York: Elsevier/North Holland. 468 p.

Horvath, S.M., and D.J. McKee. 1994. Acute and Chronic Health Effects of Ozone, p. 39–84. In *Tropospheric Ozone, Human Health and Agricultural Aspects,* ed. D. J. McKee. Boca Raton: Lewis.

Horvath, S.M., and E.C. Horvath. 1973. *The Harvard Fatigue Laboratory: Its History and Contributions.* Englewood Cliffs, New Jersey: Prentice-Hall, Inc. 182 p.

Horvath, S.M., and M.K. Yousef, eds. 1981. *Environmental Physiology: Aging, Heat, and Altitude* (Conference Procedings) New York: Elsevier/North Holland, Inc. 468 p.

Hotton, N.H. III, P.D. MacLean, J.J, and E.C. Roth, eds. 1987. *The Ecology and Biology of Mammal-like Reptiles.* 336 p.

Houdas, Y. 1981. Modeling of heat transfer in man, p. 111–120. In *Bioengineering, Thermal Physiology and Comfort,* ed. K. Cena and J.A. Clark. New York: Elsevier Scientific.

Houston, C.S. 1980. *Going High: The Story of Man and Altitude.* New York, Amer. Alpine Club. 211 p.

Houston, C.S., J.R. Sutton, A. Cymerman. 1987. Operation Everest II: Man at High Altitude. *J. Appl. Physiol.* 63:877–882.

Hoy, R.R., G.S. Pollack, and A. Moiseff. 1982. Species-recognition in the Field Cricket, *Teleogryllus oceanicus:* behaviorial and neural mechanisms. *Amer. Zool.* 22:594–607.

Hoyt, R.W., and A. Honig. 1996. Environmental influences on body fluid balance during exercise: altitude, p. 183–196. In *Body Fluid Balance, Exercise and Sport,* ed. E.R. Buskirk and S.M. Puhl. New York: CRC Press.

Hrdy, S.B. 1995. Natural-born mothers. *Natural History* 104:34–39.

Hrushesky, W.J.M. 1983. The clinical application of chronobiology to oncology. *Amer. J. Anat.* 168:519–542.

Hrushesky, W.J.M. 1985. Circadian timing of cancer chemotherapy. *Science* 228:73–75.

Hsu, S., and Li, H. 1994. Magnetism and bees. *Science* 265:95.

Huang, S.Y., J.K. Alexander, R.F. Grover, J.T. Maher, R.E. McCullough, R. G. McCullough, L.R. Moore, J.B. Sampson, J.V. Weil, and J.T. Reeves. 1984. Hypocapnia and sustained hypoxia blunt ventilation on arrival at high altitude. *J. Appl. Physiol.* 56:602–606.

Hubbard, R.W. 1979. Effects of exercise in the heat on predisposition to heatstroke. *Med. Sci. Sports.* 11:66–71.

Hudson J.W., and A.H. Brush. 1964. A Comparative study of the cardiac and metabolic performance of the dove and the quail. *Comp. Biochem. Physiol.* 12:157–170.

Hudson, J.W. 1964. Water metabolism in desert mammals, p. 21–235. *Proc. 1st Int. Symp. Thirst in Reg. Body Water.* New York: Pergamon. 308 p.

Hudson, J.W. 1965. Temperature regulation and torpidity in the pygmy mouse, *Baiomys taylori. Physiol. Zool.* 38:243–254.

Hudson, J.W. 1981. Role of the endocrine glands in hibernation with special reference to the thyroid gland, p. 33–54. In *Survival in the Cold: Hibernation and Other Adaptations,* ed. X.J. Musacchia and L. Jansky, New York: Elsevier/North-Holland.

Huey, R.B., and E.R. Pianka. 1977. Seasonal variation in thermoregulatory behavior and body temperature of diurnal Kalahari lizards. *Ecology* 58:1066–1075.

Huey, R.B., and M. Slatkin. 1976. Cost and benefits of lizard thermoregulation. *Quarterly Review of Biology* 51:363–381.

Hulbert, A.J., D.S. Hinds, and R.E. MacMillen. 1985. Minimal metabolism, summit metabolism and plasma thyroxine in rodents. *Comp. Biochem. Physiol.* 81A:687–693.

Hultgren, H.N. 1997a. *High Altitude Medicine.* Stanford, CA: Hultgren Publ. 516 p.

Hultgren, H.N. 1997b. High altitude pulmonary edema: hemodynamic aspects. *Int. J. Sports Med.* 18:20–25.

Humbolt, A. 1850. *Aspects of Nature.* (Trans. A. Sabin) Lea and Blanchard: Philadelphia. 475 p.

Hurford, W.E., et al. 1996. Splenic contraction, catecholamine release, and blood volume redistribution during diving in the Weddell seal. In *J. Appl. Physiol.*80(1):298–306.

Hutchison, V.H., H.B. Haines, and G. Engbretson. 1976. Aquatic life at high altitude: Respiratory adaptations in the Lake Titicaca frog, *Telmatobius culeus. Respir. Physiol.* 27:115–129.

Hyvarinen, J.I., S. Pyykko, and S. Sundberg. 1993. Vibration frequencies and amplitudes in the aeteology of traumatic vasospastic disease. *Lancet* 1:791–794.

I

Inbar, O., A. Rotstein, R. Dlin, R. Dotan, and F.G. Sulman. 1982. The effects of negative air ions on various physiological functions during work in a hot environment. *Int. J. Biometeor.* 26:153–163.

Ingram, D.L., M.D. Dauncey, and K.F. Legge. 1985. Synchronization of motor activity in young pigs to a non-circadian rhythm without affecting food intake and growth. *Comp. Biochem. Physiol.* 80A:363–368.

Inoue, Y., M. Nakao, T. Araki, H. Ueda. 1992. Thermoregulatory responses of young and older men to cold exposure. *Euro. J. Appl. Physiol. & Occupat. Physiol.* 65(6):492–498.

Inouye, S.T., and H. Kawamura. 1979. Persistence of circadian rhythmicity in the suprachiasmatic nucleus. *Proc. Natl. Acad. Sci.* USA 76:5961–5966.

Irving, L. 1939. Respiration in diving mammals. *Physiol. Rev.* 19:112–132.

Irving, L. 1957. *Animal adaptation to cold.* 5th Conf. on Cold Injury, p. 11–59. New York: Josiah Macy, Jr. Found.

Irving, L. 1963. Bradycardia in Human Divers. *J. Applied Physiol.* 18:489–491.

Irving, L. 1964. Maintenance of warmth in Arctic animals. *Symp. Zool. Soc. Long.* 13:1–14.

Irving, L. 1964. Terrestrial animals in cold: Introduction, p.343–347. In *Handbook of Physiology, Sec. 4. Adaptation to the Environment,* Sec. ed. D.B. Dill. Amer. Physiol. Soc. Washington, DC 1056 p.

Irving, L. 1970. Morpho-physiological adaptations in marine mammals for life in polar areas, p. 455–463. In *Antarctic Ecology, Vol. 1.*, ed. M.W. Holdgate. New York: Academic Press. 604 p.

Irving, L., P.F. Scholander, and S.W. Grinnell. 1942. The regulation of arterial blood pressure in the seal during diving. *Amer. J. Physiol.* 135:557–566.

Iverson, S.J., O.T. Oftedal, W.D. Bowen, D.J. Boness, and J. Sampugna. 1995. Prenatal and postnatal transfer of fatty acids from mother to pup in the hooded seal. *J. Comp. Physiol.* 165:1–12.

J

Jackett, M.F. 1970. Effects of sound on endocrine function and electrolyte excretion, p. 21–41. In *Physiological Effects of Noise,* ed. B.L. Welch and A.S. Welch. Plenum Press.

Jacklett, J.W. 1969. Circadian rhythm of optic nerve impulses recorded in darkness from isolated eye of Aplysia. *Science* 164:562–563.

Jacklett, J.W. 1977. Neuronal circadian rhythm: Phase shifting by a protein synthesis inhibitor. *Science* 198:69–71.

Jackson, D.C. 1988. Tolerable limits of Hypoxia: The Turtle as an Extreme Vertebrate Example, p 337–351. In *Hypoxia,* ed. J.R. Sutton, C.S. Houston, and G. Coates. Indianapolis: Benchmark Press, Inc.

Jamieson, D., et al. 1986. The relation of free radical production to hyperoxia. *Ann. Rev. Physiol.* 48: 703–719.

Jansky, L. 1966. Body organ thermogenesis of the rat during exposure to cold and at maximal metabolic rate. *Fed. Proc.* 25:1297–1302.

Jansky, L. 1971. Participation of body organs during nonshivering heat production, p. 159–172. In *Nonshivering Thermogenesis* (a symposium), ed. L. Jansky. Prague. 310 p.

Jansky, L., H. Janakova, B. Ulicny, P. Sramek, V. Hosek, J. Heller, and J. Parizkova. 1996. Changes in thermal homeostasis in humans due to repeated cold water immersions. *Pflugers Archiv-European J. of Physiol.* 432(3):38–72.

Jarvis, J.U.M. 1981. Eusociality in a mammal: Cooperative breeding in naked mole-rat colonies. *Science* 212:571-573.

Jedrzejewski, W., and B. Jedrzejewska. 1996. Rodent cycles in relation to biomass and productivity of ground vegetation and predation in the Palearctic. *Acta Theriologica* 41(1):1–34.

Jennings, B.H., and J.E. Murphy, eds. 1966. *Interactions of Man and His Environment.* New York: Plenum Press. 168 p.

Johannessen, O.M., et al. 1996. Global warming and the Arctic. *Science* 271:129.

Johansen, K. 1969. Adaptive responses to cold in arterial smooth muscle from heterothermic tissues of marine mammals. *Nature* 223:866–867.

Johansen, R. 1961. Distributon of blood in the arousing hamster. *Acta Physiol. Scand.* 52:379–386.

Johansson, B. 1960. Brown fat and its possible significance for hibernation. Proc. 1st Int. Symp. on Natural Hibernation. *Bull. Mus. Comp. Zool.* 124:232–243.

Johnson, C.H., and J.W. Hastings. 1986. The elusive mechanism of the circadian clock. *Amer. Scientist* 74:29–36.

Johnson, G.E. 1929. The fall in temperature in ground squirrels going into hibernation. *Anat. Record* 44:199–208.

Johnson, G.S. and R.S. Elizondo. 1972. Temperature regulation in *Macaca mulatta. Fed. Proc.* 31:363.

Johnson, H.D., C.S. Cheng, and A.C. Ragsdale. 1958. Environmental physiology and shelter engineering with special reference to domestic animals. XLVI Comparison of effect of environmental temperature on rabbits and cattle. Part 2. Influence of rising environmental temperature on the physiological reactions of rabbits and cattle. Univ. Mo Agric. Exp. Sta. Res. Bull. No 648.

Johnson, M.S. 1939. Effect of continuous light on periodic spontaneous activity of white-footed mice. *J. Exp. Zool.* 82:315–321.

Johnson, R.E. 1968. Doctors afield. *New Engl. J. Med.* 278:31–35.

Johnson, R.E. 1972. Some metabolic aspects of exposure to heat, p. 99–108. In *Physiological Adaptations,* ed. M.K. Yousef, S.M. Horvath, and R.W. Bullard. New York: Academic Press, 258 p.

Johnson, R.E., R. Passmore, and P. Sargent, II. 1961. Multiple factors in experimental ketosis. *Arch. Intern. Med.* 107:43.

Jürgens, K.D., and J. Prothero. 1987. Scaling of maximal lifespan in bats. 1987. *Comp. Biochem. Physiol.* 88A(2): 361–367.

Jurgens, K.D., M. Pietschmann, K. Yamaguchi, T. Kleinschmidt. 1988. Oxygen binding properties, capillary density and heart weights in high altitude camels. *J. Comp. Physiol.* B 158:469–477.

K

Kabat, H.F. 1981. Chronopharmacy: Circadian rhythms and drug dosing, p. 9–14. In *Chronopharmacy and Chronotherapeutics*, ed. C.A. Walker, et al. Tallahassee: Florida A&M Univ. Foundation, 417 p.

Kadle, R., and G.E. Folk, Jr. 1983. Importance of circadian rhythms in animal cell cultures. *Comp. Biochem. Physiol.* 76A:773–776.

Kadono, H., E.L. Besch, and E. Usami. 1981. Body temperature, oviposition, and food intake in the hen during continuous light. *J. Ap. Physiol.* 51:1145–1149.

Kaduce, T.L., A.A. Spector, and G.E. Folk, Jr. 1981. Characterization of the plasma lipids and lipoproteins of the polar bear. *Comp. Biochem. Physiol.* 69B:541–545.

Kajick, K. 1996. An esker runs through. *Natural History* 105:28–37.

Kalabukov, N.J. 1937. Some physiological adaptatons of the mountain and plain forms of the wood mouse (*Apodemus sylvaticua*) and of other species of mouse-like rodents. *J. Anim. Ecol.* 6:254–274.

Kalter, V.G., and G.E. Folk, Jr. 1979. Minireview: humoral induction of mammalian hibernation. *Comp. Biochem. Physiol.* 63A:7–13.

Kanatous, S.B., et al. 1997. Enhanced capacity for fatty acid oxidation in seals. *The FASEB Journal* 11:28.

Kang, B.S., et al. 1971. Calorigenic action of norephinephrine in the Korean women divers. *J. Appl. Physiol.* 29:6–8.

Karlsson, H., S.E. Hanel, K. Nilsson, and R. Olegard. 1995. Measurement of skin temperature and heat flow from skin in term newborn babies. *Acta Paediatrica* 84(6):605–612.

Kasal, C.A., et al. 1979. Circadian clock in culture. *Science* 203:656–658.

Kauppinen, K. 1989. Sauna and winter swimming: Health status, and physiological responses to experimental exposures to heat, cold, and alternating heat and cold. Ph.D. dissertation, Dept. Of Zoology, Physiology, Univ. of Helsinki.

Kauppinen, K., and I. Vuori. 1988. Health status of active winter swimmers. *Arct. Med. Res.* 47(2):71–82.

Kavanau, J.L., and C.R. Peters. 1976. Activity of nocturnal primates: Influences of twilight zeitgebers and weather. *Science* 191:83–86.

Keatinge, W.R. 1969. *Survival in Cold Water; The Physiology and Treatment of Immersion Hypothermia and of Drowning.* Oxford: Blackwell Scientific Publ. 131 p.

Keeton, R.W. 1924. The peripheral water loss in rabbits as a factor in heat regulation. *Amer. J. Physiol.* 69:307–317.

Keeton, W.T. 1972. Effects of magnets on pigeon homing, p. 579–593. In *Animal Orientation and Navigation*, ed. S.R. Galler et al. Washington: NASA SP-262, 606 p.

Kehl, T.H., and P. Morrison. 1960. Peripheral nerve function and hibernation in the 13-lined ground squirrel, p. 388–402. In *Proc. 1st Internal. Symp. Nat. Mammalian Hibernation.* ed. C.P. Lyman and A.R. Dawe, Bull. Mus. Comp. Zool. vol. 124. Cambridge, MA: Harvard Univ. Press.

Kendeigh, S.C. 1965. The ecology of man, the animal. *BioScience* 15:521–523.

Kenmure, A.C.F., et al. 1972. Hemodynamic effects of oxygen at 1 and 2 Ata pressure in healthy subjects. *J. Appl. Physiol.* 32:223–226.

Kenworthy, T. 1995. Arctic Oil Drilling. *The Polar Times* 2:14.

Kerkhof, G.A. 1985. Inter-individual differences in the human circadian system. *Biological Psychology* 20:83–112.

Kerr, R.A. 1995. Greenhouse report foresees stress. *Science* 270:731.

Kerr, R.A. 1996a. Millennial climate oscillation. *Science* 271:146–147.

Kerr, R.A. 1996b. The warmest year. *Science* 271:137–138.

Kerr, R.A. 1996c. Ice bubbles confirm big chill. *Science* 272:1584–1585.

Khitrov, N.K., A.V. Toloknov, T.D. Bolshakova, K.B. Vinnitskaya, and V.A. Panteleymonov. 1986. Mechanisms of adaptation to physical exertion and the effect of excess CO_2 on its formation. *Byulleten "Eksperimental" oy Biologii i Meditsiny.* CI(6):655–658.

Kholodov, Y.A. 1966. Space biology and the magnetic field. *Priroda* 4:114–115. (In Russian).

Kielblock, A.J. 1987. Heat acclimatization with special reference to heat tolerance, p. 145–152. In *Man in Stressful Environments: Thermal and Work Physiology,* ed. K. Shiraki and M.K. Yousef. Springfield, IL: C.C. Thomas. 301 p.

King, D.P. et al. 1997. A clone of a clock gene in mice. *Cell* 89:641–644.

Kirmiz, J.P. 1962. *Adaptation to Desert Environment: A Study on the Jerboa, Rat, and Man.* Washington, DC: Butterworth & Co.

Kleiber, M. 1971. Influence of environmental temperature on metabolic rate, p. 170–197. *In Environmental Requirements for Laboratory Animals.* Inst. Envir. Res. Publ. No. IER-71–02. Manhattan: Kansas State Univ. 210 p.

Kleiber, M. 1972. A New Newton Law of Cooling? *Science* 178:1283.

Kleiber, M. 1972. Joules vs. Calories. *J. Nutrition* 102:309–312.

Kleiber, M. 1975. *The Fire of Life.* New York: R.E. Krieger Pub. Co., 453 p.

Klein, D.C. 1985. Photoneutral regulation of the mammalian pineal gland, p. 38–56. In *Photoperiodisml Melatonin and the Pineal,* ed. D. Evered and S. Clark. Ciba Foundation Symp. 117. London: Pitman Publ. Ltd., 323 p.

Kleitman, N. 1963. *Sleep and Wakefulness*, 2nd ed. Chicago: Univ. of Chicago Press, 552 p.

Klinowska, M. 1972a. A comparison of the lunar and solar activity rhythms of the golden hamster. (*Mesocricetus auratus Waterhouse*). *J. Interdiscipl. Cycle Res.* 3:145–150.

Klinowska, M. 1972b. The activity of the golden hamster in relation to its meteorological environment. *Int. J. Biometeorol. Suppl.* 16:142.

Klotchkov, D.V., A.Y. Klotchkova, A.A. Kim, and D.K. Belyaev. 1971. The influence of photoperiodic conditions on fertility in gilts (in English). 10th Internat. Cong. Animal Production, Paris-Versailles, July 17–23.

Klotchkov, D.V., and D.K. Belyaev. 1977. The influence of continuous illumination of different duration on reproductive functions in the rat (Summary in English). OHTOGEHE3 8:487–495.

Kluger, M.J. 1979. Fever in ectotherms: evolutionary implications. *Amer. Zool.* 19:295–304.

Kobayashi, S., and T. Takahashi. 1993. Whole-cell properties of temperature-sensitive neurons in rat hypothalamic slices. *Proc. Royal Soc. London-Series B: Biological Sciences* 251(1331):89–94.

Koenigsberg, P.S., K.K. Martin, H.R. Hlava, and M.L. Riedesel. 1995. Sustained hyperhydration with glycerol ingestion. *Life Sciences* 57:645–653.

Konda, N., et al. 1987. Seadragon VI: a 7-day dry saturation dive at 31 ATA. IV. Circadian analysis of body temperature and renal functions. *Undersea Biomed. Res.* 14:413–423.

Kondo, N., and J. Kondo. 1993. Identification and Characterization of Novel Types of Plasma Protein Specific for Hibernation in Rodents. In *Life in the Cold: Ecological, Physiological and Molecular Mechanisms,* ed. C. Carey, G.L. Florant, B.A. Wunder, and B. Horwitz, p. 467–473. Boulder, CO: Westview Press.

Konopka, R. 1985. Genes and biological clocks. *Science* 230:1151–1152.

Kooyman, E.A., et al. 1980. Aerobic and anaerobic metabolism during voluntary diving in Weddell seals. In *J. Comp. Physiol.* 138:335–346.

Kooyman, E.A., et al. 1995. Transfer of complement from erythrocytes to endothelium. *Science* 269:89–92.

Kooyman, G.L., and P.L. Ponganis. 1997. The challenges of diving to depth. *Amer. Sci.* 85:530–539.

Kornberger, P., and P. Mair. 1996. Important aspects in the treatment of severe accidental hypothermia: the Innsbruck experience. *J. Neurosurgical Anesthesiol.* 8(1):83–87.

Koteja, P. 1987. On the relation between basal and maximum metabolic rate. *Comp. Biochem. Physiol.* 87A:205–208.

Krantz, G.S. 1970. Human activities and megafaunal extinctions. *Amer. Scientist* 58:164–170.

Krassney, J.A. 1994. A neurogenic basis for acute altitude illness. *Med. Sci. Sports Exerc.* 20:195–208.

Krauss, M. 1997. The indigenous languages of the North: a report on their present state, p. 1–34. In *Northern Minority Languages: Problems of Survival,* ed. H. Shoji and J. Jahunen. Univ. of Alaska: Senri Ethnological Studies.

Kridler, E., D.L. Olsen, and G.C. Whittow. 1971. Body temperature of the Hawaiian monk seal. *J. Mammal.* 52:476.

Krieger, E.T., ed. 1979. *Endocrine Rhythm.* New York: Raven Press, 332 p.

Kristoffersson, R., and A. Soivio. 1964. Hibernation in the Hedgehog (*Erinaceus europaens* l.). *Ann. Acad. Sci. Fennicae.* A IV. Biol., Vol. 82. Finnish Academy of Science.

Krogh, A., and M. Krogh. 1913. A study of the diet and metabolism of Eskimos undertaken in 1908, Greenland. *Medd. Om Grønland,* 51.

Kromer, W. 1980. Naltrexone influence on hibernation. *Experentia* 36:581–582.

Kuhlemeier, K.V., P.R. Fine, and F.N. Dukes-Dobos. 1978. Heart rate-rectal temperature relationships during prolonged work: males and females compared, p 71–81. In *Environmental Stress: Individual Human Adaptations,* ed. L.J. Folinsbee, J.A. Wagner, J.F. Borgia, B.L. Drinkwater, J.A. Gliner, and J.F. Bedi. New York: Academic Press. 393 p.

Kuno, Y. 1956. *Human Perspiration.* 268 p. Springfield, IL: Thomas.

Kuno, Y. 1956. *Human Perspiration.* Springfield, Ill.: Thomas. 268 p.

Kuno, Y. 1956a. The loss of water and salt by sweating, their replenishment and changes in the blood, p. 251–276. In *Human Perspiration,* ed. Y. Kuno. Springfield, IL: C.C. Thomas.

Kushner, D. J., ed. 1978. *Microbial Life in Extreme Environments.* New York: Academic Press. 465 p.

L

Laburn, H.P. 1996. The fetus and thermal challenges. *News Physiol. Sci.* 11:96–100.

Ladell, W.S.S. 1945. Thermal sweating. *Brit. Med. Bull.* 3:175–179.

Ladell, W.S.S., J.C. Waterlow, and M.F. Hudson. 1944. Desert climate: physiological and clinical observations. *Lancet* 2:491–497.

Lahiri, S., et al. 1987. Carotid body chemosensory function in prolonged normobaric hyperoxia in the cat. *J. Appl. Physiol.* 62:1924–1931.

Lambertsen, C.J., and L.J. Greenbaum, Jr., eds. 1963. *Proc. 2nd Symp. Underwater Physiol.* Nat. Acad. Sci-Nat. Res. Council Publ. 1181. 298 p.

Lambertsen, C.J., ed. 1971. *Underwater Physiology.* New York: Academic Press. 584 p.

Landau, B.R. 1960. Observations on a colony of captive ground squirrels throughout the year, p. 172–189. In *Mammalian Hibernation I,* ed. C.P. Lyman and A.R. Dawe. Cambridge: Harvard Univ. Press. 549 p.

Landau, B.R., and A.R. Dawe. 1958. Respiration in the hibernation of the 13-lined ground squirrel. *Am. J. Physiol.* 194:75–82.

Lanphier, E.H. 1987. Breath-hold and ascent blackout, p. 32–42. In *The Physiology of Breath-Hold Diving,* ed. C.E.G. Lundgren and M. Ferrigno. Bethesda, MD: Undersea Hyperbaric Med. Soc.

Lansing, A. 1986. *Endurance,* p. 102. New York: Carroll & Graf. 282 p.

Larick, R., and R.L. Ciochon. 1996. African emergence and early Asian dispersal. *American Scientist* 84:538–551.

Lasker, G.W. 1969. Human biological adaptability: the ecological approach in physical anthropology. *Science* 166:1480–1486.

Lassahn, L.K., G.E. Folk, Jr., and C.H. Seberg. 1974. Radiative heat loss from skin to cold glass windows. *Proc. Iowa Acad. Sci.* 81:85–88.

Latzka, W.A., M.N. Sawka, S. J. Montain, G.S. Skrinar, R.A. Fielding, R.P. Matott, and K.B. Pandolf. 1997. Hyperhydration: thermoregulatory effects during compensable exercise-heat stress. *J. Appl. Physiol.* 83:860–866.

Lauber, J.K., J.E. Boyd, and J. Axelrod. 1968. Enzymatic synthesis of melatonin in avian pineal body: Extraretinal response to light. *Science* 161:489–490.

Leakey, M. 1995. The Farthest Horizon. *Nat. Geographic* 188:38–51.

LeBoeuf, B.J., et al. 1986. Pattern and depth of dives in northern elephant seals, *Mirounga angustirostris. J. Zool.* A, 208:1–7.

Lee, D.H.K. 1953. Physiological Climatology as a field of study. *Ann. Assoc. Amer. Geogr.* 43:127.

Lee, D.H.K. 1965. Climatic stress indices for domestic animals. *Int. J. Biometeorol.* 9:29–35.

Lee, D.H.K. 1972. Large mammals in the desert, p. 109–125. In *Physiological Adaptations: Desert and Mountain,* ed. M.

Yousef, S. Horvath, and R. Bullard. New York: Academic Press. 258 p.

Lee, D.H.K., K. Robinson, and H. J. G. Hines. 1941. Reactions of the rabbit to hot atmospheres. *Proc. Roy. Soc. Queensland* 53:129–144.

Lefcourt, A., J. Bitman, D.L. Wood, and B. Stroud. 1982. Temperature regulation of the mammary gland. *Fed. Proc.* 41:1697.

Leithead, C.S., and A.R. Lind. 1964. *Heat Stress and Heat Disorders*. Philadelphia: F.A. Davis. 304 p.

Lemaire, C. 1987. Hydrogen narcosis, nitrogen narcosis and HPNS: a performance study, p. 579–582. In *IX Int. Symp. Underwater Hyperbaric Physiol.*, ed. A.A. Bove, A.J. Bachrach, and L.T. Greenbaum. Bethesda, MD: Undersea Hyperbaric Med. Soc.

Levine, V.E. 1937. The basal metabolic rate of the Eskimos. *J. Biol. Chem.* 119:61.

Levine, V.E., and C.G. Wilbur. 1949. Fat metabolism in Alaska Eskimos. *Fed. Proc.* 8:95–96.

Lewin, R. 1982. Food fuels reproductive success. *Science* 217:238–239.

Lewis, H.E., and J.P. Masterton. 1963. Polar Physiology: Its development in Britain. *Lancet* 1:1009–1014.

Lewis, P.R., and M.C. Lobban. 1957. Dissociation of diurnal rhythms in human subjects on abnormal time routines. *Quart. J. Exp. Physiol.* 42:371–386.

Lewis, T. 1930. Vasodilatation in response to strong cooling. *Heart* 15:177–181.

Lewy, A.J., T.A. Wehr, F.K. Goodwin, A. Newsome, and S.P. Markey. 1980. Light suppresses melatonin secretion in humans. *Science* 210:1267–1269.

Li, H., and E. Satinoff. 1995. Changes in circadian rhythms of body temperature and sleep in old rats. *Am. J. Physiolo.* 269(1 Pt 2):R208–14.

Licht, L.A., and Schnoor, J.L. 1993. Tree buffers protect shallow ground water at contaminated sites. *EPA Ground Water Currents*. December.

Licht, L.E. 1996. Amphibian decline still a puzzle. *BioScience* 46:172–173.

Lim, T.P.K., and U.C. Luft. 1965. Thermal homeostasis under hypoxia in man, p. 131–145. In *Basic Environmental Problems of Man in Space*. New York: Springer Verlag.

Lin, Y.C. 1986. Breath-hold diving: human imitation of aquatic mammals, 81–89. In *Diving in Mammals and Man*, ed. A. O. Brubakk, J.W. Kanwisher, and G. Sundnes. Trondheim, Norway: Tapir.

Lin, Y.C. 1987a. Cardiovascular deconditioning in hyperbaric environments, p. 72–92. In *Man in Stressful Environments: Diving, Hyper- and Hypobaric Physiology*, ed. K. Shiraki and M. K. Yousef. Springfield, IL: Thomas.

Lin, Y.C. 1987b. Effect of O_2 and CO_2 on breath-hold breaking point, p. 75–86. In *The Physiology of Breath-hold Diving*, ed. C.E.G. Lundgren and M. Ferrigno. Bethesda, MD: Undersea Hyperbaric Med. Soc.

Lin, Y.C. 1990. Physiological limitations of humans as breath-hold divers, p. 33–56. In *Man in the Sea*, ed. Y.C. Lin and K.K. Shida. Vol. II. San Pedro, CA: Best Publishing Co.

Lin, Y.C. 1996. Hyperbaric bradycardia, p. 57–169. In *Physiological Basis of Occupational Health*. ed. K. Shiraki, S. Saqawa, and M.K. Yousef. Amsterdam: SPB Academic Publishing, 278 p.

Lin, Y.C., D.T. Matsuura, and G.C. Whittow. 1972. Respiratory variation of heart rate in the California sea lion. *Amer. J. Physiol.* 222:260–26.

Lin, Y.C., et al. 1991. Orthostatic intolerance during GUSI-18 dive, a simulated trimix saturation dive at 46 ATA. *Undersea Biomed. Res.* 18 (suppl.):97–98.

Lindauer, M., and H. Martin. 1972. Magnetic effect on dancing bees. In *Animal Orientation and Navigation*, ed. S.R. Galler, et al. Washington, DC: NASA SP-262, 606 p.

Lipp, J.A., and G.E. Folk, Jr. 1960. Cardiac response to cold of two species of mammalian hibernators. *Ecology* 41:377–378.

Lipp, J.A., J.R. Knott, and G.E. Folk, Jr. 1960. EEG and heart rates of hypothermic cats. *Fed. Proc.* 19:179 and Dissertation Abstract: Jan. 1961, Vol. XXI, No. 7.

Lipscomb, D.M. 1974. *Noise: The Unwanted Sounds*. Nelson-Hall Company, Chicago.

Lissák, K., and E. Endröczi. 1965. *The Neuroendocrine Control of Adaptation*. London: Pergamon.

Liu, B., L.C.H. Wang, and D.D. Bulke. 1993. Effects of temperature on pH on cardiac myofilaments Ca2+ sensitivity in rat and ground squirrel. *Am. J. Physiol* 264:R104-R108.

Lockhart, A. and B. Saiag, 1981. Altitude and human pulmonary circulation. *Clin. Sci.* 60:599–605.

Lockyer, C.H. and S.G. Brown. 1981. The migration of whales. p 105–133. In *Animal Migration*, ed. D.J. Aidley. New York/Cambridge: Cambridge Univ. Press. 265 p.

Lohman, R., and T. Willows. 1991. Neuron excited by magnetic fields. *J. Exp. Biol.* 161:1–6.

Lorenz, K.L. 1957. *King Solomon's Ring*. London: Pan Books Ltd., 217 p.

Luce, G.G. 1970. *Biological Rhythms in Psychiatry and Medicine*. Washington, DC: HEW Health Serv. Publ. 2088, 183 p.

Luck, C.P., and P.G. Wright. 1962. Aspects of the body temperature and habitat of large animals, p. 334–338. In *Biometeorology* (a Symposium), Proc. 2nd Int. Bioclimatol. Congr. Pergamon Press. 687 p.

Luft, U.C., and Finkelstein. 1968. Hypoxia: A clinical-physiological approach. *Aerospace Med.* 39:105–110.

Lundgren, C.E.G. 1989. Discussion, p. 69. In *The Physiological Basis of Decompression Sickness*, ed. R.D. Vann. Bethesda, MD: Undersea Hyperbaric Med. Soc.

Lundgren, C.E.G., and W.T. Norfleet. 1987. Respiratory function in hyperbaric environments. In *Man in Stressful Environments*, ed. K. Shiraki and M.K. Yousef. Springfield, IL: Thomas. 266 p.

Lustick, S. 1969. Bird energetics: Effects of artificial radiation. *Science* 163:387–389.

Lyman, C.P. 1943. Control of coat color in the varying hare *Lepus americanus. Erxleben. Bull. Mus. Comp. Zool.* 93:393–461.

Lyman, C.P. 1958. Oxygen consumption, body temperature and heart rate of woodchucks entering hibernation. *Am. J. Physiol.* 194:83–91.

Lyman, C.P. 1963. Hibernation in mammals and birds. *Amer. Scientist* 51:127–138.

Lyman, C.P. 1970. Thermoregulation and metabolism in bats, p. 301–330. In *Biology of Bats,* Vol. 1, ed. W.A. Wimsatt. New York: Academic Press.

Lyman, C.P., and P.O. Chatfield. 1955. Physiology of hibernation in mammals. *Physiol. Rev.* 35:403–425.

Lyman, C.P., and P.O. Chatfield. 1956. Physiology of hibernating mammals, p. 80–124. In *Physiology of Induced Hypothermia,* ed. R.D. Dripps. Washington, DC Nat. Acad. Sci-Nat. Res. Council Pub. No. 451. 446 p.

Lyman, C.P., and A.R. Dawe, eds. 1960. *Mammalian Hibernation.* Proc. First Int. Symp. Nat. Mammal. Hiber. *Bull. Mus. Comp. Zool.* Vol. 124. Cambridge, MA: Harvard Univ. Press.

Lyman, C.P., and R.C. O'Brien. 1963. Autonomic control of circulation during the hibernating cycle in ground squirrels. *J. Physiol.* 168:477–499.

Lyman, C.P., and R.C. O'Brien. 1974. A comparison of temperature regulation in hibernating rodents. *Am. J. Physiol.* 227:218–223.

Lyman, C.P., R.C. O'Brien, G.C. Greene, and E.D. Papfrangos. 1981. Hibernation and longevity in the Turkish hamster, *Mesocricetus brandti. Science* 212:668–670.

Lynch, G.R. 1972. Effects of Temperature and Photoperiod on Thermoregulation in the White-footed Mouse. Ph.D. Thesis, Univ. of Iowa (May). 101 p.

Lynch, G.R., and G.E. Folk, Jr. 1970. Effect of photoperiod and cold acclimation on non-shivering thermogenesis in *Peromyscus leucopus,* p. 97–98. In *Nonshivering Thermogenesis,* Proc. of Symp. held in Prague, April 1–2.

Lyons, T.P, M.L. Riedesel, L.E. Meuli, and T.W. Chick. 1990. Effects of glycerol-induced hyperhydration prior to exercise in the heat. *Med. Sci. Sports Exerc.* 22:477–483.

Lyons, T.P., S.R. Muza, P.B. Rock, and A. Cymerman. 1995. The effect of altitude pre-acclimatization on Acute Mountain Sickness during reexposure. *Aviat. Space Environ. Med.* 66:957–962.

M

Macfarlane, W.V., K. Robertson, B. Howard, and R. Kinne. 1958. Heat, salt, and hormones in panting and sweating animals. *Nature* 182:672–673.

Machado, C.R.S., et al. 1969. Circadian rhythm of serotonin in the pineal body of immunosympathectomized immature rats. *Science* 164:442–443.

Mack, G.W., and E.R. Nadel. 1996. Body fluid balance during heat stress in humans. In *Handbook of Physiology*, Section 4: Environmental Physiology, ed. M.J. Fregly and C.M. Blatteis. 1:287–214.

Mack, G.W., H. Nose, and E.R. Nadel. 1988. Role of cardiopulmonary baroreflexes during dynamic exercise. *J. Appl. Physiol.* 65:1827–1832.

MacMillen, R.E., R.V. Baudinette, and A.K. Lee. 1972. Water economy and evergy metabolism of the sandy inland mouse, *Leggadina hermannsburgensis. J. Mammal.* 53:529–539.

Maeda, T., M. Takehawa, and H. Saito. 1995. Activation of yeast PBS MAPKK by MAPKKKs or binding of an SH3-containing osmosensor. *Science* 269:554–558.

Maeda, T., S.M. Wurgler-Murphy, and H. Saito. 1994. A two-component system that regulates an osmosensing MAP kinase cascade in yeast. *Nature Lond.* 369:242–245.

Maher, W.J., and N.J. Wilimovsky. 1963. Annual catch of bowhead whales by Eskimos at Point Barrow, Alaska. *J. Mammal.* 44:16–20.

Majumder, S.K. 1972. Vegetarianism: Fad, Faith, or Fact? *Amer. Scientist* 60:175–178.

Malan, A., and B. Canguilhem. 1989. *Living in the Cold/La vie au Froid.* London: John Libbey. 525 p.

Malconian, M.K., et al. 1993. Operation Everest II: gas tensions in expired air and arterial blood at extreme altitude. *Aviat. Space Environ. Med.* 64:37–42.

Malvin, G.M, P. Havlen, and C. Baldwin. 1994. Interactions between cellular respiration and thermoregulation in the paramecium. *Am. J. Physiol.* 267:R349–352.

Malvin, G.M. 1993. Regulation of respiratory organ blood flow in amphibians. In *Vertebrate Gas Transport Cascade,Adaptaions to Environment and Mode of Life,* ed. J.E.P.W. Bicudo. Ann Arbor: CRC Press

Malvin, G.M., and S.C. Wood. 1992. Behavioral hypothermia and survival of hypoxic protozoans *Paramecium caudatum. Science* 255:1423–1425.

Mannion, A.M. 1991. *Global Environmental Change.* London: Longman Scientific, Inc. 225 p.

Maron, M.B., J.A. Wagner, and S.M. Horvath. 1977. Thermoregulatory responses during competitive marathon running. *J. Appl. Physiol.* 42:909–914.

Martin, S.L., H.K. Stere, D. Belke, L.C.H. Wang, and H.V. Carey. 1993. Differential gene expression in the liver during hibernation in ground squirrels. In *Life in the Cold: Ecological, Physiological and Molecular Mechanisms,* ed. C. Carey, G.L. Florant, B.A. Wunder, and B. Horwitz. p 443–453.

Mathew, L., S.S. Purkayastha, R. Singh, and J. Sen Gupta. 1986. Influence of aging in the thermoregulatory efficiency of man. *Int. J. Biometeor.* 30(2): 137–145.

Matthews, G.V.T. 1968. *Bird Navigation.* New York: Cambridge Univ. Press, 197 p.

Maugh, T.H., II. 1976. The ozone layer: The threat from aerosol cans is real. *Science* 194:170–172.

Maugh, T.H., II. 1982. Magnetic navigation an attractive possibility. *Science* 215:1492–1493.

Mayr, E. 1956. Geographical character gradients and climatic adaptations. *Evolution* 10:105–108. (Reprinted in *Human Ecology*, J.B. Bresler, Reading, MA: Addison-Wesley. 472 p.)

McArdle, W.D., F.I. Katch, and V.L. Katch. 1991. *Exercise physiology.* 3rd ed. Philadelphia: Lea and Febiger.

McCook, R.D., C.N. Peiss, and W.C. Randall. 1962. Hypothalamic temperatures and blood flow. *Proc. Soc. Exp. Biol. Med.* 109:518–520.

McCutcheon, L.J., et al. 1995. Sweat composition: comparison. of collection methods and effects of exercise intensity. In *Equine Exercise Physiology 4. Equine Vet. J.* Suppl. 18:279–284. (Proc. 4th Int. Conf. Equine Exercise Physiol., Queensland, Australia 1994). Newmarket, Suffolk, U.K.: R & W Publications. 479 p.

McDonald, F., and E. Simonson. 1953. Human electrocardiograms during and after inhalation of 30% carbon dioxide. *J. Appl. Physiol.* 6:304–307.

McElhinney, T.L., et al. 1996. Patterns of behavior in the Nile grass rat. *Proc. Amer. Soc. Mammalogists* 76:14.

McGee, W. J. 1906. Desert thirst as disease. *Interstate Med. J.* 13:279–300.

McGinnis, S.M. 1968. Biotelemetry in pinnipeds, p. 54–68. In *The Behavior and Physiology of Pinnipeds*, ed. R.J. Harrison, 411 p. New York: Appleton-Century-Crofts.

McKee, W.D., ed. 1974. *Environmental Problems in Medicine,* p. 368. Springfield, Ill.: Charles C. Thomas, Pub. 860 p.

McManus, J.J., and D.W. Nellis. 1972. Temperature regulation in three species of tropical bats. *J. Mammal.* 53:226–227.

McMillan, J.P., S.A. Gauthreaux, Jr, and C.W. Helms. 1970. Spring migratory restlessness in caged birds: A circadian rhythm. *BioScience* 20:1259–1260.

McNab, B.K. 1971. On the ecological significance of Bergmann's Rule. *Ecology* 52:845–848.

McNamara, M.C., and M.L. Riedesel. 1973. Memory and hibernation in *Citellus lateralis*. *Science* 179:92–94.

Meadows, D.H., et al., eds. 1972. *The Limits to Growth*. Report for Club of Rome Project on Predicament of Mankind. New York: Potomac Assoc. Universe Books. 205 p.

Medway, W., and J.R. Geraci. 1965. Blood chemistry of the bottlenose dolphin (*Tursiops truncatus*). *Amer. J. Physiol.* 209:169–172.

Meehan, J.P. 1957. Animal adaptation to cold, p. 18–59. *5th Conf. on Cold Injury.* New York: J. Macy, Jr. Found.

Meehan, J.P., Jr. 1955. Basal metabolic rate of Eskimos. *J. Appl. Physiol.* 7:537–540.

Meier, A.H., and A.J. Fivizzani. 1980. Physiology of migration, p. 210–216. In *Animal Migration, Orientation, and Navigation*, ed. S.A. Gauthreaux, Jr. New York: Academic Press. 400 p.

Melchers, G. 1952. The Physiology of Flower Initiation. Lectures at Imperial College, London. 168 p.

Mellars, P. 1996. *The Neanderthal Legacy.* New Jersey: Princeton Univ. Press. 471 p.

Membery, J.H., and E.A. Link. 1964. Hyperbaric exposure of mice to pressures of 60 to 90 atmospheres. *Science* 144:1241–1242.

Menaker, M. 1959. Endogenous rhythms of body temperature in hibernating bats. *Nature* 184:1251–1252.

Menaker, M. 1962. Hibernation hypothermia:an annual cycle of response to low temperature in the bat *Myotis lucifugus*. *J. Cell. Comp. Physiol.* 59:163–174.

Menaker, M. 1969. Biological Clocks. *BioScience* 19:681–689.

Meurn, R.J. 1993. *Survival Guide for the Mariner,* Maryland: Cornell Maritime Press. 226 p.

Milan, F.A. 1960. *Swedish Lappland: A brief description of the dwellings and winter-living techniques of the Swedish mountain Lapps*. TR-60–7. Arctic Aeromedical Laboratory, Ladd Air Force Base, Alaska. 15 p.

Miles, L.E.M., D.M. Raynal, and M.A. Wilson. 1977. Blind man living in normal society has circadian rhythms of 24.9 hours. *Science* 198:421–423.

Miller, L.K., and L. Irving. 1963. Alteration of peripheral nerve function in the rat after prolonged outdoor cold exposure. *Amer. J. Physiol.* 204:359–362.

Miller, L.K., and L. Irving. 1964. Peripheral nerve conduction at low temperature. Proc. 15th Alaska Science Conf. *AAAS.* 15:29.

Miller, V.M., and F.E. South. 1978. Thermoregulatory responses to temperature manipulation of the spinal cord in the marmot. *Cryobiology* 15:433–440.

Mills, J.N., D.S. Minors, and J.M. Waterhouse. 1977. The physiological rhythms of subjects living on a day of abnormal length. *J. Physiol.* (London) 268:803–826.

Mills, J.N., D.S. Minors, and J.M. Waterhouse. 1978. Adaptation to abrupt time shifts of the oscillator(s) controlling human circadian rhythms. *J. Physiol.* (London) 285:455–470.

Mills, S.H., and J.E. Heath. 1970. Thermoresponsiveness of the preoptic region of the brain in house sparrows. *Science* 168:1008–1010.

Milsom, W.K., S. Osborne, P.F. Chan, J.D. Hunter, and J.Z. MacLeod. 1993. Sleep, hypothermia and hibernation: Metabolic rate and the control of breathing pattern in golden-mantled ground squirrels, p. 233–240. In *Life in the Cold: Ecological, Phsyiological and Molecular Mechanisms,* ed. C. Carey, G.L. Florant, B.A. Wunder, and B. Horwitz.

Mims, F.M. III. 1995. Smoke and rainforest. *Science* 270:1743.

Mining Survey. P.R.D. 1963. Series 13. 36 p. Chamber of Mines, P.O. Box 809, Johannesburg, South Africa.

Mittermeier, R.A., and I. Bowles. 1995. Surinam crisis: Threat to biodiversity. *Tropinet* 6:1.

Miyamoto, N., et al. 1991. Hyperbaric diuresis is associated with decreased antidiuretic hormone and decreased atrial natriuretic polypeptide in human divers. *Jpn. J. Physiol.* 41:85–99.

Moffat, A.S. 1996. Ecologists look at the big picture. *Science* 273:1490.

Molyneux, G.S., and M.M. Bryden. 1975. Arteriovenous anastomoses in the skin of the Weddell seal. *Science* 189:1100–1102.

Monk, T.H., V.C. Leng, S. Folkard, and E.D. Weitzman. 1983. Circadian rhythms in subjective alertness and body temperature. *Chronobiologia* 10:49–55.

Monod, T. 1958. Majaabat al-Koubra. Contribution a l'etude de l' "Empty Quarter" Ouest-Saharien. Memoires Inst. Franc. d'-Afrique Noire. No. 52, IFAN-Dakar. 407p.

Monro, K.M.H., and S.A. Barnett. 1969. Variation of the lumbar vertebrae of mice at two environmental temperatures. *J. Embryol. Exp. Morph.* 21:97–103.

Montagna, W.M. 1962. *The Structure and Function of Skin.* 2nd Ed. New York. Academic Press. 1454p.

Montner, P., D.M. Stark, M.L. Riedesel, G. Murata, R. Robergs, M. Timms, and T.W. Chick. 1996. Pre-exercise glycerol hydration improves cycling endurance time. *Int. J. Sports Med.* 17 27–33.

Montzka, S.A., J.H. Butler, R.C. Myers, T.M. Thompson, T.H. Swanson, A.D. Clarke, L.T. Lock, J.W. Elkins. 1996. Decline in the tropospheric abundance of halogen from halocarbons: Implications for stratospheric ozone depletion. *Science* 272:1318–1322.

Moore, F.D. 1992. Radiation burdens for humans on prolonged exomagnetospheric voyages. *FASEB J.* 6:2338–2343.

Moore, L.G., S. Zamudio, L.Curran-Everett, A. Torroni, L.B. Jorde, R.V. Shohet, and T. Drolkar. 1994. Genetic adaptation to high altitude. In *Lung Biology in Health and Disease* 76:225–262, ed. S.C.Wood and R.C. Roach.

Moore, R.Y. 1983. Organization and function of a central nervous system circadian oscillator: The suprachiasmatic hypothalamic nucleus. *Fed. Proc.* 42:2783–2789.

Moore, R.Y., and N.J. Lenn. 1972. A retinohypothalamic projection in the rat. *J. Comp. Neurol.* 146:1–14.

Moore, R., W.D. Clark, and G.S. Vodopich. 1997. *Botany*, 2nd ed. New York: McGraw Hill. 919 p.

Moore-Ede, M.C. 1986. Jet lag, shift work, and maladaption. News in Physiol. *Sciences* (NIPS) 1:156–160.

Moore-Ede, M.C., F.M. Sulzman and C.A. Fuller. 1982. *The Clocks That Time Us.* Cambridge, MA: Harvard Univ. Press.

Moore-Ede, M.C., R. Lydic, C.A. Czeisler, C.A. Fuller, and H.E. Albers. 1980. Structure and function of suprachiasmatic nuclei (SCN) in human and non-human primates. *Neurosciences Abstr.* 6:708.

Morgan, K.R., and G.A. Bartholomew. 1982. Homeothermic Response in a Scarab Beetle. *Science* 216:1409–1410.

Morgareidge, K.R., and H.T. Hammel. 1975. Evaporative water loss in box turtles: effects of rostral brain stem and other temperatures. *Science* 187:366–368.

Mori S., G. Mitarai, A. Takabayashi, S. Usui, M. Sakakibara, M. Nagatomo, and R.J. von Baumgarten. 1996. Evidence of sensory conflict and recovery in carp exposed to prolonged weightlessness. *Aviat. Space Environ. Med.* 67:256–261.

Moroff, S.V., and D.E. Bass. 1965. Effects of overhydration on man's physiological responses to work in the heat. *J. Appl. Physiol.* 20:267–270.

Morris, L., and P.R. Morrison. 1964. Cyclic responses in dormice (*Glis glis*) and ground squirrels (*Spermophilus tridecemlineatus*) exposed to normal and reversed yearly light schedules. *Proc. 15th Alaska Sci. Conf., Am. Assoc. Adv. Sci.* p. 55.

Morrison, J.B., and J.T. Florio. 1971. Respiratory function during a simulated saturation dive to 1,500 feet. *J. Appl. Physiol.* 30:724–732.

Morrison, P. 1960. Some interrelations between weight and hibernation function, p. 75–91. In *Mammalian Hibernation.* Bull. Mus. Comp. Zool., Vol. 124, ed. C.P. Lyman and A.R. Dawe. Cambridge, Mass.: Harvard Univ. Press.

Morrison, P.R. 1957. Body temperatures in aborigines. *Fed. Proc.* 16:90.

Morrison, P.R. 1962. Modification of body temperature by activity in Brazilialn hummingbirds. *The Condor* 64:315–323.

Morrison, P.R. 1962. Temperature regulation in animals native to tropical and high altitude environments, p. 381–413. In *Comparative Physiology of Temperature Regulation,* ed. J.P. Hannon and E. Viereck. Proc. 2nd Symp. Arctic Biol. and Med. 455 p.

Morrison, P.R., K. Kerst, and M. Rosenmann. 1963. Hematocrit and hemaglobin levels in some Chilean rodents from high and low altitude. *Int. J. Biometeorol.* 7:45–50.

Mosley-Thompson, E. 1996. Evidence for climate change. *Ecological Society Newsletter* 48:4.

Moss, K.N. 1924. Some effects of high air temperatures and muscular exertion upon collier. *Proc. Roy. Soc. Lond.*, B. 95:181–200.

Mosso, A. 1898. *Life of Man on the High Alps.* London: Inwin. 342 p.

Moul, D.E., C.J. Hellekson, D.A. Oren, et al. 1990. Treating SAD with a light visor: a multicenter study, p. 9. In *Second Annual Conference on Light Treatment and Biological Rhythms.* New York: Society for Light Treatment and Biological Rhythms.

Moutain, S.J., J.E. Kain, and M.N. Sawka. 1995. Reflex sweating response to drinking during exercise: effect of hypohydration level. *FASEB J.* 9:A357 (abstract)

Mouton, P., F. Woimant, O. Ille, E. Prevot, J. Mikol, M. Haguenau. 1996. Hypothermia and the nervous system (in French). *Annales de Medecine Interne* 147(2): 107–114.

Muir, A.L. 1967. Ketogenic effects of cold, p. 36–47. In *The Effects of Abnormal Physical Conditions at Work*, ed. Davies, Davis, and Tyrer. Edinburgh: E & S Livingstone. 185 p.

Muller, E.F. 1985. Basal metabolic rates in primates. *Comp. Biochem. Physiol.* 81A:707–711.

Muller, F.L. 1995. A field study of the ventilatory response to ambient temperature and pressure in sport diving. *British J. Sports Medicine.* 29(3):185–190.

Muller, J.E., and G.H. Tofler. 1991. Circadian variation and cardiovascular disease. *N. Engl. J. Med.* 325:1038–1039.

Mulvad, G., et al. 1996. The Inuit diet. *Arctic Med. Res.* 55 Suppl 1:20–24.

Muradian, KhK., and E.I. Shinkar. 1993. Effect of reduced caloric intake on circadian rhythm of gas exchange and heat production in old rats (In Russian). *Voprosy Pitaniia* May-Jun. 3:33–6.

Murdock, B.S. 1996. Mercury stockpile (Letters). *Science* 272:1247.

Murray, B.M., and D.F. Murray. 1969. Notes on mammals in alpine areas of the northern St. Elias Mountains, Yukon Territory, and Alaska. *The Canadian Field-Naturalist* 83:331–338.

Murray, R. H., and J.C. Ross. 1965. Cardiovascular effects of brief, intense thermal pulses in man. *Fed. Proc.* 24:280.

Musacchia, X.J., and D.D. Westhoff. 1964. Absorption of D-glucose by segments of intestine from active and hibernating, irradiated and non-irradiated ground squirrels, *Citellus tridecemlineatus. Ann. Acad. Fenn.* Ser. A4 71:3245–359.

Musacchia, X.J., and D.R. Deavers. 1981. The regulation of carbohydrate metabolism in hibernators, p. 55–75. In *Survival in the Cold: Hibernation and Other Adaptations,* ed. X.J. Musacchia and L. Jansky. New York: Elsevier/North-Holland.

Musacchia, X.J., and L. Jansky, eds. 1981. *Survival in the Cold: Hibernation and Other Adaptations*: Proc. Sixth Int. Symp. Nat. Mammal. Hiber. New York: Elsevier/North-Holland.

Mussalo-Rauhamaa, H., et al. 1996. Trends in the concentrations of mercury, copper, zinc, and selenium in Finnish Lapland. *Arc. Med. Res.* 55:83–91.

Myers, R.D., P.R. Oeltgen, and W.A. Spurrier. 1981. Hibernation "trigger" injected in brain induces hypothermia and hypophagia in the monkey. *Brain Res. Bull.* 7:691–695.

N

Nadel, E.R. 1977. *Problems with Temperature Regulation During Exercise.* New York: Academic Press

Nadel, E.R. 1978. Temperature regulation during exercise. P 143–153. In *New Trends in Thermal Physiology*, ed. Y. Houdas and J.D. Guieu. New York: Masson. 204 p.

Nadel, E.R., M.F. Roberts, and C.B. Wenger. 1978. Thermoregulatory adaptations to heat and exercise: comparative responses of men and women, p 29–38. In *Environmental Stress: Individual Human Adaptations,* ed. L.J. Folinsbee, J.A. Wagner, J.F. Borgia, B.L. Drinkwater, J.A. Gliner, and J.F. Bedi. New York: Academic Press. 393 p.

Nadel, E.R., et al. 1980. Effect of hydration state on circulatory and thermal regulations. *J. Appl. Physiol.* 49:715–721.

Nair, C.S., and S. George. 1972. The effect of altitude and cold on body temperature during acclimatization of man at 3,300 m. *Int. J. Biometeor.* 16:79–84.

Nakamura, K., Tanaka, M., Motohashi, Y., Maeda, A. 1997. Oral temperatures of the elderly in nursing homes in summer and winter in relation to activities of daily living. *Int. J. Biometeorol.* 40:103–106.

Nakayama, T. T. Hori, T. Nagasaka, H. Tokura, E. Taoaki. 1971. Thermal and metabolic responses in the Japanese monkey at temperatures of 5–38°C. *J. Appl. Physiol.* 31:332–337.

Nalbandov, A.V. 1970. Endocrine background of light action. In *La Photoregulation de la Reproduction chez Les Oiseaux et Les Mammifères,* ed. J. Benoit and I. Assenmacher. Montpellier Symposium. Paris: Editions du Centre National de la Recherche Scientifique. 588 p.

Neff, E. Stefansson. 1996. *Discoveries.* McGraw-Hill (in manuscript).

Nelson, R., et al. 1978. Urea metabolism in the hibernating black bear. *Kidney Intnatl.* 13 Suppl. 8:5177–5179.

Nelson, R.J., and I. Zucker. 1981. Absence of extraocular photoreception in diurnal and nocturnal rodents exposed to direct sunlight. *Comp. Biochem. Physiol.* 69A:145–149.

Nelson, R., et al. 1979. Biochemical transition from hibernation in bears. *Fed. Proc.* 38:1227.

Nelson, R.A., et al. 1983. Behavior, biochemistry and hibernation in black, grizzly, and polar bears. *Int. Conf. Bear Res. and Management.* 5:284–290.

Nestler, J.R. 1990. Intracellular pH during daily torpor in *Peromyscus maniculatus. J. Comp. Physiol.* B 159: 661–666.

Nevitt, G.A. 1996. Sniffing out sulfides. *BioScience* 45:802.

New Melleray Abbey. 1996. *Monastery Seasons.* Vol. 37, No. 1 (Spring). Peosta, IA 52068.

Newman, M.T. 1966. The application of ecological rules to the racial anthropology of the aboriginal New World, p. 149–165. In *Human Ecology*, ed. J.B. Bresler, Reading, MA: Addison-Wesley. 472 p. (Also *Amer. Anthrop.* 5.5:311–327, 1953).

New York Public Library Desk Reference. 1993. "Customary ;and Metric Systems of Measurement," p. 22–24. New York: MacMillan 944 p.

Nicogossian, A.E., C.L. Huntoon, and S.L. Pool. 1994. *Space Physiology and Medicine,* 3rd Edition, Lea Febiger, 481 p.

Nielsen, B. 1978. Sweating sensitivity and temperature regulation during exercise, p 154–156. In *New Trends in Thermal Physiology*, ed. Y. Houdas and J.D. Guieu. New York: Masson. 204 p.

Nielsen, B. 1981. Exercise and temperature regulation. p 537–544. In *Contributions to Thermal Physiology,* ed. Z. Szelenyi and M. Szekely. New York: Pergamon Press. 561 p.

Nielsen, M. 1938. Die Regulation der Korpertemperatur bei Muskelarbeit. *Skand. Arch.f. Physiol.* 79:193–230.

Nielsen, R.P. 1995. Controlling Campus Noise. *Occup. Health and Safety,* 64: No. 11, p. 65–68.

Norris, K.S. 1967. Some observations on the migration and orientation of marine mammals, p. 101–125. In *Animal Orientation and Navigation*, ed. R.M. Storm. Corvallis, OR: Oregon State Univ. Press, 134 p.

Norris, K.S., and J.H. Prescott. 1961. Whales, dolphins, and porpoises. *Univ. Calif. Pub. Zool.* 63:291–401.

Nover, L., and K.D. Scharf. 1991. Heat shock proteins. In: *Heat Shock Response,* ed. L. Nover. Boca Raton, FL CRC p 151–159.

Novick, G.E., et al. 1996. The mobile genetic element Alu in the human genome. *BioScience* 46:32–41.

Novotny J.A., et al. 1990. Xenon kinetics in muscle are not explained by a model of parallel perfusion-limited compartments. *J. Appl. Physiol.* 68:876–890,

Nunneley, S.A., and R.J. Maldonado. 1982. Head and/or torso cooling during simulated cockpit heat stress. *Aviat. Space Environ. Med.*

Nunneley, S.A., D.C. Reader, and R.J. Maldonado. 1982. Heat-temperature effects on physiology, comfort, and performance during hyperthermia. *Aviat. Space Environ. Med* 53:623–628.

O

O'Brodovich, H., et al. 1984. Hypoxia alters blood coagulation during acute decompression in humans. *J. Appl. Physiol.* 56:666.

O'Farrell, M.J., W.C. Bradley, and G.W. Jones. 1967. Fall and winter bat activity at a desert spring in southern Nevada. *Southwestern Naturalist* 12:163–171.

Obal, F., and G. Benedek, eds.. 1981. *Environmental Physiology, Vol. 18, Advances in Physiological Sciences.* New York: Pergamon Press, 371 p.

Odum, E.P. 1963. *Ecology*. Modern Biology Series. New York: Holt, Rinehart and Winston. 152 p.

Odum, E.P. 1971. Fundamentals of Ecology, p. 80. Philadelphia: W. B. Saunders Co. 574 p.

Ogle, C.W., and M.F. Lockett. 1966. The release of neurohypophyseal hormone by sound. *J. Endocrinol.* 36:281–290.

Olsson, K., M. Josater-Hermelin, J. Hossaini-Hilali, E. Hydbring, and K. Dahlborn. 1995. Heat stress causes excessive drinking in fed and food-deprived pregnant goats. *Comp. Biochem. Physiol.* 110A:309–317.

Olszowka, A.J., and H. Rahn. 1987. Breath-hold diving. In *Extreme Environments: Coping Stratagies of Animals and Man,* ed. J.R. Sutton, C.S. Houston, and G. Cowles, 417–428. New York: Praeger.

Orita, T., A Izumihara, T. Tsurutani, and K. Kajiwara. 1995. Brain temperature before and after brain death. *Neurological Research* 17(6):443–4.

Øritsland, N.A. 1969. Deep body temperatures of swimming and walking polar bear cubs. *J. Mammal.* 50:380–382.

Osadjan, M.D., A.L. DeVries, and C. Cheng-DeVries. 1996. Characterization of the antifreeze system of an Antarctic zoarcid—An evolutionary study. *The FASEB Journal,* 10:299.

Oshima, K., and A. Gorbman. 1969. Pars intermedia: Unitary electrical activity regulated by light. *Science* 163:195–197.

Ou, L.C., and S.M. Tenney. 1990. Properties of mitochondria from hearts of cattle acclimated to high altitude. *Resp. Physiol.* 8:151–159.

Ouay, W.B. 1972. Pineal homeostatic regulation of shifts in the circadian rhythm during maturation and aging. *N.Y. Acad. Sci.* 34:239–254.

P

Packer, C. 1996. Coping with a lion killer. *Natural History* 105 (June):14–17.

Page, T.L. 1982. Transplantation of the cockroach circadian pacemaker. *Science* 216:73–75.

Palmer, D.L., and M.L. Riedesel. 1976. Responses of whole-animal and isolated hearts of ground squirrel, *Citellus lateralis*, to melatonin. *Comp. Biochem. Physiol.* 53C:69–72.

Parfit, M. 1995. Diminishing Returns. *Nat. Geographic* 188:2–37.

Park, Y.S., et al. 1987. Daily energy cost of breath-hold diving, p. 57–70. In *Man in Stressful Environments,* ed. K Shiraki and M.K. Yousef. Springfield, Ill: Chas. Thomas. 267 p.

Park, Y.S., J.K. Choi, J.S. Kim, and S.K. Hong. 1993. Renal response to head-out water immersion in Korean women divers. *Euro. J. Appl. Physiol. & Occup. Physiol.* 67(6):523–527.

Passmore, R., and R.E. Johnson. 1958. The modification of postexercise ketosis (the Courtice-Douglas effect) by environmental temperature and water balance. *Quart. J. Exp. Physiol.* 43:352.

Pembrey, M.S. 1898. *Schafer's Textbook of Physiology. Vol. 1.* 838 p.

Pengelley, E.T. 1969. Influence of light on hibernation in the Mojave ground squirrel (*Citellus mohavensis*), p. 11–16. In *Physiological systems in Semiarid Environments,* ed. C.C. Hoff and M.L. Riedesel. Albuquerque: The Univ. of New Mexico Press.

Pengelley, E.T., ed. 1974. *Circannual Clocks: Annual Biological Rhythms.* AAS Symposium. New York: Academic Press, 524 p.

Pengelley, E.T., and K.C. Fisher. 1961. Rhythmical arousal from hibernation in the golden-mantled ground squirrel. *Canad. J. Zool.* 39:105–120.

Pengelley, E.T., and K.C. Fisher. 1967. Ability of the ground squirrel *Citellus lateralis* to be habiuated to stimuli while in hibernation. *J. Mammal.* 49:56l.

Pengelley, E.T., R.C. Aloia, and B.M. Barnes. 1978. Circannual rhythmicity in the hibernating ground squirrel, *Citellus lateralis,* under constant light and hyperthermic ambient temperature. *Comp. Biochem. Physiol.* 61A:599–604.

Perry, R. 1966. *The World of the Polar Bear.* Univ. Seattle: Washington Press. 195 p.

Perutz, M.F. 1983. Species adaptation in a protein molecule. *Mol. Biol. Evol.* 1:1–28.

Peterson, L.H. 1963. Cardiovascular performance under water, p. 267–269. *Proc. 2nd Symp. Underwater Physiol.* Nat. Acad. Sci. Publ. 1181.

Pettigrew, R.T., J.M. Galt, C.M. Ludgate, D.B. Horn, and A.N. Smith. 1974. Circulatory and biochemical effects of whole body hyperthermia. *Br. J. Surg.* 61:727–730.

Phillips, J.A. 1995. Rhythms of a desert lizard. *Natural History* 104:50–55.

Phillips, J.A., and H.J. Harlow. 1981. Elevation of upper voluntary temperatures after shielding the parietal eye of horned lizards (Phrynosoma douglassi). *Herpetologica* 37:199–205.

Phillips, J.B. 1986. Two magnetoreception pathways in a migratory salamander. *Science* 233:765–766.

Pickard, G.E., R. Kahn, and R. Silver. 1984. Splitting of the circadian rhythm of body temperature in the golden hamster. *Physiol. & Behav.* 32:763–766.

Pistole, D.H., and J.A. Cranford. 1982. Photoperiodic effects on growth in *Microtus Pennsylvanicus. J. Mammalogy* 63:547–553.

Pittendrigh, C.S. 1954. On temperature independence in the clock system controlling emergence in Drosophila. *Proc. Natl. Acad. Sci.* 40:1018–1029.

Pittendrigh, C.S. 1957. Perspectives in the study of biological clocks, p. 235–261. In *Perspectives in Marine Biology*. Berkeley, CA: Univ. Calif. Press, 282 p.

Pittendrigh, C.S. 1960. Circadian Rhythms and Circadian Organization. *Cold Spring Harbor Symposia* 25:159–185.

Pittendrigh, C.S. 1964. The entrainment of circadian oscillations by skeleton photoperiods. *Science* 144:565–566.

Pittendrigh, C.S. 1981. Circadian organization and the photoperiodic phenomena, p. 1–35. In *Biological Clocks in Seasonal Reproductive Cycles,* ed. B.K. Follett and D.E. Follett. New York: Wiley & Sons, 292 p.

Pittendrigh, C.S., and G.V. Bruce. 1957. An oscillator model for biological clocks, p. 75–109. In *Rhythmic and Systemic Processes in Growth*. Princeton, NJ: Princeton Univ. Press. 345 p.

Pittendrigh, C.S., and S. Daan. 1974. Circadian oscillations in rodents: A systematic increase of their frequency with age. *Science* 186:548–550.

Pittendrigh, C.S., and S. Daan. 1976. A functional analysis of circadian pacemakers. Parts I and V. *J. Comp. Physiol.* 106:223–252 and 333–352.

Pittinger, C.B. 1966. *Hyperbaric Oxygenation*. Springfield, IL: Charles C. Thomas. 113 p.

Pitts, G.C., R.E. Johnson, and R.C. Consolazio. 1994. Work in the heat as affected by intake of water, salt and glucose. *Amer. J. Physiol.* 142:253–256.

Plakke, R. K., and E. W. Pfeiffer. 1964. Blood vessels of the mammalian renal medulla. *Science* 146:1683–1685.

Pohl. H. 1969. Some factors influencing the metabolic response to cold in birds. Proc. Int. Symp. on Heat and Cold. *Fed. Proc.* 28:1059–1064.

Pohl, H. 1985. The circadian system of the turkish hamster, Mesocricetus brandti: Response to light. *Comp. Biochem. Physiol.* 81A:613–618.

Pohl, H., and J.S. Hart. 1965. Thermoregulation and cold acclimation in a hibernator, *Citellus tridecemlineatus. J. Appl. Physiol.* 20:398–404.

Ponganis, P.J., et al. 1993a. Determinants of the aerobic dive limit of Weddell seals. In *Phsyiol. Zool.* 66(5):732–749

Ponganis, P.J., et al. 1993b. Muscle temperature and swim velocity during diving in a Weddell seal (*Leptonychotes weddelli*) *Jour. Exp. Bio.* 183:341–346.

Popovic, V. 1959. Lethargic hypothermia in hibernators and non-hibernators. *Ann. N. Y. Acad. Sci.* 80:320–331.

Popovic, V. 1964. Cardiac output in hibernating ground squirrels. *Am. J. Physiol.* 207:1345–1349.

Portale, A.A., et al. 1997. Aging alters calcium regulation in healthy men. *Am. J. Physiology* 272: E139–E146.

Pough, F.H. 1980. Blood oxygen transport and delivery to reptiles. *Am. Zool.* 20:173–185.

Powell, R.A. 1982. Evolution of black-tipped tails in weasels: Predator confusion. *Amer. Naturalist* 199:126–131.

Preston-Martin, S. 1996. Breast Cancer and Magnetic Fields. *Epidemiology* 7(5):457–458.

Proppe, D.W., G.L. Brengelmann, and L.B. Rowell. 1976. Control of baboon limb blood flow and heart rate: role of skin vs. core temperature. *Am. J. Physiol.* 231:1457–1465.

Prosser, C.L. 1986. *Adaptational Biology, Molecules to Organisms.* John Wiley & Sons.

Prosser, C.L., and F.A. Brown, Jr. 1961. *Comparative Animal Physiology.* 2nd ed. Saunders: Philadelphia. 688 p.

Prothero, J.W. 1979. Maximal oxygen consumption in various animals. *Comp. Biochem. Physiol.* 64A:463–466.

Pugh, L.G.C. 1964. Animals in high altitudes: Man above 5000 meters—mountain exploration, p. 861–868. In *Handbook of Physiology, Sec. 4. Adaptation to the Environment.* Sec. ed. D.B. Dill. Washington DC: Amer. Physiol. Soc.

Pugh, L.G.C., J.L. Corbett, and R.H. Johnson. 1967. Rectal temperatures, weight losses, and sweat rates in marathon running. *J. Appl. Physiol.* 23:347–352.

Pugh, L.G.C. and O.G. Edholm. 1955. Physiology of Channel Swimmers. *Lancet* 2:761–768.

Q

Quarterman, J., A.C. Dalgarno, and A. Adam. 1964. Some factors affecting the level of vitamin D in the blood of sheep. *Brit. J. Nutr.* 18:79–89.

Qvist, J., et al. 1986. Hemoglobin concentrations and blood gas tensions of free diving Weddell seals. *J. Appl. Physiol.* 61:1560–1569.

R

Rahn, H. 1987. Breath-hold diving: a brief history. In *The Physiology of Breath-Hold Diving*, 1–3, ed. C.E.G. Lundgren and M. Ferrigno. Bethesda, MD: Undersea Hyperbaric Med. Soc.

Rahn, H., and A.B. Otis. 1949. Man's respiratory response during and after acclimatization to high altitude. *Am. J. Physiol.* 157:445–462

Ralph, C.L. 1976. Correlation of melatonin content in pineal gland, blood and brain of some birds and mammals. *Amer. Zool.* 16:35–43.

Ralph, C.L. 1980. Melatonin production by extra-pineal tissues, p. 371–378. In *Advances in the Biosciences 29: Mela-*

tonin, ed. N. Birau and W. Schloot. New York: Pergamon Press. 410 p.

Ralph, C.L. et al. 1979. The pineal complex and thermoregulation. *Biol. Rev.* 54:41–72. Printed in Great Britain.

Ralph, C.L., B.T. Firth, and J.S. Turner. 1979. The role of the pineal body in ectotherm thermoregulation. *Amer. Zool.* 19:273–293.

Ralph, C.L., H.J. Harlow, and J.A. Phillips. 1982. Delayed effect of pinealectomy on hibernation of the golden-mantled ground squirrel. *Int. J. Biometeor.* 26:311–328.

Ramanathan, N.L. 1964. A new weighting system for mean surface temperature (human). *J. Appl. Physiol.* 19:531–533.

Randall, W., J.T. Cunningham, S. Randall, J. Littischwager, and R.J. Johnson. 1987. A two-peak circadian system in the domestic cat. *J. Therm. Biol.* 12:27–37.

Ratner, P.L., M. Fischer, D. Burkart, J.R. Cook and L.P. Kozak. 1981. The role of mRNA levels and cellular localization in controlling sn-glycerol-3-phosphate dehydrogenase expression in tissues of the mouse. *J. Biol. Chem.* 256:3576–3579.

Rausch, R.L. 1961. Notes on the black bear, *Ursus americanus Pallas*, in Alaska, with particular reference to dentition and growth. Z. *Saugetierk* 26(2):65–128.

Raven, P.H. 1991. Carrying Capacity of the Globe, p. 121–138. In *Global Change and the Human Prospect, Forum Proceedings,* ed. Colwell, R.R. Published by Sigma Xi, Research Society, 294 p.

Rawson, K.S. 1959. Experimental modification of mammalian endogenous activity rhythms, p. 791–800. In *Photoperiodism*, ed. R.B. Withrow. *Amer Assoc. Adv. Sci.* Publ. 55. Wash. DC 903 p.

Rawson, K.S. 1960. An accurate recorder for animal activity. *J. Mammal.* 41:284–287.

Ray, D. and D.O. Lavallee. 1964. Self-contained diving operations in McMurdo Sound, Antarctica: Observations of the sub-ice environment of the Weddell seal, *Leptonychotes weddelli* (Lesson). *Zoologica* 49:121–139.

Rea, L.D. 1995. Prolonged fasting in pinnipedes. Ph.D. thesis, University of Alaska, 135 pages.

Reay, D. A., and F. A. J. Thiele. 1977. Heat pipe theory applied to a biological system: quantification of the role of the "resting" eccrine sweat gland in thermoregulation. *J. Theor. Biol.* 64:789–803.

Redmond, I. 1982. The Salt-Mining Elephants. *Wildlife* 24:290–293

Reed, H.L. 1995. Circannual changes in thyroid hormone physiology: the role of cold environmental temperatures. *Arctic Medical Res.* 54 Suppl 2:9–15.

Reed, J.Z., et al. 1994. Gas exchange of captive freely diving grey seals (*Halichoerus grypus*). *J. Exp. Biol.* 191:1–18.

Reeder, W. G., and R. B. Cowles. 1951. Aspects of thermoregulation in bats. *J. Mammal.* 32:389–403.

Reedy, J.R. 1964. First flight across the bottom of the world. *National Geographic* 125:454–464.

Reeves, J.T., B.M. Groves, J.R. Sutton, P.D. Wagner, A. Cymerman, M.K. Malconian, P.B. Roche, P.M. Young, J.K. Alexander, and C.S. Houston. 1987a. Oxygen transport during exercise at extreme altitude: Operation Everest II. *Ann. Emerg. Med.* 16:993–998.

Reeves, J.T., B.M. Groves, J.R. Sutton, P.D. Wagner, A. Cymerman, M.K. Malconian, P.B. Roch, P.M. Young, and C.S. Houston. 1987b. Operation Everest II: preservation of cardiac function at extreme altitude. *J. Appl. Physiol.* 63:531–539.

Reeves, R.B. 1977. The interaction of body temperature and acid-base balance in ectothermic vertebrates. *Ann. Rev. Physiol.* 39:559–586.

Regaldo, A. 1995. Oceans may be warming. *Science* 268:1436–1437.

Regan, J.D., and J.A. Parrish. 1982. *The Science of Photomedicine*. New York: Plenum Pub. Co. 637 p.

Reinberg, A., M. Smolensky, and F. Levi. 1981. Clinical chronopharmacology. *Biomedicine* 34:171–178.

Reiter, R.J., ed. 1978. *The Pineal and Reproduction.* Basel: Karger. 223 p.

Reiter, R.J. 1981. Seasonal aspects of reproduction in a hibernating rodent: Photoperiodic and pineal effects, p. 1–11. In *Survival in the Cold: Hibernation and Other Adaptations*, ed. X.J. Musacchia and L. Jansky. New York: Elsevier/North-Holland.

Reiter, R.J. 1981. *The Pineal Gland: Vol. I, Anatomy and Biochemistry.* Boca Raton, FL: CRC Press, 320 p.

Reiter, R.J. 1996. Visible and non-visible electromagnetic radiation. The good and the bad, p. 9. In *Proceedings of the 14th International Congress of Biometeorology*. Ljubljana, Slovenia.

Remmert, H. 1980. *Arctic Animal Ecology*. New York: Springer-Verlag, 280 p.

Renfree, M.B., D.W. Lincoln, 0.F.X. Almeida, and R.V. Short. 1981. Abolition of seasonal embryonic diapause in a wallaby by pineal denervation. *Nature* 293:136–137.

Rensing, L. 1971. Hormonal control of circadian rhythms in Drosophila, p. 527–540. In *Biochronometry*, ed. M. Menaker. Washington, DC: Natl. Acad. Sci., 662 p.

Rensing, L. et al., eds. 1987. *Temporal Disorder in Human Oscillatoory Systems*. Berlin: Springer-Verlag.

Reppert, S.M. 1985. Biological clocks in fetuses. *Science* 230:929–930.

Reynolds, P.E., H.V. Reynolds, III, and E.H. Follman. 1983. Responses of grizzly bears to seismic surveys in northern Alaska. *Int. Conf. Bear Res. and Manage.* 6:169–175.

Reynolds, W. W. 1979. Perspective and introduction to the symposium: thermoregulation in ectotherms. *Amer. Zool.* 19:193–194.

Rice, D.W., and A.A. Wolman. 1971. *The Life History and Ecology of the Gray Whale (Eschrichtus robustus)*. Amer. Soc. Mammal. Spec. Pub. No. 3. 142 p.

Richie, R.J., and S. Ambrose. 1996. Distribution and status of bald eagles in Interior Alaska. *Arctic* 49:120–128.

Richter, C.P. 1927. Animal behavior and internal drives. *Quart. Rev. Biol.* 2:307–343.

Richter, C.P. 1975. Deep hypothermia and its effect on the 24-hour clock and hamsters. *Johns Hopkins Med. J.* 136:1–10.

Riddles, L., and T. Jones. 1988. *Race Across Alaska: First Woman to Win the Iditarod Tells Her Story.* Stackpole Books.

Riedesel, M. L. 1977. Blood physiology, p 485–517. In *Biology of Bats, vol. 3,* ed. W. A. Wimsatt. New York: Academic Press. 651p..

Riedesel, M.L, D.Y. Allen, G.T. Peake, and K. Al-Qattan. 1987. Hyperhydration with glycerol solutions. *J. Appl. Physiol.* 63:2262–2268.

Riedesel, M.L., and G.E. Folk, Jr. 1957. Serum magnesium changes in cold-exposed mammals. *J. Mammal.* 38:423–424.

Riedesel, M.L., and G.E. Folk, Jr. 1996. Estivation, p 279–283. In *Handbook of Physiology - Environmental Physiology,* Vol 1, ed. M.J. Fregley and C.M. Blatteis. London: Oxford Univ. Press, 783p.

Riedesel, M.L., and J.M. Steffen. 1980. Protein metabolism and urea recycling in rodent hibernators. *Fed. Proc. Fed. Am. Soc. Exp. Biol.* 39:2959–2963.

Riesen, A.H. 1960. Chimpanzees raised in darkness. *Amer. J. Orthopsychiatry* 30:23–25.

Riley, J.J., et al. 1992. The response of solar radiation in Saudia Arabia to smoke from oil field fires. *Int. J. Biometeorol.* 36:176–177.

Roach, R.C. 1994. Hypoxic ventilatory response and performance at high altitude. In *Lung Biology in Health and Disease,* ed. S.C. Wood and R.C. Roach. 76:211–224.

Roberts, B., ed. 1967. *Edward Wilson's Birds of the Antartic.* New York: Humanities Press. 191 p.

Roberts, M.F., C.B. Wenger, J.A.J. Stolwijk, and E.R. Nadel. 1977. Skin blood flow and sweating changes following exercise training and heat acclimation. *J. Appl. Physiol.* 43:133–137.

Robertshaw, D. 1981. Man in extreme environments: problems of the newborn and elderly, p. 169–179. In *Bioengineering, Thermal Physiology and Comfort,* ed. K. Cena and J. A. Clark. New York: Elsevier Scientific.

Robinson, N.E., ed. 1995. *Equine Exercise Physiology 4.* Equine Veterinary J. Supplement 18. (Proc. 4th Int. Conf. Equine Exer. Physiol., Queensland, Australia 1994). Newmarket, Suffolk, U.K.: R & W Publications. 479 p.

Robinson, S., S.O. Gerking, E.S. Turrell, and R. K. Kincaid. 1950. Effects of skin temperature on salt concentration in sweat. *J. Appl. Physiol.* 2:654–662.

Robinson, S., F.R. Meyer, and J.L. Newton. 1965. Relations between sweating, cutaneous blood flow, and body temperature. *J. Appl. Physiol.* 20:575–583.

Rock, P.B. 1986. Medical problems of high terrestrial altitude. *U.S.Army Flight Surgeons' Manual.*

Rode, A., and R.J. Shephard. 1994. Physiological consequences of acculturation: a 20-year study of fitness in an Inuit community. *Euro. J. Appl. Physiol. & Occup. Physiol.* 69(6):516–24.

Roelofs, et al. 1982. Sex pheromone of the winter moth. *Science* 217:657–658.

Rolewicz, T.R., and B.G. Zimmerman. 1972. Peripheral distribution of cutaneous sympathetic vasodilator systems. *Amer. J. Physiol.* 223:939–944.

Romer, A.S. 1977. *The Vertebrate Body.* Philadelphia: Saunders. 723 p.

Rosen, L.M., S.D. Targum, M. Terman, et al. 1989. Prevalence of seasonal affective disorder at four latitudes. *Psychiatry Res* 31:131–144.

Rosenmann, M., and P. Morrison. 1971. A new method for the determination of maximum metabolism in small mammals. *22nd Alaska Sci. Conf.* 22:9.

Rosenmann, M., and P. R. Morrison. 1963. The Physiological response to heat and dehydration in the guanaco. *Physiol. Zool.* 36:45–51.

Rosenthal, N.E. 1985. Antidepressant effects of light in seasonal affective disorder. *Am. J. Psychiatry* (February) 142(2):163–170.

Rosenthal, N.E. 1993a. Diagnosis and treatment of seasonal affective disorder. In *Grand Rounds at the Clinical Center of the National Institutes of Health,* Sec. ed. S. Rosen. *JAMA* Vol. 170, No 22. p. 2717–2720.

Rosenthal, N.E. 1993b. A Decade of SAD and Light Therapy. *NIH Curr. Contents,* No 10, March 8.

Rosenthal, N.E., and T.A. Wehr. 1992. Towards understanding the mechanism of action of light in seasonal affective disorder. *Pharmacopsychiat* 25:56–60.

Roth, H. P. 1972. Ways of interpreting the thermal environment, p. 21–36. In *Studies in Environmental Physiology,* ed. G. E. Folk, Jr. and D. B. Dill. Las Vegas: Univ. Nevada Press.

Roth, J. et al. 1980. Nonpineal melatonin in the alligator *Alligator mississippiensis. Science* 210:548–549.

Ruby, N.F., J. Dark, H.C. Heller, and I. Zucker. 1996. Ablation of suprachiasmatic nucleus alters timing of hibernation in ground squirrels. *Proc. Natl. Acad. Sci.* 93:9864–9868.

Rudge, J. 1996. British weather: a serious health risk. *Int. J. Biometeorol.* 39:151–155.

Rusak, B., and G. Groos. 1982. Suprachiasmatic stimulation phase shifts rodent circadian rhythms. *Science* 215:1407–1409.

Ryan, S.B., J.P. Chiang, and D.S. Bruce. 1982. Induction of summer hibernation in thirteen lined ground squirrel, *Citellus tridecemlineatus*, through urine injection. *Fed. Proc. Fed. Am. Soc. Exp. Biol.* 41:1696.

S

Salzano, J.V., et al. 1984. Physiological responses to exercise at 47 and 66 atm abs. *J. Appl. Physiol.* 57:1055–1068.

Sampson, J.B., A. Cymerman, R.L. Burse, J.T. Maher, P.B. Rock. 1983. Procedures for the measurement of Acute Mountain Sickness. *Aviat. Space Environ Med* 54:1063–73.

Sanders, E.H., P.D. Gardner, P.J. Berger, and N.C. Negus.

1981. 6-MBOA: A plant derivative that stimulates reproduction in *Microtus montanus*. *Science* 214:67–69.

Sanderson, K. 1995. Climbing the food web. *The Pilot* 13:10–11.

Santee, W.R., and R.R. Gonzales. 1988. Characteristics of the thermal environment, p. 1–44. In *Human Performance Physiology and Environmental Medicine at Terrestrial Extremes*, ed. K.B. Pandolf, M.N. Sawka and R.R. Gonzales. Indianapolis: Benchmark Press Inc.

Sarajas, H.S.S., J.I. Hirvonen, L.K.J. Karlsson, and K.S.I. Virtanen. 1967. The hypothalamo-neurohypophyseal secretory status versus the fluid and electrolyte balance in rats undergoing cold acclimatization and/or brief exposure to severe cold. *Ann. Acad.Sci. Fennicae* Series A. IV Biologica. No. 121. 14 p.

Sargent, F., II. 1963. The nature and nurture of Biometeorology. *AIBS Bull.* 13:20–23.

Sargent, F., II. 1972. An ecological perspective for environmental physiology. *Int. J. Biometeorol.* 16:103–105.

Sarkar, S. 1996. Ecological theory and anuran declines. *BioScience* 46:199–207.

Satinoff, E. 1981. Are there similarities between thermoregulation and sexual behavior? In *Physiological Mechanisms of Motivation*, ed. D. W. Pfaff, p. 217–251. New York: Springer-Verlag.

Satinoff, E., and J. Rutstein. 1970. Behavioral thermoregulation in rats with anterior hypothalamic lesions. *J. Comp. Physiol. Psych.* 71:77–82.

Sato, K. 1977. The physiology, pharmacology and biochemistry of the eccrine gland. *Rev. Physiol. Biochem. Pharmacol.* 79:51–131.

Saunders, D.S. 1977. *An Introduction to Biological Rhythms*. New York: John Wiley and Sons. 170 p.

Savage, M.V., and G.L. Brengelmann. 1996. Control of skin blood flow in the neutral zone of human body temperature regulation. *J. Applied Physiol.* 80(4):1249–57.

Savourey, G., A.L. Vallerand, and J.H. Bittel. 1992. General and local cold adaptation after a ski journey in a severe arctic environment. *Euro. J. Appl.Physiol. & Occupat. Physiol.* 64(2):99–105.

Sawica-Kapusta, K. 1994. Heavy metals in rodent tissue. *Pol. Ecol. Stud.* 10:5–7.

Sawka, M.N. 1992. Physiological consequences of hypohydration: exercise performance and thermoregulation. *Med. Sci. Sports Exerc.* 24:657–670.

Sawka, M.N., B.J. Freund, D.E. Roberts, C. O'Brien, R.C. Dennis and C.R. Valeri. 1993. Total body water (TBW), Extracellular fluid (ECF) and plasma responses to hyperhydration with aqueous glycerol. *Med. Sci. Sports Exerc.* 25:S35.

Sawka, M.N., and J.E. Greenleaf. 1992. Current concepts concerning thirst, dehydration, and fluid replacement: overview. *Med. Sci. Sports Exerc.* 24:643–644.

Sawka, M.N., S.J. Montain, and W.A. Latzka. 1996. Body fluid balance during exercise-heat exposure, p. 139–157. In *Body Fluid Balance, Exercise, and Sport*, ed. E.R. Buskirk and S.M. Puhl. New York: CRC Press.

Sawka, M.N., A.J. Young, P.B. Rock, T.P. Lyons, R. Boushel, B.J. Freund, S.R. Muza, A. Cymerman, R.C. Dennis, K.B. Pandolf, and C.R. Valeri. 1997. Altitude acclimatization and blood volume: effects of exogenous erythrocyte volume expansion. *J. Appl. Physiol.* 81:636–642.

Schaefer, K.E. 1963. *Effect of prolonged diving training*, p. 271–278. Proc. 2nd Symp. Underwater Physiol. Nat. Acad. Sci. Pub. 1181.

Schaefer, K.E., et al. 1968. Pulmonary and circulatory adjustments determining the limits of depths in breathhold diving. *Science* 162:1020–1023.

Schmidt, K.D., and C.W. Chan. 1992. Thermoregulation and fever in normal persons and in those with spinal cord injuries. *Mayo Clinic Proceedings* 67(5):469–475.

Schmidt-Nielsen, B. 1965. Comparative morphology and physiology of excretion, p. 393–425. In *Ideas in Modern Biology*, ed. John A. Moore. Proc. 16th Int. Cong. Zool. vol. 6. Garden City, NY: The Natural History Press 562 p.

Schmidt-Nielsen, B. 1995. *August and Marie Krogh: Lives in Science*. New York: Oxford Univ. Press. 295 p.

Schmidt-Nielsen, B., and K. Schmidt-Nielsen. 1950a. A complete account of the water metabolism in kangaroo rats and an experimental verification. *J. Cell. Comp. Physiol.* 38:165–182.

Schmidt-Nielsen, B., and K. Schmidt-Nielsen. 1950b. Do kangaroo rats thrive when drinking sea water? *Amer. J. Physiol.* 162:31–36.

Schmidt-Nielsen, B., K. Schmidt-Nielsen, T.R. Houpt, and S.A. Jarnun. 1956. Water balance of the camel. *Amer. J. Physiol.* 185:185–194.

Schmidt-Nielsen, K. 1960. *Animal Physiology*. Found. Med. Biol. Ser., Englewood Cliffs, NJ: Prentice-Hall. 118 p.

Schmidt-Nielsen, K. 1964. *Desert Animals: Physiological Problems of Heat and Water*. London: Oxford Univ. Press. 277 p.

Schmidt-Nielsen, K. 1971. Marine Vertebrates—problems of salt and water, p. 205–209. In *Topics in The Study of Life*, ed. H. Ris et al. New York: Harper & Row. 482 p.

Schmidt-Nielsen, K. 1997. *Animal Physiology*. London: Cambridge University Press. 607 p.

Schmidt-Nielsen, K., S.A. Jarnum, and T.R. Houpt. 1957. Body temperature of the camel and its relation to water economy. *Amer. J. Physiol.* 188:103–112.

Schmidt-Nielsen, K., and R. Fange. 1958. The function of the salt gland in the Brown Pelican. *Auk* 75:282–289.

Schmitt, D.A., J.P. Hatton, C. Amond, D. Chaput, H. Paris, T. Levade, J.P. Cazenave, and L. Schaffar. 1996. The distribution of protein kinase C in human leukocytes is altered in microgravity. *FASEB J.* 10:1627–1637.

Schnoor, J.L., L.A. Licht, S.C. McCutcheon, N.L. Wolfe, and L.H. Carreira. 1995. Phytoremediation of organic and nutrient contaminants. *Environmental Science and Technology* Vol. 29, No. 7.

Schoene, R.B., et al. 1986. High altitude pulmonary edema and exercise at 4400 meters on Mount McKinley: effect of expiratory positive airway pressure. *Chest* 87:330–333.

Schoene, R.B., R.C. Roach, P.H. Hackett, J.R. Sutton, A. Cymerman, and Charles S. Houston. 1990. Operation Everest II: ventilatory adaptation during gradual decompressions to extreme altitude. *Med. Sci. Sports* Exerc. 22:804–810.

Scholander, P.F. 1955. Evolution of climatic adaptation in homeotherms. *Evolution* 9:15–26.

Scholander, P.F. 1956. Climatic rules. *Evolution* 10:339–340. (Reprinted in *Human Ecology*, ed. J.B. Bresler, Reading, MA: Addison-Wesley. 472 p.)

Scholander, P.F. 1958. Studies of man exposed to cold. *Fed. Proc.* 17:l054–1057.

Scholander, P.F. 1963. The master switch of life. *Scientific American* (Dec.). Reprint No. 172.

Scholander, P.F. 1964. Animals in aquatic environments: diving mammals and birds. In *Handbook of Physiology, Sec. 4. Adaptation To The Environment*, Sec. ed. D.B. Dill, 729–741. Washington, DC: Amer. Physiol. Soc. 1056 p.

Scholander, P.F. 1990. *Enjoying a Life in Science: The Autobiography of P.F. Scholander*. Fairbanks: Univ. of Alaska Press. 226 p.

Scholander, P.F., et al. 1958. Metabolic acclimation to cold in man. *J. Appl. Physiol.* 12:1–8.

Scholander, P.F., et al. 1962. Circulatory adjustment in pearl divers. *J. Appl. Physiol.* 17:184–190.

Scholander, P.F., and W.F. Scheville. 1955. Countercurrent vascular heat exchange. *J. Appl. Physiol.* 8:279–282.

Schonbaum, E. 1960. Adrenocortical function in rats exposed to low environmental temperatures. *Fed. Proc.* 19(Suppl.5):85–88.

Schottelius, B.A., and D.D. Schottelius. 1973. *Textbook of Physiology*, 17 ed. St. Louis: C.V. Mosby Co. 590 p.

Schreider, E. 1968. Ecological rules, body-heat regulations, and human evolution, p. 55–66. Reprinted in *Environments of Man*. J.B. Bresler, ed. 289 p. Reading, Mass.: AddisonWesley.

Schusterman, R. 1997. Personal communication.

Schwinghamer, J.M., and T. Adams. 1969. Cold-induced vasodilation in the footpad of the anesthetized cat. *Fed. Proc.* 28:1149–1153.

Scott, R.W. 1917. CO_2 intake and volume of inspired air. *Amer. J. Physiol.* 44:18–21.

Seals, D.R., et al. 1991. Hyperoxia lowers sympathetic activity at rest but not during exercise in humans. *Am. J. Physiol.* 260 (*Regulatory Integrative Comp. Physiol.* 31): R873–R878.

Seger, J., and N.A. Moran. 1996. Snapping social swimmers. *Nature* 381:473–474.

Sehgal, A., et al. 1994. Loss of circadian rhythms in the mutant *timeless*. *Science* 263: 1603–1605.

Seinfeld, J.H. 1996. Atmospheric Chemistry and Physics of Air Pollution. New York: Wiley. 738 p.

Selye, H. 1950. *The Physiology and Pathology of Exposure to Stress*. Montreal: Acta.

Senay, L.C. 1979. Effects of exercise in the heat on body fluid distribution. *Med. Sci. Sports* 11:42–48.

Senay, L.C., D. Mitchell and C.H. Wyndham. 1976. Acclimatization in a hot, humid environment: body fluid adjustments. *J. Appl. Physiol.* 40:786–796.

Sereno, P.C., D.B. Dutheil, M. Iarochene, H.C.E. Larsson, G.H. Lyon, P.M. Magwene, C.A. Sidor, D.J. Varricchio, and J.A. Wilson. 1996. Predatory dinosaurs from the Sahara and Late Cretaceous faunal differentiation. *Science* 272:986–991.

Severinghaus, J.W., H. Chiodi, E.I. Eger II, B. Brandstater, T.F. Hornbein. 1966. Cerebral blood flow in man at high altitude: role of cerebrospinal fluid pH in normalization of flow in chronic hypoxia. *Cir. Res.* 19:274–282.

Shapiro, Y., K.B. Pandolf, and R.F. Goldman. 1982. Predicting sweat loss response to exercise, environment and clothing. *European J. Appl. Physiol.* 48:83–96.

Shaw, R.H., ed. 1967. *Ground Level Climatology.* Washington, DC: AAAS Publications. 408 p.

Shepherd, J.T. 1963. *Physiology of the Circulation in Human Limbs in Health and Disease*. Philadelphia: Saunders.

Sherman, P.W., J.U.M. Jarvis, and R.D. Alexander, eds. 1991. *Biology of the Naked Mole-Rat*. New Jersey: Princeton University Press. 529 p.

Shiraki, K., and J.R. Claybaugh. 1995. Effects of diving and hyperbaria on responses to exercise, In *Exercise and Sport Science Reviews*, ed. J.O. Holloszy, 459–485. Baltimore: Williams and Wilkins.

Shiraki, K., and M.K. Yousef, eds. 1987. *Man in Stressful Environments: Thermal and Work Physiology.* Springfield, IL: Charles C. Thomas, Pub. 301 p.

Shkolnik, A., C.R. Taylor, V. Finch, and A. Borut. 1980. Why do Bedouins wear black robes in hot deserts? *Nature* 283:373–375.

Sicheri, F., and D.S.C. Yang. 1995. Ice-binding structure and mechanism of an antifreeze protein from winter flounder. *Nature* 375:427–431.

Simmons, E.L. 1963. Some fallacies in the study of hominid phylogeny. *Science* 141:879–889.

Simpson, S. 1912. An investigation into the effects of seasonal changes on body temperature. *Proc. Roy. Soc. Edinb.* 32:110–135.

Sinclair, D. 1967. *Cutaneous Sensation*. London: Oxford Univ. Press. 385 p.

Slater, D. 1996. Global endangered seas campaign. *Focus* 18:1.

Slijper, E.J. 1962. *Whales*. New York: Basic Books Publ. Co., Inc. 475 p.

Sluckin, W. 1965. *Imprinting and Early Learning*. Chicago: Aldine Publ. 150 p.

Smeland, E.B., et al. 1984. Modification of "diving bradycardia" by hypoxia or exercise. *Respir. Physiol.* 56:245–251.

Smith, F.A., J.L. Betancourt, and J.H. Brown. 1995. Evolution of body size in the woodrat over the past 25,000 years of climate change. *Science* 270:2012–2014.

Smith, F.A., J.L. Betancourt, and J.H. Brown. 1995. Evolution of body size in the woodrat. *Science* 270:2012–2014.

Smith, H., and H. Silvette. 1928. Note on the nitrogen excretion of camels. *J. Biol. Chem.* 78:409–411.

Smith, J. 1995. Hearing losses, which ones are recordable? *Occup. Health and Safety,* 64(5):65–68.

Smith, R.E. 1964. Thermoregulatory and adaptive behavior of brown adipose tissue. *Science* 146:1686–1689.

Smith, R.E., and D.J. Hoijer. 1962. Metabolism and cellular function in cold acclimation. *Physiol. Rev.* 42:60–142.

Smith, R.E., and J.C. Roberts. 1964. Thermogenesis of brown adipose tissue in cold-acclimated rats. *Amer. J. Physiol.* 206:143–148.

Sohol, R.S., and R.W. Weindruch. 1996. Oxidative stress, caloric restriction, and aging. *Science* 273:59–63.

South, F.E., J.P. Hannon, J.R. Willis, E.T. Pengelley, and N.R. Alpert, eds. 1972. *Hibernation and Hypothermia: Perspectives and Challenges*. Proc. Fourth Int. Symp. Nat. Mammal. Hiber. New York:Elsevier Scientific.

Spaargaren, D.H. 1994. Metabolic rate and body size: a new view on the "surface law" for basic metabolic rate. *Acta Biotheoretica* 42(4):263–269.

Specter, H. 1996. Arctic tribes unchanged for centuries. *The Polar Times* 2:15.

Sperber, I. 1944. Studies on the mammalian kidney. *Zool. Bidrag Fran Upsala* 22:249–431.

Spradling, T.A. 1996. Relative rates of evolution among rodents. *Proc. Amer. Soc. Mammalogists*. 76:3.

Spurr, G.B.. B.K. Hutt, and S.M. Horvath. 1957. Shivering, oxygen consumption and body temperatures in acute exposure of men to two different cold environments. *J. Appl. Physiol.* 11:58–64.

Steen, J.B. 1971. *Comparative Physiology of Respiratory Mechanisms*. New York: Academic Press. 182 p.

Stefansson, V. 1922. *Hunters of the Great North*. New York: Harcourt Brace. 301 p.

Stefansson, V. 1960. Tropical winter life of the polar Eskimo. Chapters XII, XIII in *Cancer: Disease of Civilization*. New York: Macmillan. 426 p.

Stefansson, V. 1964. *Discovery*. New York: McGraw-Hill. 411P.

Steffen, J.M., and M.L. Riedesel. 1982. Pulmonary ventilation and cardiac activity in hibernating and arousing golden-mantled ground squirrels (*Spermophilus lateralis*). *Cryobiology* 19:83–91.

Steffen, J.M., G.L. Rigler, A.K. Moore, and M.L. Riedesel. 1980. Urea recycling in active golden-mantled ground squirrels (*Spermophilus lateralis*). *Am. J. Physiol.* 239:R168–R173.

Stephan, F.K., and I. Zucker. 1972. Circadian rhythms in drinking and running of rats are eliminated by hypothalamic lesions. *Proc. Natl. Acad. Sci.* USA. 69: 1583–1586.

Sterba, J.A., and C.E.G. Lundgren. 1985. Diving bradycardia and breath-holding time in man. *Undersea Biomed. Res.* 12:139–150.

Stevens, E.D., and A.E. Dizon. 1982. Energetics of locomotion in warm-bodied fish. *Ann. Rev. Physiol.* 44:121–131.

Stevens, G.C. 1962. Circadian melanophore rhythms of the fiddler crab: interaction between animals. In *Rhythmic Functions in the Living Systems*, ed. Wm Wolf. Ann. N.Y. Acad. Aci. 98:926–239.

Stewart, B.S. 1996. Unique diving mammal, the northern elephant seal. *Natural History* 2:61–64.

Stewart, R.E.A., et al. 1993. *Second Walrus Workshop*. Winnipeg: Canadian Technical Report on Aquatic Science, 91 p.

Stewart, R.E.A., et al. 1993. *Second Walrus Workshop*. Winnipeg: Canadian Technical Report on Aquatic Science, 91 p.

Stones, R.C., and J.E. Wiebers. 1967. Temperature regulation in the little brown bat, *Myotis lucifugus*, p. 97–109. In *Mammalian Hibernation III,* ed. K.C. Fisher et al. Edinburgh: Oliver and Boyd.

Stout, R.W., and V. Crawford. 1991. Seasonal variations in fibrinogen concentrations among elderly people. *The Lancet* 338:9–13.

Stow, C.A., et al. 1996. Fisheries management to reduce contaminant consumption. *BioScience* 45:752–758.

Strijkstra, A.D.S. 1997. Sleep during arousal episodes as a function of prior torpor duration in hibernating European ground squirrels. *J. Sleep Res.* 6:36–43.

Strughold, H. 1967. Solved and unsolved space medical problems international status 1966–1967, p. 521–535. In *Lectures in Aerospace Medicine,* 6th Series. Brooks AFB, TX: USAF School of Aerospace Med. 535 p.

Strumwasser, F. 1959. Electrical activity of brain and temperature of brain in hibernation (*Citellus*). *Am. J. Physiol.* 196:8–30.

Strumwasser, F. 1967. Types of Information stored in single neurons, p. 291–319. In *Invertebrate Nervous Systems*, ed. C.A.G. Wiersma. Chicago: Univ. of Chicago Press. 455 p.

Strydom, N.B., and C.H. Wyndham. 1963. Effect of heat on work performance. *Fed. Proc.* 22:893–896.

Strydon, N.B., et al. 1965. Oral/rectal temperature differences during work in heat. *J. Appl. Physiol.* 20:283–288.

Studier, E.H., and J.W. Procter. 1971. Respiratory gases in burrows of *Spermophilus tridecemlineatus*. *J. Mammal.* 52:631–633.

Studier, E.H., and D.E. Wilson. 1970. Thermoregulation in some neotropical bats. Comp. Biochem. Physiol. 34:251–262.

Stump, C.S., J.M. Overton, and C.M. Tipton. 1990. Influence of single hindlimb support during simulated weightlessness in the rat. *J. Appl. Physiol.* 68:627–634.

Sturbaum, B.A., and M.L. Riedesel. 1974. Temperature regulation responses of ornate box turtles, *Terrapene ornata,* to heat. *Comp. Biochem. Physiol.* 48A:527–538.

Suarez, J, J.K. Alexander, and C.S. Houston. 1987. Enhanced

left ventriculary systolic performance at high altitude during Operation Everest II. *Am. J. Cardio.* 60:137–142

Sulman, F.G. 1980. *The Effect of Air Ionization, Electric Fields, Atmospherics and Other Electric Phenomena on Man and Animal,* p. 146. Springfield, IL: Chas. C. Thomas. 398 p.

Suomalainen, P., ed. 1964. *Mammalian Hibernation II*: Proc. Second Int. Symp. Nat. Mammal. Hiber. Annales Academiae Scientiarum Fennicae, Series A, IV. Biological, Vol. 71. Finnish Academy of Science.

Suomalainen, P., and E. Karppanen. 1956. Einflus des Winterschlafes auf das Albumin-Globulinverhaltniss des Igel-serums. *Suomen Kemistilehti* 29:74–75.

Sutherland, B.M. 1981. Photoreactivation. *BioScience* 31:439-443.

Sutton, J.R., J.T. Reeves, P.E. Wagner, B. M. Groves, A. Cymerman, M.K. Malconian, P.B. Rock, P.M. Young, S.D. Walter and C.S. Houston 1988. Operation Everest II: oxygen transport during exercise at extreme simulated altitude. *J. Appl. Physiol.* 64:1309–1321

Swan, L.W. 1961. The Ecology of the High Himalayas. *Sci. Am.* 205:68–78.

Swann, J.M,. and F.W. Turek. 1985. Multiple circadian oscillators regulate the timing of behavioral and endocrine rhythms in female golden hamsters. *Science* 228:898–900.

Swenson, M.J., and W.O. Reese, eds. 1993. *Duke's Physiology of Domestic Animals, 11th Edition.* Ithaca: Cornell University Press.

Swisher, C.C., et al. 1996. Latest *Homo erectus* of Java. *Science* 274:1870–1874.

T

Tabrah, F.L., D.L. Guernsey, S-C Chou, and S. Batkin. 1978. Effect of alternating magnetic fields (60–100 gauss, 60 HZ) on *Tetrahymena pyriformis*. Tower Internal. Institute, *J. Life Sci.* 8:73–76.

Takahashi, J.S., and M. Zatz. 1982. Regulation of circadian rhythmicity. *Science* 217:1104–1111.

Takeuchi, H. and M. Mohri. 1987. Current status of diving fishermen in Japan: distribution, age and diving method. *Jpn. J. Hyperbaric Med.* 22:227–234.

Tamarkin, L., J.B. Curtis, and O.F.X. Almeida. 1985. Melatonin: A coordinating signal for mammalian reproduction? *Science* 227:714–720.

Tapp, W.N., and F.A. Holloway. 1981. Phase shifting circadian rhythms produces retrograde amnesia. *Science* 211:1056–1058.

Taylor, C.R. 1951. Personal communication.

Taylor, C.R. 1970. Dehydration and heat: Effects on temperature regulation of East African ungulates. *Amer. J. Physiol.* 219:1136–1139.

Taylor, C.R. 1977. Exercise and environmental heat loads: different mechanisms for solving different problems? In *Environmental Physiology II. Int. Rev. Physiol., vol. 15,* ed. D. Robertshaw, p. 119–146. Baltimore: University Park Press. 246 p.

Taylor, C.R., and C.P. Lyman. 1972. Heat storage in running antelopes: Independence of brain and body temperature. *Amer. J. Physiol.* 222:114–117.

Taylor, C.R. 1966. Thermoregulatory function of the horns of the family Bovidae. TR-63–31. Fort Wainwright, Alaska: Arctic Aeromedical Laboratory.

Taylor, C.R., and E.R. Weibel. 1981. Design of the mammalian respiratory system I-IX. *Resp. Physiol.* 44:1–164.

Taylor, C.R., and J.H. Jones. 1987. Maximal oxygen consumption in mammals, p. 188–195. In *Comp. Physiol. Envir. Adaptations,* Vol 2, ed. Dejours. 8th ESCP Conf. Strasbourg: Karger Basel.

Taylor, N.A., N.K. Allsopp, and D.G. Parkes. 1995. Preferred room temperature of young vs aged males: the influence of thermal sensation, thermal comfort, and affect. *J. Gerontol. Series A, Bio. Sci. & Med. Sci.* 50(4):M216–21.

Taylor, P.A., N.A. Garrick, L. Tamarkin, D.L. Murphy, and S.P. Markey. 1985. Diurnal rhythms of N-Acetylserotonin and serotonin in cerebrospinal fluid of monkeys. *Science* 228:900.

Terman, J.S., M. Terman, D. Schlager, et al. 1990. Efficacy of brief, intense light exposure for treatment of winter depression. *Psychopharmacol Bull* 26:3–11.

Terman, M., D. Schlager, S. Fairhurst, and B. Perlman. 1989. Dawn and dusk simulation as a therapeutic intervention. *Biol Psychiatry* 25:966–970.

Terrados, N.J., C. Melichna, C. Sylven, E. Jansson and L. Kaijser. 1988. Effects of training at simulated altitude on performance and muscle metabolic capacity in competitive road cyclists. *Eur. J. Appl. Physiol.* 57:203–209.

Thalmann, E.D., et al. 1984. Accomodation to decompression sickness in HeO2 divers. *Undersea Siomed. Res.* 11 (suppl. 1): 6.

Tharp, G.D., and G.E. Folk, Jr. 1964. Rhythmic changes in rate of the isolated mammalian heart and heart cells. *Comp. Biochem. Physiol.* 14: 255–273.

Thauer, R., and E. Simon. 1972. Spinal cord and temperature regulation, p. 22–49. In *Advances in Climatic Physiology,* ed. S. Itoh, K. Ogata, and H. Yoshimura. New York: Springer-Verlag.

Thom, H. 1966. *Introduction to Shortwave and Microwave Therapy.* 3rd ed. Springfield: Charles C. Thomas. 167 p.

Thompson, D., and M.A. Fedak. 1993. Cardiac responses of gray seals during diving at sea. *J. Exp. Biol.* 174:139–164.

Thomson, K.S. 1997. Natural Selection. *American Scientist* 85:516–518.

Thornton, A. 1992. *Animal Welfare Quarterly* 41:1.

Tikuisis, P. 1995. Predicting survival time for cold exposure. *Int. J. Biometeor.* 39(2): 94–102.

Timiras, P., A.A. Krum, and N. Pace. 1957. Body and organ weights of rats during acclimatization to an altitude of 12,499 feet. *Amer J. Physiol.* 191:598–604.

Timmerman, J.C., G.E. Folk, Jr., and S.M. Horvath. 1959. Day-night differences of body temperature and heart rate after exercise. *Quart. J. Exp. Physiol.* 44:258–263.

Tipton, C.M., and A. Hargens. 1996. Physiological adaptations and countermeasures associated with long-duration spaceflights. *Med. Sci. Sports Exerc.* 28:974–976.

Tipton, C.M., J.E. Greenleaf, and C.G.R. Jackson. 1996. Neuroendocrine and immune system responses with spaceflights. *Med. Sci. Sports Exerc.* 28:988–998.

Tjandrawinata, R.R., V.L. Vincent, and M. Hughes-Fulford. 1997. Vibrational force alters mRNA expression in osteoblasts. FASEB 11:493–497.

Toda, Y. 1967. Temperature change in fingers of both hands with one hand strongly cooled in air, p. 129. In *Biometeorology, Vol. 3*, ed. S.W. Tromp and W.H. Weihe. Amsterdam: Swets & Zeitlinger. 347 p.

Tokura, H., and T. Oishi. 1985. Circadian locomotor activity rhythm under the influences of temperature cycle in the Djungarian hamster *Phodopus sungorus*, entrained by 12 hour light-12 hour-dark cycle. *Comp. Biochem. Physiol.* 81A:271–275.

Tomilin, A.G. 1950. Notes on Siberian white-sided dolphin. *Rybnoe Khozaistvo* 26:50–53.

Tonndorf, J. 1974. Personal communication.

Torbati, D., et al. 1989. Effects of acute hyperbaric oxygenation on respiratory control in cats. *J. Appl. Physiol.* 67:2351–1256.

Tosini, G., and M. Menaker. 1996. Circadian rhythms in cultured mammalian retina. *Science* 272:419–421 and 349.

Trayburn, P. 1993. Species Distribution of Brown Adipose Tissue: Characterization of Adipose Tissues from Uncoupling Protein and its mRNA, p. 361–368. In *Life in the Cold: Ecological, Physiological and Molecular Mechanisms,* ed. C. Carey, G.L. Florant, B.A. Wunder, and B. Horwitz. Boulder, CO: Westview Press.

Trefil, J. 1996. The carbon cycle. *Smithsonian* 27:30–31.

Tregear, R.T. 1965. Hair density, wind speed, and heat loss in mammals. *J. Appl. Physiol.* 20:796–801.

Tregear, R.T. 1966. *Physical Functions of Skin.* London: Academic Press. 442 p.

Triemestra, R. 1992. National Weather Service, Chicago. Personal Communication.

Tromp, S.W. 1963a. Human biometeorology. *Int. J. Biometeorol.* 7:145–158.

Tromp, S.W., ed. 1963b. *Medical Biometeorology.* New York: Amer. Elsevier Press. 991 p.

Truchot, J.P., and A. Duhamel-Jouve. 1980. Oxygen and carbon dioxide in the marine intertidal environment: diurnal and tidal changes in rockpools. *Respir. Physiol.* 39:241–254.

Tyner, J.G., and A. Hemingway. 1969. Respiration and metabolism of sheep in the cold. *Fed. Proc.* 28:791.

U

Underwood, H., and M. Menaker. 1970. Extraretinal light perception: Entrainment of the biological clock controlling lizard locomotor activity. *Science* 170:190–192.

United Nations Secretariat, comp. and ed. 1977. *Desertification: Its Causes and Consequences.* Conf. Nairobi, Kenya. Mazingira Suppl. Libr. Congr. No. 77–81423, New York: Pergamon Press.

Untersteiner, N. 1984. The cyrosphere, p 121–140. In *The Global Climate,* ed. J.T. Houghton. Cambridge University Press. 233 p.

V

Van Citters, R.L., O.A. Smith, N.W. Watson, D.L. Franklin, and R.W. Elsner. 1964. Cardiovascular responses of elephant seals during diving studied by blood flow telemetry. 15th Alaska Sci. Conf., Alaska Div. AAAS Abstract No. 106.

Van Liew, H.D. 1991. Simulation of the dynamics of decompression sickness bubbles and the generation of new bubbles. *Undersea Biomed. Res.* 18:333–345.

van Waarde, A., G. van den Thillart, and M. Verhagen. 1993. Ethanol formation and pH-regulation in fish, p. 157–170. In *Surviving Hypoxia: Mechanisms of Control and Adaptation,* ed. P.W. Hochachka, P.L. Lutz, T. Sick, M. Rosenthal, and G. van den Thillart. Ann Arbor: CRC Press.

Vauahan, M.K., G.C. Brainard, and R.J. Reiter. 1984. The influence of natural short photoperiodic and temperature conditions on plasma thyroid hormones and cholesterol in male Syrian hamsters. *Int. J. Biometeor.* 28:201–210.

Verbanck, S., D. Linnarsson, G.K. Prisk, and M. Paiva. 1996. Specific ventilation distribution in microgravity. *J. Appl. Physiol.* 80:1458–1465.

Viereck, Eleanor G. 1964. Personal communication

Virokannas, H. 1996. Thermal responses to light, moderate, and heavy daily outdoor work in cold weather. *European J. of Appl. Physiol. & Occupational Physiol.* 72(5–6):482–489.

Visser, M., P. Deurenberg, W.A. van Staveren, and J.G. Hautvast. 1995. Resting metabolic rate and diet-induced thermogenesis in young and elderly subjects: relationship with body composition, fat distribution, and physical activity level. *Am. J. Clin. Nutrition* 61(4):772–778.

Vitaterna, M.H., et al. 1994. Mutagenesis and mapping of a mouse gene. *Science* 264:719–721.

Vogel, V.B. 1969. Vergleichende Untersuchungen uber den Wasserhaushalt von Fledermausen Rhinopoma, Phinolophys und Myotis. *Z. Vergl. Physiol.* 64:324–345.

Vogt, J.J., et al. 1981. Required sweat rate as an index of thermal strain in industry, p 98–110. In *Bioengineering, Thermal phsyiology and Comfort*, ed. K. Cena and J.A. Clark, New York: Elsevier Scientific. 289p.

Vosshall, L.B., et al. 1994. Block in nuclear localization of period protein. *Science* 263: 1606–1608.

W

Wagner, P.D. 1995. The integrative biology of exercise. *The Physiologist* 38:282

Wald, G., and B. Jackson. 1944. Activity and nutritional deprivation. *Proc. Natl. Acad. Sci.* USA 30:255–259.

Walker, B.R. 1983. Inhibition of hypoxia-induced ADH release by meclofenamate in the conscious dog. *J. Appl. Physiol.* 54:1624–1629.

Walker, B.R. 1986. Role of vasopressin in the cardiovascular response to hypoxia in the conscious rat. *Am. J. Physiol.* 251:H1316–H1323.

Walkup, G., ed. 1971. Reports on fur auctions. *Fur Trappers J.* 12:18–20.

Walsberg, G.E., and B.O. Wolf. 1995. Solar heat gain in a desert rodent: unexpected increases with wind speed and implications for estimating the heat balance of free-living animals. *J. Comp. Physiol.* B 165:306–314.

Wang, L.C.H. 1978. Energetic and field aspects of mammalian torpor: the Richardson's ground squirrel, p. 109–145. In *Strategies in Cold: Natural Torpidity and Thermogenesis,* ed. L.C.H. Wang and J.W. Hudson. New York: Academic Press.

Ward, M.P., J.S. Milledge, and J.B. West. 1991. *High Altitude Medicine and Physiology,* p. 2–7. Philadelphia. University of Pennsylvania Press.

Ward, W.D. 1973. Adaptation and fatigue, p. 301–344. In *Modern Developments in Audiology*, Ch. 9, ed. J. Jerger. Academic Press.

Warkander, D.E., et al. 1990. CO_2 retention with minimal symptoms but severe dysfunction during wet simulated dives to 6.8 atm abs. *Undersea Biomed. Res.* 17: 515–523.

Warkander, D.E., et al. 1992. Physiologically and subjectively acceptable breathing resistance in diver's breathing gear. *Undersea Biomed. Res.* 19:427–445.

Warncke, Günther. 1994. Diurnal and thermal responses to changing ambient temperatures in three species of hummingbirds. In *Proc. Intersociety Meeting on Regulation, Integration, and Adaptation. The Physiologist* 37:41–42.

Warrick, R.A., et al. 1996. Sea level change. In *Climate change 1995. The science of climate change,* ed. J.T. Houghton, et al. Contribution of Working Group I to the *Second Assessment Report* of the Intergovernmental Panel on Climate Change. Cambridge: Cambridge University Press.

Wassmer, T., and F. Wollnik. 1997. Timing of torpor bouts during hibernation in European hamsters (*Cricetus cricetus L.*). *J. Compar. Physiol.* 167(4):270–279.

Watts, P.D., N.A. Øritsland, C. Jonkel, and K. Ronald. 1981. Mammalian hibernation and the oxygen consumption of a denning black bear (*Urus Americanas*). *Comp. Biochem and Physiology.* 69:121–123.

Wayne, R., and M.P. Staves. 1996. The Krogh principle applies to plants. *BioScience* 46:365–369.

Weathersby, P. K., et al. 1987. Role of O_2 in the production of human decompression sickness. *J. Appl. Physiol.* 63:2380–2387.

Weathersby, P.K., et al. 1992. Predicting the time of occurrence of decompression sickness. *J. Appl. Physiol.* 72:1541–1548.

Webb, P.I., and J.D. Skinner. 1996. Summer torpor in African woodland dormice *Graphiurus murinus* (*Myoxidae: Graphiurinae*). *J. Comp. Physiol.* B 166:325–330.

Weber, A.L., and N.T. Adler. 1979. Delay of constant light-induced persistent vaginal estrus by 24-hour time cues in rats. *Science* 204:323–324.

Weber, R.E. 1992. Molecular strategies in the adaptation of vertebrate hemoglobin function, p. 257–277. In *Lung Biology in Health and Disease,* Vol 56, Physiological Adaptations in Vertebrates, ed. S.C. Wood, R.E. Weber, A.R. Hargens, R.W. Millard. New York: Marcel Dekker, Inc.

Weems, J.E. 1967. *Peary: The Explorer and the Man.* Boston: Houghton Mifflin. 362 p.

Weesner, G. 1996. Mellowing Out. *BioScience* 46:72.

Wehner, R., and B. Lanfranconi. 1981. What do the ants know about the rotation of the sky? *Nature* 293:731–733.

Weiner, J. 1989. Metabolic constraints to mammalian energy budgets. *Acta Theriologica* 34:3–35.

Weir, E.K., and S.L. Archer. 1995. The mechanism of acute hypoxic pulmonary vasoconstriction: The tale of two channels. *FASEB J.* 9:183–189.

Wells, G.P. 1955. *The Sources of Animal Behavior.* London: H.K. Lewis. 20 p.

Welsh, J.H. 1953. Excitation of the heart of *Venus mercenaria. Arch exp. Path. Pharmakol.* 219:23–29.

Wenger, C.B. 1988. Human heat acclimatization, p. 153–197. In *Human Performance Physiology,* ed. K.B. Pandolf et al. 637 p. Indianapolis: Benchmark Press, 637 p.

West, J.B. 1984. Human physiology at extreme altitudes on Mt. Everest. *Science* 223:784–788.

West, J.R., S. Lahiri, K.H. Maret, R.M. Peters, Jr., and C.J. Pizzo. 1983. Barometric pressures at extreme altitudes on Mt. Everest: physiological significance. *J. Appl. Physiol.* 54:1188–1194.

West, N.H., and B.N. Van Vliet. 1986. Factors influencing the onset and maintenance of bradycardia in mink. *Physiol. Zool.* 59:451–463.

Wever, R. 1971. Influence of electric fields on some parameters of circadian rhythms in man, p. 117–133. In *Biochronometry,* ed. M. Menaker. Washington, DC: Natl. Acad. Sci. Publ. 1866, 662 p.

Wever, R.A. 1979. *The Circadian System of Man.* New York: Springer-Verlag. 276 p.

Whipple, H.E., ed. 1964. Thermography and its clinical applications. *Ann. N.Y. Acad. Sci.* 121:1–304.

White, F.N., and D.K. Odell. 1971. Thermoregulatory behavior of the northern elephant seal, *Mirounga angustirostris. J. Mammal.* 52:758–774.

White, R.G., R.D. Cameron, and L.K. Miller. 1971. Fresh water ingestion and water flux in a captive harbor seal (*Phoca vitulina*). *Proc. Alaska Sci. Conf.* 22:14.

Whitehead, H. 1995. The realm of the sperm whale. *Nat. Geographic* 188:57–73

Whittow, G.C. 1971. *Mammals. Comparative Physiology of Thermoregulation.* Vol 2. New York: Academic Press

Wilkins, H. 1960. The effect of light on plant rhythms. In *Biological Clocks.* Cold Spring Harbor Symposium 25:115–131.

Williams, B.A., and A. Shitzer. 1974. Modular liquid-cooled helmet liner for thermal comfort. *Aerospace Med.* 45:1030–1036.

Williams, D.D., and R.L. Rausch. 1971. Seasonal carbon dioxide and oxygen tension in the dens of hibernators, p. 32–33. In *Proc. Hibernation-Hypothermia Symp. 4*. Space Sci. Res. Ctr., Univ. Missouri, Columbia.

Willis, J.S. 1964. Potasssum and sodium content of tissues of hamsters and ground squirrels during hibernation. *Science* 146:546–547.

Willis, J.S., and M. Marjanovic. 1993. Handling of sodium ion by the plasma membrane: Differences that account for failure of cold sensitive cells. In *Life in the Cold, Ecological, Physiological and Molecular Mechanisms*, ed. C. Carey, G.L. Forant, B.A. Wunder, and B. Horwitz. Boulder, Colo.: Westview Press

Wilson, H., and J. Hansen. 1994. Global and hemispheric temperature anomalies from instrumental surface air temperature records, p. 609–614. In *Trends '93: A Compendium of Data on Global Change*, ORNL/CDIAC-65, ed. T.A. Boden, et al. Tennessee: Oak Ridge National Laboratory.

Wiltschko, R. D. Nohr, and W. Wiltschko. 1981. Pigeons with a deficient sun compass use the magnetic compass. *Science* 214:343–344.

Winfree, A.T. 1982a. Circadian timing of sleepiness in man and woman. *Amer. J. Physiol.* 243:R193-R204.

Winfree, A.T. 1982b. The tides of human consciousness: Descriptions and questions. *Amer. J. Physiol.* 242:R163–R166.

Winget, C.M., C.W. DeRoshia, and D.C. Holley. 1985. Circadian rhythms and athletic performance. *Med. and Sci. in Sports and Exercise* 17:498–516.

Winslow, R.M., M. Samaja, and J.B. West. 1984. Red cell function at extreme altitude on Mt. Everest. *J.Appl. Physiol.* 56: 109–116.

Winslow, R.M. 1984. High altitude polysythemia. In *High Altitude and Man*, ed. J.B. West and S. Lahiri. Bethesda, MD: Amer. Physiol. Soc.

Wittmers, L., et al. 1987. Cardiovascular responses of the dive reflex in human beings. *Annals of Emergency Medicine* 16:1031–1036.

Wolfson, A. 1959. Role of light in the photoperiodic responses of migratory birds. *Science* 129:1425–1426.

Wood, S.C., ed. 1980. Respiratory pigments. *Am. Zool.* 20:1–211.

Wood, S.C. 1991. Interactions between hypoxia and hypothermia. *Annu. Rev. Physiol.* 53:71–85.

Wood, S.C., and G.M. Malvin. 1991. Behavioral hypothermia: an adaptive stress response, p. 295–312. In *Strategies of Physiological Adaptations*, ed. S.C. Wood and R. Weber. New York: Dekker.

Woods, J.D., and J.N. Lythgoe, eds. 1971. *Underwater Science: An Introduction to Experiments by Divers*. New York: Oxford Univ. Press. 330 p.

Woods, R., and L.D. Carlson. 1956. Thyroxine secretion in rats exposed to cold. *Endocrinology* 59:323–330.

World Health Organization. 1969. *Health Factors Involved in Working under Conditions of Heat Stress*. Report No. 412. Geneva.

Wormworth, J. 1995. Toxins and tradition: the impact of food-chain contamination on the Inuit of northern Quebec. *Canadian Med. Assn. J.* 152 (8): 1237–1240.

Wright, H.E., Jr., and W.H. Osburn. 1968. *Arctic and Alpine Environments, Vol. 10*. Nat. Acad. Sci.-Nat. Res. Coun. Bloomington: Indiana Univ. Press. 308 p.

Wunder, C.C., and R.C. Welch. 1979. Femur strength as influenced by growth, bone-length and gravity with the male rat. *J. Biomechanics*. 12:501–507.

Wurtman, R.J., J. Axelrod, and D.E. Kelly. 1968. *The Pineal*. New York: Academic Press. 204 p.

Wyndham, C.H. 1973. The physiology of exercise under heat stress. *Ann. Rev. Physiol.* 35:193–220.

Wyndham, C.H. 1977. Heat stroke and hyperthermia in marathon runners. *Ann. N.Y. Acad. Sci.* 301:128–138.

Wyndham, C.H., and J.F. Morrisson. 1958. Adjustment to cold of bushmen in the Kalahari Desert. *J. Appl. Physiol.* 13:219–225.

Wyon, D.P., I. Andersen, and G.R. Lundquist. 1981. The effects of moderate heat stress on mental performance, p. 251–267. In *Bioengineering, Thermal Physiology and Comfort*, ed. K. Cena and J.A. Clark. New York: Elsevier Scientific.

Wyss, C. R., G. L. Brengelmann, J. M. Johnson, L. B. Rowell, and D. Silverstein. 1975. Altered control of skin blood flow at high skin and core temperatures. *J. Appl. Physiol.* 38:839–845.

X, Y

Yablokov, A.V. 1994. Whales vs. Whalers. *Animal Welfare Inst. Quarterly*, Winter, p. 1–15.

Young, S.P., and E.A. Goldman. 1946. *The Puma: Mysterious American Cat*. 358 p. Washington, DC:Amer. Wildlife Inst.

Yousef, M.K. 1988. Animal Stress and Strain. *Appl. Anim. Behav. Sci.* 20:119–126.

Yousef, M.K., and H.D. Johnson. 1967. Time course of oxygen consumption in rats during sudden exposure to high environmental temperature. *Life Sci.* 6:1221–1228.

Yousef, M.K., M.E.D. Webster, and O.M. Yousef. 1989. Energy costs of walking, in camels. *Physiol. Zool.* 62:1080–1088.

Z

Zari, T.A. 1992. Effects of Temperature on Metabolism of Geckos. *Comp. Biochem. Physiol.* 102:491–495.

Zegers, D.A., and J.F. Merritt. 1988. Effect of photoperiod and ambient temperature on nonshivering thermogenesis of Peromyscus maniculatus. *Acta Theriologica* 33:273–281.

Zernicke, R.F., A.C. Vailas, and G.J. Salem. 1990. Biomechanical response of bone to weightlessness. *Exerc. Sports Sci. Rev.* 18:157–191.

INDEX

NOMOGRAMS

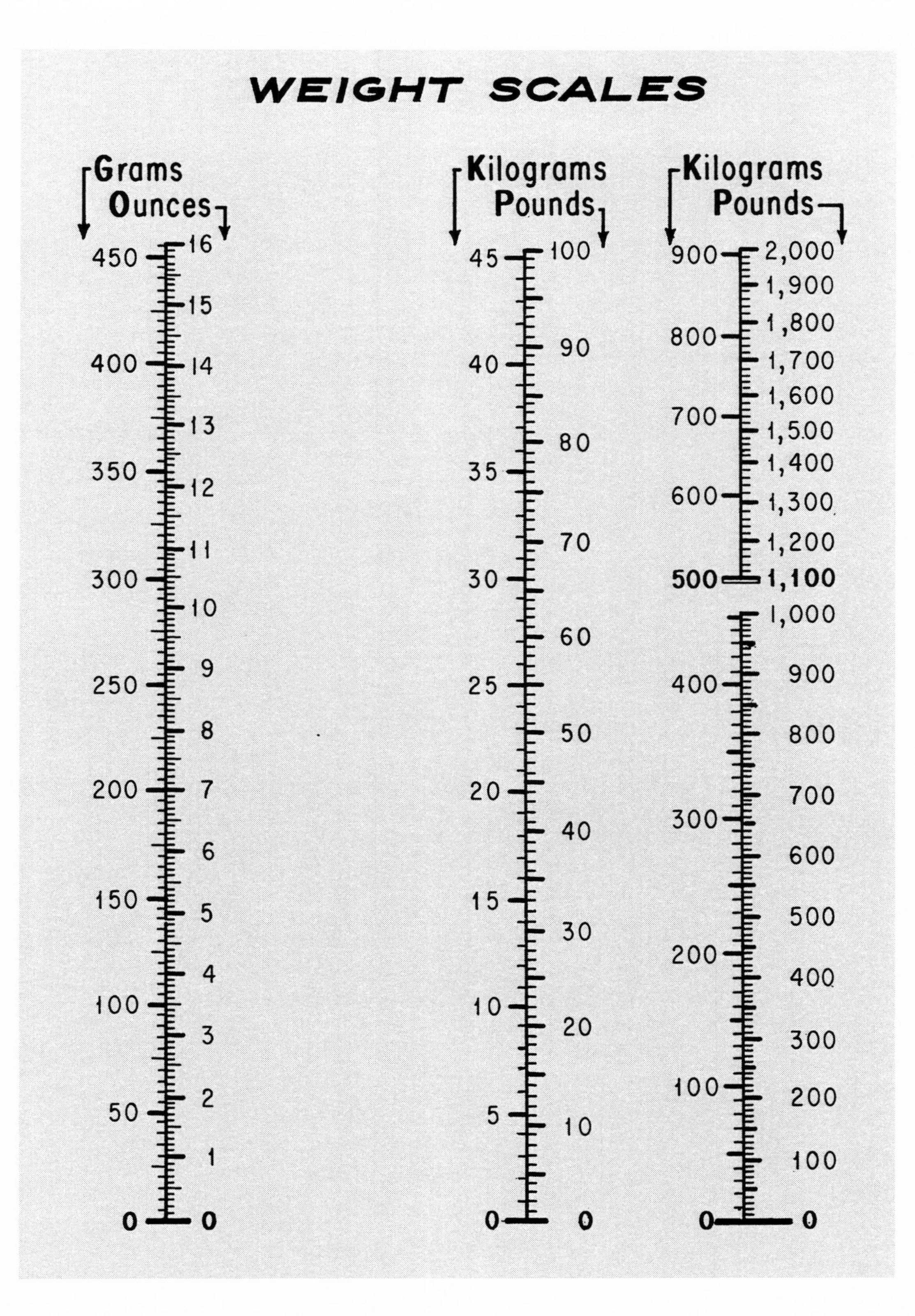
WEIGHT SCALES
Grams
Ounces
Kilograms
Pounds
Kilograms
Pounds
450
400
350
300
250
200
150
100
50
0
16
15
14
13
12
11
10
9
8
7
6
5
4
3
2
1
0
45
40
35
30
25
20
15
10
5
0
100
90
80
70
60
50
40
30
20
10
0
900
800
700
600
500
400
300
200
100
0
2,000
1,900
1,800
1,700
1,600
1,500
1,400
1,300
1,200
1,100
1,000
900
800
700
600
500
400
300
200
100
0

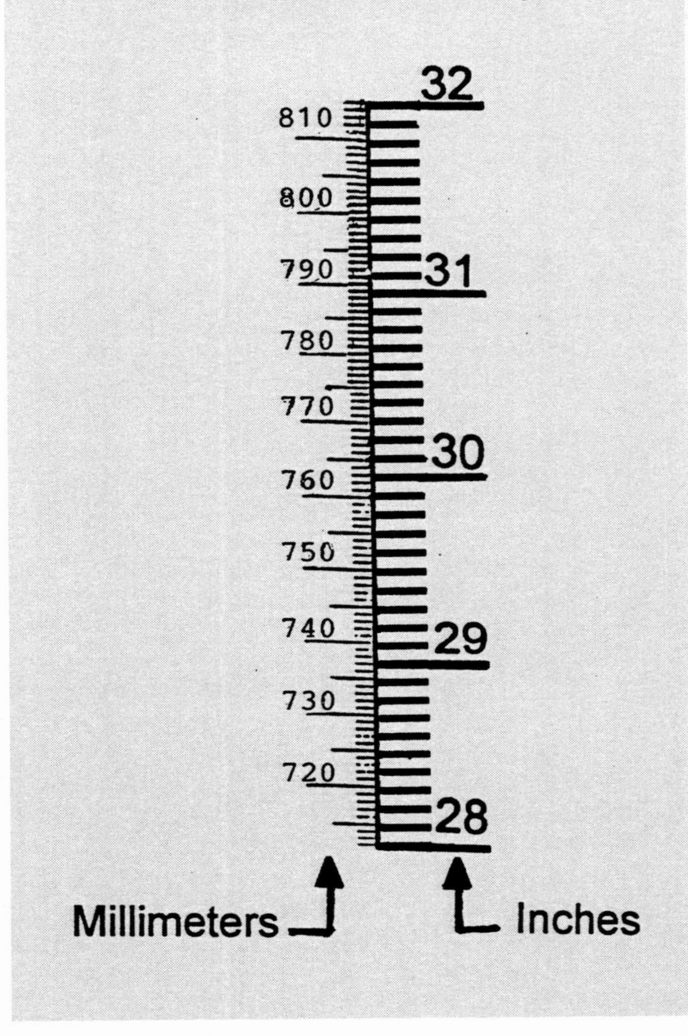
Atmospheric Pressure
in Units of Mercury
32
810
800
790
31
780
770
30
760
750
740
29
730
720
28
Millimeters
Inches

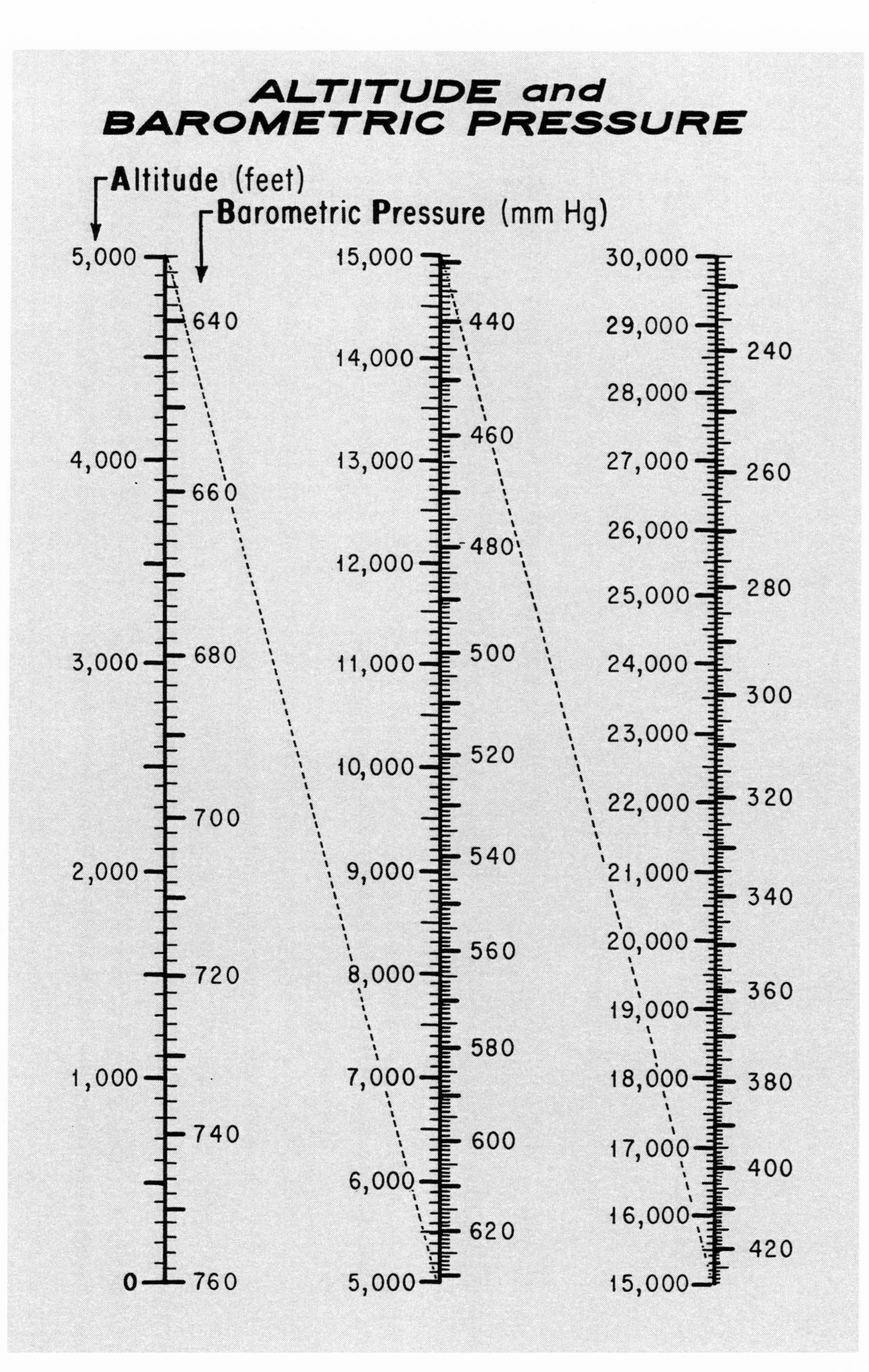
ALTITUDE and
BAROMETRIC PRESSURE
Altitude (feet)
Barometric Pressure (mm Hg)
5,000
4,000
3,000
2,000
1,000
0
640
660
680
700
720
740
760
15,000
14,000
13,000
12,000
11,000
10,000
9,000
8,000
7,000
6,000
5,000
440
460
480
500
520
540
560
580
600
620
30,000
29,000
28,000
27,000
26,000
25,000
24,000
23,000
22,000
21,000
20,000
19,000
18,000
17,000
16,000
15,000
240
260
280
300
320
340
360
380
400
420

LENGTH SCALES

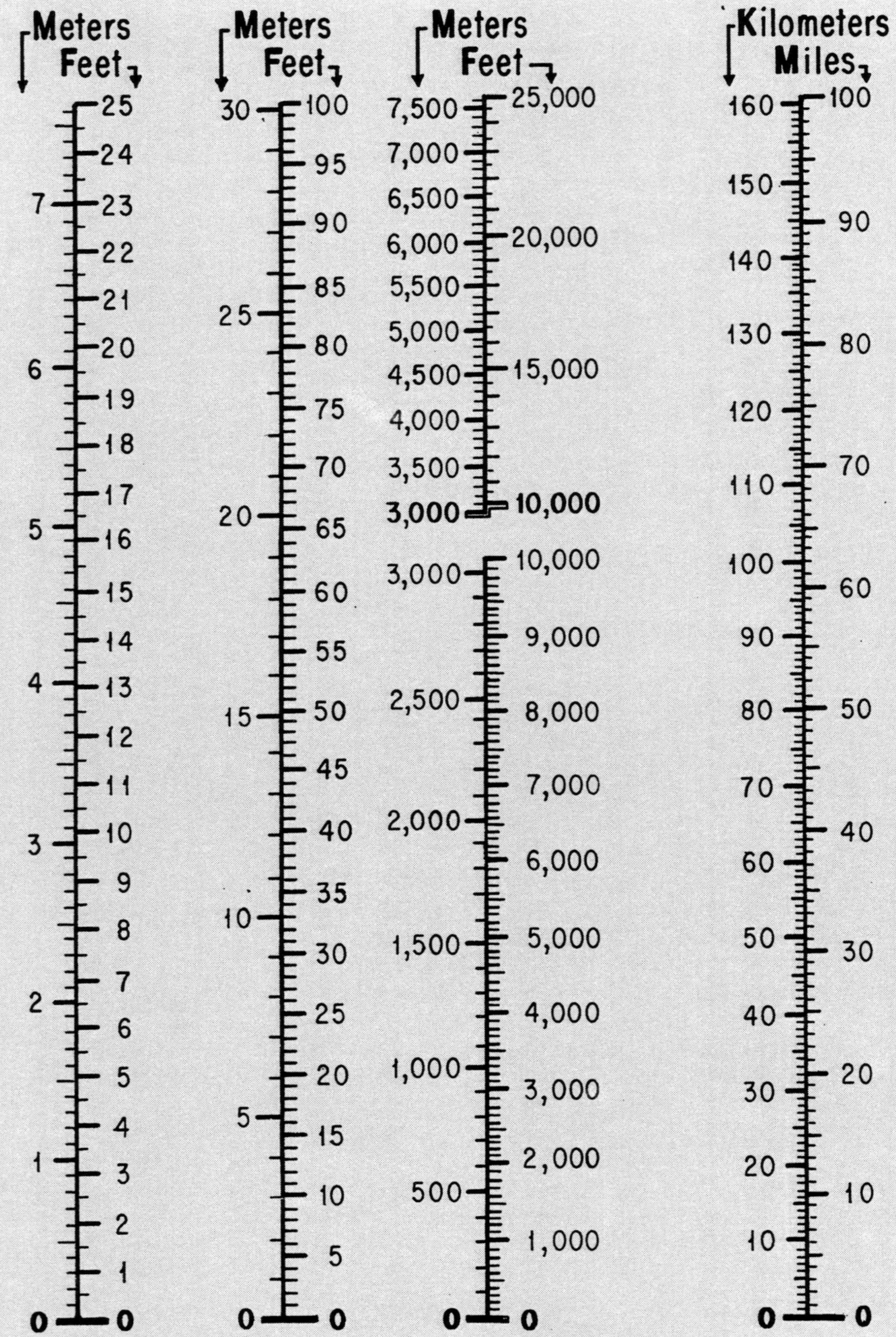

Meters / Feet
- Meters: 0, 1, 2, 3, 4, 5, 6, 7
- Feet: 0, 1, 2, 3, 4, 5, 6, 7, 8, 9, 10, 11, 12, 13, 14, 15, 16, 17, 18, 19, 20, 21, 22, 23, 24, 25

Meters / Feet
- Meters: 0, 5, 10, 15, 20, 25, 30
- Feet: 0, 5, 10, 15, 20, 25, 30, 35, 40, 45, 50, 55, 60, 65, 70, 75, 80, 85, 90, 95, 100

Meters / Feet
- Meters: 0, 500, 1,000, 1,500, 2,000, 2,500, 3,000
- Feet: 0, 1,000, 2,000, 3,000, 4,000, 5,000, 6,000, 7,000, 8,000, 9,000, 10,000
- Meters: 3,000, 3,500, 4,000, 4,500, 5,000, 5,500, 6,000, 6,500, 7,000, 7,500
- Feet: 10,000, 15,000, 20,000, 25,000

Kilometers / Miles
- Kilometers: 0, 10, 20, 30, 40, 50, 60, 70, 80, 90, 100, 110, 120, 130, 140, 150, 160
- Miles: 0, 10, 20, 30, 40, 50, 60, 70, 80, 90, 100

TEMPERATURE SCALES

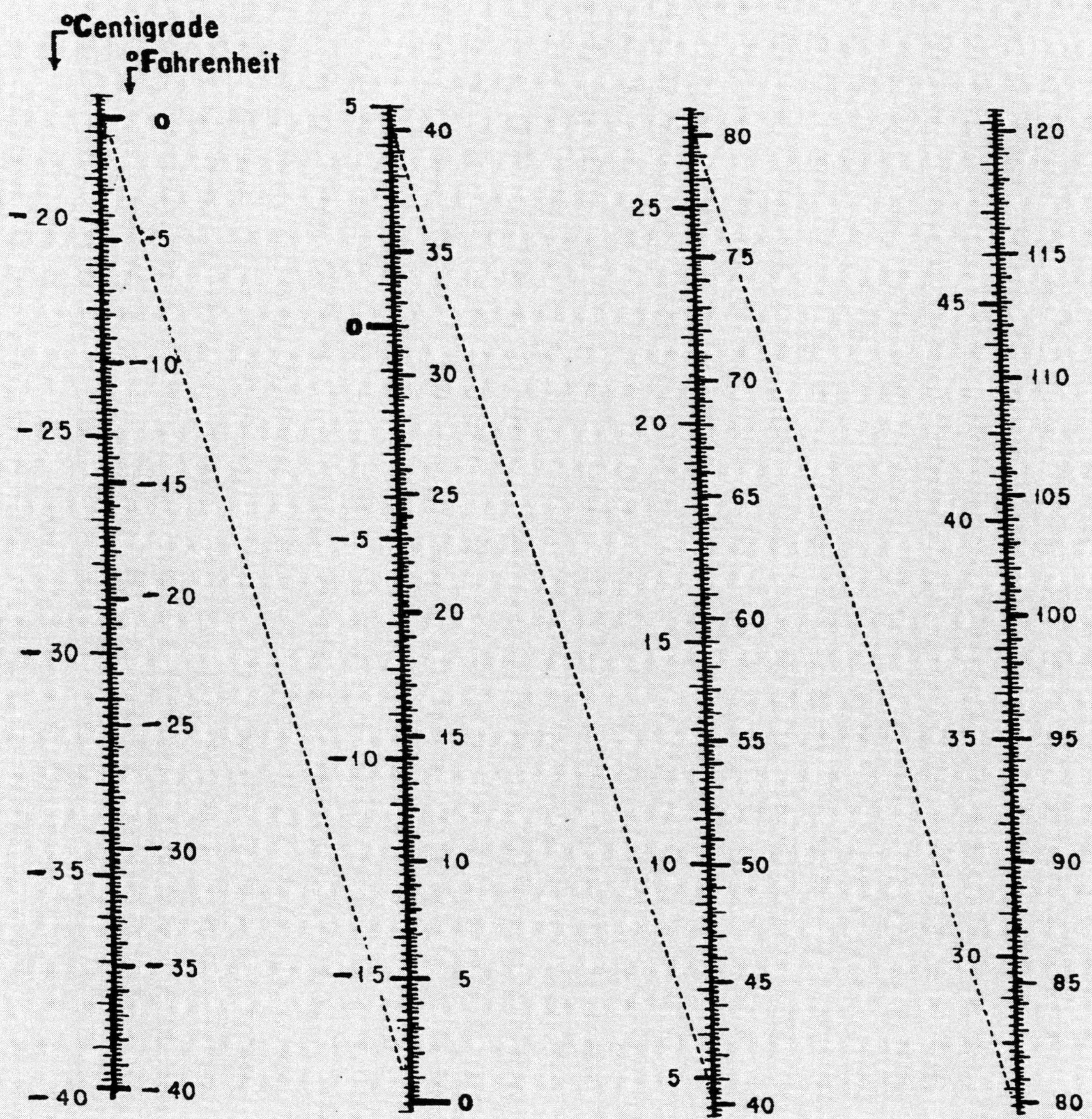